BOTANICA
NORTH AMERICA

BOTANICA
NORTH AMERICA

*The Illustrated Guide to Our Native Plants,
Their Botany, History, and the Way They
Have Shaped Our World*

MARJORIE HARRIS

*To Shirley & Sandrea
for a glorious planet
Marjorie Harris*

HarperResource
An Imprint of HarperCollinsPublishers

OTHER BOOKS BY MARJORIE HARRIS

Sciencescape, with Dr. David Suzuki

The Canadian Gardener: A Guide to Gardening in Canada

Ecological Gardening: Your Path to a Healthy Garden

Better House and Planet: Ecological Household Hints

The Canadian Gardener's Year

The Canadian Gardener's Guide to Foliage and Garden Design

Marjorie Harris' Favorite Garden Tips

Marjorie Harris' Favorite Shade Plants

Marjorie Harris' Favorite Perennials

Marjorie Harris' Favorite Annuals

Marjorie Harris' Favorite Flowering Shrubs

In the Garden: Thoughts on Changing Seasons

The Healing Garden

Four-Season Gardening (formerly *Foliage and Garden Design*)

Pocket Gardening: A Guide to Gardening in Impossible Places

Seasons of My Garden

HarperCollins books may be purchased for educational, business, or sales promotional use. For information please write: Special Markets Department, HarperCollins Publishers Inc., 10 East 53rd Street, New York, NY 10022.

FIRST EDITION

DESIGNED BY RENATO STANISIC

Printed on acid-free paper

Library of Congress Cataloging-in-Publication Data

Harris, Marjorie, 1937–
Botanica North America: The Illustrated Guide to Our Native Plants, Their Botany, History, and the Way They Have Shaped Our World / Marjorie Harris.—1st ed.
p.   cm.
Includes bibliographical references (p.).
ISBN 0-06-270231-9 (hardcover)
1. Botany—United States.   2. Botany—Canada.   I. Title.

QK115 .H365   2003
581.97—dc21
2002068720

03   04   05   06   07   WB/IM   10   9   8   7   6   5   4   3   2   1

In memory of Charlotte Erichsen-Brown
&
To Jack, for the endless love

Contents

Acknowledgments

This book was not my idea. It was dreamed of by Larry Ashmead of HarperCollins and Bruce Westwood (who became my agent). They were inspired by Richard Mabey's *Flora Britannica* and felt that North American plants deserved a similar treatment. They were kind enough to ask me to do the book. Larry Ashmead kept up the encouragement through many false starts and in the end handed me over to the wonderful Krista Stroever to edit and see this book through every one of its many stages.

I have written 13 garden books and developed a keen interest in native plants over the years, so it was with great pleasure I took it on. I had no idea what I was getting into or how long it would last (five years). A team eventually formed around me and they are now among my friends. My gratitude and thanks are boundless to: Linda Read, who spent two years gathering information and devised the plant list; Philippa Campsie, who drafted and revised much of the book, edited all of the book, and kept us going with her organization; Karen York, who drafted, researched every chapter, worked on the glossary, and then did all the botanical editing and checking; Ilone Eurchuk, who kept us all in line, started out researching and ended up with major work on the photo editing and just about every other part of the book; Tara Baxendale, who drafted, researched right up to the last minute, checked all citations, permissions, the bibliography, and the endnotes; Erik de Vries and Charlotte Masemann, who researched, copyedited, and checked citations; Juliet Mannock, who winkled out all the obscure stuff in dusty places and worked on the glossary, reread the whole manuscript and made valuable comments; Laura McCrae, who helped with citations; John Peat, who helped with California; Murray Aspden, who found people and anecdotes; Tess Fleming, who checked citations; Lorraine Johnson, who edited the first draft of the Pacific Northwest and the Prairie chapters; Wendy Greyling, who worked on the database and contacts; and Kevin Kavanagh of the World Wildlife Fund Canada, who graciously agreed to read more than half the manuscript and made invaluable comments, all of which we incorporated into the book.

I am grateful to those who kept me going: Michael Totzke and Lynda Reeves at *Gardening Life* were infinitely patient about my obsession with the book; Alix Davidson and Stephen McClare kept the garden in good shape and left me something wonderful to go back to. And, of course, my whole adorable family, especially Jack.

My deepest thanks to the people who spent time talking to us: Charlotte Erichsen-Brown for giving me the first bit of encouragement; Dr. Nancy J. Turner and Richard Hebda for interviews and the amazing work they've done in the area of ethnobotany; Arthur Kruckberg, who took time out when he was writing his own book; Neil Diboll for anecdotes and a ton of information; the wonderful people at the University of Arizona: Bob Baker, Francine Correll, Angela Powers, Jack Hill; Cecily Gill of the Tucson Botanical Gardens; Dr. Daniel Moerman, whose amazing book was an unfailing source of richness for us; Bill Terry and Kye Goodwin, who walked in the old-growth forest with me; Bob Stadyk and Julie Hrapko, who talked to Murray Aspden; Tom Thomson, who loaned me my first copy of Charlotte Erichsen-Brown's book; and Henry Kock, Ken Parker, Julie Cruickshanks, John Browne Jr., Judith

Gilley, Rolf W. Mathewes, Jim Brockmeyer of Bluestem Nursery, Andie MacKinnon, Audrey Grescoe, Bob Rodgers, E. C. Pielou, Linda Gilkeson, Dan Heims, Carolene Huesgen, Simon Watkins, Terry J. Axline, and Paul Martin Brown.

Then there were all those who not only loaned books but gave advice: Robert Fulford, Ramsay Cook, Margaret Atwood, Graeme Gibson, Ian Montagnes, Elizabeth Wilson, Katherine Ashenburg, Stan Alpern, John D. Ambrose, Phyllis Grosskurth, Diana Beresford-Kreuger, Julie Cruickshanks, Robert Deutsch, Miriam Goldberger, Paul Jenkins (who installed the database), Avrom Isaacs, Allen Mandell, April Pettinger, Carole Williams, Bret Rappaport (of The Wild Ones), Lorrey Otto, Betty Spar, Sally Wasowski, Bruce Litteljohn, Eva and Erwin Diener, Karin Shakery, Phyllis Gustafson, and Jan Watkins, who read some of the copy and was a constant cheerleader, as were Gwen Wilkes, Geraldine Sherman, and Lynda Hurst.

We contacted hundreds of people for information about their climate and plants. These are the ones who got back to us with more information, contacts, and anecdotes, and we thank those who filled in the questionnaire and didn't sign it: Richard Aaron, Acorus Restoration, Judit Adams, Penny J. Aguirre, Alberta Native Plant Council (ANPC), John D. Ambrose, American Chestnut Foundation, American Forests, Andrew Anderson, Katy and Ian Anderson, Owen Anderson, Ted Anderson, Nancy Archibald, Arizona Native Plant Society, Atlanta History Center, Tony Avent, Dr. Michael Baad, Julia Badenhope, Dr. Rene Balland, Susan Barry, Bob Bartolemy, Mark Basinger, Jim Bauml, Bayard Cutting Arboretum, Katharine Bayley, Dr. Alwynne Beaudoin, Arthur Beauregard, Christopher Beddows, Stephen M. Beckstrom-Sternberg, Dr. Ritchie Bell, Pat Bender, Berheim Arboretum and Research Forest, Bob Betz, Bickelhaupt Arboretum, Brian Bixley, Edward Blake Jr. Blue Haven Nursery, Georgia Boss, Bowood Farms, Greg

Bradenburg, Les Brake, Angela Brown, David E. Brown, John Browne Jr., Henry (Hank) L. Bruno, Robert F. Brzuszek, Dr. Ed Buckner, Naud Burnett, Frank Cabot, Gene Carr, California Native Plant Society, California Polytechnic State University, Dugald Cameron, Kathrynn Campbell, Sherwin Carlquist, Fred Case, Brenda Castango, Curtis Clark, Theodore S. Cochrane, Ken Colem, Wayne Cowles, Paul Cox, Dr. Frank Crosswhite and Carol Crosswhite, Ericka Crowell, Frank Cabot, Frank Cubot, David Culp, Roy Davidson, Bill Davit, Christie Devai, Diane Dietrich, David Drakeforde, Ken Duryer, Linda Conway Duever, Dyck Arboretum of the Plains, Margaret Eaton, Victoria Eaton, Tessa Edward, Harry Elkins, Suzanne Grew Ellis, Jamie Ellison, Larry Evers, Blanche Farley, Al and Pat Fedkenheurer, William Feldman, John Ferry, Gerald Fichtemann; Flagstaff Arboretum; Jerry John Flintoff, Folley (or Faber, in Noble, Oklahoma), BicKendrick Ford, Donna I. Ford-Werntz, Jennifer Forman, Alan Foster, Jim Fox, Charles Frasier, Colleen Gallagher, Judith Gallagher, Mary E. Gartshore and Peter J. Carson, Garvan Woodland Gardens, Heather Gerling, Barbara Glass, Don Gordon, Joyce Gould, William Grafton, Frances Kelso Graham, Greg Graham, Greg Grant, Green Gulch, Corrie Griffith, Jim Griffith, Grims Nut Nursery, Phyllis Gustafson, John Harmon, Pat Harnisch, Bonnie Harper-Lore, Jean Hartfield, Kim Hawkes, Don Heims, John Helder, Kor Hellun, Horst Hickert, Dr. Michael Hickman, Tom Hill, Dick Hillson, Wendy Hodgson, Mel Hoff, Sean Hogan, Brian Holley, Janna Homis, David Horn, Dr. Fred Hrusa, Cyril Hume, Indian River Land Trust, Iowa's Integrated Roadside Vegetation Management Program, Mary F. Irish, Elizabeth Irving, Tom Isern, Don Jacobs, Robert C. Jennings, Angelo Joaquin Jr., Baylor Johnson, Floyd Swink, Brett Johnson, Fred Johnson, Matt Johnson, Maxine Johnston, Judith Jones, Loring Jones, Gene Joseph, Helen Juhola, Bruce Keith,

Frank Kershaw, Linda Kershaw, Don and Linda Kirby, Boyd Kline, Theodore Kline, Lee Knopp, Elke Knechtel, Lesley Kreuger, Ron Lance, Dr. Harold Laughlin, Joani Lawarre and Dennis Niemeyer, John N. Lawyer, Glen Lee, Ken Lertzman, Coleman Leuthy, Bruce Litteljohn, Dawn Loewen, Jack Lord, Louise Luckenbill, Shirley Lusk, Stephen Lye, Marianne Lynch, Dan MacIsaac, Terry Maguire, Pat Malcolm, Joan Maloof, George Mannoe, Carrina Maslovat, Lee May, Ruth May, Mayo Creek Gardens, Bob McCartney, Ruth McConnell, Dr. Sidney McDaniel, Mark McDonnell, the late Paul McGaw, Donna McGlone, Bill McKibben, Bill McLaughlin, Dr. Don McLeod, Rachael McLeod, Leslie Mehrhoff, Marg Meikle, Ethan Meleg, Dr. Larry Mellichamp, Metro Parks Tacoma, Sue Michalsky, Gavin Miller, Michael Miller, Michael Moore, Dr. Susy Morales, John Morgan, Nancy Morin, Mount Cuba Centre for the Study of Piedmont Flora, Dr. Brian Mulleur, Museum of Arizona (Flagstaff), Binty and Peter Mustard, Gary Paul Nabhan, Natural Connections (Manitoba Naturalist Society), The Natural Garden, New York State Office of Parks, Recreation, and Historic Preservation (Long Island Region, Belmont State Park), Judy Newton, Wilf Nicholls, North Carolina Botanical Gardens, Northern Illinois Botanical Society, David Nowak, Cornelia Oberlander, Sharon S. Odegar, Scott Ogden, Carl Ohlsen, Bob O'Keenon, Lowell Orbison, Oliver/Osoyoos Naturalist Club, Stephen Oliveri, Ontario Herbalist Association, Karen O'Reilly, Bob Osborne, Otter Valley Native Plants, Flo Oxley, Steve Packard and Laura Balaban, Nancy Padberg, Chuck Pardimon, Dr. Jerry Parsons, Milo Payne, Peach Tree Garden Club, Tracy Penner, Yvonne Pevie, Marcus Phillips, Dr. Betsy Pierson, Pierson's Nurseries, Rita Pilgrim, Sylvia Pincott, Pinetree State Arboretum, Pinetree Garden Seeds, Dr. Dan Pitillo, Jackie Poole, Jim Porter, Richard and Barbara Porter, Doug and

Shirley Post, Verna Pratt, Diane Prorak, Milo Pyne, Nancy Quinlan, Lonny Radkin, Elfi Rahr, Dr. Gail Rankin, Ron Ratko, Steven Raven, Jim Reveal, Tony Reznicek, Helen Riggin, Richard Rindone, Roaring Brook Nurseries, Wayne Roderick, Heather and Paul Ross, Henry Ross, Lenore Ross, Gerrie C. Rousseau, J. Reed Russell, Parker Sanderson, Edward Sampson, Elizabeth Schwartz, Karin Shakery, Joanne M. Sharpe, Shaw Arboretum, Richard Shaw, Flora Skelly, Theresa Slaman, George Smith, Hugh Smith, Ruth Smith, Stephen Smith, Society for Ecological Restoration, Terry Spurgeon, Jane Stackhouse, Dr. Geoffrey Stanford, Greg Starr, State of Connecticut Department of Environmental Protection (Dinosaur State Park), Elaine Stevens, Anita Stewart, Bob and Virginia Stewart, M. Stewart, Scotty Stewart, Streamside Native Plants, Donny Strickland, Pat Taylor, Debra Teachout-Teashon, Dana Tenney, Gary Thaler, Mary Thomas, Dr. John W. Thomson, Sharon Thorne, Ron Tiessen, David Tomlinson, University of California at Davis, University of Santa Cruz (Department of Environmental Studies), John Valleau, Miriam Kritzer Van Zant, Dr. Dale Vitt, Gerry Waldron, Sally and Tim Walker, Cliff Wallis, Graham Ware, Sally Wasowski, Geraldine Watson, Linda Watson, John Weagle, Allyn Weaks, Susun Weed, John Wellwood, Michael White, Bill Whitehead and Timothy Findley, Megan Whittingham and David Malcolm, Ron Wieland, Larry Wild, Mark Wilkins, Kathleen Wilkinson, Carole Williams, Marc Willoughby, Debra Wilson, Carol Witham, Alan Woodliff, Mrs. Ingeborg Woodsworth, Jim Wye, Gustave J. Yaki, Karl Yakokor, Judy A. Yates, David Yetman, Terry Yockey, Dick Young and Wayne Lampa, Bob Zahner, and Maclovia Zamora.

Librarians without whom we would have perished: Linda Brownlee, Liz Derbecker, Luba Frastacky, Karen Graves, Richard Landon, Leslie McGrath, and Martha Scott.

PHOTOGRAPHY CREDITS

Norman Barichello viii, 535, 540; Tony Beck 76, 108, 117, 123, 171, 177, 185, 187, 191, 202, 215, 216, 219, 242, 245, 248, 263, 339, 362, 440, 466, 536, 549, 550; Paul Martin Brown 239; Barbara Collins 372, 420, 421; Janet Davis 9, 17, 20, 27, 37, 42, 43, 48, 55, 56, 64, 88, 90, 119, 125, 140, 165, 166, 175, 183, 188, 256, 303, 311, 349, 352, 353, 361, 374, 509, 573; Leslie Degner ii, 115, 180, 253, 272, 273, 330, 331, 356, 436, 457, 463, 485, 492, 502, 515, 529, 532, 533; Mark Degner 190, 261, 267, 276, 328, 443, 451, 521, 528, 538, 539, 547, 557; Ken Druse 45, 66, 70, 291, 407, 433; Derek Fell xxii, 46, 150, 155, 390, 404, 432; John Glover viii, 28, 29, 34, 63, 95, 112, 128, 161, 250, 319, 397, 402, 439, 445, 506; Pamela Harper 53, 135, 143, 167, 172, 209, 367, 383, 427, 446, 461, 479, 481, 499, 508, 523, 562; Jane Huber 425; Rosemary Ann Kautzky 210; Glen Lee 113, 186, 324; Andrew Leyerle 71, 83, 89, 100, 102, 107, 120, 127, 163, 192, 260, 280, 285, 286, 293, 297, 300, 305, 308, 309, 313, 314, 317, 321, 326, 333, 465; Bruce Litteljohn xiv, 7, 38, 73, 85, 103, 111; Janet Loughrey 50, 52, 57, 145, 194, 301, 316, 467, 498, 510, 513; Allen Mandell 470; Marilynn McAra 174, 524, 542, 545; Scott Millard 336, 377, 379, 385; Hugh Nourse 61, 79, 93, 197, 320, 370, 392; Cheryl Richter 289, 345, 394, 417, 459; Royal Botanical Gardens 258; Eleanor Saulys 99; Patricia Taylor 137, 544; Mark Turner 15, 77, 146, 169, 200, 204, 469, 474, 489, 496, 501; John Valleau 86, 122, 288, 306, 386; Paddy Wales 68, 505, 516, 518, 519; Andy Wasowski 369, 388, 430; Dudley Witney 11, 131, 139, 149, 152, 196, 208, 220, 224, 227, 229, 232, 233, 235, 237, 241

Introduction

Plants give us a sense of territory. We like to make our mark on them whether it's carving an initial on the trunk of a tree or moving them with us as we've wandered over the planet. We've done this for the past 40,000 years with little thought except for our own needs and pleasure. The question in our time has become: is this transporting of plants willy-nilly a good thing to be doing at all? And should we not know more about plants in their ecosystems and how they ultimately affect our lives? These questions will become increasingly important for the survival of our very much stressed-out planet. But the most important question is, why should we care about native plants at all?

A native plant, in this book, is a plant that can be documented to have been in North America before European contact—about A.D. 1450. These are portraits of the plants that evolved along with the animals of the continent and eventually with the Native Americans who moved among them and modified them for thousands of years. These are the plants the first Europeans would have seen, and the very same ones which gave them economic success beyond their wildest dreams and changed the history of the world.

But they have another, infinitely touching, quality. They managed to survive, in many cases, millennia of catastrophic forces, glaciation, and dramatic changes in the land around them. All the while they developed internal systems and relationships that made them among the hardiest plants on earth. But they couldn't take the onslaught of the invading aliens of the past 400 years. Nonnative plants may have been blown over the oceans or dropped by birds, but the most ruthless of them came inadvertently with Europeans—hanging on to pant cuffs or the bottom of a shoe, or even escaping from their gardens. We cannot tell how much these exotic aliens changed the virgin soil of this amazing landscape as settlers traipsed across the continent, but we do know they flourished so rapidly that many age-old systems were ravaged. These new plants could whip through a habitat and alter it within a generation, elbowing aside insects, changing soil structure and water tables.

Through the pollen records of such invaders as ragweed, bluegrass, and white clover, scientists now can spot the precise date when newcomers arrived in any given place. We can see how native plants were clobbered by the new plants and then finished off when annual grasses and crops took hold. Native plants had no time to adapt to many of the new animals such as cattle and horses (which hadn't been on the continent in thousands of years). We know many plants have been lost to us. We can tell by the faint, almost ghostly traces they left behind as they lurk in ragged patches.

No plant lives in isolation. There are always other living creatures or plants nearby that are companions and there is always a reason for them to be together. Native plants evolved symbiotic relationships with insects, bacteria, and other minute creatures living in the soil. These, in turn, keep trees, other plants and animals—and eventually people—healthy. This fragile web of life took millions of years to develop. When people disparage "tree huggers" because they want to save the trees for a small bird or even large butterfly, it's not some wacky spaced-out concept. Every organism depends on other organisms for survival. For example, without

lupines there would be no Karner blue butterflies, and without them to pollinate the lupine, the plants would become inbred and eventually die out. You can't save a plant or an animal in isolation, you have to save the whole habitat.

One of the major ecological disasters of our time is the shrinking of natural habitats and with that comes a reduction in biodiversity. Biodiversity, or biological diversity, is the rich mix of flora, fauna, and their interactions within an ecosystem. Major ecological disasters (burning of the forests in Indonesia, chopping down the rain forest) are felt around the world. We also know we must preserve genetic diversity (seed banks are just one example) for future generations.

We need to learn as much about these native plants as possible before it is too late. If we begin to understand how everything is connected, from the microscopic bacteria that live and farm in the soil to the giant redwood trees which are an ecosystem unto themselves, then we might become a little smarter in how we treat plants. Remove any one element and the whole system starts to falter. Oh, not immediately, but slowly, painfully over years. That doesn't mean we can't work with nature or natural systems or even use them to our own advantage. Not at all—but first we must understand them before we do drastic things to them.

Our problem has always been the lack of ability to see these systems as a whole. We compartmentalize because it is easy and convenient economically. The interconnectedness of all things on this planet seems to be too large for most human brains to absorb. For instance, salmon has always been out in the ocean, free for the taking. We know the collapse of the salmon stocks is an economic disaster. But the impact stretches much further. Now Dr. Tom Reimchen, a biologist with University of Victoria, has found the link between the Pacific coastal forest and salmon. The tiny salmon head out to the open ocean where, as they mature, they develop high levels of the heavy isotope of nitrogen, N-15, traceable through DNA. The fish come back to the estuaries to spawn. Bears catch them (eating up to 700 in a season, a major source of protein). First they eat the brains and then drag the rest of the fish into the forest. What's left of the fish carcasses attracts dozens of species of birds, and millions of microbes that eventually break it down. What Dr. Reimchen found to his astonishment is that this whole process provides the forest with 40 percent of its total nitrogen needs, and in cases of a particularly large salmon run, up to a staggering 85 percent. "The open ocean and insects at the top of those trees, the animals, the birds . . . are linked in a series of incredibly beautiful sets of interactions in which there is no separation at all."[1] But with no spawning fish, what will happen?

It's not news that we live within a closed system where a catastrophe, either natural or man-made, in one part of the globe will affect all other parts. But we are now finding, to our dismay, tropical insecticides in trees near the Arctic; that even underarm deodorant spray from the south makes its way into core samples of arctic ice. Everything we do has a global impact. What most of us worry about is not change—nature has always changed. Rather it is the speed of change caused by human activity that is terrifying. We struggle to understand the present and yet we are still learning about this continent's past.

North America hosted some of the world's largest cities hundreds of years ago, all gone now and most, we think, disappeared through ecological degradation. Many great cultures arose and then fell long before white man's diseases destroyed almost 90 percent of the Native Americans. There were the Cahokia with their huge mounds (we are still not sure of their function); the Hohokam with large cities and a desert culture that included farming; the great cedar culture of the Pacific Northwest.

When people first entered North America following the big game, it's estimated that it took them 1,000 years to spread throughout North and South America, and only about 100 years to learn the plants. They used every known medicinal plant and passed on their information from generation to generation. They moved slowly in small groups, breaking up and going several ways when the groups got too large—too many people meant starvation, and too many people meant overtaxing resources, as those lost civilizations warn us. For the most part, however, these people lived within the systems of nature they found. Native Americans say you must look at a plant for seven generations before you can understand or use it. They observed animal behavior and chose the plants animals chose. If it was bitter, it usually became a medicine, if sweet, a food.

The European newcomers were following big game of another kind as they entered the continent. We must remind ourselves that what they saw was a bounty unknown to them before: fish knee-deep in the rivers at spawning time; birds by the thousands that could be slaughtered every week and would still come on; wild animals all through the woods needing only to be hunted. And the woods themselves—complex places—glorious as well as frightening.

The reaction to the new was to make it familiar: bring the plants and animals (which happened after 1620) that would make this foreign land look like home. The unfamiliar was discarded or ignored. No one was thinking ecologically in those days. A big tree was there to cut down and burn, make a cabin (and keep it warm), or ship home. Quite often it was just an inconvenience. If we have not been observant of our landscape and its fragility, the habit has been part of our heritage for 500 years.

"What was merchantable in New England," writes William Cronon, "was what was scarce in England. . . . Only if this was true would it make sense to pay the cost of transporting it across the ocean. . . . Seeing landscape in terms of commodities meant something else as well: it treated members of an ecosystem as isolated and extractable units."[2] In the 1500s and 1600s there was little widespread knowledge of how plants worked.

We can't blame them. They had come with two visions in their heads: religious and economic freedoms. A way to live where hard work would always be rewarded and all men were equal—except for those pesky Indians living in the woods. And though many people today spend their lives trying to understand plants, generally we don't know a lot more about them than our forefathers did.

We have allowed the clear-cutting of old-growth forests to go on unabated for over a hundred years. We replant trees in tidy rows but we aren't replacing forests. It might take from 45 to 70 years to grow commercial, marketable wood which is not nearly as good as the old-growth (hence the hunt for old-growth wood). Tree planting doesn't grow a forest, it's farming. To make a forest would require at least 400 to 500 years just to let the trees bulk out. Another 500 years to let it find its various levels of succession. With concentrated effort during all those years, we could study the forest intensely and perhaps we'd understand what's going on. For all our technology, nature continues to be elusive.

Though we have a long tradition of revering wilderness and we love the *idea* of wilderness, we have lost our sense of living in nature. We still live to exploit it, to find new resources within it, to make money from it. We are perfectly capable of setting aside a nature reserve and then running a logging, mining or gas road through it. We don't take nature very seriously and, it seems, we don't understand what these actions do to it. We save little patches of natural habitat and expect it to function as usual. We never really seem to feel the weight or power of our incursions into nature.

We have to change to ensure the survival of biodiversity, which is to the natural system what a healthy body is to a human. We must be more sensitive to our plants, especially within their natural habitats. It's very difficult to transplant native plants because they are so completely tied into their ecological systems. Move them a few feet and they might not grow; move them a few miles and they may never recover.

The point of this book is to honor some of the thousands of native plants left in North America. I am a gardener and I use native plants in my garden, but I became aware of the plight of native plants many years ago when a local ecological group took over a vacant city lot. They installed a native garden—no petunias, no impatiens. The neighbors hated it with a passion. "A garden of weeds," they scoffed. "Ruin the real estate values on the street," others sneered. I watched with fascination over the space of two years as the plants grew and showed their glories in different seasons. Epithets such as "waste site," "dump," and "weeds" were forgotten. People went out of their way to go through the little park, to listen to insects, watch butterflies, and feel refreshed. It was a beginning.

When I take my grandchildren to a nearby public park and walk through the "untended" sections, where no herbicides are used, we watch with fascination what's going on around us. Then over to the grass of the soccer pitch. "What's the difference?" I ask. "No noise." Absolutely right: no insects, just green cement. Okay for soccer, but not a place to live in if you're a butterfly or a bird. I like to think of this as a small lesson in the difference biodiversity makes. The kids seem to pick it up pretty quickly. They've watched the plants change from year to year, seen bugs, especially caterpillars, plump up. Their ambition is to watch a monarch butterfly emerge someday. And they just might have that opportunity if we leave this little bit of land alone.

And that's what we have left of our native plants: little patches. If the point of this book is to honor native plants, it is also to honor their history, the secrets they shared, the role they have played and must continue to play in the survival of our species. E. O. Wilson says in *The Future of Life* that "It is exquisitely human to search for wholeness and richness of experience."[3] What we find in nature is precisely that richness. Who wants to live in a world with 10 plants? We want to be surprised and amazed, and that's precisely what native plants still do for us. We are the stewards of this land and if we do our best not just to save a plant here or there from the juggernaut of development, but to save as much of their habitat as possible, then we will be giving these plants the respect they deserve.

Apologia

Our choice of plants was confined to those used by people and those crucial to a specific plant community. We apologize if some favorites are left out or some important ones missed. We wanted the plants with a history to show not just their botanical complexity but their personal history and ours as well. We do vouch that the information is as accurate as we could possibly make it given the contradictory nature of botany

and taxonomy—and the constant variation of the plants themselves. They are endlessly adapting, changing, and out there confounding us at every turn.

Organization of This Book

We divided North America (the continental United States and Canada) into 10 major plant communities. This requires a leap of imagination because almost every square mile of this complicated continent has a great many plant communities that will change with the height above sea level, the terrain, and exposure to light. We've looked for the common native plants that have been useful to people or are critical to a given biotic system. This means ignoring what was brought in (by accident or design), or blown in by the wind, in the past few hundred years.

All sections (chapters) have the important plants arranged historically, ecologically, or economically to start off with. They are then succeeded by plants gathered together in their botanical families. In most cases they go from the larger to the smaller, but there are, like nature itself, exceptions. Each plant entry has information about its botanical structure, its ethnobotany, and natural history, as well as its usefulness today.

PLANT REGIONS OF NORTH AMERICA

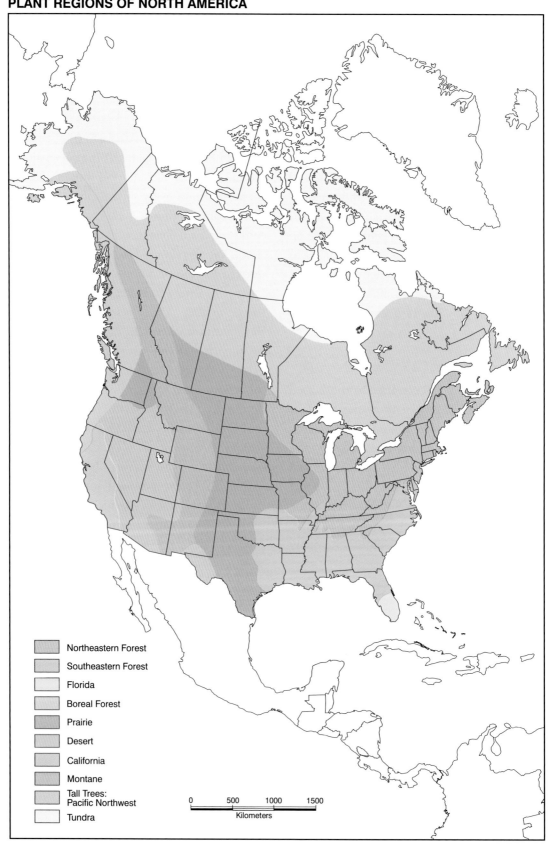

Northeastern Forest

Southeastern Forest

Florida

Boreal Forest

Prairie

Desert

California

Montane

Tall Trees:
Pacific Northwest

Tundra

0   500   1000   1500
Kilometers

BOTANICA
NORTH AMERICA

The Eastern Forests

When naturalists look at a field or a forest, they see trees, insects, and flowers, hear sound, and sense energy—a system at work, a habitat. When developers look at a field or a forest, they see potential commercial value—a place to generate wealth. When the Pilgrims stepped on shore more than 400 years ago, they came, not just as religious refugees seeking freedom from persecution, but as developers. They believed themselves appointed by God to possess this new place and make money from it.

What confronted these new arrivals was

A stream tumbles through an open glade in the Eastern forest.

astonishing. Animals and fish in abundance. Trees of a size that hadn't been seen in Europe for hundreds of years. But it wasn't a giant wall of green or a dark and gloomy forest that greeted them. It was something like a garden: wide-canopied trees, backed by huge evergreens, light shifting from dappled brilliance to deep shade, meadows filled with shrubs covered with berries in autumn, and a profusion of flowering shrubs and perennials—most of them unfamiliar to the European eye. No wonder they referred to the land as the New Eden.

It was not just God who had wrought this beauty. This was a forest that, in places, had been managed for almost 12,000 years by the people who lived within it and along its edges. It seemed to Europeans to be empty, though of course they knew it was populated—they could hardly miss seeing the villages, fishing traps, and farms. But since the Pilgrims considered the indigenous peoples to be "savages," they did not see them as a barrier to their God-given mission to exploit the land.

The Indians had controlled the size and extent of the forest by burning the underbrush at regular intervals. This left the almost parklike setting referred to so often by the Europeans. When Giovanni da Verrazzano visited New England in 1524, he described open areas and forests that could be traversed "even by a large army."[1] William Wood, 100 years later, observed "in many places, divers Acres being cleare, so that one may ride a hunting in most places of the land, if he will venture himselfe for being lost."[2]

Selective burning, which had been going on for thousands of years, kept the Eastern forest growing as a mosaic: it opened up areas for crops, and created edges that drew game attracted by the berry-laden shrubs which inevitably sprang up after a fire. Strawberries, blackberries, raspberries, and many other food plants grew along these edges. Fire recycled elements back to the forest floor, allowing grasses and shrubs to grow luxuriantly. Plant diseases were kept under control, as were a lot of other pests such as fleas.

The native peoples had farms or, more accurately, summer camps where they planted corn, squash, and other vegetables. Their winter camps were closer to hunting grounds and sources of fuel wood. When a field became infertile, they moved on and left it fallow with the intention of returning several years down the road.

Just as the Europeans had not seen a forest like this before, so they had not seen a civilization like this before and dismissed it as unimportant. Indians kept no animals; their only farm technology seemed to be hoes made with clamshells; and the women apparently did most of the work. This is mentioned again and again. The men had only to hunt (considered by the newcomers as a sport, not as work). The women tilled the fields, provided food and clothing, and looked after children. Although the native peoples knew and kept to their own boundaries, these boundaries were invisible to the Europeans, who allowed their cattle to run loose in the forest, setting in motion a cycle of destruction.

What happened over the next 200 years is a tragedy that still makes the heart cry out. Natives died by the hundreds of thousands because of diseases they had never been exposed to before; villages were filled with so many bodies that no one was left to bury them. Entire forests were cut down and shipped back to Great Britain, leaving behind an ecological disaster in the New World. From Newfoundland to the tip of Florida, inland to the Mississippi River, the mighty forest fell to the settlers' axes. And when these industrious people finally looked up, what was once considered virgin forest was gone.

This East Coast forest reveals the history of North America in a fascinating way partly because it was exploited so ruthlessly. The Pilgrims were very good at keeping accounts, making laws, and conducting trade. If they weren't very observant about the plants that surrounded them, they could certainly quote what price that stand of trees over there would fetch on the market. They were serious, hardworking people who came to the unknown with no wilderness skills and found themselves in a terrifying forest peopled with strange beasts and exotic-looking natives. They were driven partly by fear and partly by a strong ambition to succeed. They found themselves in an ecology so complex as to be incomprehensible to them, even as it is to us today. They saw their job as subduing nature, not celebrating it.

The last ice age left a varied and nutrient-rich soil that supported a hugely diverse range of plants and animals. Thousands of years later when the settlers let their own animals loose in the forest, the trampling destroyed a delicate relationship between trees and the plants of the understory that had been built up over millennia. With their great weight, cattle compacted the soil, depriving it of oxygen and giving European plants more chance to supersede the oxygen-needy native species. Trees were hacked down not just for wood to burn but to create pastures. Plowing prevented native seeds from germinating and each action led to erosion that became more extreme as the forest was felled and watercourses were diverted or destroyed. The settlers brought invasive weeds with them: dandelions, chickweeds, bloodworts, mulleins, mallows, nightshades, and stinging nettles. So readily did plantain follow in their tracks it was called "Englishman's foot."

The native peoples were aware of the biological invasion going on around them, but they were so busy fighting for their own survival, they were unable to stop it, let alone curb the worst practices. Settlers did everything the natives did, but did it more intensely (cropping, burning, hunting) and they changed the forest irrevocably in a very short time. Today the Eastern forest is almost entirely second-, third-, fourth-, or later-generation growth.[3]

A few settlers became concerned about the rate of destruction, and the first conservation laws were passed in 1640. By 1875, the ravaged forests and a few hundred years of wholesale clear-cutting prompted citizens to create non-profit conservation organizations. New groups sprang up to protect specific landscapes or even individual trees.

And the battle goes on. The eastern United States contains more than 47 million acres of national forest, but more timber was sold from these areas in 1999 than from any other forest in the country. Massive industrial logging means that 1.2 million acres of land are cleared every year to keep about 140 chip mills running.

Clear-cutting has led to invasions by insects, fungi, and pathogens, because trees have been weakened and stressed. Pollution and erosion have wreaked havoc among trees endangered by suburban sprawl, and now off-road vehicles go where only hikers once strode.

How does nature reward this incredibly destructive behavior? Left alone, the land will always astonish us: nature revives, it regenerates—even where the soil is poor, growth comes back thick and lush in less than a century. Despite our destructive ways, the Eastern forest remains one of the most glorious, transfixing forests in the world.

In this chapter, we've divided this complex forest into three distinct parts: the Northern hardwood forest, the understory, and the Southern hardwood forest. (Swamp forests are treated in the next chapter.)

NORTHERN HARDWOOD FOREST

The Northern hardwood forest is a mix of deciduous and evergreen trees spreading west in a band 100 to 500 miles wide. The varied canopy covers a richly diverse understory. This forest includes several distinct ecosystems, including much of the Appalachian forests and the Pine Barrens of New Jersey. In this ecoregion, yellow birch and sugar maple dominate a forest made up of their own saplings, with ferns and mosses beneath. The higher the elevation, the more closely the vegetation resembles the boreal forest that once covered the area, when glacial ice sheets came as far south as New Jersey.

The richness of this forest is most obvious in autumn. From a bird's-eye view, it is breathtaking—a forest so widespread and dense there seems to be no possibility it could have any problems. It has a longer growing season than the boreal, and though subject to strong winds and powerful cold fronts, it receives frequent soakings with rain and heavy dew. The soil is rich and deep, covering a wide variety of rock formations; but there is also alluvial soil with stone and sand, small lakes, bogs, marshes, and swamps. Now, only a few isolated specimens of the original trees are left, such as a white willow in New Hudson, Michigan, and, until recently, the famous Wye Oak in Wye Mills State Park, Maryland, which fell over in a storm in 2002. They are but symbols of what was once the true prowess of these trees.[4]

Pinaceae

Eastern White Pine

Pinus strobus, Northern white pine, Canadian pine, hard pine, Weymouth pine.

Pinus strobus

grows to 50 to 80 feet and, in favorable conditions, to 150 feet or more. It has a short taproot, and the roots spread outward. The bark is grayish-green when young, dark grayish-brown on older specimens. The bluish-green needles come in bundles of five, each 3 to 5 inches long. The cones hang down and grow up to 8 inches long; they mature in the second year. Found in fertile, moist, well-drained soils.

**   *Range:* from Newfoundland to Manitoba, south to the Appalachian Mountains and northern Georgia, west to Illinois and Iowa.**

**   Official tree of Maine, Michigan, and Ontario.**

Remember Bert Lahr as the Cowardly Lion in *The Wizard of Oz* singing "If I Were King of the Forest"? Sorry, Cowardly, but the forest already has a king, and it is the Eastern white pine.

When the Europeans arrived, what they saw before them represented a forest of ships' masts. The average height of those white pines was 150 feet, about the height of a 14-story building. Each stood perfectly straight, with 80 feet free of branches. The wood was white yet strong, light, and fine-grained. Its growth was swift—it could stretch by 16 inches a year—and it seemed in endless supply. This was the tree that would help create the greatest navy the world had ever seen.

The white pine's wind-battered profile is helped by large ascending limbs that give the tree a distinctive, almost feathery appearance. The cone shape of its youth changes in old age to a horizontal branching pattern. White pine is monoecious, which means both the male and female flowers are on the same plant, although they are found in separate parts of the tree. The female flowers start as small cones (or *strobili,* giving the tree its scientific name) at the ends of the main branches right at the top. When they are ready to be pollinated, they turn green. The yellow-green male seed cones ripen into a light brown several weeks before the females are ready. They release pollen into the wind in such enormous clouds of yellow dust that, as natural historian Donald Culross Peattie put it, "Great storms of pollen were swept from the primeval shores far out to sea and to the superstitious sailor it seemed to be 'raining brimstone' on the deck."[5]

Once the pollen is shed, the male cones drop off. The females stay on the trees for two years. Every five years or so, the females produce 400 cones and up to 18,000 seeds, spreading 700 feet in all directions. Most of the seed, however, ends up as food for wildlife—a red squirrel can strip a cone of its 45 seeds in a matter of minutes.

During the seedling stage when growth is slow, the white pine is vulnerable. If it's competing for sun, rain, and space with aspens, oaks, and maples, the pine will lose out. It does best among its own kind in pure stands.

Pines produce resins (water-insoluble terpene polymers) used to make tar, pitch, turpentine, and rosin. These compounds have been used for centuries for caulking and waterproofing the wood, rope, and canvas of ships (the reason why sailors have been referred to as "tars" since the 1600s). The first inhabitants of the Eastern forest, including the Algonquin, Iroquois, Ojibwe, and Mi'kmaq, used pine resin not only to seal the cracks of canoes and boats, but also as a medicine. They applied the antiseptic resin to wounds to help them heal, and boiled it up with water to make a tonic said to cure everything from coughs to smallpox. Pine needles, which are high in vitamins A and C, were used to make an anti-scurvy tea.

Native Americans also used wood from young trees for construction and in making canoes, but their tools of stone and shell were not heavy enough to cut down the really massive trees. When the European settlers came with their metal axes and saws, the Eastern forest had something like 3.4 billion cubic feet of *Pinus strobus*. Most of these big trees were 200 years old, but many were as much as 450 years old. The pines grew conveniently near waterways where the settlers erected their earliest large buildings—sawmills. Europe was pretty much logged out by this time and was almost panting with desperation for such wood. The English, constantly at war, had depended on the Swedes and Russians for masts which were often pieced together out of the biggest and best the Baltic forests had to offer. But nothing there could equal the Eastern white pine of North America.

The first shipment left Portsmouth, New Hampshire, in 1634, in a ship specially designed to hold 100 white pine masts laid out horizontally in the hold. From the 1630s, the center of the export lumber trade was in Maine and New Hampshire. By 1682, there were 24 sawmills operating in Maine alone.[6]

With masts from the New World fetching up to 100 pounds apiece, Americans grew rich and were able to bring slaves back to work on their estates and in their fine homes built of white pine. It was the perfect wood. It didn't warp, it was strong and easy to work, and took stains and finishes well. It became everything to the colonists: from their cabins to their coffins, from church pews to dining tables. They even put the white pine on their earliest coins. Only fur was the pine's rival in commercial greatness for the dollars that accrued to the colonies.

The North American pine became essential to prosperity in England as well. Therefore, the British Parliament under William III, outraged by settlers who would dare cut down large pine trees merely to clear space for a farm or to burn for warmth, passed laws in New Hampshire and Maine reserving trees for the Crown. All trees with a diameter of 24 inches, measured a foot from the ground, belonged to the king and a fine of 100 pounds awaited anyone with the temerity to take one illegally. Burning, which the Native Americans had done to manage the forest and clear away the undergrowth, was strictly forbidden.

As a result, the forest became darker and denser than it had been for thousands of years. Ancient trees more than 200 feet high were marked with a special blaze, known as the King's Broad Arrow—one of the king's trees. In Europe, the land and its woods had belonged to the monarchy and the elite. Here in North America, they were supposed to be free and available to all. The settlers found history repeating itself—the monarchy and elite were taking over once again.

Rebellious settlers, occasionally disguised as Indians, worked through the night to take down a tree, in spite of the possibility of being caught, flogged, and fined. Sometimes, to spite the British, they replicated the King's Broad Arrow on inferior trees. But spies were everywhere in the forest. If one of the woodsmen was nabbed, the punishment might even include burning down the sawmill that had processed the wood.

The struggle over the ownership of the trees was only part of the settlers' grievances leading up to the American Revolutionary War, but it was a significant part. The first flag of the Revolutionary forces used the white pine as an emblem. It enraged the Americans that the same British warships threatening their colonies were fitted with white pine masts from their own forests. In 1774, Congress stopped the export to Britain of the pines. After gaining their independence, Americans cut down trees at a furious rate, as though driven by the fear that someone else might get to them first. François André Michaux, who traveled through the States in the 1790s, observed the devastation and noted that "for a space of 600 miles from Philadelphia to a distance beyond Boston, I did not see a single stock of the white pine large enough for the mast of a vessel of 600 tons."[7]

The cutting was indiscriminate. Some get-rich-quick types cut blocks from standing trees and used them to make shingles, leaving the rest of the tree to die. By the end of the nineteenth century, once-vast stands had been pretty much logged out. Formerly virgin forest was now in its second and third growth, and the white pine had changed from the predominant species to just one species among many in the mixed forest.

White pine is still used in building and furniture making and even to make wooden matches. It is also used for landscaping and reforestation, especially in former mine areas, because it can grow in contaminated soils. Christmas tree farms sell young white pines. But few trees will ever again have the chance to reach the heights of their ancestors. Only a few giants remain. The current champion in the Eastern forest is a white pine in Marquette, Michigan, that is 201 feet tall with a 16-foot circumference.[8]

Today, a few woodworkers reclaim original white pine from old barns and houses and make it into flooring, paneling, furniture, and architectural details. Furniture maker Fred Savage of Hillsburgh, Ontario, who has been using reclaimed wood since the early 1990s, tries to recapture the "mood or spirit" found in antique furniture "by choosing boards that already have that look to them." He sometimes muses on the forests that were once home to this wood: "It must have been an awesome sight: hundreds of thousands of square kilometres of old-growth forest. These buildings [which yield the reclaimed wood] are the last remnants."[9]

Pinaceae

Pitch Pine
Pinus rigida

Pinus rigida
grows 30 to 60 feet high, depending on the site. It has slender green needles 4 inches long, in groups of three, and cones 3 inches long with rigid prickles. Tufts of needles may grow right out of the thick plates of the brown bark on the trunk. Pitch pine is a monoecious tree. The pistillate (female) green flowers, tinted rose, are borne on stalks 8 inches long high up in the tree. The staminate (male) flowers are yellowish, 1 inch long, and grow lower down. It can grow in almost sterile soil, in rocky or sandy locations.

***Range:* Northeastern and Mid-Atlantic states.**

Light was a luxury for the colonists in their small dark homes. They soon discovered, by observing how the Indians made fires, that the rich resins in pitch pine burned perfectly. Small faggots made from the heartwood and known as "splint lights" lit colonists' homes up and down the eastern coast for decades.[10]

Pinus rigida, pitch pine

The pitch pine is remarkable in other ways as well. For instance, the cones release their seeds at different times, depending on the tree. Sometimes the cones open immediately; in other cases, they remain closed on the ground for years, and open only after a fire. Cone behavior is thought to be inherited, so trees growing in areas that were routinely burned evolved a way of surviving fire and regenerating afterward (such trees are called serotinous). The Native Americans set fires around the base of the trees and planted crops among the standing dead trees. An area could be used for 10 years before

the soil lost its fertility and the people moved elsewhere.

Once a tree is on fire, the resin in the cone makes the scales adhere to one another and close, preserving the seed in good shape for many years. The pitch pine also has unusually thick bark and its shoots sprout quickly from the trunk after a burn. It can endure other forms of stress as well. Deer can browse on the lower branches and the tree will still produce a thick green canopy. Tolerant of salt spray, it grows equally well in sandy soil, salt marshes, or on sand dunes. Because it thrives in areas such as Cape Cod, it was probably one of the first trees settlers saw when they arrived.

In the early days of the colonies, when shipbuilders were major employers, calls went out for men who were skilled in the art of making pitch. To extract the tar, they built kilns that let in as little air as possible, then fed the kiln with the pitch pine wood. The surrounding ditch caught the ooze running out as the wood burned. This practice was much less ecologically sound than the traditional method—an Iroquois would expertly whack a channel in a tree and let the pitch drip out into a vessel. The tree might have been slightly damaged but it was not destroyed outright. The Iroquois used the pitch not only for fires and torches, but also as a treatment for burns and a poultice for boils. Fleas and bugs were a constant bother to them and they burned pine needles as an insecticide.

The colonists chopped pitch pines down so quickly that, in 1650, 30 years after the founding of Plymouth, a halt was called on cutting down trees within six miles of the Connecticut River in Massachusetts. By 1715, conservation measures were enacted to control the amount of pitch pine that could be cut down. Even so, the work carried on well into the eighteenth century. Eventually, the pitch-making industry moved south from New England. It became so important to North Carolina that it became known as the Tarheel State and its citizens "Tarheelers."[11]

The pitch pine can reach 100 feet and live up to 200 years with some individuals enduring for 350 years. There is one in New York state and another in Newberry, New Hampshire, more than 90 feet tall, but most grow to about 60 feet and add a foot a year until they are 50 to 60 years old. In spite of its adaptability and durability, pitch pine is now threatened by acid rain and air pollution, which inhibit its growth.

Pinaceae

Virginia Pine
Pinus virginiana, alligator pine, bastard pine, black pine, cedar pine, hickory pine, New Jersey pine, North Carolina pine, old-field pine, poor pine, poverty pine, river pine, scrub pine, short shucks, shortleaf pine, Virginia tall

Common names of plants are like a mini-history of each plant. Virginia pine's many names suggest not only its importance as a commercial lumber tree, but also how widespread it was. It was cut down to make mine props, railroad ties, rough lumber, fuel, tar, and charcoal. At the same time, it was considered such a weed that scrub pine was its epithet.

The tree's ability to survive in harsh environments makes it a good choice for reforestation projects; and its quick growth makes it a popular Christmas tree. The sapwood is yellowish, the heartwood is orange to brown, and both are light, soft, and coarse-grained. Most Virginia pines have many side branches, so the wood tends to be very knotty. Because of its elongated fibers it has become an important pulpwood species.

Pinus virginiana grows to 50 feet with a trunk 20 inches in diameter. It has an erect shape and a flattened crown. The paired needles are up to 2 inches long. When the tree is young, the bark is

quite smooth, then fissuring as it ages, turns into rough plates. The cones are 1½ to 2½ inches long, conical or oval, with red-brown scales. The tree seldom lives more than 150 years.

Range: **southeastern New York (Long Island) and New Jersey, west to Indiana, south to Alabama and Georgia.**

Ulmaceae

American Elm
Ulmus americana, white elm, water elm, soft elm

Ulmus americana
can grow up to 80 feet high. The trunk is tall and straight with rough bark; the branches emerge in a fountain shape; and the twigs seem to zigzag from bud to bud. The shallow root system spreads widely around the tree. Its distinctive leaves are oval, pointed at the ends, with notched edges and a lopsided (asymmetrical) base. Small reddish flowers appear in spring, followed by seeds that are individually enclosed in a flat, papery sac.

Range: **Cape Breton Island, Nova Scotia, west to Saskatchewan, south to Wyoming and central Texas, east to central Florida, and along the entire Gulf coast.**

State tree of Massachusetts and North Dakota.

The American elm was once a common sight throughout eastern North America. It was easy to recognize from a distance, because of its lovely upswept branches, which form a graceful Y shape. In forests, fields, and along urban streets, the elm grew quickly, tolerated moderate shade, and adapted to all kinds of environments.

Who could have known that it would prove so vulnerable?

Before the settlers arrived, the indigenous peoples found many uses for the waterproof elm bark. The Huron made it into containers for water or maple sap, while the Iroquois and Chippewa, who often lived in areas where birch was uncommon, used it to make canoes. They toppled trees by burning the base, and poured boiling water on them to loosen the bark, which was then stripped off in large sheets. The inner bark was used to make rope. Settlers, noticing the unusual properties of the bark, employed it to waterproof the roofs of their log cabins. In his poem *The Song of Hiawatha*, Longfellow wrote:

> With his prisoner-string he bound him,
> Led him captive to his wigwam,
> Tied him fast with cords of elm bark
> To the ridge-pole of his wigwam.

Ulmus americana, American elm

John Tanner, author of *A Narrative of the Captivity and Adventures of John Tanner . . .* (1830), explains the term *prisoner-string:* "These cords are made of the bark of the elm-tree, by boiling and then immersing it in cold water. . . . The leader of a war party commonly carries several fastened about his waist, and if, in the course of the fight, any one of his young men takes a prisoner, it is his duty to bring him immediately to the chief, to be tied, and the latter is responsible for his safe-keeping."

The elm's wood, however, proved much more difficult to exploit, since it possesses an interlocking grain which makes it well-nigh impossible to split. "They are the most useless piece of vegetation in our forests," wrote one elm opponent. "They cannot be used for firewood because they cannot be split. The wood cannot be burned because it is full of water. It cannot be used for posts because it rots in a short time. It can be sawn into lumber, but it warps and twists into corkscrews, and gives the building where it is used an unpleasant odor for years."[12] The "unpleasant odor" of slippery elm was reminiscent of urine; and some colonists called it "piss elm." Nevertheless, the contrariness of the wood makes it shock-resistant and well able to hold nails securely. It was used to make chair seats and ships' keels.

The main use of elm, however, was as an ornamental. By the mid–nineteenth century, there was so little shade in the towns and byways of the Eastern seaboard that drastic measures had to be taken. Since elms were among the most common trees in the wild, they were ripped out of their natural sites and planted along roadsides and the streets of towns across the country. Obligingly, the elms grew quickly and put up with the stress of urban life. Their graceful shape made them endearingly familiar, and hugely popular. Henry David Thoreau claimed that he found elms superior to people; Walt Whitman was inspired with "large and melodious thoughts" when he walked under elm trees.[13]

Even before elms were being deliberately planted, they were witnessing historic events in town and country. "If you want to be recalled for something that you do," wrote Donald Culross Peattie, "you will be well advised to do it under an Elm."[14] In the 1940s, when Peattie offered his advice to the future famous, historic elms outnumbered historic oaks in the United States by a ratio of two to one. William Penn's treaty with the Indians was signed beneath an ancient elm. Boston's original Liberty Tree was a huge elm, chopped down and burned by General Gage's soldiers in 1776. Daniel Boone meted out justice to his fellow citizens beneath the Judgment Elm, which grew next to his house in Defiance, Missouri. John Quincy Adams and Woodrow Wilson planted elms on the grounds of the White House. Ebenezer Webster, whose son Daniel so soundly defeated the Devil with his verbal pyrotechnics, planted an elm on the grounds of the family homestead in Franklin, New Hampshire. The Osage in Oklahoma sold off the oil leases on their reservation under a tree that became known as the Million-Dollar Elm. In 1943, physicists at the University of Chicago met beneath a campus elm to discuss the experiments paving the way to the first atomic bomb. The elm became known as the Scientists' Council Tree.[15] More recently, an elm survived the bombing of the Alfred P. Murrah Federal Building in Oklahoma City in 1995. It has now become known as the Survivor Tree, and is preserved in a walled memorial on the site that was dedicated in April 2000.[16]

The most famous elm of all was the one underneath which George Washington formally took command of the Continental Army on July 3, 1775, in Cambridge, Massachusetts. The site is now on the campus of Harvard University. The story does not end there, however. In 1902, the University of Washington in Seattle, which collected Revolutionary memorabilia, asked Harvard for a piece of the original elm. Harvard sent a cutting from the tree, which was planted in front of Lewis Hall. In 1923, when a massive storm hit the Atlantic seaboard, the original Washington elm in Cambridge was uprooted. The University of Washington returned a cutting to reestablish the original tree. Then in 1963, the elm at the University of Washington was struck by lightning. Harvard sent a cutting from its regenerated tree to replace it.[17]

Everyone (except perhaps a few grumpy woodworkers) loved the elm. Then in the 1930s, disaster struck in the form of a fungus known as Dutch elm disease. The disease originated in Asia but was first officially recorded in France in 1918. It spread to Holland, Belgium, and Germany and reached England in 1927. Initially, it was suggested that fumes from tons of exploding gunpowder during the Great War had poisoned the sensitive elm trees. But investigations by Dutch botanists revealed the true cause: a pathogenic fungus (*Ceratocystis ulmi*) carried by the elm bark beetle.

A few years later, the disease had crossed the Atlantic—in imported logs of English elm, used in making veneer. It occurred first in Ohio in 1930. In 1933, 3,800 diseased elms were found in New Jersey, and 23 in Connecticut. The disease spread quickly because of the tradition of monoculture—so many elms had been planted along the roads, the fungus simply followed them from one community to another.

A few isolated trees escaped, and researchers are studying the genetics of elm varieties to select those with resistance (if not immunity) to the disease. Although the days of boundless elm-lined streets are gone forever, the sight of a fountain-shaped elm in a field or park will, it is hoped, once again be something to gladden the heart.

Cupressaceae

Eastern Red Cedar
Juniperus virginiana

Juniperus virginiana

grows in almost any kind of soil, from dry rocky outcrops to swamps. It can reach 50 feet in moist, well-drained sites. The narrow, dark green leaves have overlapping scales and are about ¼ inch long. The berrylike fruits are dark blue and covered with a kind of pale powdery wax.

Range: **every state east of the 100th meridian, as well as Ontario and southern Quebec. Range has been extended into the Great Plains by natural regeneration from planted trees.**

Eastern red cedar is the most widely distributed tree-sized conifer in the eastern United States. It can survive as an understory tree, even though it is classified as shade-intolerant, because it does most of its photosynthesizing when bigger trees are leafless in winter. Its wide range is partly explained by its ability to grow in poor soils and dry conditions.

This is not a true cedar; it is a juniper with little blue berries (strictly speaking, they are cones). Yet it has the wonderful fragrance of the Western cedar and, like the Western cedar, was enormously important among Native Americans as a source of wood for buildings and canoes, as well as a medicine and a red-brown dye.

In large amounts, the essential oil in the red cedar is toxic, but in small amounts it was a popular healing substance. The Ojibwe ate the crushed leaves and berries to relieve headaches, the Chippewa boiled the twigs to make a remedy for rheumatism, and the Dakota and Omaha burned the twigs and inhaled the smoke when they were suffering from colds. The Pawnee inhaled the smoke for nervousness and bad dreams. "In the year 1849–50 Asiatic cholera was epidemic among the Teton Dakota. . . . Many of the people died and others scattered in a panic. Red Cloud, then a young man, tried various treatments, finally a decoction of cedar leaves. This was drunk and was used also for bathing, and is said to have proved a cure."[18] This report gives us an idea of the experimental nature of so much medicinal use of plants. The Indians were constantly trying out plants to protect themselves from new diseases.

Juniperus virginiana, Eastern red cedar

The settlers quickly learned to use the wood, and by 1749, Swedish botanist Peter Kalm noted:

> *Of all the woods in this country this is without exception the most durable, and withstands weathering longer than any other. . . . The best canoes, consisting of a single piece of wood, are made of red cedar; for they last far longer than any other and are very light. In New York I have seen quite large yachts built of this wood. . . . I saw a parlor . . . wainscotted many years ago with boards of red cedar. . . . Some people put the shavings and chips of it among their linen to secure it against being worm eaten.*[19]

Settlers used it for fences and posts. The highly aromatic and durable wood was so prized that from 1660 on, cedar furniture was listed in people's wills as an important asset. "In 1699, plans for the 'Capitole, now erecting in the City of Williamsburg' were revised to specify that the porches stand on cedar columns."[20]

Cedar remains a source of medicine today. The leaves and twigs contain a substance known as podophyllotoxin, which inhibits the growth of tumors and has been used in the treatment of cancer. The wood oil, known as cedrol, is also used in medicines and perfumes. Cedarwood is also highly prized. The pink heartwood becomes reddish-brown over time; and the creamy white sapwood darkens to a tan color with age. Because the wood smells pleasant to humans but deters moths and woodworm, the tradition of using cedarwood to make or line chests and wardrobes for storing clothes and linen endures. Woodworking shops advertise patterns for making "cedar hope chests." Somewhere, presumably, there are still girls who are "saving for their hope chests." And cedar chests and blanket boxes remain antique-store prizes.

Salicaceae

Eastern Cottonwood

Populus deltoides, Eastern poplar, Carolina poplar, necklace poplar, cotton tree, whitewood

For many people, huge clouds of cottonwood seeds herald the arrival of summer. The yellow cottony fluff billows out in clouds, giving this tree its common name.

Cottonwood, a cousin to the quaking aspen, is the fastest-growing native tree in North America, getting to 190 feet by age 60 if it's growing in the East. In open areas, its large trunk divides into branches close to the base, and it has a wide, spreading crown. In a closed stand, however, the trunk grows tall and straight with few branches and the crown is small and rounded. While pure stands of *Populus deltoides* do occur, it is most commonly found in mixed stands containing black willow, sandbar willow, peachleaf willow, green ash, box elder, hackberry, bur oak, and American elm. Eastern cottonwoods can survive for up to 150 years or more.

It is a dioecious tree, and though occasionally male and female flowers are found on the same tree, a male and a female tree are needed to reproduce. The male catkins are reddish and the female are slender and greenish-yellow. Once a tree is between 10 and 15 years old, the female catkins start to make greenish-brown, ¼-inch-long, elliptical capsules containing the fluffy seeds, which are dispersed by the wind. The tree can also reproduce vegetatively, sprouting from root collars or stumps, or sending out suckers from the roots, and is particularly resilient after a fire. Because it grows so quickly, *P. deltoides* is very useful in reclaiming disturbed sites such as former strip mines. Although it is a popular urban tree because it grows so quickly, its aggressiveness can cause problems when the roots strangle sewer pipes.

The ability of cottonwood to regenerate is remarkable, as the following story illustrates:

When I was a seven-year-old boy, my father told me that cottonwood trees had a unique characteristic: If you break off a branch and stick it in the mud, it will sprout. . . . A few weeks later I was walking through the arboretum when a huge storm blew in. . . . A strong gust of wind snapped off a branch from a cottonwood tree and it stuck in the mud. I thought, "Aha! This is my chance to prove my father wrong." I came back each day for five days to gather evidence. Sure enough, after the third hot summer day it started to wilt. . . . By the fifth day the leaves were curled. This branch was dead. I went home and told my father he was wrong, and I had the proof. My father calmly listened to my interpretation of the facts. He said, "Son, you're jumping to conclusions. You need to collect more data." He told me to go back to that tree every day for the next 10 days, write down what I saw, and then tell him what I thought. . . . Sure enough, after five more days the leaves started to uncurl. After seven days they started to plump up, to fill with fluid. By the 10th day the stick was indeed alive. . . . I carefully dug down around one side of the stick. I saw the small sprouting roots that had begun to grow. So, my father was right.[21]

The alternate, roughly triangular leaves can be up to 6 inches in length, and from 3 to 6 inches wide. (*Deltoides* comes from the Latin word for "triangular.") They are pointed at the tip, with coarse, rounded teeth along the edges, a shiny green on top, paler below. The leafstalks, which are often yellow and flattened, can be up to 4 inches long. According to an Indian legend, the first tepee design was conceived by someone who twisted a cottonwood leaf into a cone shape, forming a miniature tepee. Native children used to make toy tepee towns with cottonwood leaves.

When the tree is young, the bark is smooth and yellow-green to gray. As the tree matures, the bark becomes deeply furrowed, corky, and ash brown to gray. The Lakota called Eastern cottonwood *canyáh'u*, meaning "peel-off wood," because it is so easy to strip from the tree.

Native Americans had many uses for cottonwood. It was considered a spring treat to make a slash in the bark and extract the cambium, which could be eaten raw or turned into a soup. The inner bark was first cut into strips, dried, and ground into a meal. The Dakota did this and also used the leaf buds to make dye. The Ojibwe used the buds, stewed in bear fat, to treat earache, bronchitis, or coughs and used the down from the seeds like cotton to cover and protect wounds. The Chippewa made a syrupy mix from the buds to treat colds and respiratory problems, heart ailments, sprains, and strains. The Potawatomi cooked buds in tallow to make an ointment for eczema and sores. Canoes were sometimes made from large cottonwoods 4 to 6 feet in diameter, though the Omaha used cottonwood to make the Sacred Pole that was an important part of their ceremonies. According to Jenifer Morrissey, cottonwoods figured prominently in certain creation stories:

> *The Arapaho Indians of present-day Colorado believe that cottonwoods are fundamental to the creation of the star-studded night skies. When the night spirit needs stars, it asks the wind spirit to blow, and blow hard it does, until cottonwoods shed some of their branches. At the broken places is left a star-shaped pattern in the wood, where a new star was born into the sky.*[22]

The wood is relatively light, weak, and soft, with a uniform texture and a straight grain. Set-tlers used it to build barns and houses, not because it was the best wood for construction, but because there was so much of it. Huckleberry Finn knew its virtues as well. It was a cottonwood he found growing beside the Mississippi: "When the first streak of day began to show we tied up to a towhead in a big bend on the Illinois side, and hacked off cottonwood branches with the hatchet, and covered up the raft with them so she looked like there had been a cave-in in the bank there. A towhead is a sandbar that has cottonwoods on it as thick as harrow-teeth."[23]

Today, cottonwood is used for lumber, veneer, plywood, particleboard, fiberboard, pulpwood, pallets, food storage crates, and small items like ice cream sticks and strawberry boxes. High-grade book and magazine paper is sometimes made from the pulp of Eastern cottonwood.

At Washington State University, experiments are currently being conducted to create a hybrid between Eastern cottonwood, *P. deltoides*, and the native black cottonwood, *P. trichocarpa*, to allow the pulp and paper industry to combine the benefits of black cottonwood's exceptional height and Eastern cottonwood's exceptional girth.[24]

Conservationists are trying to save the historic cottonwood of Balmville, New York. The tree is 85 feet tall, and is thought to be at least 300 years old. The town was named after the tree because it was originally misidentified as a Balm of Gilead (*P. balsamifera* var. *candicans*). The tree was close to George Washington's headquarters in 1783, and provided shelter for Matthew Vassar, who supposedly slept beneath it on his journey to Poughkeepsie, where he established Vassar College and a brewery. On visits to his uncle, Colonel Frederic Delano of Balmville, Franklin Delano Roosevelt used to park his Ford Phaeton on the opposite side of the road so he could take a moment to admire the venerable tree. The tree is aged and decaying, yet it is so important to the town that a 65-foot steel column has been installed on a concrete pad at the base of the tree to prop it up.[25]

Range: **southern Quebec and Ontario south to Georgia, west to North Dakota, and in eastern Nebraska, Kansas, Oklahoma, and Texas. Cultivated extensively throughout North America.**

State tree of Kansas and Nebraska.

THE NUT TREES

Fagaceae

American Chestnut
Castanea dentata

Castanea dentata

once grew to 70 to 90 feet, but now the sprouts from stumps reach only 10 to 30 feet. The bark is smooth and brown when young, furrowed into flat ridges as the tree ages. The alternate, simple, pinnately veined leaves are 5 to 8 inches long. Small, creamy yellow flowers are borne in bisexual catkins 6 to 8 inches long, appearing April to May in the South, June in the North. The brown nut, which is ½ to 1 inch in diameter, and flattened on one side, matures in August. The blight *Cryphonectria parasitica*, which infects this tree, is an orange fungal growth.

** *Range:* southern Ontario, New England states, south to Georgia.**

It is impossible now to imagine that the American chestnut, *Castanea dentata*, was once considered a weed tree. It just grew too easily. And though it is a relative of the American beech and

Castanea dentata, American chestnut

oaks, it is horse chestnut *(Aesculus hippocastanum)*, which is a European import (and related to our native buckeyes) that was once considered a more desirable species, probably because it was imported.

At one point, American chestnut was the most common hardwood forest tree in the Eastern woodland. Native Americans ate the sweet-tasting nuts, which they often dried. The Iroquois of New York state ground the chestnut kernels in wooden bowls. The crushed nuts were boiled until the oil rose to the top. The oil was then skimmed off to use as a condiment and the nut meal was dried and pounded into a starchy flour.[26] The Iroquois also roasted the nuts to make a coffeelike drink. The Seneca and other groups used the leaves in a medicine to treat whooping cough.

The settlers learned to love the chestnuts, too.

> *In central West Virginia . . . the woodlands of which were well stocked with chestnut, it was the good practice of farmers to leave groups of scattered trees in certain fields as chestnut groves. These were the delightful resorts of young and old when October frosts had opened the burs and showered the nuts to the ground. . . . The choicest of the nuts were gathered and eaten fresh, roasted or boiled, or were hoarded for winter consumption.[27]*

The easiest way to gather the nuts was to spread a sheet on the ground and get a young boy to climb up into the tree and knock the nuts down with a pole.

Elsewhere, the tree was prized for the lovely rich grain of its wood, not unlike oak but with a deeper tone of brown. It was used for everything from cribs to coffins. The heartwood is rot-resistant, so pioneers used it for shingling their houses and barns and as fence posts and mine timbers. The amount of chestnutwood used in fence posts alone was staggering. Rebecca Rupp estimates it took about an acre of woodland to fence in 10 acres of farmland, and by the 1880s there were about 6 million miles of wooden fence in the United States.[28] The railroads also used chestnutwood in vast quantities. It took 75 feet of standing timber to produce an 8-foot tie and 2,500 ties to lay a mile of track. Another 500 ties were needed to maintain that mile. By 1890,

the United States Forest Service was replacing 70 million ties every year and using about 19 million ties to create new rail lines.[29]

As early as 1860, the chestnuts were becoming depleted. Henry David Thoreau noted in his journal for October 17, 1860: "It is well known that the chestnut timber of this vicinity has rapidly disappeared within fifteen years, having been used for railroad sleepers, for rails, and for planks, so that there is danger that this part of our forest will become extinct."[30]

The chestnut did become virtually extinct, but it wasn't the fault of the railroads. Its extinction was caused by a fungus brought to North America in a shipment of **Chinese chestnut** (*C. mollissima*) imported from Japan in 1886 and advertised in nursery catalogs as a valuable ornamental tree. *C. mollissima* carried the fungus *Cryphonectria parasitica*, but was unaffected by it. The native American chestnuts, however, were not so lucky. They were unadapted to the invader, and quickly succumbed. In 1904, the fungus was found on chestnuts in the Bronx Zoo. By 1907, it had moved on to the New York Botanical Garden, as the spores were carried by animals or the wind. Even the slightest lesion in the chestnut bark gave the spores entry. The fungus attacked the internal conductive tissue, creating sunken cankers, killing the tree in a single season. By 1940, about 4 *billion* American chestnuts had perished. Those that remain are branches sprouting at ground level from stumps, fed by root systems that survived the blight. They seldom get bigger than 30 feet, and often die before bearing fruit.

This illustrates, a little too vividly perhaps, the problem with importing plants. The Chinese chestnuts appeared to be healthy because they had coevolved with the fungus, and achieved a symbiotic relationship. But the poor old American chestnut had never been exposed to the fungus and had never developed any defenses. This is the plant-world version of the epidemics which killed so many Native American peoples. They had no immunity to the common viruses carried by Europeans—viruses which made Europeans ill, but not fatally so.

Some dead trees remained upright for years, like ghostly reminders of human folly. The loss was an economic disaster as well as an environmental tragedy with untold side effects (for instance, seven moth species that fed exclusively on American chestnuts are now extinct). Today, groups such as the American Chestnut Foundation of Bennington, Vermont, are trying to breed a strain of chestnut with the fungus-resistance of the Chinese or Asian chestnut and the qualities of the native American chestnut.[31] The Committee on the Status of Endangered Wildlife in Canada has also formed a "recovery team" to try to return the chestnut to its northern range. If these groups succeed, then future generations may once again look up into the spreading canopies of giant chestnut trees.

Fagaceae

American Beech
Fagus grandifolia, beech, red beech

Fagus grandifolia

grows to 60 to 80 feet with wide spreading crown; gray smooth bark; alternate simple leaves 3 to 5 inches long, glossy dark green, unlobed, leathery (sun) to waxy (shade) in texture, persisting into winter, yellow to golden bronze in autumn. Monoecious—male and female parts exist on the same tree; triangular nut reddish-brown in October; reddish-brown wood.

Range: **Nova Scotia to Ontario south to eastern Texas, east to north Florida.**

Beeches, oaks, and chestnuts are all members of the same family, but there is only one species of beech, Fagus grandifolia, native to North America. All beeches are distinguished by their flowers.

Fagus grandifolia, American beech

The males appear in dangling yellow-green clusters. The female flowers bloom in pairs on short stalks. But it is the thin smooth bark that seems to attract the attention of almost everyone, especially any penknife-toting youngster. For centuries, people have written on the bark, and some ancient inscriptions remain visible on trees in North America and Europe. "Such arboreal record-keeping is made possible by the rapid formation of wound cork, the tree's equivalent of the scab that forms over human skinned knees. Tears, cuts, or incisions in the bark are quickly sealed over . . . leaving distinctive scars visible for decades or even, in the case of the beech tree, centuries."[32]

Some ancient beeches are historic artifacts simply because of their graffiti. The Presidents' Tree of Takoma Park, Maryland, is marked with all the presidents' names from George Washington to Andrew Johnson, along with the signature of the carver, Samuel Fenton, a soldier stationed nearby during the Civil War. The Old Benchmark Beech near Covington, Louisiana, is a boundary marker with two centuries' worth of surveyors' notations carved on its bark.[33]

The beech is equally noted for beechnuts. The taste is similar to hazelnuts and they can be roasted like chestnuts. The Algonquin, Ojibwe, Menominee, Chippewa, Potawatomi, and Iroquois ate the nuts whole, or crushed them to make meal. Animals are equally fond of them. Red squirrels and chipmunks hide the nuts for winter food. This works out very nicely for the locals, who, in turn, raid the caches for their own forage.

Beechnuts were once the staple food of the passenger pigeon. These birds landed on beech trees by the hundreds of thousands. Aldo Leopold described their coming as a "biological storm." James Audubon once witnessed such a storm:

As soon as the Pigeons discover a sufficiency of food to entice them to alight, they fly around in circles, reviewing the country below. During their revolutions, on such occasions, the dense mass which they form exhibits a beautiful appearance, as it changes its direction, now displaying a glistening sheet of azure, when the backs of the birds come simultaneously into view, and anon, suddenly presenting a mass of rich deep purple. . . . The air was literally filled with Pigeons; the light of noon-day was obscured as by an eclipse; the dung fell in spots, not unlike melting flakes of snow; and the continued buzz of wings had a tendency to lull my senses to repose. . . . Before sunset I [had traveled] . . . fifty-five miles. The Pigeons were still passing in undiminished numbers, and continued to do so for three days in succession.[34]

After they had passed, there was not a beechnut to be seen anywhere. Audubon often counted 50 to 100 nests in a tree, with an attendant 2-foot pile of droppings at its foot. It's impossible to think that a bird with such a profound connection to the ecology of the country could be knocked off so easily. But the passenger pigeon was extinct by 1900. Millions upon millions of birds were slaughtered for food and sport (which included letting pigs run loose into the dead and dying pigeons to finish them off). The other factor in their demise was the staggering number of beeches felled by settlers for their farms, depriving the birds of both food and lodging. Of course without the birds, the soil changed as well. Those droppings and the beeches had evolved together. The very last bird was shot as a trophy.

The settlers also ate beechnuts, and even made a type of coffee by roasting and grinding the kernels. The nuts could also be pressed to release oil, which was comparable in many ways to olive oil, suitable for frying and salad dressings. Farmers fed beechnuts to their pigs and turkeys to fatten them. This is presumably why a New York firm selling hams and bacon adopted the name Beech-Nut Company in 1891 (it introduced Beech-Nut gum in 1911).

Beechwood was not generally used for construction. It tended to warp during seasoning and rotted if left buried in the soil. It was extremely useful as barrels, bowls, butter boxes, and kitchen utensils, because it had no smell or taste to leach into the food. It was also a sturdy wood that could withstand friction and scouring so was used for handles of carpenters' tools, thresholds, clothespins, and the soles of clogs.

A type of creosote was made by distilling the tar from beechwood. Once used to treat coughs and as a disinfectant, it is rarely used for these purposes today. Modern beechwood is now used for flooring, furniture, veneer, plywood, railroad ties, baskets, pulp, charcoal, and rough lumber. It is popular as a fuel because it burns well.

Betulaceae

American Hazelnut
Corylus americana, American filbert

Corylus americana is a smooth-barked, multistemmed shrub growing up to 20 feet tall. The simple leaves are rounded or heart-shaped, with toothed edges. The twigs are hairy. In early spring, before the leaves unfold, drooping yellow male catkins and tiny female flowers appear. The latter have no petals but their red pistils emerging from the ends of the buds give a dash of contrasting color. The nuts are surrounded by two hairy, lacy leaflets.

Most of the hazelnuts you can buy in a store are from the imported **European filbert** *(C. avellana)*, but according to naturalists Steve Brill and Evelyn Dean, the native nuts taste even better.[35] Indigenous peoples such as the Iroquois and Menominee ate the nuts raw or ground up and added to other foods. Today, some hazelnuts are grown commercially. Hazelnuts are often called filberts, the name given to European hazelnuts because they were usually ready for harvest on Saint Philbert's Day, August 20.

The hazelnut has been cultivated as an ornamental tree since the eighteenth century because it grows quickly and provides dense shade—too dense, sometimes. Without pruning, it can turn into a small thicket and crowd out other plants.

C. cornuta, **beaked hazelnut,** is the other native filbert species. It is also widespread in the northern forest, but is being replaced by American hazelnut in its southern range.

Range: **southern Ontario and Manitoba, New England, and the Eastern states as far south as Georgia and as far west as Oklahoma.**

Juglandaceae

HICKORY
Carya spp.

Hickories belong to the walnut family. *Carya* comes from the Greek word for nut-bearing tree. The word "hickory" comes from the Indian term *pohickery* or *pawcohiccorra*. The "pawco" seems to have been dropped by the end of the seventeenth century, and the tree was known as hiquery, hickery, hiccory, or hackerry. Each species seems to have dozens of common names. What's even more confusing is that many different species bear the same common name.

Juglandaceae

Shagbark Hickory

Carya ovata, bird's-eye hickory, Carolina hickory, curly hickory, little pignut, mockernut hickory, red hickory, redheart hickory, scaly-bark hickory, shagbark, shagbark walnut, shellbark, shellbark hickory, skid hickory, small pignut, Southern shagbark hickory, Southern shellbark, sweet walnut, true hickory, upland hickory, white hickory, whiteheart hickory, white walnut

    Carya ovata grows 60 to 90 feet. The gray bark is smooth in youth and shaggy in age. The alternate, pinnately compound leaves (with five leaflets) are 8 to 16 inches long. It is monoecious—both male and female flowers grow on the same tree. The males are three-branched catkins, 3 to 4 inches long, coming from the tips of the previous year's wood or leaf axils; the females, on short terminal spikes, are on shoots from the current year. They bloom in spring just as the leaves unfurl completely.

    Range: southern Maine, west as far as Nebraska, south to Georgia, Alabama, Mississippi, Louisiana, and eastern Texas.

Carya ovata, shagbark hickory

Shagbark hickory is one of North America's hardiest, most historically fascinating trees. Ages ago, it grew all over Europe, until glaciation scraped that continent clean of much of the forest cover. From then on it grew only in North America and China. In the United States alone, from the edge of the prairies eastward, this tree covered 16 million acres. It is seldom found in pure stands and lives on dry slopes or well-drained lowland areas, where it mixes with oaks. The evolution of the forest never stops. Even in an old-growth forest, hickory will replace oak, black walnut, and honey locust, only to be replaced in turn by the more shade-tolerant sugar maples, hop hornbeams, or basswood.

    The distinctive feature of *C. ovata* is the bark, which is shaggy and peels off in long strips. The effect is so satisfying there's a great temptation when wandering through the woods to spend time playing with these trees. When the tree is young, the bark drops off in strips, creating a litter in which seeds can germinate safely. During the winter, the dormant seeds lie on the forest floor; in spring they sprout vigorously.

They are survivors throughout their lives. If a fire hits a hickory stand with trees over 10 inches in diameter, chances are that the stumps will either sprout or send out suckers.

Hickories can survive, and in some cases bear seeds, for up to 300 years, though the best seeds are produced between the ages of 60 and 200. The fruit, actually a nut, is 2 inches in diameter, and varies from 1 to 2½ inches in length. The husk is green until it ripens in September or October, when it turns brownish-black and splits open to expose the brownish-white nut.

A year when a hickory produces an unusually large quantity of nuts is called a mast year. During a mast year, a single tree can produce 70 pounds of nuts. This may happen every two or three years, and when it does black bears tend to produce more young the following year. Many animals not only eat the nuts but also distribute them, including red fox, gray fox, mice, chipmunks, ducks, woodpeckers, and yellow-bellied sapsuckers. In its day, the passenger pigeon ate the nuts, distributing the seeds over their extraordinary range.

The Iroquois used the hickory to treat arthritis, worms, and headaches by placing small shoots on hot stones and inhaling the steam. They also used the nutmeat oil on their hair or mixed it with bear grease to rub on their skins. This had two functions: it was a protection against mosquitoes which proliferated in the estuaries where many Eastern woodland natives lived; and it was a good protection against the ravages of the sun. To get the oil, they shelled the nuts, pounded the innards, and steeped them in boiling water. The Dakota, Pawnee, and other peoples made soup with the nuts and sugar from the sap.[36]

While the close-grained, sturdy heartwood is a reddish-brown, the sapwood is nearly white. The colonists used the wood to make long-lasting fences, furniture, door hinges, ramrods

for guns, and hoops for pork barrels. Settlers also found that the dense wood of shagbark hickory made excellent fuel wood and charcoal, as well as a pleasant smoke for curing meat. And, of course, there was the use of the hickory stick so dreaded by students. In the early twentieth century, hickorywood was even used to make the first automobiles.

Today, along with pignut hickory wood, another important commercial species, shagbark hickory is used for furniture, flooring, tool handles, ladders, and sporting goods. The hickory is so sturdy it can be used to revegetate unpromising areas such as abandoned strip mines. It can even be tapped like a sugar maple in the springtime and the sap made into syrup.[37]

Hickory's reputation for toughness was legendary and has entered the language. In 1812, when General Andrew Jackson was given the order to demobilize his troops near Natchez, 500 miles from their Tennessee homes, he refused, and marched back all the way to Tennessee with them. He ate and slept with the soldiers, and let his horse be used to transport the wounded. His men adored him. "He's tough," they said, "tough as Hickory." They nickname him "Old Hickory." Jackson took his nickname all the way to the White House.[38]

Jackson's toughness had a much darker side, however. He hated Indians and fought them with a savage fury that now seems astounding. He killed so many that he kept count by cutting off their noses. He had them skinned after they were dead and used the skins for ammunition bags. Jackson isn't admired today as he was in the past, but nonetheless he is buried beneath six tall shagbarks, which is quite a memorial in itself.

Juglandaceae

An extraordinarily tough hickory, the pignut was another legendary hickory and also entered the language: people of a certain age still talk about strapping on the old hickories when they go skiing. The wood absorbs vibration so well that early settlers used it for ax handles and hubs on the covered wagons that took them westward. Later, hickorywood was used to make textile looms, which sustain great vibrations when the shuttles are in motion. According to Donald Culross Peattie, "Hickory is stronger than steel, weight for weight, more elastic, less brittle, less heat-conductive."[39] The Omaha used to weave thongs over a hickory stick form to make snowshoes.[40]

This species was also called broom hickory because the early settlers made brooms with it. Dr. Daniel Drake, one of the first physicians in Ohio and an important medical pioneer, wrote:

A small hickory sapling was the raw material. The "splits" were stripped up for eight or ten inches with a jackknife pressed by the right thumb, bent back, and held down with the left hand. When the heart was reached and the wood became too brittle to strip . . . the splits [were] turned forward and tied with a tow string made for the purpose on the spot. It only remained then to reduce the pole above to the size of a handle.[41]

People made scrubbing brushes the same way with a smaller sapling.

Carya glabra, with smooth light to gray bark which becomes deeply furrowed in age, grows 60 to 70 feet high. The alternate leaves are pinnately compound and 6 to 12 inches long. Male catkins are 2 to 3 inches long, female flowers appear at the end of branches, and the fruit, which ripens in September, has a distinct pear shape. In autumn, the leaves turn a dull gold and leave behind the leafstalk once they fall off. They remain only a short time but give the tree an interesting spiky look for a while.

Range: **Massachusetts to southern Ontario, southeastern Iowa, Mississippi, Arkansas, Louisiana, central Florida, and west to Texas.**

Juglandaceae

Bitternut hickory got its Latin name (*cordiformis*) from its heart-shaped nuts, and its common name from their bitterness. They are so bitter that only a squirrel will eat them. But that little "tree rat" (as some people uncharitably think of squirrels that nibble on their on favorite tulips) is the linchpin to keeping the Northern forest alive and healthy. It is hard to think of the squirrel as a noble creature, but it does perform a vital function in the forest: it eats three out of four nuts and buries the fourth, usually in a place it's most likely to sprout.

Carya cordiformis grows up to 80 feet high with slate gray to light brown, smooth bark which becomes fissured with age. The alternate, compound leaves, composed of seven to nine leaflets, are 6 to 10 inches long. In spring, the emerging leaflets are shiny green on top and slightly pale and hairy underneath. As they twist and turn in the wind, they change the entire look of the tree. The tree is monoecious with male and female catkins on the same tree.

The wood isn't as hard as shagbark or pignut, but it is quicker-growing though much shorter-lived at 200 years. Its main claim to fame is its popularity as the finest hickory for smoking food. The Cherokee knew this and used the wood ash along with salt and pepper to cure pork. The ash was also used to make lye.

Settlers used bitternut hickory oil to light their lamps, and even today people make hickory bark torches. Benjamin Pressley of Stanley, North Carolina, writes: "I have seen this torch burn for as long as two hours. This torch is a combination of green and dry strips of hickory bark and a core of an easily combustible material that keeps a constant supply of tinder furnished to the torch."[42]

Range: **Maine west to southern Quebec, southern Ontario, and Minnesota, south to Texas, east to Florida.**

Juglandaceae

Pecan
Carya illinoensis

Pecan, the opposite of the bitternut, has the sweetest nuts of all. The name *illinoensis* is probably a botanical error, since the trees more than likely come from Texas and Mississippi, rather than one of the more northerly states. And though it originally grew only in the South, it was, like many other foods, brought north by Indians who planted them along portages as a convenience on their travels. It does best in warm climates and needs a long hot summer to help the nuts develop fully. Pecan trees are also the largest of the *Carya* species, often reaching 100 feet high. One giant in Mississippi grew 130 feet tall.

Thomas Jefferson liked hickory and planted a specimen of each species at Monticello, his estate in Virginia. He imported what were then called "paccans" from the Mississippi Valley and shared them with his friend George Washington.[43] Today, pecans are widely cultivated and are an integral part of Southern and Texan cooking. Eighty percent of the world's pecan crop is produced by the United States, mostly in Georgia and Texas, and the demand increases yearly.[44] Pecans fetch up to one dollar a pound, and a pecan grove can yield up to 1,000 pounds an acre if well managed.[45]

Carya illinoensis has dark, reddish brown bark with vertical ridges. The alternate compound leaves have 11 to 17 leaflets, and the whole leaf is 12 to 30 inches long. The nuts are 1 to 2 inches long and grow in groups of three to six, with thin, smooth shells. Found in rich, well-watered soils.

Range: **throughout the Mississippi Basin and Southeastern states.**

Fagaceae

OAK
Quercus spp.

The oak is one of the most commanding of all North American native trees. It has all the aspects of greatness: it lives for hundreds of years, and ages into a dignified form with a gnarled trunk and a breathtaking sweep of canopy. There are hundreds of species around the world; estimates run from as few as 300 to as many as 600. The fact that different species of oak hybridize freely on disturbed sites such as abandoned farms may explain the confusion about the actual number. In general, however, oaks are divided into three groups:

White oaks produce acorns that mature in autumn, and are considered annual oaks because they shed their leaves each year. They have rounded leaf tips and lobes; the acorns are quite sweet. This group includes **swamp white oak** *(Quercus bicolor)*, **bur oak** *(Q. macrocarpa)*, post oak *(Q. stellata)*, **chinquapin** or **yellow oak** *(Q. muehlenbergii)*, and **chestnut oak** *(Q. prinus)*.

Red oaks are biennial (taking two years to produce a ripe acorn) and may also be called black oaks. Their leaves have pointed lobes tipped with fine bristles; the acorns mature two years after formation. This group includes **scarlet oak** *(Q. coccinea)*, **Northern pin oak** *(Q. ellipsoidalis)*, and **black oak** *(Q. velutina)*.

Live oaks are evergreen or semi-evergreen, typical of the Southern hardwood forest. *Q virginiana* and *Q. laurifolia* are two examples.

One of the oak's survival techniques is the capriciousness of its acorn production. It is confusing enough to baffle the many predators and parasites that attack the tree. In a mast year, a single tree can produce up to 3,000 acorns, after which production drops off drastically. The populations of the animals that eat the acorns also fluctuate with the crop size.

Oaks are so tough that they will survive lightning strikes. If lightning hits the upper branches of an oak and the tree is rain-soaked, the charge races along the surface of the bark. If the tree is dry, lightning goes right into the sapwood on one side of the tree and zaps straight through to the ground. The sap boils instantly and strips of bark are literally torn off. Nevertheless, the oak will often live on.[46]

Native Americans used oak in every imaginable way. The acorns were important food and medicine for treating diarrhea and toothaches.

The Chippewa chewed the fresh root and applied it as a poultice to wounds. Several groups added soil to the bark to make a black dye; and when cedar ashes were mixed with the inner bark—*voilà*—a red dye. A hugely important part of the natives' lives was devoted to dedication and purification rituals, and they would use only important plants in the ceremonies. An oak tree would qualify not only because it is imposing but also because the wood is extremely hard. The Navajo inserted oak sticks in the crevices above the door of a hogan during ceremonies. But oak was also employed to make war clubs, bowls, mortars, and paddles, and the curled twigs were used as drumsticks in the War Dance.[47]

Fagaceae

White Oak
Quercus alba

White oak is a massive tree opening up into a haunting shape, with branches 50 feet long or more springing at right angles from the trunk. Equally impressive is the tree's underground structure. As Donald Culross Peattie puts it, "So deep is the taproot of such a tree, so wide the thrust of the innumerable horizontal roots, that if one could see its whole underground system this would look like a reflection, somewhat foreshortened, of the giant above ground."[48] Those same deep taproots make the oak invulnerable to the wind; and shoots grow back from a fire-ravaged stump within a few years. Puts one in mind of Tara and the survival of the South in *Gone With the Wind*.

Quercus alba adopts a pyramidal shape during its youth and rounds out with age. It will grow to 100 feet, even more in the wild, with a spread of 50 to 80 feet with grayish-pink leaves in spring turning to deep green in summer, then transformed to a rich red, brown, or purple in autumn. The alternate simple leaves are 5 to 8 inches long, dark green above, almost gray beneath, and, on younger trees, may hang on all winter. The bark turns a light gray with age and becomes scaly and plated; the reddish-brown twigs are covered with an almost white bloom. Brown-green male and female flowers appear as catkins in spring. The acorns ripen in September or October.

This tree stands alone in every way, although in some forests is it codominant with shagbark hickory. A white oak can live up to 600 years, producing acorns when it is about 50 years old, and has a mast year every 4 to 10 years. The white oak acorns, however, start to spoil the minute they hit the ground and rot if they aren't eaten immediately. Talk about fast food. Squirrels perch tensely on the twigs, waiting for the acorns to drop, and then descend for an instant meal. Other animals also depend on the white oak acorn: black bears, raccoons, Eastern chipmunks, white-footed mice, and many birds. Although they are relatively sweet when they are fresh, the acorns become even sweeter after they've been boiled up, and were a favorite of Native Americans, who ground them into a meal used in mush, soups, or to make hard dry bread.

The oak is a host to many other plants. A small, brownish-yellow parasite called *Conopholis americana* may grow on the oak's root, rising above the soil to become a spike 2 to 8 inches high. The tiny scales are its leaves. This strange plant survives by pulling all its nutritional needs from the oak. It's called squawroot because it was used by some Native American women to relieve symptoms of menopause; or cancerroot because it presumably helped in the control of the disease. Honey mushrooms (*Armillaria mellea*) are a less benign parasite. This is an edible fungus which causes white rot and usually indicates the tree is under great stress or may be dying.

The inner bark of the white oak contains a high concentration of tannins that were used as a medicine by Indians. It was boiled into a syrup to make an antiseptic as well as to soothe a sore throat, to control diarrhea, or to treat, in the hesitant language of male ethnobotanists of a certain generation, "diseases peculiar to women." It was also applied like a mustard plaster (dried, ground up, and dampened) to hemorrhoids. But it wasn't just a powerful medicine tree. The Cherokee, for instance, used the wood in everything from wigwams to cooking utensils. Oak was an ideal firewood and even the fibers from the interior of the tree were gathered to make baskets.

The Europeans recognized the tree immediately. Oak was a national British symbol (the British throne was made of oak) and oakwood had long been used for building houses and ships in England, but by the seventeenth century, it had been pretty much removed from the landscape. Imagine their delight at seeing this wealth of oaks (in every sense of the word) on arriving in North America. Settlers went to work chopping and sawing their way through the New World forests to build ships for the Old. The frigate *Constitution* ("Old Ironsides") had a gun deck of solid white oak from Massachusetts and a keel made from white oaks growing in New Jersey. According to Peattie, these oak ships carried New England sea captains around the world right up to the Second World War, when the keels of minesweepers and patrol boats were still made of white oak.[49]

At least one oak survived this massive harvesting. The biggest oak tree in America began its life in the 1500s. It was there when Native Americans used the trail around Chesapeake Bay, which later became a road for settlers. The land it stood on was bought by Thomas Williams (1665) and then by Richard Bennett in 1705. As it changed hands over the centuries, the parcel of land the tree was standing on became smaller and smaller until it was on a 2½-acre property owned by Johannes Arants. By the 1830s, it was beginning to be recognized and referred to as the Russum Oak (for its owner Dr. S. T. Russum). By the 1900s, it was known as the Wye Oak, after the village of Wye Mills, Maryland. And in 1919, it was entered into the *American Forestry Magazine*'s Tree Hall of Fame.[50] It was the biggest white oak in the country until it toppled over in a storm in 2002 at the age of 460.

What set the Wye Oak apart from other oaks were the massive "knees" or buttresses that supported the huge tree. This unusual formation is a matter of some speculation. One account says an old country store was once located at the site of the tree, and the stomping hoofs of the customers' horses tethered to the tree bruised the roots, resulting in a malformation which eventually developed into large burls.

Around 1900, houses were being built and floored with "golden oak" named for the high varnish laid on quartersawn boards. These wide boards come from trees 100 to 300 years old. Today, brand-new houses are still being advertised with "oak floors," but the wood comes from younger, narrower trees and it is coated with polyurethane. Kevin Jackson, who installs hardwood floors for a living, loves oak. He doesn't even mind the splinters. With oak, he says, "you wait a day or so and the splinter will swell up and is easy to remove. With other woods it can take weeks."[51]

However, perhaps the man who loved white oak more than anyone else was Colonel W. H. Jackson of Athens, Georgia, who in 1820 deeded a piece of ground to a white oak tree in his town, with the words, "For and in consideration of the great love I bear this tree and the great desire I have for its protection for all time, I convey entire possession of itself and the land within eight feet of it on all sides." Although the original tree, known as the Tree That Owns Itself, died in 1942, one of its acorns was planted on the same ground and survives to this day.[52]

Range: **Maine west to southern Quebec, southern Ontario, and Minnesota, south to eastern Texas, and east to northern Florida.**

State tree of Connecticut, Illinois, and Maryland.

Fagaceae

Red Oak
Quercus rubra, Northern red oak

Quercus rubra

is distinguished from white oak by its leaves, which are pointed at the ends, dull green, 5 to 8 inches long, and 4 to 5 inches wide. The red oaks can grow to over 100 feet, but are usually about 70 feet high. The bark develops flattened ridges as it matures. The slender green twigs become dark red before turning brown. Once the pointy-tipped leaves have started to open up, the flowers appear. The green females, usually found in the lower part of the tree, may go almost unnoticed because of the flashy quality of the slender male catkins.

Range: **Nova Scotia to Minnesota, eastern Kansas, and northern Arkansas, south along the Appalachians to Georgia, northern Alabama, and central Mississippi.**

State tree of New Jersey.

Red oak erupts like a flame in autumn and it must have dazzled the Pilgrims who were landing when it was in its most furious flush. In those days, red oak was a major inhabitant of the Eastern forest and one of the fastest-growing of all oaks. The shape of each tree reflects its own light conditions. Out in the open, it will have a short thick trunk and a wide crown. In the middle of the forest, it grows straight and tall with a narrow crown. Any-

Quercus rubra, red oak

where an oak finds itself, the roots plunge into the soil with a huge branching system, anchoring the tree firmly. This is a quick-growing tree that does well in very cold temperatures but is susceptible to drought. It will sprout from its root if fire-damaged. The trees of the red oak group (such as black oak, scarlet oak, and pin oak) have a tendency to hybridize partly because they are wind-pollinated.

The bitter acorns, which take two years to mature, are large (1 to 1½ inches long) and flat with a shallow cup. They begin to appear when the tree is about 50 years old and take up to 16 months to mature; a hairy lining surrounds the yellow meat, which is an excellent source of energy, protein, carbohydrates, and calcium. A mature red oak can produce up to 5,000 acorns in a mast year, and can keep fruit on the tree for 16 months after pollination. The acorns survive the winter and mature in autumn (black, pin,

and Northern pin oaks have the same characteristic, as do pines and autumn-flowering witch hazels).

In spite of the bitterness caused by tannic acid, the high fat content (18 to 25 percent) of the acorns provides food for insects, animals, and people. The Ojibwe and Potawatomi made the acorns more palatable by steeping them in wood ash, or by scooping out the insides, putting them in a basket, and leaving them in a clear stream for several days to rinse away the bitterness. They dried or baked the acorns slowly and then crushed or ground them to make flour.

Hundreds of insects and diseases prey on red oaks. One of the worst is the gypsy moth *(Lymantria dispar)*. Once again, the tragedy was caused by foreign imports. A French scientist named Étienne Leopold Trouvelet living in Medford, Massachusetts, in the 1880s decided to start a silk industry in North America. He wanted to breed a hardier version of the silk moth *(Bombyx mori)* and intended to cross it with the gypsy moth, which was believed to be

related. He brought his breeding stock into the country in 1868. Within weeks, his caterpillars had escaped from the lab. Pretty soon, his neighbors noticed great buff-colored patches of egg masses appearing on the trunks of their oaks as the leaves dropped off. The infestations traveled quickly as the caterpillars floated on the wind from one stand of oaks to another in a process called "ballooning." Trouvelet fled the country when his neighbors realized how their troubles had been started. He went on to become a well-known astronomer, a much wiser choice of profession, but the damage was done. Millions of acres of oak forest continue to be defoliated every year.

Though oaks can survive a few years of being chewed up by gypsy moths, it makes them weaker and more susceptible to disease and infestation by other insects. The trees use up vast stores of energy to produce phenol compounds intended to stunt the growth and reproduction rates of the invader. The gypsy moth in its turn has population highs and lows, increasing in number for 10 years and staying numerous for 3 until disease and predators once again beat them into submission. On the other hand, when gypsy moths are numerous, they become a source of protein for young chicks, and so give some bird populations a boost.

The red oak's bark, with its high tannin content (6 to 10 percent, compared to white oak's 2 percent), makes it useful for medicinal purposes. The Chippewa mixed the powdered inner bark of red oak, quaking aspen, and balsam poplar, plus the ground root of Seneca snakeroot, for a powerful heart medicine. Mild tannic solutions also provided a remedy for sore throats, skin irritations, cold sores, and (taken internally) for diarrhea and hemorrhoids. The bark also provided a dye, and the interior fiber was used for basketmaking.

The settlers quickly found uses for the red oak, although it was considered more porous than white oak. Barrels made from red oak were fine for transporting dry goods such as flour, sugar, and nails, while white oak made the best casks and hogsheads for holding liquids (wine actually improved when stored in white oak barrels). Red oak with its strength and hardness is still much used for flooring, furniture, and veneer, among other items.

Perhaps the most interesting use of oakwood was in the building of roads. Henry Kock of the University of Guelph, Ontario, describes one still visible: "I grew up near Sarnia, Ontario, not far from the Plank Road. Never curious enough for the details when I was young, it wasn't till a few years ago that I found out that it consisted of 4-by-12-inch oak timbers. The planks were laid diagonally in three alternating layers—a 12-inch-thick, oak road running a straight 28 kilometers from Oil Springs, where Canada's first crude oil was found, to Sarnia, where the refineries were built."

Most red oaks live about 300 years, but some survive even longer. Kock says, "I heard of a red oak growing in Niagara-on-the-Lake [Ontario] that was determined by arborists to have been a seedling at the time Marco Polo met Kublai Khan! When I saw the majestic old tree (then the oldest in Ontario) in 1978, it was in decline and it's now (apparently) just a stump after more than 700 years of growth."

Aceraceae

MAPLE
Acer spp.

"The scarlet of the maples can shake me like a cry / of bugles going by," wrote Bliss Carman. The familiar leaves have become the symbol of the Canadian nation; the running sap is a sign of spring and of youth, and the flaming colors of its autumn foliage mark the passage of time. There are 150 species of maple worldwide, 13 of them native to North America.

Aceraceae

Sugar Maple
Acer saccharum, hard maple, rock maple, black curly maple, bird's-eye maple, sugar tree

This tree with its distinctive chevronlike markings is codominant with American beech *(Fagus grandifolia)* when the Eastern forest reaches the most stable part of its life—the climax stage. This stage means stability in the ecosystem. The sugar maple can live for 300 or 400 years under these circumstances.

It grows in pure stands or with other species such as basswood, birch, black cherry, red and white spruce, beech, Eastern white pine, and Eastern hemlock. It grows just about anywhere, except in places of spring flooding. Its seedlings hang on until there's a break in the canopy large enough to let in light and rain. At that moment, the litter of the forest floor becomes an ideal nursery, easy to penetrate as the seedlings form roots. The tragic death of the American elm benefited the sugar maple in the competition for space and light.

Acer saccharum grows from 90 to 120 feet tall with a 3- to 4-foot circumference. The bark on old trunks is deeply furrowed, brown or gray. The leaves, palmate, opposite, and five-lobed, are dark green above, paler green beneath. The reddish-brown twigs are covered in lenticels. These are special cells that allow for the exchange of gases inside and outside the tree. The long vertical plates of the bark curve slightly outward as they age, forming a chevron

pattern. The small greenish-yellow flowers are borne in tassel-like clusters or racemes about 1 inch long. Male flowers outnumber females about 50 to 1. In fall, the little maple keys called samaras ripen and then take off, hitching a ride on every breeze that comes along.

In autumn, the tree seems to catch fire. In his wonderful essay on maples, John Eastman writes:

> *Without a microscope, you probably won't notice the tough sheaths that enclose the leaf veins or the tiny sparkles of glasslike crystals scattered throughout the leaf. These mechanical features plus tannic compounds are probably defences that give the leaf a fighting chance to survive its many feeders. The leaves usually turn blinding yellow or orange-yellow in the fall, a brilliant explosion of carotene pigments unmasked by the shut-down of leaf factories and the subsequent decay of green chlorophyll.*[53]

The sugar maple is under attack from all sides. Acid rain is taking a terrible toll. Armies of scale insects, thrips, aphids (and their attendant ants) feed on buds and leaves. Maple seeds are also eaten by songbirds (red-breasted nuthatches, purple finches, and grosbeaks), squirrels, chipmunks, and white-footed and deer mice; porcupines, cottontails, and mice gnaw the bark in winter, leaving behind distinctive tooth marks. Some trees are completely girdled by foxes and gray squirrels, which strip the bark to get at the cambium (the layer below the bark). And then there is the whole maple syrup industry.

As much as 15 percent of the tree's carbohydrate reserves are removed by tapping the tree. Though we've been exploiting maples for hundreds of years, we still know very little about the sap we find so desirable. The sap starts to move through a series of vessels in the xylem (the woody tissue which both carries the sap and gives the plant support) once daytime temperatures reach 40° to 50°F but nights are still below freezing. The warming increases air pressure, compressing the vessels in the tree, and the sap moves upward. Every day the sap runs, 200 to 250 gallons move through a sugar maple, which has a greater proportion of sapwood to heartwood than other species. Since the sugar content is only 2 or 3 percent, it takes 35 to 40 gallons of sap to make 1 gallon of maple syrup.

Native Americans used the sap in a variety of ways. The Iroquois combined maple sap, thimbleberries, and water to make a drink for longhouse ceremonies. In the early days, they put hot rocks in their sap containers to get it boiling. The Ojibwe allowed the sap to turn sour to make a vinegar and then mixed it with maple sugar. They cooked venison in the vinegar, which must have tasted a bit like an Asian sweet-and-sour dish. The Potawatomi and Meskwaki used maple sugar instead of salt as a seasoning. While salt is commonly used today, many of the older people still prefer to season their food with maple sugar.

The Iroquois conducted a thanksgiving ceremony for the maples in early spring. The rising sap, says Eastman, was a renewed survival covenant for another year. Maple was considered a gift of the Creator for its syrup and sugar.[54]

Frances Densmore, in her book *Uses of Plants by the Chippewa Indians*, has left an invaluable portrait of how Native Americans moved with ease over their land and just how hard they worked. Maple syrup was one of the two most important foods in their diet (the other was wild rice). She recounts the great pleasure they had, young and old, of going to their temporary camps for sugaring off.[55]

The camps had small huts which housed kettles and fire on one side with sleeping platforms covered with cedar boughs on the other. They left their utensils from year to year: the paddles, bowls, sap buckets, and *mukuks* (foot-long bark containers to hold the sugar). The first run of sap was the best, with snow still on the ground, and freezing temperatures at night. A diagonal cut was made on each tree 3 feet from the ground and a spile (a type of spout) inserted to catch the drip but not split the wood. Sap dishes were set out in the morning and gathered late in the afternoon to be taken back to camp kettles. All night long the fires were tended and the kettles kept boiling. If the boil came on too rapidly, the women would swish a spruce branch through the froth. By morning, the sap was thick enough to strain. In the old days, they used narrow strips of basswood to strain it; later they used burlap or an old sheet.

Then came the sugaring off. "Sirup," Densmore writes, was put into scrupulously clean kettles.

A maple-wood paddle was used in stirring the sirup, and when it had thickened to the proper consistency it was quickly transferred to the granulating trough, where it was again stirred with a paddle, and at the proper time "rubbed or worked" with the back of the granulating ladle, or in some instances pulverized by hand. . . . Granulated sugar, however, was not the only form into which maple sap was converted. . . . It was customary to pour some of the thick sirup into small containers where it hardened solidly. Little cones were made of birch bark and fastened together with strips of basswood bark so that the group resembled a cluster of berries. These cones filled with sugar were a favorite delicacy among the children. The upper mandible of a duckbill was similarly filled, several of these being fastened together in a row by a little stick.[56]

Europeans were fascinated with sugaring off. In 1691, Christien LeClercq wrote of the

Mi'kmaq: "A Thing which seemed to me very remarkable in the maple water is this, that when by virtue of boiling it is reduced to a third, it becomes a real syrup, which hardens to something like sugar, and takes on a reddish colour. It is formed into little loaves which are sent to France as a curiosity, and which in actual use serve very often as a substitute for French sugar."[57]

Between 1785 and 1795, as French botanist François André Michaux explored New England, he observed: "Wild and domestic animals are inordinately fond of maple juice, and break through their enclosures to sate themselves; and when taken by them in large quantities it has an exhilarating effect upon their spirits." Perhaps all it took thousands of years ago was some smart observer who saw a squirrel lap up the sap from a wound in the tree and become "exhilarated."

The settlers found this new confection very much to their liking. In *Roughing It in the Bush* (1852) Susanna Moodie recalls: "I sugared off the syrup in the house; an operation watched by the children with intense interest. After standing all day over the hot stove-fire, it was quite a refreshment to breathe the pure air at night. Every evening I ran up to see Jenny in the bush, singing and boiling down the sap in the front of her little shanty."[58] Today, almost half of the functioning sugar bushes in North America are in the province of Quebec.

The settlers also loved the maple for its wood, which takes on a shimmer under the right hands. The most highly prized of all were the pieces of bird's-eye, an unusual grain caused sometimes by knotting, sometimes by a fungus feeding on the cells under the bark. Treasured for its beauty and strength, it has been fashioned into musical instruments (violins, violas, cellos, guitars, harps) of great value, as well as gunstocks and knife handles of enduring quality. During the 1970s Japan went through a major bowling alley boom and used sugar maple for lanes and pins.[59]

Alas, highly prized woods can bring out the worst in certain people. The popularity of bird's-eye maple led to the rise of bird's-eye rustlers: people who go into the woods mutilating trees to find a bird's-eye maple—as if trees didn't have enough troubles already.

Range: **eastern Canada to Georgia, Alabama, Mississippi, and Texas.**

Aceraceae

Red Maple

Acer rubrum, scarlet maple, soft maple, swamp maple, Carolina red maple, Drummond red maple, water maple

Acer rubrum

grows from 30 to 90 feet tall, with a diameter of up to 4 feet. Large leaves come out with a reddish tinge and turn to green (paler underneath) with red veins and leafstalks. Small, fragrant, drooping clusters of red and orange flowers hanging from red twigs are among the first to come out from March to May, depending on elevation and latitude. Grows in low rich soil near water.

Range: **Newfoundland, Nova Scotia west to southeastern Manitoba, Minnesota, Wisconsin, and Illinois, south through Missouri, eastern Oklahoma and Texas, east to southern Florida.**

State tree of Rhode Island.

It's easy enough to confuse red maple and silver maple, but if you smell the twigs the distinction is clear (silver maple smells slightly unpleasant). Red maple is all red: red flowers in April, red seeds in May, red leaves in autumn, and little red twigs at the end of the branches in winter. "Thoreau," John Eastman says, "wrote admir-

Acer rubrum, red maple

ingly of red maple in one of his last essays, 'Autumnal Tints.' He wondered about the Puritans' reaction 'when the maples blaze out in scarlet' and slyly suggested that they erected meeting houses to avoid worshipping under red maple's indecent 'high colors and exuberance.' "[60]

The commonest tree in the eastern part of North America, red maple tolerates waterlogged soils and usually grows along with black ash, American elm, and pin oak in swamp country. It survives not only flooding but also fires, and this fire resistance increases as the bark thickens with age.

The red maple flowers so early it has become one of the most welcome sights in spring. The orangy-red flowers expose themselves when most other trees are still bare. Male and female flowers grow separately, although both appear on the same tree on different branches (in other words, the tree is monoecious). The winged fruits hang in clusters on long red stalks, standing out against the pale green leaves. Autumn brings another spectacular change because it is one of the first trees to change color.

By four years of age, it will produce seeds and yield a good crop every second year, sending a million seeds off into the wind. If they land anywhere suitable, they germinate within 10 days; otherwise, they wait until the following year.

In 1898, Canadian poet Bliss Carman wrote:

> Let me have a scarlet maple
> For the grave-tree at my head,
> With the quiet sun behind it,
> In the years when I am dead.

Carman died on June 8, 1929, and his wish was granted on May 13, 1954, when historian A. G. Bailey and others planted a scarlet maple beside the poet's grave in Forest Hill Cemetery outside Fredericton, New Brunswick.

Oddly, the wilted leaves of the red maple—not the fresh leaves, nor indeed the leaves of any other maple—form chemicals that are toxic to horses.[61]

Ever adaptable, the red maple is a "supergeneralist" that thrives in all kinds of sites. It can develop wide-reaching roots in a wet site, or a deep taproot in drier locations. It can also adjust its gender, producing mostly male or female flowers, depending on local needs. Today, red maple is replacing trees of higher economic value (black cherry, oak, and walnut), along with the wildlife that depend on the nuts and acorns of other species. This can be devastating for the health of the forest, which needs up to a dozen different species for healthy survival, since any large group of single species could be hit with one disease or bug and be wiped out completely.

Red maple was first cultivated in 1656 and used for timber and pulpwood. The close-grained, light brown wood was discovered to be ideal for furniture, veneer, pallets, cabinetry, and plywood. The tree could even be tapped for maple syrup, like a sugar maple.

In 1748, the intrepid Swedish botanist Peter Kalm wrote from Delaware:

The red maple, or Acer rubrum, *is plentiful in these places. . . . Out of its wood they make plates, spinning wheels, feet for chairs and beds, and many other kind of turnery. With the bark they dye both worsted and linen, giving it a dark blue color. For this purpose it is first boiled in water and some copperas . . . is added before the stuff (which is to be dyed) is put into the boiler. This bark likewise yields a good black ink. . . . In Canada they make both syrup and sugar of its juice. There is a variety of this tree which they call the curly maple, the wood being as it were marbled within . . . utensils made of this wood are preferable to those of any other kind in this country, and are much dearer. . . . The tree is there-fore cut very deep before it is felled, to see whether it has veins in every part. . . . The curled maple is a species of the common red maple, and likewise very difficult to obtain. You may cut down many trees without finding the wood which you want.*[62]

Of course this is precisely what the people kept on doing until the legendary virgin forest was no more.

In 1894, Rhode Island schoolchildren were asked to vote on their favorite tree. Of the 16,776 ballots cast, silver maple won with 5,720 votes (elm came second with 5,260). However, the state government chose red maple as the state tree instead of the silver maple. Jimmy Carter also had a fondness for red maples, and planted one on the grounds of the White House.

Aceraceae

Silver Maple

Acer saccharinum, soft maple, river maple, swamp maple, water maple, white maple

With its great trunk, the gentle slope of its long lower branches, and a deeply etched trunk of silver and gray, the silver maple is a stunningly attractive tree. The shape is like a gnarled upturned hand. This is the urban tree of choice, and no matter what dreadful things have been perpetrated on it by city pruning crews, it still keeps a lovely shape because of those weeping twigs.

Acer saccharinum grows 70 to 120 feet high, with silvery gray bark that becomes deeply furrowed as the tree ages. The deeply cut pale green leaves turn a light yellow but aren't really spectacular in autumn. Its most compelling quality is the greenish-yellow flowers which open before the leaves come out in spring. Small red buds swell and then turn yellow as the tiny flowers emerge. The females mature quickly into winged fruits, germinate quickly, and can be established seedlings by autumn.

Native Americans used the silver maple sap, bark, and wood in much the same way as they used sugar maple. But the silver maple's real place in history is connected with the regreening of America. By 1830, there was so much deforestation around small towns throughout New England that settlers realized the damage inflicted on this once-beautiful land. In a remarkably forward-thinking move, the townspeople of Woodstock, Vermont, decided they would replant the lost forest. They held meetings, raised money, and appointed each village boy to look after a specific tree on the green. Other towns followed. Thousands of silver maples were planted in Washington in 1833 and the Dartmouth Ornamental Tree Planting Society set out to establish maples and elms in the spring of 1844.

In *Garden and Forest* in 1891, Charles Sprague Sargent wrote that the silver maple was "one of the largest and most beautiful trees [but] . . . a valuable tree in ornamental planting only when it can be placed in deep, rich, and moist meadow-land or by the banks of streams or lakes."[63] He planted one in such a location in 1881. It is now the largest tree (180 feet high) at the Arnold Arboretum. (*A. saccharinum* usually grows 40 to 100 feet high.) Frederick Law Olmsted, the great landscape architect (a term he coined), planted the tree wherever he could, because it grows several feet a year.

Silver maple fell out of favor as a city tree by the end of the nineteenth century. The huge brittle branches tend to come tumbling down in storms, and smaller branches litter the ground after a high wind. Its aggressive roots also heave up sidewalks and infiltrate sewer lines. As trends change, other trees will take its place in the urban forest.

Range: New Brunswick, west as far as South Dakota, south to Oklahoma and Louisiana, east to northwest Florida and Georgia.

Aceraceae

Striped Maple
Acer pensylvanicum, moosewood, moose maple, goosewood, whistlewood

Acer pensylvanicum

grows on average 8 to 15 feet high, but may get to 30 feet. It has smooth green bark with whitish stripes that eventually darken. The large leaves, which have three sharply pointed lobes, smaller than other maples, turn yellow-gold in autumn.

Range: **New Brunswick, Prince Edward Island, Quebec, Ontario to Michigan, south to Ohio, Tennessee, and northern Georgia.**

Here is the prettiest of all little understory trees. With such a firm presence and startling appearance, it makes a walk in the woods worthwhile to look for it alone. The elegant white striations running vertically along the green bark are completely different from anything else in a dark forest. The Baily Arboretum in New York has an astounding example—a rare 77-footer.

Stripled maple is related to species found mostly in China and Japan. Geologists call plants like this "stranded species," which means that when the continents separated millions of years ago, these plants were left alone on this continent, while their closest relatives lived on another continent.

The spelling of its Latin name is both peculiar and confusing. Back in 1753, Carolus Linnaeus, in his determination to name all living things, spelled *pennsylvanicum* as *pensylvanicum* and his error has caused arguments between botanists (who spell it as Linnaeus did) and editors (who love to add a second '*n*') ever since.

Striped maple is commonly known as moose-

Acer pensylvanicum, striped maple

goosewood because of its odd-shaped leaves, which have a faint resemblance to goose feet. The name whistlewood comes from the fact that children used to make whistles out of the bark, which slips off easily, as any hungry porcupine can attest. Porcupines, however, eat only small patches, perhaps saving the rest for future meals.

Native Americans took many cues from animal behavior. They observed moose chewing away at this tree and believed it was because of its healing qualities; therefore, they made a tea from the bark as a cold remedy, for respiratory trouble, and for kidney problems. The Penobscot made a poultice of the bark which they applied to swellings. They also used the wood to make arrows and burned it for firewood.[64]

Today, it is widely planted as an ornamental under the shelter of taller trees. The wood is used only occasionally by cabinetmakers for decorative inlay. One new use may bring moosewood to more prominence: an active antitumor substance has been isolated from striped maple, and is currently being tested to determine its value.

THE BIRCHES

Betulaceae

BIRCH
Betula spp.

Birches with the sun or moonlight shining behind them can evoke the vast peacefulness of the early Eastern forest. The white bark shimmers in the cool green forest. They beckon one to move closer to see and touch.

Ah, such fleeting beauty: by 60 a birch is elderly, and by 100 positively Methuselan. Its

wood, because moose eat the sweet young branches. They nibble at the leaves in summer, and the shoots in winter. It's also known as

Betula spp., birch

Birches are part of a family that includes alders, hornbeams, hop hornbeams, and hazels. Altogether there are six genera in the Betulaceae family, and more than 100 species, 15 of which are found throughout the temperate region of the northern hemisphere; one-third grow in New England. North America has seven native birches and they all have the familiar serrated edges on their leaves and the instantly recognizable bark. However, an imported white birch from Europe, *Betula pendula*, sometimes called silver birch, can also be found in North American forests. "Calvin Coolidge, in a shameful display for a Vermonter born in the heart of *B. papyrifera* country, planted a European birch on the grounds of the White House."[65] Birches grow best in moist woodland soil. The further south they grow, the greater elevations they can tolerate.

Birch has several negative associations. Since the seventeenth century in England, the term "birching" meant getting hit with a birch switch on the hands, legs, or backside. Witches were said to ride a birch broom, or besom, a large whisklike bundle of twigs bound around a handle. (Harry Potter on his Nimbus 2000 uses a more modern variety.)

Perhaps the worst crime ever perpetuated against a tree by a human being was performed upon an 11-foot birch in September 1980 by Jay Gwaltney of Chicago, Illinois, when a radio station asked: "What would you do for $10,000?" Over a period of 18 hours, he ate the whole thing.

Betulaceae

Yellow Birch
Betula alleghaniensis, gray birch, curly birch

dogged resilience can be observed in any winter storm as the birch bends over like a croquet hoop and then gamely stands upright again in spring. Supple birches were immortalized by Robert Frost, that cold and crotchety poet, who certainly knew his trees ("One could do worse than be a swinger of birches").

In autumn, the yellow birch turns a gold so luminous it outshines all the trees around it.

Only eastern North America is graced with yellow birch. It grows in deeper shade and in a wider variety of soil types than white birches, though, like them, it seldom grows in pure stands. It is often accompanied by other birch species or by striped maple or mountain maple.[66] The long horizontal markings on its bark are the lenticel cells that allow gas to pass between the inside and outside of the tree.

Mostly associated with the climax forest (the most stable stage of a forest's evolution), the yellow birch is also known as a fugitive species, one which takes advantage of any break in the forest canopy to seed. When an old tree falls or is knocked over by the wind, yellow birch seeds germinate quickly. It is monoecious—separate male and female catkins grow on the same branch. The tiny greenish male catkins appear in spring along with the leaves. During late summer, the females form at the end of long shoots in clusters of two or three. When they become receptive, the males shed pollen for three to five days. By August or early September, little winged nutlets begin to ripen.

Betula alleghaniensis grows to 100 feet high. The leaves are dark green above and light yellow-green below with serrated edges. It has pale gray-brown bark with a slight yellow cast. Either the bark or the yellow fall color gives this birch its common name.

Birches produce seeds by the thousands for at least 40 years. As they fall to the ground in cold October weather, the wind scatters them in all directions. Few seeds survive since most are gobbled up by birds and animals. The remainder must find a receptive place, but not in leaf litter, which will smother them. Once the warmth of June arrives, they germinate in the most unlikely spots on mossy logs, decayed wood, rotten stumps, or cracks in boulders. Driven by a gust of wind, they can even perch on the face of a cliff. Some birches appear to grow on stilts because the seeds originally landed on stumps. The stump

eventually rots away, leaving behind the thick roots, which seem to leap out of the soil. These above-ground roots are known as "prop roots."

Red-shouldered hawks use the birch as an important nesting site. During the winter, many other birds will feed on the brown seed catkins and, in spring, white-tailed deer can be seen browsing on pale saplings. As a food source for humans, however, the birch was limited. The Algonquin of Quebec mixed the sap with that of maples to extend the syrup they had acquired from other species and the Ojibwe used the combination of both saps as a drink. Since the wood is soft, it was used for lumber or in making birchbark canoes. The Ojibwe used the bark to make storage containers, sap dishes, rice baskets, buckets, trays, and winnowing dishes as well as ordinary eating vessels.

Birch was also a medicinal plant. Many native peoples used the bark as an emetic, to remove bile from the intestines, as a blood purifier, or to treat other internal blood diseases. Cleansing rituals and acts of purification, extremely important to these peoples, were conducted regularly. The Iroquois boiled up bark and water to treat what they called "Italian itch." The Mi'kmaq heated the bark and used it as we use hot water bottles. They also made trumpets for calling game, and torches for night fishing, an important part of their food-gathering routine. Finally, as a cleansing ritual, Native Americans placed birch bark in coffins when they buried their dead.[67]

Huron Smith, an anthropologist of the Eastern woodland Native Americans, spent a lot of time observing how they used the plants of the forests around them. In 1933, he noted:

*The [birch] twigs are aromatic . . . [and]
gathered . . . to extract fragrant oil which is used
as a seasoner for other less pleasant medicines. . . .
The Forest Potawatomi recognize the strength of
Yellow Birch and it is a preferred material in its*

sapling stage for wigwam poles. These poles are set up in a circle and then bent down at the tip to meet and overlap in the center where they are tied together in the form of a hemisphere which makes the framework for the wigwam or medicine lodge. It also endures for a fair length of time and when the family moves it is left in position for it is but a matter of half a day to throw together another wigwam.[68]

François André Michaux, in 1803, had the distinction of being the first European to describe the tree in detail. At that time there were still astoundingly large birches in existence. In Maine, he acquired several quart-sized containers full of the minute seed of white and yellow birches. *Betula alleghaniensis* was the name given to the tree by Nathaniel Lord Britton (1859–1934), first director of the New York Botanical Garden and a specialist in North American flora, indicating it comes from the Allegheny Mountains.

As a commercial wood, birch came into its own in 1840, when John Dresser of Stockbridge, Massachusetts, invented a hand-cranked veneer lathe. This tree, considered one of the finest veneer trees, has a slightly pink overtone in the yellow wood and a beautifully figured grain with lustrous sheen.[69] In lumber camps, yellow birch made yokes for oxen and walls for bunkhouses. It was often used for the hubs of cartwheels because its fine grain meant the spokes seldom worked loose.

Birch furniture is usually made with this species because, once it is stained and varnished, the nearly white sapwood deepens to pale gold. In Nova Scotia, shipbuilders were particularly fond of the wood, and yellow birch was used in the construction of the famous "ghost ship," the *Mary Celeste*. A brigantine built in Nova Scotia in 1861, she was found mysteriously sailing across the Atlantic in 1872, with not a soul on

board. There was no sign of damage or violence; everything was eerily in its place. She was returned to service and ended up sinking off a coral reef near Haiti in 1885. Modern searchers combing the ocean floor 116 years later discovered her watery grave when some wooden fragments were analyzed—and found to be yellow birch specific to certain vessels built in Nova Scotia. However, the true story of that day in 1872 will likely remain one of the great mysteries of the sea.

The birch was chosen as the Mother Tree of America in the 1920s by the American Forestry Association. The idea was that all over the nation, everyone would plant a tree on Mother's Day. Today it is marketed as Canadian silky wood and is much prized for cabinets, paneling, and marquetry. Much of the birch harvested nowadays, however, ends up as toothpicks, Popsicle sticks, tongue depressors, dowels, and clothespins.

Range: **Newfoundland to Wisconsin and Minnesota, south through Pennsylvania and the Appalachian Mountains to Georgia.**

Cherry Birch
Betula lenta, sweet birch, black birch

The bark of the cherry birch resembles a wild cherry tree with colors ranging from smoky gray-black to dark red to dark brown. The most distinguishing feature of cherry birch is its catkins, those cheekily erect fruiting bodies poised at the tips of the branches. They remain on the tree over the winter and are most obvious in the spring, when interest in the cherry birch once ran at its highest.

The copious flow of sap occurs later in the season than it does in maples, so the settlers would transfer their spiles and buckets from the maples to the birches in late spring. Birch sap, however, is less than half as sweet as maple sap. It takes 10 gallons or more of sap to make a mere pint of birch syrup, which Euell Gibbons pronounced, "is good, with a flavor resembling that of sorghum molasses." The sap was also used to make a beer which Gibbons (who must have tried everything that grows in the forest) did not recommend for children, as he said it "has a kick like a mule."[70]

Birch twigs have a wintergreen scent and taste, and at one time were the source of wintergreen flavoring. Before the nineteenth century, creeping wintergreen (*Gaultheria procumbens*), also called checkerberry or teaberry, was harvested for wintergreen oil, but later cherry birch was used instead. The people living in the Appalachian Mountains, where the tree flourished, gathered huge quantities of the twigs to supply the demand. They nearly wiped out the species, because it took 100 birch saplings to derive 1 quart of essential oil. Fortunately, the development of synthetic wintergreen oil saved the tree.[71]

Betula lenta has dark bark marked with horizontal lines and vertical cracks and grows up to 60 feet tall. The finely toothed oval leaves are 2½ to 5 inches long, with tufts of white hair on the undersides. Range: Maine to Alabama and west as far as Ohio.

B. occidentalis, **water birch**, occasionally called black or red birch, is a smallish shrubby tree bordering the cold streams of the Rocky Mountains, while the **river birch**, *B. nigra*, also called black birch, ranges from Massachusetts to Florida and then west in a wide swathe along the Gulf Coast to Texas, extending north through Kentucky, Tennessee, and Missouri, making it the most southerly growing birch in the world. The Carolina planters once used it to make hoops for their rice casks.

Caesalpiniaceae

Kentucky Coffee Tree
Gymnocladus dioicus, king tree

Gymnocladus dioicus

grows 60 to 80 feet high and spreads out 50 feet wide; the double compound leaves may be up to 2 or 3 feet long made up of leaflets 2 inches long. The inflorescences are almost invisible, but do appear in May; male and female flowers occur on separate trees. The 8-inch-long fruits are purplish-brown pods that mature in autumn on the female trees; the seeds are buried in the sweet pulp of the fruit.

Range: southern Ontario (where it is now on the rare list), south to Michigan, Wisconsin, southeast Minnesota, east South Dakota, south to Alabama, Arkansas, and Oklahoma.

State tree of Kentucky.

Imagine seeing a tree unlike anything you've ever seen before in your life. Questions pop into the mind: What is it? Why does it look this way? Questions that no doubt occurred to the first settlers when they saw the Kentucky coffee tree: a tall slender tree with massive leaves giving it a strange ferny look, and great long exotic-looking beans.

Gymnocladus means "naked branch" (in winter, a young tree looks like a dead stick) and *dioicus* means it is dioecious; that is, there are separate male and female trees. The thick gray bark has scaly ridges; the twigs, hairy in youth, turn dark brown and have heart-shaped leaf scars as they age. The gigantic double compound leaves are the last leaves to come out in spring in the Eastern forests. Its only competition for double pinnately compound leaves in our native

Gymnocladus dioicus, Kentucky coffee tree

rifically hard to spot. Jan Watkins of Milton, Ontario, recounts her experience with this floral enigma and the difficulty of telling a plant's sex:

I read that it was dioecious. Two sexes, yes, but which sex did I have? My graceful tree showed no sign of puberty as it cast a dappled shade on the lawn. Huge doubly-compound leaves clothed the tree, yet, at age 16 and 16 feet tall, never a bean nor flower nor catkin had graced her. Him? Was she sick for romance? Would I ever see the huge bean pods which lent the tree its name?

Finally one snowy Christmas Day, a friend noticed two large beans hanging from the gaunt naked tree and, each year thereafter, the bean crop increased steadily. Yet, I saw no flowers in summer. At last, one sunny June 22, 18 years after planting the tree, my heart missed a beat. I saw one blossom nestled at the base of the long-stemmed leaves, pointing straight at me and looking like an enlarged spike of lily-of-the-valley. Each year I search, but I have never seen her flowers again, nor ever found her suitor.

The Dakota, Oto, Ponca, and Winnebago used an infusion of the roots as a remedy for constipation and turned the root bark into a type of snuff to revive unconscious patients. The Meskwaki fed the wax of the pods to those suffering from lunacy; the Omaha fed the powdered root mixed with water to women during childbirth, to ease labor pains. The roasted seeds were popular as snacks: the Pawnee ate them like chestnuts. The Winnebago used the seeds as counters or tally checks in gambling.[72]

Early settlers noticed that the seeds resembled coffee beans and roasted them to make a coffeelike drink, but it never tasted quite like the real thing. The great botanist Thomas Nuttall considered it "greatly inferior to *cichorium* (chicory)."[73] The name "coffee tree" appears in

forest is the honey locust *(Gleditsia triacanthos)*— but even that tree is nothing like the scale of the Kentucky coffee tree with its 4-foot leaves. The greenish-white flowers produce brown pods 4 to 8 inches long and 1½ to 2 inches wide containing seeds that look like coffee beans. It is a strange and wonderful sight to behold anywhere, but in comparison with the other trees of this forest, it is exotic.

The flowers aren't just subtle, they are ter-

George Washington's diary in the late 1700s. He was given some seeds to plant at Mount Vernon, but does not mention what he thought of the drink.

Today the Kentucky coffee tree is used for lumber and woodworking. The sapwood is narrow and yellowish-white, and the heartwood is light red to reddish-brown. The wood is hard and heavy, with a coarse, straight grain. As interest in native plants increases, it is being used more and more as an ornamental and street tree.

Magnoliaceae

Tulip Tree

Liriodendron tulipifera, yellow poplar, tulip poplar, tulip magnolia, whitewood

Liriodendron tulipifera, tulip tree

Liriodendron tulipifera

grows 60 to 90 feet high, sometimes taller, with a spread of 45 feet. The smooth gray to gray-brown bark gets furrowed as it grows older. It blooms in late May or early June, depending on the location. The yellow-green tuliplike flowers (up to 2 inches wide) have three sepals and six greenish-yellow petals in two rows, with an orange corolla. It grows in full sun in moist soil.

Range: southern New England, west through southern Ontario and Michigan, south to Louisiana, east to north-central Florida.

State tree of Indiana, Kentucky, and Tennessee.

Backlit by watery spring sunshine, the leaves of the tulip tree unfold in an astonishing array, vividly bringing to mind its long history. It is a primitive plant and, according to the fossil record, back in the Cretaceous period (100 to 70 million years ago), there were several *Liriodendron* species spread throughout North America and Eurasia. This is one of only two to survive the great Pleistocene ice age, and they are the only two species in the genus (rare in itself). For a long time, people thought that *Liriodendron* in North America was unique, but in 1875 another species was discovered in western China.

It grows relatively quickly (up to 3 feet a year) with fleshy roots reaching deeply into the soil, and there are some specimens with huge thick trunks which may be as much as 500 years old. Although the tulip tree is among the most primitive of all seed plants, the duck-billed shape of the leaves seems more exotic than ancient. As John Eastman puts it, "successions of bills within bills uncurl and unfold, revealing a marvel of leaf packaging."[74]

The tree doesn't begin to flower until it's 20 years old, but once started, it will continue for the next two centuries. Even though only about 10 percent of the flowers are ever fertilized, nothing seems to disrupt its pattern of bloom.

The blossoms are more like magnolias than

tulips (it's forgivable to confuse the two) and they emerge after the leaves. But they are difficult to get a good look at because their real splendor is high in the canopy. The copious nectar is a magnet for honeybees, and the honey produced is highly prized. Though the blooms are usually pollinated by beetles, flies, or bees, it's not unusual to see tiger swallowtail butterflies lurking in the canopy so their caterpillars can feed on the four-lobed leaves. Pollination takes place during the 12 to 24 daylight hours after the flowers open, when the stigmas are light colored. Once they darken, the window of opportunity slams shut for another year.

The seeds are held in conelike clusters which produce a huge amount of seed, especially when the weather is hot and dry during pollination. The winged samaras holding the seed can be flung by the wind four to five times the height of the tree in distance. They can also hang on to the tree over the winter.

The inner bark of the root and trunk is the source of the bitter alkaloid, tulipiferine, a heart stimulant. Native Americans sometimes used the powdered bark and seeds for rheumatic and digestive problems and as a remedy for worms.

The wood is a soft creamy white (hence whitewood), with a grain straight and free of knots, and is resistant to splitting. The Cherokee used it for making dugout canoes and cradles. Because of the immense girth of the trunk, it was possible to whack out enormous pieces of whitewood. Woodworkers in the nineteenth century made great use of it. John Tradescant the younger is credited with taking this tree back to England and introducing it to gardeners there in 1654.[75] Specimens are now found in many European cities. It's an astonishing sight to visit one of the major Italian gardens, the Boboli in Florence, and see a massive tulip tree in its full May bloom. Today, the wood is mainly used for paneling and veneers.

Pinaceae

Eastern Hemlock
Tsuga canadensis, Canadian hemlock, hemlock spruce

Tsuga canadensis

grows 60 to 90 feet tall, with a conical crown. The young trees have red flaky or scaly bark, becoming furrowed with age. The terminal twigs droop to one side. The needles are ¼ to ¾ inch long, shiny green above and pale green below, with two white bands. The needles are actually arranged spirally but lie flat. The cone is ½ to ¾ inch long with light brown scales.

Range: Nova Scotia to southern Ontario, south to northern Alabama and Georgia.

State tree of Pennsylvania.

The shadow of the Eastern hemlock is so dark and so dense it is referred to as casting a "blue shade" in contrast to the more delicate green shade of deciduous trees. It is one of the great nurse logs of the whole Eastern forest, playing as important a role in death as in life. Nurse logs are crucial to a healthy forest, providing food and habitat to millions of fungi and microorganisms that slowly break down the wood and fiber. (See page 469–470 and 477–478 for more on nurse logs.)

Eastern hemlocks start to produce pollen and seed cones when they are about 20 years old. And though they need moist ground free of deep leaf litter and shade in order to germinate, the seeds won't do anything at all in the shade of the parent plant, which is way too dark. In an opening in the canopy, these seedlings shoot up rapidly to bask in full sunlight. The shallow root system, however, makes those that get to the sapling stage vulnerable to drought, fire, wind, and disturbances such as road clearing. Left undisturbed, an Eastern hemlock can live to be 1,000 years old.

This is not the hemlock that Socrates made so famous. That was *Conium maculatum*, a member of the carrot family (Apiaceae), a completely different plant. Why this tree acquired the same name is anyone's guess. The Onondaga called it *On-e-tah*, and the Iroquois *haneta*, both names apparently meaning "greens on a stick."

Native Americans had a wide variety of medicinal uses for it. The bark contains a high concentration of tannic acid, making it similar to tea, and the Mi'kmaq used it to treat colds and coughs. The Ojibwe crushed and powdered the bark as a remedy for diarrhea. Because the tannic acid had a disinfectant quality, the bark was applied to cuts and wounds. The Delaware used an infusion of the roots in a steam treatment for aching joints and swollen muscles. The bark also produced a red-brown dye and the wood was used for lumber or fuel. As Campbell Hardy explains: "The sojourner in the woods seeks the dry and easily detached bark which clings to an old dead hemlock as a great auxiliary to his stock of fuel for the campfire; it burns readily and long, emitting an intense heat, and so fond are the old Indians of sitting round a small conical pile of the ignited bark in their wigwams, that it bears in their language the sobriquet of 'the old Grannie.'"[76]

The early settlers copied some of these traditions. In 1853, Samuel Strickland, one of the many travelers of the period who kept diaries, recounted: "I have camped out, I dare say, hundreds of times, both in winter and summer; and I never caught cold yet. I recommend, from experience, a hemlock-bed, and hemlock-tea, with a dash of whiskey in it merely to assist the flavour, as the best preventive."[77] Of course, Strickland may have underestimated the medicinal qualities of whiskey, but he was certainly not alone in extolling the qualities of hemlock tea.

Tsuga canadensis, Eastern hemlock

Catherine Parr Traill, writing in 1855, noted that, while not the most palatable beverage, "As a remedy for a severe cold, I believe a cup of hemlock-tea, drunk quite warm in bed, is excellent, as well as suderific [causing a sweat]."[78]

The Europeans didn't pay much attention to the hemlock until they discovered its use as a tanning agent. Hundreds of tanneries were established in eastern Canada and the United States in the late nineteenth century, and whole forests were cut down and stripped of their bark for the tannic acid. The industry disappeared as new synthetic tanning agents were developed in the twentieth century.

The coarse, somewhat brittle wood of Eastern hemlock used in roofing, subflooring, boxes, crates, and pallets, was considered inferior to other lumber species for more sophisticated uses. Today, most hemlock winds up as newsprint and wrapping paper.

Recently, an insect pest from Asia, the unpleasant-sounding hemlock woolly adelgid, has started to attack the species in the southern

Appalachians, giving rise to the fear that another important species may go the way of the American chestnut.

Tiliaceae

Basswood

Tilia americana, linden, American linden, lime tree, whitewood

Tilia americana

grows to 60 to 80 feet on average but can reach 100 feet. The dark green, heart-shaped leaves are 4 to 8 inches long, serrated with long teeth, paler underneath with tufts of hair in the vein axils. They turn pale yellow in autumn. The bark is gray to brown with long, narrow, scaly ridges and some smooth patches. The fragrant pale yellow flowers are borne in clusters 2 to 3 inches wide on drooping stalks fused to leaflike bracts. The fruit is a nutlike drupe ⅓ to ½ inch long. Light brown wood is tinged with red, with a straight grain.

Range: southern Manitoba and Ontario, Quebec, and New Brunswick, south to Virginia and Alabama, west to North Dakota, Kansas, and Texas.

Tilia americana, basswood

This must be one of the very few trees known for its sound. It's sometimes called the "humming tree" or the "bee tree" because of the hordes of honey- and bumblebees drawn to its blooms.[79] The creamy-yellow scented flowers emerge after the leaves unfurl in June when the tree hits the age of 15 and it produces seeds nearly every year thereafter. The seeds are like tiny nutlets with samaras which, like little propellers, pilot the seed away from the parent tree into fertile ground for germination. Though the seeds may not sprout immediately, they remain

viable for several years. When the leaves drop, they are so rich in minerals they provide a good humus-filled soil ideal for seed growth. If there is a dry summer, the leaves turn brown. Even so, the brightly colored buds can be seen all winter long.

All *Tilia* species have bracts. In some species, the bracts look like flowers, but in this one they are green leaflike structures. In either case, they emerge just before the flowers.

Basswood has an interesting survival strategy. As one stem matures, another forms just below or next to it, so that when the older one dies, the newer one can replace it. As a result, an individual basswood plant may survive for centuries, constantly renewing itself.

Basswood grows beside other hardwoods such as sugar maples and white ash. A compact symmetrical tree, it is considered the strongest of all North America's forest trees because of its soft, light, tough wood. Indians used to watch where the bees would alight so they could find "bee trees" easily. For them, the inner bark's woody fibers (called the bast) were a source of rope, thread, and twine for lines and nets. Settlers soon learned the techniques of making rope and twine, and eventually the North American fishing industry came to depend on the tree (which became known as the bastwood tree).

Native Americans also used infusions and decoctions (rather syrupy brews) of the bark to make medicines, including an antiseptic eyewash, a poultice for burns, and a remedy for worms. A jelly made from the bark became known as an aid for consumption among the Cherokee. The colonists must have been cheered by the sight of this tree when they arrived in North America because it looked so much like the European linden trees (*T. × vulgaris*, a smaller species). Today basswood is still a popular tree along urban boulevards in both Europe and North America.

Europeans used the dried blossoms of linden to make soothing teas. This was the infusion or *tisane* into which Marcel Proust dipped his madeleine at the beginning of his seven-volume opus, *À la recherche du temps perdu (Remembrance of Things Past)*. The North American species, although useful as medicine, can cause nausea and was never used for infusions in the same way.

The wood is light as well as strong, and was used to make boxes for colonial exports, from chocolate and candy to pickles and fish. "Bee-keepers used basswood for beehives, bee boxes, and honey crates," which seem such an appropriate use for the humming tree. Farmers made chicken coops and incubators as well as basswood cartons for dressed poultry and eggs; and "dairymen once used basswood butter churns and cheeseboxes."[80] Basswood, its value unchanged, is still used in box making, as well as for yardsticks and chair bottoms.

In 1794, Matthew Lyon of Vermont developed a way to make paper from a combination of bark and rags. By the mid–nineteenth century, pulpwood was developed for commercial use, and basswood is still used for it.

White basswood (*T. heterophylla*), is a related species growing throughout the Northeast and as far south as Alabama and Tallahassee, Florida. The almost imperceptible blooms come out in June and July but are so small they are obscured by its leaves. The great humming of bees and the scent are still unmistakable here as they are in its relatives.

Fabaceae

Black Locust
Robinia pseudoacacia, yellow locust, false acacia

Robinia pseudoacacia

grows up to 70 feet with several large ascending branches and an open crown. The deeply furrowed bark is reddish-brown to nearly black; the orange inner bark becomes obvious in old trees. The pinnately compound leaves have 7 to 19 bluish-green oval leaflets. The leaves emerge later in spring than those of many other trees. The fragrant white flowers appear in racemes 4 to 8 inches long. Flat brown-black 2- to 4-inch-long pods mature in October.

Range: **Pennsylvania to Georgia, west to Iowa, Missouri, and Oklahoma. Widely naturalized beyond its range.**

Robinia pseudoacacia, black locust

Michael A. Dirr, author of definitive books on American woody plants, calls the locust an "alley cat" of a tree,[81] because it grows anywhere under just about any circumstance. It is a member of the pea family—therefore it has roots with nodules where nitrogen-fixing bacteria live and work, enabling the tree to grow in very thin or poor soils. For this reason locusts are sometimes planted in areas that have been strip-mined, to regenerate the soil and bring the land back to life.

During the great scouring of the land in the Ice Age, the only surviving black locusts lived in a small area of the southern Appalachians. The modern trees, all offspring of those trees, were subsequently planted throughout North America and Europe.

The scientific name honors Jean and Ves-pasian Robin, who were among the first to culti-vate the trees in Europe, in the sixteenth century. (It is now one of the commonest American trans-plants in Europe.) *Pseudoacacia*, "false acacia," sug-gests that the plant was once mistaken for an acacia, which also has compound leaves. It is also similar to the acacia in the way the leaflets fold and the leaves droop on rainy days and in the evening.

It is so tough, durable, and rot-resistant, it became a favorite wood for railroad ties, grapevine stakes, and fences. "Captain William Fitzhugh of Bedford, Virginia, enclosed his orchard with locust fencing which, according to a letter he wrote in 1686, was 'as durable as most brick walls.'"[82] Colonial shipbuilders used black locust to make dowels to fasten parts of a ship together. They were called treenails because they would swell up and make the hull tight and waterproof. Fishing boats especially were made with treenails because they were even more resistant to decay than boats made with oak.[83]

The fragrant flowers attract bees, which produce an excellent honey from the nectar. According to naturalist Steve Brill, the flowers are edible, and he adds them to salads and soups. The leaves and bark, on the other hand, are highly toxic. Although the native peoples used them as an emetic and laxative, modern herbalists avoid them. The toxins can be used to produce a fungicide for treating wood.

Hippocastanaceae

Buckeye
Aesculus glabra, Ohio buckeye

Ohio is called the Buckeye State and this tree is its state symbol. The tree got its common name because the split seed capsule resembles the half-opened eye of a deer. This poisonous relative of the horse chestnut was sometimes called "stinking buckeye" by pioneers because of the unpleasant smell of the flowers and the bruised bark. The nuts and roots were poisonous, and Native Americans sometimes threw them into ponds to stun the fish, which they could then grab with their hands—a practice now illegal. The Cherokee pounded the nuts to make poultices to soothe aches and pains. Back in the early days, Ohio farm wives made the nut kernels into a kind of laundry soap which would foam up when it was swished around in water. It contains saponins which have a molecular structure very similar to that of steroid hormones.[84]

A shade-tolerant tree, it is found in wet soil on streambanks and in bottomlands. The dark green leaves consist of five leaflets which turn a soft pumpkin color in fall.

A near relative, **sweet buckeye**, *Aesculus flava* (also known as *A. octandra*), grows from 60 to 90 feet high with gray or dark brown bark which becomes furrowed with large smooth flat plates

and covered by large scales as it ages. Michael Dirr considers it the most beautiful of the large-growing buckeyes. The yellow-green flowers grow in spikes and produce a nut encased in a spiky husk.

Elizabeth Lawrence in *Gardening for Love*, her book about the market bulletins of the Southeast, writes:

> *The mountain people [of Georgia] also advertise buckeyes. They consider them lucky. An extract made from them is believed to cure rheumatism. Even carrying a buckeye in your pocket, it is thought, will keep you free from rheumatic pains. One species,* Aesculus octandra, *is used to give the appearance of age, a caramel tone, to the clear raw whiskey some moonshiner has distilled in some deep ravine. In the Georgia mountains, buckeyes are sold by the gallon.[85]*

Range: **throughout Pennsylvania to Tennessee and northern Georgia, west to Ohio and Illinois. State tree of Ohio.**

Cornaceae

Flowering Dogwood
Cornus florida, American dogwood

Cornus florida
grows from 15 to 50 feet high, depending on its location. It has yellow flower clusters surrounded by large white bracts that open in spring. The bark and the hardwood are both a reddish-brown; the wide sapwood is a creamy color. The glossy green, simple leaves are 3 to 6 inches long and turn red in autumn. The clusters of red fruit ripen from September to October and remain hanging on the tree.

Range: **Maine to New York, southern Ontario, Michigan, Illinois, Mississippi, south to Kansas, Oklahoma, and Texas, and east to Florida.**

State tree of Virginia.

This was Thomas Jefferson's favorite tree. He planted it all around his home at Monticello, Virginia.

Virginians loved it enough to name it their state tree. Explanations of its common name vary. Some say it cures mange in dogs, others that dogs like chewing on it. John Eastman says that the smell of the fresh-cut wood is like dog feces.[86] The Latin name, *Cornus*, means "horned" and describes the way it grows from a forked trunk close to the ground. The branches grow horizontally, giving it a graceful layered look.

In the record of his travels through North America (1777), William Bartram recorded the following:

We now entered a very remarkable grove of Dog wood trees (Cornus florida), *which continued nine or ten miles unalterable, except here and there a towering* Magnolia grandiflora; *the land on which they stand is an exact level: the surface a shallow, loose, black mould, on a stratum of stiff, yellowish clay. These trees were about twelve feet high, spreading horizontally; their limbs meeting and interlocking with each other, formed one vast, shady, cool grove, so dense and humid as to exclude the sun-beams, and prevent the intrusion of almost every other vegetable, affording us a most desirable shelter from the fervid sun-beams at noon-day.*[87]

Cornus florida, flowering dogwood

The flowering dogwood has a graceful profile with a rounded canopy, and glossy green leaves. What look like spring blooms are actually creamy white bracts (this is a structure beneath a flower that can easily be mistaken for a petal when the flowers themselves are minuscule). The berries or drupes (a fleshy fruit containing a single stone) have the high fat content so important for migrating birds, which use gigantic amounts of energy on their long flights. The robin, for instance, depends on this plant in getting from its winter grounds in the South to the northern breeding grounds of summer. In return, birds help spread the seeds far and wide on their travels.

In early times, the Indians used this as an indicator plant: when the dogwood bloomed, it was time to plant the corn. Both this plant and its near relative, the alternate-leaved **pagoda dogwood** *(C. alternifolia)*, were also valued as medicinal plants. The highly astringent bark was simmered in water and used as a tonic to relieve sore muscles, or as an antiseptic. Chewing on the bark was recommended for relieving a headache or soothing a sore throat. People also gave it to children suffering from colic, worms, measles, or diarrhea. A tonic made from the roots was taken as an all-around preservative of good health and to purify the blood. Frayed twigs were sometimes used as toothbrushes.

The amount of time Native Americans spent gathering plants to use for personal hygiene was remarkable. Morning and evening purification rites involved washing all parts of the body. It was, therefore, a great shock for these people to come upon Europeans who thought nothing of staying in the same clothes and going without anything but a cursory wash for months on end. The olfactory clashes must have been immense for both, because part of survival in the bush for the Indians was covering themselves with bear grease against insect bites—not exactly an agreeable smell to their neighbors.

Dogwood berries in large quantities are poisonous to people though not to animals. In spite of this, they could be used as a form of bitters with brandy or whiskey without drastic effect. During the Civil War, when American ports were closed, supplies of cinchona bark (quinine) from Peru, the main remedy for treating malaria, were cut off. Dogwood bark, however, was found to contain the same substance and be just as effective.

The wood is strong and heavy, with a fine grain. Almost completely shock-resistant, it has been widely used over the years to make the shuttles of looms, tool handles, wheel cogs, hay forks, golf club heads, and knitting needles.

Today, the flowering dogwood is threatened in the wild. It is being attacked by an anthracnose fungus called *Discula destructiva*, which makes the plant susceptible to borers, acid rain, and other types of environmental stress. The fungus is believed to have been brought into North America on **Japanese dogwoods** *(C. kousa)*. It arrived simultaneously in Seattle and New York and is ravaging many native dogwoods in the Eastern states. Flowering dogwoods in cultivation are generally well enough cared for to resist the infection.

Not all dogwoods are trees. **Bunchberry**, *C. canadensis*, is a ground cover with a lovely white bloom made up of four white petal-like bracts, each with a cluster of tiny yellow flowers. It rarely grows more than 1 foot high. The berries are edible, and Philip Henry Gosse, in *The Canadian Naturalist* (1840), described them as "farinaceous and agreeable." A more modern writer, Peter Landry of Dartmouth, Nova Scotia, dismisses them as "tasteless," but adds offhandedly that they can be "thrown into a pudding for color."[88]

Red osier dogwood, *C. stolonifera*, grows into a 9-foot-high shrub. It is sometimes known as "kinnikinnick" (a name also used for *Arctostaphylos uva-ursi* or bearberry), because the Chippewa used it for smoking. The word *kinnikinnick* was Algonquin for "mixture"—describing the variety of substances often mixed to make a satisfying smoke.

Styracaceae

Halesia carolina

is a small tree that grows to 30 to 40 feet high with gray to brown bark developing into scaly plates. The alternate simple, dark lime-green leaves are 2 to 5 inches long. The white flowers are bell shaped in two- to five-flowered nodding clusters on year-old wood in spring. The fruit is flattened with four wings, which come away from the stone containing one to three brown seeds.

Range: North and South Carolina, Georgia, Alabama, Tennessee, to eastern Texas.

The slender white bells of the *Halesia* are among the most delicate in a forest full of exquisite spring nuances. This small graceful tree is part of the climax forest stage of the Appalachian Cove forest, along with tulip trees, white basswood, and magnolias. *Halesia* was named after one of the greatest English scientists of the seventeenth century, Stephen Hales (which is why it's pronounced hales-ia and not haleesia). He conducted rigorous experiments on plants and is considered the founder of experimental plant physiology. He figured out how to measure transpiration (water being drawn from the roots of plants and given off by the leaves) and found out that plants draw carbon dioxide from the air. This revolutionary finding was pretty much ignored in his time, but many of his other discoveries were not. He proved that sap ran upward, that fresh air (a.k.a. good ventilation) was crucial to health. But one of his sensible contributions to everyday life was the practice of

Halesia carolina, Carolina silver bell

putting an inverted teacup into a pie to keep the pastry from getting soggy.

Lumbermen used to call the little tree "bell-wood." "Its beautiful figures do not appear when the wood is cut like ordinary lumber, but when the log is turned against the rotary veneer knives, a beautiful bird's eye figure . . . may emerge."[89] Though few use it today for lumber or woodworking, it is one of the finest and most charming of ornamentals used in the garden.

Lauraceae

Sassafras

Sassafras albidum

Sassafras albidum, sassafras

Sassafras albidum

is a medium-sized tree in the southern part of its range, growing to 80 feet tall; a 10-foot shrub in the north. It had rough gray bark and smooth green twigs. The bright green leaves, which turn vivid colors in fall, may be oval, two-lobed (mitten-shaped), or three-lobed. The flowers, which appear before the leaves, are greenish-yellow on racemes; the pealike drupe is bright blue. Grows quickly on old fields and open woodlands, where there isn't too much shade, and even does well in urban parks or along city streets.

Range: southern Ontario, New England, and the Eastern states as far west as eastern Oklahoma and eastern Texas and as far south as central Florida.

Legend says the odor of sassafras blowing offshore first alerted Columbus to the nearness of land. It was the smell of the New World coming to meet him.

European travelers to and settlers in North America were united in their praise of the plant.

Fanny Trollope, mother of novelist Anthony Trollope, who found much to criticize in the United States, was completely taken with the sassafras. "The sassafras is a beautiful shrub," she wrote in *Domestic Manners of the Americans,* "and I cannot imagine why it has not been naturalized in England, for it has every appearance of being extremely hardy. The leaves grow in tufts, and every tuft contains leaves of five or six different forms. The fruit is singularly beautiful; it resembles in form a small acorn, and is jet black; the cup and stem looking as if they were made of red coral."[90]

Long before Mrs. Trollope noticed it, other Europeans had taken an interest in the tree's healing qualities. Nicholas Monardes of Seville, who wrote the exuberant *Joyful News Out of the*

New Found World in 1577, considered it a cure-all, but especially good for the ague (malaria, really). He promoted it so enthusiastically that it also became known as the ague tree. Others were equally impressed. The Jesuit Paul Le Jeune wrote in the *Jesuit Relations* (1656–1657): "But the most common and most wonderful plant in those countries [of the Iroquois] is that which we call the universal plant, because its leaves, when powdered, heal in a short time wounds of all kinds; these leaves which are as broad as one's hand, have the shape of a lily as depicted in heraldry; and its roots have the smell of the laurel."[91] He also noted that its roots produced a wonderful red dye. The Iroquois, Seneca, Onondaga, and others made a tonic from sassafras roots. It was used to reduce fevers, purify the blood, relieve dysentery, and treat syphilis.[92] (It is still being argued today whether venereal disease was brought to the New World or if it was a gift from the Americas, though most recently 1,000-year-old bones from southern Africa and 2,500-year-old bones from a Greek skeleton in Italy showing evidence of venereal disease have been found.[93]) The important medicinal ingredient in the roots was safrole, a volatile oil that is also present in nutmeg. Modern science has found not only that it is an antiseptic and a stimulant, but also carcinogenic to rats in large doses.

Sassafras quickly became one of the most important exports from the colonies—picked as an excellent commodity right from the beginning of their tenure. It was as important as beaver and cod, and became one of the main reasons for exploring the Eastern forest. "As for Trees the Country yeeldeth Sassafras a plant of sovereigne vertue for the French Poxe, and as some of late have learnedly written good against the Plague and many other Maladies," wrote Martin Pring of the plants he saw in Martha's Vineyard.[94] The local Indians were employed to help cut down these trees. And

Pring himself, after finding a huge market in London for the trees, came back with two ships to load up with sassafras as booty.[95] It was the wonder drug of its time and sold even better than tobacco. It gained an almost magical reputation, as Rebecca Rupp notes:

> *The odor of the wood alone was held to ward off illnesses, evil, and vermin. A ship with a sassafras hull as deemed safe from shipwreck. Cradles, spoons, and Bible boxes were fashioned of sassafras; hen roosts were built of it, in the belief that it repelled chicken lice; bedsteads were built of it, in the belief that it repelled bedbugs.*[96]

The Choctaw were probably the first to grind the leaves into a powder; the Louisiana settlers, who learned the technique from them, incorporated it so completely it has become an essential part of Cajun-Creole cooking. Today people simply enjoy sassafras for the flavor. It was once used as an ingredient in root beer (replaced by artificial flavorings). It is also still used to make gumbo filé, the Louisiana spice powder added to soups and other dishes.

Betulaceae

Eastern Hop Hornbeam
Ostrya virginiana, ironwood, leverwood, American hop hornbeam

Eastern hop hornbeam, though slow to develop and with little commercial value, has an admirable ability to flourish almost anywhere. It's rather like one of those people who keeps popping up in the most unexpected places. It grows at an altitude of as high as 5,000 feet in the Appalachian Mountains, and can take 100 frost-free days in some areas, 290 in others. It even thrives in Central America, which you'd expect to be right out of its range. It is usually found along with sugar maples and beech trees, but cohabits with all kinds of larger trees, including evergreens. It blooms at the end of March in the South, mid-May in the North.

Ostrya virginiana is deciduous, grows to about 35 feet, and has oval-shaped alternate leaves with serrated edges. They are dark greenish-yellow above, pale below. The grayish-brown bark is smooth, but the twigs and leaf-stalks are hairy. The male catkins each have 3 to 14 stamens and stay on the tree all winter. The female flower has a closed calyx with hairy inflatable scales, and the fruits, containing a single nutlet, are triangular and papery.

Its scientific name *(Ostrya)* derives from the ancient Greek term *ostreon*, meaning "oyster" or "shell," because the seeds are as closely protected as if they were in a shell. The name was conferred by Philip Miller (1691–1771), an English gardener, curator of the Chelsea Physic Garden, and author of *The Gardener's Dictionary*. It's called "hop" hornbeam because the little fruits resemble hops. "Hornbeam" is a name coined long ago, when related European trees were used to make ox yokes. It's also called ironwood because the wood is strong, hard, and heavy (not to be confused with *Olneya tesota*, the ironwood that grows in desert areas). The trees are small and rather scattered. It is not usually harvested for its timber, but the wood can be used for posts, golf club handles, tool handles, mallets, and woodenware. In the nineteenth century, it was used for sleigh runners, farm wagon-wheel hubs and axles, mauls, cogwheels, and levers—giving it the name leverwood.

It also had medicinal uses. According to Frances Densmore, "The inner bark of the chokecherry, the root of the hazel and white oak and the heart of the wood of the hop hornbeam

Ostrya virginiana, Eastern hop hornbeam

were steeped together and the liquid drunk for hemorrhages from the lungs."[97] Strong syrupy mixes were made to treat kidney trouble and rheumatism, and hop hornbeam was an ingredient, along with white cedar, in cough syrup. Today, it is rather overlooked and there is little demand for its wood. It has, however, been cultivated as an ornamental, although it grows very slowly.

Range: **Nova Scotia to southeastern Manitoba, south to western Wyoming, Texas, and Florida.**

Rosaceae

American Plum

Prunus americana, American wild plum, wild yellow plum, wild red plum, goose plum, hog plum

Prunus americana, American plum

Prunus americana

is a small, many-branched tree that grows up to 30 feet tall. The leathery single leaves are finely toothed, about 3 to 4 inches long. The white, five-petaled flowers smell unpleasant. The fruit, which ripens in late summer or early fall, has a bright red skin and yellow flesh inside. Found in wet soils near streams or swamps or in fields in the East, on dry uplands and slopes in the West.

Range: southwestern Quebec west to Saskatchewan, south to Arizona and Florida.

The American plum (*Prunus americana*) and the **chokecherry** (*P. virginiana*) are related and both grow wild over a huge range. The fruits of the American plum are slightly sour. It reproduces sexually, depending on insects to pollinate the early-blooming flowers which open before the leaves unfurl. It also reproduces asexually by creeping rhizomes that form dense thickets. A fire can rampage through one of these pockets, leaving behind devastation. But some buds survive on each plant and immediately start to grow. The fruit attracts birds and insects, making it a good edge plant in the area between the forest and an opening or glade.

The drupes or seeds, in this case, are relatively large and have been found on archeological sites dating back 1,800 years. The Huron of eastern North America used to bury the fruit for a few weeks to soften them, then dig them up and dry them out. By then, the fruit was quite edible.

Frances Densmore reported that the Native Americans used a combination of the roots of **black cherry**, *P. serotina,* and wild plum to get rid of worms. They were boiled, strained and the woody material pounded into a pulp. She also noted that the inner bark could be used for gangrenous wounds. Combined with bloodroot, it made an attractive red dye for the porcupine quills used as decorative elements on clothing and artifacts such as boxes.

Jacques Cartier, like so many of the explorers, looked for familiar plants. In 1534, he noted that the forest was full of beauties "of the same kind as those in France, such as . . . plum trees."[98] The settlers made the fruits into jelly with plenty of sugar. Catherine Parr Traill, who wrote *The Backwoods of Canada* (1836) about her pioneer experiences, noted: "Among our wild fruits we have plums, which . . . make admirable preserves, especially when boiled in maple molasses."[99] Traill may also have been describing the **Canada plum** *(P. nigra),* which is reddish-orange, larger and sweeter than the American plum. It grows in southern Ontario and throughout the eastern states as far west as Iowa and Texas.

Today, Tom Isern, professor of history at North Dakota State University, likes to make the fruit into plum butter. He says that "it was known to my grandparents' generation as the shelterbelt plum, as they came to know it through the shelterbelt programs of the 1930s. *P. americana* is still commonly distributed by public agencies for planting in plains states and is highly rated for wildlife value."[100]

Caesalpiniaceae

Eastern Redbud
Cercis canadensis

Cercis canadensis

a small deciduous tree that grows 15 to 20 feet high. The flowers, which may be pink to magenta, come out before the leaves. The heart-shaped leaves are about 4 inches by 4 inches,

Cercis canadensis, Eastern redbud

bronze when they first emerge, and light to dark green when fully open. Found in moist, well-drained sites.

Range: **New Jersey west to Pennsylvania, southern Michigan, southern Ontario, and Nebraska, south to central Oklahoma, central Texas, east to central Florida.**

State tree of Oklahoma.

This is a tree of the edges. It grows in that zone of small trees and shrubs which makes the transition from the forest into an opening or glade or perhaps onto a meadow or prairie. This is where the sun can shine in and the trees and shrubs of the understory get a chance to bloom and fruit. Animals are attracted to these edge zones (ecotones or transitions between two distinct areas), because this is where the best forage is likely to be found.

Stretching its branches toward the light, the redbud takes on a graceful fountain shape enhanced by large heart-shaped leaves. The taproot reaches deep into the soil and, when it encounters clay subsoil, starts to grow horizontally. The bisexual, self-pollinating flowers appear like tiny bits of magenta ribbon along the naked branches, exposed as they are in early spring. They are produced from small buds that develop on the previous year's growth of twigs, branches, and trunk. After two or three weeks, as the leaves begin to appear, the flowers fall off, scattering like drops of blood on the soil. The ovaries of some flowers (which are pollinated by bees) develop into flat reddish-brown pods. Like other members of the pea family, the redbud has nodules on the roots that fix nitrogen from the air by turning it into a form the roots can easily absorb. And despite the name *canadensis* (of Canada), it isn't common in the wild in Canada, although fossil evidence suggests that it was once more widely dispersed in the North than it is now.

Because of its early flowering, it was much loved by Native Americans. Cherokee children liked to eat the blossoms and took flowering branches into their homes to drive out winter. Medicinally, the inner bark was used to treat just about anything that had to do with the lungs, from congestion and colds to whooping cough and flu. The twigs boiled in water produce a yellow dye.

The related European species, *C. siliquastrum*, is known as the **Judas tree** because Judas Iscariot, as legend has it, hanged himself from it after betraying Jesus. It has blood-red buds in spring.

Ericaceae

Rhododendron
Rhododendron spp.

Rhododendron arborescens

the sweet azalea, is a deciduous, loosely branched shrub that grows 8 to 20 feet high and equally wide. The white or pale pink flowers, which contain a reddish style and pink to rose stamens, are 1½ to 2 inches in diameter. They come out after the leaves unfurl in late May or early June. It's sometimes described as the smooth azalea because of the texture of the stems (other azaleas have hairs on them). It has a strong fragrance similar to heliotrope. Like all other native azaleas, it attracts butterflies and hummingbirds. Found in light, moist, acid soil.

Range: **New York and Pennsylvania south to Alabama and Georgia.**

Rhododendrons and azaleas are among the most attractive and popular plants in North America. There are 15 native species, spread throughout the continent. The native azaleas have so many common names it's confusing to the nonconnoisseur. Moreover, the various native species are hard to distinguish because they naturally hybridize with each other.

The word *azalea* comes from the Greek word *azaleos* or "dry" (because the plant often flourished in dry ground). Native azaleas range in color from white, yellow, and orange to scarlet and crimson, in many vivid shades and hues. Some have conspicuous color blotches in the throat of the flower and many are very fragrant. *Rhododendron*, also from the Greek, means "rose tree" (*rhodos*, rose, and *dendron*, tree). Native wild rhododendrons have been called deertongue, big leaf, bush honeysuckle, or wild honeysuckle.

The beautiful bushes certainly cheered the heart of Fanny Trollope who, with four children in tow, struggled across country in 1832.

As soon as we had wearily dragged to the top of one of these [hills], we began to rumble down the other side as rapidly as our four horses could trot; and no sooner arrived at the bottom than we began to crawl up again. . . . The latter part of the day, however, amply repaid us. At four o'clock we began to ascend the Alleghany mountains. . . . The whole of this mountain region, through ninety miles of which the road passes, is a garden. The almost incredible variety of plants, and lavish profusion of their growth, produce an effect perfectly enchanting. . . . The magnificent rhododendron first caught our eyes; it fringes every cliff, nestles beneath every rock, and blooms around every tree. . . . All that is noblest in nature was joined to all that is sweetest.[101]

According to Donald Culross Peattie, "Tradition has it that rhododendron hides the moonshiner and his still, and tradition is frequently right. Rhododendron may even play its part in the excessive isolation of mountain life; too high

to see over, yet low enough to form an impenetrable twiggy thicket of crooked stems."[102]

At first, most settlers regarded rhododendrons as something of a nuisance. "In those parts of the country where it flourishes most luxuriantly, veritable rhododendron jungles, termed 'hells' by mountaineers, are formed. The branches reach out and interlace in such a fashion as to be almost impassable."[103] Once exported to England and Europe, "rhodies" became part of the rage for what was known in the 1860s as the American garden—native plants from the Eastern forest, especially those broad-leaved evergreens. According to one source, "The money spent on rhododendrons during the twenty years in this country would nearly suffice to pay off the national debt."[104] At the Centennial Exhibition in 1876 in Philadelphia, an English nurseryman put on a display of hybrid rhododendrons (crosses of American and Asian species) that so dazzled the Americans they stopped taking their native bushes for granted and began to cultivate the flowers.

Today, some native species are under serious threat. Captain Dick Steele, rhododendron breeder and founder of the Rhododendron and Native Plant Foundation, crusades to save native species. He writes in the *Rosebay Newsletter*:

While searching in the wild near Hopewell, Virginia I located a truly significant form of the azalea then known as nudiflorum; *. . . I called on the lady who owned the farm and took her to that beautiful plant, photographed the two together and arranged permission to return that fall to collect seed and a section from the plant. When I returned that October, I found the whole area bulldozed level and that plant lost forever.*

In 1954 en route to take up duties in the Canadian Embassy in Washington, I stopped for an hour break in driving, near Pocologan, on the shores of the Bay of Fundy in New Brunswick,

Canada. and went off into shrubland to look at the masses of Rhododendron canadense *that were in full bloom. About a half mile back from the road I found a rich deep purple form of the plant which I captured for posterity with my camera.*

I returned to Canada 2½ years later to collect both seed and the plant. To my chagrin a new Super Highway had been built through that area and the plant had been destroyed. We have the photo but the plant is gone. Similarly a new highway in Nova Scotia destroyed our only colony of the very rare white form of Rhododendron canadense.[105]

Captain Dick (as everyone calls him) created the foundation to establish a living collection of plants, by exploring for them, collecting them, and breeding them. The dues are a few hours of work each year.[106]

Ericaceae

Rosebay Rhododendron
Rhododendron maximum

Rhododendron maximum

grows 4 to 15 feet tall in the northern part of its range and up to 30 feet in the most southerly parts. It has large, dark green, evergreen leaves 4 to 8 inches long. The flowers, which emerge in June, range from rose and purple pink to white; they have olive-green to orange spots and are about 1 to 2 inches across.

Range: New England and New York, south to Georgia, Alabama, and Ohio.

The first European to spot a rosebay rhododendron was Dr. Jacob Bigelow. He saw one in a swamp near Medfield, Massachusetts, and later discovered a red-flowered variety growing "abundantly on the banks of the Charles River, a dozen or fifteen miles from Boston." He reported his find in 1820 in *American Medical Botany, Being a Collection of the Native Medicinal Plants of the United States.*[107]

Although the plant had a reputation for being poisonous, Bigelow ate a rhododendron leaf and suffered not at all. The stories about poisoning by rhododendrons go back thousands of years. Xenophon (once a student of Socrates) led an army of 10,000 Greek soldiers through the territory of Colchis, where they noticed masses of beehives. They collected the honey and then, Xenophon wrote, collapsed by the thousands, unable to move for a couple of days. About 300 years later, in 67 B.C., Mithridates had his allies place toxic honeycombs along Pompey's route. The soldiers ate the honey, fell ill, and were easily slaughtered. Honey made with nectar from Mediterranean rhododendrons became known as "mad honey."

The first report of "mad honey" in the United States appeared in 1802 in the *Transactions of the American Philosophical Society* (volume 5). One J. Grammar described having blurred vision, nausea, and loss of muscle control from eating honey derived from rhododendron nectar. In 1891, a P.C. Plugge "was able to isolate the toxic compound in the honey from Trebizond and identified it as andromedetoxin, now known as acetylandromedol. . . . [The same toxins] occur in Mediterranean oleander, members of the heath family (Ericaceae) which includes the rhododendron, azaleas of the Black Sea and Caucasus area, and the mountain laurel of the Eastern United States and Pacific Northwest."[108]

John Muir, writing in 1869, noted what happened when sheep ate the plant:

June 7—The sheep were sick last night, and many of them are still far from well . . . cough-

ing, groaning, looking wretched and pitiful, all from eating the leaves of the blessed azalea. . . . Having had but little grass since they left the plains, they are starving, and so eat anything green they can get. Sheep men call azalea "sheep-poison," and wonder what the Creator was thinking about when He made it.[109]

Swamp laurel *(Kalmia angustifolia)*, with its similar effects, is often called lambkill.

Native Americans used rosebay rhododendron to make liniments and poultices for aches and pains. The Cherokee threw the leaves on a fire and danced around it to bring on cold weather. They also used the wood to make spoons or pipes.

By 1850, the rosebay was being cultivated in gardens, as Thoreau noted in his journal, June 4, 1853: "The date of the introduction of the *Rhododendron maximum* into Concord is worth preserving, May 16th, '53. They were small plants, one to four feet high, some with large flower-buds, twenty-five cents apiece; and I noticed the next day one or more in every front yard on each side of the street, and the inhabitants out watering them. Said to be the most splendid native flower in Massachusetts. . . . I hear to-day that one in town has blossomed."[110]

Rhododendron maximum, rosebay rhododendron

Ericaceae

Pinxter Flower
Rhododendron periclymenoides syn. *R. nudiflorum*, pink azalea

Pinxter is the Dutch word for Pentecost, the Christian feast day which occurs 40 days after Easter; Dutch settlers named the flower pinxter because it blooms around the time of Pentecost. In New Hampshire, the color of the blooms was called "election pink" because the flowers open at the time when elections for governor used to take place.

Found in dry areas and open woods, *Rhododendron periclymenoides* grows 4 to 6 feet tall and 8 to 10 feet wide with gray-brown twigs and alternate, oval, papery-textured leaves. It has fragrant white to pale pink to deep violet flowers which open in mid-April. The fruits are oblong capsules ¼ to ½ inch long that release tiny seeds in late summer.

Fred Galle of Callaway Gardens in Georgia says this dwarf azalea, which forms large colonies, is the parent of many of the garden hybrids used in the North. It closely resembles the **rose-shell azalea** (*R. prinophyllum*), which is found through northern Ohio and northeastward into New England.

In Massachusetts, residents used to gather an edible fungus that grew on the plant, which they called mayapple (not to be confused with *Podophyllum peltatum*), swamp apple, or swamp cheese. They pickled the fungus in spiced vinegar.

Range: **Massachusetts to Ohio and North Carolina.**

Hamamelidaceae

Witch Hazel
Hamamelis virginiana

Hamamelis virginiana **is a shrub that grows up to 15 feet tall, occasionally larger. The leaves are oval or almost round, 2 to 6 inches long, with wavy edges and often hairy underneath. The flowers have narrow yellow petals less than 1 inch long. The flowers appear in the fall at the same time as the fruits, which are little brown woody capsules containing two black seeds.**

Range: **Nova Scotia to southern Ontario and southeastern Minnesota, south to Texas and northern Florida.**

In February all through its range some species of witch hazel brighten up the whole winter scene by bursting forth with intensely colored flowers, like a surprise visit of butterflies in the snow. They hang on for weeks and weeks, sometimes the only color around in the North. This native species, however, is different—the flowers appear in autumn.

Despite the name, this shrub has nothing to do with witches and is not a hazel. The first part of the common name actually comes from *wych*, meaning a pliable branch—they either were used or looked as if they could be used by water diviners for dowsing. Opinions seem to be divided on this point. The second part of the name refers to the leaves, which resemble those of the hazelnut tree (*Corylus americana*).

The scientific name comes closer to the mark. *Hamamelis* is a compound of two Greek words meaning "apple" (or fruit) and "at the same time," because the fruit and the flower

Hamamelis virginiana, witch hazel

bottles of vile-smelling liquid found in many medicine chests. It can be used as an astringent to improve the look of the skin, or applied to bruises or mosquito bites to ease the swelling. Its healing properties have been known for hundreds of years. The Osage used the bark to treat sores and ulcers of the skin much as we do today. The Potawatomi and Menominee used the twigs, either placed on hot rocks in a sweat lodge or boiled in water, to soothe aching muscles. The Mohegan used a decoction of the leaves for cuts, bruises, and insect bites.

In the 1840s, Theron T. Pond of Utica, New York, started marketing a distilled liquid made with the bark and twigs of witch hazel. The industry he founded (Pond's skin care products) is still flourishing.

Today, most of the witch hazel sold in North America is harvested and processed in Connecticut. However, the commercially available witch hazel does not contain all the active substances present in the tree. The bark contains tannin, which contributes to its anti-inflammatory and analgesic qualities (and may even be antiviral), yet the tannin is largely lost during distillation. The European form of witch hazel, which is created by steeping the bark and leaves in a combination of water and alcohol, retains more tannin. Some herbalists make their own extract from the bark and consider it more efficacious than the drugstore variety.

Ericaceae

Highbush Blueberry
Vaccinium corymbosum

Vaccinium corymbosum
is a shrub that grows up to 12 feet tall with a similar spread. It is multistemmed in the wild, but

appear together on the plant in autumn. Although this sounds biologically impossible, in fact, the previous year's seed capsules mature as the current year's flowers emerge. *Virginiana*, is, of course, a reference to the state of Virginia—a rather arbitrary choice, since the plant is found all through eastern North America.

One of the interesting features of witch hazel is the way its seeds disperse with a resounding pop. The mature capsules fire off seeds like miniature catapults, up to 30 feet away, depending on the prevailing winds.

Most people associate the name with those

usually more compact in cultivation. The twigs are bright red when young. The leaves are smooth, dark green, and shiny, in elliptical shapes 1 to 3 inches long. The pink bell-shaped flowers appear in a corymb (a flat-topped inflorescence) in May. The blue-black berries appear in July or August.

Range: **Nova Scotia to Wisconsin, south to Louisiana and Georgia.**

The young bright red twigs of highbush blueberry are cheery reminders of what an important source of food this plant is. The related **lowbush blueberry**, *V. angustifolium*, grows close to the ground, usually less than 1 foot high in densely packed masses. It is found in eastern Canada and New England, and occasionally farther south. The highbush blueberry, however, can grow to 12 feet high and grows as far south as Louisiana and Georgia.

Blueberries have been an important source of food for humans and animals for thousands of years and today they are grown commercially for desserts and baking. In 1643 Roger Williams, in his *Key into the Language of America*, noted that the Native Americans called the blueberry "sautaash" and the berries were "dried by the natives, and so preserved all the year; which they beat to powder, and mingle it with their parched meal, and make a delicate dish which they call *sautauthig*, which is sweet to them as plum or spice cake to the English."[111]

Collecting the blueberries was the responsi-

Vaccinium corymbosum, highbush blueberry

bility of women and children. Families had their own territory where they had a "right to collect"; this right was often passed from one generation to the next. "On a cool, clear morning in late July, women and children would rise early, line special collecting baskets or bark pails with basswood leaves to protect their harvest from the heat that would come later in the day, and set off for the barrens to gather the first blueberries of summer."[112] The berries contained vitamin C, needed to stave off the scurvy.

Settlers soon learned the value of the flavorful berries and, by the early 1900s, Elizabeth Coleman White of Whitesbog, New Jersey, offered local prizes for highbush blueberries bearing the largest fruits. She handed out aluminum gauges with ⅝-inch holes; only the larger berries qualified. She also distributed questionnaires requesting information about plant vigor, cold and disease resistance, flavor, texture, productivity, and the time of ripening. By 1909, working with USDA plant breeder Dr. Frederick V. Coville, she had crossed many plants selected from the wild in the Pine Barrens of New Jersey. White named the varieties after their finders (such as Rube Leek—a berry that became the Rubel, a key plant of their breeding). In 1916, they had the nation's first commercial crop of blueberries marketed under the name of Tru-Blu-Berries. White's farm yield peaked at 20,000 barrels of berries a year. She also introduced the use of cellophane to package blueberries. By Coville's death in 1937, there were 30 large-fruited, named varieties of highbush blueberry being grown commercially—in less than 25 years, the wild blueberry had become a mass-produced fruit. White died in 1954.[113]

Today, the blueberry industry includes both cultivated and wild varieties. Wild blueberries are generally smaller and sweeter than the cultivated ones. Blueberries are grown in more than 30 states and provinces, from Florida to British Columbia, and grow best in areas with fairly acidic soil. The blueberries from the Northern states and Canada tend to have a stronger flavor and higher sugar content than those from the Southern states. Blueberries are sold dried, canned, frozen, or fresh, or made into jams, jellies, and syrups. They are available fresh for about eight months of the year, because they are harvested at different times in different latitudes.

Researchers at Tufts University who study phytochemicals (plant chemicals) have found that blueberries contain high levels of antioxidants, which help in preventing cancer and heart disease. A blueberry a day . . . well, not quite, but health food advocates tell us that half a cup a day might keep cancer, along with most other diseases, at a distance.

Ericaceae

Cranberry
Vaccinium macrocarpon

Vaccinium macrocarpon
grows about 6 inches high on wiry, creeping stems with small, oval, leathery leaves. In May, pink flowers appear, giving way to bright red berries in fall. Found in bogs and sandy meadows.

Range: Atlantic provinces of Canada, Quebec, Ontario, New England, and the Eastern states as far south as North Carolina, west to Wisconsin, Ohio, Illinois, and Indiana.

The pink flowers with their swept-back petals look a bit like the head of a bird, and this supposedly inspired the early settlers of North America to call the plant "crane berry." The settlers quickly came to appreciate the taste of the

FOLLOWING PAGE: *Vaccinium macrocarpon,* cranberry

tart berries. Of course, the local peoples had known about them for years, and ate them fresh, dried, or cooked into a jamlike conserve. The Narraganset called them *sasemineash* (very sour berry), the Wampanoag called them *ibimi* (bitter berry), and the Huron called them *toca* or *atoca* (good berry).

The berries contain vitamin C and helped to ward off scurvy. They can remain all winter on the plant, and provide food for animals at a time when pickings are slim. Campbell Hardy observed bears tempted out of their hibernation by the "genial warmth of a spring day" feasting on "the bright crimson berries still clinging to their tendrils . . . and rendered tender and luscious by the winter's frost."[114] They are pollinated by bees. The seeds are dispersed by birds and mammals and, most important, by water.

Cranberries grow well throughout eastern North America, and as far north as Hudson Bay. They were used extensively by Native Americans, and in his journal, explorer Samuel Hearne gives us an idea of what its northern range looked like in the eighteenth century: "Cranberries grow in great abundance near Churchill, and are . . . as common on open bleak plains and high rocks as among the woods. When carefully gathered in the Fall, in dry weather, and as carefully packed in casks of moist sugar, they will keep for years, and are annually sent to England in considerable quantities as presents, where they are much esteemed."[115]

Probably the first person to grow cranberries commercially was Henry Hall, of Dennis, Massachusetts. Around 1810, he noticed how much better his cranberries grew when sand blew across the bog. He transplanted the cranberries, fenced them in, and spread sand on them. The plants thrived and so did Hall's new business.[116] Soon he and his neighbors, who caught on quickly, were shipping their berries to Europe, making the cranberry one of the first native plants grown commercially for export.

The technique of covering the cranberry fields with sand before planting persists. Growers often harvest the berries by flooding the fields and skimming off the ripe berries which bob to the surface.

Today, the cranberry business is booming, aided by reports that drinking cranberry juice regularly can help prevent urinary tract disease and kidney problems. It adds tang to soda, wine, or champagne in bar drinks, and who can imagine Thanksgiving turkey without a dollop of ruby-red cranberry sauce?

Lauraceae

Spicebush

Lindera benzoin, common spicebush, wild allspice, feverbush

The sweet lemony fragrance of the spicebush attracts bees and ladybugs. The caterpillars of the swallowtail butterfly find succor in the leaves, which they fold up around themselves, and in winter you can see their delicate chrysalises hanging on the twigs.

Spicebush, *Lindera benzoin*, is a deciduous shrub that can grow to 10 feet tall and nearly as wide. The tiny scented yellow flowers bloom in April, and the bright red berries ripen from August to September. It is dioecious, needing both a male and a female plant for reproduction, and thrives in a variety of soils in both sun and shade.

Native Americans used just about every part of the plant in their vast repertoire of plant medicines. Bark, leaves, and berries are all aromatic and new bark was considered a delicacy to chew on. They also steeped leaves, twigs, and bark in boiling water to make a mildly astringent tea to soothe menstrual cramps or induce sweating. The berries contain an aromatic essential oil used to treat arthritis and rheumatism. The leaves contain camphor in very small quantities and were used in much the same way we use insect repellents—a property very important to Native Americans, who usually lived near water, where mosquitoes and other insects abounded. The leaves also made a good poultice for bruises and rashes. The plant quickly found its way into the medicine chests of the early settlers. Contemporary herbalists have suggested the plant should be studied further for its remarkable healing properties.

Spicebush is related to such widely diverse plants as avocado and sassafras. Settlers gave the plant its common name because its fruit, which is about the size of an olive, could be dried and made into a substitute for allspice in baking. Steve Brill, in *Identifying and Harvesting Edible and Medicinal Plants*, says, "Since spiceberries are ripe in apple season, they often find themselves in the same pot. I love compotes with sliced apples, walnuts, orange rind, and spiceberries,

simmered about 15 minutes." He recommends freezing rather than drying the berries, because they can go rancid at room temperature.[117]

Range: **eastern North America from Maine and Ontario to Kentucky, Missouri, and Kansas.**

Anacardiaceae

Sumac
Rhus spp.

Rhus spp.

are shrubs distinguished by their compound leaves, which turn bright red in the fall; densely packed conical flower heads of white or greenish-yellow; and red hairy fruit. Some species, such as the fragrant sumac (*R. aromatica*), tend to be under

Rhus spp., sumac

6 feet. Oddly enough, dwarf or prairie sumac *(R. copallina)* is one of the taller species, and can reach 30 feet. Grows in fairly dry soils in sunny areas.

Range: southeastern Canada, New England, and throughout the eastern United States south to Texas and Florida.

Sumac is the unmistakable herald of fall when it is all decked out in its scarlet foliage. Old-fashioned nature poets were forever writing of "sumac torches burning" and "hills on fire." They blaze away in autumn with berries remaining on the shrub throughout the winter, providing forage for animals in less urban areas.

Sumac is related to poison ivy *(Toxicodendron)*, and one species, *Rhus vernix* or **poison sumac**, is to be avoided, but other species, such as **staghorn sumac** *(R. typhina)*, **smooth sumac** *(R. glabra)*, and **dwarf** or **shining sumac** *(R. copallina)*, have edible berries with a lemony taste and were once used to make a drink called "Indian lemonade." Even the leaves smell faintly lemony. The poison sumac is, fortunately, fairly rare, and grows in bogs and swamps, unlike most other species, which are found in much drier soils, on hillsides, and along roads and railway embankments.

The nonpoisonous sumacs had a variety of medicinal uses for Native Americans. A drink made from the berries was used to lower a fever, and boiled-down syrup made from the bark was used to treat diarrhea, dysentery, sore throats, gum disease, and cold sores. The berries were also mixed with tobacco to enhance the aroma of the smoke. The main use of sumac, however, seems to have been in dyeing. Different parts of the bush yielded different color dyes: light yellow from the stem pulp; orange or yellow from the roots; brown from the autumn leaves; and red, brown, or black from the boiled fruit, depending on the species. The berries of some species, such as staghorn sumac, even yielded a kind of black

ink. The plant also contains tannins, useful in treating leather.

After the Civil War, in an attempt to rebuild the economy, some people in the devastated Southern states harvested and marketed sumac for the dyeing and tanning industries, to replace imported dyes from Europe. This industry was eventually superseded by chemical dyes and tanning agents. Today, the sumac has no commercial use, although it is sometimes planted on hillsides to control erosion or cultivated as an ornamental.

Caprifoliaceae

Viburnum
Viburnum spp.

Viburnum spp.

are shrubs or small trees with horizontal spreading branches. The leathery green leaves are simple and opposite, and are brightly colored in fall. The bisexual flowers grow in branched clusters, with five stamens and a stigma with three lobes. In many species, the flowers have a pleasant scent. The red or purple fruit is berrylike, containing a flattened stone which in turn contains a reddish-brown seed.

Range: eastern Canada and northeastern United States.

There are 120 species of viburnum and it's hard to imagine a forest or a garden without these marvelous plants. Viburnums were probably among the first shrubs the settlers noticed along the waterways and throughout the land cleared by the Indians. They were considered weeds because they were so common. The settlers, though not ardent observers of plants, eventually

Viburnum spp., viburnum

viburnum (*V. cassinoides*), and **highbush cran-berry** (*V. trilobum*) are edible. Steve Brill and Evelyn Dean recommend the black haw particularly, "an autumn fruit heavy, warming, and full of carbohydrates, very different from the light, juicy, cooling fruits of summer." They simmer them for 20 minutes in a combination of orange and lemon juices, strain out the seeds with a food mill, and use the resulting purée in strudels, pastry, or muffins or poured over pancakes.[118]

Ericaceae

Dwarf Huckleberry
Gaylussacia dumosa, black huckleberry, bush huckleberry, gopher berry

Gaylussacia dumosa

is a small, deciduous, erect, many-branched shrub which grows 12 to 30 inches high. Many stems ascend from the base, forming a low, dense, rounded crown. The twigs are covered with short curly hairs. The simple oval leaves are leathery, with a fringe of hairs around the edge. The bell-shaped white flowers, which appear in May and June, are borne on racemes at the end of the branchlets. The fruit is a black berry with 10 nutlets, ½ inch long, each carrying one seed. Grows in rich soil, often in association with pitch pine or scrub oak.

Range: **Newfoundland and Nova Scotia south to central Florida and east to western Mississippi and central Tennessee.**

began to appreciate the plants that surrounded them and viburnums started to appear as garden ornamentals in the eighteenth-century.

Native Americans had a number of medicinal uses for these interesting shrubs. The Iroquois used an infusion of the plant as a contraceptive, and the Menominee and Chippewa took infusions of inner bark for stomach and menstrual cramps. It was later found that the inner bark contained an antispasmodic glucoside called viburnin, which accounted for its effectiveness.

The berries of many viburnum species, including the **black haw** (*V. prunifolium*), **nannyberry** (*V. lentago*), **wild raisin** or **witherod**

Huckleberry Finn, Huckleberry Hound, and my huckleberry friend from "Moon River" show how deeply this plant made its way into the American psyche. Though it is sometimes confused with blueberry and may even be called highbush blueberry, it is another species altogether. The scientific name *Gaylussacia* honors the eminent French scientist Joseph Louis Gay-Lussac (1778–1850), who uncovered what became known as Gay-Lussac's law: the fact that gases expand equally with the same increase in temperature.

If you hold a leaf up to the light, you can see it is covered with tiny glands full of resin. This is the only huckleberry with little leaflike bracts that persist when the fruit is mature. Small animals feed on the berrylike fruits and so do quail, wild turkey, and ruffed grouse, all of which help disperse the seed.

Thoreau wrote on August 8, 1858, about berries growing near Ledum Swamp:

I see there, especially near the pool, tall and slender huckleberry bushes of a peculiar kind. Some are seven feet high. They are, for the most part, three or four feet high, very slender and drooping, bent like grass to one side. The berries are round and glossy-black, with resinous dots, as usual, and in flattish-topped racemes, sometimes ten or twelve in a raceme, but generally more scattered. Call it, perhaps, the tall swamp huckleberry.[119]

The berries of this plant are edible, but the sweetest fruits are found on the **black huckleberry** (*G. baccata*). The Iroquois ate the berries fresh, dried them for winter use, made them into drinks, and used them in a tonic to purify the liver and the blood. They also smoked the huckleberry leaves, along with many other plants. Smoking was important not only for purification rituals but also for communication and just plain fun.

Another relative, **box huckleberry** (*G. brachycera*), was lost to American gardens until it was reintroduced by the Arnold Arboretum. One particular stand in central Pennsylvania, which covers about 300 acres, is believed to have originated from a single plant, and may be more than 12,000 years old.[120]

EASTERN FOREST UNDERSTORY

Trilliums, ferns, and bluebells carpet the forest floor.

The understory, a place of soft light under the overhanging canopy of the trees, is critical to the Eastern forest's health. Though it may appear as serene as the moon, it is actually teeming with life above and below the surface.

Since the canopy grabs most of the direct rays of the sun, the lower layers make do with filtered light. As understory plants co-evolved with big trees, many of them developed large leaf surfaces to intercept as much light as possible for photosynthesis. Even more action goes on in an exquisitely balanced way under the ground. Each tree develops its own microhabitat of fungi and insects which depend on each other.

Many of the plants found in the forest floor are spring ephemerals. They pop up early when there are few leaves on the trees, and the light on the forest floor is at its greatest intensity. To make their little shoots less tempting for hungry animals, they produce toxins that taste unpleasant or burn the mouths of unwary foragers. They photosynthesize quickly and attract wakening insects with their showy blooms. The object, of course, is to make sure the right insects get the pollen moving around so fertilization can take place. And so ephemerals have persisted—attracting, repelling, then going dormant in summer as other plants emerge.

The forest floor fills up with duff: a combination of fallen leaves, needles, animal droppings, and dander that form a rich mulch keeping the soil both cool and moist. Duff also adds nutrients to the soil and feeds the millions of microscopic creatures leading subterranean lives. This rich tapestry of life is also one of its most fragile ones. Each time we move these plants, they are unlikely to survive elsewhere. Trying to save them is an arduous business not taken on lightly. They are essential to a healthy forest.

Araceae

Jack-in-the-Pulpit

Arisaema triphyllum, wild turnip, arum, Indian turnip, devil's-ear

Arisaema triphyllum

grows 1 to 3 feet tall in open places or semishade. It has three leaflets up to 6 inches long and 1½ to 3 inches wide; the greenish spathe often has purple stripes. Inconspicuous flowers produce fruits that ripen into bright red berries. The corm is turnip-shaped.

***Range:* New Brunswick and Nova Scotia east to Manitoba, south and east to Kansas, Missouri, and South Carolina.**

Jack-in-the-pulpit is such a strange plant, yet so compelling it's hard not to want to touch it. Perky little Jack (the spadix) sits straight up in his pulpit (the spathe). Jack-in-the-pulpit in the wild is found in the company of bellworts and wild geraniums in a shady forest in rich damp soil. The word *triphyllum* means three-leaved. *Arisaema* comes from a Greek word *(aima)* meaning "blood red"—a good description of the amazing red berries.

It is one of six members of the arum family, which includes 150 species throughout the world with relatives such as skunk cabbage, philodendron, breadfruit, and calla lily. Jack-in-the-pulpit is either tuberous or rhizomatous, a herbaceous perennial living in swamps, bogs, and damp woods. There are three forms: *A. triphyllum*, which has a three-part leaf and is found in wetter habitats; the woodland *A. atrorubens*, which is dark red with a two-part leaf and is considered a separate species; and the northern Jack-in-the-pulpit, *A. stewardsonii*, which is found in bogs in Pennsylvania and New Jersey.

The spathe is the sheathlike membrane with dark purple or brown stripes. It curls around the clublike spadix (Jack). Hidden beneath the sheath are two kinds of flowers: sterile and fertile. These inconspicuous flowers lurk deep inside the plant at the base of the spadix. The females are tiny green berries; the males are threadlike shedders of pollen. In autumn, the bright red berries are one of the identifying marks of the plant. As the berries ripen, the sheath shrivels up, opening the seed to heat and light.

Jack-in-the-pulpit has a remarkable way of

adapting to its environment. The spadix can develop flowers of either sex. The root, a fleshy corm slightly bigger than that of a crocus, is bisexual—either a female or a male plant can rise from it—and each corm produces a single stem with distinctive points on the basal leaves. Which sex the plant will be depends on how much food is available. The female, because it develops the seeds, needs more food than the male. If there's too little food, only one leaf comes out in spring, and the plant keeps its major energy for root survival. Somehow, over the following year, about equal numbers of male and female flowers emerge.

Although this isn't one of the insect-eating plants, the lower chamber, where the sex organs are located, acts as an insect trap. The unsuspecting creatures can get in, but can't get back up the slick walls and past the flap to escape. The insects, especially fungus gnats, are attracted by the plant's rather erotic fungusy smell. Jack also has fierce chemical defenses. Both the foliage and the corm produce calcium oxalate, a chemical so caustic most insects avoid eating its foliage.[121] Humans and animals should also avoid eating this plant, which at best burns the mouth, and at worst causes such violent irritation to the stomach that it can be fatal.

Timothy Coffey, who collects wildflower lore, has a prodigious collection of common names for plants from different parts of North America. Among his list for *A. triphyllum* are Adam's apple, bog onion (Maine, Massachusetts), brown dragon, cooter-wampee, cuckoopint, cuckoo plant, devil's-ear, hopnis, Indian cherry (Texas), Indian cradle (New York), Iroquois breadroot, lady-in-a-chaise (Massachusetts), lord-and-lady, marsh turnip, memory root (Massachusetts), *oignon sauvage* (Quebec), parson-in-the-pulpit (South), *petit prêcheur* (Quebec), pepper turnip, plant of peace, preacher-in-the-pulpit, priest's-pintle, starchwort, swamp cherry (Texas), swamp turnip, thrice-leaved arum,

Arisaema triphyllum, Jack-in-the-pulpit

thrice-leaved Indian turnip, tuckahoe, wake-robin, wampee (South Carolina), wild pepper, and wild turnip (Vermont).[122]

Each name suggests a different use, or perhaps a different attitude toward the plant. Many evoke the little-person-in-an-enclosure. Some refer to the calcium oxalate—"memory-root"—alluding to a favorite schoolboy trick of tempting others to bite into the blistering corm, an experience that burns itself in the victim's memory forever. "Lord-and-lady" is thought to derive from the fact that some plants are purple (lords) and some pale green (ladies). *Hopnis* is a Seneca word also used for *Apios americana* (groundnut), because both have similar kinds of underground root. "Starchwort" comes from the practice of making starch from the roots: it was a nice, white starch, but tended to burn the hands of those who used it, and create a rash on the skins of those wearing clothes starched with it. *Tuckahoe* in the Algonquin language meant "a bulbous plant used for food."

According to some legends, Jack-in-the-pulpit was the plant at the bottom of Christ's cross and the striations on its petals were from drops of His blood.

> Beneath the cross it grew;
> And in the vase-like hollow of the leaf,
> Catching from that dread shower of
>   agony
> A few mysterious drops, transmitted thus
> Unto the groves and hills their healing
>   stains,
> A heritage, for storm or vernal shower
> Never to blow away.[123]

While some people looked at the plant and saw either a little person or drops of blood, others saw a male appendage and gave it the name "priest's-pintle" (i.e., penis). They therefore assumed that the plant would be valuable for problems related to that part of the body. This thinking was typical of the doctrine of signatures, an idea that became popular through a book called *Signatura Rerum: The Signature of All Things* by Jakob Böhme (1575–1624). He was a German shoemaker who had a mystical vision of the relationship between God and man through nature. By studying nature, he concluded, you could divine the intention God had for each plant. For example, the leaves of liverleaf, *Hepatica*, resemble the liver; ergo, it holds a treatment for the liver. So the interestingly shaped *Arisaema* clearly had something to do with the penis.

It's a mystery, though, how people ever found uses for this poisonous plant. Swedish clergyman and amateur botanist Peter Kalm, who traveled through North America from 1748 to 1751, was amazed that anyone would think of using it.

> *It is remarkable that the arums . . . are eaten by men in different parts of the world, though their roots, when raw . . . are almost poisonous. . . . How can men have learned that plants so extremely opposite to our nature were eatable, and that their poison, which burns on the tongue, can be conquered by fire?[124]*

Kalm assumed that starvation probably drove people to eat the seemingly inedible. He found the Indians of his acquaintance roasting the roots, or drying them for at least six months, to counteract the sharpness. Catherine Parr Traill in *Canadian Wild Flowers* (1868) noted:

> *The Indian herbalists use the Indian Turnip in medicine as a remedy in violent colic, long experience having taught them in what manner to employ this dangerous root. The juice of [Arisaema] triphyllum has been used, boiled in milk, as a remedy for consumption.[125]*

Native peoples were very aware of its toxic effect. They made very dilute infusions for a sore throat or as an eyewash. The roots were pounded into poultices for headaches, or boiled and mixed with a meal to help bring boils to a head or heal abscesses. Iroquois women used the rhizomes in an infusion as a form of contraception. Later, when horses were introduced to North America, the boiled plant was ground up and added to a mare's feed to induce pregnancy and reduce listlessness.

The native peoples attributed special powers to the plant. Huron Smith, writing in 1928 about the Meskwaki, noted: "This [Indian turnip] is one of the diagnostic medicines. The central part of the seed, divested of pulp, is dropped into a cup of water. If it goes around four times clockwise, before dropping to the bottom, the patient will recover. But if it goes down the fourth time, or fails to float at all, the patient will die." In some cases, the plant's poisonous qualities were deliberately used to kill people. "Charles Keosatok told us that the Meskwaki used to chop this root fine and put it in the meat they fed to their Sioux enemies and others. A few hours after eating, this would cause them much pain and they would die."[126]

Anacardiaceae

Poison Ivy
Toxicodendron radicans

Toxicodendron radicans
is a vine that grows up to 100 feet long, with compound leaves of three leaflets each, glossy green in summer and bright red in fall. The tiny white flowers appear in clusters in summer, giving way to white berries in fall. Found in woods, fields, and hedgerows.

Range: **throughout the continental United States (except Alaska) and southern Canada.**

Here's the plant everyone loves to hate, including the botanists who classify plants, because they can't agree on its subclassifications. Some consider it a relative of the sumac *(Rhus);* others make a distinction between the plants containing the toxin *(Toxicodendron)* and those that don't (true sumacs). Distinctions were once made between poison ivy and poison oak subspecies (the latter is shrubby rather than viny), but when cuttings of both were grown in different places, botanists found those planted in moist forests became the ivy type, and those planted in drier, sunnier spots became the shrubby oaky ones. (To confuse the matter further, there is a different species called **poison oak**, *T. diversilobum,* which grows only on the West Coast. See pages 429–430.) Such is life in the botanical underbrush.

Real fear and loathing, however, are reserved for the poison ivy sap, which causes allergic contact dermatitis in humans. The active agent is urushiol, which is contained in ducts in most parts of the plant. When the plant is touched, the ducts break and the toxin is released. The toxin can even become airborne when the plant is burned. The degree of toxicity depends on the time of year and the age of the plant. Though people react differently to urushiol, hardly anyone is completely immune. A nanogram (billionth of a gram) can cause the dreaded rash. About 15 percent of the people who are sensitive to urushiol have extreme reactions, and may need corticosteroid injections to bring down the swelling.

Home remedies abound for dealing with a poison ivy rash, including a theory about building up immunity by eating a little poison ivy for a few days each spring. Euell Gibbons, the famous forager of wild plants, tried it with good effect—or so he claims.[127] It sounds dangerous; no one else should try it. Home recipes for washes to take away the stinging include strong catnip tea mixed with olive oil, goldenrod leaves in cold water, cider vinegar followed by cornstarch paste, or a recipe called "Five Buddies in a Blender," containing jewelweed, plantain, comfrey, burdock, and chickweed mixed with water and witch hazel. Once

Toxicodendron radicans, poison ivy

blended, the mixture is frozen and then used to make a cooling potion.[128] Even coal rubbed on a rash helps, or so some people believe.

Native Americans knew poison ivy, and treated it with respect. Anna Jameson, who traveled through Canada in the 1830s, learned about it from a man who lived near Georgian Bay, Lake Huron.

Old Solomon asked me once or twice how I felt; and I thought his anxiety for my health was caused by the rain; but no, he told me that on the island where we had dined he had observed a great quantity of a certain plant, which, if only touched, causes a dreadful eruption and ulcer all over the body. I asked why he had not shown it to me, and warned me against it? And he assured me that such warning would only have increased the danger, for when there is any knowledge or apprehension of it exist-

ing in the mind, the very air blowing from it sometimes infects the frame.[129]

Other settlers were not warned in time, as a letter written in 1847 from Quebec, by one George Stacy, shows:

A very strange thing happened to us all during the summer. Frederick, during the haying season, poisoned himself in the legs and feet, due to touching what they call here "poison ivy." It was six weeks before he could wear his boots, and even now he is not free from lameness. Eliza, myself and the baby are now suffering. It first shows itself by a small white pimple, increasing to that of a half crown. It discharges all the time, and is very painful. I am now writing sitting on the bed, for I am not able to put my legs to the floor without great pain.[130]

Nature, of course, does not produce any plant in vain. Poison ivy is indestructible: Katherine Carter Ewel, an ecologist with the USDA Forest Service in Florida, has found in her research on the cypress swamps that poison ivy survived even when it was regularly deluged with waste water over several decades. This may only sound like perversion on the part of this tiresome plant, but to people who are concerned with sewage treatment and the preservation of wetlands, it is something of a revelation. Poison ivy may have positive environmental uses not even thought of and yet to be explored.[131]

Apiaceae

Sweet Cicely

Osmorhiza longistylis, aniseroot, sweetroot, licorice root, scented root, paregoric root, sicily root, sweet anise, sweet chervil

Osmorhiza longistylis

grows 2 to 3 feet tall, with compound leaves made up of toothed leaflets and a big carrotlike root which is sparsely hairy. The flower head is made up of umbels with one to six primary rays and small white flowers that appear in April or May. The flowers give way to brown seedpods in summer. Found in moist soil in woods, on shaded slopes.

Range: Quebec to Saskatchewan, south to Georgia and Texas.

Osmorhiza longistylis, sweet cicely

Spring in the woodlands brings the joy of discovering the lacy green leaves and tumbling white flowers of this fragrant perennial herb. It is related to carrots, parsnips, parsley, and Queen Anne's lace.

The aromatic roots, which smell like licorice, are considered both an antiseptic and an aphrodisiac. Every part of the plant is useful. The Cheyenne pulverized the leaves, stems, and roots to make an infusion for stomach problems such as bloating, or for kidney trouble. The Omaha pounded the roots to make a poultice for boils, and the Pawnee used a root decoction for weakness and lassitude. The name paregoric root suggests another medicinal use: paregoric is an opium-based remedy for diarrhea.

Many people steeped the leaves in water to make an eye medicine. Eye infections were a common problem among Native Americans because of exposure to smoke. Samuel de Champlain described the native lodges, in which two dozen families might live together and a dozen fires might be kept alight, yet there were no windows and nothing but a small hole in the roof for the smoke to escape. "The smoke inside is thick and blinding," he wrote, "and diseases of the eyes are common."[132]

The effect of sweet cicely on animals was apparently extraordinary. The Omaha and Ponca had merely to hold the plant in their hands and, says their lore, wild horses would trot right up to them and be caught. Rather like equine catnip. Other animals enjoyed it too, including sheep and cattle as well as deer. The part above ground gets killed off by frost, but the fleshy root holds food reserves for the plant. For this reason it survives all manner of grazing.

The leaves are still grown and collected as a spicy addition to salads, chicken dishes, or to give fruit salad a bit of a kick. Naturalist Steve Brill also uses the roots of sweet cicely, but warns, "I've become very cautious about walking through the aniseroot when I gather nuts in autumn. Otherwise, the hard, sharp seedpods catch on to and pierce my clothing like botanical needles." He notes the roots are "hard to clean, but a little of this strongly flavored treat goes a long way."[133]

Apocynaceae

Indian Hemp
Apocynum cannabinum, dogbane

Apocynum cannabinum

grows 3 to 5 feet tall. The erect stem is brown, with oval or lance-shaped leaves in pairs. Tiny white flowers appear in summer and grow in clusters at the top of the stem; the flowers give way to long, thin, curved seedpods. It is found in fields, marshes, thickets, and waste places.

***Range:* throughout the continental United States (except Alaska) and southern Canada.**

The scientific name *Apocynum*, loosely translated from the Latinized Greek, means, "Go away, dog!" because the plant is poisonous to dogs—and other animals. *Cannabinum* means "hemp-like." (This plant is no relation to *Cannabis sativa*, also known as Indian hemp and a member of the family Cannabaceae, although it shares certain properties. Rather, it is a relative of the oleander and the Florida rubber vine.)

This plant is best known for the long, strong fibers in the stem, which native peoples used to make fishnets, rope, twine, bow strings, and strong cloth. They processed the fibers by soaking the stems in water until the soft parts rotted away, then beat them to separate the long thin fibers. The Iroquois and Mohawk believed any cloth made from this hemp would protect them from arrows and ax wounds. The plant was considered so valuable that Algonquin women sometimes took *Apocynum* seeds with them when they got married, rather like a dowry.

The bark is toxic, but like many other toxic plants, it served its purpose as an emetic when used in very small quantities. The Chippewa also dried the root and pulverized it to produce a kind of snuff they inhaled for headaches. The Ojibwe boiled the leaves to make a poultice to soothe poison ivy rashes. Since dogbane itself exudes a milky sap that causes rashes in some people, this sounds a bit like a Native American example of homeopathy (like treating like).

Today, dogbane is pretty much considered a weed, and a tough one, because of the unbelievable underground root system which will reach down 14 or 15 feet in short order. According to Jerry Doll of the University of Wisconsin, who has studied dogbane, "single plants arising from seed and growing without competition or disturbance could invade an area nearly 40 feet in diameter in two seasons." He recommends a number of strategies to keep dogbane under control in agricultural areas, including planting alfalfa, which competes successfully with dogbane for space.[134]

Araliaceae

American Ginseng
Panax quinquefolius

Panax quinquefolius, American ginseng

Panax quinquefolius

grows 8 to 27 inches tall, rising from a short gnarled rhizome with a parsnip-shaped root. It thrives in heavy shade and humusy but well-drained soil. It has five leaflets with small greenish-white flowers at the end of the stalks. It develops bright red berries.

Range: southern Ontario and Quebec west to Manitoba, from Maine west to Minnesota, and in higher elevations in some Southern states.

Every culture looks for a cure-all that will prolong life, enhance sex, improve memory, and probably keep away vampires. Well, *Panax* means "panacea"—very much how ginseng has been viewed historically and is still viewed today. In particular, it is said to improve energy and sexual prowess, accounting for its widespread popularity. Native Americans have used it medicinally for thousands of years, just as it has been used in Asia, and there has been a resurgence of interest in its medicinal and tonic properties. This is not entirely good news: the demand for herbal medicines has pushed some of our finest native plants right onto the endangered species list.

The power of the ginseng lies in its root. The aboveground greenery is inconspicuous, buried under leaves or hidden by larger plants. "One must burrow deep, like the rabbits, to find its round, pungent, sweet, nut-like root, measuring about half an inch across, which few have ever seen," wrote naturalist Neltje Blanchan in 1925.[135]

The ginseng root is thick, fleshy, and wrinkled, with fibers emerging from one end. The upper part has scars from the annual shoots which grow from it. When dried, it has yellowish-white bark with a slight scent and a bittersweet taste. According to Chinese tradition, the roots shaped like a human body contain the most powerful magic. The word *ginseng* is believed to be a corruption of the Chinese *schin-seng*, or *jun-shen*, meaning "manlike" or "man-plant." Chinese emperors paid fortunes for human-shaped roots. Chairman Mao, it is said, drank ginseng tea at least three times a week, made from roots that cost $100 an ounce.[136] He lived to a ripe old age and never lost his interest in very young women.

The Native Americans were also familiar with its powers. Uses varied from the Meskwaki of Wisconsin, who made a love potion from ginseng, to the Penobscot women of Maine, who steeped the root as a fertility drink.

In Quebec in the early eighteenth century, Jesuit priests who had served in Asia recognized the plant immediately. Although the fur trade was lucrative, the trade in native ginseng quickly became the second most important export leaving New France. By 1750, however, most plants with large roots had been shipped to China. The Canadian trade petered out, but exports from the United States continued. Get-rich-quick books written in the nineteenth and early twentieth centuries promised maximum profits for minimum labor by selling the plant into the Asian market. Alas, it's neither an easy nor a quick business: the plants take three years for seed to germinate, another three years to produce fruit, and another five years before the roots are ready for market. Few greedy entrepreneurs had the patience to cultivate the plant and they stuck to ripping it out of the wild.

Modern studies suggest that though it's not a miracle cure, it does contain vitamins, minerals, and saponins that make the body more resistant to stress. Some evidence also suggests it might help in treating type 2 diabetes. A recent study noted the blood sugar level of patients with adult-onset diabetes who took three grams of ground ginseng before or during a meal dropped an average of 20 percent. The lead author of the study, Dr. Vladimir Vuksan, nutritional scientist at the University of Toronto, says if this reduction "could be sustained long-term—and we don't know that yet—[it] would save people with diabetes from many complications" (including damage to the nerves, kidneys, and eyes). Dr. Vuksan used *P. quinquefolius* because it is commonly grown in Ontario, and the provincial government has a program to encourage farmers to start planting ginseng as an alternative to tobacco.[137]

A century ago, ginseng was said to be disappearing from the woods. Today, as the demand for the plant increases, more and more people are learning how to cultivate it in North America. About 2 million pounds are harvested each year. Ginseng growers' societies have sprouted up to share information about techniques and markets.

Ginseng has its personal fanatics who insist the best plants grow in the wild, because cultivated ginseng is often contaminated with pesticides and herbicides. In remoter parts of the United States, such as Appalachia, people dig for the wild roots, most of which are exported to Asia. Although it is protected by law and in most states may only be picked from September 1 to November 30, digging still goes on at other times, supplying a profitable black market.

It takes 400 plants to make 3 pounds of dried roots. A forest might have as many as 100 plants, but the norm is 20 or even fewer. If poachers find them all, wild ginseng may disappear completely. Few people now have the attitude of a wise old black man who made his living around the turn of the nineteenth century. Old Ned took his grandson to a mountainside near Sewanee. " 'I have been getting my herbs here for 30 years,' he said. 'I always take some and leave some to grow more. This you must learn—not to destroy the growth of anything. You can make a living in the mountains if you learn the value of its gifts.' "[138] Advice that poachers would do well to observe.

Aristolochiaceae

Wild Ginger

Asarum canadense, Indian ginger, colicroot, heartleaf, heart snakeroot, Vermont snakeroot, Canada snakeroot, false coltsfoot, asarabacca

Asarum canadense

grows 12 inches high in rich moist soil. Its creeping perennial root grows into a hairy stalk that divides into two; each stem bears a large heart-shaped leaf and a cup-shaped brownish-purple flower. It is creamy colored inside and is pollinated by ground insects.

Range: **New Brunswick east to southeast Manitoba, south to Kansas, Tennessee, and North Carolina.**

Wild ginger, with its singular heart-shaped leaves, has nothing to do with the spice from tropical countries—a completely unrelated tropical plant called *Zingiber officinale*. The roots of wild ginger have a similar but much milder taste. In spice-deprived areas in the eighteenth and nineteenth centuries, however, wild ginger was powdered and used to flavor food. *Asarum europaeum* was a familiar wild herb in Europe, cultivated as a remedy for headache and deafness. Many people swore that a dab of wild ginger in tea would keep one healthy. It was also a well-known remedy for seasickness and is still viewed as such today.

Ginger, along with Dutchman's-pipe (*Aristolochia macrophylla*)—an insect-trapping machine—is a member of the birthwort family. *Asarum* is a small genus, with only 70 species found in temperate zones around the world.

Often obscured beneath woodland foliage, the solitary three-lobed purplish-brown flowers huddle in the leaf litter for warmth. The heart-shaped leaves grow on long hairy stalks to capture whatever sunlight filters through to the forest floor. The caterpillars of the pipevine swallowtail butterfly (*Papilio philenor*) feed on the leaves. The dark cup-shaped flowers provide food and shelter for pollinating flies such as March flies (Bibionidae), fungus gnats (Mycetophilidae), and syrphid flies (Syrphidae).[139]

The female pistil matures first and has sticky lobes. As the lobes wither, 12 male stamens bursting with pollen develop. The roots are rhizomes and each one a single unisexual plant, assuring cross-fertilization and therefore the health of the species. The mature seeds explode out of the leathery capsule and scatter about the parent plant.

Wild ginger can form colonies quite rapidly—sometimes too much of a good thing. Garden writer Lewis Gannett deplored it in *Cream Hill* (1949):

Asarum canadense, wild ginger

Why the hills about us are not all carpeted with wild ginger is a deep mystery. Three small plants which we set out fifteen years ago have each spread over square yards; one of them reached down through a stone wall and reappeared rods away from its base of operations. We like wild ginger. Its rusty hidden flower makes a lovely design, though you have to stoop to see it. . . . But enough is enough. For years now we have been weeding wild ginger and trying to give it away to friends who complain that nothing will grow in their dry soil.[140]

The analgesic properties of wild ginger were well known to Native Americans, who used it as a remedy for coughs and colds and to relieve heart pain. They also discovered that the ground-up dried root added spice to almost any meal and could protect meat and fish from going off. If a piece of meat spoiled, a bit of wild ginger guarded against ptomaine poisoning. The Meskwaki and

Ojibwe used the roots as seasoning for bottom feeders such as mud catfish, to disguise the muddy taste. Women used wild ginger to relieve swollen breasts or alleviate painful menstruation. It was a contraceptive as well as an anti-inflammatory, and effective in treating children with convulsions or a fever.[141] The plant was also called snakeroot—perhaps it was used to treat someone who had been bitten by a snake and word got around, though a great many plants of a wide variety of species were dubbed snakeroot, indicating this was a serious concern.

Ginger had many other uses as well. It was dried to perfume clothing. And it was also prized for warding off evil (rather as Eastern Europeans thought garlic would ward off vampires). Diamond Jenness, one of the great early anthropologists, reports on a fear of witchcraft among the Ojibwe he encountered near Lake Huron in Canada:

So potent is this fear of witchcraft that every Parry Islander takes counter-measures for his own protection, and for the protection of his family. . . . Since a sorcerer may visit a house by night and place evil medicine in dishes prepared for the following day, some of the Indians regularly add a little wild ginger to their food. In the earlier days, they say, warriors always mixed this wild ginger with their war-rations of dried berries and dried meat, for it prevented the contagion of the food from several sources, from the touch of a little baby, of a woman in her seasons, and of a sorcerer. . . . My foster-mother put wild ginger in all our food to prevent any ill effect, and she gave me wild ginger to chew.[142]

Catherine Parr Traill was a remarkable woman who not only managed to raise a family on a pioneer farm but also found time to draw the plants around her and describe them grace-

fully. Writing in the 1830s, she noted that wild ginger was often used by people who didn't have the money for a doctor and relied on ginger tea to rid themselves of the ague.[143]

By the 1950s, laboratory evidence confirmed that the root has antibacterial properties. By the 1960s, scientists had isolated the antitumor agent aristolochic acid, an active agent against bacteria and fungi.

Some people just enjoyed the flavor. The nineteenth-century plant scientist George Washington Carver considered the *Asarum* leaf the "acme of delicious, appetizing, and nourishing salads." Pieces of root were candied by being cooked in sugar water for long periods (some recipes call for cooking up to four days); the leftover syrup was poured over flapjacks and fruit. The fragrant oil from the plant was also used as an ingredient in expensive perfumes. Today, it is chiefly known as an attractive plant for shady gardens.

Asteraceae

New England Aster

Aster novae-angliae, farewell-summer, hardy aster, last rose of summer

Aster novae-angliae

has hairy stems and grows 4 to 6 feet tall, reaching 8 feet in a damp thicket or moist meadow. The blue to purple to pink heads have a yellow disk which contains the inner tubular yellow flowers held up by green bracts; the petals form a raylike halo around the center. The fruit is a tiny 1/6-inch achene.

Range: New England, Quebec, and Ontario west to Alberta, south to Wyoming, New Mexico, Arkansas, and Alabama.

When the meadows and verges have given up and turned to dusty colors, along comes the richest palette of the year. Fields come alive with asters, the most charming and familiar of all the late-summer plants, splashing bright pinks, purples, and whites across the landscape, attracting dozens of insects in a final fling before the cold weather begins.

The New England aster belongs to one of the largest families of plants in the world—the Asteraceae, with at least 1,300 genera and more than 21,000 species. There are more than 250 species of asters alone, about 120 native to North America and perhaps 54 or more indigenous to the eastern part of the continent. Asters crossbreed prolifically, sharing their wealth of beauty as they go, giving each part of the continent its own asters, each species adapted to the local insects. Heights vary from 1 to 8 feet tall, but most have hairy stems and rough, lobed leaves clasping the stem on each side.

The word *aster* comes from the Greek word for star, which describes the flowers perfectly. The flower head has two kinds of florets to appeal to as many pollinators as possible. A fringe of ray flowers directs the pollinators into the flower's central disk. This disk is made up of hundreds of minute flowers crammed together so that each emerges at a different time from its neighbor. An insect can buzz around an aster again and again, coming back relentlessly for more nectar, and still be satisfied. The timing also ensures successful cross-pollination.

Some early users felt that rubbing the flower heads produced a smell of camphor or turpentine, which explains why asters were widely used as a fumigating agent. They were burned to produce smoke thick enough to drive away what the Potawatomi and others considered "evil spirits" and what the Europeans colonists considered the "harmful vapours" of the sick. What evil spirits avoided, deer liked—Frances Densmore reported in the 1920s. The Chippewa smoked

Aster novae-angliae, New England aster

the root in a pipe "to attract game." Apparently, the smoke was a little like the smell of the deer themselves, and they would come to the familiar scent, expecting to find their own ilk.[144]

Asters are such a heartening sight, and now that the ban on spraying insecticides and herbicides along the roadsides in some states and provinces has allowed them to return, they are renewing their traditional autumn tapestry.

Wood aster (*A. divaricatus*) is one of the few asters to thrive in dry shady places. It grows 1 to 3 feet high; the flower heads are small, ¾ to 1 inch across, with the typical yellow center surrounded by white petals. This plant is endangered in some areas, as it is driven out of its habitat by garlic mustard (*Alliaria petiolata*). The latter was brought here by the settlers as a medicinal plant and, unbeknownst to them, was to become a major pest all over this continent.

Heart-leaved aster or common **blue wood aster** (*A. cordifolius*) is also known as beeweed, blue devil, fall aster, and stickweed in West Virginia, and as tongue in Maine. The names suggest it was not much appreciated in West Virginia, but in Maine the greens were considered edible. The common name is apt: the leaves are broadly heart-shaped. The numerous small flower heads come in pleasant shades of violet or blue, occasionally white. It grows in open woods and thickets. It was also smoked in pipes to attract game, and had a reputation as a "an excellent aromatic nervine, in many cases preferable to Valerian,"[145] according to C. S. Rafinesque (1783–1840). He was an American naturalist who devoted his life to discovering plants, and named more species and genera than almost any other American scientist of his time. He was also a great observer of the settlers' habits. Since valerian was a sedative, perhaps this admirable nervine was their answer to Valium. Given how difficult and stressful the settlers' lives were, they probably needed something to soothe shattered nerves at the end of a hard day.

The flat-topped white or **parasol aster** (*A. umbellatus*) got its name because the top of the whole plant looks flat or umbrellalike. The Mohegan used the dried leaves to make tea, which was considered good for the stomach. The Potawatomi also burned the flowers to repel evil spirits from a room in which a patient was recovering from illness.

Balsaminaceae

Spotted Jewelweed

Impatiens capensis, touch-me-not, balsamweed, horn of plenty, brook celandine

Impatiens capensis

grows to 5 feet, with succulent, translucent, hollow stems bearing opposite leaves on the lower branches, alternative leaves on the upper shoots. The leaves are oval with rounded teeth. The orange-yellow flowers are 1 inch long or less, trumpet-shaped, with red, dark orange, or brown spots. They bloom from midsummer to early fall, before giving way to seed capsules less than 1 inch long. They dry out and disperse their seeds when they explode.

Range: throughout the continental United States and southern Canada.

Impatiens capensis became jewelweed because tiny jewel-like beads form on the surface of the waterproof leaves when rain falls or dew forms on them. The whole plant glitters. "Touch-me-not" is descriptive of what happens if they do get touched: the dried-out seed pods explode, spraying seed everywhere. Kids in the past used to play games bursting them, or trying to carry them about without detonating them—an even more difficult task.[146]

For sufferers from poison ivy rash, the modern name should be "touch-me-a-lot," because the anti-inflammatory and fungicidal compounds in its juice soothe the miseries of the rash. Better still, the plant grows in places near poison ivy (except for dry, sandy, or coastal areas), so help is often close by. Indians said that the Great Spirit was kind enough to put the remedy next to the poison. A more modern

Impatiens capensis, spotted jewelweed

wild-plant lover calls it "the forager's American Express, because I never leave home without it."[147] The juice soothes all kinds of irritations, from nettle stings to athlete's foot. Native peoples made up infusions of the whole plant, which they drank to cure a cold or made into a liniment for aches, bruises, and sprains. They also obtained a yellow dye from the juice.

Euell Gibbons writes of eating jewelweed sprouts in cream sauce on toast, suggesting that they are every bit as good as asparagus, although the flavor is quite different.[148] Steve Brill and Evelyn Dean liken the taste of the seeds to walnuts.[149] Apart from all these virtues for humans, jewelweed is a major attraction for hummingbirds. Mathew Tekulsky, a major expert on hummingbird gardens, cites a 1906 report in which two naturalists at Point Pelee, Ontario (the southernmost tip of Canada), saw hundreds of hummingbirds hovering over the patches of jewelweed creating "a low hum that arose from the vibrations of many little wings."[150]

Berberidaceae

Blue Cohosh

Caulophyllum thalictroides, squawroot, papooseroot, blue ginseng, blueberry root

Caulophyllum thalictroides

grows 1 to 3 feet tall from a long thick root. It has a smooth stem sheathed at the base. The single leaf halfway up the stem consists of three stalked, lobed leaflets, each about 2 to 3 inches long. The small yellowish or dark green-purple flowers bloom before the leaves are open, followed by dark blue berries. It thrives in rich soil.

Range: **Nova Scotia, New Brunswick, southern Quebec, Ontario, southeastern Manitoba, and south to South Carolina, Alabama, and Missouri.**

The graceful elongated naked stems of the blue cohosh push up out of the soil looking much like a stalk of purple asparagus. The tiny yellow flowers open up before the elegant ferny leaves unfold. They are so inconspicuous that they are easy to miss. There is a bluish cast all over the leaves called a "bloom" which makes it all the more mysterious.

After insects pollinate the flowers, the ovary bursts open as the seeds enlarge and develop outer coats, also blue, looking somewhat like

Caulophyllum thalictroides, blue cohosh

berries. They turn into the plant's major attraction as something to observe, but given how poisonous they are, looking is the only thing one should do.

Cohosh is an Algonquin word used by indigenous peoples for several different and unrelated species, including black cohosh (*Cimicifuga racemosa*) and white cohosh (*Actaea alba* syn. *A. pachypoda*).

All three species had medicinal uses. Blue cohosh was and is particularly important in matters of the uterus. Because it induced childbirth it was called papooseroot. In 1828, American naturalist C. S. Rafinesque wrote:

This is a medical plant of the Indians, and although not yet introduced into our official books, deserves to be better known. I have often found it used in the country and by Indian doctors. . . . The root is the only part used: in smell and taste, it partakes of Ginseng and Seneca root, and is sometimes mistaken for both. It is sweetish, a little pungent and aromatic. . . . It is used . . . for rheumatism, dropsy, cholic, sore throat, cramp, hiccup, epilepsy, hysterics, inflammation of uterus, &c. It appears to be particularly suited to female diseases, and . . . the Indian women owe the facility of their parturition to a constant use of a tea of the root for two or three weeks before their time. . . . It promotes delivery, menstruation, and dropsical discharges. It may be used in warm infusion, decoction, syrup or cordial.[151]

By the mid-1850s, it was being sold in drugstores in powdered form or as the raw root. It was eventually included in the list of *Canadian Medicinal Plants* and the *Dispensatory* of the United States. In the 1950s, a French Canadian pharmaceutical researcher, J. Auguste Mockle, confirmed that a saponin, caulosaponin, provokes strong uterine contractions.[152]

Today, blue cohosh is still a popular herbal medicine, used to bring on menstruation, ease painful menstruation, and hasten childbirth. But knowledgeable herbalists warn about its powerful alkaloids, which can cause heart problems and dangerously stress the kidneys. Don't even think about trying it without reliable professional advice, and never combine it with other medication.

Berberidaceae

Mayapple

Podophyllum peltatum, devil's-apple, hog apple, Indian apple, American mandrake, American mayapple, raccoonberry, wild lemon, umbrella leaf

Podophyllum peltatum

is a herbaceous plant found in moist woods on well-drained soil. When it has grown to about 12 inches high, the stalk opens into an umbrella shape with two large, dark green, long-stemmed, palmate, lobed leaves. In the node between the leaves, a pretty white nodding flower with six to nine petals emerges in early spring. The yellow fruit starts to set in summer and ripens by early fall.

Range: **southern Quebec to Minnesota, south to Texas and Florida.**

When the mayapples gently unfurl their great leaves they make a protective canopy over their delicate flower buds. *Podo* means "foot" and *phyllum* means "leaf," and the plant does look a bit like a webbed foot stretching upward. Rather eccentrically, insects gather the pollen even though the flower has no nectar. The mayapple is a perennial and may form circular colonies from a single plant. An average-sized colony is probably about 45 years old. The yellow, egg-shaped fruit which appears in the fork of the stem comes out naturally in May and is edible when ripe. It smells like and has the taste of lemon. Unripe mayapple fruit, however, can cause vomiting and every other part of the plant is extremely toxic.

Mayapple fruit was traditionally eaten ripe or mashed and dried in cakes for use as a portable food for hunters or travelers. The dried fruitcakes could be soaked in warm water and cooked as a sauce or mixed with corn bread. They were also made into a type of preserve. In the American South, some people make mayapples into a drink like lemonade.

Catherine Parr Traill, in the spare time she had from raising several children and supporting a feckless husband, wrote a book of advice for other women immigrants. Here's her take on mayapple:

Gather the fruit as soon as it begins to shew any yellow tint on the green ring: lay them in a sunny window for a day or two; cut them in quarters and throw them into a syrup of white sugar in which ginger slices, and cloves, have been boiled: boil the fruit till the outer ring is tender: take the fruit out and lay them in a basin, sift a handful of pounded sugar over them and let them lie till cold. Next day boil your syrup a second time, pour it over the fruit, and when cold put it into jars or glasses, and tie down. It should not be used till a month or six weeks after making.[153]

For many people such as the Delaware or the Iroquois, the important part of the plant was the root, which acted as a laxative, tonic, and purgative. Being cleansed was very important to their well-being, and purification rituals were essential to a long and healthy life. This plant

Mayapple's wide use as a medicinal plant is reflected in the common name given by European settlers, American mandrake. The European mandrake *(Mandragora officinarum),* an unrelated species, was known for the magical powers in its roots, which were often shaped like the human form. According to legend, the root would scream when it was pulled from the ground, causing anyone hearing it to go mad.

The Indians were astute about the plant because, as it turns out, the mayapple root contains alkaloids which are the plant's way of storing nitrogen, maintaining a healthy pH balance, and protecting itself from parasitism or infection. Alkaloids make good purgatives as well as pain relievers, pesticides, and herbicides. The Cherokee took the ooze from a root and soaked their corn seeds in it before planting to keep off crows and insects, and the Menominee sprinkled it on potato plants to kill bugs. Integrated pest management was a cornerstone to the way native peoples managed their land long before it was

Podophyllum peltatum, mayapple

was a way of getting rid of worms, one of the plagues of life in the forest. As the ever-observant William Bartram noted:

The [mayapple] root . . . is the most effectual and safe emetic and . . . efficacious in expelling worms from the stomach—the lives of many thousands of the people of the southern States are preserved, both of children and adults. In these countries it is of infinitely more value than the Spanish

And will any poet sing
Of a lusher, richer thing
Than a ripe May-Apple, rolled
Like a pulpy lump of gold
Under thumb and finger tips,
And poured molten through the lips.

James Whitcomb Riley, the "Hoosier poet," in *Rhymes of Childhood* (1890)

introduced into modern agriculture. They observed the life cycles of insects very closely and understood the controls necessary to keep pests off their crops.

Today, several important drugs come from mayapple. The podophyllotoxins produced in the leaves and stems and stored in the rhizome interfere with cell division, including cancer cells and the cells in venereal warts. Alas, mayapple is one of the few known sources of podophyllotoxins. As we cut down forests and destroy its habitat, the plant is becoming increasingly rare. Scientists at the University of Mississippi are trying to develop the plant as an alternate crop for American farmers, but so far no cultivar suitable for commercial farming has been found.

Campanulaceae

Cardinal Flower
Lobelia cardinalis

Lobelia cardinalis, cardinal flower

Lobelia cardinalis

grows up to 4 feet tall on an erect stem, with alternate lance-shaped, toothed leaves. At the top of the stem the scarlet flowers appear in a spike; each flower is 1 to 1½ inches long, with a tubular base, flaring into two lips—the upper with two lobes and the lower with three. Found in marshes, meadows, streambanks, and low-lying open woods.

Range: **throughout the eastern and central United States and southeastern and south-central Canada.**

The cardinal flower stands ramrod straight like a scarlet grenadier. It is a magnet for hummingbirds, which land on the lower lobes and dip their long bills into the nectar all the while beating with their tiny wings to maintain pur-chase on the bloom. It is a ravishing sight.

And ravishing is what the Meskwaki had in mind when they used this plant as a love medicine (they also used the **great blue lobelia**, *L. siphilitica*, for the same purpose). When lovers quarreled, friends or family chopped up cardinal flower roots very finely and put them in the couple's food without their knowing. The fractious pair would somehow find themselves patching things up and falling in love all over again. According to a slightly different version, minced roots were sprinkled on the bed of an estranged couple to rekindle their love. Like the great blue lobelia, a decoction of the roots was used to cure syphilis.

According to anthropologist Huron Smith, the Meskwaki ground up the entire plant into fine dust and stored it in a pouch. If bad weather was approaching, they threw this dust into the air to dispel the storm. It was also tossed into an open grave after a funeral.[155]

Today it is grown as an ornamental plant, and is especially recommended as an attractor for hummingbirds.

Commelinaceae

Spiderwort
Tradescantia ohiensis, dayflower, flower-of-a-day, snotweed

Tradescantia virginiana, spiderwort, a related species

Tradescantia ohiensis

is a clump-forming herbaceous perennial which grows up to 3 feet tall with dark bluish-green, arching, grasslike leaves up to 18 inches long and 2 inches wide. Clusters of blue (occasionally pink), three-petaled flowers bloom from late May into early July. Each flower opens up for only one day.

Range: throughout the eastern and midwestern United States and southeastern Canada.

I always thought of this as an alley plant because it's about the only ray of beauty you'll see near a parking lot. *Tradescantia* can grow in hidden places, down back alleys, practically under the exhaust pipes of cars. Nothing the modern world dishes out seems to bother this plant. It looks a bit like a chunk of thick grass before it breaks into bloom. Then the long skinny leaves rise up and droop down at the ends, giving it the look of a squatting spider.

The sap may also have to do with its name. When the stems are broken, the gooey sap stretches out in threads like the filaments of a spiderweb. This also accounts for the name "snotweed." "Dayflower" refers to the fact that the flowers bloom for only a single morning. The petals contain an enzyme that autodestructs within a few hours. This curious habit also means its cobalt blue flowers open only when there are clouds about. The bisexual flowers (this is also called having a perfect flower—it's got all the sexual parts on one plant), have petals enclosed by pairs of boat-shaped bracts.

The scientific name honors John Tradescant, the seventeenth-century gardener who served Charles I of England. He was also a celebrated plant collector. His son, also called John, came to North America on a plant-collecting expedition and found this plant on one of his trips in 1637. John Jr. collected many other plants in 1642 and 1654 and this one quickly became a favorite in English gardens. The best known of all the spiderworts is *T. virginiana*, another species found by John Tradescant the Younger and brought back to England. Linnaeus

BOTANICA NORTH AMERICA

named the species in honor of this family, whose collection of curiosities, artifacts, plants, and stuffed animals at Tradescant's Ark in Lambeth, England, forms the basis of the Ashmolean Museum in Oxford.

The stems, leaves, and flowers of spiderworts are edible and may be eaten raw or added to stews. However, the most fascinating role of this plant is that of radiation detector. The blue stamen hairs, if exposed to low-level radiation, turn pink. If the stamens are put under a microscope and the pink cells are counted, the level of radiation can be determined fairly accurately. For this reason, *Tradescantia* is often planted near nuclear power plants and regularly examined. It is also the parent of many hugely successful garden hybrids now sweeping the country.

Convolvulaceae

Wild Potato Vine

Ipomoea pandurata, manroot, manroot morning glory, man-of-the-earth, wild sweet potato

Ipomoea means and sounds wormlike, and describes this plant's writhing stems. The vine can weave its way around fence posts, oftentimes stealing light from sun-loving plants in its way as it snakes around dry fields, plains, and woods. It can be a real nuisance in certain agricultural areas, when it coils around corn stalks or soybean plants. *Pandurata* means "shaped like a pandura," a stringed instrument somewhat like a violin, and alludes to the shape of the leaves. The trumpet-shaped flowers remind people of morning glories except that they are white and pink instead of blue, and they bloom throughout the summer. Morning glories are a close relative and, like them, the potato vine

blooms open shortly after dawn and close again in a few hours.

The edible tubers were well known to Native Americans, who called them *mecha-meck.* The tubers must be thoroughly cooked (baked, boiled, or roasted), and the tough skin peeled off. Some people find they taste like sweet potatoes, others are convinced that there is little but a bitter taste, though the smaller the root the tastier it is likely to be. Uncooked, they are a powerful laxative, and therefore had a medicinal use.

I. pandurata grows up to 12 feet long from a single massive root, hence the "manroot" name. "I once dug up a root of the wild potato." wrote Oliver Medsger. "It extended nearly vertically to a depth of three feet or more. The great fleshy root, resembling a large sweet potato, was two and a half feet long, and weighed 15 pounds. Friends . . . unearthed a plant in southern New Jersey that weighed more than 30 pounds."[156] Doug Elliott, another writer on wild plants, also reported on a go at hauling a wild potato out of the ground:

As I started to dig, I could see that the stems all originated from a central area about two inches thick, but as I carefully dug deeper around the root, I came to the narrow area around the neck and started to become disappointed because I thought the root was tapering off. Two hours later, as I was lying on my belly midst piles of dirt, the trowel and my hand stretched deep down into the hole, carefully removing dirt from around the root that was still going deeper, I certainly had to chuckle at my pre-judgement. The root was three feet long where I cut it off, and about as big around as my thigh.[157]

Range: Connecticut west to southern Ontario, Michigan, and Kansas, south to Florida and Texas.

Dennstaedtiaceae

Western Bracken
Pteridium aquilinum, pasture brake

The name *Pteridium* is an echo of this plant's ancient past. It looks like an escapee from the Carboniferous era, 345 million years ago—and that is exactly what happened. It isn't a great leap of the imagination to picture these as gigantic plants 20 feet high, but today they grow to 3 feet or so from an underground rhizome with feathery leaflets forming a triangular shape. When the plant is mature, the stems are brownish and the backs of the fronds are covered with brown sori (the structures that produce spores). They can make very large colonies.

Western bracken is tougher than most ferns and will grow wherever there's an opening in an abandoned field, on the edge of a hill, or in the middle of a parking lot. It can defeat just about anything, including asphalt and grass.

Native Americans used bracken in many ways: to line steaming pits, cover berry baskets, wrap dried foods for storage, or wipe fish. The fronds made bedding and baskets, sunshades, mulch, tinder, and torches. It was a handy, all-purpose plant. As the groaner goes, with fronds like these, who needs anemones?

Now we come to a tricky question: Is it or is it not the source of fiddlehead greens? The fiddleheads are the new shoots, which emerge tightly coiled up, like the tips of violins. Some books on edible wild plants say *P. aquilinum* is the fern of fiddlehead fame; others say that's not true and fiddleheads are actually from the **ostrich fern** (*Matteuccia struthiopteris*), which grows in eastern Canada, New England, around the Great Lakes, and through the north-central states, and on the west coast of Canada. The ostrich fern believers state that *Pteridium* is car-

cinogenic. Mature bracken *is* toxic, but like pokeweed, the new shoots may be harmless.

Bracken does produce fiddleheads, and apparently some foragers eat them, but the commercial producers of fiddleheads favor the ostrich fern, which is even taller (up to 4 or 5 feet tall) with a much larger, more oval frond. Fiddleheads are sold fresh or frozen, and are usually steamed and served with butter or added to stir-fries.

In fall, the bracken turns brown or bronze colored, and then fades away. Thoreau described the look of dying bracken in autumn:

Brakes are fallen in the pastures. They lie flat, still attached to the ground by their stems, and in sandy places they blow about these and describe distinct and perfect circles there. The now fallen dark-brown brake lies on or across the old brake, which fell last year and is quite gray but remarkably conspicuous still. They have fallen in their ranks, as they stood, and lie as it were with a winding-sheet about them.[158]

Range: **throughout the continental United States, and southeastern and southwestern Canada.**

Ericaceae

Trailing Arbutus
Epigaea repens, mayflower, winter pink, mountain pink, gravel plant, ground laurel, shadflower, crocus

Epigaea repens

grows on woody stems covered with long reddish, bristly hairs trailing along the ground and may be

Epigaea repens, trailing arbutus

up to 5 feet long. It roots wherever it touches the ground. The alternate, evergreen, hairy leaves are 1 to 4 inches long and leathery green. The flowering stalk is 1 to 2 inches high holding a cluster of fragrant white to pinkish flowers. The fruit develops in a seven-sided orangish capsule, which appears in August. Thrives in sandy or rocky acid soil.

Range: Newfoundland, Nova Scotia west to Saskatchewan, south to Florida, Alabama, Kentucky, and Iowa.

State flower of Massachusetts; provincial flower of Nova Scotia.

Legend has it that this was the first flower the Pilgrims saw in the spring after their initial winter in the New World. They immediately gave it the name of their ship, the *Mayflower*. The Pilgrims obviously had sharp eyes, since mayflowers can be difficult to spot, nestling as they do

among dead and decaying leaves. This habitat provides the mycorrhizal fungi vital to its survival.

Trailing arbutus is a perennial subshrub, but the stems die back leaving a remnant shoot system lying on the ground. This is called being a hemicryptophyte. It is a dioecious plant—male and female flowers are on separate plants.

The Algonquin of Quebec used the leaves for kidney disorders, as did many other Native Americans. It became famous as a gravel plant because it purged kidney stones. The Cherokee gave an infusion of the leaves to children with diarrhea, and used it to treat indigestion. The Iroquois used it for rheumatism as well as kidney ailments, and made it into a tea to aid digestion. Many different groups considered its flowers to come directly from their divinity, which anthropologist Huron Smith called *kitcimani-towwiwin*.[159]

The mayflower's spicy scent has been a compelling lure to plant lovers, but perhaps it has become too well loved—it may be in danger from its own fans. In their desperate anxiety to put winter behind them, people tend to go out and grab the first flowers of spring. In New England, it used to be customary to have a "mayflower picnic" to see the first of these spring flowers. It is now illegal to pick them in Massachusetts.

Coin collectors also valued the mayflower. Nova Scotia, which minted some of its own money in the nineteenth century, produced a beautiful penny in 1856, with the head of Queen Victoria on one side and the mayflower on the other. Later, Nova Scotia became the first province in Canada to adopt a floral emblem, choosing the mayflower. In 1918, it also became the official flower of Massachusetts.

Ericaceae

Wintergreen
Gaultheria procumbens, checkerberry, boxberry, partridgeberry, mountain tea, teaberry

Gaultheria procumbens
is a shrubby plant creeping along the ground, sending up erect branches 3 to 6 inches high. The glossy, evergreen, pleasant-smelling, oval leaves are usually clustered at the summit of the branches. They are dark green on top, paler underneath. In midsummer, tiny bell-shaped white flowers appear in the leaf axils, singly or in pairs. These give way to bright red berries about ⅓ inch across in fall, which remain on the plant through the winter.

Range: throughout the northeastern United States and southeastern Canada.

Wintergreen, a glossy green aromatic plant and close relative of salal of the Pacific coastal regions, is a cheerful autumn sight with its bright red berries. According to Lenape Indian legend, the berries represent the blood shed when the mastodons fought the other animals and lost. The mastodons sank into the bogs, where their bones are still found from time to time, and the Great Spirit transformed their blood into red berries to compensate people for the loss of this source of meat.

This is the kind of plant that seems to reek of cleanliness and health, as anyone who has sniffed wintergreen will attest. The berries are edible and the leaves have long been used to make tea. Native Americans added the leaves to their tobacco for a more pleasant smoke and chewed the roots to keep their teeth clean and healthy. Wintergreen oil, which is made from the leaves, contains methyl salicylate, a compound close to salicylic acid, the pain-suppressing ingredient in aspirin. Appropriately, the scientific name honors a Quebec physician, Dr. Jean-François Gaultier (1708–1756). Gaultier was a botanist as well, and showed the plant to his friend Peter Kalm. Kalm took it to Sweden to Linnaeus, who gave it the good doctor's name.

Wintergreen tea was a popular remedy for rheumatism, arthritis, lumbago, gout, and just about anything else causing aches and pains. The settlers quickly caught on and by the 1880s were distilling wintergreen oil in factories. They used it to make patent medicines and liniments, as well as candy and root beer. Wintergreen oil was also added to cosmetic lotions, because it softened the skin, and was even used to make leather easy to work in bookbinding. However, it takes a ton of wintergreen leaves to make a pound of oil, so today most commercial winter-

Gaultheria procumbens, wintergreen

green oil is a synthetic form made using birch twigs, which also contain methyl salicylate. (The undiluted oil is highly toxic, and only tiny amounts are used as flavoring or in medicine.)

Wintergreen is still used for tea, but there is a special technique involved. Euell Gibbons tried it the usual way with fresh or dried leaves, with boring results. One day he filled a quart jar with fresh leaves, added water, and left it. After two days, he noticed it had fermented slightly and, once heated, the liquid was a delightful tea with a noticeable wintergreen flavor. He also experimented with jam made from the berries, and found heating destroyed the flavor, but that uncooked jam with some of the wintergreen tea

thrown in was a success. He even made wintergreen wine—a pink, sparkling, pleasant beverage.[160]

Because the berries last through the winter, wintergreen is a good substitute for holly at Christmas—as it was in the days before imported holly was available in the colder regions of North America. The ever-buoyant Catherine Parr Traill wrote:

I remember the first Christmas Day I passed in Canada—being laughed at because I wandered out on to the plains near Peterboro', and brought a wreath of the boxleaved trailing wintergreen (which with its scarlet berries reminded me of the varnished holly with which we were wont to garnish the old house at home), and hanging it over the mantelpiece, and above the pictures of my host's parlor, in honor of the day. It seemed to me these green branches might be held as emblems to remind us that we should keep faith bright and green within our hearts.[161]

Fabaceae

Groundnut

Apios americana, glycine, hopniss, wild bean, Mi'kmaq potato, bog potato, ground pea, trailing pea, potato pea, pig potato, Dakota potato, Indian potato, white apple, traveler's-delight

Early in the history of European settlement, the Indians taught the settlers how to harvest these underground tubers, which are packed with proteins. They became an essential food for the newcomers. They could be boiled or roasted like potatoes, or dried and ground into a substance like flour. The tuberous roots, which were collected in autumn, would last in an edible state until spring. Groundnuts kept many people alive over a hard winter. Alas, the settlers became possessive about these plants. By the 1650s, some towns had passed laws prohibiting the Native Americans from digging for groundnuts on the settlers' land, on pain of whipping or incarceration.

The groundnut is a deciduous vine that grows up to 10 feet long from white tubers or rhizomes, 1 to 4 inches in diameter. The vine has compound leaves with five to nine lance-shaped leaflets, each 1 to 2 inches long. The fragrant purplish-brown flowers appear in summer and give way to seedpods 2 to 4 inches long, containing edible seeds that look a little like lentils. It grows in moist woods and in thickets with light sandy to loamy soil.

The taste of groundnuts has been variously described as being like mushrooms (Samuel de Champlain), artichokes (Marc Lescarbot), parsnips (Father Theodat Gabriel Sagard), prunes (Nicolas Perrot), truffles (Pierre Biard in the *Jesuit Relations*), chestnuts (Campbell Hardy), and potatoes (Peter Kalm). The European reports exaggerated the vegetation of North America. The trees were always bigger (that was certainly true) and the food was always more delicious than anything in Europe. As social historian Sheila Connor writes: "The descriptions of the edible plants approached the inflated prose of a supermarket advertisement. The fruits, nuts, and berries were more plentiful, juicier, sweeter, larger, and better than any that had been seen before. Groundnuts . . . [were as] 'big as hens eggs and good as potatoes.'"[162]

Just as there were varying descriptions of the flavor, there were dozens of names for the plant. The name has even survived in some place-names, as Campbell Hardy explains:

Paddling down a picturesque Nova-Scotian stream called the Shubenacadie some ten years since in an Indian canoe, it occurred to me to ask the steersman the proper Micmac pronunciation of the name. He replied "We call 'em 'Segeeben-acadie.'" Plenty wild potatoes—segeeben—once grew here." "Well, 'acadie,' Paul, what does that mean?" I inquired. "Means—where you find 'em," [he] said.[163]

Thoreau waxed eloquent about groundnuts, and looked to the day when the indigenous plants would come into their own again.

Digging one day for fish-worms I discovered the ground-nut (Apios tuberosa) *on its string, the potato of the aborigines, a sort of fabulous fruit, which I had begun to doubt if I had ever dug and eaten in childhood, as I had been told, and had not dreamed it . . . Cultivation has well nigh exterminated it. It has a sweetish taste, much like that of a frostbitten potato, and I found it better boiled than roasted. This tuber seemed like a faint promise of Nature to rear her own children and feed them simply here at some future period. In these days of fatted cattle and waving grain-fields, this humble root, which was once the* totem *of an Indian tribe, is quite forgotten, or known only by its flowering vine; but let wild Nature reign here once more, and . . . [perhaps the groundnut will] resume its ancient importance and dignity as the diet of the hunter tribe. Some Indian Ceres or Minerva must have been the inventor and bestower of it; and when the reign of poetry commences here, its leaves and string of nuts may be represented on our works of art.*[164]

Thoreau was overoptimistic about the future of the groundnut. The tuber tastes slightly turnipy and requires a lot of butter to make it palatable. Wild foragers recommend it highly, although most suggest great caution in gathering the plant, since several species of poisonous wild plants resemble the groundnut.

Range: **throughout eastern Canada and the eastern and midwestern United States.**

Fumariaceae

Dicentra spp.

There are 20 species of *Dicentra*. The most familiar, *Dicentra spectabilis*, or **bleeding heart,** with its fountains of bright pink hearts edged in white, graces spring gardens around the world. Originally from the Far East, the species was brought to England in 1847 and caught on immediately among the gardeners there. Many immigrants to North America brought it along to their new country, carrying it westward in their wagons. This explains why clumps of bleeding hearts still grow around old farmhouses all across the continent.

The native herb, *D. canadensis*, is quite different. Smaller and more delicate, with white flowers and lacy foliage, it is one of the glories of spring. *Dicentra* means "two spurs," referring to the double-spurred flowers. These plants have a symbiotic relationship with bumblebees, which have the long tongues necessary to get into the nectaries in the spurs. *Dicentra* species contain alkaloids known to be toxic, but can be useful as medicinal plants.

Fumariaceae

Squirrel Corn
Dicentra canadensis, bleeding heart, colicweed, ghost corn, Indian potato, lady-and-gentleman, lyreflower, staggerweed, turkey corn, turkey pea, white heart, wild hyacinth

Dicentra canadensis
grows 6 to 12 inches high from small yellow bulblets the size of peas. The compound, highly

divided leaves spring from the base of the stem. The stalk bears racemes of four to eight white, heart-shaped flowers with short rounded spurs. Lives in rich, moist woods.

Range: New England, southern Quebec, southern Ontario, Minnesota, to southern Missouri, Tennessee, and North Carolina.

――――――――――――――――――――――

The usual common name for this plant, squirrel corn, refers not to the flower, but to the underground yellow bulblets, which are about the size of corn kernels. They are often eaten by squirrels. The name "wild hyacinth" refers to its lovely scent, and "staggerweed" to the toxic juices of the plant—poisonous to cattle. The Onondaga gave it the name "ghost corn," believing it was food for the spirits.

Squirrel corn flowers in spring (as early as March in some areas). The tubers (or roots) contain the medicinal element of this plant and must be collected when the flowers are in bloom. When fresh, the tuber is dark yellow throughout; dried, it takes on a light grayish-yellow color and has a faint odor.

Nineteenth-century America was a time of medical free enterprise. Many kinds of healers, some more scrupulous than others, offered their services as physicians. During this period a group calling themselves "Eclectic Practitioners" came to prominence. Dr. John King (1813–1893) was one of their pioneers. He fervently believed in the power of herbs to heal and backed up his belief with scientific research. His partner was pharmacist John Uri Lloyd. They analyzed the specifics of American plants, performed many experiments, and wrote up a great deal of anecdotal information gathered from every source available to them. They combined the best medical knowledge of the day, along with the most complete information about plants they could find, in the book *King's American Dispensatory.*

In their book, King and Lloyd reviewed the use of *D. canadensis* in the treatment of syphilis:

In all syphilitic affections, *it is an excellent tonic and alterative; and will likewise be found valuable in* scrofula . . . *As a tonic, it possesses properties similar to gentian, calumba, or other pure bitters; its alterative properties, however, render it of much value. In* syphilis, *especially in the constitutional form, when occurring in debilitated or broken-down constitutions, its efficacy is not equaled by any other agent as an alterative tonic; but from considerable experience with it, I am by no means satisfied that it exerts any real influence as an antisyphilitic, properly so-called, as has been heretofore believed.*[165]

John D. Gunn, author of an 1859 guide to home health, wrote: "The only obstacle to its more general use is the difficulty of procuring it, as the top disappears so soon that there is but a short time, in the early spring, when it can be found. In consequence of this, but little is gathered, and it is generally difficult to find in the drug stores."[166]

Its near relation, **Dutchman's-breeches** (*D. cucullaria*), has white, yellow-tipped flowers with double spurs that look like a pair of breeches hanging out to dry. Indians were great runners and they used a compound infusion of the leaves as a liniment to strengthen their legs. Chosen when they were small children, runners trained with and without moccasins over every possible terrain. Running was a part of the rites of passage into manhood. Access to a good liniment was important.

D. cucullaria was also considered a very important love charm by the Menominee. A young man would throw the flowers trying to

Dicentra canadensis, squirrel corn

Hydrophyllum virginianum, Virginia waterleaf

Hydrophyllaceae

hit his intended with them. Huron Smith wrote in 1923 that a young man would chew the root, exhaling its scent toward her. He then circled around the girl until she caught the perfume; the idea was that she would follow him wherever he went. Smith failed to note whether this charming system really worked in practice.[167] Today, most of us are just content to have such an enchanting plant in the garden.

Virginia Waterleaf

Hydrophyllum virginianum, appendaged waterleaf, brook flower, burr flower, Indian salad, Iroquois greens, John's-cabbage, Shawnee salad

Hydrophyllum virginianum

grows on smooth stems with a slightly hairy space near the tip, up to 2 feet high. The alternate, pinnately lobed, hairy, toothed leaves are divided into three to seven segments which can be lance-shaped or oval and are covered with splotches of gray or pale green. The flowers are up to 1/3 inch long, with five petals, in clusters, and are purplish to white. Flowers bloom from May through August. Grows in moist humusy soil in dappled shade. The roots are fibrous rhizomes.

Range: Quebec and New England to Virginia, west to Tennessee and Kansas.

Virginia waterleaf got its common name because to some observers the splotches on the leaves look like watermarks on paper; to others the leaves themselves appear stained with drops of water. The scientific name *Hydrophyllum,* which also means "waterleaf," was bestowed because a small quantity of water is held in the cavity of each leaf.

Native Americans ate the young leaves, usually boiled up. Huron Smith, in his book on the Menominee, wrote: "The leaves are eaten as greens. First they are wilted like lettuce in vinegar made from the last run of maple sap, then simmered in a kettle. The first water is thrown away. They are then boiled with pork and fine meal until ready to serve."[168] Many of the common names of the plant refer to this practice.

Ethnobotanist Daniel Moerman reports that the Ojibwe fed the chopped-up roots to their ponies to make them fat and have glossy hair. They also employed the astringent root to treat "flux," the name of a complaint similar to dysentery, one of those age-old complaints found in every culture. The Iroquois boiled the roots to make a liquid medicine which they used as a wash for cracked lips and mouth sores.

Waterleaf is still used as a food plant today. The young plants are cooked, seasoned with salt, pepper, and butter, and eaten immediately. It can also be used as a pot herb when boiled for five minutes with one or two changes of water and served with vinegar. According to wild-food

advocates Steve Brill and Evelyn Dean, "The flavor is like parsley, only more delicate."[169]

This plant, like many others needing rich, moist, woodland soil, is becoming a rarity in places like Connecticut because its preferred habitat is being destroyed.

Liliaceae

Bluebead Lily

Clintonia borealis, yellow clintonia, corn lily, yellow lily of the valley, dogberry, heal-all

---

Clintonia borealis

grows up to 16 inches tall, usually with two to five widely oval leaves springing from the base, and a leafless flower stalk producing a cluster of bell-shaped yellow flowers at the top in summer. The flowers are 1/2 to 3/4 inch wide, and give way to blue, sometimes white, berries in the fall.

Range: throughout the eastern and midwestern United States north of South Carolina, and throughout southern and central Canada.

Clintonia was named after DeWitt Clinton, who lived from 1769 to 1828. He was mayor of New York City for three terms, a state senator, then lieutenant governor; he ran for president but lost to James Madison, and ended up New York state governor. He was instrumental in getting the Erie Canal built and promoted the abolition of slavery. Somehow he found the time to write books on natural history.

The plant bearing his name has little blue berries just as enchanting as its tiny yellow flowers. According to Ojibwe and Menominee legend, dogs chewed the roots to coat their teeth with

FOLLOWING PAGE: *Clintonia borealis,* bluebead lily

Erythronium americanum, trout lily

poison so they would kill anyone they bit. To counteract this poison, the same root had to be found. Dogberry roots aren't very toxic, so this story joins the annals of plant legends rather than plant facts. The roots contain diosgenin, which is a chemical precursor of the female hormone progesterone; native peoples made a tea from the roots which was then given to women in the throes of painful labor.

The young leaves are edible, and can be added to salads or boiled like spinach. Given that the berries are slightly toxic, however, wild-food lovers should take care.

Liliaceae

Trout Lily

Erythronium americanum, adder's tongue, dog's-tooth violet, fawn lily, yellow trout lily

Erythronium americanum **grows from underground stolons up to 10 inches high. Solitary flowers have upswept tepals in a turbanlike shape on a stalk with two leathery, mottled leaves emerging from the stem base. The tepals are yellow or white with anthers 2 inches long; the fruit is in a capsule.**

Range: **throughout eastern North America.**

The yellow nodding heads of the trout lily seem to define the term "spring ephemeral." Although the term sounds romantic, it is, in fact, an ecological strategy to ensure survival. Spring ephemerals pop up from barely thawed soil, and bloom quickly in the early spring under deciduous trees—everything must happen before the unfurling canopy blocks out sunlight. The trout lily has to grow rapidly in a few weeks, store up enough food for the whole season, as well as produce flowers easily seen by potential pollinators. As the plant's leaves wither, the pistils ripen with seed. After this heroic effort, its life retreats underground and to all intents and purposes it's no longer part of the understory. But not so. During the spring rains, nutrients leach into the soil. The trout lily absorbs phosphorus and transfers it to its leaves to develop enzymes and fat compounds. Once the leaves have died, the phosphorus is transferred back to the soil by bacteria and other microorganisms.

There are about 20 species of trout lily. On the East Coast, the yellow varieties predominate; further inland, the white ones become more common. Each plant produces a single stalk with one nodding flower containing six long stamens and anthers. The petals and the sepals look so much alike they are called tepals. It has small storage organs with transparent scales around the corms. The base of the stem has two leaves with troutlike mottling. These leaves act like a sheath around the growing stem. The larger of the two leaves forms a point, and the other fits against it so snugly that between the two of them they can push up through the soil and surface litter and still protect the flower.

The nodding yellow bell-like flower keeps the rain from washing away any of the precious pollen on the anthers. Trout lilies are pollinated by bees and large flies, which need nectar and pollen to survive when they emerge from hibernation or pupae. One bee, the *Andrena erythronii*, depends completely on the trout lily's pollen to feed its larvae. The bee hangs upside down with one pair of legs clinging to the tepals, while the second pair scrape pollen from the anthers. The pollen is packed into the hollows on the bee's hind legs, so that it looks as though it is carrying excess luggage. The nectar does double duty—it gives the bee energy and sweetens the pollen for the babies. Bee larvae fed this way mature in time to pollinate clover, alfalfa, and other early crops.

Once the capsules ripen, the seeds are spilled on the ground (very onanistic). The ephemerals go into dormancy and the stalk collapses. The plant may lie there for five months in a little pile. During this time, hungry insects take away the seeds to eat the meaty caruncles (tips of the seed), usually leaving the remains some distance from the mother plant.

Some evolutionary ecologists speculate that trout lilies evolved from plants that produced seeds, but other methods of reproduction became more important and the ovules took on a much reduced role. Most trout lilies are "droppers." A dropper is a tubular fleshy bud that grows from the base of the old corm. Threadlike stems from the dropper drive deep into the soil and new corms develop at their ends. When this happens, the old dropper withers. The offsets produced in this way will form a new colony.

This explains why it's so difficult to dig up wild trout lilies. Once you reach the storage organ, the plant snaps off and the corm is still safely hidden.

Native Americans crushed warmed trout lily leaves and poured the juice over wounds. They also chewed the root and spat it into the river to make fish bite (another reason to call it a trout lily).

Catherine Parr Traill wrote admiringly:

The dog's-tooth violet, Erythronium, *with its spotted leaves and bending yellow blossom, delicately dashed with crimson spots within, and marked with purple lines on the outer part of the petal, proves a great attraction in our woods. . . . There are two varieties of this flower, the pale yellow, with neither spots nor lines, and the deep yellow with both; the anthers of this last are reddish-orange, and thickly covered with a fine powdery substance.[170]*

The homesick settlers called it dog's-tooth violet after the European species, which is somewhat different. John Burroughs, the nineteenth-century naturalist and writer, successfully promoted the name trout lily because of its mottled leaf and because it grows on the banks of trout streams—the name has stuck.

Liliaceae

Cucumber Root
Medeola virginiana, Indian cucumber, cushat lily

Medeola virginiana

grows 6 to 18 inches high. The pointy leaves grow in whorls of 5 to 11 partway up the stem. Another whorl of three to five smaller leaves emerges at the top of the stem, from the center of which the flowers emerge in summer. The stalks from this whorl bear yellowish-green flowers with red stamens, followed by purple-black berries.

Range: Nova Scotia and Quebec, north of Lake Superior and Minnesota, south to Virginia and Missouri, the mountains of Georgia and Alabama, and Florida.

Cucumber root has a whorl of green leaves partway up the stem and another group of leaves out of which the yellow-green flowers emerge. They have all the energetic look of a helicopter about to whirl off into the distance.

The crisp white root of the *Medeola* gives the

plant its name. It is about the size of a small finger and has a taste similar to, though even milder than, cucumber. The botanical name comes from Medea, the sorceress of Greek myth, and suggests the plant's almost magical medicinal properties. The name "cushat lily" is a bit of a mystery: a cushat is a pigeon, but it indicates that birds like to eat the berries.

The plant was mainly used as food by the indigenous peoples and does not seem to have been a major medicinal plant, although it had some diuretic properties. According to Daniel Moerman, the Iroquois made an infusion of the crushed dried berries and leaves to give to colicky babies. They also chewed the raw root and put it on hooks to "make fish bite."[171]

Cucumber root is still enjoyed by those who love wild foods. Paige Peters says:

While on an extended backpacking trip I came upon a campsite and was greeted by seven adventurous women, ages 52 to 70 years. They all had bright smiles and invited me to eat dinner with them. . . . To my surprise, on the table was a huge Ziploc bag full of fresh greens. Having been living on dehydrated food and energy bars, my mouth watered as I asked what it was and where it came from. Ellie, the 70-year-old, replied that she had been doing some gathering on her hike today and prepared a salad . . . composed of young dandelion leaves, blue violet leaves, Indian cucumber root and day lily shoots. We topped it off with raspberries found behind a tent just 20 feet away from our dinner table.[172]

This plant needs care. Cucumber root has become rare in many places, and endangered in others—uprooting all the tubers kills off the plant. Unless a site is going to be ripped out, always leave enough behind so the plant can grow stalks again the following year.

Liliaceae

Solomon's Seal

Polygonatum biflorum, wild lily of the valley, Saint-Mary's-seal, lady's-seal, sealwort

One of the extraordinary sights of early May is the luminous, almost phallic, tips of Solomon's seal peeping through the dark soil. Out of the thick-jointed rhizomes come the graceful stems. To add glory, pairs of white heart-shaped bells with a decorative green tinge around the edges dangle in pairs under the slender arches, protected by the leaves from harsh weather.

Poly means "many" and *gonu* means "joint" or "knee" in Greek. This probably refers to the jointed rhizome. Naturalist Mary Durant, author of a wonderful book about plant names, believes the name "Solomon's seal" was introduced in the Christian era because its six-petaled flowers resemble the "Star of David, which in early days was always known as Solomon's seal."[173] Nature writer Doug Elliott quotes a story suggesting "Solomon, who knew the diversities of plants and the virtues of the roots, had set his seal upon it as a testimony of its value as a medicine for all humanity."[174] The seal is the scar left on the root. Each year the root produces a new stem which withers in the summer, leaving one scar. Count the number of scars and you can tell how old the plant is.

Polygonatum biflorum grows 1 to 6 feet tall, with arching stems and oval, parallel-veined leaves that turn yellow in autumn. Tubular or bell-shaped, white or cream flowers with green markings hang below the stems singly or in clusters. A blue or black berrylike fruit follows. It is found in woods and on riverbanks.

The bisexual flowers have a short female style and longer pollen-bearing anthers, which help the plant avoid self-fertilization. The seeds are spread by the wind and insects, but it's not a major food plant for animals and birds. It has, however, been used by humans for many centuries as both food and medicine.

The Cherokee beat the dried roots into a flour to make bread, or cooked the leaves as greens. The stems and leaves could be parboiled, rinsed, fried with grease and salt until soft, and eaten as a pot herb.[175] The young shoots can be eaten like asparagus.

The Cherokee also made a poultice from the roots to apply to boils and carbuncles. (It contains a substance called allantoin, which is used in modern medications for the external treatment of wounds and skin ulcers.) The Chippewa boiled the roots until they were almost syrupy, sprinkled them on hot stones, and inhaled the steam to relieve a headache. According to Huron Smith, the Menominee dried and pulverized the root and mixed it with cedar (the twigs and leaves), which they "burned as a smudge to revive one who has become unconscious. If they suppose the patient is about to die, then smoke of this smudge is blown into his nostrils to bring him back to life."[176]

John Gerard, the English herbalist, writing in 1597 noted the root "taketh away in one night, or two at the most, any bruise, black or blew spots gotten by falls or womens wilfulnesse in stumbling upon their hasty husbands fists, or such like."[177] Either Gerard was being sarcastic about the treatment of women in his day, or he had an extraordinarily one-sided view of domestic relations.

Perhaps because of its name, Solomon's seal has always been considered to have magical power. It was once believed to be an aphrodisiac and many practitioners of the black arts used it to ward off evil. Steve Brill and Evelyn Dean mention "Dr. John Moore, a self-taught herbalist who believed the plant could increase a person's psychic ability." According to Moore, the plant helped people "hear the trumpets of Jeri-

Polygonatum biflorum, Solomon's seal

cho."[178] A modern Wiccan recipe for "spirit oil" includes Solomon's seal, along with orris root, rosemary, powdered jade or turquoise, sandalwood oil, mint oil, and safflower oil. The oil is supposed to be used to "anoint candles for exorcisms, seances, counterspells, purification rituals, protection against evil influences and spells to increase clairvoyant powers."[179]

Range: Massachusetts, New Hampshire, New York, Connecticut, southern Quebec, eastern Ontario, and southern Manitoba, south to Texas, east to Florida.

Liliaceae

Large-Flowered Trillium

Trillium grandiflorum, white trillium, common trillium, showy trillium, snow trillium, trinity flower, white wood lily, wake-robin, birthroot

Trillium grandiflorum

grows to 18 to 24 inches high and has solitary upright flowers rising from the leaf axils; three outer sepals surround three petals 2 inches long; and there are three pale green leaves whorled around the stalk. It carries berries in early summer and needs a moist deciduous forest to grow in. The white flowers turn pale pink with age.

Range: Tennessee, Georgia, North Carolina, north into Canada, west to Manitoba.

Provincial flower of Ontario.

A forest carpeted with brilliant white trilliums and their intensely green leaves is a great temptation to stalkers of spring flowers. Many people believe picking the flower kills the plant. This isn't true, but the flower won't come up for another year. Of

Trillium spp., trillium

course, if it is picked or gets hit by frost regularly, the plant will die. Even clipping the leaves and flowers can kill the rhizome, which takes years to become productive. So it's best to leave trilliums alone. In some states and provinces, trilliums are protected by law. Elsewhere, one hopes it is protected by common sense.

The growth cycle of trillium is complicated. The seed needs 83 days of cold to germinate and then two winters in the soil. The roots start to grow in the first winter and, by the second, the initial shoots are released. For the first five or six years of its life, the plant produces only leaves. When it does flower, the flowers emerge as white, then slowly turn pink. The goal among trillium growers is not just to collect the seed, but also to get the plant to flower within five years. If the habitat suits them (deciduous forest conditions, early warmth, and then shade), they will continue to grow along rhizomatous roots.

All trilliums are myrmecochorous, which means they attract ants. Ants love the elaiosomes, or fatty seeds. About a quarter of each ripe seed is taken up by a great blob of oil (the strophiole). When the seeds fall out of the capsules in clusters, the ants rush in. They carry the seeds away to their nests, or eat the oil on the spot, leaving the seeds on the ground. Fred Case, a trillium collector from Saginaw, Michigan, says, "I have ants that crawl up the stem even before the seed is ripe, so they can get all the seed out. They carry the seeds down to the tunnel, then simply cut off the oil organ (which contains unsaturated fatty acids they need for valuable food) and ignore the seeds."[180]

Trilliums are moved ant by ant throughout the forest. Case gives an idea of how much time it takes and what distances ants can transport these seeds. He moved into his home in 1957. Since then, the trilliums have traveled 900 feet down and across the road toward the woods closest to his house.

The main trillium pollinators are bees and butterflies. The goldenrod or flower spider, *Mis-umena vatia*, hangs out nearby, preying on the insects visiting the flower (in summer, it migrates to goldenrods). A moth caterpillar, *Clepsis melaleucana*, likes to eat the leaves.

Native Americans used a solution made from the rhizome for various ailments, including ear infections and rheumatism. The solution, which is astringent and antiseptic, was used as an external dressing for inflammation or to control bleeding during childbirth, hence the name "birthroot." Very young leaves are edible as salad greens, but the older leaves are bitter.

Catherine Parr Traill described the trillium's simple beauty in spring in the countryside of what is now Ontario:

> *Nature has scattered with no niggardly hand these remarkable flowers over hill and dale, wide shrubby plain and shady forest glen. In deep ravines, on rocky islets, the bright snow white blossoms of the Trilliums greet the eye and court the hand to pluck them. The old people in this part of the Province call them by the familiar name of Lily. Thus we have* Asphodel Lilies, Douro Lilies, &c. *In Nova Scotia they are called Moose-flowers, probably from being abundant in the haunts of Moose-Deer.*[181]

A related species, **painted trillium** (*T. undulatum*), has a flower that looks as if someone has dipped a fine pen in red ink and sketched a hint of a V in the center. From 6 to 12 inches high, it grows in moist but well-drained acidic soils. In the Algonquin culture, pregnant women were believed to be able to bring on labor by eating the flowers and leaves of the painted trillium. *T. ovatum*, **coast trillium,** is one of the most widespread native species on the West Coast (along with Western flowering dogwood and Pacific rhododendron). *T. ovatum* has large white three-petaled flowers that gradually fade to red. They stand erect over three graceful leaves in early spring. The clump-forming perennial has red-green stems and stalkless, pointy, dark green leaves and thrives in the moist forest floors.

Like those of its cousin *T. grandiflorum*, the seeds of *T. ovatum* take their time to germinate, and grow and bloom even more reluctantly. It can take up to 15 years for this trillium to flower. The blossoms appear in April or May, and start out white, but once they have been pollinated they change to pink, to warn would-be pollinators that they should look elsewhere for food.[182]

T. ovatum was first reported by the Lewis and Clark expedition in 1806 and by European travelers going up the Columbia River, who were impressed with the masses of white flowers gracing the landscape.

According to ethnobotanist Erna Gunther, the Makah name translates as "sad flower" because the Makah believed that picking the flower brought on rain. The Ditidalt of Vancouver Island also called it "sad flower" because they believed to pick it would bring fog.

Many First Nations used this plant in their medicinal practices. The juice was extracted from the roots or a poultice made from the bulbs was applied to boils to bring them to a head. An infusion of the juice was used to soothe sore eyes. Among the Quinault, it was common for a woman to cook the bulb and drop it into the food of a man she desired as a lover.

Range: British Columbia, south and west as far as Montana.

Monotropaceae

Indian Pipe
Monotropa uniflora, corpse plant, ice plant, death plant, ghost flower, bird's nest

Monotropa uniflora

grows on 4- to 8-inch stems. No part of the plant is green, the leaves are scaly, and a nodding odorless white flower appears in summer. Upright seedpod. Grows in rich, moist, woodland soil and is dependent on nearby plants for nourishment.

Range: throughout the continental United States (except Alaska) and southern Canada.

Indian pipe looks like an upside-down clay pipe sticking out of the leaf litter. This mysterious plant has no color and no chlorophyll, making it different from anything else in the Eastern forest. It turns black or brown if it is picked (probably accounting for its name of corpse plant) and is utterly dependent on other plants for its nutrients (making it a saprophyte). Although it grows amid decaying plants and looks like a fungus, it isn't. At one time, it was lumped into the same family as wintergreen and kalmia, the Ericaceae, but now it is classified as another family altogether, the Monotropaceae, which includes parasitic and epiphytic plants.

Annora Brown in *Old Man's Garden* calls this plant a miscreant and one with an individuality all its own: "Cold and clammy, it is the ghost that walks the forest aisles. The straight stem and nodding bell-shaped flower imitate the shape of a pipe."[183]

Mary Thatcher Higginson (1844–1941) takes a warmer view, calling them "a spotless sisterhood . . . [of] forest nuns . . . heads . . . bent

Monotropa uniflora, Indian pipe

Mary Chiltoskey and Paul B. Hamel, in the book *Cherokee Plants*, tell how Indian pipes came to be:

Before selfishness came into the world . . . the Cherokee people were happy sharing the hunting and fishing places with their neighbors. All this changed when Selfishness came into the world and man began to quarrel. The Cherokee quarreled with tribes on the east. Finally the chiefs of several tribes met in council to try to settle the dispute. They smoked the pipe and continued to quarrel for seven days and seven nights. This displeased the Great Spirit . . . [who] decided to do something to remind people to smoke the pipe only at the time they make peace. The Great Spirit turned the old men into greyish flowers we now call "Indian Pipes" and he made them grow where friends and relatives have quarreled. He made the smoke hang over these mountains until all the people all over the world learn to live together in peace.[185]

in prayerful mood" making "shy devotions near the singing brook."[184]

Indian pipe grows in the deepest, densest, shadiest woods. The waxy nodding flowers come from the rootlets, which draw nutrients from fungi living in the soil. This feeds the transparent stems and the scales, which are vestigial leaves. The fungi act as a kind of pipeline between soil and plant since they draw their own nutrients from green plants. The flowers are pollinated by insects, and turn to dry capsules enclosing the seeds.

Native Americans used the plant as a sedative, and gave it to children for epilepsy or convulsive fits. The juice of the plant was watered down to make an eyewash, and the Cree chewed the flower for toothaches. The Thompson believed if there was an abundance of this plant, there would be many mushrooms in coming seasons.

Perhaps because these plants are so strange, there is a great temptation to try and cultivate them in the garden. But it's almost impossible to set up the complicated system they need to survive. As saprophytes, they depend on certain fungi to grow, and it's almost impossible to tell just how much of anything this rather vampirish plant needs.

Onagraceae

Oenothera biennis

grows up to 5 feet tall from a fleshy taproot, with reddish hairy stems, a rosette of lance-shaped, opposite leaves, and a branching spike of bright yellow flowers at the top. The flowers, which appear in summer, have four petals and a cross-shaped stigma in the center. They open in the evening, remaining closed during the day. The seed capsule is woody, containing many tiny, hard, reddish seeds. Found in fields and waste ground.

Range: throughout the eastern and central United States, and southeastern and south-central Canada.

The yellow flowers of evening primrose are like crushed silk. In the brightness of the sun the dashing petals fade, and they appear as limp as laundry on a line, but at night they gleam. Their phosphorescent glow invites moths to come and visit.

In her book *Healing Plants,* Ana Nez Heatherley tells us: "When the Old Ones of the Ani-Tsalagi tell the story of the beginning of the world, they speak of the two brothers who were made by the Father. These two brothers knew that people lived in the depths of the underground world in darkness and filth. The two descended into the earth and led the people up. The sun was so bright it made

Oenothera biennis, evening primrose

them cry, and the tears which fell grew into flowers of the sun [and] became the evening primrose and the sunflower."[186]

These pretty flowers are not related to the true primroses *(Primula)*, and presumably got their name simply because they are yellow. Their scientific name is even stranger: *Oeno* comes from the Greek *oinos*, meaning "wine"; *thera* means "hunt" or "chase." Apparently this was the name given to a related European plant that was used to flavor wine. *Biennis* means it's a biennial—putting out a rosette of leaves one year and blooms the next.

Evening primrose is edible. It may be hard to identify the first year's leafy rosette, but this is when it tastes almost radishy and when the edible root was gathered by early users. By the second year, it was considered too tough to eat. Even the first year's root required at least two hours of boiling. Though it wasn't grown as a food plant by North American settlers, it was taken to Europe in the seventeenth century and became widely cultivated there. Annora Brown was amused to see North American farmers uprooting evening primrose to make way for other crops such as potatoes or carrots: "The farmer classes them as weeds and so we see, by a queer twist of fate, a man digging out and casting aside a vegetable which has for centuries helped to feed the Indian race, and replacing it with a vegetable which has for centuries helped to feed the European race."[187]

Native Americans have long been familiar with the plant's medicinal reputation. They used a poultice of the leaves to heal wounds and made a tea from the leaves to soothe stomachaches. Today's scientists have discovered that the oil from the seeds contains compounds preventing blood clots; it also speeds up healing, and contributes to the formation of certain human hormones. Herbalists recommend it for PMS, problems in menopause, rheumatoid arthritis, and the prevention of heart disease.

Even now its use in the treatment of multiple sclerosis is being studied.

Orchidaceae

Showy Lady's-Slipper

Cypripedium reginae, showy moccasin flower, whippoorwill's-shoe, shepherd's purse, nerveroot

Cypripedium reginae

grows from rhizomes. The parallel-veined oval alternate leaves clasp the stem. The flowers, which usually appear in June, are borne in terminal racemes and have three white tepals fanning

Cypripedium reginae, showy lady's-slipper

out around a larger, slipper-shaped, pink lip. It is found in bogs and moist woodlands.

Range: throughout eastern Canada, New England, central eastern and midwestern states. State flower of Minnesota.

Paul Martin Brown has hunted for orchids from Alaska to the wilds of Newfoundland. He has identified more than 60 orchid species in New England and New York alone, despite the urbanization of the area. Along with the asters, the orchid family is another huge family of flowering plants, with more than 17,000 species. Most are tropical or epiphytic (air-rooted plants) but some of the showiest species are found in the northern temperate zone.

Cypripedium is perhaps the favorite of all wild orchids because it is so easy to identify. All it requires is a sharp eye to spot the toe of the slipper, called the labellum, hanging over its greenery. The labellum is a fused petal, which may be white, yellow, pink, or purple, depending on the species, surrounded by ribbonlike petals. The slipper is actually a benign little trap for insects. Although it contains no nectar, it smells as if it does. Insects crawl inside, hoping to get a nourishing drink. As they crawl out again, no doubt feeling a bit cheated, they brush up against the back of the pouch, where the pollen is stored. The attraction of the lady slipper is so beguiling that the insects just never learn. They repeat the process with other lady's slippers, thereby cross-pollinating the plants. Sometimes larger insects such as bees get their revenge by tearing open the flower in their anxiety to get out.

Once the fruits ripen, they split open to release hundreds of thousands of fine seeds to be scattered by the wind. Most are doomed to die if they don't fall on precisely the right site. They have no food stored up for the future seedling, which depends entirely on mycorrhizal fungi to absorb nutrients from the soil. Even if it does

find the right spot, it may take years to flower. Anyone who has tried to transplant lady's slipper finds that within a few years the exact mycorrhizal balance will be off and suddenly the plant disappears altogether. This is one of the problems we have in transplanting native plants from their original sites: they are so delicately tuned to their habitat, and it's almost impossible for us to replicate it perfectly. Many plants become extinct once their habitat is destroyed.

The curious shape of lady's slipper has intrigued people since early times. Medieval botanists named this plant for the Virgin: Lady's-slipper—*Calceolus marianus,* "the shoe of Mary." When Linnaeus set out to name or rename every plant family known to his time, he reasoned that Venus rises from the sea off Cyprus bringing her promises of glory and fertility, so he named the flower *Cypripedium,* from the Greek *Kypris* (one of the names for Venus and probably the source of the name of the island Cyprus) and *podion,* "slipper" or "little foot." The French, who had called this the *sabot de la Vierge,* had to rename it *sabot de Vénus.*[188]

Lady's-slipper root, like the aster root, was considered a "nervine," a sedative or tonic for the nervous system, inducing calm and tranquility. It was often given to women in the throes of childbirth and sometimes used as a drug to induce sleep. The Menominee even used the root to induce dreams of the supernatural. The settlers quickly learned to use the root, too, making infusions by steeping the powdered root in boiling water. Some people took doses on a regular basis as a tonic. By 1859, the powdered root was found everywhere in drugstores.

The flower itself was also sought after. According to Mrs. William Starr Dana in *How to Know the Wild Flowers* (1893):

Near Lenox, Mass., there is one locality where the showy lady's-slipper can be found. Fortunately, one would suppose, this spot is known only to a few; but one of the few who possess the secret is a country boy who uproots these plants and sells them by the dozen in Lenox and Pittsfield. The time is not distant when the flower will no longer be found in the shadowy silences of her native haunts, but only, robbed of half her charm, languishing in stiff rows along the garden-path.[189]

Cypripedium pubescens, yellow lady's-slipper

Orchidaceae

Yellow Lady's-Slipper

Cypripedium pubescens (or *C. calceolus* var. *pubescens*), moccasin flower, golden slipper

Cypripedium pubescens

grows up to 24 inches tall, with similar leaves and structure to the showy lady's-slipper. However, the slipper part of the flower is bright yellow, with brown or greenish surrounding petals.

Range: throughout North America.

Catherine Parr Traill was a patient observer, and given the difficult life of a pioneer, she must have longed for the first sight of the lady's slipper after the harsh northern winters. In her book *The Backwoods of Canada,* she wrote:

The moccasin flower or lady's-slipper (mark the odd coincidence between the common name of the American and English species) is one of our most remarkable flowers; both on account of its beauty and its singularity of structure. Our plains and dry sunny pastures produce several varieties; among these, the Cypripedium pubescens, *or yellow moccasin, and the* C. arietinum *are the most beautiful of the species. The color of the lip of the former is a lively canary yellow, dashed with deep crimson spots. The upper petals consist of two short and two long; in texture and color resembling the sheath of some of the narcissus tribe; the short ones stand erect, like a pair of ears; the long or lateral pair are three times the length of the former, very narrow, and elegantly twisted, like the spiral horns of the Walachian ram: on raising a thick yellow fleshy sort of lid in the middle of the flower, you perceive the exact face of an Indian hound, perfect in all its parts—the eyes, nose, and mouth.*[190]

The roots of the yellow lady's-slipper were used by the Cherokee for neuralgia and other pain, and an infusion taken for "rupture pains," worms, "spasms," and "fits." The pattern of its use as a narcotic might explain why it is mentioned so often as taken for "female" problems, which could have been anything from gynecological problems to stomach cramps to nervous agitation. Dried powdered root could also be moistened and applied to decayed teeth for pain.[191]

Pink lady's-slipper (*C. acaule*) was used, at least by the Menominee, for "male disorders," which generally meant problems with the urinary tract. It was also used as a love potion and a sedative (presumably it put people in the relaxed frame of mind necessary for love). In another sense, the flowers may have contributed to romance when women wore them in their hair, a rather fanciful use noted by Rafinesque in his 1828 book on medical botany. Pink lady's-slipper is the provincial emblem of Prince Edward Island.

Oxalidaceae

Wood Sorrel

Oxalis montana, mountain wood sorrel, wood shamrock, white wood sorrel, sour grass

The shamrock-shaped leaves make this an easy plant to identify in the woods. Several species grow in North America—some have yellow flowers, such as *Oxalis stricta*, others, such as *O. violacea*, have purple flowers. The leaves fold downward at night or in the cold, like little butterfly wings.

O. montana grows up to 6 inches tall from an underground rhizome. Three-part, shamrock-shaped, dark green leaves spring from the base of the plant. The plant has no main stem, only a cluster of flower stalks. The five-petaled flowers, which bloom from May to August and attract bees and butterflies, are white with red or purple stripes and yellow centers. It is found in swampy lowlands and moist woodlands. *Oxalis* comes from the Greek *oxys*, which means "sour," and the plant contains oxalic acid, which is indeed extremely sour. In small quantities, the tartness is quite pleasant and, indeed, sorrel is well known as an edible plant. In 1612, John Smith wrote that the Indians of Virginia enjoyed them in spring and, 300 years later, the anthropologist Huron Smith observed the Potawatomi cooking the leaves with sugar to make a type of dessert.[192] According to Bernard Assiniwi, the Algonquin even considered it an aphrodisiac.[193]

The sharp-tasting leaves are diuretic and astringent, and herbalists recommend sorrel as a tonic to purify the blood. People with kidney stones, rheumatism, or gout should avoid the plant, since the oxalic acid could worsen their condition.

Sorrel has long been a common vegetable in Europe, where a number of species grow, but it is not as popular in North America, perhaps because it is something of an acquired taste. With the movement toward eating wild foods, however, this may change.

Range: **Labrador to Manitoba, south to Minnesota, North Carolina, and Tennessee.**

Papaveraceae

Bloodroot

Sanguinaria canadensis, red puccoon, red Indian paint, redroot, pauson, snakebite

Sanguinaria canadensis, bloodroot

Sanguinaria canadensis

grows 6 to 9 inches high and blooms in early spring with snow-white, 3-inch, cup-shaped petals with gold anthers. The leaves are blue-gray to medium green, smooth with a subtle scallop, and 6 to 12 inches across. Found in rich woodland soil, this ephemeral self-sows and will form large colonies in time.

Range: Nova Scotia to Florida, west to Manitoba and Nebraska.

The "blood" of the bloodroot is the scarlet juice of the stem and root. Its pure white blossoms are among the earliest to emerge in the woodlands in spring. Like all other ephemerals, they are fleeting. By summertime, the last few leaves have disappeared. Bloodroot is in the same family as poppies and yellow celandine. Although there are more than 200 species in 23 genera (most of them in the northern temperate zone), bloodroot is the only member of its genus in the world.

Because its nature is evanescent and it blooms in an unreliable and cold time of year, when insects may not be active, bloodroot must self-pollinate to survive. Therefore it has a bisexual flower. Bumblebees and honeybees, once they shake off their lassitude, collect the copious pollen from the flowers. This is another ant-dependent, or myrmecochorous, plant. Its seeds attract ants, which move them about and bury them.

The flower, which usually lasts only two days, is so exquisite, so elegantly delicate, it's hard to believe anyone would touch the plant. Early garden books and manuals of native plants, however, are full of warnings not to pick the flower because the scarlet juices stain anything from hands to white pinafores—indelibly. "In childhood," recalls Alice Morse Earl in 1901, "I absolutely abhorred Bloodroot; it seemed to me a fearsome thing when first I picked it. I remember well my dismay, it was so pure, so sleek, so innocent of face, yet bleeding at a touch, like a murdered man in the Blood Ordeal."[194]

The Algonquin called bloodroot *puccoon* or *paucon.* The word *puccoon* came from *pocan,* which meant "blood red" and was used for any red-juiced plant used for staining and dyeing. H. W. Youngken, who studied Native Americans' use of plants in the 1920s, wrote:

The fresh rhizomes and roots containing a red milky juice were used for decorating their skin, while wearing apparel was often boiled with these parts. Bachelors of some of the tribes, after rubbing some of the red milky juice on their hands, would contrive to shake hands with girls they desired; if successful in this, after five or six days, these girls are said to have been found willing to marry them.[195]

The Chippewa dug up the roots in autumn to make their brilliant red dyes. The green or dried roots were pounded into a mash, then steeped in water to make a dark yellow dye. Shredded roots of wild plums were added to make the dye even darker. The Iroquois used the rhizomes as an orange or yellow dye for fabric. Others used bloodroot dye to decorate baskets or to stain the face for specific clan identification, decoration, or for important ceremonies.[196] The European settlers quickly caught on, and the French even imported bloodroot to dye wool in the early years of North American settlement.

Timothy Coffey tells how the men on the William Byrd expedition to Virginia and North Carolina in 1728 were caught fraternizing with the women of the Nottaway:

Our chaplain observ'd with concern that the ruffles of some of our fellow travelers were a little discolor'd with pochoon, wherewith the good man had been told these ladies us'd to improve their invisible charms.[197]

Bloodroot was also used to make a tonic and a cough medicine. The Delaware took pea-sized bites of the root for 30 days to counteract general weakness. As a cold remedy, the roots were boiled and mashed up or made into an infusion. The mashed roots were also used on cuts, to counteract poison ivy, to draw out thorns and slivers, for leg

sores, for gonorrhea and syphilis, and even to cause abortions. As with many plants, even the smoke made from burning it was useful: it was wafted over someone who had seen a dead person.

The newcomers complained constantly about how they were plagued with insects. If they had observed their native neighbors a little more carefully, they would have seen bloodroot used as an insect repellent.[198] They did, however, catch on to its uses as a tonic, emetic, and expectorant. By 1852, extracts of bloodroot were widely available in commercial cough syrups, and marketed as an emetic and narcotic. It was common to take a few drops on a lump of sugar to relieve a sore throat. Sugar was necessary because the taste was awful, apparently, reinforcing the belief that what tastes bad must somehow do good. The active element, sanguinarine, is poisonous in large quantities, but effective in small amounts.

By the mid–nineteenth century, a doctor in London was using it as part of a treatment for skin cancer. The treatment fell out of favor until the 1960s, when bloodroot was reconsidered as a treatment for minor cancers of the nose and ears. Today, sanguinarine is used in toothpastes and mouthwashes to kill plaque and oral bacteria.

Phytolaccaceae

Pokeweed

Phytolacca americana, inkweed, poke, scoke, pigeonberry

This startling plant can grow to 8 feet tall with striking purplish stems. Pokeweed manages to be both edible and poisonous. The poison, which is a powerful and potentially fatal alkaloid, is found in the roots. When the new green shoots come up in springtime, they are free of poison, and are considered a delicacy. Later in the growing season, however, the poison from the roots moves up into the leaves, stems, flowers, seeds, and berries. This is a strategy to protect the plant during the time of pollination, fertilization, and seed production because foragers steer clear of it. At this time it can cause a rash that is just as devastating as a dose of poison ivy.

The name "pokeweed" might have stuck because each berry is slightly indented, as if it had been poked. A much more likely explanation is that it comes from the Algonquin word *pocan*, which means "blood red." (This word also gave rise to *puccoon*, one of the common names for bloodroot, *Sanguinaria canadensis*.) "Inkweed" is easy to understand: the purple berries produce a dark ink or dye. *Phyto* means "plant" and *lac* means "crimson." (Just because of the sound of the name, presidential candidate James K. Polk and his supporters used pokeweed leaves during the 1844 campaign as their emblem.) One of the old-fashioned names for it was "pigeonberry," because the now-extinct passenger pigeons used to eat the berries—presumably they had developed an immunity to everything but man's rapacity.

Phytolacca americana is herbaceous and has oval, pointed, bright green leaves up to 12 inches long and 5 inches wide growing alternately on the reddish-purple stem. The tiny flowers (¼ inch wide or less) have green centers and five white sepals, but no true petals. They grow in a spike with a red central stalk, and emerge in summer and fall, giving way in late autumn to dark purple berries, each about ⅓ inch in diameter. It is found in fields, on the edges of woods, and in waste ground.

As they did with many poisonous plants, Native Americans used pokeweed medicinally in minute doses. They made pokeweed poultices to "burn out" certain cancerous growths and pokeweed teas to induce vomiting. Some people in the southern United States still consider "pokeberry tea" a remedy for arthritis or simply a

Phytolacca americana, pokeweed, inkweed

tonic to "purify the blood," and there are stories of people drinking the stuff every day for years, but don't do this, it's much too dangerous.

In the Southern states, many people grew it in their backyards so that they could harvest the new shoots for "sallets," or mixed boiled greens (not, as many assume, a salad green). As Angela Gillaspie puts it:

This vegetable cannot be purchased from a grocery store, but it can be picked from most any backyard, [or] bartered from a friend or neighbor. . . . This plant is called poke salad, or poke salet, as my family says it. . . . The leaves are carefully and thoroughly washed then boiled until they are tender. The liquid (which is also

believed to be poisonous) is drained and the leaves are rinsed again. When poke salet is cooked, it resembles spinach and tastes like asparagus. It is a very nutritious greens dish.[199]

Range: Maine, west to Minnesota, south to Florida and Texas.

Portulacaceae

Spring Beauty

Claytonia virginica, wild potato, fairy spud, grassflower, good-morning-spring, miskodeed

Claytonia virginica

grows 4 to 6 inches high from a corm with many eyes. Two grasslike, slightly succulent leaves rise up, with a slender stem supporting a cluster of flowers. Each pale pink flower is ½ inch wide, with five petals, and delicately striped with a deeper shade of pink. The flowers close at night. It is found in the rich humusy detritus of the woodland floor, often in partly cleared woods of beech and maple.

Range: throughout eastern North America and the Midwest.

The first carpet of spring is the delicate pink and white of spring beauty—so perfectly named. In some quarters, where the soil is rich, warm, and humusy and the plant gets a chance to mass, it creates a breathtaking mist of pink. It is usually the first plant to open in spring, but delays blooming if the weather remains too cold. The blooms open just as insects are warming up and looking for a place to alight and eat. It comes

Claytonia virginica, spring beauty

up at the same time as bloodroot, bellwort, and violets.

The name honors John Clayton (1694–1773), an amateur botanist who held the post of clerk of Gloucester County, a job sufficiently well paid without being too onerous, giving him time to travel throughout Virginia collecting plants.

The corms were eaten by the indigenous peoples. Raw, they are said to taste like radishes; boiled, they taste like chestnuts. They are so small, however, that the settlers called them "fairy spuds"—it takes a long time to collect enough for a meal. The Iroquois made cold infusions with the powdered root to give to a child having convulsions.

In *Song of Hiawatha*, Longfellow calls the plant "miskodeed":

And the young man saw before him,
On the hearth-stone of the wigwam,
Where the fire had smoked and smouldered,
Saw the earliest flower of Spring-time,
Saw the Beauty of the Spring-time,
Saw the Miskodeed in blossom.

Newcomers to North America were often filled with an ache for home and a longing for the gentle English springtimes they'd left behind. The chill of a short, late North American spring was a terrible shock. Catherine Parr Traill consoled herself with the sight of the pretty flowers: "Our Spring Beauty well deserves its pretty poetical name. It comes in with the Robin, and the song sparrow, the hepatica, and the first white violet; it lingers in shady spots, as if unwilling to desert us till more sunny days have wakened up a wealth of brighter blossoms to gladden the eye."[200]

We know now it self-sows in a natural habitat but that it's a very difficult plant to propagate. Never take it from the wild, and don't buy it from nurseries that do.

Ranunculaceae

White Baneberry

Actaea pachypoda, syn. *A. alba*, doll's-eyes, chinaberry, white cohosh

Actaea pachypoda, white baneberry

Actaea pachypoda syn. *A. alba*
grows up to 3 feet tall with many branches and compound toothed leaves. It can get quite bushy and form waist-high canopies. The long, thick, dark red flower stalk produces a cluster of tiny, white, many-petaled flowers that give way to white berries, each with a black spot at the end, like little eyes. Found in rich woods and along streambanks.

Range: maritime provinces, Quebec, and Ontario, New England, south to Louisiana and Georgia, west to Oklahoma.

In spring, white baneberry is a puff of white flowers over bright green leaves with a slight scent, all of which is very charming. But in autumn, there's something slightly unnerving about the white, sinister, eyelike, poisonous berries of this little plant. Eyeballs on a stick. Its close relative, **red baneberry** (*Actaea rubra*) is similar, except it has red poisonous berries. Occasionally the two hybridize to produce a pink-berried variety.

"About the first of September," Thoreau wrote, "the white cohosh startles the intruder into moist and shaded grounds with its remarkable spike of ivory white berries—which contrast singularly with the greenness around, as if they contained a pearly venom. The berries are wax-white tipped with a very dark brown or black spot—imp-eyed—on stout red pedicels. The red variety is rarer hereabouts."[201]

Some accounts of Native American uses of the plant suggest that the white baneberry was reserved for the treatment of women's health problems (a tea made from the roots was used to treat dysmenorrhea or to relieve the pain of childbirth) and the red baneberry was used for men's problems. Other accounts suggest that both species were used to treat women and were sometimes called squawroot as a result.[202] Other plants were also called squawroot and were used for the same symptoms.

Today, some people cultivate baneberry bushes in woodland gardens, often along with ferns and other woodland plants as companions.

Ranunculaceae

Wild Columbine

Aquilegia canadensis, rock lily, honeysuckle, clucky, bell, meeting house

Aquilegia canadensis

is a perennial, growing up to 3 feet tall, with compound blunt-lobed leaves in groups of three. The nodding flowers, which are usually 1 inch wide, have scarlet sepals, yellow petals, and bushy yellow stamens. Blooms in spring or early summer on cliffs, and in clearings or semishady woods.

Range: Nova Scotia, south to Florida, and west to Minnesota.

The wild columbine has an exquisitely refined scarlet and yellow nodding flower. It's quite unlike the bigger hybridized forms. Both the common and scientific names are a little hard to understand. *Canadensis* doesn't necessarily mean the plant grows only in Canada, only that it might have been first spotted in Canada. Although *aquilegus* in Latin means "water drawer," *aquila* means "eagle"; strangely enough, "columbine" comes from the Latin word *columba* or "dove." Perhaps the name has something to do with the fact that the flower has spurs pointing backward, which might have reminded an observer of the talons of a bird. At one time, it was a candidate for the national flower of the United States, simply because of this reminder of the eagle.

Native Americans considered an infusion of the columbine an antidote to poison ivy rash, or

Aquilegia canadensis, wild columbine

any other skin itch. The Iroquois also believed it detected bewitchment—an important quality in a forest redolent with light and shade, unknowable in so many ways. The Meskwaki used the seeds to perfume their smoking tobacco—its scent held strong powers of persuasion when the pipes were smoked at councils or trade meetings. The roots were chewed for stomach and bowel troubles. It was also used both as a love medicine and a perfume when seeds were spread among clothing, or chewed into a paste and spread in just the right places. It was even considered effective in increasing stamina when smeared on the legs of both humans and horses before a race.[203] Recent laboratory work, however, has shown the plant to contain strong alkaloids, and may be poisonous if eaten in any quantity.

John Tradescant the Younger, who trekked through the Eastern forest on several different trips, sent columbine plants from Virginia to Hampton Court in 1640. Tradescant named at least 40 North American plants in his garden list of 1634, including *A. canadensis*. English gardeners crossed the red columbine with the European columbine, *A. vulgaris*, to produce the hybrids that we're familiar with today.

By the middle of the nineteenth century, the flower was being cultivated as a pretty garden plant, and it is still planted for its splashes of color, which attract hummingbirds to its nectar.

Ralph Waldo Emerson listed it among the things he found most soothing:

All my hurts
My garden spade can heal. A woodland
    walk,
A quest of river-grapes, a mocking
    thrush,
A wild-rose, or rock-loving columbine,
Salve my worst wounds.[204]

Ranunculaceae

Round-Lobed Hepatica

Hepatica americana, liverleaf, heart liverwort, liver moss, mouse-ear, spring beauty, crystalwort, golden trefoil, ivy flower, herb trinity, squirrel cup, snow trillium, mayflower, flue anemone, kidneywort

Hepatica americana

grows 4 to 6 inches high. Three-lobed leaves unfurl from the base of the hairy stem once the blooms are fully open. The lilac, blue, pink, or white flowers are ½ to ¾ inch across. Found in dry woods rich in organic soil.

Range: Nova Scotia to Minnesota, south to Missouri, and east to Florida.

Hepaticas are always a surprise. They launch themselves early in the year and are often concealed under the dried-up leaves of the beech-maple woodlands they favor. Their new buds are hairy, perhaps to discourage insects until the exquisite ½-inch anemonelike flowers emerge. The hepaticas belong to the buttercup family and are related to columbines, anemones, and baneberries. There are two native hepaticas: *H. americana* (which has round-lobed leaves and grows in acid soil) and *H. acutiloba* (which has sharp-lobed leaves and grows in limestone soil). If they grow in the same woods, they often hybridize.

The flowers come in clusters with up to 12 sepals (rather than petals); they close at night and on cloudy days. New leaves replace the old after flowering to make the plant evergreen. Bees and flies do the pollinating, while insects spread the one-seeded fruits. The effusive naturalist John Burroughs wrote of hepatica:

What an individuality it has! No two clusters are alike; all shades and sizes; some are snow-white, some pale pink, with just a tinge of violet, some deep purple, others the purest blue. . . . Then . . . there are individual hepaticas . . . that are sweet-scented. The gift seems as capricious as the gift of genius in families. You cannot tell which the fragrant ones are till you try them . . . the odor is faint, and recalls that of the sweet violets.[205]

Hepatica americana, round-lobed hepatica

The scientific name comes from the Greek for "liver": *hepar*. The lobed leaves were thought to resemble the shape of the liver; and as the leaves fade, they do take on a recognizable liverish tone. According to the doctrine of signatures popular in Renaissance Europe, if it looks like liver, it must be good for the liver. Although this idea was eventually discredited, the name remains.

Native Americans made no claims for the use of hepatica in liver complaints. The Cherokee chewed the root to relieve a cough and the Chippewa used a decoction (a boiled-down syrup) of the root to treat convulsions in children. The Potawatomi used the root and the leaves as a tea to relieve vertigo. The roots alone were considered an excellent dye.

The plant also had some rather odd qualities attributed to it. According to James Mooney, who wrote about the Cherokee in 1885, "Those who dream of snakes drink a decoction of this herb and the walking fern to produce vomiting, after which the dreams do not return." Huron Smith, writing about the Meskwaki in 1928, noted: "When the mouth gets twisted or crossed and the eyes get crossed, this root is brewed into a tea, and the face is washed with it until it returns to normal. At the same time the patient has to take two teaspoonfuls daily."[206] Somebody should tell Jim Carrey.

Mooney also wrote that in the Carolina mountains, "a girl can infallibly win the love of any sweetheart she may desire by secretly throwing over his clothing some of the powder made by rubbing together a few heart leaves which have been dried by the fire."[207]

Urticaceae

Stinging Nettle
Urtica dioica

Urtica dioica

grows 2 to 6 feet tall on a hairy, erect stem that is square in cross section. The hairy, dark green, toothed leaves grow in pairs up the stem. The inconspicuous green flowers sprout from the leaf axils in summer, growing in small drooping clusters underneath the leaves. The plant is dioecious, as its scientific name suggests: the male and female flowers grow on separate plants. Found in wetlands, meadows, and at the edges of woods.

Range: throughout the eastern and central United States and Canada.

It is said that Roman soldiers deliberately rubbed themselves with nettles to keep themselves warm when they were on duty during cold nights. Or perhaps they did it to keep themselves awake on duty in dismal outposts. It's just extreme enough to be a legend, but the scientific name comes from the Latin verb *urere,* "to burn."

The common name may come from the Old English word for "needle." Those nasty stinging hairs are actually tiny hollow tubes which exude an irritating sap containing ammonia. The end of the tube pierces the skin slightly and injects the sap into the wound. Like poison ivy and jewelweed, nettle often grows alongside its own traditional cure: the dock plant. Jewelweed is another cure for a hit of nettle irritation, though plain old white toothpaste may help out just as effectively.

How anybody got close enough to this plant to find a use for it is a mystery, but Native Americans certainly ate it—quite sensibly, since it's full of vitamins. The leaves lose their sting when they are cooked, and taste like spinach (before spinach was imported to North America, nettles were the green equivalent). Nettle fibers were also used to make hemp (along with species such as milkweed, dogbane, and basswood). Father Theodat Gabriel Sagard, who worked among the Huron in the seventeenth-century, took note of how the women gathered the plants and the men made them into fish nets. He doesn't explain how they dealt with the stinging sap, but said they spun it "on their thighs," which must have required considerable skill. The Huron carried on a thriving trade in twine with neighboring tribes.[208]

The Europeans copied their example and, by the eighteenth-century, they were weaving the nettle twine into a material rather like linen. The pioneer in this business seems to have been Dame de Repentigny of Quebec City, who sent her cloth to the French king for his approval.

The nettle has some medicinal uses as a diuretic and a cure for urinary ailments. According to ancient herbal lore, the stinging of a nettle was said to reinvigorate paralyzed limbs.

Nettles produce a huge number of seeds (20,000 seeds from a single shoot) in a sunny location and can spread quickly. Today, most people see them as unpleasant weeds and try to eradicate them. But they are essential to compost. They supply useful trace minerals and nutrients when chopped up and thrown on the pile. Nettle tea is great for plants as well. Simply pour hot water over nettles and leave it in a covered bucket for a week. Strain it and dilute it with 10 parts water to 1 part nettle goop. It's a terrific fertilizer.

Uvulariaceae

Bellwort
Uvularia grandiflora, merrybells, haybell, strawflower, wild oat

---

Uvularia grandiflora

grows up to 30 inches tall from a white tuberous rhizome. Pointy-tipped oval leaves clasp the stem. The yellow flowers, which emerge in April and May, hang down from the tips of the leafy branches. Each flower is 1 to 2 inches long. The seeds are borne in capsules. Grows in rich humusy woods.

Range: southern Ontario and Quebec, New England, south to Georgia, west to Oklahoma and Arkansas.

Uvularia grandiflora, bellwort

Merrybells seems the proper common name for this cheerful woodland plant. The golden bell-like flowers hanging down over the cool spring soil are a promise of what is to come in the forest.

The name comes from the uvula, that odd-looking fleshy lobe hanging down at the back of the throat. It seems a strange name to be applied to something so sprightly. It's also interesting to know why some plants are dubbed "worts" as in bellwort. The *bell* part is obvious, but the *wort* part comes from *wyrt*, an Old English word for root. Eventually *-wort* connoted a plant with an edible stem and root.

The flowers of bellwort are made up of six deeply divided petal-like sepals, which point downward at the end of the stems. They look like elegant drooping lilies and, indeed, they are related to lilies. At one point, *Uvularia* was taken to be a species of Solomon's seal, because its smooth stalks seem to grow through the base of the leaves. Now it is recognized as being in its own family, Uvulariaceae. There are five species growing north of South Carolina, including *U. perfoliata* and *U. sessilifolia*, and one, *U. floridana*, which is native to the Southern states and Florida.

The rhizomes were once collected and eaten in much the same way as asparagus. The Cherokee boiled or fried them in fat, or ate them as greens. The Iroquois mashed them up to make a poultice for boils or for broken bones. The Potawatomi made an infusion to treat backache and mixed the boiled root with lard to rub on sore muscles.[209]

U. perfoliata is endangered in New Hampshire and rare in Canada and New York. It is a popular shade plant now and increasingly showing up in nurseries. If you buy some from a nursery, make sure they have been propagated vegetatively and not hauled out of the wild. An important principle when buying any kind of native plants: know your dealer.

Violaceae

Violet
Viola spp., Johnny-jump-up, heal-all

Viola spp.

include dozens of species in North America. In some, the leaves and flower stalks emerge directly from the roots and the plant stays close to the ground; others have stems bearing both leaves and flowers and may grow to be 1 foot high. Some have heart-shaped leaves; others have lance-shaped leaves; still others have finely divided leaves. The purple, blue, white, or yellow flowers

have five petals in a distinct arrangement: two at the top, two at the sides, and a striped petal at the bottom.

Range: **temperate areas of North America.**

State flower of Illinois, New Jersey, Rhode Island, and Wisconsin.

We are inclined to take violets for granted because they are so common, so easy to acquire, and so hard to get out of the garden. But if you think of them as specially evolved landing pads for bees, you may view them in a whole new light, as one of nature's amazing inventions. The stripes on the lowest petal are lines for guiding the bees into the heart of the flower.

The showy blooms of spring aren't the only ones violets produce. Later in the season, a second, different flower emerges, growing close to the ground or even underground. It remains closed and fertilizes itself. The scientific term for this ability is cleistogamous.

Native violets include the common **woolly**

Viola spp., violet

or **blue violet** (*Viola sororia*), with heart-shaped leaves; **bird's-foot violet** (*V. pedata*), with claw-like leaves; the **downy yellow violet** (*V. pubescens*), with hairy leaves; and the white **Canada violet** (*V. canadensis*), which can grow up to 18 inches high. One variety of the common blue violet which is grayish-white with a lavender-colored center is known as the **Confederate violet,** because it is the same color as the Confederate Army uniforms.

It's practically impossible to find anyone who doesn't like violets (and their sisters, the pansies). Many of the popular species are imports from Europe, including the sweet-smelling *V. odorata*, which is used in French perfume and was Napoleon's emblem on his return from Elba. Most native species, however, have no smell but they do have knobby tenacious roots.

Violets contain salicylic acid, one of the ingredients in aspirin, hence the name "heal-all." However, the medicinal use of the plant in North America seems to stem from the knowledge of immigrants, who were familiar with the European species. The Native Americans made less use of them.

Writer Jack Sanders tells us: "It is said that John Bartram, the first famous American botanist, was inspired to pursue botany instead of farming by one day finding a colony of bird's-foot violets at the edge of a field. That night he dreamed of the flower and the next day announced that he was hiring a manager for his farm and heading for Philadelphia to study botany. . . . He mastered Latin in only three months so he could deal with the nomenclature."[210]

Though a swath of violets is a pleasure to look at and recalls the dozens of poems written about their fleeting beauty, and the flowers are considered a valuable addition to the salad bowl, one of the chores of post-bloom production is to get rid of them before they take over completely. To spend a day rooting them out, always leaving one

behind of course, is one of the pleasures of gardening. They are too much a part of our psyche, way too charming, and too important to bees to eradicate altogether.

Vitaceae

Virginia Creeper

Parthenocissus quinquefolia, American ivy, five-fingered ivy, five-fingered poison ivy, woodbine, false grape

This climbing vine clings to stone and brick with sticky little pads which are actually modified root structures. It belongs to the same family as grapes, but has poisonous berries. A fast grower, it can put on 6 to 10 feet in a single season and eventually reach 150 feet. The leaves are divided into three or five stalked, toothed leaflets, green in summer and red in fall. The tiny green flowers grow from June to August in inconspicuous clusters, and give way to dark blue berries in fall. It has magnificent autumn color, which is one of the main reasons people allow it into gardens and up the sides of houses. The bright leaves invite birds to come in and get fat eating rich berries on their flight paths south.

The first part of the scientific name is a curious roundabout. *Parthenocissus* is a Latinization of the Greek translation of the common name, Virginia creeper. *Partheno* means "virgin" (which is the source for *Virginia*) and *cissus (kissos)* means "vine" or "ivy." *Quinquefolia* is Latin for "five leaves." (**Boston ivy** is an Asian relative, *P. tricuspidata*, with three-lobed leaves.)

According to anthropologist Frances Densmore, the Chippewa cut the stalk into short lengths, peeled it, and boiled it. "Between the outer bark and the wood there was a sweetish substance which was eaten somewhat after the manner of eating corn from the cob." Herbalists list it as an antidiarrheal, antiseptic, astringent, diuretic, expectorant, and tonic, and there are records of its use by Cherokee, Iroquois, and Meskwaki, but it does not seem to have been one of the really important medicinal plants.[211]

It crawls around on the forest floor, sometimes growing underground and emerging to climb up and drape itself companionably over a tree—sometimes not so companionably cutting off all of its light. Michael Dirr says the famous ivy-covered walls of universities should more properly be called creeper-covered since it is this plant and not true ivy *(Hedera)* clothing them in luxuriant green and brilliant fall red.[212] Should we be calling them the Creeper League Colleges?

Range: **throughout the eastern and central United States, and southeastern Canada.**

SOUTHERN HARDWOOD FOREST

The Southern hardwood forest is home to some of the loveliest trees on the planet. It must have seemed just as miraculous to those who arrived here thousands of years ago, as those who came only a few hundreds of years ago. Towering pines, magnificent live oaks the size of ten-story buildings, an understory that included magnolias and persimmons, filled with all kinds of animals—an overwhelming richness. Four hundred years ago, there were more flowering trees and shrubs than had ever been seen anywhere else on this continent or in Europe.

The Southern forest, which stretches 600 miles from east to west and 400 miles north to south, has no dominant tree. It is a mosaic of different forests, each tailored to its specific geography. A pine-dominated forest in Georgia is very different from one in Mississippi, for example. Every biological community of the warm temperate climate is here: maritime forests on sand dunes, hammock forests on rises; scrub forests, swamp forests, bogs, bayous, and broad-leaved forests.

A very different kind of culture arose in southern North America after European contact, compared to that in the Northern settlements. Although the Indians didn't fare any better here than in the North, they managed to hang on to larger areas for their own use, because swamps and bogs were considered of no economic importance to the newcomers. The culture was also very much influenced by the arrivals of boatloads of slaves from Africa who not only brought African customs but also secretly brought seeds with them and searched for plants that reminded them of home.

The forests were cut down to grow tobacco and cotton in many areas, and those who grew these products for the European market amassed great riches. Without the slave trade, however, there would have been no commerce. No one could have worked the land in the heat and humidity as strenuously as those who were forced into it by slavery.

The Southern hardwood forests were pretty much left alone until after the Civil War, when the South began to rejuvenate itself and logging started in earnest. It took only a few decades to remove the biggest, oldest, and best trees. By 1909, the Southern forests were supplying 21 billion board feet of lumber—half the nation's output.

Once Virginia and areas to the south had been pretty much logged out, the lumbermen lit out for the Pacific Northwest. Conversion of the land to farms slowed, leaving behind second-, third-, and fourth-growth forests. None of the big animals were left, except bears and the occasional large cat. The canopies are still hand-

some, but much diminished. On plantations, trees grow in stately rows like tulips, but elsewhere the growth is bushy and crowded, not at all the way the forest was only 400 years before.

Fagaceae

Live Oak

Quercus virginiana, Virginia live oak, Southern live oak, sand live oak, bay live oak, Texas live oak, West Texas live oak, scrub live oak, plateau oak, plateau live oak, escarpment live oak, encino

Quercus virginiana

grows 40 to 80 feet high, with a 60- to 100-foot spread, in sandy, low, coastal areas. The simple, alternate leaves are elliptical and up to 3 inches

Quercus virginiana, live oak

long—when young they look a bit like holly, but end up with flat margins. The acorns are on a long stalk in an elongated form. The red-brown bark has shallow grooves and becomes dark, almost black, as it ages.

> ***Range:* Virginia to Florida, west to Mexico. State tree of Georgia.**

Though "live oak" seems to imply all other oaks are dead, it only means this species is evergreen, or, in Walt Whitman's words, can be found "uttering joyous leaves all its life."

The live oak is the quintessential image of the Old South, with Spanish moss draped from every limb, evoking an age of gentility mixed with a midnight-in-the-garden-of-good-and-evil aura of mystery. It is magnificent, often massive, "wide as a boulevard, with limbs that look like transcontinental oil pipes."[213] Along windswept

(a threatened species that nests only in live oaks). The acorns don't last long in two senses: they are so sweet that animals and people are attracted to them, and the taste diminishes as soon as they germinate. Live oaks also provide a home for other plants, such as Spanish moss and mistletoe. Although Spanish moss is usually harmless, large swags can block the light to the oak's lower branches, seriously inhibiting photosynthesis. **Mistletoe** (*Phoradendron* spp.), on the other hand, is a parasitic plant which draws both moisture and nutrients from the tree and can cause great damage. Live oaks also have a merciless enemy in the form of *Ceratocystis fagacearum*, the dreaded oak fungus. It attacks the sapwood and survives in dead wood for at least a year.

Healthy live oaks can live 200 to 300 years. Ancient trees such as these can be seen at Wormsloe, the famous plantation near Savannah, Georgia. The fortified house was built in the 1740s by Noble Jones, one of Georgia's first 114 British colonists. His passion was horticulture. Unlike his neighbors, he resisted the urge to chop down the cypress and oak forest. He planted cotton and rice on his 86-acre plantation but left the trees, shrubs, and watercourses much as he had found them. To him, it was a reminder of England. His son, Noble Wimberly Jones, planted the 400 live oaks lining the driveway, and today, their gnarled, spreading limbs form a cool canopy overhead.

Many of Savannah's live oaks, planted together so many years ago, are aging. And, because the city's canopy is made up almost exclusively of live oaks, they are all beginning to die. As Michael Dirr notes: "Savannah is a beautiful city . . . [and] . . . is looking for suitable substitutes to maintain the biological architecture but, alas, the Live Oak, like the American Elm, can never be adequately replaced."[214]

dunes, though, it takes on a sheared look and becomes stunted; it holds fast against hurricane-force winds with taproots thrusting deeply into the soil. These trees are survivors: not only do they produce an annual crop of acorns, they can also reproduce vegetatively, sending up sprouts from the roots—sprouts that are a favorite snack of foraging animals.

Live oaks feed many species of birds: from the Northern bobwhite to the Florida scrub jay

The Indians used to eat the acorns, or feed them to livestock. Originally, Native Americans had no use for domesticated animals, given the bounty of the wild meat in the woods around them. But once their habitat as well as the animals they depended on became threatened, they took up the European habit of keeping animals for both work and food. They also used live oak saplings as markers for trails. They'd bend them in the direction of the trail and the trees consequently grew into odd but useful shapes.

The wood, though strong, did not produce the straight lumber needed for most practical purposes. However, in the early nineteenth century, the American navy used the "knees"—the wood bracing up the huge branches as they grew outward from the trunk—to strengthen the joints of warships.

Today, live oaks are still planted for their appearance, and are being used to revegetate areas cleared for mines or farms, especially in the Mississippi Valley.

Fagaceae

Laurel Oak
Quercus laurifolia, Darlington oak, diamond-leaf oak

Laurel oaks were once part of the landscape known as the oak savanna—open areas of grasses and scattered stands of trees. They stayed open because of occasional fires, and flourished because of occasional flooding. This was a practice the Indians maintained for many thousands of years. But the European farmers and loggers did their best to prevent fires, dammed the rivers to prevent flooding, and tore up the grass to plant cotton or let their animals graze on it. The oaks were cut for lumber and firewood, then replaced with loblolly pines. The delicate balance that made up the oak savanna vanished.

Despite some attempts at restoring the natural landscape, the lands those first settlers saw have largely disappeared.

The most beautifully formed laurel oaks are found growing as ornamentals in north Florida and in Georgia. This is partly due to the graceful leaves, but also because the tree grows quickly, maturing to a height of 150 feet in only 50 years with a large deep taproot. The alternate 3- to 5-inch leaves are not lobed as they are in other oaks. The brown-to-black acorns are ½ to 1 inch long. Cut or burned, this tree generates a mass of sprouts from the base.

This is a semievergreen oak which grows in the moist woods of the Southeastern coastal plain. On a floodplain it forms almost pure stands. Though not long-lived, it is one of the top ten food trees for wild animals.

At 15 to 20 years old, the tree produces acorns every second year in September and October, when squirrels collect and bury the fallen acorns. Spring germination of its acorns is one of the qualities of the black oak group, of which this tree is a member. (See page 24 for oak groups.)

Range: **Atlantic and Gulf coastal plains from southeastern Virginia to southern Florida, and westward to southeastern Texas.**

Fagaceae

Post Oak
Quercus stellata, iron oak, box oak, brash oak

When the American railroad conquered the West, this is the tree that supplied many of the crossties. The railroad was a monumental undertaking which was possible only because there was so much wood conveniently at hand.

The wood is close-grained and hard, and since it rots slowly in soil, it was ideal for fence posts, hence its common name. Post oak grows in dry, sandy, even gravelly soils and can reach heights of 40 to 50 feet with gray or brown fissured, scaly bark. The leaves have gray or yellow hairs on the underside. The squarish lobes on either side of the leaves form a cross shape; the acorns, with a deep, bowl-shaped cap, are ½ to 1 inch long, occurring singly or in pairs.

Post oak's leathery leaves are deep red when they first appear, turn dark green for the summer, and finally fade to yellow or brown. They hang on the branches with a fierce tenacity until forced off by the following spring's new growth. The bark is full of tannin capable of poisoning cattle, sheep, and goats, particularly in dry weather. Because it's a drought-tolerant plant, it is sometimes the only green thing available to grazing animals. The acorns are edible, though, and an important source of food for white-tailed deer, wild turkeys, and squirrels.

Post oak was one of the species dominating the Cross Timbers area stretching from Texas north through Oklahoma to Kansas. In this area, two belts of forest are surrounded by prairie, marking the western extension of the great Eastern forests. Indians used the area first, then the Europeans turned it into grazing range and it has been built on since the nineteenth century. (No one knows why it's called Cross Timbers—maybe because people had to cross it as they went west to the Rockies.) In 1841 it appeared as an apparition to William Kennedy:

This belt of timber varies in width from five to fifty miles. Between the Trinity and Red Rivers it is generally from five to nine miles wide, and is so remarkably straight and regular, that it appears to be a work of art. When viewed from the adjoining prairies on the east or west, it appears in the distance as an immense wall of woods stretching from south to north in a straight line, the extremities of which are lost in the horizon.[215]

Getting through Cross Timbers was quite a feat for anyone and baffling to newcomers. "The ground was covered with a heavy undergrowth of briars and thorn-bushes, impenetrable even by mules, and these, with the blackjacks and post oaks which thickly studded the broken surface, had to be cut away, their removal only showing, in bolder relief, the rough and jagged surface of the soil which had given them existence and nourishment."[216]

Native Americans roasted, boiled, or ground post oak acorns for food. Settlers found them generally unpalatable, and fed them to livestock. The tough, close-grained wood was used for fuel, rough construction, and in making barrels for storing and shipping liquids in the days before plastic and glass. Today, it is often planted in urban parks as a shade tree.

Range: **throughout the southeastern states as far north as Massachusetts and as far south as Florida; also found as far west as Iowa, Kansas, Oklahoma, and Texas.**

Fagaceae

Southern Red Oak
Quercus falcata, Spanish oak, water oak

Quercus falcata

grows to 80 feet, with a rounded open crown. The dark gray bark is furrowed into plates. The leaves are alternate, elliptical, deeply divided into a long narrow end lobe and curved side lobes (*falcata* means "scythelike"), 4 to 8 inches long, shiny green above, hairy gray below, with bristles on the tips. Small acorns ½ inch long, with a shallow cup, mature in the second year.

Range: **New York, south to northern Florida, west across the Gulf states to the valley of the Brazos River in Texas, north to Illinois and Ohio.**

The Southern red oak is one of the hardiest and fastest-growing of the oaks and usually lives for about 150 years. Time enough for it to accumulate a long and important history, according to naturalist Donald C. Peattie:

On the piedmont and coastal plain of Virginia, where every foot of ground, it sometimes seems, is steeped in memories of two wars and two great eras of peace, the Colonial and the Federal, the Spanish Oak is commonly the Chief Inhabitant and Oldest Settler. It is a poor country church . . . that is not shaded by at least one great old Spanish Oak.[217]

Quercus falcata, Southern red oak

Though it has monumental qualities, the Southern red oak is also an important nesting site for birds and mammals. The acorns provide food for waterfowl, wild turkeys, blue jays, common grackles, white-breasted nuthatches, deer, and squirrels. People found them inedible when raw, but the Cherokee leached out the bitterness of the tannins and other phenolic compounds by boiling them in water, burying them for a time, or leaving them to soak in streams. When it was suitably flavorless, they mixed this material with fat to make cakes and bread or a kind of mush. They also used the bark as an antidiarrheal treatment for chronic dysentery; "astringent bark [was] chewed for mouth sores . . . infusion of bark [was] applied to sore, chapped skin"; and it was useful as an antiseptic and an emetic. The bark was also woven into baskets.[218]

A variant called **cherrybark oak** (*Q. falcata* var. *pagodifolia)* produces strong and heavy wood ideal for furniture, interior finishes, and veneer. Wood of the regular Southern red oak variety is strong and heavy, but it also tends to be rough, coarse-grained, and have insect and stain damage. Now used mainly for factory lumber, railroad ties, and timbers, it was once used for what was known as slack cooperage—barrels for dry goods such as flour. Like the post oak, this industry declined with the rise of modern cardboard and plastic packaging.

Fagaceae

Blackjack Oak
Quercus marilandica, scrub oak

Lichens form on the rough bark of this weed tree—as blackjack oak is unfairly known. It's long been an indicator of poor soil, so the sight of it growing in any area discouraged most settlers. By the same token, land that would support little else could grow this oak. The dark brown wood is used mainly for charcoal and fuel.

The common name refers to the leaf shape, which is similar to the weapon of cops and thugs. *Marilandica* means "Maryland," where this species was originally identified. Blackjack oaks grow 20 to 40 feet tall on dry, sandy flatlands or barren rocky slopes, developing thick black bark with deeply furrowed plates. The simple alternate leaves are leathery and shiny, up to 6 inches long with three very shallow lobes, making them look like duck's feet. The acorn is ¾ inch long, elliptical, with reddish-brown scales and a yellow seed.

The Comanche used the leaves as cigarette wrappers. The Choctaw made an infusion of the tree bark coal "to remove the afterbirth and ease cramps." An infusion of the bark coal was used to aid in childbirth as well. Though animals enjoyed the acorns, humans considered them strictly starvation food.[219]

Range: **Long Island, west to Nebraska and Kansas, south to Texas and Florida.**

Fagaceae

Turkey Oak
Quercus laevis

Turkey oak is often found in the understory of longleaf pine forests, on infertile, sandy soils. In fact, without turkey oaks in a forest managed for wildlife habitat, several species of the longleaf pine community will not prosper.

Turkey oak has such a thick bark that it acts as a firebreak in the sand hills and dunes where it usually grows, then resprouts from the stump after a fire. It is not as fire-resistant as the long-leaf pine. If fires occur frequently, longleaf pines will dominate; if they are rare, the turkey oak will take over. Turkey oak grows to 65 feet with dark gray bark and has five-pointed leaves, giving it the common name.

As fodder for deer, these oaks are especially important, but they are not at all palatable to humans in spite of being high in digestible fat. In central Florida, whenever a square mile of turkey oaks is removed, the deer population drops drastically.

Because it is small and of little commercial value, it is frequently considered a weed tree. The wood is sometimes used locally for firewood and sometimes for rough construction. At one time, the bark was used to tan leather.

Range: **Southeastern coastal plain from Virginia to central Florida, and west to Louisiana.**

Pinaceae

Loblolly Pine
Pinus taeda, old-field pine, bull pine, Indian pine, long-straw pine, rosemary pine

Pinus taeda

grows to 90 feet tall in a loosely pyramidal form. The tree loses its lower branches as it ages. It is one of the fastest-growing of all Southern pines. The scaly bark is almost black when the tree is young, turning to a reddish-brown as it ages. The needles are dark green, 6 to 9 inches long, and occur in bundles of three. The oval or cylindrical cones are 3 to 6 inches long and covered with spines to protect the seeds from animals and birds.

Pinus taeda, loblolly pine

Range: **eastern Texas, southern Arkansas, to central Florida, north through the Carolinas to Virginia, Delaware, and southern New Jersey.**

A loblolly is a clownish person, according to the dictionary, but in North Carolina a loblolly is a natural pocket or depression in the ground or rock filled with sandy soil—an ideal spot for the loblolly pine to grow. It's often found in old fields or on abandoned farmland. Another name, "bull pine," refers to the sturdy and thick trunk, which is often 5 feet in diameter. The name "rosemary pine" is from the soft scent of the tree.

Loblolly pine forms stands in moist depressions, swampy areas, by the sea, and also in drier inland soil. Over the years, it has been cut and cut and cut, and what we see now is probably third or fourth growth. It's considered ideal for commercial plantations because it grows a phenomenal 55 to 65 feet in 25 years. Better still, the long straight trunk has no knots for at least the first 30 or 40 feet, so it can be used for both lumber and pulpwood.

François André Michaux, the great eighteenth-century French botanist and plant collector, noted that "three-fourths of the houses in Virginia were built of loblolly pine, which provides better timber than scrub pine. Pine was frequently specified for construction. It was designated for the flooring of the capitol in Williamsburg in 1700."[220]

Its dark brown, deeply furrowed bark is now ground up as mulch for acid-loving ericaceous plants such as rhododendrons rather than being wasted, as it was by earlier lumber barons. The needles are also collected to make mulch.

The loblolly pine is subject to attack by pine beetles and fungi. When this happens, the tree defends itself by producing oleoresins, phenolic compounds, and tannins in the inner bark, or phloem. This discourages the beetles from making active colonies, and the tannin acts as an antimicrobial substance.

These trees have now been widely exported, especially to countries such as Brazil and South Africa, where reforestation is desperately needed. Loblollies do very well, because neither country harbors any of their natural predators. Because it grows so swiftly and responds to plantation conditions, loblolly pine is now one of the most important commercial lumber trees in the Southern states.

Pinaceae

Shortleaf Pine
Pinus echinata, shortleaf yellow pine, Southern yellow pine, old-field pine, short straw pine, Arkansas soft pine

Shortleaf pine is one of the four most significant commercial conifers of the Eastern forest and second only to loblolly pine as a commercial softwood. It has the widest range of any pine in the Southeast, growing in 22 states. The pollen records show it once grew as far north as Michigan. Its oval crown is smothered with dark blue-green leaves and it grows up to 80 feet tall. Flowers appear in early spring, followed by winged seeds. The wood of old shortleaf pine trees is heavy, hard, fine-grained, and less resinous than the other Southern pines. It is used for interior and exterior finishing, construction, veneers, and in the making of pulp.

Range: southeastern New York, New Jersey, west to Illinois, Missouri, and eastern Oklahoma, south to Texas and northern Florida.

Pinaceae

Longleaf Pine
Pinus palustris, lone star pine, yellow pine, Southern yellow pine, swamp pine, Georgia pine

Longleaf pine is the most striking of the Southern pines. Huge stands once covered the Atlantic and Gulf coastal plains, occupying perhaps 60 million acres; by 1985, less than 4 million acres were left, and very few of those acres

Pinus palustris, longleaf pine

were in a truly natural state. This fire-resistant species also benefits from periodic burnings to clear away competing species. With the declining frequency of fires because of both public attitudes and improved fire-fighting techniques, its habitat has gradually been taken over by slash pines, loblolly pines, and shortleaf pines. Related trees vie for space like siblings battling over an inheritance.

Longleaf pine grows 80 to 120 feet tall, with an open, irregularly shaped crown, supported by a massive taproot thriving in moist soils (*palustris* means "of the marsh"). The long, bright green needles can grow from 8 inches to as long as 18 inches, in groups of three, with large reddish-brown cones 6 to 10 inches long. This is in sharp contrast with the orange-brown, rather rough and scaly bark. Winged seeds are dispersed by wind from October to November, most landing within 75 feet of the parent tree. Germination usually occurs within a week—if

the seeds aren't scooped up by birds or animals. Unlike other pine species, longleaf pines produce seedlings that remain stemless for up to five or six years. This is known as the "grass stage" of the plant. During this period, while the root is elongating and establishing itself underground, the tree barely grows and is especially vulnerable to fire.

The wood is heavy, hard, tough, and durable with a high content of amber-colored heartwood and the tight pattern unique among pines. It has been used for poles, piling posts, plywood, and pulpwood. During the nineteenth century, millions of board feet were exported to England, where good lumber continued to be scarce. Today, pretty much the only old-growth longleaf pine available is from companies such as Appalachian Wood, which buy old wood from demolition outfits and resaw the timbers for fine flooring and cabinetry.

The resin is used to make tar, pitch, rosin, and turpentine. Large plantations of longleaf pine and **slash pine** *(P. elliottii)* are grown specifically for this purpose. The sap is distilled to separate volatile essential oils from the nonvolatile diterpene residue called rosin. The volatile oils end up as turpentine (basis for paint thinners and organic solvents); rosin is used in making varnishes and oil-based paints.[221]

Range: Atlantic and Gulf coastal plains from southeastern Virginia to east Texas and south Florida, and the Piedmont Ridge, Valley, and Mountain ranges of Alabama and northwest Georgia.

Magnoliaceae

Southern Magnolia
Magnolia grandiflora, bull bay, large-flowered magnolia, big laurel

Magnolia grandiflora

grows 60 to 80 feet with a spread of up to 50 feet. The alternate simple evergreen leaves are 5 to 10 inches long. The creamy-white flowers, 8 to 12 inches in diameter, have a strong scent. The rose-red fruit is 3 to 5 inches long, splitting open to expose red seeds in October and November.

Range: North Carolina, south along the Atlantic coast to central Florida, west to southeast Texas.

State flower of Louisiana, state tree of Mississippi.

Magnolia grandiflora, Southern magnolia

Once called "worthy to be trees of heaven,"[222] the magnolia seems to symbolize the beauty and graciousness of the Old South. Margaret Mitchell's famous heroine Scarlett O'Hara had "magnolia-white skin," and novels set in the Old South invariably feature avenues lined with magnolias leading to elegant, white-columned mansions. It is said that Southern belles used to write billets-doux to their sweethearts on magnolia petals. The tree positively drips romance, mystery, and old-world charm.

It's easy to see why. Southern magnolia, with its huge fragrant white flowers and evergreen leaves, is arguably one of the most striking forest plants anywhere. The magnificent leaves fall to the ground and slowly disintegrate—glorious even in death. Collectors and flower arrangers patrol the forests, picking up their skeletal remains.

In order to thrive, Southern magnolia needs 210 to 240 frost-free days, moderate temperatures in winter, and an average of 80°F in a wet summer. It grows along rivers and near swamps, where the soil is moist and fire is rare.

The large, white, highly scented flowers come out from April to June and the fruit ripens all autumn. Once mature, the fleshy cone-shaped fruits open to reveal ½-inch-long seeds held by gossamer threads. The seeds eventually fall to the ground. Every few years, the trees produce a bumper crop. Birds and small mammals feed on the seeds, distributing them widely. If the seed falls into a rich, moist soil protected by tree litter, it will produce leaves fairly quickly.

William Bartram was impressed by the magnolias he saw on his trip to Florida on the St. John's River in 1765. He called them laurel magnolias, but in fact he was talking about Southern magnolias.

How majestically stands the Laurel, its head forming a perfect cone! its dark green foliage seems silvered over with milk-white flowers. They are so large, as to be distinctly visible at the distance of a mile or more. The Laurel Magnolias, which grow on this river, are the most beautiful and tall that I have any where seen. . . . Their usual height is about one hundred feet, and some greatly exceed that. . . . [The] berries possess an agreeable spicy scent, and an aromatic bitter taste. The wood when seasoned is of a straw colour, compact, and harder and firmer than that of the poplar.[223]

Today, the Southern magnolia is usually found in the company of other hardwoods. The wood is harvested and marketed as magnolia lumber and, along with other magnolia species, is used to make furniture, pallets, and veneer. Mostly, however, magnolias are planted as shade trees and ornamentals. Several U.S. presidents have planted magnolias on the White House grounds. Andrew Jackson brought a couple from Tennessee in 1830 and planted them on the west side. They used to be visible on the 20-dollar bill, which showed the southern view of the house, before the bills were redesigned in 1996 to show the northern façade.

The name honors Pierre Magnol, who was professor of botany and director of the botanical gardens of Montpellier, France, during the late seventeenth and early eighteenth centuries.

Magnoliaceae

Sweet Bay

Magnolia virginiana, swamp magnolia, beaver tree

Another glorious Southern magnolia, sweet bay is intoxicating in its beauty and scent, and captivating with its bright red seeds. It grows to 20 feet tall in the northern part of its range, but up to 50 to 90 feet in the southern. In the North, it is a multistemmed, open-topped tree that drops its leaves in winter; in the South, it is an evergreen with a large pyramidal shape. The oval leaves are shiny bright green on top and white underneath, 3 to 5 inches long. The winter buds are covered with fine hairs and the large, creamy-white, fragrant flowers, appearing in May or June, have 9 to 12 petals on a smooth stem with a brilliant yellow center. The oval fruit produces dark red, smooth, brilliant red seeds ¼ inch long.

It was the first magnolia to be cultivated in English gardens when a missionary sent sweet bay seeds to Henry Compton, the Bishop of London from 1675 to 1713. Compton, whose diocese included the American colonies, was an avid plant collector and created a botanical garden at the bishop's residence, Fulham Palace. The sweet bay was even more sought after in the early nineteenth century, when botanizing became a fashionable pastime for the leisured middle classes of America. The trend was boosted by the hugely popular *Florula Bostoniensis*, one of the first major works about North American flora, written by Dr. Jacob Bigelow. While the book made people appreciate the native plants around them, it also led to a certain amount of destruction, as plant enthusiasts seized prize specimens for their collections without a thought for the future.

Sometimes things got out of hand, as they often do with fads. A Boston judge named Theophilus Parsons was on his way from Manchester to Gloucester, Massachusetts, in 1806 when he was surprised to see a sweet bay slightly out of its usual range. He contacted the Reverend Manasseh Cutler and showed the flowers he'd collected to Judge John Davis. Word got out and, for 80 years, plant lovers traveled to the site so regularly its name was changed from Kettle Cove to Magnolia. Some unscrupulous types, not content with observing the plants in their natural habitat, ripped out the magnolias and carted them off. By 1916, only two little plants a few feet high had escaped the magnolia hunters of Massachusetts.

It was the scent of *M. virginiana* along streambanks that drew the plant hunters to their quarry, but its attractions go well beyond the ornamental. The fleshy roots were used to bait beaver traps (hence the name beaver tree), and deer consider its protein-rich leaves and twigs a great delicacy. The seeds are much loved by gray squirrels and to a lesser extent by white-footed mice, quail, and songbirds. Sweet bay's flowers are used in perfume, while its aromatic, easily worked wood is a popular choice for veneer, boxes, and containers.

Range: **Massachusetts to Florida and Texas.**

Ebenaceae

Persimmon

Diospyros virginiana, common persimmon, possumwood, boawood, date plum

Diospyros virginiana

grows to 20 to 60 feet tall. The oval leaves are dark green and leathery, pale green and hairy on the underside, turning golden or red in autumn. It has a long taproot and grows in both moist and dry sites. The bark is brown to black, with deep fissures, broken into rectangular checkered sections. The twigs are dark with orange lenticels and black lateral buds. The inconspicuous, greenish-white, fragrant flowers emerge between March and June; male and female flowers are usually borne on separate trees. It fruits every two years between the ages of 25 and 50. The fruit is an orange or brown berry about 1 inch in diameter.

Range: **southern Connecticut and Long Island to southern Florida, west to southeast Iowa, eastern Kansas, and Oklahoma, and in the valley of the Colorado River in Texas.**

Here is the fruit of the gods—or perhaps the wheat of the gods, since *diospyros* comes from two Greek words, *dios* meaning "divine" and *pyros*, "grain." The most northerly member of the ebony family, it is one of the most useful plants of southeastern North America for both people and animals.

Persimmon trees are sometimes called "possum trees" because opossums are said to climb the tree to get at the fruit. According to one old song:

> 'Possum up the 'simmon tree,
> Raccoon on the ground.
> Raccoon says to the old 'possum,
> "Won't you throw them 'simmons
>   down?"

Persimmon thrives in the bottomlands of the Mississippi River and in the south Atlantic and Gulf states. Once a space opens up in the Southern forest, persimmons move in quickly with their shrubby growth, surviving wherever the humidity is high. Once cut, it sprouts rapidly from the stump.

Ripe persimmons are soft, with a sticky orange center with the consistency of a date, hence one common name: "date plum." The fruit has unattractive skin which turns brown when it ripens, but the taste is sweet and the fruit is filled with vitamins A and C, as well as calcium and potassium. The Cherokee and Comanche ate the fresh berries or dried them to a prunelike consistency to be made into cakes. The unripened fruit is high in tannins and astringent. Captain John Smith of Jamestown, Virginia, remarked, "If it be not ripe, it will draw a man's mouth awrie with much torment."[224] Astringency and a slightly antiseptic quality were precisely the ingredients to make the unripe fruit a useful topical dressing for burns. Ripened fruit eaten to excess can cause problems. The bark was chewed to relieve heartburn and other digestive problems, so perhaps the plant provided its own remedy. An infusion of the

Diospyros virginiana, persimmon

bark was also used by the Cherokee and Rappahannock for throat infections.

According to garden writer Elizabeth Lawrence, "All parts of the persimmon are useful: the inner bark is a tonic, small articles are manufactured from the wood, the fruit can be made into persimmon beer, and the seeds have been roasted as a substitute for coffee."[225]

Dense, heavy, and strong, the rich black heartwood is highly prized for making billiard cues, flooring, and veneers. At one time it was used to make the heads of golf clubs; in the early textile industry, shuttles made from persimmonwood lasted for thousands of hours without needing replacement. The fruit even yielded an indelible kind of ink.

In the seventeenth century, cuttings were taken to England, where they were cultivated by gardeners, despite the transplanting difficulty caused by the enormous taproot. Today, the persimmon tree is often used in Europe for erosion control, and in North America as a landscape plant.

The persimmons generally found in today's supermarkets are the Oriental persimmon, *D. khaki*, grown in California from imported cuttings. Their fruits are larger, have fewer seeds, and less tannin in the pulp.

Theaceae

Franklinia
Franklinia alatamaha

Franklinia alatamaha

grows 20 to 25 feet high and up to 15 feet wide. It branches out in the lower part of the plant with striated marks on the bark. The twigs are green, the older bark is gray. The bright green alternate leaves are 4 to 8 inches long with serrated margins. They turn orange-red in autumn. Fragrant white flowers appear in late summer (August to September), 3 inches in diameter, followed by small rounded fruit.

***Range:* once native to southeastern Georgia, but now propagated around the world.**

Owners of this tree belong to a small but exclusive club. There are only a few thousand franklinias in the world, according to a census conducted by the Historic Bartram's Garden near Philadelphia. It is a striking plant with glorious white flowers in August and spectacular fall color. But part of its appeal is its history. Those who own a franklinia own a piece of North American history.

John Bartram and his son William came upon it in 1765 along the Alatamaha River in Georgia. The Bartrams, who were well-known botanists and nurserymen of Philadelphia, named the tree after John Bartram's close friend, Benjamin Franklin. William Bartram included the story in an account of his travels he wrote in 1791:

I set off early in the morning for the Indian trading-house, on the river St. Mary, and took the road up the N.E. side of the Alatamaha to Fort Barrington. . . . On drawing near the fort, I was greatly delighted at the appearance of two new beautiful shrubs, in all their blooming graces. One of them appeared to be a species of Gordonia, *but the flowers are larger, and more fragrant . . . the seed vessel is also very different.*[226]

Father and son collected the seeds and took them home to Philadelphia, where they grew. The franklinia hasn't been found in the wild since 1803, and all franklinias today are descendants of those seeds collected by the Bartrams over 200 years ago.

William's father, John Bartram (1699–1777), came from a Quaker farm family who lived on the banks of the Delaware River. As he grew up, he acquired a great deal of knowledge about medicinal herbs. In 1736, James Logan, the owner of a great scientific library, lent him a copy of *Systema Naturae* (1735) by Linnaeus. Fascinated, Bartram became the first American-born botanist, and indeed the founder of American botany. He settled near Philadelphia, grew native plants on his farm on the Schuylkill River, and exported seeds and plants from America to England. Though not the first successful hybridizer of plants, he was the first in America to make a deliberate experiment and to describe it, in Latin, accurately enough to be replicated. In Europe a theory about the sexuality of plants had been argued for years. Bartram proved it was true in the New World.

An outspoken conservationist, he criticized his fellow farmers for their continual cropping of corn and tobacco, which depleted the soil. He let his fields lie fallow from time to time, and fertilized them with a mixture of spring water, gypsum, horse manure, and garbage. His farm was 50 percent more productive than his neighbors'. As a Quaker, he rejected slavery; he freed his slaves, educated, clothed, and housed them.

In 1734, Peter Collinson, an English Quaker, asked Bartram to supply American plants to his nursery in England. Bartram did so for the next 35 years, sending hundreds of plant species, and became famous for the quality of his seeds. He wrote thousands of letters describing the country he lived in and his life in it. As the king's botanist in America, he was paid 50 pounds a year, and earned another 10 pounds a year from his seed sales, providing him and his large family with a good living and allowing him to become one of the first to do so strictly from raising plants.

William, his fifth son (1739–1823), went on to found the first American nursery and publish the first plant catalog. In the 1770s, Bartram junior spent four years traveling through the Carolinas and Georgia, gathering important specimens, sending seeds and drawings back to England. His book *Travels of William Bartram* became an instant classic and his descriptions of the landscape inspired English poets such as Coleridge and Wordsworth.

The franklinia is now a cherished ornamental grown in gardens around the world. Sadly, the loss of a plant in the wild is not a rare occurrence in contemporary botany. With the rate of building, paving, and cementing going ever faster, small populations of plants existing nowhere else on earth can be easily wiped out without anyone noticing. For example, in January 1999, *Science News* reported: "Cheyenne botanist Robert Dorn

Franklinia alatamaha, franklinia

discovered another oddball, now in the newly created genus *Yermo*, while riding his dirt bike along the proposed route of an oil pipeline in Wyoming back country. The knee-high, yellow-flowering member of the sunflower family is known only from the population he located." It's frightening to muse on how many plants that once existed in North America are no longer here.

Arecaceae

Saw Palmetto
Serenoa repens, sawtooth palm, windmill palm

Serenoa repens

is a shrub growing 3 to 6 feet tall. The blue-green to yellow-green leaves are like palm leaves, fan-shaped, slit more than halfway to their bases, and up to 3 feet wide. Clusters of white flowers bloom among the fans in early summer. The dark blue to black, olivelike fruit is less than 1 inch long, maturing in the fall. It grows in dry, well-drained soil with little mineral or organic content.

Range: Georgia, Florida, Alabama, and Louisiana.

Saw palmettos may have sharp edges, but they provide succor for birds and rodents, from the Florida grasshopper sparrow to the Florida woodrat. In spring, the palmettos send up prickly green shoots which become a refuge for hogs and rattlesnakes. The plant often grows in dense, bushy clusters under pine trees, with stems creeping along the ground. The common name comes from the sharp, sawtoothlike spines along the leaf petioles.

Saw palmetto can be a small shrub or small tree up to 20 feet high. Either way, it ends up in a tangled mass. Between April and July, it pro-duces perfect white flowers on stalked panicles from the leaf axils. The flowers are a source of nectar for bees (the honey is considered delectable), while the fruit is eaten by many animals which then distribute the seed widely. The seeds themselves are covered with a coat impermeable to oxygen. There is no germination until the covering disintegrates.

Saw palmettos grow quickly on burned-over sites. Bears find the new sprouts very tasty. The plant was once considered something of a pest,

Serenoa repens, saw palmetto

though it did have a few uses: the Seminole of Florida ate the fruit, made dance fans and rattles from the leaves, and wove the leaf stems into baskets or used them for thatching material. The stems are a source of tannic acid extract and can be processed into a cork substitute.

Saw palmetto was not really valued until the early twentieth century, when certain substances in its fruit were discovered to be highly effective in treating benign enlargement of the prostate gland and chronic urinary tract infections. The extract from the dried berries, which had been used in Europe for years, was finally made a dietary supplement in the United States in 1994. When word got out that saw palmetto could also return sexual function to men with prostate gland trouble, "It was like a gold rush," rancher Everett Loukonen recalls. Anyone hoping to make a quick buck went in search of saw palmetto through Florida and Georgia. "You had people climbing over fences, cutting locks, trespassing," said Loukonen, a ranch manager for Barron Collier Partnership, which owns about 70,000 acres of land in south Florida. In the summer of 1995, there was a poor crop, and the prices for the berries skyrocketed. Some people made a fortune; others weren't so lucky. "There were people drowned, mauled by bears, bit by rattlesnakes, poked in the eye with the sharp leaves," he said. "It was bad."[227]

With about 2,000 tons of berries harvested from south Florida exported to Europe each year, the weedy little plant has grown into what some estimate is a $50-million-a-year crop. Though there is a movement afoot to have it declared an agricultural crop, so far the industry has operated quietly, with immigrant workers earning cash for berries, no questions asked, no records kept.

Bromeliaceae

Spanish Moss
Tillandsia usneoides, old-man's-beard, Florida moss, Southern moss, long moss

Tillandsia usneoides

is an epiphytic plant that lives and photosynthesizes above the soil by supporting itself on a host plant, such as a live or evergreen oak. The tiny gray-green stems appear as swags up to 25 feet long, covered with small scalelike leaves about 1 inch long. It has small, inconspicuous, fragrant yellow-green flowers. The pods contain windborne seeds.

Range: Southeastern states, Virginia to Florida, west to Texas.

Spanish moss drapes itself decoratively around trees, fence posts, and anywhere the wind blows these minute plants. Epiphytes, of which this is a really good example, are plants that seemingly live on air. They need a host, but they aren't parasitic. In other words, they don't kill what they use. Spanish moss, which isn't really a moss, is a member of the bromeliad family—along with pineapples. *Tillandsia* is the largest genus within this family with more than 500 species that we know of (new species are being discovered all the time). Spanish moss is the most widespread *Tillandsia* species, and grows from the Southern states right down to Argentina and Chile. Some species grow in the humidity of the Amazon rain forest; others grow on mountaintops in freezing weather. It usually uses a live tree as a host, but can be seen clinging to telephone poles and other structures as well, driven there by the wind.

What appears to be a delicate silvery-gray

veil or garland is actually a collection of thousands of individual plants. Though it is a much larger plant, it resembles lichen in that it collects water and nutrients from whatever is around, and carries out photosynthesis. If Spanish moss gets too heavy or dense, it can break a branch or block sun completely from the host plant.

The flowers are so minuscule they are hard to spot. After fertilization, the blossoms produce seedpods. As these dry out, they burst open, launching seeds into the wind. Each seed has a hairy tip called the coma (the plural is comae), which acts like a sticky parachute. Once the seed lands on a branch or rough surface, the coma holds it in place until it germinates. Then rootlets form, keeping it attached to the host.

At this early, vulnerable stage, most of the plantlets die, but the survivors put out a stalk with a wide shield to collect any drop of rain or fog. The water goes through the stalk into the body of the leaf and is absorbed by osmosis into the plant. Dust from rocks, the host, or the air makes its way into the plant, which then extracts calcium, nitrogen, phosphorus, and other available nutrients.

Animals such as snakes and bats live in the shelter of Spanish moss. Egrets and mockingbirds use the moss for bedding, and, like Edward Lear's man with a beard, the Parula warbler nests right in the moss itself. Insects burrow around inside it—one spider, *Pelegrina tillandsiae*, knows no other home.

Linnaeus named *Tillandsia* after the Swedish botanist Elias Tillands (1640–1693), who once got so seasick on a voyage he vowed he'd rather walk home (a stroll of some 1,000 miles). Forever after, Tillands was known to hate water. So it was a natural leap of Linnaeus's imagination to name a plant that doesn't seem to need water after this professor. Linnaeus also chose the name *usneoides* because the plant looked like lichen (*Usnea* is a type of lichen; see pages 555–556).

There are many variations on the story of its common name, all with much the same theme. According to Dennis Adams of the Beaufort County (South Carolina) Public Library. Rufus Joseph Tillman, who lived on the Gulf Coast (near Mobile, Alabama), created this romantic version to entertain his great-granddaughter, Sherry Hicks:

A Spanish soldier fell in love at first sight with an Indian chief's favorite daughter. Though the chieftain forbade the couple to see each other, the Spaniard was too lovestruck to stop meeting the maiden in secret. The father found them out and ordered his braves to tie the Spaniard high up in the top of an ancient oak tree. The Spaniard had only to disavow his love to be freed, but he steadfastly refused. Guards were posted to keep anyone—the chief's daughter above all—from giving food or water to the poor Spaniard.

He grew weaker and weaker, but he still would not renounce his love. Near the end, the chief tried to persuade him once more to stay away from his daughter. The Spaniard answered that not only would he refuse to disavow his love, but that his love would continue to grow even after death. When at last the Spaniard died, the chief kept the body tied up in the tree as a warning to any other would-be suitors.

Before long, the Indians noticed that the Spaniard's beard only continued to grow. The Indian maiden refused ever to take a husband—unless the Spaniard's beard died and vanished from the tree. As the years went by, the beard only grew stronger and longer, covering trees far from the Indian maiden's village. Legend says that when the Spanish Moss is gone, the Spaniard's love will have finally died with it.[228]

The Native Americans of the Southeast, including the Houma and the Seminole, twisted the fibers to make ropes and mats, and used a boiled-down mixture of the plant to reduce fevers. The early settlers mixed it with mud to create a mortar in their log cabins and, in some places where the homes have been left undisturbed, the Spanish moss has been found still in good condition after a passage of 150 years. Dried, it was used as a kindling and a mulch.

Because it allows cool air to circulate within its dense mass, Spanish moss was once in demand as a mattress stuffing. This was so common that when German Duke Karl Bernhard visited South Carolina in 1825, he noted that it was being exported to European mattress makers. Henry Ford even stuffed the upholstery of his Model Ts with Spanish moss. It is still used for crafts, mulch, or packing material.

Harvesting Spanish moss is a thriving industry. Pickers gather the moss with long poles during the winter, taking up to a ton of material from one tree. In the old days, they dried it on wires strung like clotheslines, reducing it to about 20 percent of its weight. Curing was done by various methods, including heaping up the moss, soaking it, and then burying it in pits. This loosened the scales and left the fluffy strands. Then it needed to be "ginned," like cotton. The moss was placed on a conveyor belt and moved toward a toothed cylinder partly enclosed in a drum. Fans blew out the dirt and straightened the strands. Finally, it was assembled into a bale of about 150 pounds. Today, it's usually dried in a microwave oven, which kills any insects lodged in it.[229]

Spanish moss has acquired a new use, as the plant equivalent of a canary in a coal mine: it is used to test for air pollution. Samples collected along highways such as I-95 in South Carolina showed high concentrations of lead. Moss collected from Highway 17 south of Charleston had high levels of copper, sodium, nickel, and manganese. Researchers are also investigating Spanish moss for medicinal uses: it may prove to be valuable in the treatment of diabetes.

Tillandsia usneoides, Spanish moss

Swamps and Wetlands

THE SWAMP FOREST

When I would recreate myself, I seek the darkest wood, the thickest and most interminable and, to the citizen, most dismal, swamp. I enter a swamp as a sacred place, a sanctum sanctorum. There is the strength, the marrow, of Nature.

Henry David Thoreau[1]

The ever-mysterious swamp forest is now a threatened landscape.

For hundreds of years after the arrival of the Europeans, anything that was swampy, wet, soggy, or even somewhat moist was viewed by settlers with extreme distaste. They considered the swamps and lowlands as breeding places not only for mosquitoes and disease but also of sin. Abandoning people in swamps as punishment is a common literary device, and a whole raft of movies features them as terrifying places of darkness and mystery (think *Cape Fear*). As a consequence, we have now spent 400 years draining anything that looks wet or mucky.

These woefully misunderstood wetlands (defined simply as places where water is at or near the surface all the time) are found from the tundra to the southernmost tips of South America, and in every continent except Antarctica. Wetlands come in many forms: as sloughs or potholes on the prairies, in alpine meadows in the mountains, as bottomland hardwood swamps in the South, at the edges of lakes and ponds, as groundwater saturating the soil (vernal pools), or as wet meadows (filled with herbaceous plants). Wetlands are classified according to the behavior of the water. Marshes are wetlands through which water flows steadily. In fens, water seeps slowly through decaying vegetation. Swamps have water flowing very slowly through them and come in two versions: those with shrubs and those with trees (living and dead). Bogs are stagnant and the water has a high acid content.

A swamp forest has water in it for most of the year. It is dominated by trees with shallow intertwining roots supporting each other. The forest wetland, on the other hand, is flooded for only a few months. The water table is high, sometimes as close as 6 inches from the surface. These areas are sensitive to winds which can blow the shallow-rooted trees over easily, but the stumps create nesting sites for a vast number of birds and insects.

Almost every squelchy area is now under serious stress, mainly because people have been so slow to realize the critical role they play in the local ecosystem. Wetlands are some of the most productive areas on the planet. They produce microbes on a gargantuan scale, feeding species from both land and sea—from plants, crustaceans, and mammals to amphibians, fish, and reptiles—and ultimately people.

Plants that adapt to a constant state of wetness are called hydrophytes. Their detritus decays into a rich soupy mix that forms a web of life for hundreds of species. These special habitats not only support an incredible number of species, but also prevent erosion and filter out soil and water pollution. Riverbanks and shorelines are protected from eroding by sedges, bulrushes, reeds, and cattails, which screen out pollutants such as heavy metals.

Wetlands also store vast amounts of carbon within their plant communities, moderating the climate around them, and offsetting some of the impact of global climate change. Inland, wetlands on flood plains and along streams act as sponges, storing up water, braking its flow, and slowly releasing it back into the system. Once upon a time, the Mississippi River could store water for 60 days, but that has now been reduced to 12 days because of the loss of wetlands.

The hydrology of saturated soils is complex, and every time these soils are drained for another, more "useful" purpose than nature intended, problems ensue. Birds' nesting sites are destroyed, places where fish and crustaceans spend their lives or part of them are wiped out, and food on which animals depend is removed.

It is thought that, in the 1600s, there were more than 220 million acres of wetlands in North America; almost half of these are now gone. Between 70,000 and 90,000 acres are still lost every year because of drainage, dredging, landfill, damming, and diking, making wetlands one of our most endangered habitats. And as the wetlands go, so goes the North American ecology. If wetlands are threatened, so are we.

Taxodiaceae

Bald Cypress

Taxodium distichum, common bald cypress, swamp cypress, Southern cypress, Gulf cypress, Tidewater red cypress, deciduous cypress

Taxodium distichum

grows 80 to 130 feet high. The leaves are bright yellow-green in spring and reddish-brown in autumn. The cones are globular or ovoid, 1 inch across, green to purple, brown at maturity. Bald cypress is found in swamps, but can survive in dry soil when planted as an ornamental.

Range: Texas to Florida, west to southern Illinois, Missouri, Arkansas, and Louisiana, north to Maryland, up the Mississippi River Valley.

State tree of Louisiana.

Bald cypress is the perfect symbol of the Southern wetlands. It belongs to the redwood family, which includes the coast redwood *(Sequoia)* and the giant redwoods *(Sequoiadendron)* and is sometimes called the redwood of the East. Though the others are evergreen, bald cypress drops not only its leaves, but even its twigs in winter—hence the epithet "bald." The scientific name *Taxodium* comes from *Taxus* (yew) because the leaves are similar to yew leaves. *Distichum* means "two-ranked": the leaves grow in two rows.

Bald cypress once had a massive range and some specimens grew to enormous girth and height. The few that survived the pillaging of the new settlements are hundreds of years old. One bald cypress in the Cat Island Swamp (south of St. Francisville, Louisiana) has a circumference of 53 feet, and is thought to be at least 1,000 years old. There are even rumors of trees 4,000 years old. If they exist, no one will reveal their location, because we know from experience there will be someone who wants a souvenir from the oldest or rarest or last of anything.

Taxodium doesn't produce seeds every year and, when it does, conditions have to be ideal for germination to occur—the land must be fairly dry but moist enough for the seedlings to thrive. Therefore several years may pass without any seedlings sprouting.

Bald cypresses living in watery areas often send up strange-looking protuberances known as "knees," which grow 3 or 4 feet above the level of the water. They were once thought to aerate the roots, but research has disproved this theory. They may provide stability, although pruning them doesn't seem to harm the tree. Theories abound, but nobody is quite sure what their real function is.

The trees growing in swamps provide excellent feeding areas and cover for waterfowl. Nearly all the nesting pairs of bald eagles in Louisiana select old-growth cypress for their huge nests.

When the Europeans first settled in southeastern North America, you couldn't see the

Taxodium distichum, bald cypress

forests for the bald cypress, according to William Bartram, writing in 1772:

On the west side it [the St. Juan River] was bordered round with low marshes and invested with a swamp of cypress, the trees so lofty as to preclude the sight of the highland forests beyond them; and these trees, having flat tops and all of equal height, seemed to be a green plain, lifted up and supported upon columns in the air.[2]

The bald cypress soon began to disappear under the Europeans' axes. At first, loggers cut only the easily accessible trees near waterways, but in the late nineteenth century, the industry was stepped up. After the Civil War, Southerners, hoping to rival the industrial might of the North, looked to their forests as a vast resource to be ruthlessly exploited. Southern leaders lobbied Congress to pass the Timber Act in 1876, which paved the way for the sale of large parcels of federally owned forests and provided tax breaks to lumber barons. An estimated 5.7 million acres of federal lands were sold off by 1888, not including millions of acres of swamplands.[3]

Using the "pullboat," a steam engine mounted on a barge, which allowed loggers to clear-cut in the swamps, huge tracts of land were denuded. In the South, logging went on all year round, and "skidder" towns grew up on barges moored near cypress-harvesting sites. About 1 billion board feet a year were processed in the South from 1905 to 1913.

The loggers became very efficient at removing 1,300-year-old trees. Bald cypress was used in building construction, fences, boats, river pilings, outdoor furniture, flooring, shingles, and greenhouses. Certain bald cypress trees were attacked by a fungus that gave them an unusual appearance. Their wood was known as "pecky" cypress and prized for use in paneling or furniture.

Bald cypress was called the "wood eternal" because it will endure in spite of being in contact with water. Water pipes in New Orleans made of this wood lasted from 1789 to 1914, and bald cypress shingles have been known to hold up for 250 years. An oil called cypressene is believed to give old-growth bald cypress its powerful resistance to decay and rot. Alas, second-growth bald cypress, which is virtually all we have left, lacks this quality. It is not known at what age or size this decay resistance develops—something we will never be able to find out.

By the 1920s, what had once appeared to be a limitless resource was logged out.

Timber scouts from the eastern portion of the Atchafalaya Basin [Louisiana] combing the area for suitable timber began meeting scouts from the western part of that enormous swamp, which people had long thought inexhaustible. Because loggers had used pullboats and skidders, the mucky forest floor had been scoured and disrupted to such a degree that young trees could not germinate in the place of the felled ancients. . . . Despite the timeless qualities of its "wood eternal," over 1.6 million acres of cypress bottomlands in Louisiana alone were liquidated in a matter of decades. The natural riches that had taken hundreds of years to create vanished. And with the primal forests went the black bears, the panthers, the ivory-billed woodpeckers, and countless other creatures of the swamps.[4]

Only a few virgin stands of bald cypress are left in state parks, the largest of which is Francis Beidler Forest in Four Holes Swamp, South Carolina. They call their 1,800 acres "original growth," but even there, some timbering was done back in the 1600s. The tree is protected in places such as Big Cypress State Park, Tennessee, and the Louisiana Nature Conservancy's

Cypress Island Preserve. Second-growth stands are valued for their ecological contribution, because bald cypress swamps soak up floodwaters, reducing flood damage, act as sediment and pollutant traps, and provide critical habitat for wildlife. As the bald cypress has proved, no natural resource is "inexhaustible," no matter what economic authorities like to tell us.

Platanaceae

Sycamore

Platanus occidentalis, American plane tree, buttonball tree, buttonwood

Platanus occidentalis

grows 80 to 150 feet high. The alternate simple shiny palmate green leaves have large irregular teeth, and are pale and hairy along the veins on the underside. The tree is monoecious: male flower clusters grow on short stalks of previous year's growth and the female flower clusters on short stalks of older branches. The fruit is a ball of closely packed, long narrow fruits (or achenes); each achene has a hair-thin hard seed coat. It grows in rich alluvial soils.

Range: from southwestern Maine west to southern Wisconsin and eastern Nebraska; south to Texas; east to northwestern Florida and southeastern Georgia.

Platanus occidentalis, sycamore

The Old World and the New World come together in this tree. In 1637, 1642, and 1654, John Tradescant the Younger, who worked at his father's famous nursery near Lambeth, traveled to the New World to look for fresh plant species to sell to English and European gardeners. On his first trip, he collected a sycamore (either the seeds or some cuttings) and took it back to En-

gland. He was the first to grow it as an exotic species. Thousands of trees were propagated from Tradescant's specimens.

It turned out that the North American *Platanus* hybridized easily with the Oriental plane tree *(P. orientalis)* that was already growing in London. By 1670, this new hybrid, *P. × acerifolia*, had become known as the famous London plane tree, distinguished from its American parent by the fact that it has pairs of little buttonballs rather than single ones.[5] The sturdy hybrid tolerated coal dust, industrial smoke, and compacted city soils. The mature trees rose to more than 100 feet, and became a favorite shade tree because of their vigor, immunity to fungal disease (anthracnose), and swift growth. One given in the mid–seventeenth century to the newly appointed custodian of the Oxford Botanical Garden, Jacob Bobart, stands today in the garden of Magdalen College, Oxford.[6] Its astounding rate of growth is legendary. Older trees have been known to grow around and engulf fences or benches that lean against them, creating sculptural shapes of their own.

Although it has the familiar paint-by-numbers patterning, the American sycamore is not quite as flamboyant and is not so much associated with city streets. (American sycamore also has nothing to do with the sycamore of the Bible; that's *Ficus sycomorus*.) *P. occidentalis* is more likely to be found on a "biomass farm" (a.k.a. tree plantation) in the Southeastern states, because it can grow a phenomenal 70 feet in 17 years. It is also the quintessential commercial tree in these businesslike times.

American sycamore has a massive trunk, larger in diameter than any other tree of the Eastern forest. It can live for 500 to 600 years, making it a treasure. The biggest known sycamore is in Jeromesville, Ohio, roughly halfway between Cleveland and Columbus. It is 129 feet tall, and 49 feet in girth.[7] Huge branches develop into a large open crown which, in the case of the Jeromesville tree, stretches to 105 feet.

As the tree ages, the thin red-brown outer bark falls off in patches to reveal the white, pale green, and ecru shades of its interior (a bit like skin with stretch marks). It thrives in rich soil along streams and in bottomlands, even if they are occasionally flooded. It was favored by settlers because it was an indicator of what they termed "sweet" or fertile soil.

Sycamores have such shallow roots that those in wetland areas can easily tumble over in a heavy wind. Their rapid growth also means that they tend to have divided trunks and enormous branches, which don't help their stability in stormy weather. They produce so many seeds, however, that with sun and good soil, they can generate masses of seedlings.

P. occidentalis is a monoecious tree: the dark red male and greenish female flowers grow on the same tree and pop out where the leaf emerges on separate stalks. An achene (a one-seeded fruit) is produced on the female and a distinctive buttonball is developed. The little buttonballs are actually hairy balls of fertilized achenes all clumped together clinging to the tree all winter long. As the balls disintegrate, each single-seeded nutlet comes loose. The little hairs act like propellers when the wind comes along and spins them off to a new home well away from the parent tree.

Sycamores cannot tolerate shade so it becomes necessary for them to secrete a toxin in their leaves and fruit—a herbicide that kills off local competition. This is called allelopathy. When the huge leaves drop to the ground and decompose, the toxin leaches into the soil, discouraging any other plants from invading the tree's space. As it matures, the sycamore develops large holes. Some become almost completely hollow. The Native Americans sometimes used hollow trees to make dugout canoes, and the colonists used them as pigpens, stables, or even temporary shelter, when they weren't amusing themselves with sycamore stunts:

Colonial record-makers crammed into sycamore trunks like Guinness-*bound fraternity brothers packing into telephone booths. Ohio, which seems to have held a monopoly on amazing sycamores, boasted specimens that could hold fifteen men on horseback, or forty men off.*[8]

Some pioneers, however, blamed the tree for giving them diseases such as malaria. No one, at that time, had made the connection between malaria and the mosquitoes that bred in the swampy lands favored by the sycamores.[9]

Sycamores appealed to Thomas Jefferson, who wrote to Martha Randolph from Philadelphia in July 1793: "I never before knew the full value of trees. My house is embosomed in high plane trees, with good grass below, and under them I breakfast, dine, write, read, and receive my company."[10]

As the colonies started to rebel against their British masters and British taxes, they designated a certain tree in every colony as a Liberty Tree. This was a spot where the Sons of Liberty could count on meeting each other to plot against the British. Many citizens, including those of Newport, Rhode Island, chose a sycamore as their Liberty Tree. Once the British soldiers were on to this, they retaliated by hacking down or setting fire to many of the designated trees.

In 1971, Stuart Allen Roosa, astronaut and former Forest Service smoke jumper, took hundreds of seeds (including sycamore seeds) into space with him on *Apollo 14*. In spite of the fact that his canister blew open on reentry, the seeds germinated. These "Moon" trees are now growing back on earth. One lives in Washington Square in Philadelphia, across from Independence Hall.[11]

Sycamores may have been tapped for their sap, but it was most likely a desperate measure in time of drought, as Euell Gibbons reveals in *Stalking the Wild Asparagus*:

On reading that certain Indian tribes used the sweet sap of the Sycamore (Platanus occidentalis) *for making sugar and sirup, I tapped a large, double-trunked sycamore . . . with one tap in each trunk, I soon had the two gallons of sap I wanted for the experiment. But when I boiled it down I discovered that I would have done about as well had I merely dipped up two gallons*

of water from the nearby creek. It gave off a maple aroma as it boiled, but the entire amount produced little over a tablespoonful of very dark-colored sirup which tasted like a poor grade of blackstrap molasses.[12]

Sycamore wood is extremely hard to work. Wood grain is created by the way the xylem cells line up. If all the cells lie in the same plane parallel to the vertical axis of the tree, the wood is referred to as splittable. If they are irregular, it is considered incredibly tough. With both elm and sycamore the growth alternates from one direction to another in succeeding years. This interlocking arrangement makes them among the most formidable woods to chop up. As a result, sycamore was seldom used as fuel.

Instead of kindling, the creamy white wood is used to make butchers' blocks, barber poles, rolling pins, and shipping crates. Some of its veneer has become the backs of violins. It often contains what is known in the wood biz as "lacewood," which is used for decoration in furniture. The interiors of Pullman cars were once exclusively decorated with this wood.[13]

Salicaceae

WILLOW
Salix spp.

Salix spp.

branch out to create shrubs and trees that can reach over 100 feet in height with a spread of 50 to 100 feet in some species; leaves are elongated, mostly pale yellow-green; they are dioecious with showy male (the pussy willows) and female catkins.

The weeping willow is a sure sign of an underground stream and a high water table. It has two moments of glory: the pale yellow new growth in spring, which makes a transitory golden background to the brief beauty of ephemerals around them; and as the last tree to turn golden yellow in autumn. *Salix* comes from the Celtic *sal-lis*, meaning "near water." Willows range from a few inches high (the **Arctic willow,** *S. arctica*), to behemoths (*S. nigra* at more than 100 feet).

The weeping willows we see most often are *S. babylonica*, an Asian import which has hybridized successfully with native species. "They hybridize freely," says Michael Dirr, "and it is often difficult to distinguish hybrids from species."[14] Around the world there are about 300 willow species, from the Arctic to the southern parts of South America and from Europe to Asia. About 250 species grow in the northern hemisphere. Of those, between 75 and 90 species, maybe more, are native to North America.

With water at their feet, they grow so rapidly it's possible to notice the difference in only a year and they may get to 80 feet in 15 years. They race for the sky in areas with at least 50 inches of rainfall a year, and thrive in soil that's often flooded or even below water level. They transpire huge quantities of water into the air, acting as a conduit from the water in the soil to the winds and the air around them.

This is a plant that reproduces ferociously between the ages of 25 and 75. The trees may be either male or female, but you can't tell male from female until they bloom. The male catkins emerge just before the leaves appear. Though the flowers have nectar and are usually insect-pollinated, the pollen can also be blown by the wind. Water carries off the seeds of those near riverbanks, taking them far from the parent trees. The light brown capsules split open to shed tiny green seeds with a hairy cover. If the trees are particularly stressed, they shed even more seeds than usual and the long silky hairs cover the ground like a thick fluffy blanket or are moved great distances by the slightest breeze.[15]

Willows have shallow roots, especially if they grow in clay soil. A sudden gust of wind can send twigs dive-bombing to the ground and a really stiff wind can break off whole branches. If the fallen twigs or branches sink into the soil, they will root. This is an incredibly tenacious plant, as those who live near them (I'm one) can attest.

Willow bark has been known as a medicine for hundreds of years. The Greek physician Dioscorides knew that ashes from willow steeped in vinegar would remove corns. In 1633, the English surgeon (and England's most famous herbalist) John Gerard wrote: "The leaves and barke of Withy or Willowes do stay the spitting of bloud whatsoever in man or woman, if the said leaves and barke be boiled in wine and drunke. The greene boughes with the leaves may very well be brought into chambers and set about the beds of those that be sicke of fevers, for they do mightily cool the heate of the aire."[16]

Because the branches bend so easily, willow withes have been used for thousands of years in basket weaving, the making of bows and arrows, and for the construction of furniture. The Plains Nations made smoking lodges of willow draped with buffalo skins. The Ojibwe in the North cured moose hide for their moccasins over a willow fire, and, in cold weather, fashioned willow snowshoes. The multifaceted willow could also produce water jugs, bags, pouches, and fishing nets.

Willow has traditionally been used by water diviners or dowsers to make their Y-shaped divining rods. Willow furniture fashionable once again, though it's never possible to tell how long a trend like this will last and it may pass out of fashion as quickly as it arose. Wattling, that is, weaving willow into fences, a very old craft transported from the old country, is also experiencing a comeback. And everywhere, it seems, willow sculptures can be found composting in an ever-so-chic manner. There is even the American Willow Growers Network, an association that shares information on the use and development of willow species and cultivars in different climates and growing conditions. They advocate the use of willow in timber production, windbreaks, and other ecological uses.[17]

Salicaceae

Black Willow
Salix nigra

Salix nigra

grows on average to 80 feet, but can reach 140 feet in ideal sites, with one to four trunks each 20 inches thick. The long narrow leaves have fine incurved teeth, green on both sides and hairy when young, although they get smoother as they age. In this species, the catkins, which come out in April to May, mature with the leaves and are slender and loose, about 1 inch long. The seed capsule is smooth on the outside.

Range: New Brunswick, southern Quebec and Ontario, northwest of Lake Superior, central Minnesota, south to Mexico, Texas, and northwest Florida.

The black willow is the largest of the native willows and one of the largest willows in the world. It is called a black willow because of its deep brown, almost black, bark. Thickets of black willow are colloquially known as willow bats or willow slaps, referring to their habit of doing the same to any interloper.

The Native Americans put dense willow thickets to good use. "Prince Maximilian zu Wied-Neuweid, sightseeing along the Missouri in 1833, noted nervously in his diary that the local Indians habitually lay in ambush among the river willows."[18] Today, the hunter in a duck blind may also use black willows to screen himself from his prey.

A major advantage of riverside willows is their effectiveness as erosion-controllers. The wide-spreading root systems can hold unstable riverbanks in place and keep obstreperous rivers within bounds. (This same invasive root system makes willow the bane of city water commissioners: willow is famous for wreaking havoc on septic systems and strangling water pipes.) In 1878, Captain Oswald Ernst of the Army Corps of Engineers, looking at the destruction of the banks of the Mississippi around St. Louis and the constant overflow, initiated the first erosion control program by planting willows along the banks.[19]

Always more of a medicinal plant than a serviceable timber plant, the bitter bark was used as a tonic in spring—its advocates believed it purged the blood of impurities. Willow bark was steeped and drunk in quantities for colds, to relieve asthma, and applied to a cloth on the forehead for headache. Again and again it is mentioned as a substitute for tobacco; a cure for sprains and bruises and vomiting; a treatment for rheumatism; and a gargle for sore throats. The substance in the bark is salicin, which is the precursor of acetylsalicylic acid, also known as aspirin. Did early healers recommend patients take two pieces of willow bark and call them in the morning?

Although most willows are too weak to be of value as timber, black willow is the exception. It does not split, yet it is lightweight. Today it is used for pulpwood, flooring, veneer, boxes, and crates. At one time, artificial limbs were made from lightweight, easily tooled willow. It makes excellent charcoal for medical or chemical uses, including gunpowder.

Modern willow artists include Heather Sanft of Nova Scotia, an anthropologist and craftsperson who re-creates historic willow objects, such as eel baskets and lobster pots (many of which are used in historic sites such as Fortress Louisbourg), develops contemporary willow sculptures, and builds living willow fences.[20] Other artists follow the tradition of willow basketmaking of the Native American peoples of the southwestern United States. These baskets are formed on a foundation of willow strips, and other fibers, such as sedge, bulrushes, and fern roots, are coiled around to form the sides. Many have intricate geometric patterns.

Salicaceae

Pussy Willow
Salix discolor

The first time children feel the velvety touch of the pussy willow's paws, they are hooked. According to a Polish legend, a cat whose kittens had fallen into a river when they were chasing butterflies was crying loudly at the edge of a river. Hearing her cries, the nearby willows swept their graceful branches into the waters to rescue the tiny kittens. The kittens held on tightly to their branches and were safely brought to shore. Each springtime since, goes the legend, the willow branches sprout tiny furlike buds at their tips where the tiny kittens once clung.[21]

Salix discolor is a multistemmed shrub or small tree which grows 12 to 25 feet tall with thick reddish to dark brown twigs. When young the leaves are hairy. Mature, they have raised veins beneath, are a shiny green on top, and about 2 to 4 inches long. The stalkless catkins grow 1 to 3 inches long, appearing before the leaves from March to May. The furry gray pussy paws are actually tiny male catkins. Flower shops sell the forced branches to be used in fresh or dried flower arrangements destined to gather

dust as the year advances and the willows stay intact. Children use them to make little mice or glue them to construction paper in countless kindergarten classes.

Like black willow, pussy willow contains salicin. Anthropologist Huron Smith tells us that among the Forest Potawatomi, the bark of the pussy willow is a universal remedy. "The root bark is boiled down to make a tea, which is used in stopping a hemorrhage." He also notes another intriguing use: "The buds have been considered anti-aphrodisiac."[22] The European settlers seized on this quality and, right up to the 1930s, some physicians recommended it to quell overly libidinous youth, or those who had frequent night emissions. This use may either reflect the sexual strictures of the Victorian period, or it may indicate an attempt to stem the increase in sexual disease at that time.

Range: **Labrador east to Hudson Bay in Ontario to British Columbia, south to Delaware, Kentucky, Missouri, South Dakota, and Montana.**

Hamamelidaceae

Sweet Gum
Liquidambar styraciflua

Liquidambar styraciflua
grows up to 100 feet tall (and sometimes up to 140 feet), with a diameter of 3 feet and a canopy up to 40 feet wide in a pyramid shape. The alternate leaves are starfish-shaped, 3 to 6 inches long, shiny green with rusty hairs below; they turn brilliant red or magenta in fall. The tree is monoecious and flowers in April or May; the flowers are an insignificant greenish color. The small winged seeds are carried on the wind.

Liquidambar styraciflua, sweet gum

Range: Connecticut south to central Florida and eastern Texas, as far west as Missouri, Arkansas, and Oklahoma, and north to southern Illinois.

Sweet gum oozes a sticky sweet substance just asking to be picked off and stuck in the mouth for chewing. Anyone who grew up in the Southern states knows that it was pointless to resist temptation. Like chestnuts in Northern states, the seedballs seem designed as toys. To a child, the leaves look like starfish.

It is known as a pioneer tree—the first to move into an abandoned field or logged-out area in uplands and coastal plains. As one of the most important hardwoods of the Southeastern states, sweet gum is amazingly adaptable. It can live in many different kinds of soils and sites, though it is happiest in moist alluvial clay and loamy soils. It

has very thin bark which burns easily, making it way more susceptible to fire than most other trees.

Sweet gum is a prime example of stranding—this is a species that was widely distributed, then almost completely wiped out by massive geological forces, and now grows only in a few locations and usually in small numbers. *Liquidambar* once grew all over the planet and even shows up in the Eocene rocks (55 million years old) of Greenland, which, at the time, was subtropical. Given what we know from the fossil record, during the various epochs of glaciation, at least 20 species of this tree alone became extinct. What remains are two species besides our North American form—one in Turkey, the other in Formosa.[23]

Liquidambar styraciflua grows in Mexico and South America and was mentioned in the account of the meeting between Cortez and Montezuma in the sixteenth century. According to Don Bernal Díaz del Castillo, "After he had dined, they presented to him three little canes highly ornamented, containing liquidambar, mixed with a herb they call tobacco, and when he had sufficiently viewed and heard the singers, dancers, and buffoons, he took a little of the smoke of one of these canes."[24]

Sweet gum exudes a yellow sticky resin called storax or styrax or copalm balsam, which has long been known for its medicinal properties. Francisco Hernández, an herbalist whose work was published in Europe in 1651 and who worked in Mexico from 1571 to 1575, wrote: "Added to tobacco, it strengthens the head, belly, and heart, induces sleep, and alleviates pains in the head that are caused by colds. Alone, it dissipates humors, relieves pains, and cures eruptions of the skin. . . . It relieves wind in the stomach and dissipates tumors beyond belief."[25]

Native Americans of the South used storax to treat wounds, warts, and other skin problems. The resin also made an effective expectorant to relieve congestion caused by colds. Mixed with tallow or lard, it was transformed into an ointment used to treat hemorrhoids and ringworm. Like Montezuma, many Native Americans flavored their tobacco with the gum.

The settlers quickly realized there was a market for storax and an industry was born. The resin was used to make medicinal syrups and ointments and chewing gum. In spring the trees were peeled and after a few days the highly flammable gum was scraped off, gently heated, and then canned. For three to five years the healthy young trees would survive this treatment and grow back new bark. Later they'd be cut for lumber.[26] During the Civil War, physicians made a preparation with storax to treat dysentery. The whole business waned as cheaper resin from Asia became available, but when supplies from Asia were cut off during the Second World War, the industry was temporarily revived.

The thick white sapwood hasn't much to recommend it, but the heartwood, which develops over 60 or 70 years, becomes pink and can be stained to look like black walnut. It is used for veneer and furniture, as well as for boxes, crates, and plywood.

The trees make wonderful ornamentals. Frederick Law Olmsted introduced sweet gum into Central Park, New York, and in Philadelphia it was planted along Independence Mall. Alexander Hamilton was so fond of sweet gum that he tried to have it declared the national tree of the United States, but the idea never caught on. Today, landscape architects and urban foresters are divided: some recommend the tree for city streets because it is attractive; others shun it because they feel its seedballs create a nuisance on the streets when they fall.

Rubiaceae

Buttonbush

Cephalanthus occidentalis, riverbush, buttonball, pond dogwood, pinball, little snowball, honey bell

Cephalanthus occidentalis
is a shrub that grows 3 to 6 feet high and spreads equally wide. The shiny green leaves are lance-shaped or ovate and 2 to 6 inches long. The flowers are bright white and spherical, about 1½ inches in diameter; they come out in July and August. The fruit turns into a brown ball in September and stays on the shrub all winter. Grows on riverbanks or in wet areas.

***Range:* Ontario and Quebec, southern Nova Scotia, and New Brunswick, south to the Gulf Coast and west across the Great Plains states.**

Cephalanthus occidentalis, buttonbush

Cephalanthus means "head flower," and the white flowers are spherical enough to be considered a head shape, though bristling with antennae (the stamens). *Occidentalis* means "western" as designated by Linnaeus. "Western" was applied to several North American species. To him, it seems, anything across the pond from his native Sweden was western. Butterflies, bees, and hummingbirds swarm around the creamy-white flowers ensuring pollination and, later, seeds for the dozens of birds who flock to it for forage. It's also a good nesting cover for water birds.

The Meskwaki and other native peoples traditionally used buttonbush bark for making laxatives, emetics, and diuretics, and for curing skin, bronchial, and venereal diseases. The early settlers realized its ameliorative effects and used it as a substitute for quinine in treating malaria. This makes sense since it is a member of the coffee family, the same family as the plants that produce quinine *(Cinchona)*. Sometimes, however, the cure must have been worse than the disease, because the bark also contains cephalathin, a poison that can induce vomiting, paralysis, and convulsions.

Nyssaceae

Black Tupelo
Nyssa sylvatica, black gum, sourgum, pepperidge, tupelo, tupelo gum

Nyssa sylvatica
grows 30 to 50 feet with a spread of 20 to 60 feet. The leathery leaves, which turn bright red in fall, are smooth-edged and alternate, and cluster together at the end of its twiggy branchlets. It flowers from April through June. The fruit is blue-black, about ½ inch long, ripening in September or October. It grows in well-drained, light-textured soils.

Range: southwestern Ontario, the Atlantic coastal states from Maine to Florida, west to eastern Texas, and up the Mississippi River Valley to Arkansas and Tennessee.

With springtime's lavish bounty, it's easy to overlook black tupelo's subtle flowers, but in the fall its scarlet leaves are riveting. The pigment anthocyanin is primarily responsible for most fall color. It is found in the cell sap and is a brilliant red. For most of the growing season, the bright green of the chlorophyll in the leaves masks the anthocyanin. But in the fall, when temperatures drop, the leaves stop making chlorophyll and the tree's scarlet beauty is revealed.

Black tupelo has small, greenish-white male and female flowers borne on separate trees but it also has perfect flowers—those that contain both male and female parts on each tree as well (this is known as being polygamodioecious, to be botanically precise). The brilliant autumn color signals to birds that the fruit is ripe and ready. Passing birds scoop up the blue-black fruits quickly and excrete the seeds far from the parent plant.

As the tree ages, it dies from the top down, getting shorter and shorter as it grows older and older but developing long taproots and a swollen base.

Decay attacks the heartwood first, so hollow trees are common. Southerners used to cut such trees into short sections, then stand them on boxes with a board on top as beehives. As a result, in some places, beehives were known as bee gums. The flowers of black tupelo are a source of nectar for bees and in certain locations, such as the Apalachicola River area of west Florida, tupelo honey is produced in commercial quantities.

The early colonists probably created the name "tupelo" for the tree from the Creek

Nyssa sylvatica, black tupelo

Nation words *ito*, "tree," and *opilwa*, "swamp." Linnaeus chose *Nyssa*, the name of a water nymph in classical mythology, to make the connection with the water, although the trees really grow better beside, not right in, swamps and wetlands. Other common names, such as "black gum" and "sour gum," refer to the color and taste of the berrylike fruits. No one seems to know where the common name "pepperidge" comes from. Some sources suggest it is old English dialect for "barberry," but that's a completely different plant, of the genus *Berberis*. Tupelo is, of course, the name of the town in Mississippi where Elvis Presley was born in 1935; the name was also made famous by blues singers such as the late John Lee Hooker, who were forever coming from or going to Tupelo.

The bark is an emetic, and the Cherokee gave it to children to get rid of worms. Liquid obtained from the roots was also used as an eye medicine. The edible berries are still made into preserves. The wood is light and strong, but hard to work because of its interlocking grain. It is the grain, however, that makes it durable and shock-resistant. It was once used to make plows, gunstocks, and chopping bowls, but today is generally used for lumber, veneer, railroad ties, and pulp and paper. When stained, it makes a pretty fair substitute for mahogany.

The black tupelo's near relation, **water tupelo** (*N. aquatica*), can be found in wetland and bottomland areas and prefers a warm, humid climate where summers are long and hot and winters are short and mild. Louisiana once had vast swamps in which bald cypress and water tupelo were the main species, but many of the trees were felled for timber and little remains of the cypress-tupelo swamps today.

Clethraceae

Clethra alnifolia

is a deciduous shrub that grows up to 8 feet tall and 8 feet wide. The bark is reddish-orange or gray. The alternate oval leaves are dark green, 1½ to 4½ inches long, turning yellow in fall. In late summer it produces fragrant white or pale pink flowers in clusters 4 to 6 inches long. The fruits are little brown capsules that look like peppercorns.

Range: Nova Scotia, southern Maine, southern New Hampshire, Massachusetts, southeastern New York, eastern Pennsylvania, and Florida to east Texas.

Clethra is one of the forest's late-summer secrets. It blooms in August and the sweet-smelling flowers attract bees, which in turn encourage beekeepers to cultivate the plants. There are two native species: *C. alnifolia*, a wetland plant, and *C. acuminata*, which grows along watercourses in montane areas. The scientific name of *C. alnifolia* seems redundant, since the genus name comes from *klethra*, the Greek name for alder (the leaves look like alder leaves), and *alnifolia* is Latin for "alder-leaved."

One of the common names is "poor man's

Clethra alnifolia, sweet pepperbush

soap." When rubbed together, the flowers can create a sort of soapy lather with a bit of elbow grease. Today the plant is used in landscaping because it is attractive and fairly easy to care for.

The pepperbush's mountain relative, *C. acuminata*, is a slightly larger version of *C. alnifolia*, with a cinnamon-colored bark. It is sometimes known as **cinnamon clethra**. According to Daniel Moerman, the bark of this species was used by the Cherokee in a decoction to induce vomiting or, in combination with wild cherry bark, to break a fever.[27]

Aquifoliaceae

American Holly
Ilex opaca

Ilex opaca

grows up to 50 feet tall with shiny, spiky, dark green leaves 2 to 4 inches long, small white flowers in summer, and bright red berries in fall. It can survive in thin soil with little humus or even in sand, but the biggest grow in the rich bottomlands and swamps of the coastal plain. It is a shade-tolerant understory plant.

Range: **New England, south throughout the coastal plain, Piedmont and Appalachian Mountains, south to mid-Florida, and west to Texas and southeast Missouri.**

State tree of Delaware.

The American holly was yet another sign to the Pilgrims upon landing that this wild and savage place did have some possibilities. They arrived before Christmas in 1620 and were cheered by the sight of familiar-looking holly. The dark evergreen leaves and bright red berries were used then, as they are now, for Christmas decor.

There is still a tradition that if you plant holly near the entrance to your house, the house will be protected and a place of good luck.

Depending on where it grows, holly may be evergreen (in Florida) or may shed some of its leaves in winter (in the Appalachian Mountains). These long-lived trees are dioecious: the female bears the bright red berries used for Christmas decoration, but only when fertilized by a male of the species. Male and female flowers are produced on separate plants, so holly needs bees, wasps, yellow jackets, and night-flying moths to move the pollen around. The berrylike fruits (drupes) contain four seeds that take 16 months to three years to germinate. Birds, squirrels, deer, and, especially, wild turkeys love the drupes (another way to stuff a turkey).

Ilex opaca, American holly

Holly has both white sapwood and white heartwood, making it one of the palest woods known. It is so finely grained it's almost impossible to see the growth rings, which makes it such a luxurious furniture inlay. Best-suited for carved scrollwork, scientific instruments, engravings, and handles, it is also useful in many other forms of fine woodworking. When dyed black, it looks like ebony, and is used for piano keys, violin pegs, and fingerboards.

Roasted holly leaves, although they contain no caffeine, were used as a tea substitute during the American Civil War. (Another holly, *I. paraguayensis*, is used in South America to make the popular tealike drink *maté*.) The tea was also used as a laxative, an emetic, and a diuretic. Tea made from the bark was used to treat malaria and epilepsy or as a wash for sore eyes and itchy skin. Peter Kalm, who found holly in wet places in the forests of New Jersey, noted that the Swedes who lived there boiled the dried leaves for a pleurisy medicine. A relative, *I. verticillata*, was known as **feverbush** because a decoction of the bark was used by many native peoples to reduce fever. The Iroquois used the bark as an antiseptic wash.

Elizabeth White, most famous for developing the cultivated blueberry, was also active in rescuing the native American holly from obscurity. In 1947, she helped found the Holly Society of America. Holly plants are protected in Maryland and Delaware, to save them from unscrupulous marketeers who whack away at plants before Christmas and sell off the branches. Proper nurseries buy from holly orchards, where it is grown plantation-style to stem the vandalizing of the wild populations. It has been cultivated since the eighteenth century (both Washington and Jefferson planted holly bushes) and today there are more than 1,000 cultivars available in North America.

Annonaceae

Pawpaw
Asimina triloba, custard apple, false banana

Asimina triloba

grows 20 to 30 feet high and is distinguished by large, long, droopy, 12-inch leaves that look almost tropical and turn golden in autumn. The flowers are 1 to 2 inches across and emerge between February and May. The fruit, which looks like a long green potato, develops between July and September. It grows on the slopes of valleys and on floodplains, in deep rich, damp soil.

Range: eastern Pennsylvania west to Michigan and Illinois, eastern Kansas south to Florida and Texas, and north into southern Ontario.

"Picking up pawpaws, put 'em in a basket . . . way down yonder in the pawpaw patch." According to some versions of this song, it's "put 'em in your pocket," but you'd need huge pockets. Pawpaw is one of the largest fruits native to North America.

A dense deciduous tree or shrub, it tends to form thickets or colonies by cloning itself with suckers from a single trunk. It's a pretty strange-looking tree: the hairy young twigs have enormous leaves—up to a foot long—and rather smelly, meat-colored flowers that appear before the leaves in midspring. Pawpaws are pollinated by flies or beetles. Few diseases or pests bother this plant—even the gypsy moth spurns it. But that gorgeous black-and-white butterfly, the zebra swallowtail, has larvae that eat only pawpaw leaves.

Pawpaws grow in forests filled with black tupelo, Ohio buckeye, honey locust, and Kentucky coffee tree. Although pawpaw is generally a shade-tolerant plant, once the canopy becomes too dense in an old-growth forest, it tends to die out. It breeds either by sexual reproduction, requiring a male and female plant to cross-pollinate, or by cloning.

The fruit is a lumpy greenish-yellow thing about 6 inches long, borne in clusters, a bit like bananas. When they ripen, they smell slightly sweet, but once on the ground, they make a squishy mess. If chewed, the seeds are poisonous, but swallowed whole, they pass through the digestive tract without a trace. Birds, raccoons, gray foxes, opossums, squirrels, and black bears seem to enjoy the fruit, while deer and beavers stick to eating the bark.

Fans of the pawpaw say they taste something like a mix of banana, mango, and pineapple. Euell Gibbons, author of *Stalking the Wild Asparagus*, quotes a "Hoosier lad" who described pawpaws in this way: "They taste like mixed bananers and pears, and feel like sweet pertaters in your mouth." Gibbons finds that those who have never before tasted the fruit like it very ripe and very soft, but warns that the heavy fragrance becomes cloying after a few days and cautions against ripening it in the house.[28]

The Cherokee and Iroquois are among the native peoples who cultivated the pawpaw and ate the fruit. They selected the best to make sure their trees produced the most succulent fruit. The seeds moved with them as they traveled from summer to winter camps, so the new plants are often found in unlikely locations. For example, the Ohio Valley would seem right out of the tree's usual range, but at the end of the nineteenth century, fishermen in the Ohio Valley were using the bark to string fish on.

The Iroquois used the mashed fruit in small cakes, which they dried. Raw or cooked fruit was dried in the sun or close to a fire and used as a portable food by hunters. Alternatively, they soaked the dried fruitcakes in warm water and cooked them into a sauce or mixed them with

Asimina triloba, pawpaw

corn bread. The Cherokee also used the inner bark to make strong ropes and string.

Pawpaw fruits saved Lewis and Clark from starvation in 1806, and Daniel Boone and Mark Twain were reported to have been pawpaw fans.[29] Perhaps they suffered from constipation, pawpaw being something of a laxative. Early explorers sent specimens to England, but somehow it never caught on there. Even in North America pawpaw cultivation languished during the nineteenth century, and picked up in the Great Depression, when people once again began growing native fruit trees. The astonishingly brittle roots make transplantation from the wild almost impossible. Botanists since 1900, however, have been selecting the best clones from the wild and grafting their way to an improved pawpaw hybrid.

Neal Peterson, who is considered the foremost pawpaw expert in the world, founded the PPF (Pawpaw Foundation) in 1988. The Foundation oversees two orchards at the University of Maryland with 1,900 pawpaw trees, gives advice to gardeners, and even holds taste testings. Its members hope that the pawpaw may regain its popularity and become an important fruit crop. Pawpaw recipes abound. They can be used like bananas in most banana bread recipes. *Leo Zanelli's Home Winemaking from A to Z* even includes a recipe for pawpaw wine.

It may one day be grown, not as fruit, but as a medicinal plant. Pawpaw's seeds contain an alkaloid called asimicin, which can kill insects (the powdered seeds are excellent for treating head lice) and has been used in chemotherapy against cancer. Dr. Jerry McLaughlin, professor of pharmacognosy at Purdue University, says, "One molecule of asimicin is enough to kill a cancer cell."[30]

WETLANDS

Alismataceae

Arrowhead
Sagittaria latifolia, broad-leaved arrowhead, wapato, duck potato, swamp potato, water nut

Sagittaria latifolia
grows up to 4 feet tall. Large, three-pointed leaves arise from the base of the plant on stalks. The flower stalk emerges in the middle of the leaves, with whorls of small white flowers, each with three petals and yellow centers, about 1½ inches wide. The roots and lower leafstalks are usually submerged in the water of a pond, stream, or swamp.

Range: throughout the continental United States (except Alaska) and in all Canadian provinces.

Sagitta means "arrow" in Latin, and with its three-pointed, arrow-shaped leaves, this plant is unmistakable. "Wapato" is the name given to the plant by the Chinook people of the Pacific Northwest, who ate the potatolike tubers.

European settlers brought their animals with them when they first came to this country. This in itself had a serious impact on the local flora. Peter Kalm noted the plant in the 1740s when hogs brought over by the European settlers rooted them up, much to the annoyance of the Indians. The hogs weren't the only ones competing for the plants. Muskrats and beavers also dug them up and stored them in caches. These animals, however, were extremely useful to the Ojibwe, who let them do the work of digging, then raided their stashes. Eating the tubers with deer meat was as common as eating potatoes with roast beef is today.

In the early nineteenth century, Lewis and Clark observed the peoples of the Pacific Northwest harvesting the arrowhead:

This bulb, to which the Indians gave the name of wappatoo, is the great article of food, and almost the staple article of commerce, on the Columbia [River]. . . . It is collected chiefly by the women, who employ for the purpose canoes from 10 to 14 feet in length, about 2 feet wide, and 9 inches deep, and tapering from the middle, where they are about 20 inches wide. They are sufficient to contain a single person and several bushels of roots, yet so very light that a woman can carry them with ease. She takes one of these canoes into a pond where the water is as high as the breast, and by means of her toes separates from the root this bulb, which on being freed from the mud rises immediately to the surface of the water, and is thrown into the canoe. In this manner these patient females remain in the water for several hours, even in the depth of winter.[31]

Sagittaria latifolia, arrowhead

When cooked, the tubers taste a bit like sweet chestnuts. The Indians made them into a gruel, or dried and ground them into a powder and mixed them with flour to make bread. One delicacy was to slice the boiled roots into thin sections and string them on ropes to dry in the sun, a bit like apples. Sometimes they were cooked with maple syrup to make a dish similar to candied sweet potatoes. A poultice made of the roots was used to treat wounds and sores.

Araceae

Wild Calla

Calla palustris, water arum, swamp robin, water dragon

Calla palustris

grows in ponds, bogs, and swamps. The roots lie beneath the surface of the water and the stem rises up to 1 foot above. The large, glossy, dark green leaves are heart-shaped. The flower, which appears in late spring or summer, is a white spathe surrounding a yellow spadix. In late summer and fall, the flower gives way to a cluster of red berries.

Range: all Canadian provinces, south to Colorado, Texas, and Florida.

This relative of the Jack-in-the-pulpit pokes up through the water of a bog. If the plant looks familiar, it's because it has more than a passing resemblance to the calla lilies beloved of florists (*Zantedeschia*, which come from South Africa).

Calla palustris, wild calla

Its spadix of tiny yellow flowers is surrounded by a large white, petal-like spathe. Unlike most bog plants, this one has shiny leaves which make it stand out. It has submerged rhizomes and the roots pop up at the nodes. If you peer at it closely, you might see a snail crawling up and down the plant, attracted by the somewhat stinky smell. Snails fertilize the plant as they travel slowly from plant to plant.

The fruit is a cluster of bright red, attractive berries containing calcium oxalate crystals, which, if eaten, can cause lips to swell and the mouth to feel as if it's on fire. The leaves, however, are harmless, and have been used to make a tea to ease the symptoms of flu, colds, and arthritis. The Potawatomi used them to make a poultice to soothe swellings.

The rhizomes also contain calcium oxalate, but Native Americans discovered that if they were roasted, the toxic effect was neutralized. The ever-observant Peter Kalm noted:

When they are fresh, they have a pungent taste, and are reckoned a poison in that fresh state. Nor did Indians ever venture to eat them raw, but prepared them in the following manner: they gathered a great heap of these roots, dug a great long hole; they made a great fire above it, which burnt till they thought proper to remove it; and then they dug up the roots and consumed them with great avidity. These roots, when prepared in this manner, I am told, taste like potatoes. The Indians never dry or preserve them, but always dig them fresh out of the marshes, when they want them. . . . It is remarkable that the arums . . . are eaten by men in different parts of the world, though their roots, when raw, have a fiery pungent taste, and are almost poisonous in that state. How can men have learned that plants so extremely opposite to our nature were eatable, and that their poison, which burns on the tongue, can be conquered by fire?[32]

This is one of the questions that fascinates anyone interested in the plants of North America: how did people ever discover that otherwise poisonous plants were edible? Apart from observation of animals, what brave soul would try something completely unknown and then proceed to experiment on its various virtues? Hunger was perhaps one driving force, but ethnobotanist Nancy Turner says that the native peoples of North America were true scientists. They tested plants over and over again until they fully understood their properties.

Cyperaceae

Bulrush
Scirpus spp., tule

Scirpus spp.
Some *Scirpus* spp. have triangular cross sections in the stems; other species have round stems. The tall stems stand up like spears, and the brown flowers emerge from one side or the top of the stem. They range in height from 1 to 5 feet. Usually found standing in water in a pond or swamp.

Range: Various species found in all American states and all Canadian provinces and territories.

Different species of *Scirpus* grow in different parts of North America, and it's hard to find a bog or swamp that doesn't have some representative of the genus. These are not the bulrushes of Moses-in-the-bulrushes fame, which were a type of papyrus *(Cyperus papyrus)*. Instead of finding babies in baskets, you are more likely to find birds' nests, or the haunts of otters and muskrats in North American bulrushes.

Bulrushes don't depend on seeds to invade new territory. The tubers extend throughout an

Scirpus spp., bulrush

area and more shoots come up each spring, interlocking with each other to form a dense mat. These plants are famous as land builders, anchoring soil in place.[33]

Native Americans used bulrushes such as *S. acutus* (**hardstem bulrush**) or *S. validus* (**soft-stem** or **great bulrush**) to make mats and baskets, and ate the rhizomes. Iroquois in the Lake Ontario area boiled the rhizomes to make a syrup that they ate with cornmeal. The Wood Cree of east-central Saskatchewan ate the stem bases fresh, and the Chipewyan of northern Saskatchewan also ate the top of the rhizome just below the stem. "It has been gathered at Black Lake and Riou Lake, and people traveling by canoe down the rivers used to stop and pull them up to eat. If it was red instead of white then it was too old and tough, and not good to eat."[34] The rootstocks of *S. maritimus* and other species were ground into flour.[35]

Children's uses for bulrushes and cattails are endless—from sword fights to torches to competing to see how far the seeds will blow away—but it probably doesn't occur to most kids to eat the plant. However, wild-food foragers recommend bulrushes. Melana Hiatt explains how to prepare them:

Harvest the roots, peeling and cutting the older roots into sections. Crush them and boil in water until you have a white gruel. Separate out any fibers and dry the gruel for flour. When camping you can add oil and water to the flour with a bit of baking powder and salt until you reach a consistency of biscuit dough. Wrap some around a stick and roast in your camp fire. I add wild garlic and other wild herbs to the batter and serve with butter. Or mix your dough and wrap around hamburger and cheese or fish and a thin lemon slice, cover with foil and lay on the edges of your coals until cooked or in flames.[36]

Ericaceae

Swamp Laurel

Kalmia polifolia, bog laurel, pale laurel, Alpine laurel, Western mountain laurel, bog kalmia

Kalmia polifolia

is a low, evergreen, spreading shrub up to 20 inches tall. The dark green, glossy, leathery, elliptical, opposite leaves emerge directly from the stem without a leafstalk and curl under slightly at the edges. The pink or purple, five-petaled, cup-shaped flowers appear at the top of the stem in clusters. Each flower is less than 1 inch across. Found in bogs, sometimes in coniferous forests.

Range: Alaska to Newfoundland, south to Oregon, Minnesota, and New Jersey.

Kalmia latifolia, mountain laurel, a related species

Kalmia is named for Pehr (Peter) Kalm, who is one of the true heroes of North American botany. He was a Swedish professor of natural history and economy who was the first to collect and record the flora of Canada. He was a student of Linnaeus and eventually taught at Abo in Sweden. In 1748, the Swedish Academy of Sciences sent him to North America to collect the seeds of economically useful hardy plants that could possibly thrive in Sweden's northern climate. He spent four years in eastern North America and married the widow of a Swedish clergyman, whom he met in New Jersey. His account, *Travels into North America*, was published in Sweden from 1753 to 1761, and later translated into English. It is a wonderful, down-to-earth record of plants and people. He brought back hundreds of specimens to show Linnaeus and, in gratitude, Linnaeus named a whole genus after him.

Swamp laurel is a plant of distinction and beauty, which has a fascinating method of polli-nation. The pretty pink flower has 10 stamens, each one tucked into a kind of pouch. As they grow, the tension within the pouch increases. When an insect lands on the plant, the stamens explode from their pouches, flinging their pollen from all angles toward the center and onto the insect's body. The insect heads off toward another flower, carrying the pollen to a waiting pistil.

The plant contains andromedotoxin, which can cause vomiting, low blood pressure, convulsions, paralysis, and, occasionally, death. Even honey made from the flowers can be toxic. According to English writer Maude Grieve in her book *A Modern Herbal*, some native Indian tribes used it to commit suicide. The leaves, however, were used externally in poultices and skin washes, to treat skin diseases, sores, and wounds. In minute doses, an extract of the leaves has also been used by qualified practitioners to treat internal hemorrhages and diarrhea.

K. latifolia, also known as **mountain laurel** or **spoonwood**, is a close relative. "The spoon tree, which never grows to a great height, was seen today in several places." wrote Kalm. "The Swedes here have named it thus, because the Indians used to make their spoons and trowels of its wood."[37]

Kalmia species are favorite garden plants in Europe, stemming, no doubt, from the fad for the American Garden taken up by English gardeners in the late nineteenth century. This style used native plants of North America, usually large-leaved evergreen plants. The fashion led, in some cases, to the eradication of plants in certain areas, as plant collectors grabbed too many specimens, but also to the introduction of many new cultivars developed by nurseries in Europe.

Iridaceae

Blue Flag

Iris versicolor, flag lily, poison flag, water flag, liver lily, snake lily, American fleur-de-lys

Iris versicolor

grows up to 3 feet tall. The stalk and leaves arise from a rhizome. The pale green, slightly fleshy leaves are shaped like swords. The flowers, which appear from May to August, are blue or violet, similar in shape to garden varieties of irises. The fruit is a three-angled capsule. Found in freshwater or brackish marshes, wet meadows, and forested wetlands.

Range: Manitoba to Newfoundland, south to Minnesota and Virginia.

Iris was the goddess of the rainbow, and irises were said to spring up on the earth wherever she walked. Irises (not lilies) are also the flower that are stylized as the fleur-de-lys emblem, first of the French monarchy, and today of the province of Quebec.

The root of the iris contains iridin, which is a toxic chemical, but also a medicinal drug in very small doses. In the southern United States, an extract from the root was used as a cathartic and an emetic by the Native Americans, including the Ojibwe, who also carried pieces of the root as a charm. They believed that the scent of the iris kept snakes away.

Iris roots can also be used to make a poultice to relieve pain and swelling from injuries. According to Peter Kalm, writing in 1749,

The Indians make use of the iris root as a remedy for sores on the legs. This cure is prepared as follows. They take the root, wash it clean, boil it a little, then crush it between a couple of stones. They spread this crushed root as a poultice over the sores and at the same time rub the leg with the water in which the root is boiled. . . . It is the blue iris, which is extremely common here in Canada, that is used for this purpose.[38]

William Bartram mentions a dramatic incident in 1775, where the iris root figured prominently.

Iris versicolor, blue flag

The town [of Attasse, a Creek Indian town] was fasting . . . taking medicine to avert a grievous calamity of sickness. [They had] laid in the grave an abundance of their citizens. They fast seven or eight days, during which time they eat and drink nothing but a meagre gruel, made of a little corn flour and water, taking at the same time by way of medicine or physic, a strong decoction of the roots of Iris versicolor, *which is a powerful cathartic. They hold this root in high estimation, every town cultivates a little plantation of it, having a large artificial pond just without the town planted and almost overgrown with it.*[39]

In his groundbreaking book *American Indian Medicine,* Virgil Vogel says the pulverized rootstock of blue flag when mixed with saliva was used as an earache remedy or an eyewash and for sores and bruises. Among the Penobscot, blue flag was a panacea for all kinds of ailments. The roots were kept in every home and steaming them was believed to keep disease away.[40]

I. versicolor and *I. virginica* found their way into the United States Pharmacology from 1820 to 1895 as blue flag. Resins were named irisin or iridin and used as a diuretic and purgative. These days irises are confined to gardens, where they attract plenty of bees.

Acoraceae

Sweet Flag

Acorus calamus, calamus, flagroot, myrtle flag, muskrat root

The flower, which takes the form of a rather phallic yellow-green spadix, seems to emerge from halfway up the leaves; in fact, the green part below is a flattened stalk and the green part above is a flattened, leaflike spathe. It grows up to 6 feet tall, with slender, swordlike, light green leaves with a rigid midvein from the base of the plant, which sits in the mud. It is found in bogs, marshes, streams, lakes, and ponds.

The leaves, when crushed, have a pleasant, faintly orangy smell. In medieval times, since *Acorus* also grows in Europe, the leaves were used as a floor covering. No doubt the first settlers were glad to see this familiar plant growing in the New World.

The sweet-tasting root acts as a slight stimulant, so it was popular in tonics. It was also used to settle the stomach, and cold sufferers could take the powdered root like snuff to clear the sinuses. The Native Americans knew sweet flag well, and the Penobscot even had a popular story about it:

A plague of sickness was sweeping the Indians away. There was no one to cure the people. One night a man was visited by a Muskrat in a dream. The Muskrat told him that he was a root and where to find him. The man awoke, sought the muskrat root, made a medicine of it, and cured the people of the plague. Sections of the dried root are cut up strung together, and hung up for preservation in nearly every house.[41]

The plant is sometimes known as muskrat root, because muskrats are fond of eating the root—explaining partly why a muskrat was part of the dream.

The Dakota had an original use for the plant. Before a battle, they chewed the root to make a paste, then smeared it all over their faces to give them courage and help them remain steady when they faced the enemy.[42] The Cree chewed large amounts of the root; the stimulant in the plant, known as asarone (it is chemically related to mescaline), gave them hallucinations, which were considered gifts of the spirits.[43]

The settlers used the roots as a pleasantly sweet foodstuff which they candied like ginger (perhaps with more of a kick). This was a specialty of Shaker communities. Euell Gibbons, who went looking for a recipe, found one that recommended boiling the roots for "two or three *days*."[44] Perhaps the Shakers were the only people with sufficient patience to persist. It was also used to make aromatic vinegars, or added as a flavoring to drinks like gin and beer. Some people simply chewed the root to sweeten the breath. Nature writer Doug Elliott says that chewing sweet flag in small quantities helps reduce the desire for tobacco, which suggests an interesting modern use for the plant.[45]

Range: **throughout North America, except for the Far North and the Southwest.**

Liliaceae

Janet Lyons and Sandra Jordan, authors of *Walking the Wetlands*, write:

> *In early spring the somber gray-brown hue of the marsh is dotted with an array of tiny, bright green spears. Within a few weeks as the weather warms, these rigid miniature totems unfurl to form conspicuous masses of vivid green foliage. Often attaining a height of six feet, the swamp hellebore displays a coarse beauty. The mature plant flaunts a majestic presence in an otherwise bleak landscape.*[46]

Veratrum viride is a perennial with large leaves 6 to 12 inches long clasping the thick stem. The yellow-green flowers emerge at the top of the stem on branched flower stalks bearing many small flowers, each about ¾ inch wide. The fruit is a three-lobed capsule. It grows in swamps, wet woodlands, and wet meadows.

False hellebore has one thing in common with the Eurasian hellebore (*Helleborus*): it is poisonous. Although false hellebore seldom kills humans because it causes immediate vomiting, it may kill animals. The smaller the animal, the more lethal the effect. *Veratrum* makes a good insecticide and parasiticide, hence the name "bugbane."

The first settlers were amazingly unobservant about the wild land around them, no doubt confused by the plethora of unfamiliar plants. Their attitude to the "savages" often prevented them from learning the ways of the woods from the local peoples. But seventeenth-century writer John Josselyn was that rare person who spent his time examining what was going on around him and watched how the indigenous peoples used plants. He was a keen observer and recorder of everything he saw. He wrote of the white hellebore, as he called it:

> *The Indians cure their wounds with it; annointing the wound first with raccoon's greese or wildcat's greese, and strewing upon it the powder of the roots: and, for aches, they scarify the grieved part, and annoint it with one of the foresaid oyls; then strew upon it the powder. The powder of the root, put into a hollow tooth, is good for the toothache. The root, sliced thin and boyled in vineagar, is very good against herpes milliaris.*[47]

The Indians of the East Coast used the poison as a test in certain rituals. Josselyn noted that they downed a boiled syrup made from the root and "he whose stomach withstood its action the longest was decided to be the strongest of the party, and entitled to command the rest."[48] The young men prepared for this ritual by eating hellebore in small amounts, learning to keep it down as long as possible and get their stomachs accustomed to the poison.

The settlers eventually learned to use it as an insecticide. According to Peter Kalm, "When children are plagued with vermin, the women boil this root, put the comb into the decoction, and comb the head with it, and this kills the lice most effectually."[49]

In the language of the *King's American Dispensatory*, the rhizomes are "inodorous, but strongly sternutatory when powdered"; in other words, they make you sneeze and were used to make

snuff. The *Dispensatory* goes on to say, "Applied to the skin, *veratrum* is rubefacient"—which, being translated, means that it makes the skin red (hence the name "itchweed").[50] In the nineteenth century *V. viride* was used in medicines to slow the action of the heart and reduce blood pressure, but by the twentieth century more reliable drugs replaced this rather dangerous remedy. It is still considered an effective natural insecticide.

Range: **Quebec, New Brunswick, Minnesota, east to Tennessee and Georgia in upland areas.**

Nymphaeaceae

Yellow Pond Lily
Nuphar lutea, spatterdock

Nuphar lutea

is a floating aquatic plant with large, lung-shaped leaves up to 15 inches long that lie flat on the water surface, and pale, wavy-margined leaves that are submerged in the water. The yellow, bulbous, malodorous flowers also sit on the water and emerge between May and September. The thick underwater stem emerges from a rhizome.

Range: **all Canadian provinces, south to South Dakota, Nebraska, Iowa, Illinois, Indiana, Ohio, Maryland, Delaware, and the New England states.**

The exquisite flower of the yellow pond lily belies its aggressive qualities. One plant alone can colonize an entire pond. Its rhizomes grow year after year, branching out, then sending up more flowers like little yellow fists raised above the water in support of Wetland Power. What look like the petals on its flowers are actually yel-

Nuphar lutea, yellow pond lily

low sepals, enclosing numerous small petals and stamens and a disk-shaped compound pistil. They are sometimes tinged with red or green.

The plant once had a slightly mystical reputation because it sprouted from the water. In 1923, Huron Smith noted that "this plant is described by the Menomini as belonging to the 'Underneath Spirits' and is accounted a great medicine. . . . The Menomini say that this plant makes the fogs that hover over the lakes."[51]

The roots are supposed to produce an edible starch, but this is emergency food only. Naturalist Doug Elliott, who has eaten all sorts of things that most people wouldn't dream of trying, says, "When I have tried this, and no matter how I cooked it, the taste was of boiled, baked, or roasted SWAMP! I wonder how much personal experience is involved in these modern reports. There probably is a method of making the root palatable, but so far I have not been able to find it."[52]

The Mi'kmaq, Ojibwe, and Menominee made poultices from the roots to put on bruises, cuts, and swellings, which sounds like a much more sensible use. The Quinault saw the figures of men and women in the shape of the roots and chose the appropriate shape before using one as a pain remedy.[53]

Polygonaceae

Smartweed
Polygonum spp., jointweed

Polygonum means "many joints," a distinguishing feature of the genus. The jointed look of the plant accounts for the name "jointweed," not because it's something you stick in a pipe to smoke and get high. The name "smartweed" refers to the sharp taste of the leaves and other parts of the plant, caused by an acrid oil contained in tiny oil glands. In England, smartweed was called "arsmart," according to John Minsheu, writing in 1626, "because if it doth touch the taile or other bare skinne, it maketh it smart, as often it doth, being laid into the bed green to kill fleas."[54]

Polygonum species may float in the water or grow in wet soil. Horizontal stems grow either on the ground or in the water, with leaves emerging alternately from prominent joints or nodes. Tiny flowers (less than ¼ inch across) appear in clusters at the ends of stalks rising above the water or standing up from the stem.

Several different species are native to North America, including **water smartweed** *(P. amphibium)*, which has bright pink flowers in a dense cone-shaped cluster and floats on the water; **dotted smartweed** *(P. punctatum)*, with greenish flowers loosely clustered on the flower stalk; **lady's thumb** *(P. persicaria)*, which has pink flowers and a dark patch like a thumbprint on the leaves; and **Pennsylvania smartweed** *(P. pensylvanicum)*, which has pink or white flowers and grows up to 3 feet tall.

Every part of the plant has some use: sprouts can be harvested in spring and cooked like asparagus; the leaves, which have a peppery taste, can be added to other dishes to make them spicy; and the rootstock peeled and boiled is eaten like a potato. Smartweed contains vitamins C and K (a coagulating substance). The Native Americans knew to use smartweed to treat wounds, hemorrhoids, and mouth and stomach ulcers. The Reverend John Clayton, writing in 1687, described the Indians of Virginia as treating a wound by chewing smartweed and spitting the liquid into the wound.[55] This sounds painful not only for the sufferer, but also for the person doing the treatment.

Range: **Newfoundland to Alaska, south to Utah, Nebraska, and New Jersey.**

Ranunculaceae

Three-Leaved Goldthread

Coptis trifolia, goldthread, cankerroot, mouthroot, yellowroot

Coptis trifolia

is a perennial plant that emerges from a slender yellow rhizome. The compound leaves have three leaflets each, with toothed edges, and emerge on stalks from the rhizome. They are dark green on the upper surface and lighter green below. The white star-shaped flowers also emerge from the base of the plant on a stalk 3 to 6 inches high in early to midsummer. What appear to be the petals are actually showy sepals; the petals are small and inconspicuous. Found in conifer swamps and wet forests.

Range: **Newfoundland to Alaska, south to British Columbia, Iowa, and North Carolina.**

The common names suggest two of the main uses of the plant: the roots yield a bright yellow dye, and Native Americans chewed the root to treat mouth ulcers or canker sores. An infusion made from the roots was also used as a gargle to treat mouth infections and thrush (a throat infection) and as an eyewash to treat eye infections. It tastes bitter, and contains an alkaloid, berberine, which is anti-inflammatory and antibacterial.[56] Jonathan Carver, who traveled through North America in the 1770s, wrote, "It is also greatly esteemed by the Indians and the colonists as a remedy for any soreness in the mouth, but the taste of it is exquisitely bitter."[57]

The Shakers of Rennselaer County in New York state used goldthread in dyeing, and bought as many plants from all local families as they could find.

The women and children of several families living in the vicinity of this piece of woods [where Coptis grew] spent most of one summer in these woods (which were mostly of spruce and balsam trees), digging Goldthread to sell to the Shakers. They dug it out with their hands only, sitting down by a smudge [fire] built to keep off gnats and mosquitoes, while they pick out the "roots" from among the sphagnum and other mosses among which they crept.[58]

French Canadian names for the plant were *fil d'or* or *savoyane*, which came from a Mi'kmaq word meaning "dye for skins," *tisavoyane*. It was accepted in the U.S. Pharmacopeia as a tonic and "stomachic."[59]

Amelanchier, spp., serviceberry

Rosaceae

Serviceberry
Amelanchier spp., sarvisberry, saskatoon, shadbush, Juneberry, sugar pear, Indian pear, grape pear

Amelanchier spp.

About 15 different species of *Amelanchier* grow in North America, mostly in wet woodlands and thickets, bogs, streambanks, and dry slopes. They range from low shrubs to small bushy trees. The leaves are simple, elliptical, sometimes pointed, toothed around the edges. The white flowers have five petals and appear in clusters in spring. The fruits are dark blue or black berries, each containing 5 to 10 white seeds.

**   *Range:* throughout Canada, New England, and around the Great Lakes. Different species have different ranges.**

This elegant little tree is one of the most useful native plants. It serves all who see and use it: people, insects, birds, and animals. In early spring, like a definition of the season itself, it becomes a fountain of pure white flowers gracing landscapes throughout the upper part of the continent.

The distinctive striped gray bark of this vase-shaped shrub or small tree would make it attractive even without the bright red buds that burst forth early in the season. They are followed by racemes of white flowers which develop into edible red berries darkening to purple-black as they ripen. Birds and just about any local animal love them. Raccoons clamber up the trees, clumsily breaking branches just to get at the berries.

Everywhere these trees and shrubs grow, people dub it with an affectionate local name. Many early settlers named it "serviceberry" because the flowers bloomed in mid-April, when religious services, halted for the winter, began again as the snow melted and roads became passable once more. Those who didn't go to church went fishing to catch the shad as they migrated upstream in the spring, hence the name "shadbush." The name "Juneberry" alludes to the fact that the berries ripen in June. The name "saskatoon" comes from the Cree word for the berries: *misaskwatoomina*. The name was also given to the city in Saskatchewan because the local people used to gather the berries and walk through the camp on the original town site, crying out their wares.

The juicy berries have a strong flavor, a little like pears, with nutty seeds. They were perhaps the most important fruit of western and central North America, and large quantities were harvested and dried for winter use. Eaten fresh or in oil, they improved the taste of any less palatable berries. Methods to preserve them included spreading them out in the sun to dry. They were also turned into a jamlike mixture, or made into

cakes or loaves. "Sometimes the berries were smoke-dried over a slow fire, and some people strained off the juice from the drying berries and drank it, used it to marinate other foods, or simply poured it over the berries as they dried."[60] Saskatoon berries' most important use was in the making of pemmican, a portable meal of dried meat, animal fat, and berries that native peoples carried on hunting trips or ate in winter when fresh food was scarce. The antioxidants in the berries helped to slow down the spoilage of the dried meat. A few groups still make pemmican today.

The berries have remained a favorite for jams, pies, desserts, and other dishes. Widely available in western Canada, where commercial berry production is increasing, they are less well known elsewhere. The most commonly grown species is *A. alnifolia*, but *A. canadensis* and cultivars from this parent are also popular. Saskatoons are picked using a coarse-toothed comb to rake off the berries, which are then immersed in water to remove the twigs and leaves—as the detritus floats to the top, it's easily skimmed off.

The berries weren't the only useful part of the plant. The bark and root were once used to make an infusion that native women drank to treat excessive menstrual bleeding or prevent miscarriage after an injury. The infusion was also administered to expel parasitic intestinal worms. *Amelanchier* wood is strong, close-grained, and hard, and has been used to make arrow shafts, the handles of tools, and other small implements.

Scrophulariaceae

Chelone glabra

is a perennial wildflower that grows 1 to 3 feet high in swamps, wet thickets, streambanks, and ditches. The toothed leaves are narrow, up to 6 inches long, opposite, and pointed at the ends. The white or pinkish flowers are tubular, with a concave upper lip and a three-lobed lower lip. The fruit is a smooth brown capsule containing many flattish, winged seeds.

Range: **Nova Scotia to Ontario, south to Minnesota, Missouri, New England, and Georgia.**

All the way up the stiff erect stems, the little flowers open up, looking very much like turtles poking their heads out of their shells. The flowers emerge in sequence: the ones at the base of the flower spike come out first, followed by the ones higher up. When the flower opens, only the stamens are mature. Later, the pistil matures. Insects crawl right into the newly opened flowers in their search for nectar. When they push their way out, they are covered in pollen. If they visit a mature flower farther down the flower spike, they carry the pollen to a waiting pistil.

Turtlehead leaves are food for the checkerspot or Baltimore checkerspot butterfly in its caterpillar stage. The caterpillars munch away all summer, hibernate over the winter, wake up in spring and eat some more, and then form a chrysalis. When they emerge, the butterflies

Chelone glabra, turtlehead

mate and lay new eggs on the leaves, thus perpetuating the cycle.

"Leaf teas of turtlehead, exceedingly bitter, were staples of folk medicine introduced by similar native usage," writes John Eastman. "Healers of various stripe often presented them for jaundice and liver ailments, as well as for worms and as an appetite stimulant and laxative. Decoctions of the plant were applied externally to tumors, herpes, and hemorrhoids."[61]

Scrophulariaceae

Veronica americana

is a perennial herb that often creeps along the ground, with fleshy stems 1 or 2 feet long, found in swamps and ponds or along streambanks. The

Veronica americana, American speedwell

The impression of his face appeared afterward on the cloth (*Veronikon* means "true image" in Greek). Some people believed that they could also see the face of Christ in the little flower.

Europeans attributed healing powers to the plant. It contains vitamin C, and so prevents scurvy, but it also acts as an expectorant, diuretic, and stimulant. Modern foragers look for the edible speedwells, though some people find them bitter-tasting because they contain so much tannin. Naturalist Janice Schofield suggests that "the upper stems and leaves of veronica can be nibbled as a snack, added to salads, or steamed lightly as a potherb. . . . The leaves and stems (add flowers if you wish) can be steeped for tea . . . the taste is reminiscent of Chinese green tea."[62] Speedwell can be eaten along with watercress in a sandwich: just add sour cream and soft cream cheese.[63]

Typhaceae

Cattail

Typha latifolia, common cattail, broad-leaved cattail, soft flag

Common in wetlands, and along the margins of rivers and lakes, cattails grow so quickly and make such crowded colonies that plants are in danger of choking themselves to death. In some wetland areas, to prevent this, people whack back cattails severely when they threaten to take over. It's an aggressive plant that can grow to 6 feet or more from thick rootstocks.

Typha latifolia has long, narrow leaves emerging from stiff, unbranched stems. The flowers are in two parts: a sausagelike female part that eventually turns brown, and a spiky yellow male flower above it that falls off after it has shed its pollen.

The cattail has a few surprises in store for those who have never been intimate with the plant. In 1840, naturalist Philip Henry Gosse wrote:

leaves are opposite, lance-shaped, and toothed along the margins. The pale blue- or lilac-colored flowers have four unequally sized petals.

Range: Alaska to Newfoundland, south to California and North Carolina.

There are several theories about the origin of the name "speedwell." According to one, the bright blue flowers, growing along roadsides and riverbanks, seem to speed the traveler along. Other people think that the name refers to the fact that the plant grows very quickly. The third is English in origin. English peasants used the leaves of some species for an expectorant and cough medicine and called it "spit-well." Yet another theory is that if the flowers are picked, the corolla soon falls off, hence the name speedwell or "good-bye."

The scientific name recalls Saint Veronica, who, according to legend, wiped the face of Christ with a cloth as he carried his cross toward Calvary.

The thick cylindrical head appears like a fine, but very closely set brush, radiating from the axis or stalk, which it covers for about six inches. On picking out a lump of what we may call the bristles of this brush, we are surprised to see we have a handful of the softest down, that which before was not bigger than one's thumb, now on being freed from the stalk, filling one's hand . . . the whole head is composed of this very expansive down; and I am told that poor persons sometimes collect quantities of it to make beds, which are said to be soft and elastic.[64]

Settlers not only used cattail down as stuffing for quilts and pillows, but also put it in mitts and socks for extra warmth. No matter what Gosse said, this wasn't just something that the poor did. Anybody who lived in such isolated conditions needed insulation, and the practice was common right into the twentieth century. Today, the down is sometimes used as a filling for children's toys or ornamental pillows.

During the Second World War, everyone believed in collecting for the war effort. People were comforted by the idea that all should pitch in and help their fathers and brothers who were serving overseas. Along with metal and paper, people gathered huge numbers of cattail spikes and milkweed pods from abandoned lots and nearby ditches. The fluff in the cattails and milkweed pods was used in life preservers and sleeping bags. This no doubt made a dent in the local ecology but it certainly made those of us collecting the material feel important and useful.

The leaves had long been used by Native Americans, who wove them together to make mats and baskets. Settlers used them to cane chairs. The leaves were hung up to dry, then soaked until pliable. They are so tough that some of these cattail-caned chairs have survived a century of use. The flower heads also make spectacular outdoor torches when they are soaked in kerosene or diesel fuel.

Most parts of the cattail are edible. Doug Elliott in *Wild Roots* says: "It has been said that one could not starve living next to a Cattail

Typha latifolia, cattail

Angelica atropurpurea, angelica

Apiaceae

Angelica
Angelica atropurpurea, archangel, bellyache
root, masterwort

Angelica atropurpurea

grows up to 12 feet tall. It has a thick purple stem and an underground taproot. The double compound leaves grow from a bulbous sheath that wraps around the stem. The leaves are divided first into three parts, and then again into three to five toothed leaflets. The greenish-white flowers form little globes clustered together at the top of the stem. Found in swamps, bottomlands, and beside streams.

***Range:* throughout eastern North America, except for the Far North and the southernmost regions.**

marsh. I think this is close to the truth."[65] The rhizomes can be peeled, pounded, and dried to make a type of flour; the sprouts on the rhizome can be used in salads and stir-fries; the inner core of the stem and the leaves can be eaten raw or cooked; and a syrup can be extracted from the roots. Even the long brown sausage-shaped female flowers can be roasted and eaten like corn on the cob; the yellow pollen from the male flowers can be added to breads, muffins, or pancakes. The famous down mixed with tallow makes a kind of chewing gum—something to remember if you are stuck out in the wilderness with nothing else to relieve tension.

***Range:* throughout North America, except for the Far North tundra regions.**

One year you'll find angelica and be thrilled by its beauty; the next year it seems to have disappeared. Angelica is a biennial. It grows foliage the first year and produces its flowers and seeds the following. This relative of carrots, celery, fennel, dill, and parsley grows in the same way—from a taproot. Pull out that long root and you'll notice a definite carroty smell.

Atropurpurea means "dark purple." The Europeans were familiar with *A. archangelica,* as a herb used in cooking. According to legend, the archangel Michael appeared to a monk in a dream, and told him to instruct people to chew the root to ward off the plague, hence the name *Angelica.*

The Reverend John Clayton, who traveled through Virginia in the 1680s, recounted a story of Indians using the plant. A woodsman was traveling through the forests with a native guide who spotted some angelica. He gathered part of the root, but left the rest to resprout, saying that the plant was scarce, and highly esteemed. Later in the trip, they spotted some deer. Upwind of the deer, the Indian sat down on a tree stump and rubbed the root between his hands. The deer got wind of the scent and walked toward the two men. The Indian was then able to easily shoot one of the bucks. Some people called it "hunting root."[66] Pieces of angelica root were carried as talismans for good luck in hunting (or in gambling, depending on the need).

The plant was also valued for treating stomachaches and indigestion, giving it the name "bellyache root." Although the raw root is toxic, decoctions and tinctures are made from it, and are still found in stores selling herbal remedies. Among other properties, it is said to help treat addictions such as anorexia and alcoholism. The fresh shoots are edible, and can be boiled to make a pleasant green vegetable, while the seeds can be used like caraway and are quite tasty.

Hypericaceae

Marsh St. John's Wort
Hypericum virginicum syn. *Triadenum virginicum*

Hypericum virginicum syn. *Triadenum virginicum*

grows up to 2 feet high. The pale green leaves are opposite, about 2 inches long, oval, with pinkish veining, heart-shaped at the base, clasping the dark pink erect stalk. The leaf is dotted with glands that look like translucent spots. The flowers grow at the tops of the stalks and in the leaf axils. They are pink with five sepals and petals, and three groups of three prominent yellow sta-

mens. **The fruit is an oblong reddish capsule.**

Range: **Labrador to Saskatchewan, south to Nebraska and Georgia.**

St. John's wort has become one of the most famous of all medicinal plants. There are about 400 species of *Hypericum* worldwide, including annuals, perennials, and deciduous and evergreen shrubs. The **common St. John's wort,** *H. perforatum,* which has yellow flowers, is a species introduced from Europe, where the genus got its name and its reputation. Most species have glands on the leaves, which were said to look like wounds, and so, according to the doctrine of signatures, the plant should be good for treating wounds.

Hypericum means "above an image": the flowers were hung over religious pictures to ward off evil. There are competing theories about the common name. According to most sources, the plant usually came into flower around St. John's Day, June 24, close to the summer solstice. Another story, however, says the plant was used by the Knights of St. John to treat the Crusaders.

The Native Americans used this and other species such as **great St. John's wort** (*H. pyramidatum*) to treat tuberculosis. Today, St. John's wort is an increasingly popular herbal remedy for depression. It contains a substance called hypericin, which has tranquilizing and sedative properties. Dr. Jim Duke, an economic botanist with the U.S. Department of Agriculture who had collected many specimens of *H. virginicum* and other *Hypericum* species for the National Cancer Institute in Frederick, Maryland, says this plant "is one which I predict will have significant anti-retroviral activity . . . that is, action against the AIDS virus."[67]

CARNIVOROUS PLANTS

Carnivorous plants evolved because there weren't enough nutrients in the soil to satisfy their needs. Insects became a good source of food and the plant's digestive systems adjusted accordingly. Something draws us to these plants. Perhaps we identify with them; maybe we have a morbid fascination with watching something else destroy and eat its prey. Whatever the psychology attached to these carnivores, we like to make pets of them, which is a curious way to treat a plant. Venus' flytrap and all the other popular carnivorous plants should never be collected in the wild, but bought only from a greenhouse where they have been grown from seed. There is some illicit trafficking in these plants, most of which are protected by law in the states and provinces in their range.

Dionaea muscipula, Venus' flytrap

Droseraceae

Venus' Flytrap
Dionaea muscipula

Dionaea muscipula

is an insect-eating plant with two-part, hinged fleshy leaves, 1 to 6 inches long, fringed with bristles; the leaves are green outside and sometimes orange inside. The leaves are on leafstalks that rise from the base of the plant. The flower is white, about 1 inch wide, with five petals, rising on a flower stalk from the base of the plant. The whole plant may be only a few inches high or up to 1 foot tall. Found in moist sandy areas.

Range: the Carolinas.

This plant is to the world of flora what the spiderweb is to the world of fauna: a killer. When an insect lands on the leaf and disturbs the three sensitive hairs in the center, the two halves of the hinged leaves snap shut. The bug is trapped by the bristles on the leaf margins and the plant's digestive juices start to flow, turning the insect into a mush of nutrients that can then be absorbed into the plant's body. End of insect. The leaves open for business once again.

Venus' flytrap has become such a popular houseplant that there are dozens of books and Web sites devoted to their care and feeding (for example, they all say never, ever feed it hamburger; on the other hand, the lengths they suggest owners should go to find suitable bugs is a bit alarming). Oddly enough, the word *muscipula* means mousetrap, not flytrap (which is *muscarium* in Latin), but there is no evidence that the plant can manage anything as large as a mouse. Dione was the mother of Venus.

Drosera rotundifolia, sundew

Droseraceae

Round-Leaved Sundew
Drosera rotundifolia, dew plant

Drosera rotundifolia

is an insect-eating plant with fleshy leaves on stalks that tend to hug the ground. The leaves, which are about ½ inch long, are covered with hairs, at the tip of each of which is a sticky drop of fluid. The white flowers emerge in a cluster on a stalk up to 9 inches high. Each flower is about ¼ inch wide, with five petals, sometimes tinged with pink. Found in bogs.

Range: **Alaska to Newfoundland, south to California, Louisiana, Alabama, and Florida.**

The sundew takes a different approach to catching and eating insects. It has glandular secretions at the ends of hairs which cover the surface of its leaves. An insect lands on the leaves, sticks to the little drops, and in the throes of struggling to get free, gets stuck to more and more hairs. Eventually, it gives up and is slowly digested, drop by deadly drop. This may be the ultimate in mosquito traps, since mosquitoes are what *Drosera* eats.

Most of what we know about sundews we know from Charles Darwin, who wrote the book *Insectivorous Plants* in 1875. He was fascinated by the sundew, and apparently once said, "At this present moment I care more about *Drosera* than the origin of all the species in the world."[68]

The plant is protected in many American states, since it is tiny and fragile and suffers terribly when its environment is threatened. In the past, however, it was gathered for use in medicines for lung and respiratory ailments. It also had something of a reputation as an aphrodisiac, and the dew applied to the skin was said to cure warts, corns, and pimples.

Lentibulariaceae

Common Bladderwort
Utricularia vulgaris, greater bladderwort, bladder-snout, popweed, hooded water milfoil

Bladderwort traps tiny aquatic animals in its network of feathery leaves and underwater bladders.

The bladders have a membranous opening at one end, surrounded by hairs. When the hairs are disturbed by a passing bug or tadpole, the membrane snaps open and sucks the creature inside like a vacuum cleaner. Then its juices go to work.

The underwater branching structure may be as large as 7 feet across. Yellow flowers shaped a bit like a snapdragon rise on stalks that are from 2 to 10 inches above the water's surface.

Thoreau wasn't much impressed, and in his *Journal* called the bladderwort "a dirty-conditioned flower, like a sluttish woman with a gaudy yellow bonnet."[69] Maybe it was the thought of all that underwater carnage that put him off.

Range: **throughout North America, except for the Arctic.**

Sarraceniaceae

Pitcher Plant
Sarracenia purpurea, Northern pitcher plant, hunter's cup

Sarracenia purpurea
is a carnivorous plant with tubular leaves that grow from the base of the plant and are open at the top with lips that flare outward. The interior of the leaves is covered with stiff, downward-pointing hairs. The flower stalk bears a single, red, drooping flower with five petals and numerous sepals. The whole plant may be 8 to 24 inches tall and grows in sphagnum bogs and marshes.

Range: **Saskatchewan to Labrador and Newfoundland, south through the Eastern states to Texas and Florida.**

Provincial flower of Newfoundland.

Sarracenia purpurea, pitcher plant

The pitcher plant drowns its prey. The nectary lies at the bottom of its tubular leaves. Bugs sniff out the nectar, dip into the plant, then find they can't escape because they are caught on the downward-pointing hairs lining the leaves.

Some creatures which find themselves in this predicament make do and remain in the pitcher plant, unaffected by the otherwise lethal environment. These are tiny aquatic mites that swim about in the fluid and contribute their wastes to the plant's nutrition. A rather nasty-sounding fly called the *Sarcophaga sarraceniae,* or Eastern flesh fly, lays its eggs on the lip of the leaf. The maggots hatch and slide down the inside of the leaf, where they feast on the remains of the plant's victims. They finally chomp through the wall to get out, where they transform into adult flies. Presumably, this drains the fluid out of the leaves, which then lose the ability to attract and kill other insects.

Spiders also make use of the plant, by constructing webs across the mouths of the tubular leaves. Once again, the pitcher plant does not gain from this tactic, since the insects are prevented from getting into the liquid at the bottom of the leaves.

Anthocyanin is what gives *Sarracenia* its vermilion color. There are extremely rare populations of all-green forms which have all been poached so thoroughly that they are no longer found in the wild. This means we have lost whatever information their natural systems might have held for us.

The plant was named by Linnaeus for Michel Sarrazin de l'Étang (1659–1734), a French physician in New France (Quebec), who used the plant in the treatment of smallpox, a treatment he claimed to have learned from the Indians. He made an infusion with the roots and used it to bring out the spots, then followed up with second and third doses. This apparently caused the pustules to subside, the patient to recover, and Sarrazin to become famous.[70]

Darlingtonia californica, cobra plant

Sarraceniaceae

Cobra Plant
Darlingtonia californica

Darlingtonia californica is a Western relative of the pitcher plant that consumes insects in a similar manner. It can be a monster in another way: it grows over 3 feet tall with hooded tubular leaves that look like a rearing cobra.

It was named for William Darlington (1782–1863), a botanist from Pennsylvania who neither found the plant nor cataloged it. It was first collected by a naturalist called William

Brackenridge. The name was assigned by botanist John Torrey in the 1840s, as something of a botanical thank-you present, after Darlington wrote *A Plea for a National Museum and Botanic Garden to Be Founded on the Smithsonian Institution at the City of Washington*. At the time, the institution was languishing for a lack of funding and lack of public interest. The original collections amassed by James Smithson and willed to the United States were housed in the Patent Office. They were not even accessible to the public. Torrey named the plant in gratitude for Darlington's public-spirited intervention.[71]

Range: **California and the West Coast.**

THE EDGE OF THE SEA: BEACHES AND DUNES

For millennia beyond computation, the seas' waves have battered the coastlines of the world with erosive effect, here cutting back a cliff, there stripping away tons of sand from a beach, and yet again, in a reversal of their destructiveness, building up a bar or a small island. Unlike the slow geologic changes that bring about the flooding of half a continent, the work of the waves is attuned to the brief span of human life, and so the sculpturing of the continent's edge is something each of us can see for ourselves.[72]

Rachel Carson

In the midst of this constant push and pull, plants survive with miraculous tenacity. Seaweed and the plants of the intertidal zones and dunes look like the ancient plants they are—having just left the primordial sea of their origin—perhaps not even quite fully developed. People have eaten and used them in various ways since the beginning of human time.

There is a theory that *Homo sapiens* started its evolution by spending 10,000 years living by the sea and developing brains before marching upright into the savanna to hunt big game. Perhaps this is why people love to be by the sea: we want to return to our origins. Today, however, we are becoming aware just how fragile the seashores and their attendant dunes have become. The sight of massive, dramatic erosion has become commonplace. We can all observe the dangers. If it's dangerous for people, it doesn't take much of a stretch of imagination to understand how difficult it is for plants living amid apparent chaos and hostility all the time. During extremes in weather, such as hurricanes, life becomes even more stressful.

When there are wild winds and heavy seas, the sand is scoured from the beach and dumped offshore to make shoals and sandbars. As calm is restored, the sand returns to the beach through the rolling motion of waves. Gullies form in the intertidal area where the sand is drier, and gradually dunes are built up. An equilibrium is established in which plants and sea work their magic together and plant communities change as protected areas are developed by the action of sand and wind.

The intertidal beach is the expanse between high and low tide. As the tides rise and fall, life on the beach goes underground. At the high-tide line, marsh wrack is left behind, wind blows sand in, seeds are trapped, and the possibility of plant life increases.

On the primary dunes, the ones closest to the sea, plants must be able to deal with salt spray. On the sea side of the primary dune are the pioneers like sea rocket and beach orach; when the humus from the dying plants has built up enough, the grasses eventually move in. The areas behind the dunes are as dry as the desert, and plants adapt by developing succulent leaves to store moisture and reduce their leaf surfaces to combat damaging evaporation. They can also grow marvelously long taproots that run for yards in search of water. In the meantime, sea spray provides the plants with most of their essential nutrients.

There has been such a long tradition of eating from the sea that we often don't hesitate to help ourselves. But there are two cautions: always check to see if a plant is rare or endangered and sample it very carefully for unknown toxins. Consult an expert to be on the safe side.

The World Below the Brine

The world below the brine,
Forests at the bottom of the sea, the
    branches and leaves,
Sea-lettuce, vast lichens, strange flowers
    and seeds, the thick tangle, openings,
    and pink turf,
Different colors, pale gray and green,
    purple, white, and gold, the play of
    light through the water,
Dumb swimmers there among the rocks,
    coral, gluten, grass, rushes, and the
    aliment of the swimmers,
Sluggish existences grazing there
    suspended, or slowly crawling close to
    the bottom,
The sperm-whale at the surface blowing
    air and spray, or disporting with his
    flukes,
The leaden-eyed shark, the walrus, the
    turtle, the hairy sea-leopard, and the
    sting-ray,
Passions there, wars, pursuits, tribes,
    sight in those ocean-depths, breathing
    that thick-breathing air, as so many do,
The change thence to the sight here, and
    to the subtle air breathed by beings
    like us who walk this sphere,
The change onward from ours to that of
    beings who walk other spheres.[73]
                              Walt Whitman

Uniola paniculata, sea oats

Poaceae

Sea Oats
Uniola paniculata

Uniola paniculata

grows from 3 to 6½ feet high. It is a coarse, perennial grass with a hollow stem that grows from a rhizome. The leaf blades grow to 24 inches with an inflorescence of condensed panicles. In late summer, it produces oatlike, straw-colored flowers in flattened spikelets with 8 to 20 florets.

Range: on the Atlantic and Gulf Coasts from Virginia south to Texas, and on the Pacific Coast in southern California.

Sea oats are so critical to the ecology of the seashore that it's now illegal to either pick or uproot them. This plant is a major stabilizer of dunes which, in turn, attract other plants. In areas of Florida where dunes have been lost, sand is brought in, machine-sculpted into dunes, then planted with sea oats and other grasses to keep them in position.

Sea oats dominate the prime dunes, in one of the most difficult of all locations, facing the ocean. They live under circumstances no other species can take—the constant onslaught of salt spray—with such tolerance that even in the most extreme cases, they thrive. When the dune retreats and the distance between spray and grass lengthens, the sea oats are eventually replaced by other species. Sea oats spread from 2 to 6 feet a year, but researchers are finding even this rate isn't fast enough to keep up with the rapidly subsiding shoreline.

Sea oats are named for the large plumes they produce during the summer. Though they look like regular oats, they are actually perennials. The seeds are blown every which way by the wind. Those that hit sand are quickly picked up by small rodents and birds. Wind, ocean currents, and animals can take the seeds from mainland to islands and back again. If the seeds happen to land where sand is building up, they will probably sprout. According to Stephen Facciola, the seeds can also be cooked to make a hot cereal with a pleasant flavor.[74]

The rhizomes produce roots wherever the nodes are buried by sand. When the leaves are developed and there is a dry period, the stomata close, and growth stops. It restarts as soon as there is rain. If it is inundated for more than a few days, the plant will be killed.

At one time, florists used to gather the plumes, dye them bright colors, and sell them for dried flower arrangements. Now the plants are protected and collecting the seed plumes is illegal. Even so, it is still a plant seriously under assault. The use of off-road vehicles, pollution from raw sewage still being dumped along some shores, and fertilizer runoff all threaten its habitat. On North Padre Island in Texas, the dunes are still recovering from grazing, which was allowed between 1850 and 1971. There is no instant fix for the seashore.

Researchers in Florida are trying to select the best strains of *Uniola* to plant on dunes as stabilizers. According to Mike Kane, a Florida Sea Grant researcher at the University of Florida, and Dave Crewz, of the Florida Department of Environmental Protection: "In the future, the sea oats planted to restore Florida's storm-ravaged beaches may look just like their silky-stalked neighbors who survived the pounding surf. In reality, these new plants may be superior in many ways to their cousins growing in the wild." Modern techniques of DNA fingerprinting, plant tissue culture, and comparative field tests are all employed to determine the range of genetic variability in natural sea oat populations. What scientists are searching for is a plant that has even better shoot and root growth, because sea oats are critical to preserving dunes.[75] Other researchers in North Carolina are looking at the plant not only to stabilize shifting sand dunes, but also as an alternative crop for some tobacco farmers.

Brassicaceae

Sea Rocket
Cakile edentula

Cakile edentula

an annual, has narrow, alternate fleshy leaves about 3 inches long, and small pink, lilac, or white four-petaled flowers in clusters at the tip of the stem. It blooms between July and September. The plant can grow up to 12 inches tall when it is

Cakile edentula, sea rocket

not creeping along the sand dunes and beaches where it grows.

Range: Atlantic Coast from Labrador to Florida, around the Great Lakes, and on the Pacific Coast from Alaska to California.

The foredune, where the beach meets the dunes, is a forbidding place. Ninety percent of the surface of coarse sand has nothing growing over it, but is strewn with the remains of marine organisms, other flotsam, and scattered inorganic debris. Wind sculpts the sand in spring, ice freezes over it in winter, and people trample over it all summer long. This is where sea rocket survives. This tough little plant can grow quite close to the sea on tidal wrack—the vegetative debris pushed up a beach by the waves. In sheltered areas, it grows upright, but in exposed places it flattens itself against the sand. Some sources suggest that *Cakile* is nonnative, others list it with native plants. Thoreau spotted it on Cape Cod in 1849, although he called it *Cakile americana*.[76]

Sea rocket stores water in its thick, fleshy leaves and stems, germinating each spring from a floating, water-resistant seed which produces a bushy plant by late summer. The elongated seedpods are divided into two segments to double its chances of reproduction. The top half breaks off and floats to a new site while the bottom half stays attached to the plant and germinates in place.

You can find sea rocket not just on ocean beaches but also around the Great Lakes, particularly Lake Erie and Lake Huron. Conditions on the Great Lakes can be just as harsh as on the Atlantic Coast, with violent storms and winter ice.

The leaves are edible and taste a little like horseradish (it's a member of the mustard family). Foragers eat the leaves, flowers, and seedpods when they are young and tender. The leaves are often mixed with other greens in salads or cooked as a vegetable.[77]

Wild-food fan Christopher Nyerges says:

While hiking on the beach at Point Reyes National Seashore (north of San Francisco), my friends and I spotted an old, partially dried-up sea rocket plant from the previous season. The plant had already produced and dropped hundreds of seeds into the sand. Under the plant, we saw hundreds of tiny sea rocket sprouts, each no more than two inches long. We carefully collected a bag of the tender sprouts for our dinner. That evening, we washed all the sprouts to eliminate the sand. We used half of the sea rocket sprouts mixed with miner's lettuce for our salad. The rest of the sprouts we added to our clam chowder (made with fresh clams).[78]

Eating from the edges of the sea has been going on for hundreds of thousands of years. And it is still a possibility in areas not touched by pollution. Szczawinski and Turner tell us,

Sea-rocket leaves can be harvested from May through the summer months. If you live near a patch of sea-rocket, you can maintain a continuous supply of fresh, young leaves simply by harvesting the growing tips, which will regenerate within one or two weeks. If you stagger your picking over many plants, you will not deplete the populations. Be sure to allow some plants to flower and produce seed before the summer finishes, however, as this plant is an annual and must produce seed to grow the following year.

They recommend adding a cup of sea rocket leaves to six large eggs and ½ cup of cream to make scrambled eggs.[79]

Chenopodiaceae

Seabeach Orach

Atriplex arenaria syn. *A. pentandra,* crested saltbush, beach spinach

Atriplex arenaria syn. *A. pentandra* is an annual that grows up to 2 feet tall. Its alternate leaves are silvery gray with rounded edges, and the stem is tinged with red. It grows on dunes or beaches or in salt marshes. From July through October, it has small green flowers that grow in spikelike clusters at the ends of the stems.

Range: New England to Texas.

Seabeach orach tolerates salt spray and grows from the mean high-tide line up to dune or beach crests and on beach slopes. It is often found growing in the same areas as *Cakile edentula.*

Native Americans ate orach, including seabeach orach and **common orach** (*A. patula*) and several other species. They also beat orach roots in water to produce a soaplike substance, and pounded the seeds to make flour. According to Janice Schofield, the mashed leaves can be applied to insect bites and stings to reduce swelling.

The leaves and stems of orach can be prepared like spinach: simmered in water until limp. The seeds are also edible and can be used in soups, cereals, or breads.[80]

Cistaceae

Beach Heather

Hudsonia tomentosa, beach heath, false heather, poverty grass

Beach heather conserves water in two ways. First, it has tiny, almost scalelike leaves flattened against the stem which reduce its surface area. Second, it is covered with a dense layer of insulating hairs. By forming mats on the seashore, it traps sand to create more insulation. Little hillocks develop around each cluster of plants, making ground-covering colonies. This elevates the stems and in its flowering season, bright yellow flowers can be seen flouncing about the dune.[81] The yellow flower has five petals with oval seedpods, surrounded by sepals, which contain one seed, sometimes two. It is a creature of the dunes but can be found on sandy flats as well.

In 1865 Thoreau wrote in his book on Cape Cod: "The sand by the roadside was partially covered with bunches of a moss-like plant, *Hudsonia tomentosa,* which, a woman in the stage [coach] told us, was called 'poverty grass,' because it grew where nothing else would."[82]

Hudsonia was named not after the explorer, but after a London apothecary named William Hudson (1730–1793), the prefect of the Chelsea Physic Garden and author of *Flora Anglica.*

Range: **Nova Scotia to North Carolina.**

Fabaceae

Beach Pea

Lathyrus japonicus, purple beach pea, seaside pea, sea pea, sea vetchling, raven's canoe

Lathyrus japonicus

grows up to 2 feet high. The showy flower is like other flowers of pea plants, with reddish outer petals and bluish-purple inner petals. At the ends of the stems are curling tendrils. The leaves are opposite, oval, and may be slightly hairy. The pods are smooth and unjointed, up to 2½ inches long, very much like those of garden peas. Grows on beaches and dunes.

Range: **East, West, and northern coasts, from California around to the Mid-Atlantic states.**

Beach pea is another tough survivor that helps to stabilize sand dunes. It sends out runners ending in tendrils over the surface which coil themselves around tufts of grass. It is this kind of nitrogen-fixing legume with nodules that helps to create fertile soil from sterile sand. Another survival mechanism is the ability of its seeds to survive in extremely cold seawater for up to five years and still germinate.

The Mi'kmaq and Wabanaki of the East Coast and the Dena'ina of Alaska often ate beach peas, which contain healthy amounts of vitamins A and B as well as protein. In summer, they ate the peas fresh, but gathered extra peapods to fry and store for the winter. These fried peas were boiled in stews or ground up to make pudding. The Dena'ina people called this plant "raven's canoe" because of the shape and color of the ripened seedpod.[83]

Lathyrus japonicus, beach pea

Here is a brief retelling of an origin myth recorded from an old Unaligmiut man who had learned it as a boy from an old man living in the Bering Strait region.

Raven created beach peas which grew on the naked earth, and from one pod a full-grown man burst forth. Raven approached, pushed up his beak, and also became a man. He told the first man that while he had created the [beach] peas, it was without knowing that a man would emerge from them. From clay Raven formed various pairs of animals at different times and gave life to them. Mountain sheep were created first, followed by reindeer and then caribou; a woman was formed next, to become the wife of the first man. Raven went on to create certain fish as well as other creatures and to teach the human couple how to live in their emerging environment. The woman bore a son and then a daughter, who were to marry. Raven turned back to the original pea pod from which the first man had been born and found that three other men had emerged from the same pod. Raven led the first man inland, but the other three were taken to the coast . . . where they were taught to exploit the resources of the sea.[84]

Among the European explorers and early settlers who ate the peas was Captain Thomas James, wintering in James Bay in 1632. He fed beach peas to his scurvy-stricken men, who revived. In 1609, Marc Lescarbot in *Acadia* wrote: "Peas in great quantity along the seashore, the leaves whereof we took in the springtime and put among our old peas, and so it did seem unto us that we did eat green peas."[85]

Thoreau also noted the popularity of a variety of beach pea in Cape Cod. "He [an old Wellfleet oysterman] liked the Beach-pea . . . cooked green, as well as the cultivated. He had seen it growing very abundantly in Newfoundland, where also the inhabitants ate them, but he had never been able to obtain any ripe for seed."[86]

Beach peas are popular with modern foragers. Inexperienced persons, however, should be very careful. Some poisonous plants look exactly like the peapods. Although beach pea seeds are nutritious in small quantities (it takes an age to gather them anyway), they contain a toxic amino acid which causes lathyrism, a serious disease of the nervous system.

Biologists at Memorial University in St. John's, Newfoundland, along with staff of Agriculture Canada, "are trying to turn *[L. maritimus]* which naturally grows along the shorelines of Newfoundland, into a cultivable cold-climate crop." What tipped off researchers Chinnasamy Gurusamy and Dr. Arya K. Bal, who are studying the beach pea, were the historical texts which told of stranded sailors eating the seeds as survival food. Their research found that the seeds are rich in nutrients and minerals much like soybeans. They analyzed plants for the toxin that causes lathyrism and found that the native beach pea is 40 times lower in the level of neurotoxins than plants found in Asia and Africa.[87]

Gurusamy and Bal's research is focused on Salmon Cove, Newfoundland, where they gathered most of their samples. In 1998 the whole town took on the job of protecting and promoting the beach pea, and in 1999 they had their first "Beach Pea Festival." The townsfolk declared it the official plant of Salmon Cove and have organized a major cleanup at the local beach, now declared home of the beach pea.

Ammophila breviligulata, dune grass

Poaceae

Dune Grass
Ammophila breviligulata

Ammophila breviligulata
sends out rhizomes under the surface of the sand which shoot up from every node to about 5 feet high. Aboveground leaves come from the base of the plant. Spiky flowers appear in summer.
Range: **Newfoundland to North Carolina.**

Dune grass is sand-loving as well as sand-needy—it must be covered with up to 3 inches for most of the year. Its name *Ammophila* comes from *ammos* (Greek for "sand") and *philos* (Greek for "loving"). The abrasiveness of the sand particles stimulates the underground rhizome system into growth and this traps the sand like a net, helping to keep the dunes in place. Dune grass can survive burial by up to 3 feet of sand a year. The decaying organic matter it collects helps, in turn, to create a microclimate for other plants. Taproots plunge as deep as 30 feet, gathering moisture from sources far underground. Shallow surface rootlets also collect the moisture available from rain. Because it is a pioneer species (a plant that moves in as soon as a niche is opened to it), dune grass can colonize the foredune area.

Environmental groups use dune grass to stabilize beaches. In 1998, volunteers from a group called the EveryDay Angels Foundation, along with Clean Ocean Action, a marine advocacy group, helped place 60,000 dune grass plants along the shoreline at Gateway National Recreational Area in Sandy Hook, New Jersey. This millennium project began with a walk along the beach for naturalist Kirby Tomkins and Lisa Bernstein. They were discussing what to put into a millennium time capsule.

" 'A plug of dune grass,' was his reply. 'Without dune grass, none of this would be here,' he said, his arm sweeping towards the beach in front of us. 'Ten thousand years ago, this spot was fifty miles from the ocean. As the sea level rose, the grass helped stabilize the land at the edge of the water.' "

Each year now, a few volunteers plant more and more grass, keeping the sand where it should be—at the edge of the ocean.[88]

Poaceae

Sandspur
Cenchrus tribuloides, sandbur, dune sandbur, grassbur

The name "sandspur" describes the sandpapery feel of the leaves and the spurs or burs on this bristly plant. The burs have tiny, backward-pointing spines that stick to clothing and—painfully—to skin. The burs are there partly to protect the plant against disturbance, and partly to help with seed dispersal to other areas. Sandspur, like many other seashore plants, helps to stabilize sand dunes.

An annual grass, it grows up to 18 inches tall and has fibrous roots and grayish-green stems with ribbonlike alternate leaves. The spiky, yellow-brown flowers are borne in clusters at the top of the stem. Seeds, contained in bristly spherical burs, are produced by midsummer and are green first, then turn to tan or purple. It grows on dunes and beaches, and in desert grasslands.

In some places, such as Brunswick County, North Carolina, sandspur has been deliberately planted to keep people off the dunes so the dunes can develop undisturbed.

Range: **East and West Coasts, from Nova Scotia south to Texas, and from British Columbia south to California; also found in sandy areas throughout the United States.**

Zosteraceae

Eelgrass
Zostera marina, sea grass, grass wrack

Eelgrass looks like a seaweed, but is actually a flowering plant growing in seawater. As the stems creep horizontally along the sea bottom, the leaves emerge vertically above the water level, which is why it's called an emergent plant. The tiny green flowers hide in sheaths at the base of the leaves. Other marine plants and small marine animals attach themselves to the leaves. Eelgrass tends to grow in shallow water (less than 4 feet deep) just below the low-tide line. If the water is clear enough to allow sunlight to filter through, and the high tide isn't too extreme, eelgrass will grow in water up to 100 feet deep. It can also form dense mats, and can be found in bays, shoals, and protected seashores.

Great stands of eelgrass support large populations of fish with food and shelter; marine birds eat eelgrass. It also produces oxygen, helps to filter out pollutants, and stabilizes the sediments in which it grows. Eelgrass is a linchpin for the whole shore ecology. Unfortunately, it is also threatened by industrial pollution and dredging, especially along the Atlantic. When eelgrass is stripped away, the creatures that use it as a nursery are abandoned and the shoreline is lost to erosion.

The stems and leaves are edible, and the native peoples of British Columbia often used eelgrass as a spring vegetable, steamed or cooked with shellfish. The sweet root was either dried for winter fare or used as a special food at feasts. The Tla-o-qui-aht dyed and bleached eelgrass

along with other materials to make articles of clothing such as hats.

According to Captain John Walbran, author of *British Columbia Coast Names, Their Origin and History* (1909), "At Hesquiat village [southeast of Nootka Sound on Vancouver Island] a salt water grass called 'segmo' drifts on shore in large quantities, especially at the time of the herring spawning, which the Indians are in the habit of tearing asunder with their teeth to disengage from the grass or weed the spawn, which is esteemed by them a great delicacy."[89] If you run a piece of eelgrass through your teeth, it makes a sound like "hsh." Thus Hesquiat Harbor got its name from a Nootka word meaning "people of the sound made by eating herring eggs off eelgrass."[90]

Eelgrass was once used commercially as insulation and stuffing. One of the centers of this industry was New Jersey, near Island Beach.

Around 1910, squatters, attracted by a new "seaweed" industry, erected shacks along the shore [of Island Beach] and began collecting eelgrass. During summer months this grass was gathered by boat from Barnegat Bay to the south, and in fall

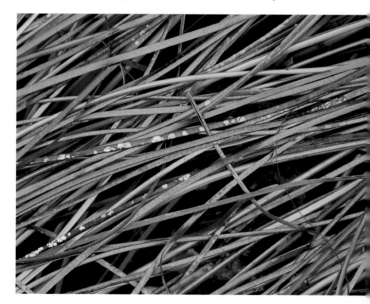

Zostera marina, eelgrass

and winter the men raked what had washed ashore. The eelgrass was then hung on racks to dry, picked clean of clinging crabs and shrimp, and air-dried to remove moisture and odor. This was bailed and shipped by railroad freight to the mainland and used as insulation, mattress stuffing, and as upholstery stuffing in the Model T Ford. However, this new endeavor came to a halt when disease wiped out the supply of eelgrass.[91]

The blight affected the fish and shellfish that were also part of the eelgrass's habitat, and this, in turn, affected the folks who gathered them all along the coast of New England. From 1830 to 1931, they had lived by collecting a cornucopia of food from the sea: crabs, clams, quahogs, and bay scallops along the west end of Cape Cod. In 1931, however, a parasitic fungus, *Labyrinthula*, wiped out 90 percent of the eelgrass along the Eastern Seaboard. Although it is coming back slowly, it has never been found in the same abundance again. Because it was the habitat for bay scallops, they declined as well.

In 1920, poet Edna St. Vincent Millay began what was the first of several years acting in and writing plays for the Provincetown Players. When she returned to New York, she expressed her longing for the seashore in an elegiac poem about eelgrass.[92]

Eel-grass

No matter what I say,
All that I really love
Is the rain that flattens on the bay,
And the eel-grass in the cove;
The jingle-shells that lie and bleach
At the tide-line, and the trace
Of higher tides along the beach:
Nothing in this place.

Edna St. Vincent Millay

Today, fishermen still pay attention to eelgrass because it's where fish can be found in between the tidal shifts. Living in a current stimulates them to bite. "As the time approaches either the high tide or low tide, the current will slow. This is the time to watch the Eel Grass. During a fast current, the Eel Grass will lay way over. As the high or low tide approaches, the Eel Grass will start to stand straighter. This is the time to lighten up your swim bait."[93]

Range: **Pacific Coast from Alaska to California, along the coast of Hudson Bay and northern Quebec, and Atlantic Coast from Labrador to Florida and Texas.**

SALT MARSHES

Salt marshes are places of intense turmoil. One hour they are underwater, the next they are exposed to the air. Twice a day they are affected by tides gently washing in and out. Each time this happens, particles float slowly to the marsh floor and produce an anaerobic (oxygen-free) sediment. It is not just a natural filtration system, it's also a way of adding organic nutrient-rich matter.

The unusual circumstances of salt water flowing in twice a day can be handled by only a few plants: two of the major ones are *Juncus* and *Spartina*. They are called emergent plants because the roots are in the water and the leaves are in the air. What these roots are sitting in can produce a definitely foul odor. The stems of the plants get covered with microalgae, and there are bacteria, protozoa, and metazoa all working away to trap particles as the water moves between the marsh plants.

Other filterers such as mussels, barnacles, and oysters also remove particles from the water as it floods over the marsh. Even though the sediment is anaerobic, there is a load of plant biomass—plant material and organic particles—that doesn't fully decompose. This is compacted into peat which then gets broken up into bits of debris. This, in turn, is consumed by the animals that inhabit these places.

Poaceae

Cordgrass

Spartina alterniflora, smooth cordgrass, saltwater cordgrass

Spartina alterniflora

grows 1 to 8 feet high on stiff, hollow branching stems. The narrow, tapering leaves sheathe the stem alternately. The flower is also narrow, emerging at the top of the plant in a cluster of scaly spikes, arranged alternately on the flower stalk. Found in salt and brackish marshes.

Range: **Newfoundland to Texas.**

Cordgrass covers vast areas of land. The roots of cordgrass hold the soil of the salt marsh in place. On the East Coast this is one of the most important species in seaside habitats. The thick complicated roots build up matted detritus that becomes surprisingly firm. In the upper marshes, where the tide just barely sweeps away plant litter, this is even more obvious. The whole structure of the plant encourages this dense quality of the marsh floor.

Smooth cordgrass is a halophyte, which means it can tolerate the salt water that would kill most plants. It has special glands that excrete salt to the edges of its leaves to get rid of the excess. When it has been particularly dry, look closely and you'll see the crystals on the plant's leaves.[94] The hollow stems, leaves, and rhizomes

allow gases to exchange so that even when it's standing in fetid water, the plant can take up oxygen from the atmosphere and transport it to the roots. Without oxygen, plants can't use nutrients efficiently. Cordgrass has been called one of the most productive land plants in the world, rivaled only by agricultural crops like sugar cane, a plant needing huge amounts of fertilizer and an intense amount of work to harvest. Cordgrass goes it alone.[95]

The submerged roots send out new shoots during the winter when the growth cycle begins. At this time they can spend 12 hours a day underwater. They may have to go for several days without direct sunlight. By the time the summer solstice comes about, cordgrass is about 3 to 4 feet tall; at this time high tides are after dark and there's plenty of light. This is the true meaning of power growth.[96]

Not all invasive alien plants come from other countries. *Spartina* is an extremely useful native plant on the East Coast, but back in the 1890s it was used for packing oysters to be shipped from the East to the West Coast. Seeds from the grass packing escaped and it became one of the most prolific and baffling exotics along West Coast marshes. Not realizing that it would go completely haywire in this new environment, some well-meaning folks planted *Spartina* in the 1970s in the San Francisco Bay Area, hoping to restore the local salt marshes. They had no idea it had no known enemies in the West, or that it would produce 20 times the pollen of the local *S. foliosa*, or that it would become an expensive threat to the resting sites of migrating birds.

Today, Adopt-A-Beach programs try to halt the relentless march of this East Coast alien. There have also been desperate attempts to knock it out using certain insects (*Prokelisia* spp., a.k.a. leafhoppers) that attack it.[97] Naturalists worry about the "genetic pollution" of the species on wildlife in the Bay Area. Because the hybrids spread farther down the mud flats than the natives do, less open land is available for foraging shorebirds and families of harbor seals. The invaders and the hybrids also choke flood-control channels, causing problems for navigation.[98]

S. patens is a shorter version of *S. alterniflora*. It is known as **salt-meadow hay, salt-meadow cordgrass**, or **marsh hay**. The light-green grass forms thick mats in marshes. It was once mown for hay, hence the name. This was particularly important to very early settlers, who often had no other fodder for the animals they had brought with them.

Juncaceae

Black Needlerush
Juncus roemerianus, needlerush, needle grass

Salt tolerance is the pivotal element of the needlerush's existence, making it one of the major plants of the salt marsh. It has two distinct features: a lateral-appearing flower and sharp pointy "stems" that are strong enough to puncture skin. They are actually cylindrical dark brown leaves furled tightly into a needle-sharp tip. Male and female flowers grow above a rigid, coarse stem reaching a foot or so in salty areas and up to 7 feet high in less salty areas. The rhizomes from which it grows may form colonies in coastal tidal marshes and along estuaries as far as 15 miles inland.

Black needlerush, one of the dominant species in the salt marshes of the southern Atlantic states and along the Gulf Coast, is usually found at higher elevations than other salt-marsh plants, where salt water completely covers the land only during unusually high tides. Quite often, between the needlerush stands and the sea there is a band of **smooth cordgrass** (*Spartina alterniflora*). The two species don't intermingle and the boundary

between the two is usually very distinct and sharp. It makes a wonderful pattern along the shore.

The *Juncus* deals with salt by transporting it to the cells at the tips of the leaves, adding to its spikiness. Naturalist Jack Rudloe recalls an encounter with needlerush during a trek through one of northwest Florida's wetlands:

The tips of its spiky blades came up to our waists and continually pricked our fingers. And in Juncus, *one doesn't dare bend over for fear of getting stabbed in the eyes. Although wetter, it was easier to walk through the creek beds, following the winding watercourses that wove in and out through the marshes. Despite its prickliness, its uniform long thin reedy stalks, needlerush is extremely fragile. Deer and raccoon leave permanent trails snaking through the wet grasslands. At one point we crossed a freshly flattened wide line of grass, an alligator trail—a big, fresh alligator's trail.*[99]

Early settlers used the sharp spines for needles. Rush leaves have also been used in basketmaking. Today, black needlerush is being planted in an attempt to restore coastal wetlands.

The wetland at Galveston Island State Park in Texas covered 1,000 acres in the mid-1950s. Today there is less than half due to draining of oil, gas, and groundwater from below the surface. The major plants being used to restore this site are black needlerush, along with smooth cordgrass, bulrushes, and cattails. This will "reproduce the ecological zones characteristic of natural coastal marshes that show a gradation from . . . the low marsh, to brackish areas."[100]

At the University of Delaware, black needlerush is being considered as a substitute for the common reed, *Phragmites australis*, which is invading wetlands. Where once there was rich diversity, this plant is going berserk, ruining the amount of food available to wildlife, and blocking the sun available to other plants. Under Jack Gallagher, a team is at work identifying varieties of marsh plants to form a natural barrier to the common reed. These plants are affectionately called "*Phrag* blockers." "They include plants such as black needlerush and three-square sedge, which have rigid stems; deep, dense root systems; and other features that would stop *Phragmites* from reaching the 'end zone.' "[101]

Range: **Maryland to Texas.**

Asteraceae

> ### Groundsel Tree
> *Baccharis halimifolia,* groundsel bush, sea myrtle, silverling, cottonseed tree, salt-marsh elder

Baccharis halimifolia is the only member of the family Asteraceae that grows to the size of a tall shrub and it is very tall—to 12 feet. Its many branches are covered with alternate pale gray-green oval leaves, 3 inches long with scalloped edges. It is dioecious: male (pale yellow) and female flowers (white) appear on separate plants, clustered at the end of the stems in late summer. The fruits have lustrous white bristles, giving the whole tree a silvery tinge. It's called cottonseed tree because of the white hairs surrounding each seed and can be found in salt marshes, brackish marshes, and in fields along coastal areas.

According to Timothy Coffey in *The History and Folklore of North American Wildflowers*, the presence of *Baccharis halimifolia* was once thought to indicate places where oil could be found.[102]

The groundsel tree's tolerance of salt spray and rapid colonization of disturbed sites may make it useful for rehabilitating disturbed sites. Mary Parker Buckles describes a hardy survivor on Long Island Sound. It was one of the few plants left after a shattering drought in 1995.

> *The drought was even more evident in plants on flatter, maximally exposed bedrock. . . . Like the prematurely fallen leaves, these woody plants looked sere. . . . The only prominent color anywhere around belonged to goldenrod, bronzy poison ivy foliage, and a two-foot-tall groundsel tree, a shrub often associated with salt marshes. The groundsel tree, with white fruits and iridescent underpinnings on its branches, grew in a protective V of stone.*[103]

Range: **Massachusetts to Texas.**

Salicornia spp., glasswort

Chenopodiaceae

Glasswort

Salicornia spp., beach asparagus, pickleweed, false samphire, chicken-claws, pigeonfoot, leadgrass

Salicornia spp.

grow 20 inches high, and are leafless. It is an emerald-green succulent plant that turns a translucent pink or red in fall. The thick, fleshy stem grows from a rhizome and has many opposite branches. Inconspicuous green flowers appear in late summer growing in the joints along the stem.

Range: Atlantic Coast from Quebec and Newfoundland to Florida, Pacific Coast from Alaska to California.

Glasswort, according to some sources, got its name from the scrunching sound it makes when stepped on, as if one were walking on broken glass. This is one of those neat bits of botanical information that sounds like an urban myth. The scientific name *Salicornia* means "salt horn," because it grows in salty environments and has small branches that look like little horns. In the fall, it carpets the salt marshes in red and bronze.

"Glasswort, and some other seaside plants, have a simple way to cope with their salty environment," says Steve Brill. "If you can't lick 'em, join 'em. They contain enough internal salt to balance the sodium around them."[104] Each bladder has one hairy seed that falls separately from the fruit.

The common names all point out that this is a food. The edible stems can be added to green salads or steamed lightly before being eaten.

The Kaigani Haida of Alaska eat them fresh, canned, or pickled. Some Native Americans on Vancouver Island sell it at markets. It's wise to be careful tasting glasswort for the first time—it can irritate the throat. "Glasswort is at its best eaten raw, eating only the upper and more tender parts of the plant," Christopher Nygeres advises. "Glasswort can be used also as the main salad ingredient, mixed with other salad greens (wild or domestic), or even as a garnish with any seafood dish (same as restaurants use watercress and parsley sprigs). Mixed with stews and soups, glasswort provides all the salt needed."[105]

In the past, the ashes of glasswort were used to make glass and soap. Mary Ann Spahr of the Rainier Audubon Society says: "The plants were dried, and then burnt in large heaps. The crude ash was fused with sand to make a rough glass, or leached with lime water to make a solution of caustic soda. On evaporation this would make a clearer glass, or be used with animal fats for making soap."[106]

Malvaceae

Seashore Mallow

Kosteletzkya virginica, pink mallow, salt-marsh mallow

Seashore mallow is the showiest plant in the salt marsh. It looks like the ordinary garden mallow, *Althaea officinalis*, but is quite a different species. Its name commemorates Vincent Franz Kosteletzky (1801–1887), who lived in Prague and wrote on medical botany.

It grows up to 4 feet tall, with triangular, hairy alternate leaves, larger on the lower part of the plant and smaller on the upper part. The large pink flowers each have five petals emerging from the axils of the leaves and in a cluster at the top. Look for them in salt and brackish marshes.

Kosteletzkya virginica, seashore mallow

Plumbaginaceae

Sea Lavender

Limonium carolinianum, L. nashii, marsh rosemary, cankerroot, ink root

Limonium carolinianum and *L. nashii*

are very similar, except that the flowers of *L. nashii* have fine hairs at their bases. The narrow oval leaves of the sea lavender grow from the base of the plant and may be up to 6 inches long. From the center emerges a single branching flower stalk bearing many tiny, lavender-colored, tubular flowers. Found in salt and brackish marshes.

Range: Newfoundland and Labrador to Texas. *L. carolinianum* grows south of Long Island (its name suggests that it is associated with the Carolinas), *L. nashii* from Labrador south.

Seashore mallow resembles the rose mallow, which is actually a type of hibiscus. It also looks very much like the old-fashioned mallow settlers brought with them for their New World gardens. They ground up the roots of this mallow to make a sticky substance that was sweet and gooey. If it sounds familiar, it's because it was dubbed marsh mallow. (Today's packaged marshmallow no longer has ground-up mallow roots in it and is made with other ingredients.)

Range: **Long Island to Florida and Texas.**

Limonium carolinianum, sea lavender

"Who can laugh at my Marsh Rosemary," wrote Sarah Orne Jewett in 1886, "or who can cry, for that matter? The gray primness of the plant is made up of a hundred colors, if you look close enough to find them. This same Marsh Rosemary stands in her own place, and holds her dry leaves and tiny blossoms steadily toward the same sun that the pink lotus blooms for, and the white rose."[107]

This attractive plant is almost its own worst enemy. It was a fairly common plant at one time but has been so overcollected by dried-flower arrangers and commercial enterprises it is now rare to find it in its natural habitat. Once a familiar sight all along the Massachusetts coast, it is now endangered there. This is a highly specialized plant, growing in mud just above the high-tide mark in marshes. When the flowers are picked, the roots usually come along with them, which is why the species is in so much trouble. It is now illegal to remove sea lavender from the national seashore.[108]

Deborah Rose of Plymouth, Massachusetts, remembers collecting sea lavender as a child. "At the Little Beach we watched a huge rock, that was nicknamed 'The Tide Rock' to determine if it was high tide. . . . We were allowed to swim to the other side and pick sea lavender, and we would swim back with the lavender in our mouths or hold it with one hand and swim the side stroke."[109]

Limonium is derived from the Greek word for "meadow." In some areas, sea lavender and other seashore plants can indeed make a salt marsh look like a meadow with masses of yellow salt-marsh grass, sea lavender, and goldenrod.[110]

The thick, reddish, perennial root of sea lavender is recommended for use in medicine: it is used to heal ulcers and sores since it contains tannin and is astringent. Doug Elliott in *Wild Roots* says: "Sea Lavender was used on the battlefields of Britain in ancient times as a means to stop bleeding. The plant earned the name canker root because of its use in healing canker sores."[111]

SEAWEEDS

When Christopher Columbus churned his way across the Atlantic Ocean, he encountered the mysterious Sargasso Sea and found himself amid a vast platform of seaweed. Columbus thought he was so close to shore that he could smell the land. He was wrong, of course. He still had 1,400 miles to go. The Sargasso Sea and its massive kelp bed is an ecosystem unto itself as it floats along the Gulf Stream near the Canaries. The gas-filled bladders of the kelp hold it up and provide a home for the sea creatures that inhabit the bed.

Seaweeds are placed in three categories: green, red, and brown. They share some common traits: all lack true stems; have anchors called holdfasts; absorb nutrients directly from the water; and manufacture their food by photosynthesis. The red and green seaweeds have pigments that mask chlorophyll. Green seaweed is part of the upper shore along the high-tide zone and is able to tolerate pollution even from sewage. Red seaweed can dry out and still survive near the low-tide line. Brown seaweed, which is found in the middle zone (between high- and low-tide lines), can survive in air for long periods. These are the largest algae in the world—they have no roots, no stems, no leaves, and yet can grow 200 feet long. There are more than 50 species of brown seaweed and more than 200 kinds of red seaweed.

As soon as they stepped from the small boats taking them from ship to these new shores, the early colonists would have been surrounded by red and brown seaweed. At that time the algae were far more common than they are now. This would have been a familiar sight to those who had come from maritime towns, though they would not have recognized the exact species nor such a huge variety. Sean and Robert Manley imagine the scene: "The seaweed clung to the rocks, sometimes to other algae, but any storm cast them up along the sand beach, and they glistened so on the rocks of the rocky coasts that they must have seemed like the skins of great seals lying upon the rocks."[112]

Laminariaceae

Horsetail Kelp
Laminaria digitata, tangle, kombu, black kelp, finger kelp, seastaff, sea girdle, ribbon weed

Laminaria digitata
is an olive-brown seaweed with a broad blade that splits into numerous straplike fingers (*digitata* means "fingers"). The holdfast is yellowish and the stalk is stiff and almost woody.

Range: Atlantic Coast from Newfoundland to New England, Pacific Coast from British Columbia to California.

Horsetail kelp, a brown seaweed, grows along the East Coast shoreline. The long slender stalk and leaflike blades manufacture food through photosynthesis. It grows on a long stem with a holdfast, which is like a root except it attaches itself to rocks rather than soil. The stem is normally about 5 feet, but can reach 10 feet in length. Among the wondrous tales about the size and strength of this seaweed is this one by Henry David Thoreau, written in 1865:

The long kelp-weed was tossed up from time to time, like the tails of sea-cows sporting in the brine. . . . There was but little weed cast up here, and that kelp chiefly, there being scarcely a rock for rock-weed to adhere to. Who has not had a vision from some vessel's deck, when he had still his land-legs on, of this great brown apron, drifting half upright, and quite submerged through the green water, clasping a stone or a deep-sea mussel in its unearthly fingers? I have seen it carrying a stone half as large as my head. We sometimes watched a mass of this cable-like weed, as it was tossed up on the crest of a breaker, waiting with interest to see it come in, as if there was some treasure buoyed up by it; but we were always surprised and disappointed at the insignificance of the mass which had attracted us. . . . This kelp, oar-weed, tangle, devil's-apron, sole-leather, or ribbon-weed,—as various species are called— appeared to us a singularly marine and fabulous product, a fit invention for Neptune to adorn his car with, or a freak of Proteus. All that is told of the sea has a fabulous sound, to an inhabitant of the land, and all its products have a certain fabulous quality, as if they belonged to another planet, from sea-weed to a sailor's yarn, or a fish-story.[113]

Red is the sea-kelp on the beach,
 Red as the heart's blood,
Nor is there power in tide or sun
 To bleach its stain.
 It lies there piled thick
 Above the gulch-line.
It is rooted in the joints of rocks,
 It is tangled around a spar,
 It covers a broken rudder,
 It is red as the heart's blood,
 And salt as tears.

E. J. Pratt, from "Newfoundland," in *Complete Poems*

Japanese doctors treat high blood pressure with horsetail kelp or *kombu* tea, which is available in most health food stores. A small piece of this kelp can also be added to soups and stews like a bay leaf—once it has flavored the dish, it is removed. Adding *kombu* to bean dishes makes the beans easier to digest. This nutritious seaweed is also a good source of vitamins A, B, and C, iron, zinc, and many other minerals.

Any kind of kelp makes a terrific fertilizer. It should be left in a large container with water for a few months and then diluted 10 to 1 with water before being poured around plants. It is also used in animal feed.[114]

Lessoniaceae

Bull Kelp
Nereocystis luetkeana, bullwhip kelp, bladder kelp, ribbon kelp

Bull kelp is one of the largest of all marine algae. It is golden to dark brown and grows in slender, unbranched stalks up to 100 feet long. It has a thick holdfast like a suction cup, which clamps it to the sea bottom in water as deep as 65 feet. At its base, the stalk is solid and fairly narrow, but it widens and becomes hollow toward the top. At the very top is a spherical float up to 6 inches in diameter, with two clusters of flat, elongated blades attached to its upper surface.

Children can turn almost anything into playthings and these giants of the sea are no exception. For uncounted generations, kids have chased each other up and down beaches trying to hit each other with the strands, or raced around stomping and popping the bladders as fast as possible. Nuu-chah-nulth children on the West Coast dried the floats and put them in the fire to explode like firecrackers. Haida children liked to squirt water out of the sacs like milk from a cow's udder.

Their parents also had many uses for kelp. The Comox lined steaming pits with bull kelp to flavor the food and generate steam. The Kwakwaka'wakw used the hollow tubes in steaming pits to add water to the hot rocks at the bottom during cooking. Sea wrack and other seaweeds were piled over hot rocks to generate steam for cooking, bending and molding wood, or medicinal sweat baths. Coastal peoples used giant kelp to collect herring eggs during the spawning season. A great delicacy was to eat the fronds with the eggs.

Virtually all West Coast native peoples made fishing lines, nets, ropes, and harpoon lines from bull kelp. They dried and cured the long stalks, then spliced or plaited them together. Kelp was also used to make anchor lines, and fishermen often fastened their boats to living kelp plants anchored on the bottom.

Native Americans also sucked the long salty strips of kelp to combat a sore throat and to clear mucus out of the system. The leaves could be dried, ground into a mulch, and then rubbed into a child's hair to make it grow long.

Range: **Alaska to California.**

Fucaceae

Knotted Wrack
Ascophyllum nodosum, knotted rockweed, sea whistle, Norwegian kelp

In the mid–tidal zone of rocky seashores you can clearly see a brown seaweed called knotted wrack. The botanical name gives a vivid description of its qualities: *Ascophyllum* is composed of two Greek words: *askos,* meaning "a leather water bottle," and *phyllon,* meaning "leaf." *Nodosum*, from the Latin *nodosus*, means "to be covered in lumps or knots." Slow-growing and long-lived, knotted wrack reaches 3 to 4 feet in length, sometimes even longer, varying in color from yellow to olive to dark brown, depending on the time of year. The smooth flattened fronds branch in two along the stem. Every few inches, a frond enlarges to form an air bladder. Beds of knotted wrack form complicated ecosystems with many other species, such as goose barnacles or epiphytes, attached to them. The dense mats cover submerged rocks in this zone and provide shelter for marine invertebrates and a canopy to hide young fish.

Since the days of the early settlers in northern New England, knotted wrack has been gathered as a fertilizer containing large amounts of iodine and potassium. Today it is used to produce specialty paper, printing dyes, and welding rod coatings. Knotted wrack is also used in the East Coast fishing industry as packing material for shipping live lobsters, clams, and sandworms (used for bait).

All brown algae are sources of algin, a substance used as a thickener and stabilizer in ice creams, puddings, cough medicines, and laxatives. The algin in knotted wrack has been used in air fresheners and explosives.

Range: **Newfoundland to New Jersey.**

Gigartinaceae

Irish Moss
Chondrus crispus

Irish moss, a red seaweed, grows in curly clumps, anchoring itself on rocks, wood, or even shells, and surviving even in exposed rocky areas where the waves pound the shore. It needs light and suffers in areas where the water is not clear. Water pollution and dredging are disasters for this plant. Sediment adhering to rocks also prevents it from becoming attached. Irish moss is also threatened by intense harvesting.

The Greek word *chondros* means, among other things, "grains" as in grains of salt, while the Latin word *crispus* means "curled." Though *Chondrus crispus* is a branching red algae, it may also appear as yellowish-white, green, purple, or brownish-red, up to 6 inches tall. It usually grows in clusters of 10 to 20 fronds from a single holdfast on a rock surface in rock pools and the lower intertidal zone of rocky seashores.

Irish moss is the source of carrageenan, and is grown and harvested commercially. Carrageenan, a gummy, mucilaginous substance, acts as a thickener, stabilizer, and gelling agent in foods such as Jell-O, ice cream, salad dressing, chocolate, and instant puddings, as well as in pharmaceuticals, toothpaste, cosmetics, paints, and textile sizing. Long a folk remedy for tuberculosis, coughs, bronchitis, asthma, and intestinal problems, it has also been used to clarify beer and tan leather. As someone once said, "It puts the gleam in your beer and the frosty glint in your ice cream." Opening a pint of ice cream really means getting a whiff of the sea, rather than of mountain frost.

The word *carrageenan* comes from County Carragheen in Ireland, where the properties of this seaweed were discovered about 600 years ago. Fifteen percent of the world's supply of hydrocolloids comes from carrageenan. In the nineteenth century, gathering moss was an important industry for many North American coastal communities.

An Irish immigrant, Daniel Ward, and his friend Myles O'Brien came to Scituate, Massachusetts, in search of mackerel and found to their surprise that an abundant supply of Irish Moss

Chondrus crispus, Irish moss

grew on the ledges of Scituate and could be easily harvested at low tide using long-handled rakes. At certain locations a good mosser could harvest between 500 and 2,000 pounds a day. The moss could be harvested twice daily at low tide. It was spread on the beaches to dry out for up to two weeks, then packed in tubs each night. Next day they'd be filled with brine and spread out again.

In the early days, some 100 to 200 families made their living mossing from about mid-May until August. The height of sea mossing was between 1847 and 1860. The price went from 50 cents a pound to a post civil war depression price of 4 cents a pound.[115]

Salome MacLeod of Beach Point, on the south coast of Prince Edward Island near Murray Harbour, wrote down her memories of mossing:

For many years there were thousands of dollars worth of moss collected and sold here.

After a big wind storm, the shore around the Cape would be piled high with this moss, and people of all ages and sexes were there, even the children would run down at low tide and earn most of their spending money. . . .

It was very healthful work in the warm sunshine and the salt water sea breeze. The moss was brought home, and spread out of doors to dry in the sun, then it was hand screened to remove kelp and other foreign matter, weighted in, then sold, then pressed into bales and shipped to New Bedford, Massachusetts. . . . Now the moss is extinct here.[116]

Range: **Newfoundland to New Jersey.**

Fucaceae

Rockweed

Fucus vesiculosus, bladderwrack, bladder rockweed

Fucus vesiculosus or rockweed is an olive-brown seaweed with flattened stems and a raised midrib which branches in repeating Y shapes. Paired bladders occur on either side of the midrib. The fronds end with swollen areas which are the reproductive receptacles. This plant lives in the intertidal zone, where there is so much stress due to wave action that it must attach itself to a rock or it will be torn away and washed up on shore. Many things gather themselves into rockweed as described by Julie Hodgkins of Trescott, Maine: "I find all different types of things in the rockweed—periwinkles, crabs, eels, limpets, sand fleas—each piece of seaweed is home to something. The fish come in to feed in the rockweed when the tide is high, and I see eagles and loons and other birds follow them in. When I'm clamming, I find more clams beneath rockweed beds. And there are animals tracks all over the beach in the morning. Many of these animals feed on the crabs and snails living in the rockweed—it's all connected."[117]

This plant can also be used as garden mulch, as Margaret Eaton recalls:

When I was growing up in the fifties in Portland, Maine, our family had a lovely perennial flower garden at the front of the house, which my mother tended, and a vegetable garden at the back, which my father planted every spring. The fifties and early sixties were a time of technological "advances," including commercial fertilizers, which my father leaned toward as being convenient and relatively cheap, but my mother would have none of them. She came from a family of growers in Guernsey, Channel

Fucus vesiculosus, rockweed

BOTANICA NORTH AMERICA

Islands, UK, who had used natural fertilizer in the form of seaweed, in particular bladder wrack, or as she called it, "Vraic."

And so, every year, late in August or early in September, we climbed into our trusty old Studebaker and drove north along the Maine coast, collecting the common brown seaweed, which when torn by storms from the rocks which it carpeted, formed floating mats and washed up on shore, lying in tide lines along the beach. We bundled it into garbage bags, stowed it in our trunk and drove back to the city to spread it on the garden as mulch in the early fall, tilling it into the garden in the spring. The biological breakdown returned nutrients to the soil and I'm sure now that the veggies we ate were all the better for it; the tomatoes were plumper and redder, the lettuce curlier and greener. At the time, however, I used to be embarrassed by the rather strange smell emanating from the trunk of our car.

Range: Hudson Bay to New England and as far south as North Carolina.

Rhodymeniaceae

Dulse
Palmaria mollis syn. *Rhodymenia palmata*, Neptune's girdle

Dulse summons up particular memories for me. When as a child I spent summers beside the Bay of Fundy, it was one of the sea plants that we were forced to collect, mainly to sell it to people inland. I got to hate the taste of dulse, but recently someone gave me a small bag of it. Just opening it lifted the top off a box of memories of the sea and the cliffs and the 50-foot tides. We would terrify ourselves following the tide out to pick up the dulse, then racing back to the weir as the sound of the tide turning or rain pounding down echoed across the water. I am older now and wiser, but I still hate dulse. Some people swear by it as a tonic for long life and actually like its cloying salty taste.

Dulse is a reddish-purple seaweed with broad, often branching blades, and the best place to find it is in tidal areas. It grows on rocks near the low-water mark, usually along with other seaweeds such as *Laminaria*.

Some Inuit peoples used to boil dulse with fish. Today, dried dulse is a salty snack sold in health food stores and some supermarkets because it is rich in calcium, iron, phosphorus, potassium, and vitamins A and B. It is harvested commercially in Nova Scotia, New Brunswick, Maine, British Columbia, and Washington state. According to Alaskan naturalist Janice Schofield, "Sea vegetables, dulse included, are said to contain every trace mineral known, in the same concentration as human blood."[118] Herbalists recommend tea made with dulse for menstruating and lactating women. For a cold remedy that fisherfolk swear by, soak a handful of dulse for 10 to 15 minutes in enough hot water to cover. Then cook the dulse in that water, at very low heat, for 15 to 20 minutes. Strain. Save seaweed; add to soup or compost. To the leftover liquid add honey and lemon juice. This tastes pretty awful unless you put a lot of honey and lemon in it (and maybe a bit of rum or brandy), but it actually does work well.

Rosemary Hill of Massachusetts recalls her first introduction to dulse as an alternative snack:

While the rest of our family were munching on chocolate bars, chips and popcorn, my Aunt Sarah, who was diabetic, used to snack on dulse. I think it was when she was visiting the international site of the summer home of FDR on neighboring Campobello Island that she first became acquainted with it, although the small cellophane bags of dried dulse were and still are shipped to towns and cities along the Atlantic seaboard to be enjoyed by people like my aunt who were denied the dubious "pleasure" of high sugar, high cholesterol snacking.

Perhaps I shouldn't have been so hasty forty years ago in turning my nose up at this healthy alternative. Just one handful of dulse contains 100% of the recommended daily allowance of Vitamin B6, along with generous amounts of potassium, magnesium, calcium, phosphorus, iron, and iodine!

Range: **Pacific Coast from Alaska to California, Arctic coast from Nunavut to Labrador, Atlantic coast from Newfoundland to Maine.**

Ulvaceae

Sea Lettuce
Ulva lactuca, water lettuce, green laver

Ulva lactuca with its thick, translucent, lettuce-like sheets up to 18 inches long, often with ruffled edges, comes in such a bright green it looks almost fake. It has no stem, only a small holdfast, and is found mainly in the upper intertidal zone, in rock pools, mud flats, estuaries, and tidal areas, anchored to rocks and jetties, or clinging to something as small as a shell or even a live crab.

These fast-growing seaweeds tolerate a wide range of salt and temperature conditions. In sheltered areas, sea lettuce may occupy entire tide pools with individual plants measuring more than 2 feet. They can rapidly colonize any part of the rocky seashore if the conditions are right. After a disturbance, *Ulva* is one of the first seaweeds to reappear on the seashore.

This edible seaweed is best harvested in spring, when it is tender. It gets tougher later in the year. Washed of sand, boiled briefly, it can be eaten like spinach, chopped, and added to salads and soups, or dried and used to season soups and casseroles. Raw, it doesn't have much taste, but the dried form is said to be more flavorful—though an acquired taste. *U. lactuca* is wonderfully nutritious, high in iron, as well as high in protein, iodine, aluminum, manganese, and nickel, and contains vitamins A, B_1, and C, sodium, potassium, magnesium, calcium, soluble nitrogen, phosphorus, chloride, silicon, rubidium, strontium, barium, radium, cobalt, and boron.

At one point during all the excesses of the Victorian era it was fashionable to collect seaweeds and press them. They look like fantastic pieces of fabric.

Range: **all North American seashores.**

Ulva lactuca, sea lettuce

Florida

Florida's landscape is unique in North America. Moody everglades with silent alligators lurking in their depths, mangrove swamps and hammocks that are not quite land yet not fully water, and tropical vegetation make it seem more Caribbean than continental. It has been called a land bridge to the tropics—the Keys are virtually Caribbean islands, hopscotching their way toward Cuba. Tropical and temperate plants mingle here in an entirely new way. Cooling breezes blow and there are daily rains during the summer.

Florida's rich landscape ranges from swamps like this to inland dunes.

Next to California, Florida has the most diverse native flora of any state, with many endemic plants—plants that require a very specific habitat and grow only in certain restricted areas.

Florida is also living proof of the movements of continents over billions of years. The basement rock, the very foundation of this state, is the same as that of northwest Africa. The two places were once close, when the whole planet was one vast continent called Pangaea. As the continents pulled away from each other, many plant communities were stranded in one continent or the other. Plants were also carried by fierce winds such as hurricanes and tornadoes, ocean currents, and by birds flying from one landmass to another. The great ice sheets that lowered the land with their massive weight thousands of years ago, drowned the forests of the Florida plateau. As the ice retreated north, the melting left Florida with a system of lakes and wetlands, a high water table, interior sand dunes, and the indelible marks of ancient seas. By the time people arrived here 10,000 to 12,000 years ago, the north was pineland and sandhills with broadleaf forests, the middle was pine scrub and oak woodlands, and the far south had cypress swamps, sawgrass marshes, and the Everglades.

Among the earliest human inhabitants of the peninsula were the Calusa, who built canals and left evidence of a sophisticated culture along the bays and estuaries of the coasts. They must have been an amazingly attractive and healthy-looking lot. Ponce de León, who was the first Spaniard to set foot on their shores, believed he had discovered the fountain of youth when he encountered them.

Not much has changed; Florida still attracts people who want to prolong their lives in its balmy climate. However, the popularity of the state has led to heavy, often crass development, some of which threatens Florida's fragile environments. The chronicler of all things Floridian, novelist Carl Hiaasen, writes of the sharp contrasts between urban Florida and natural Florida:

> Later as he walked along the dike, . . . Keyes marveled at the contrast: to the western horizon, nothing but sawgrass and hammock and silent swamp; to the east, diesel cranes and cinderblock husks and high-rises. Not a hundred yards stood between the backhoes and the last of South Florida's wilderness.[1]

In spite of the sea of cement surrounding them, plants still grow in astonishing profusion.

Citizens are finally beginning to recognize the damage done by overdevelopment and managed water systems, and there is an increasing respect for native grasses that hold down the seashore. Floridians are slowly coming to realize that they must save what attracted them in the first place: the exotic beauty that makes this place unique in all of North America.

MANGROVES

Florida's mangrove swamps cover 1,000 square miles. The word *mangrove* can refer to one of three plant species or to an ecosystem. In the latter case, mangroves consist of several salt-tolerant plant species that live in the intertidal zone of a subtropical or tropical region. In certain places along the coast, the mangrove swamps are 3 or 4 miles wide. The word *mangrove* is thought to be derived from *mangue*, a Senegalese word for "tree" taken up by the Portuguese, elided with the English word *grove*. In other words, a stand of trees, which hardly conveys the singularity of this plant and the landscape it creates. It can build new land and islands by stabilizing the sea floor off the coast and creating soil by trapping sediment and plant material in its tangled branches.

Mangrove swamps have survived because they resist close inspection by humans. No boat can navigate through the dense mazes of stems, prop roots, and, in some places, aerial roots, standing in water or mud, and it's impossible to walk through them, although boardwalks have been built in some places.

Christopher Columbus described them as trees growing in the water. Sir Walter Raleigh noticed oysters in the trees, and thought it seemed rather odd. In fact, oysters and other sea creatures can become attached to prop roots, which are exposed at low tide. Birds, crabs, lizards, and snakes also make their home in mangrove swamps.

Large numbers of invertebrates and fish depend entirely on mangrove swamps as their habitat. Snook, snapper, tarpon, jack, sheepshead, red drum, oyster, and shrimp feed and thrive in the swamps. Florida's recreational and commercial fisheries would dwindle without its fetid mangroves. Why the smell? Because decaying plant matter is an important part of the thick mud that builds up in the swamps, and this produces the revolting odor of hydrogen sulfide.

Plant communities start with a pioneer—one of those intrepid plants that will get things started. In southern Florida, the pioneer of the mangrove swamp is the **red mangrove** (*Rhizophora mangle*), which establishes itself in the sand or mud in shallow water at the edge of the sea. As the plant takes hold, silt collects in the prop roots, nurturing the growth of new seedlings. The sediment builds up over time, preventing erosion and extending the land out to sea. This creates further shallow areas where more mangroves can sprout. Eventually, conditions become suitable for the **black mangrove** (*Avicennia germinans)* to take over sediment collecting from the red mangrove. The former doesn't have the prop roots that allow red mangrove to grow in the sea, and exploits the sediment already built up by its predecessor. As former seashore becomes land, the **white mangrove** (*Laguncularia racemosa*) gains a foothold. Finally, the buttonwood tree (*Conocarpus erectus)*, the gumbo-limbo (*Bursera simaruba*), and other trees take over, and the area is no longer simply a mangrove swamp, it's a tropical forest. Meanwhile the red mangrove is out there, creating more land, bridging the gaps between islands or keys.

All three kinds of mangrove grow in southern Florida, but in the northern part of the state, only the black mangrove can survive. Mangrove growth also differs depending on the areas: in northern Florida, most are less than 20 feet tall, but farther south on the Gulf Coast, monster mangroves 80 feet high and 7 feet wide can be found.

Florida's mangroves have existed for about 60 million years, according to the fossil record. Why, then, have mangroves not filled in the entire Caribbean? Probably because mangroves grow in shallow water (parts of the Caribbean are very deep), and because the level of the water has risen and fallen over the eons, drowning mangroves at times, leaving them high and dry at others. At the moment, the sea level is rising about ½ inch every 50 years.

Mangrove swamps also protect the shore from waves and erosion by dissipating the energy of the waves as they hit the dense tangle of branches. During tropical storms, when the waves are high, animals hide in the safest places around—the mangroves. Hurricanes, however, can rip up mangrove swamps, and forest fires can wipe out thousands of acres at a time. The mangroves' deadliest enemy is, of course, tourism and development. Former mangrove swamps now lie under shopping-mall parking lots, trailer parks, and amusement grounds. Elsewhere they have been cleared and dredged to make room for yachts.

Rhizophoreaceae

Red Mangrove
Rhizophora mangle

Rhizophora mangle

is usually a small bushy tree 15 to 30 feet high, although it can reach 80 feet, with spreading branches and arching roots that prop up the stems. The leathery, oval, evergreen leaves have black dots on the undersides. The flowers are pale yellow, each with four petals, and grow in

Rhizophora mangle, red mangrove

clusters of two or three in the leaf axils. They give way to long, spearlike seedpods that hang down and eventually fall off into the water or mud under the tree. The red mangrove grows in salt water, in coastal bays or swamps.

Range: southern Florida.

"Though they may look like a vegetative ballet troupe dancing on tiptoes along the fringes of island waterways, mangroves are actually more like a construction crew," says one guide to Florida wildlife.[2] Red mangroves are known as "walking trees" since the plants appear to be either standing in or walking on water. Their stiltlike prop roots reach down into the water. Hence the "tiptoe" effect.

The seeds can actually sprout in water rather than on land. Once it falls from the parent plant, the seed may stay afloat for months, at first lying horizontally on the surface of the water, but continuing to grow. As its center of gravity shifts, it slowly becomes vertical with roots shooting out the lower end. When the roots touch the sea bottom, they cling and, almost immediately, the seedling starts to send out prop roots to stabilize itself and settle in. Its other specialty is that it can grow in either salt- or freshwater.

The roots can block salt and extract fresh water from salt water as the tide flows back and forth. Eventually mangroves form a barrier between land and sea. The deep roots become like anchors since mud has no oxygen to supply nutrients to the plant. To absorb oxygen from the environment, the plant has developed absorptive root hairs close to the surface. Openings known as lenticels in the bark of the prop roots allow gases to exchange inside and outside the plant and take in necessary oxygen.

Marjory Stoneman Douglas in her book *The Everglades* gives this description of what appears to be a miraculous feat of nature:

> *The headwaters of many of those wide, slow, mangrove-bordered fresh-water rivers [are] like a delta or an estuary into which the salt tides flow and draw back and flow again.*
>
> *The mangroves become a solid barrier there, which by its strong, arched and labyrinthine roots collects the seepage of the fresh water and the salt and holds back the parent sea. The supple branches, the oily green leaves, set up a barrier against them. There the fresh water meets the incoming salt, and is lost.*[3]

The hard, heavy, dark reddish-brown mangrovewood is streaked with a lighter brown and is surrounded by lighter sapwood. It has been used for cabinetry, construction, piling, poles, posts, shipbuilding, and wharves. The bark has been used in folk medicine for a variety of complaints, from dyspepsia to scrofula.[4] Writer J. F. Morton "even describes a wine made from mangrove leaf and raisin. Dried hypocotyls [the stalks] have been smoked like cigars."[5]

However, the most important role played by mangroves is as a shore stabilizer. They are so important to the shorelines that it is illegal to cut them down. According to many Floridians, they are the best place to sit out a hurricane, because they absorb the shock of the waves.

Avicenniaceae

Black Mangrove
Avicennia germinans

Avicennia germinans

is a large, many-branched tree up to 60 feet tall, with dark brown, scaly bark. Many roots are visible at the base of the tree. The leathery leaves are yellow-green, 2 to 4 inches long, and coated with fine hairs on the undersides. The white, fragrant flowers grow in clusters at the ends of twigs and give way to pale green, hairy, podlike fruits. Found along the coast and on islands offshore, often in mangrove swamps.

Range: Florida, south Louisiana, and Texas.

Black mangrove is the hardiest of the three mangrove species and is found farther north than the other two types along the Gulf Coast. In its most northern range it is small and shrubby and dies off in cold winters. As the red mangrove advances out to sea, the black mangrove establishes itself on the newly formed land.

To help the plant breathe, the roots send up vertical, pencil-like extensions called pneumatophores. The roots are buried in soil and these small foot-long extensions help in the exchange of gases. Because these swamps are usually near salty water, black mangrove adapted by drawing salt into the plant and then excreting it through the leaves. Licking the leaves is one way to distinguish a black mangrove.[6]

The wood is not as strong as that of the other mangrove species, and is not much used. The resin, however, seems to repel insects, and when the wood is burned, the smoke keeps the mosquitoes away. It was also used for smoking fish.[7] The bark has been used in tanning and the resin has been used in folk remedies to treat rheumatism.

Combretaceae

White Mangrove
Laguncularia racemosa, white buttonwood

At the seeming end of the world, otherwise the end and jumping-off place of the United States of America, there lies a wide dreaming country, flat as a painted floor. It is touched here and there with strange blanched thickets of buttonwood and of white mangrove. It is spread with coarse salt grass and patched broadly with amazing thick saltweeds, which in the intense summer turn orange and canary and shell pink and scarlet.

Marjory Stoneman Douglas[8]

White mangroves move in after the red and black mangroves become established. They form a third zone in the mangrove swamps, inland, behind the other two species. White mangroves have no prop roots, although they may have a few aerial roots, because they live above the high-tide mark.

L. racemosa grows up to 60 feet tall and is many-branched with reddish-brown, flaky bark and leathery, dark green leaves up to 3 inches long. Fragrant white flowers grow on a spike, each flower about ¼ inch long, with five round petals. The fruits are hard pods, with 10 ribs, containing a single dark red seed. It is not as common as other mangroves and can be found in water in bays and lagoons along the coast.

The hard, heavy wood is used for carpentry, construction, posts, and tool handles. The bark is used for tanning and for dyeing fishnets. The bark is used in a folk remedy for dysentery.

Range: **southern Florida and the Keys.**

THE PALMS

Linnaeus called palms the princes of the plant kingdom. They are the defining plants of the tropical world, useful to people as food, shelter, clothing, and fuel. Palms are not only agreeable plants, they have also proliferated around the world with more than 2,600 species in more than 200 genera. Some are trees, others shrubs, and a few are vines.

The palms that most people would recognize are tall, unbranched, with a cylindrical trunk up to 200 feet high, crowned by a tuft of large leaves. Some palms, however, have no trunk to speak of and the leaves seem to shoot straight out of the ground. The leaves emerge like pleated or feathered fans from stout sheaths and tough stalks. The trunks of some palms show scars where the bases of the stalks of previous leaves grew. In other species, the trunk is covered with stiff, spiny fibers that are left when the leafstalks decay and fall off. The small flowers emerge in clusters on simple or branching spikes and the fruit may be dry or fleshy.

Arecaceae

Royal Palm
Roystonea elata

Roystonea elata
is a tall, imposing tree, capable of growing to heights of more than 120 feet. The lower part of the trunk is the color and texture of concrete; the upper part is bright green, and the leaves emerge from the crown. The leaves are large fronds up to

Roystonea elata, royal palm

10 feet long, and may weigh as much as 25 pounds each. The individual leaflets are strap-shaped, narrowing at the tips. Yellow flowers appear in spring in dense clusters about 2 feet long; each individual flower is about ¼ inch across. The fruits are dark blue, oval, each containing a single seed. It grows naturally in sandy areas, but also lines the streets of many urban areas in Florida.

Range: Florida.

This is the quintessential palm, the instantly recognized cliché from thousands of cartoons, hokey tropical movies, and neon signs. The huge trunk is smooth and the large leaves fan out majestically from the top. During the 1920s and 1930s, many specimens of a related species, **Cuban royal palm** (*R. regia*), were brought to Florida, and the two may be confused. *R. regia* is much like *R. elata*, but has a slightly more curvaceous trunk.

Originally, the genus name was *Oredoxa*, but it was changed to *Roystonea* in honor of General Roy Stone, an army engineer who served in the Caribbean at the end of the nineteenth century.

R. elata was a native of the cypress swamps of south Florida originally, but today it has largely disappeared from the wild and only a few stands remain in the Everglades National Park, the most accessible being one near the Royal Palm Visitors Center. The palm became endangered in the wild because its seedlings were taken by nursery owners to sell to gardeners. The royal palm is also subject to a fungal disease with the unpleasant name of ganoderma butt rot, which produces white lumps or woody protrusions on the trunk. Once a tree is infected, it cannot be healed and must be cut down and burned.

Today, the palms have become ornamental trees, lining city streets throughout southern Florida. Many of them are found in indoor atriums and malls as part of the urban jungle.

Arecaceae

Cabbage Palmetto
Sabal palmetto, blue palmetto, cabbage tree

Sabal palmetto

may rise up to 90 or 100 feet high. The top half of the trunk has a wickerlike pattern made from the bases of dried and fallen leafstalks. The fan-shaped leaves, each 5 to 8 feet long, sprout from the crown in dense clusters. Each leaf has a stout midrib with sawlike teeth; pointed segments branch off from it. The tree tolerates salt well, and can be found in sandy soil, in hardwood hammocks, or planted along streets in cities.

Range: South Carolina, Georgia, Florida, Alabama, and Louisiana.

State tree of Florida and South Carolina.

The cabbage palmetto or *Sabal palmetto* has become another popular urban tree in tropical cities. This one, however, grows where it's too hot or stormy for other trees to thrive. It is hardy, it is handsome, and it tolerates salt spray, all of which explain its widespread use. Though its native range is along the East Coast, it has been successfully transplanted in places such as Las Vegas and Phoenix.

Cabbage palmettos can grow in a variety of sites: from sand dunes to salt flats, from wet prairies to cactus thickets to Florida "hammocks"—tropical forests of mixed hardwood and shrubs on an elevated area. Cabbage palmetto is also the most wind-resistant tree in southern Florida and it seems to fend off damag-

Sabal palmetto, cabbage palmetto

ing insects and other pathogens. Florida exports mature cabbage palmettos for landscaping.

Harriet Beecher Stowe, the author of *Uncle Tom's Cabin*, wrote a wonderful book called *Palmetto Leaves* in 1867, in which she explained:

The palmetto-tree appears in all stages—from its earliest growth, when it looks like a fountain of great, green fan-leaves bursting from the earth, to its perfect shape, when, sixty or seventy feet in height, it rears its fan crown high in air. The oldest trees may be known by a perfectly smooth trunk; all traces of the scaly formation by which it has built itself up in ring after ring of leaves being obliterated. But younger trees, thirty or forty feet in height, often show a trunk which seems to present a regular crisscross of basketwork—the remaining scales from whence the old leaves have decayed and dropped away. These scaly trunks are often full of ferns, wild flowers, and vines, which hang in fantastic draperies down their sides, and form leafy and flowery pillars. The palmetto-hammocks, as they are called, are often miles in extent along the banks of the rivers. The tops of the palms rise up round in the distance as so many hay-cocks, and seeming to rise one above another far as the eye can reach.[9]

The wood is light, soft, pale brown, and fibrous. Early American settlers built their forts with this wood because it was tough enough to withstand arrows, bullets, and cannonballs. South

Carolina adopted the cabbage palmetto as its state tree after a fort built of palmetto logs on Sullivan's Island resisted an attack by the British fleet. It also withstands sea worms, and can be used for pilings. The trunks have been sawed into disks for ornamental tabletops. Baskets, mats, and hats are made from the fronds, and brushes from the fibers in the sheaths of the young leaves. American poet Sidney Lanier wrote:

No traveler of proper sentiments in Jacksonville neglects to have all his womankind furnished with a braided palmetto-hat, trimmed with wild grasses; and this particular writer, with a profound ignorance of all millinery, declares without hesitation that some combinations of these lovely grass-plumes with richly-woven palmetto plaits form quite the most beautiful coverings he has ever seen on a female head.[10]

The trunks and leaves were once used by the Seminole and Miccosukee in Florida to make shelters. One of the earliest Europeans to see these structures was Jonathan Dickinson, who was shipwrecked on the coast of Florida with his wife and baby in 1696. The three of them, with their shipmates, made their way through Florida over the next seven months.

After we had traveled about five miles along the deep sand, the sun being extreme hot, we came to an inlet. On the other side was the Indian town, being little wigwams made of small poles stuck in the ground, which they bended one to another, making an arch, and covered them with thatch of small palmetto-leaves. . . . At length all our people were brought over, and afterwards came the Cassekey. As soon as he came to his wigwam he set himself to work, got some stakes and stuck

them in a row joining to his wigwam and tied some sticks whereon were these small palmettos, tied and fastened them to the stakes about three foot high; and laid two or three mats of reeds down by this shelter. . . . The Cassekey went into his wigwam and seated himself on his cabin cross-legged having a basket of palmetto berries brought him, which he ate very greedily.[11]

These houses were called chickees. Local cypress was used for the posts and rafters and cabbage palmetto for the thatching. Chickees look somewhat like shaggy open huts, the thickness of the thatch far outweighing the slender posts. They were traditionally clustered in a circle and could measure 20 by 8 feet for a communal chickee, or 10 by 16 for a sleeping chickee. The packed dirt floor and sides open to cooling breezes made them ideal for the hot, humid climate. The only threat to a chickee was a hurricane. But if a chickee blew over, it didn't take long to put it back up. Two men could build one in a few days and, when the roof was replaced every few years, it would stand for a decade. This is a skill fading with time, though at one point in Florida's history, it was considered *de rigueur* for a man to know how to build one. Apart from a demand for chickees in resort areas, few people live in them now.

Today, some young men still learn the skill by helping older male relatives. With a growing demand for recreational chickees by non-Indians, a number of Seminole and Miccosukee have established successful chickee contracting businesses. Since some designs requested by outsiders are Hawaiian or Polynesian, native builders have expanded their repertoire. Due to government restrictions on obtaining cypress and palmetto grounds, many journey a great distance for materials.[12]

The Seminole also made palmetto dolls, which are now an item for the tourist trade in a few places. The dolls consist of a simple palmetto-fiber trunk with arms, legs, and feet made of cardboard covered with palmetto fiber. The male dolls are often clothed in the style of the 1930s, with shirt, turban, and neck scarf.

Everglades trader George Storter is sometimes cited as being the first to promote the making of Seminole palmetto dolls for the tourist market. By 1900, Seminole women and girls were making wooden dolls dressed in the traditional Seminole patchwork. Presumably these early dolls were made for young Seminole girls as playthings. However, early on, dolls became one of the best-selling Seminole souvenirs, and were in great demand from the [tourist] market.[13]

The tree acquired the name "cabbage palmetto" because of its edible terminal bud, which has a cabbagey taste. The bud, also called swamp cabbage or heart of palm, is eaten raw or cooked and can be found in the gourmet sections of supermarkets. Because cabbage palmettos grow upward from the terminal bud, removing the bud kills the tree and these delicacies now come strictly from palmetto plantations. The bud is removed from a year-old plant, then a new plant is seeded in its place. In Florida, where palmetto is the state tree, it is protected by law and hearts of palm may not be gathered in the wild without written permission. The cabbage palmetto flowers are a source of a strong but delicious dark-amber honey and the fruits are also edible, although they are said to be an acquired taste to humans, if not to wildlife, who devour them greedily. There is an annual Swamp Cabbage Festival held in LaBelle, Florida.

Traditionally, young cabbage palmetto fronds were collected and shipped worldwide in spring for use in Christian churches on Palm Sunday. Each year whole families set out to cut palmettos in the thousands, packing them in bundles of 25 or 50 and sending them off to churches.[14]

Pinaceae

Slash Pine
Pinus elliottii, Southern pine, yellow slash pine, swamp pine, pitch pine, Cuban pine

Pinus elliottii
grows about 100 feet tall, with a straight trunk crowned with many small branches forming a round head. The reddish-brown bark forms plates on the surface of the trunk, which can be 3 feet in diameter. The shiny dark green needles are 4 to 11 inches long, appear in twos or threes, and cluster thickly on the branches. The glossy brown cones are 3 to 6 inches long, with thin prickly scales. Grows naturally on sandy soil, but is planted as a timber tree throughout the South.

***Range:* South Carolina, Georgia, Florida, and Louisiana.**

The tall, scrawny slash pine is such an individual that each tree seems to have its own personality. The needles of the slash pine are long and the branches start more than 20 feet off the ground. Due to the tightly compacted thick bark, this is a highly fire-resistant tree, though young trees without these layers are still susceptible.

In her book of stories from Florida, Marjory Stoneman Douglas describes running through a pineland:

Pinus elliottii, slash pine

The original inhabitants of Florida used the wood of the slash pine in construction and in certain medicines. The Seminole boiled up slash pine wood or bark to use externally as an analgesic, as well as in baths for aches and pains, or for sores and cuts or hemorrhoids.[16]

In *South Florida Folklife*, the authors write:

The wood used to make most of South Florida's original Cracker houses was virgin pine, known as "Dade County pine," [P. elliottii var. densa] in South Florida, but called "lighter" or "heart" pine in north Florida. The wood that was cut from these trees is full of resin. It could be worked easily while it was green, but when it aged the resin hardened, making it almost impossible to hammer a nail through the wood. The positive side to this is that the termites won't eat the wood: the negative side is that if there is a fire, the resin-filled wood is extremely flammable—thus the name "lighter."[17]

They were like no pine trees he [Larry Gibbs] had ever seen in his life, these Caribbean pines. Their high, bare trunks, set among palmetto fans that softened all the ground beneath them, rose up so near the road that he could see the soft flakes of color of their scaly bark, red and brown and cream, as if patted on with a thick brush. Their high tops mingled gray-green branches, twisted and distorted as if by great winds or something stern and implacable in their own natures. Their long needles were scant, letting the sky through. They were strange trees, strange but beautiful. The brilliance of the sun penetrated through their endless ranks in a swimming mist of light. They were endlessly

The Dade County slash pine variety has pretty much been destroyed by clearing for agriculture and housing, or burned by fire.

Slash pine got its common name from the slashes that were made to collect the sap used in the making of turpentine. George M. Barbour, in his 1882 book *Florida for Tourists, Invalids, and Settlers*, reported that "'Turpentining' has become quite an industry, and there are several large turpentine farms in the country that are reported to be very profitable."[18] Today, fast-growing slash pines are used for lumber and pulp and paper, though sometimes sold as "longleaf pine" for general building and construction uses.

Burseraceae

Gumbo-Limbo
Bursera simaruba

Bursera simaruba, gumbo-limbo

Bursera simaruba

is a many-branched tree up to 60 feet high, with a stout trunk and massive, widely spreading branches. It has red, peeling bark. The leathery, compound leaves, which grow at the ends of the branches, are shed each year. Each leaflet is 2 to 4 inches long and 1 to 2 inches wide. The greenish flowers grow in clusters, with male and female flowers on different trees, appearing at about the same time as the new leaves. The dark red fruit is triangular in cross section, containing a ¼-inch triangular seed. It grows in coastal hammocks.

Range: southern Florida and the Keys.

The name gumbo-limbo is thought by some etymologists to be a Bantu phrase meaning "slave's birdlime." This term refers to a sticky glue that slaves made from the sap and spread on tree limbs to catch songbirds. This practice continues to the present and its adherents will tell you birdlime sticks better than duct tape. Others believe the name comes from the Spanish *goma-elemi* meaning "gum resin." In Florida, locals call it "the tourist tree," because it is always red and peeling.

This salt-tolerant, fast-growing species is easy to propagate. Stick a branch in the ground, and it will root. It's sometimes used to create living fences. The soft, spongy, light brown wood is especially good for carving, and generations of young boys have used it to make toy sailboats, racing them against one another in the shallow seashore waters. Carousel horses were also made from gumbo-limbo, before the days of plastic.[19]

The tree exudes an aromatic resinous sap smelling a little like turpentine. It has been used as glue, varnish, water-repellent coatings, incense, and liniment. Not surprisingly, it is a relative of the Middle Eastern trees that produce the resin used to make frankincense and myrrh. Some Florida settlers even used the resin to mend broken china. The leaves make a substitute for tea. According to popular lore, the bark will treat a toothache, or soothe the rash caused by touching the tropical poisonwood tree *(Metopium toxiferum).* Some people swear that the touch of the gumbo-limbo is cooling.

The gumbo-limbo also grows in Central and South America and in many of the Caribbean islands. Haitian drums used in voodoo rituals were made from the wood.

Zygophyllaceae

Lignum Vitae

Guaiacum sanctum, Brazil wood, gum guaiacum, pockwood, wood of life, tree of life

Guaiacum sanctum

is a slow-growing tree that may rise to 15 to 50 feet high. The trunk is usually crooked, with many knobby branches and grayish, furrowed bark. The leaves are evergreen, opposite, appearing along the axis in leaflets, and are leathery in texture. When the flowers open, they are bright blue or purple; over time they fade to white. The five-celled fruit is about 1 inch long, yellow, and heart-shaped.

Range: southern Florida.

Lignum vitae, like so many of Florida's plants, is not only endangered, it's on the list of plants heading for extinction. In Central America and Florida, remaining populations are confined to postage stamp–sized areas threatened constantly by development. What's left can be found in lowland tropical to subtropical dry forest scrub, near the coast or at lower elevations inland. The trees can live to a great age, and a specimen discovered in the 1960s was estimated at more than 1,000 years old.

An intensive international trade in *Guaiacum sanctum* began in the sixteenth century when the Europeans learned that the Arawak made a decoction from it to treat syphilis. For the next two centuries, there was strong demand for the heartwood, which was used to prepare the extract, and it was well known as a major treatment for the disease. Though this was no cure-all, at least the symptoms were arrested noticeably. The resin could also be used to treat rheumatism, constipation, and skin diseases, but its use in treating syphilis gained the most attention. There was even a runup in the price of the wood in the 1520s and 1530s, when a pound of the wood sold for the enormous price of seven gold crowns—phenomenal at the time.

Because the brownish-green wood was so hard and resilient, it was used in die-cutting and to make hard-wearing devices such as ship propeller shafts, bearings, caster wheels, rollers, mallet heads, pestles, and even bowling balls. The interlocking grain makes it almost impossible for the wood to split, and though it is too heavy to float, the wood was virtually waterproof because of its high resin content.

The Woodex Bearing Company collects anecdotes about the uses of lignum vitae for bearings in everything from washing machines to ships' propellers. John Harrison, who solved the problem of longitude, created a friction-free clock in 1722 using lignum vitae bearings. The clock still works. In some places, hydroelectric turbines have been run on lignum vitae bearings: one such plant at the Rock Island Arsenal was in operation from 1919 to 1998.[20]

Moraceae

Strangler Fig

Ficus aurea

Ficus aurea

is an epiphyte, often found growing on *Sabal palmetto.* The seeds germinate in the tops of the palm tree and send air roots down to the ground. These roots grow thicker until they are as big as the tree's trunk. They eventually crush the life from the host tree. The bark on the air roots is smooth and gray. The yellow-green leaves are 2 to 5 inches long, thick, and leathery. The flowers are reddish-purple, and grow in the leaf axils. The rounded fruit is red, brown, or yellow.

Range: southern Florida and the Keys.

Strangler figs may be essential to Florida's forest ecosystems, since they provide food for many wildlife species, but it's hard not to impute some murderous intent to these lethal epiphytes. They can live on air and once their sticky seeds are dropped by birds in the tops of other trees, they will germinate in the tiniest crevice given the slightest amount of moisture. They start their rapid growth by sending thin rootlets downward in search of soil. These long coils of snakelike roots wrap themselves around the host, cutting off sunlight by sending out a canopy of leaves. The host plant stops growing and may eventually die, leaving an upright skeleton under the fatal clutch of its strangler's root system.

A mature strangler fig can grow to more than 60 feet tall with multiple trunks reaching down

Ficus aurea, strangler fig

for water in every direction. The strangler fig flowers almost continuously, producing small, spherical fruits that are eaten by birds (important to the birds and to the strangler), which then spread the seeds, and start the process all over again.

According to ethnobotanist Daniel Moerman, the Seminole used the strangler fig in several ways. A poultice of mashed bark was applied to cuts and sores, and the root bark was twisted and used to bind together the frames of houses. The plant was used for food and chewing gum and to make arrows, bowstrings, and fishline.[21] Nowadays most people avoid the strangler fig because the broken twigs exude a milky sap that can cause a bad skin reaction.

Zamiaceae

Florida Coontie

Zamia floridana, syn. *Z. pumila*, Florida arrowroot, comfort root, comptie, conti, Seminole bread

---

Zamia floridana

is a fernlike perennial plant with dark green, feathery leaves, often growing in clumps up to 6 feet in diameter. The compound leaves are usually narrow, sometimes curling along their length. Male cones can be up to 6 inches long; female cones are slightly larger and rounded. The salt-tolerant coontie is found in sandy soil in pine hammocks; occasionally on coral.

Range: **Florida.**

The Florida coontie seems to come in three species, although botanists disagree about the distinctions. The first, *Zamia floridana* or *Z.*

pumila, is a salt-tolerant plant with thin leaves, which grows in pine hammocks on the eastern coast in south-central Florida, and also on the western coast, north to the Panhandle. On the west coast, some grow on coral.

The second, *Z. umbrosa*, has wider leaves, and grows in small pockets along rivers in sandy soil along the west coast, in coastal hammocks of northeastern Florida and southeastern Georgia.

The third, *Z. silvicola*, has the widest leaves of all, and grows in northeastern Florida, mostly in sand and under pines and oaks. In a few areas, wide-leaved coonties can be seen growing on land that is occasionally flooded by salt water. These plants are preferred for ornamental use. A variation of the wide-leaved form grows in northern Florida and is called the "Palatka giant" by the locals—its leaflets are wider, its cones are larger, and its leaves are up to 7½ feet long.

The coontie's cylindrical male cone is black or brown or a kind of reddish color; the female cone, which is larger and more oval, grows 2 to 7 inches long and is full of seeds. Beetles and weevils pollinate the plant, attracted by its starch-rich nutrients. When the seeds have been on the ground just long enough to begin germination, birds and animals hone in on them. Mockingbirds, grackles, and blue jays all like the seed; mice and rats will also hive them off to make a stash. It's also a primary food source for the larvae of the Atala butterfly.

Coonties, once common in Florida, are now rarely seen in the wild but are planted as ornamentals in tropical gardens. The decline of the coonties in the wild came about in part because of commercial exploitation. The Calusa and Timuca were among the first known groups to use coonties as a food. In certain areas there were vast colonies of these plants, mainly in south Florida near what is now Miami and Ft. Lauderdale. There is a reference to huge quantities of the plant around

Zamia floridana, Florida coontie

New River in Ft. Lauderdale, so many that the Indians called this place "Coontie Hatchee."

The Seminole moved into Florida in the mid–eighteenth century and this plant became a very important source of food for them. The coontie was their name for it, which roughly means "flour root." They cut up pieces of the stems, pounded them close to powder, then put them through several washes, letting the starch sink to the bottom. The paste was fermented, and then dried to a powder. When the white men came to Florida, they also used the stems for food. Their name for this plant was "arrow root."

By the middle of the nineteenth century, factories all over southern Florida produced starch from coonties, some of them up to 10 or 15 tons a day. They did this right up to 1925 and the once-vast populations of coonties are now reduced to small remnants.[22]

Polypodiaceae

Resurrection Fern
Polypodium polypodioides syn. *Pleopeltis polypodioides*

Polypodium polypodioides syn. *Pleopeltis polypodioides*
is a vascular epiphyte found in trees, on rocks, or on the ground. The leathery, midgreen fronds are about 6 inches long and have up to to 17 pairs of well-spaced, oblong, scaly leaflets.
Range: Delaware to Florida to Texas.

The resurrection fern is the most abundant and common of all the small ferns of the hammocks. The scientific name *Polypodium* comes from the

plant withstands drought by curling up to conserve moisture and reduce evaporation. It looks dead in this state, but bounces right back to life uncurling its leaves the minute there is a hint of moisture in the air.

In the southeastern United States, resurrection fern is most often found on large spreading branches high up in live oak trees, and is sometimes associated with Spanish moss, another epiphyte. It can also appear on fence posts, in palm trees, and in pine forests. It is often deliberately grown on a fallen oak as a focal point in Southern gardens—very much the same as it does in nature.

Moerman notes that the Seminole used an infusion of the leaves for "chronic conditions," and that an infusion of the whole plant was used to bathe people suffering from insanity. The Houma used a decoction of the fronds to treat bleeding gums and mouth sores, or for headaches and dizziness.[23]

Orchidaceae

Butterfly Orchid
Encyclia tampensis syn. *Epidendrum tampense*

Encyclia tampensis syn. *Epidendrum tampense*

grows up to 30 inches tall, with green, straplike leaves springing from a gray-green pseudobulb. The long thin flower stalk bears numerous fragrant flowers, each about 1 inch wide, with five dull green petals splayed out behind a scoop-shaped white lip with purple markings.

***Range:* southern Florida.**

Polypodium polypodioides, resurrection fern

Greek for "many feet," because the roots form footlike structures. They are long, thin, creeping rhizomes effective in working their way into any nook and cranny including between the furrows of tree bark, even spreading over rocky cliffs. Though it's an epiphyte, it is not a parasite—it doesn't kill its host, drawing any water and nutrients it needs from rain and the atmosphere.

The common name is understandable: the

This orchid is an epiphyte growing on tree trunks, in branches of cypress hammocks and pinelands, on live oaks, or even on rocky out-crops. At the base of the stem there is what is called a pseudobulb which collects nutrients to feed its protruding roots and allows it to live in diverse places.

Orchids have spread from the tropics, where they originated, around the world because they produce millions of seeds. Any dramatic weather system such as a hurricane can carry the minute seeds for thousands of miles. They survive because they can find spots almost anywhere that will provide a home—in the top of a tree, on a branch, or in an opening on a log. They have the ability to absorb food from the rain and the air whether their roots are grounded or not. What they need is light. They have other survival strategies as well. Look closely and you'll find that the flowers have a well-defined insect shape to draw in pollinators. It takes about seven years after a major storm to see where orchid seeds have landed, but you can bet that orchophiles will be out scouting likely locations looking for them. There are so many of these possessed plant collectors that a book called *The Orchid Thief* by Susan Orlean became a surprising best-seller in the late 1990s.

Orchids seem to drive people crazy. Those who love them love them madly. Orchids arouse passion more than romance. They are the sexiest flowers on earth. The name "orchid" derives from the Latin orchis, *which means testicle. This refers not only to the testicle-shaped tubers of the plant but to the fact that it was long believed that orchids sprang from the spilled semen of mating animals. . . . In Victorian England the orchid hobby grew so consuming that it was sometimes called "orchidelirium"; under its influence many seemingly normal people, once smitten with orchids, became less like normal people. . . . "The bug hits you," a collector from Guatemala explained to me. "You can join A.A. to quit*

drinking, but once you get into orchids you can't do anything to kick the habit."[24]

Orlean describes the extraordinary lengths to which orchid collectors will go to find rare specimens. In the nineteenth century, it meant enduring hardships in the South American rain forest; in the late twentieth century, it appeared to include housebreaking and theft.

The butterfly orchid, which is not as rare as other species and is widely cultivated throughout Florida, is often used in orchid breeding. Once

Encyclia tampensis, butterfly orchid

crossed with other genera, new species are created with a similar attractive shape but a brighter color. However, like many native Florida species, it is declining in numbers because of development and because selfish plant collectors haul the plants out of their native habitat. Of the thousands that once grew at the base of cabbage palmettos in the pinelands of southern Dade County, almost all have disappeared, according to Ruben P. Sauleda, a Florida orchid grower. He says, however, that the occasional *E. tampensis* has survived in some of the small pockets of pinelands.[25]

Butterfly orchids still grow in abundance in the protected Everglades National Park along the waterways, and can best be seen from a boat. A few Florida orchid collectors tell of sneaking out after dark to areas slated for development and gathering orchids illegally (they are protected by law in Florida) because a day or so later the orchids would be legally bulldozed into oblivion. Furious debates on the ethics of how best to save Florida orchids can be found on the Internet, where there is a great deal of interesting and serious discussion.[26]

Cyperaceae

Sawgrass
Cladium jamaicense

Cladium jamaicense
grows on hollow stems 6 to 10 feet tall in shallow water. It has tough, stiff leaves with serrated edges ending in brown flowers.
Range: Florida.

Sawgrass, a sedge, is one of the primary plants of the Everglades, which has been evolving here for more than 4,000 years. It probably covered 2 million acres of the Everglades in its heyday and is one of the oldest green plants in this part of the world.

The leaves rise directly from the base and can reach 3 feet in length. Sawgrass gets its name from the fact that the midrib and the margins of the leaves have sawlike teeth. At the center is an edible white base much sought after by deer. Tiny snail eggs often cling to the stems just above the water's surface. Alligators like to hide in sawgrass, and muskrats use it to build nests.

The name "Everglades" now has a romantic ring to it: a place of mystery. But, in fact, the area got the name because an eighteenth-century mapmaker managed to get "river glades" transcribed incorrectly. It is a massive area of low, flooded land covered with grasses and related species—in other words, a wet prairie. A third of it now comprises a national park including such other ecosystems as pinelands, cypress swamps, and hammocks.

The Seminole called the area *Pahayokee*, which means "river of grass." The rivers of grass have declined over the years as the water of the Everglades becomes increasingly salty. Today, the sawgrass glades (or prairies, as they are called by the locals) extend south from Lake Okeechobee toward Florida Bay. Most of the year, the area is flooded by fresh water from Lake Okeechobee, but it is drier in winter and spring.

Marjory Stoneman Douglas, in her evocative book *The Everglades*, writes:

> In the times of high water in the old days, the flood would rise until the highest tops of that sharp grass were like a thin lawn standing out as blue as the sky, rippling and wrinkling, linking the pools and spreading and flowing on its true course southward . . .
>
> The water moves. The saw grass, pale green to deep-brown ripeness, stands rigid. It is moved only in sluggish rollings by the vast push of the winds across it. . . .

Nothing less than the smashing power of some hurricane can beat it down. . . . Even so, the grass is not flattened in a continuous swath, but only here and over there, as if the storm bounced or lifted and smashed down again in great hammering strokes or enormous cat-licks.

Only one force can conquer it completely and that is fire. Deep in the layers of muck there are layers of ashes, marks of old fires set by lightning or the early Indians.[27]

The Mewuk, Pomo, and Yokut peoples used sawgrass to make baskets,[28] and it may sometimes have been used for thatching instead of palm leaves. It has also been used to make paper and newsprint.

Cladium jamaicense, sawgrass

The Boreal Forest

The dark, seemingly impenetrable vastness of the boreal forest has moments of incredible light. During the late summer nights, the aurora borealis flashes and crackles across the blackest sky on earth. No artist has ever captured its ethereal colors, no composer its eerie music.

The woods are crisscrossed with pathways left by animals traveling from breeding ground to summer forage to winter protection. Bears, caribou, and moose move through the deciduous areas, beating highways into the soil that have lasted 1,000 years. In the Mealy Mountains of

The stillness of the boreal forest belies its intense life.

Labrador, caribou were so plentiful only a 100 years ago that it would take days, sometimes weeks, for a herd to pass completely through a given spot.

Although this community is usually called the boreal forest after Boreas, the Greek god of the north wind, it's sometimes known as the taiga, from a Russian word meaning "little sticks." It sweeps in a band more than 600 miles wide from Alaska east to Newfoundland and the Atlantic Ocean and south through the Rocky Mountains.

These are the lands of long and intensely cold winters (often far colder than the Arctic, which is moderated by the ocean). The frost-free season is only 90 days long, and its arrival is always amazing. One moment the land is frozen solid, mute, and then, overnight it seems, spring glides in and an even more glorious beauty is revealed. The woods explode with the singing of birds, the buzzing of insects, and a carpet of wildflowers, ferns, sedges, and low-growing vines. Drifts of ephemerals pop up to attract the insects; once their job is over, they disappear, leaving the ground to the mosses.

Forest makes up 28 percent of the North American continent north of Mexico, and more than 60 percent of what is classified as forest is boreal.[1] The boreal's importance is shown in the statistics: there are estimated to be 140,000 kinds of living organisms in the Canadian forest, half of which are not even classified. Plants and vertebrate animals make up 5 percent of those 140,000 organisms; invertebrates, fungi, and microorganisms make up over 90 percent.[2] When this forest is at its maximum growth in spring and summer, worldwide levels of carbon dioxide fall and levels of oxygen rise. The plants of this vast area remove millions of tons of carbon dioxide from the atmosphere every year.[3]

When the ice sheets melted 12,000 years ago, the conifers took over, and they still predominate. They can survive in poor soil and photosynthesize even in cold temperatures. It can take them up to 200 years to reach the climax part of their cycle. The trees live in what are known as podzols (from another Russian word)—ash-colored acidic soils produced beneath any canopy of evergreens.

The boreal forest is not a uniform region. It is bordered to the north and south by broad ecotones—areas in which plants from two distinct plant communities combine. In its most northern extremities the tundra makes thrusting advances into the boreal. This is known as the tundra-taiga ecotone, and it is characterized by open coniferous forests. On the southern edge is an ecotone dominated by broad-leaved deciduous trees. The really dense, quintessentially boreal coniferous forest is smack-dab in the center.

The boreal regions also contain many wetlands, including bogs, fens, swamps, and marshes, known as muskeg, which are essential to the boreal ecology. The insects in the muskeg provided food for the birds that at one time flew over by the millions to their northern nesting grounds. For the people who drifted over these lands, the acidic, mushy muskeg were areas of pestilence and horror. Legends persist of people becoming lost and being eaten, not by the animals but by the insects.

For millions of years, regular forest fires set the boreal's ecological balance. The conifer forest developed its survival strategies to accommodate fires every 50 to 100 years.[4] For example, in many coniferous species, it takes a fire to open up the cones and spill the seeds. The flames also stamped out diseases and stimulated the trees' reproductive cycles, regenerating the forest in a continuous cycle. Fire created open spaces where aspen, poplar, and birch could grow. Moose browsed in the open forests of these trees. Heavy browsing lowers the nitrogen and carbon content of the soil, which then becomes poor and thin—ideal for conifers. Gradually, the spruce displaced the deciduous trees, and the forest grew darker and denser. As the forest aged, caribou moved in, using it for shelter, eating lichens and conifer needles, until a fire started the whole cycle over again.

Because of its relationship with spruce trees, the moose is known as a keystone species for the boreal spruce forest. Get rid of all the moose and the balance of the boreal forest would be lost forever.

Big animals have been critical to this forest since the Clovis people came from Asia 12,000 years ago. They were pursuing the mastodon, which chewed up the leaves and twigs of conifers and left extremely large, fertile deposits behind. As the hunting bands broke up when they became too numerous to live on a given hunting ground, the Clovis people formed new communities which eventually flourished as the rich cultures of North American native peoples.

For the Europeans, who first came via Greenland 1,000 years ago, the boreal forest was a forbidding landscape. Later arrivals built a whole literature on the terror of the taiga, the place of the wolf, the dark menacing forces deep in the forest.

The world seems destined to change its climate more quickly in the next few hundred years than it has in the past 10,000. This means that the boreal forest is threatened by more than clear-cutting, mine projects, oil and gas exploration. Levels of carbon dioxide are predicted to double, which will mean that the forest would need to shift 600 miles to the north to find the same climatic conditions. It's impossible. The farther north the forest is pushed, the closer it gets to water, and a completely different type of climate. The boreal forest, which once seemed to have an eternal quality, is a fragile, disappearing ecosystem.

Pinaceae

Jack Pine

Pinus banksiana, scrub pine, Northern scrub pine, gray pine, black pine, blackjack pine, bull pine, Banksian pine, Hudson Bay pine, Labrador pine, Banks pine, princess pine

Pinus banksiana, jack pine

The lonely majesty of a wind-swept jack pine has inspired generations of poets and painters. These trees endure in spite of terrible weather because they have developed taproots penetrat-ing as deep as 10 feet that anchor the trees securely.

Jack pines can grow 70 feet tall and live for 200 years, though decay may attack the top of the tree when it is more than 75 years old. A jack pine forest has a dense, closed canopy with an understory of cherry, blueberry, hazels, bracken, and sweet fern along with trailing arbutus. A jack pine savanna, on the other hand, has grasses such as big bluestem, little bluestem, Indian grass, and blazing star as companions.

The seed is stored for several years in closed yellowish-brown cones that hang down from the crown. After 5 to 10 years, the seed is ready to sprout.[5] However, the cones are glued shut with resin, which means that jack pines depend on fires to release the seeds. Once the resin melts in the heat of a forest fire, the cones open and the minute winged seeds fall to the ground or are carried off by the wind. They germinate on land that has been newly cleared by fire. The seedlings grow quickly in the full sun. Pyramidal when young, they spread irregularly with age. Natural fires every 50 years will make a healthy forest, but stands burned more frequently than every 5 to 10 years become pine barrens or jack pine savannas (a savanna is a transition zone between forest and grassland).

The lichen on jack pine is an important source of food for the woodland and barren ground caribou that trot across the top of North America. The tree also provides cover and food for moose. Rodents and birds eat the seeds. White-tailed deer, snowshoe hares, and caribou browse on the branches. Kirtland's warbler, an endangered species, breeds only in homogenous stands of jack pine that are 20 to 40 years old in Michigan's lower peninsula.

People, too, have long depended on the jack pine. Pine bark tea mixed with crowberries (*Empetrum nigrum*) is a traditional remedy for coughs and also has a diuretic effect. Pine oil and pine tar are well-known antiseptic and insecticidal cleansers. The Cree used to burn lumps of pine pitch to fumigate the living quarters of sick people. It was also a common practice to remove splinters by warming a lump of resin over a fire and applying it to the skin. When it hardens, the resin and splinter peel off in one fell swoop.

The Pines

I heard the pines in their solitude
    sighing,
When the winds were awakened, and
    day was dying;
And fiercer the storm grew, and darker
    its pall,
But the voice of the pines was louder
    than all:
"We fear not the thunder, we fear not
    the rain,
For our stems are stout and long;
Or the growling winds, though they
    blow amain,
For our roots are great and strong.
Our voice is eternal, our song sublime,
And its theme is the days of yore—
Back thousands of years of misty time,
When we first grew old and hoar! . . .
"Sublime in our solitude, changeless,
    vast,
While men build, work, and save,
We mock—for their years glide away to
    the past,
And we grimly look on their grave.
Our voice is eternal, our song sublime,
For its theme is the days of yore—
Back thousands of years of misty time,
When we first grew old and hoar."

Charles Mair (1838–1927),
Dreamland and Other Poems,
Tecumseh: A Drama

The Tête-de-Boule Algonquin used pine gum to caulk canoes; Algonquin, Ojibwe, and

Potawatomi made the roots into cords for sewing birchbark canoes. The male cones were edible if they were boiled, roasted, fried, or pickled.[6]

In 1932, ethnobotanist Huron Smith reported that there was no part of the tree that wasn't used by the Ojibwe. "The leaves are used as a reviver. . . . Jack pine roots have ever been esteemed by all Ojibwe as fine sewing material for their canoes and other coarse and durable sewing. They dig the roots with a grub hoe . . . and often find them fifty or sixty feet long. These are split lengthwise into two halves starting at the tree end, and are wrapped in coils. . . . They are then sunk in the lake which loosens the bark and enables them to be scraped clean, as well as adding to their flexibility. They are ivory white when used and very tough and flexible."[7]

The indigenous peoples, however, thought that evil spirits lurked in the jack pine, causing infertility among women and animals. The native suspicion of jack pines seems to have been picked up by French Canadian lumberjacks, who believed that a woman who passed within 10 feet of one would become sterile (like the closed cones of the tree itself). Because it was so highly feared, nobody would cut it down, either. Early settlers considered the jack pine an evil or unlucky tree, because very little grew in its vicinity. This is probably because it grew on the infertile soils other plants could not tolerate.[8]

The name *banksia* honors Sir Joseph Banks (1743–1820), former director of the Royal Botanic Gardens at Kew in London. He was a plant explorer and botanist and was so influential that Linnaeus thought that Australia should be named Banksia after him. Cooler heads prevailed, but he is remembered wherever he traveled—and he traveled extensively with captain James Cook on the voyages of the *Endeavour*. On his first trip to Newfoundland and Labrador in 1766, he took back dozens of species to be part of the botanical collection at Kew.

Pinus banksia has white sapwood, light brown to almost orange heartwood, and is relatively lightweight, very knotty, and regarded as rather coarse. Today it is an important commercial timber for telephone poles, fence posts, and railway ties. It is used as pulpwood and grown on farms as Christmas trees.

Range: **from close to the Arctic Circle south to northern New York and Minnesota.**

Pinaceae

Tamarack

Larix laricina, larch, tamarack larch, Eastern larch, Alaskan larch, American larch, black larch, hackmatack

Larix laricina

a deciduous conifer, grows 40 to 80 feet high. It is upright with a straight trunk and a pyramidal crown and lives to about 180 years. Needles are 1 inch long, in clusters of 10 to 20, and turn yellow in fall before they drop. Cones grow about ½ inch long. The tiny seeds have wings and are dispersed by the wind or by animals.

Range: **Alaska and northwestern Canada from the treeline south to Ontario, Quebec, the Atlantic Provinces, the northern United States around the Great Lakes, and New England.**

For most of the year, it's entirely possible to ignore this inconspicuous tree, which looks like an ordinary evergreen. But when autumn arrives, tamaracks turn an eye-catching gold. Then, like deciduous trees, they drop their needles. They grow in areas that have permafrost

Larix laricina, tamarack

and in climates too dry and cold for spruce or fir. To prevent water loss by transpiration, which would otherwise go on all winter, tamaracks get rid of their needles.

The tree is monoecious—both sexes are on the same tree—with small yellow male flowers and red to red-brown female flowers. Tamarack usually grows in peaty soil, but it will also grow on dry bedrock because of the shallow root system. In the very northern limits of its range, at the edge of the treeline, it usually reproduces by layering (branches dip down to touch the forest floor and take root).[9] Bald eagles and ospreys often nest in tamaracks.

The native peoples of Alaska and northern Canada used the wood for dogsled runners, boats, and fish traps. In northern Alberta, the Cree shaped the branches into duck and goose decoys. The Ojibwe used the roots to sew together the edges of birchbark canoes. Henry Wadsworth Longfellow in *The Song of Hiawatha* mentions this:

The Chippewa and Wood Cree also used its wood to make toboggans, snowshoes, drums, and paddles. Because of its hardness, the Europeans used it for shipbuilding. In Quebec, the tree is known as *violon* or violin wood, perhaps because it was used in making musical instruments.

Tamarack has many medicinal uses, especially the bark, which was used to make poultices for wounds or skin ailments. The gum was chewed for indigestion, or kidney and liver problems. A tea made from the needles was used to treat stomach and lung problems or excessive menstrual bleeding. The young shoots of the tamarack were sometimes eaten as emergency food, and the gum could be dried and pounded into a kind of baking powder.[11]

Settlers were certainly aware of the benefits of the tamarack. In his book *New England Rarities*, John Josselyn writes in 1672, "I cured once a desperate bruise . . . with an unguent made with the leaves of a larch tree and hogs grease, but the larch gum is best."[12]

In 1938, A. R. M. Lower, in his book *The North American Assault on the Canadian Forest*, noted that the roots had once been used to make the curved sections in ships, but that most larches had been killed by an insect infestation that had swept across the country some years previously. He noted that new growth was occurring, and today the larch population is back to normal.[13]

Before creosoting was developed, tamarack was the principal source of railway ties because of the hard straight quality of the wood—the same virtues that made it ideal for telegraph poles.[14] Today, tamarack is used in the pulp and paper industry, in particular, to make the transparent windows—a substance known as glassine—in the business envelopes used for checks and bills. It is sometimes used as rough lumber for boxes and crates or fuel wood, and because it doesn't rot easily, as railroad ties, telephone poles, fence posts, or timbers in mines.

Betulaceae

Paper Birch
Betula papyrifera white birch, canoe birch

Betula papyrifera

grows to 70 to 80 feet high, though it tends to be short in the northern part of its range. It is a single- or multistemmed, deciduous tree with a slender trunk and a narrow crown in forests; a wider crown in the open. Multistemmed trees are relatively common. It is shallow-rooted with pale, papery bark, which peels easily off the slender trunk (less than 1 foot in diameter). The bark is golden when the tree is young, but turns white when the tree is more than 10 years old. The leaves are alternate, shaped like the spades on playing cards: rounded at the base, pointed at the end, 2 to 4 inches long. Male and female flowers occur in separate catkins on the same tree. Fruits are winged nutlets dispersed by wind. Grows in moist areas.

Range: Alaska, all Canadian provinces, and New England.

Provincial tree of Saskatchewan and state tree of New Hampshire.

The birch and its bark are among the most romantic wild icons of North America. The delicate tracery of its leaves and the brilliant white of the bark light up an otherwise dark part of the forest. It is not a long-lived tree, usually stopping its growth after about 60 or 70 years. Most birches die before they reach 140 years old.

The male and female flowers grow on the same tree in separate catkins. The brilliant yellow of the male catkins makes spring a luminous experience. The winged nutlets take off into the wind once they are produced.

Birches are important for browsing moose, which munch away at them all winter when not much else is available. This forces the roots to produce new shoots, creating a tree with multiple stems. Snowshoe hares, porcupines, and beavers all take a turn at the feast birch supplies. Seeds are eaten by birds, voles, and shrews. Ruffed grouse gobble up both the catkins and buds. Yellow-bellied sapsuckers drill holes to get at the sap, and then they are followed by hummingbirds and red squirrels, which like to take a sip as well. There is at least one caterpillar that devotes itself to the birch and spins itself a silk nest by folding the leaf and then sewing it up. It's a wonder the tree survives. But it does, and will even regenerate from root crowns after a fire.

Because there is both a lush understory and a great deal of moisture in the canopy, pure stands of birch are unlikely to succumb completely to fire. They may even live to become seed trees once again, ensuring that there will be regeneration.[15]

Birch trees grow well on disturbed sites, and often appear on land that has been burned or sites where vegetation has died because of air pollution from nearby mining and smelting operations. Near Sudbury, Ontario, the site of nickel mines and smelting operations, for example, paper birches and red maple dominate in the forests.

FOLLOWING PAGE: *Betula papyrifera*, paper birch

But as recently as the late 1930s and the 1940s, large stands of birch in eastern North America were killed or damaged by a condition called birch dieback. The twigs and roots died and the trees languished; most of those affected died within six years. The condition has not occurred on such a large scale since then.

The almost froufrou quality of the birch is pointed out by Arthur Plotnik. "In more sexist times, male poets portrayed the white-barked birch tree as meek and demurely feminine. . . . James Russell Lowell called it 'the most shy and ladylike of trees.' They saw femininity in the silkiness of the tree's bark, the grace and delicacy of its branch structure, the fluttery lightness of its foliage. The bark was often virgin white. And when exposed to oppressive heat, most species soon fainted away."[16] In 1923, the American Forestry Association designated the paper birch as a tree to be planted on Mother's Day. Robert Frost, however, in his poem "Birches," celebrated the tree's strength.

Just about every part of the birch has been used at some time for some purpose. The native peoples of North America used the waterproof bark of paper birch to make wigwams and baskets, and used it as paper for drawing. The bark is remarkably durable: drawings and writings on birch bark from the sixteenth century are still legible.

The most significant use of birch bark has been the making of canoes. Is there another image of the North more evocative of this wild land than Indians or voyageurs in their long canoes paddling through the wilderness? These craft could be anywhere from 10 to 20 feet long. Natives used a patchwork approach, stitching sheets of bark together with roots, and sealing them with pitch. The image in paintings is of pure white, but it was more likely that the craft were reddish-brown. These flexible waterproof vehicles could survive almost anything and were

used for at least 10 years. They were often decorated with porcupine quills.[17]

Samuel de Champlain in 1603 described the canoes: "Their canoewes . . . are made of the barke of a Birch tree, strengthened within with little circles of wood well & handsomely framed and are so light, that one man will carry one of them easily. . . . Their cabins are low like their tents, covered with the said barke of a tree. . . . The men sat on both sides of the house . . . [each] with his dish made of the barke of a tree."[18]

George Heriot in his book *Travels through the Canadas* wrote in 1807 about the nomadic Algonquin: "They generally carry with them large rolls of the bark of the birch-tree, and form the frames of the cabins of wattles or twigs stuck into the earth in a circular figure, and united near their upper extremities. Upon the outside of this frame the bark is unrolled, and thus affords shelter from rain and from the influence of the sun."[19]

Birchwood was also used to make sleds and snowshoes. This work was mostly done in the winter when the wood was frozen, which made it easier to split in a straight line, and when there was less sap in it. It was also easier to steam and bend the wood without it splitting. Birchwood was used for all sorts of purposes: wooden nails, bows, arrows, drums, ax handles, hammers, spoons, snowshoe webbing, dog whip handles, and grease lamp bowls. Rotted birch was used to smoke meat and fish; mixed with white spruce and jack pine cones it could be used to tan hides.[20]

Birch bark was used by the peoples in Northern areas as "sunglasses" against snow blindness. They would tear the bark into strips about 2 inches wide, which they would wrap around their eyes, using the lenticels as eyeholes. The Dakota shredded the bark and bound it in bundles for torches. Samuel Strickland observed that the native women "have a curious method of forming patterns upon this bark with their teeth, producing very elegant and elaborate designs. They double a strip of bark many times into angles, which they bite at the sharp corner in various forms. Upon the piece being unfolded, the pattern appears, which is generally filled in very ingeniously with beads, and coloured porcupine quills. They perform this work in the dark quite as well as the daylight."[21]

When Native Americans chewed the gum of the tree, they knew what they were doing. Birch leaves, twigs and buds contain methyl salicylate (a component of aspirin) and the bark contains betulinic acid, an anticancer, anti-inflammatory, and antiviral compound. Leaves and twigs were often brewed to make a kind of medicinal tea or an infusion for treating skin and scalp problems. Following an outbreak of contagious disease, people used to burn birchwood to clear the air of contagion. Lengths of bark were used as casts for broken limbs. First of all, a soft cloth was put next to the limb, then well-soaked birch bark was wrapped and tied about the limb, and when it was gently heated, the bark shrank to fit the limb perfectly.

Birch sap can be used to make vinegar, wine, beer, and syrup. Birch syrup is a bit like maple syrup or molasses and birch beer is a little like root beer, but not as sweet. Birch sap contains only 0.9 percent sugars, compared to 2 to 3 percent for maple, and is more acidic.

In their book on traditional foods, Harriet Kuhnlein and Nancy Turner write:

In May, before the leaves appeared, the trees were tapped. A V-shaped flap was cut in the outer bark, and propped out to make a "spout" by placing a small stick underneath it horizontally. Several cuts were made in the bark directly above, and a birch-bark basket was placed on the ground beneath it. The sap oozed from the upper cuts, dripped from the tip of the V and was caught in the basket. It was drunk as a beverage or added to soups.[22]

Today birchwood is used commercially to make toothpicks, Popsicle sticks, pulp and paper, and fuel wood. Birchwood is famous as a veneer and is employed in making furniture. After a good sweat, Scandinavians and other sauna enthusiasts beat their bodies with birch switches and then jump into an icy lake.

The birch tree now battles for its range and its life. Acid rain has damaged the whole of the boreal forest, but birch damage is particularly noticeable. But the birch also has a remarkable relationship with the mycorrhizae in the soil. Scientists in Northern areas are studying the birch to see what beneficial microorganisms keep certain plants free of the birch borer and diseases that afflict other species.

Children and campers still collect the bark to make drawings or to use as tinder. However, tempting as it is to take the bark, it should be stripped off so carefully that the inner layer, or cambium, is not touched. Stripping it off in one big chunk or taking bark from right around the trees will seriously harm the tree. A better solution is to take bark from a fallen branch.

Pinaceae

White Spruce

Picea glauca, Western white spruce, Alberta white spruce, Canadian spruce, Adirondack spruce, Nova Scotia spruce, Quebec spruce, St. John's spruce, Black Hills spruce, Porsild spruce, skunk spruce, cat spruce

Picea glauca

grows to 78 feet but will be shorter at its northern limits. The stiff blue-green needles are ¾ inch long. It is a coniferous evergreen with a long straight trunk and a narrow crown. The thin bark can be either scaly or smooth and is grayish-brown. The root can grow 35 inches into the ground, but taproots tend to be deeper. In dense stands the lower branches are shed. The light brown cones are up to 2 inches long and produce tiny winged seeds. White spruce grows in moist, well-drained soils, usually in mixed forests.

Range: Alaska, throughout most of Canada, northern New England, Minnesota, Wisconsin, and Illinois. Isolated populations grow in the Black Hills of Wyoming and South Dakota.

Provincial tree of Manitoba and state tree of South Dakota.

Spruces are different from all the other conifers: they have square needles and their cones, which are about 2 inches long, hang down from the branches rather than sit up perkily erect. The prickly spruce needles grow in whorls all around the twigs.

The tree is monoecious: the male and female flowers grow on the same tree. The males start out light red, then turn yellow. The females are purple. Both are almost inconspicuous, unless you happen to be at the tree with a magnifying glass to see the females in the tops of the whorls of needles. The fruit is a pendulous cone of light green which later turns pale brown. Once the male produces pollen, the female becomes receptive and pollen is shed over five days in May, June, or July, depending on where the tree grows. The seeds fall in autumn and are wind-borne, but cones can remain on the tree for a few years after seed dispersal is supposed to take place. Because boreal forest fires usually occur earlier than the period when seed matures, the seeds can survive the fire. If they fall in places that have been burned, they are more likely to germinate. But the tree's thin bark and shallow roots usually mean that the trees themselves are killed by fire.

It's tough being a spruce. The rodents, especially red squirrels, cache the seeds. And the seeds that evade red squirrels have a hard

Picea glauca, white spruce

time getting started if the litter of dead leaves and needles on the forest floor is too deep. Sometimes the tree reproduces itself by layering: the branches dip down to touch the forest floor and take root.

Spruce are "plastic," that is, able to survive under highly variable conditions.[23] For instance, in some parts of its range, the winter temperatures are as low in January as −65°F; in other parts the summers may be as hot as 94°F in July. The only place spruce can't live is in standing water.[24]

Spruce trees, both **black** (*P. mariana*) and **white,** are common in the boreal forest and were immediately apparent to the newcomers who arrived on the shores of Labrador and Newfoundland to fish and, ultimately, to take possession of the land. The most famous encounter with the spruce was Jacques Cartier's in 1536, when it saved the lives of his crew. Cartier's men were so stricken with scurvy that their gums were bleeding, their teeth were falling out, and they could barely stand. He described how they were rescued:

[Two women] showed us how the bark must be pulled off the leaves of these branches, and all put into water to boil; then the water drunk once in two days; and the juice of the leaves and bark pressed out and the water put on the swollen and sick limbs. . . . They call the said tree in their language, annedda. *Immediately afterwards, the captain had the drink made, so the sick could drink, of these however no one wanted to try it, only one or two, who decided to try it. As soon as they had drunk, they felt better, which they found a true and evident miracle; for of all the sicknesses they had suffered from, after having drunk two or three times, they recovered their health and were cured, so that some of the company who had had syphilis for more than five or six years before getting this sickness [scurvy] . . . were completely cured.[25]*

By the seventeenth century, boiling the tips of spruce boughs in water or in beer had become a well-known treatment for scurvy. The crews of Captain Cook were required to drink spruce beer every other day to ward off the disease. Sometimes spruce beer would be mixed with stronger spirits to produce a drink called callibogus. Thomas Anburey, writing in 1789, recorded in his *Travels:* "The Soldiers, not only at the *Isle au Noix,* but likewise at St. John's, have been very subject to the scurvy, not having any other than salt provisions, but by drinking plentifully of spruce beer, they are now all in perfect health, which clearly proves that liquor to be a powerful antiscorbutic."[26] In the eighteenth century, Christopher Middleton, wintering in Churchill on Hudson Bay, reports that, while celebrating Christmas, the men ran out of English beer and turned to "spruce beer and brandy, the only means used here to prevent scurvy."[27]

Spruce is still an ingredient in certain root beers, usually sold in health food stores. Elizabeth Goudie in her book *Woman of Labrador* wrote, in the early twentieth century: "Father boiled spruce and juniper and we had to drink that with our dinner meal. It was a good drink. It was made of spruce, molasses, yeast, raisins and rice and tasted something like Coke. We had to drink that for a month to clean our blood. That was done every spring all the time we were growing up."[28]

Janice Schofield, a modern forager, recommends spruce tea, made by taking a handful of the green shoots which grow from the tips of spruce branches in spring, steeping them in hot water, and sweetening the mix with honey, orange slices, cinnamon, cloves, or a nip of brandy. The tea also contains lots of vitamin C. She also makes a jelly from spruce tips, which is good with lamb or venison. The soft centers of young spruce cones can be roasted in campfire coals until they are soft and syrupy.[29]

You who have made good, you foreign
  faring;
You money magic to far lands has
  whirled;
Can you forget those days of vast daring,
There with your soul on the Top o' the
  World?
Nights when no peril could keep you
  awake on
Spruce boughs you spread for your
  couch in the snow;
Taste all your feasts like the beans and
  the bacon
Fried at the camp-fire at forty below?

Can you remember your huskies all
  going,
Barking with joy and their brushes in air;
You in your parka, glad-eyed and
  glowing,
Monarch, your subjects the wolf and the
  bear?
Monarch, your kingdom unravisht and
  gleaming;
Mountains your throne, and a river your
  car;
Crash of a bull moose to rouse you from
  dreaming;
Forest your couch, and your candle a star.

Robert Service, from
"Men of the High North"

The Native American uses of the white spruce were imaginative and varied. The Dene, for instance, split the tops of spruce, heated the branches next to a fire, and the resin that oozed out was applied to the eyes for snow blindness. The resin was used by many other groups against rheumatism, swollen and sore joints, or as a salve for the skin, to soothe cuts, scratches, boils, and bites. To prevent feet from cracking in the extreme cold, some people rubbed on this salve as a preventative measure. The Huron boiled spruce up to make a poultice for burns— an important remedy, given how close most peoples lived to their fires and how likely they were to get hit by a flying ember. The Chippewa used spruce pitch to make a caulk for canoes, to waterproof and preserve babiche (strips of hide) in ropes and snares, and for chewing; the teeth of Chippewa women were said to be especially white because they chewed spruce gum. The Penobscot chewed spruce gum just because they liked the taste. Spruce gum has even become popular with some American athletes as an alternative to chewing tobacco.[30]

When the Europeans came, they looked at the spruce and saw ships' masts. Cartier, ever practical, reported that the spruce at Bay de Chaleur in 1534 was, "for making masts, sufficient to mast boats of 300 tons or more."[31] The hunt for ships' masts went on for almost 200 years, right up to the time when just about every big tree with a long straight trunk had largely disappeared.

The word *spruce* comes originally from *Prusse*, the French word for Prussia. England used to import sprucewood from Prussia in the eighteenth and nineteenth centuries. It is light and splits easily into straight pieces. During the Second World War, England imported Canadian spruce to make Mosquito bombers and other light airplanes. Today, sprucewood is used in making gliders and rowing skiffs because it is very light. The wood is fine enough to use for musical instruments, but tough enough for construction.

Abies balsamea, balsam fir

Pinaceae

Balsam Fir
Abies balsamea, balsam, Canadian balsam, Canadian fir, Eastern fir, bracted balsam fir, blister fir

Abies balsamea

grows 40 to 80 feet tall. It has smooth gray bark with resinous blisters and, as it ages, the bark tends to be brown and scaly. The root system is shallow, penetrating to only 30 inches below the surface of the ground. The needles are flat, blunt-tipped, and resinous and have two pale green lines on the undersurface. The cylindrical cones sit upright on the branches. It can be found on mountain slopes and alluvial flats, in forests and wetlands.

Range: Alberta to Newfoundland, around the Great Lakes in Minnesota, Wisconsin, Michigan, and in Pennsylvania and northern New England. Provincial tree of New Brunswick.

One of the sounds of the boreal forest in winter is the swoosh of snow falling off trees. The pyramidal shapes of evergreens are as practical as they are graceful, since snow slides off them easily and keeps the boughs from breaking under its weight. It doesn't protect them from animals, however. Moose with their wide feet are able to plod through the snow to reach balsam fir, which is their favorite food and about the only forage around when the snow is deep.

Balsam twigs are mostly opposite, with round or conical buds hidden by the leaves. The male cones are tiny and reddish, right through to orange at pollination. The females, which are 2 to 4 inches long, are resinous, with stems, and are blue-gray to green or purple. The cones produce tiny brown winged seeds.[32]

Robust as it may look, balsam fir falls victim to spruce budworm, which can kill a mature tree in three to five years. Because of its thin bark, balsam fir is also the least fire-resistant of all the conifers in the Northeast. If a stand is wiped out, it is unlikely to regenerate easily because the fire kills not only the tree but the seeds as well.

Native peoples used the pleasant-smelling resin to make the seams of their canoes watertight. Henry Wadsworth Longfellow described Hiawatha doing just that:

> "Give me of your balm, O Fir-tree!
> Of your balsam and your resin,
> So to close the seams together
> That the water may not enter,
> That the river may not wet me!"
>
> And the Fir-tree, tall and sombre,
> Sobbed through all its robes of darkness,
> Rattled like a shore with pebbles,
> Answered wailing, answered weeping,
>   "Take my balm, O Hiawatha!"[33]

The resin from balsam fir is also considered therapeutic and antiseptic, and has been used in folk remedies for everything from bronchitis to earaches to warts. Different native groups had different uses: the Chippewa used the gum as an analgesic; the Kwakiutl as a laxative; the Menominee found it useful for colds and sores; the Ojibwe for colds and venereal diseases; the Penobscot for cuts; and the Caughnawaga for skin cancers. The Cree used the resin to soothe insect bites, boils, scabies and other infections, and skin sores; it was applied directly or mixed with oil to make an ointment. Other groups considered it food: the Mi'kmaq used the bark to make a type of tea, and the Upriver Halkomelem of British Columbia found it effective to simply eat the bark or the gum on its own. The Ojibwe made a warm liquid of the sap and drank it to cure gonorrhea.[34] The bark can even be used to make a kind of bread.[35]

By 1609 the commercial value of the tree was noticed in Acadia by Marc Lescarbot and Pierre Erondelle: "Good profit may be drawn from the fir and spruce trees, because they yield abundance of gum; and they die very often through overmuch liquor. This gum is very fair, like the turpentine of Venice, and very sovereign for medicines. I have given some to churches of Paris for frankincense, which hath been found very good."[36]

These days balsam firs are grown on Christmas tree farms. It's popular because the needles cling to the branches even in overheated houses, and the shape is attractively conical—very Christmas card–like. Wreaths are made from the boughs and the highly scented needles are stuffed into souvenir pillows and sold in New England stores. Balsam fir wood is a creamy white to light brown and is also used for both lumber and pulpwood. It is lightweight, not very strong, and not very good at holding nails, so it tends to be used to make plywood or things like packing crates.

The blisters on the bark contain oleoresin ("Canada balsam"), which is used as a medium for mounting microscopic specimens in laboratories, a cement for glassware, and in the manufacture of spirit varnishes. It is the provincial tree of New Brunswick, although spruce budworm has denuded large areas of Canada's maritime provinces.

Rosaceae

Pin Cherry

Prunus pensylvanica, Northern pin cherry, fire cherry, bird cherry, wild red cherry, pigeon cherry, Pennsylvania cherry

Prunus pensylvanica **grows up to 45 feet high with red-brown bark. Leaves are alternate, ovate to lanceolate, to 4 inches long, and dentate. The white flowers appear in flat-topped clusters at the same time as the leaves; the fruit consists of bright red, sour-tasting cherries. The tree will colonize disturbed areas and burned-over areas as well as dry and moist woodlands.**

Range: **British Columbia to Newfoundland, the Great Lakes, New England, and the Rocky Mountains as far south as Colorado.**

Prunus pensylvanica, pin cherry

In the grand scheme of things, the pin cherry serves an important purpose in its short 40-year life-span—its role is to move in quickly after a disaster such as a fire has hit an area. No one is quite sure why, but the seeds (which can be dispersed by either gravity or animals) may wait for years to break dormancy. They can lie in the soil for 100 years waiting for ideal circumstances to hit. Perhaps they need to become more permeable to water and oxygen, maybe it's more light, or that the soil conditions must change dramatically. Then the pin cherry becomes the pioneer plant par excellence.

The roots of the pin cherries protect the soil by holding valuable nutrients in place while other, larger, trees become established. It acts as a nurse tree for other trees' seedlings, giving them the cover they need to grow. However, once these seedlings take hold and grow taller than their nurse, the shade-intolerant pin cherries die out.

Pin Cherry Stumps

Our cherries have gone down the beaver
 run,
The beavers care not for our cabin's
 charm.
Food they must have for now, and when
 the sun
Goes south, a snugger house to keep
 them warm.

The cabin and I grieve. But brave small
 sprouts
Of the gone cherries, staunching our
 distress,
Stretch, trying to be trees. We squelch
 grave doubts
And humor them for mutual
 cheerfulness.

Stellanova Osborn, *Beside the Cabin*

The pin cherry has seductive warm, shiny, reddish-brown bark marked by lenticels (the horizontal lines that let gases pass from the inside to the outside of the plant). The charming white flowers bloom between March and July and then turn into drupes, or one-seeded fruits.

An eighteenth-century traveler noticed the berries: "The cherries are small, and extream red; and though their taste is not good, yet the Roe-bucks like 'em so well, that in the Summer time they scarce ever miss to lye under the Cherry-trees all Night long, especially if it blows hard."[37] The stone is poisonous, since it contains hydrocyanic acid, as do the leaves and bark, all of which can be lethal to any livestock and wild animals that eat them by mistake. Most animals, however, simply vomit them out. To reproduce, pin cherries don't need seeds since they can form colonies by sending out shoots to take root and form new plants.

The cherries are used to make jams, jellies, and even wine (if birds leave enough behind). Indigenous peoples from all parts of Canada and the northern United States ate pin cherries fresh from the trees, cooked, or dried in the sun. Dried cherries were often powdered and stored as a winter food.

The Cree used the inner bark of pin cherry to make a medicinal tea for sore eyes.[38] An infusion of the bark was used by the Algonquin to counteract blood poisoning. Sores and ulcers were treated with a wash made from root bark by the Cherokee. The Ojibwe used an infusion of the inner bark as a cough remedy; the Cherokee used the same for laryngitis.[39] Various other nations used the bark in basket decorating.[40]

Onagraceae

> ### Fireweed
> *Epilobium angustifolium,* willowweed, willow herb, wild asparagus, blooming Sally, great willow herb

Epilobium angustifolium

is a tall herb that can grow up to 9 feet high, but is usually about 3 to 4 feet high. The stout stem rises straight up from the woody roots, with many dark green narrow leaves branching off alternately. In midsummer, the top of the stem produces a large spike of 15 to 50 flowers, usually pink or red, but sometimes white. Each flower emerges on a short stalk with four petals and a long narrow ovary that turns into the seedpod. Each seed has a tuft of silky white hairs that help the wind disperse it. Fireweed grows in open woods, on roadsides, and on disturbed or burned land.

Range: throughout North America, except for the Deep South and desert areas.

Territorial flower of Yukon.

As its name suggests, this is another pioneer plant that grows in the aftermath of fire. After the Second World War, bombed-out areas of London sprouted fireweed, something not seen there in generations.[41] Fireweed survives even the most intense of fires because the fine roots and rhizomes are in the top 2 inches of the soil. They can sprout within a month and bloom the same year. Another survival strategy is to produce copious amounts of seed which waft along on the wind on their silky tufts. Chances are pretty good that if they fall on a mineralized soil exposed by fire, they will grow.[42]

Beverley Skinner, a wildlife biologist and

Epilobium angustifolium, fireweed

All parts of the plant are useful. Bees make a famously rich honey from the nectar, and the young shoots can be peeled, steamed, and eaten as a vegetable ("wild asparagus"). The leaves can be tossed in a salad or dried and made into tea, said to be good for stomachaches and constipation. The interior pith of the stem is sweet and gelatinous, and various native peoples ate it with oil or sugar. The long stalks can be dried and woven into mats or baskets. Even the fluff from the seeds has been used as tinder for starting campfires or woven with other fibers into clothing.

The native peoples of the Northwest used the tough stem fibers to make twine and fishnets. They ate the young leaves and shoots, as well as the pith. The Flambeau Ojibwe, who call it "slippery root" or "soap root," use the outer rind of the root to make a poultice against boils and carbuncles by pounding it and then soaking it in water.[44]

The Dena'ina of Alaska also ate young fireweed stems for food, sometimes with fish eggs. They also used the raw stems to treat pus-filled boils or cuts.[45] The Chippewa boiled fireweed plants to make a remedy for worms. The Wood Cree chewed the inner root or the stems to make poultices for boils or wounds, and treated bruises with the fresh leaves.[46]

Today, fireweed is mainly admired as an attractive wildflower. Marianne Lynch of Neustadt, Ontario, tells the following story about the fireweed: "My aunt, Mrs. Viola Lemay, was a great naturalist in her time. (She died in 1989.) She lived most of her adult life in South Porcupine, northern Ontario, and she and her husband spent their free time wandering through the bush and swamps looking for wildflower species. She loved to dry and press flowers. One of her favourites was fireweed. When Viola retired to Grey County, Ontario, she continued her hobby. For years she looked in this area for fireweed. . . . [Then] five years

broadcaster in Alaska, says: "When I first moved to McGrath, I was told that when the last flower opens on the top of the stalk, summer in the interior of Alaska is officially over. Since the flowers open one at a time from the bottom, it's a plant worth keeping an eye on."[43]

The *epi* part of the scientific name comes from the Greek for "upon"; *lobos* means "foot" or "pod" and it describes the placement of the petals on top of an elongated ovary, which later becomes the slender seedpod. *Angustifolium* means "narrow-leaved."

after her death, I found a patch of this elusive plant growing along the side of a bush in Normanby Township. The area had been used to burn old fence posts which creates the habitat for fireweed, hence the name. I dug up some plants and have them growing in my garden in her memory."[47]

Ericaceae

Labrador Tea

Ledum groenlandicum, Hudson's Bay tea, trapper's tea, marsh tea, muskeg tea, moth herb

Ledum groenlandicum

is an erect, fragrant evergreen shrub that grows 2 to 4 feet tall. The leathery leaves roll under at the margins. They are shiny on top and have orange fuzz below, up to 2 inches long with a fairly narrow, drooping shape. The umbel-shaped clusters of white flowers come out in spring. Reproduces mainly vegetatively, but can reproduce from seed.

Range: **Alaska and across northern Canada to Newfoundland, and around the Great Lakes.**

Ledum groenlandicum, Labrador tea

When I was growing up in Labrador, we saw masses of Labrador tea. It was pivotal to all our games about survival in the bush. We made tea with it, though none of us actually drank any. We'd heard all the local lore about how trappers used it when they were desperate in the bush, and the Montagnais were reputed to drink it as well. But we were much too chicken to try it. Still, we'd pretend, since it was all part of the game of being lost in the woods and surviving. We'd scare ourselves half to death every time we actually saw an animal on the trail. I can still smell the spicy scent and recall those endless perfect few days in between the blackfly and the mosquito season, when we could wander at will without being covered with netting. We saw the Naskapi elders, who still moved their summer camp near the Grenfell Mission in North West River, brewing and sipping away on the tea, which no doubt gave us the inspiration for our games.

Labrador tea is easy to identify by its pretty white flowers with downward-protruding stamems. After pollination, the plant releases tiny seed capsules complete with hairs to carry them away on the winds. It grows on rough rhizomes which also help it to reproduce vegetatively. Since it grows in boggy areas and moist forest, it isn't usually subject to fire.

The leaves of Labrador tea have a strong smell and are still used to repel and kill insects, hence the name "moth herb." Farmers used to sprinkle the leaves around barns and granaries to keep away rodents. In the nineteenth century, it was used to repel fleas, bedbugs, lice, and mosquitoes.[48]

For some aboriginal peoples, Labrador tea was one of the most important and powerful of all healing plants. In their book *Voices from Hudson's Bay*, Flora Beardy and Robert Coutts recorded the words of a Cree from York Factory:

If you cut your finger or accidentally injured yourself with an axe, the person would chew on kâkikê-pakwa *[Labrador tea] leaves. Then the leaves are put on the wound and bandaged. The wound would heal. The person got well. . . . The most powerful plant for healing was the one called* kâkikêpakwa. *I once heard an elder talk about this plant, the* kâkikêpakwa, *and what it can cure. If a person had a sore chest, it's applied to the chest. If a young woman had problems with menstruation these leaves were boiled to make a drink and this made the person well. This plant cured them.[49]*

The Wood Cree made a salve for burns by mixing the leaves with grease. They also made a medicinal tea to wash burns, bites, and sores.[50] Others made a powder of the roots to prevent ulcers or formed a poultice of leaves and grease to relieve burns. Burn remedies were often necessary for these families who spent their lives around open fires. An infusion of leaves was used to relieve poison ivy and as an aid for the kidneys, while a strong decoction was taken to ward off colds.[51]

The Dena'ina Athabascan used Labrador tea against distresses such as colds, heartburn, and hangovers, as well as diseases such as tuberculosis and arthritis; as a wash it was used to heal sores. The Yupik natives of western Alaska boiled the whole plant to make a remedy for food poisoning.[52]

The white settlers rapidly became aware that it was highly effective in warding off scurvy. The leaves contain ascorbic acid (vitamin C), in addition to tannic acid and gallic acid. Henry Ellis in 1748 wrote:

The Plant, by the Indians *called Wizzekapukka, is used by them, and the* English *as a Medicine, in nervous and scorbutick Disorders; its most apparent and Immediate Effect, is promoting Digestion, and causing a keen Appetite. To this Plant, the Surgeons residing at the Factories, ascribe all the Qualities of Rhubarb; it is a strong Aromatick, and tastes pleasantly enough when drank as a Tea, which is the common Way of using it.[53]*

Turner and Szczawinski believe that it was eventually taken up by indigenous peoples as a drink (rather than strictly for medicinal purposes) when they saw traders doing this, since they had not yet developed the tea-drinking habit.[54]

Labrador tea is made by mixing dried leaves with those of black tea, since pure Labrador tea contains a substance called ledol, which can be toxic in large doses, and andromedetoxin (the same substance found in rhododendrons), which causes headaches and stomach cramps. Steeping the leaves for longer than a few minutes increases the potency of these chemicals.

In the 1770s, explorer Samuel Hearne wrote that Labrador tea was "much used by the lower class of the [Hudson's Bay] Company's servants as tea; and by some is thought very pleasant. But the flower is by far the most delicate, and if gathered at the proper time, and carefully dried in the shade, will retain its flavor for many years and make a far more pleasant beverage than the leaves."[55]

During the American Revolution, when citizens boycotted imported British tea, Labrador tea was considered a good substitute. It was also popular in Canada: "The Labrador or country tea . . . was used very extensively in the country before our entrance into Confederation. In the hay and harvest fields it was considered by many of the old settlers of the Red River colony to be superior to any other beverage in allaying thirst."[56]

The leaves can also be added to stews like bay leaves to impart a pleasant spicy taste. Sometimes the Ojibwe smoked Labrador tea leaves instead of tobacco.

Labrador tea makes an appearance in the memoirs of Mina Benson Hubbard. She was an intrepid woman who decided, after her husband was killed trekking through Labrador in 1903, that she would follow his footsteps and complete his journey. She undertook this amazing trip in 1905 in record time and completed it in good health (her husband had starved to death). Her notes and observations were very detailed. "The river's course was now cut deep into the plain, the banks being from thirty to forty feet in height, and the current very swift. The plain had once been sparsely wooded but was burned over and very desolate looking now. Huckleberries, cranberries, and Labrador tea grew in profusion and were in blossom, while patches of reindeer moss were seen struggling into life where we made our camp."[57]

Cyperaceae

Sedge
Carex spp.

Carex spp.

may grow from 2 inches to 2 feet tall. They are divided into six groups and have one characteristic throughout the genus: the ovary and seedlike fruit (achene) are encased in a membranous sac (perigynium) in the axil of a single bract. They may be loosely tufted or spiky. Flowering stems are triangular and solid without nodes. Flowers are either male or female.

Range: **throughout Canada and the northern United States.**

"The first thing one notices [about a sedge meadow], other than the stillness, is the total lack of any brushy undergrowth—there's a carpet and a canopy, but not much else. In very early spring, the sedge meadow can be rather

Carex spp., sedge

gray . . . but by late spring, it's a lush place full of dappled sunlight and shifting shadows,"[58] writes Monique Reed. No matter where they are found (and this one was in Texas), sedge meadows have haunted everyone who has ever walked through them, from poets to children.

Thoreau described the autumn landscape when he was out walking in 1859. "Surely russet is not the name which describes the fields and hillsides now, whether wet or dry. There is not red enough in it. I do not know a better name for this (when wet) yellowish brown than 'tawny.' On the south side of these warm hills, it may perhaps be called one of the fawn-colors, i.e., brown inclining to green. Much of this peculiar yellowish color on the surface of the Clamshell plain is due to a little curled sedge or grass growing at short intervals, loosely covering the ground (with green mosses intermixed) in little tufts like curled hair."[59]

Sedges are grasslike plants mostly found in moist places. What distinguishes *Carex* from other species are the ovary and seedlike fruit (achene), which are enclosed in a membranous sac (perigynium) in the axil of a single scalelike bract.

Ducks and other waterfowl are particularly fond of sedges, and disperse them freely. Each seed is covered by a protective membrane, which survives the duck's digestive system, and may be carried for miles by migrating birds. Other sedges are eaten by wildlife and domestic livestock in winter.

Native Americans wove baskets from the stems of sedges or used them to produce yellow and brown dyes. The younger the plant, the brighter the color. The stems of *C. lacustris* (**lakeshore sedge**) can be chewed raw, like celery. Inland Dena'ina eat the bulb of the *C. gigantea* (**large sedge**) for food. The Chipewyan used the roots of *C. aquatilis* (**water sedge**) to make a medicine for women whose menstrual period was delayed.[60] The leaves of *C. nebrascensis* were

used by the Blackfoot in the sun dance ceremony: they tied the leaves around the horn of the sacred buffalo skull used in the ceremony.[61] Dry sedge leaves were also used by the pioneers as stuffing for boots and mittens to protect hands and feet from winter cold.

According to John Eastman, "Up until the last fifty years, sedge meadows provided ample gratuitous sources of livestock feed and bedding, especially for pioneer farmers. Tussock sedge, especially, was laboriously hand-harvested as

No wind there is that either pipes or
   moans;
The fields are cold and still; the sky
Is covered with a blue-gray sheet
Of motionless cloud; and at my feet
The river, curling softly by,
Whispers and dimples round its quiet
   gray stones.

Along the chill green slope that dips and
   heaves
The road runs rough and silent, lined
With plum-trees, misty and blue-gray,
And poplars pallid as the day,
In masses spectral, undefined,
Pale greenish stems half hid in dry gray
   leaves.

And on beside the river's sober edge
A long fresh field lies black. Beyond,
Low thickets gray and reddish stand,
Stroked white with birch; and near at
   hand,
Over a little steel-smooth pond,
Stand multitudes of thin and withering
   sedge.

          Archibald Lampman,
      from "Autumn Landscape,"
   in *The Poems of Archibald Lampman*

'marsh-hay'—not as nourishing as pasture grasses, but a useful supplement to them. Marsh hay also provided efficient insulation for storing ice in the summer and 'to keep Milwaukee beer cold until it reached Chicago,' as one observer wrote. Though no longer used for such purposes, this sedge is still sold—in bags as garden mulch."[62]

John Muir, one of the fathers of the American conservation movement, is given the credit for starting up the national park system, and sedges started it all. First there was a small bit of wet sedge on Fountain Lake Farm in Wisconsin—his first attempt at wilderness preservation. He told a 1896 meeting of the Sierra Club:

The preservation of specimen sections of natural flora—bits of pure wilderness—was a fond favorite notion of mine long before I heard of National parks. . . . On the north side of [Fountain] lake, just below our house, there was a Carex *meadow full of charming flowers— cypripediums, pogonias, calopogons, asters, goldenrods, etc.—and around the margin of the meadow many nooks rich in flowering ferns and heathworts. And when I was about to wander away on my long rambles I was sorry to leave that precious meadow unprotected; therefore I said to my brother-in-law, who then owned it, "Sell me the forty acres of lake meadow, and keep it fenced and never allow cattle or hogs to break into it, and I will gladly pay you whatever you say. I want to keep it untrampled for the sake of the ferns and flowers; and even if I should never see it again, the beauty of its lilies and orchids is so pressed into my mind that I shall always enjoy looking back at them in imagination, even across seas and continents, and perhaps after I am dead."*

Though his brother-in-law refused to part with the land, the Fountain Lake Farm has been protected since 1957, when Marquette County acquired the adjoining land to form the John Muir Memorial Park. "The county continued to acquire parcels of land until all the land immediately surrounding the lake was included."[63]

A sedge marsh left an equally strong impression on Kurt Buhlmann, a conservation research scientist, although his meadow was far from the boreal forest. "It was an older 1950s-era development in the northwest, mountainous, and rural part of New Jersey . . . the area closely resembles the Blue Ridge mountains and valleys of Virginia, rather than being the typical asphalt and concrete image many people have of New Jersey. The development surrounds a lake where I took swimming lessons during the summers of my youth. . . . [But] best of all was a sedge meadow marsh (perhaps a 'swamp' to some) that was accessed by a wooded trail a few hundred feet behind my parents' house. We all spent some time there, catching turtles and frogs or playing hide and seek. In fact, playing in that marsh led to my interests in ecology and my present profession as a conservation research scientist."[64]

There are more than 100 species of sedges growing throughout the circumpolar boreal region. The species are classified into six groups.

The first group consists of sedges with a single spike (e.g., *C. leptalea*, **bristle-stalked sedge,** a tufted perennial with slender stems 4 to 20 inches tall that is found in moist woodland and thickets, and in boggy areas across the boreal forest). All the other groups have several spikes.

The second group has stalkless side spikes, usually monoecious, with the female flowers above the male; the achenes are lens-shaped, and the flowers have two stigmas (e.g., *C. interior,* **inland sedge,** a densely tufted perennial with thin wiry stems, that grows 6 to 20 inches tall in fens, seepage areas, and wet meadows).

The third group is like the second except that the male flowers appear above the female (e.g., *C. siccata,* **hay sedge,** a perennial that grows 6 to 30 inches tall in dry sandy, open areas).

The fourth group has a male end spike and female stalked side spikes, the achenes are lens-shaped, and the flowers have two stigmas (e.g., *C. aurea*, **golden sedge,** a loosely tufted perennial with slender stems that grows less than 1 foot high and is found on streambanks and in wet meadows).

The fifth group is distinguished by its triangular achenes and the fact that the flowers have three stigmas and the perigynia are hairy (e.g., *C. lasiocarpa*, **hairy-fruited sedge** or **slender sedge,** a loosely tufted perennial with stems up to 3 feet tall which are reddish at the base and grow at the edges of lakes and ponds and in fens).

The sixth group is like the fifth except that the perigynium is hairless (e.g., *C. media*, **Norway sedge,** a densely tufted perennial that grows up to 18 inches high in moist open forest and wetlands, or along streambanks and pond edges).

Equisetaceae

Horsetail
Equisetum spp., joint grass, shave grass, scouring rush, horse pipe, snakeweed

Equisetum spp.

grow 6 inches to 4 feet tall from rhizomes, with jointed, hollow, green stems. The leaves are tiny whorls that fuse to form a sheath around the stem. The stems contain chlorophyll, the leaves do not. Spore-bearing cones grow at the top of the stems. Horsetail grows in moist woods, along riverbanks and lake shores, and in ditches and marshes. *E. hyemale* is perennial, growing from creeping rhizomes, 1 to 4 feet tall, found in moist sandy sites, widespread in the circumpolar boreal forest.

Range: **Alaska to Newfoundland.**

This plant looks like something left over from the set of a dinosaur movie. As a matter of fact, horsetails have been around for millions of years. It is thought to be one of the most ancient of all plants, and seems to have stopped evolving. The corpses of dead horsetails deposited over millions and millions of years created the coal seams that underlie large areas of the Eastern forest. During the coal-making period of geological history, some horsetails grew up to 25 feet high with stems 3 feet thick. They covered huge expanses of the planet. So if you find this strange plant, now much diminished in size, treat it with respect: it has an honorable history.

Horsetails survive because they are remarkably persistent. People who are trying to eradicate them from an area have their work cut out. Kem Luther wrote about his struggles with the plant: "My most recent candidate for the weed from Hell is the horsetail, *Equisetum arvense*. I have been trying to establish a perennial garden in a place where horsetail has taken hold. . . . Though I hoe and hoe, I'm not sure that I'm making any progress. I pull out as much of the root system as I can get, until it breaks off, but a week later a new shoot comes up from the same root. . . . There is a documented case, described in *The World's Worst Weeds*, of horsetail erupting through a meter of silt after a Vermont flood in 1936."[65]

The plant got its common name "scouring rush" because the stems contain silica, which makes them feel as rough as sandpaper or an emery board. Horsetails have long been used for cleaning. The Southern Kwakiutl, Bella Coola, and Thompson peoples polished their canoes with fistfuls of horsetail; the Okanagan and Thompson used horsetails for sharpening bone tools, and native peoples throughout the boreal region used them to clean pots.[66]

"Romans always used horsetail to clean

Equisetum spp., horsetail

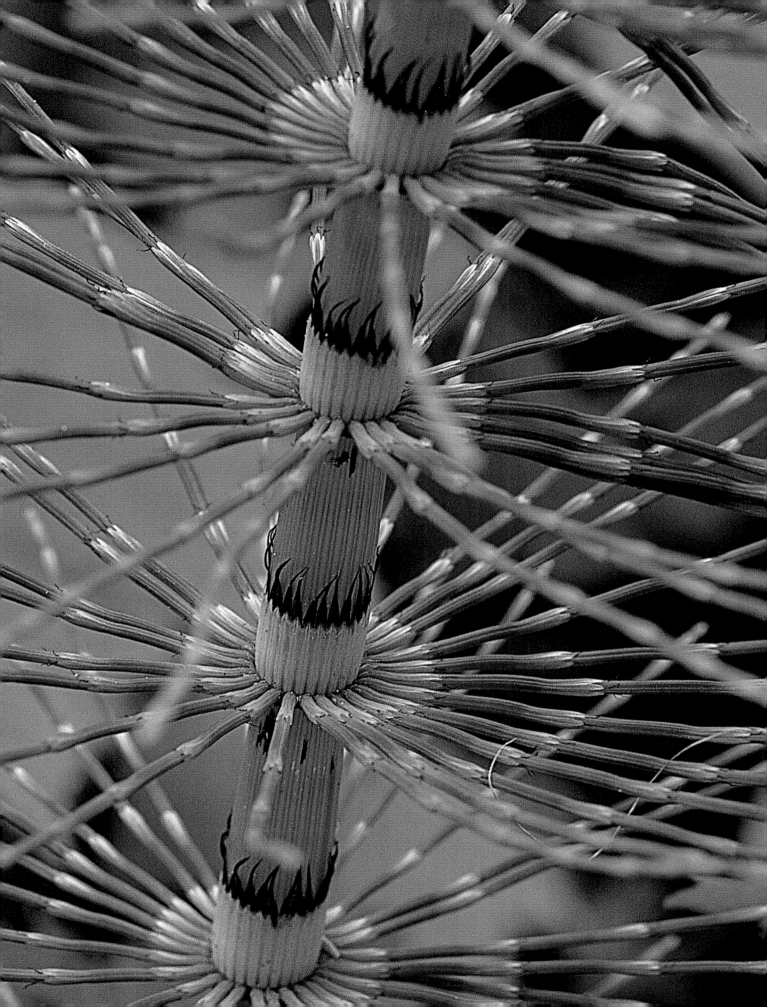

their pots and pans," says Carla Vine, who recently experimented to see how effective this would be on a plate covered with dried lasagna. "It performed admirably," she says, "but be prepared to strain out all kinds of little pieces of horsetail from the wash water."[67] The silica in the horsetails also allowed people to shine brass buttons, or remove rust from a shotgun. As Charles Millspaugh wrote in 1892: "It is gathered into bundles by many housewives and used to brighten tins, floors, and woodenware, and in the arts of polishing woods and metals."[68] Today many campers still use horsetails to do outdoor dishwashing.

Unlike most cleaning tools, horsetails are edible. The shoots, when young, can be boiled in two changes of water and consumed just like asparagus. The roots of the **common horsetail** (*E. arvense*) were ground up and eaten with berries or in soups by the Inuit. The Dena'ina used the tubers of the horsetail for food, and referred to them as berries, owing to their sweetness and juiciness early in the spring; they become inedible later.[69] The Hesquiat took special trips to collect the roots, up to 50 pounds at a time, which they would eat at a special feast.[70]

Captain John Palliser (1817–1887), the explorer who crossed the prairies north of the 49th parallel in the mid-1850s, fed horsetails to his horses. "More than once, as [explorer] Sir James Hector made his perilous way through the forest of the Kicking Horse Pass, the only food the horses had was a few blades of this same plant."[71]

It must have done the horses good, because it also has a reputation as a medicine. Herbal suppliers still sell dried, sometimes powdered horsetail and recommend it for a variety of purposes, from the prevention of osteoporosis to strengthening brittle nails. But don't try to pick your own: some *Equisetum* species are poisonous. Horsetail extract is also used in various nail polishes. In the early twentieth century, it is said,

horsetails were burned as a disinfectant to purify the air in houses where someone was suffering from a contagious disease.

Settlers' children used them as playthings: "Children pulled stems apart at the joints like tinker toys and then challenged each other to see who could reassemble the longest stem."[72] Kids can also make panpipes from them. "Pull a horsetail apart and blow into a section of stem; the tone is related to the length of the section. After you master a single stem, try three at once and you'll have a chord. Finally, figure a way to hold 6 to 10 stem sections in a row arranged from longest to shortest, and you'll have a panpipe."[73] However, some herbalists caution against allowing the children to play with the hollow stems of the plant, as the juice may cause adverse reactions, such as itching or rashes.[74] And watch out for sudden movements in the grass: legend has it that a whistle made from the stems will give the player the ability to summon snakes.[75]

Music stores sell pieces of common scouring rush to clarinet and oboe players, who use them to shape the reeds of their instruments. Roy Lukes, writer of the "Nature-Wise" column in Wisconsin's *Door County Advocate*, tells a recent story about horsetails:

It was Robert Marcellus, principal clarinetist with the Cleveland Orchestra, and also with the Peninsula Music Festival orchestra in Door County during the summer, who approached me one day with an unusual request when I was working at the Ridges Sanctuary at Baileys Harvor.

"Do you happen to know the whereabouts of some rather thick robust scouring rushes?" Naturally my interest was immediately aroused as I asked him why he was looking for these unusual plants. "I use the individual dried sections as miniature cylindrical sanding devices for shaping my clarinet reeds. The natural abrasives in

these plants are as fine and perfect for this job as any manufactured item."

Fortunately I was aware of a small patch of tall scouring rushes, also referred to as horsetails, and one two-foot-tall plant was all he needed. It was exactly the species he had hoped we would find.[76]

Another modern-day use for horsetail is to make a concoction to prevent fungal diseases on household or garden plants. Pick a few in the woods, let them dry, and put them in about a gallon of water from the dehumidifier, then let it simmer for about 20 to 30 minutes. Cool and strain, then use it in a dilution of 1 to 10 parts water and spray on plants affected by fungus every few weeks.

Fabaceae

Wild Sweet Pea
Hedysarum mackenzii syn. *H. boreale,*
Mackenzie's sweet vetch

Hedysarum mackenzii **syn.** *H. boreale*
grows from a woody taproot and may be 2 feet tall. The compound leaves are narrow and hairy, somewhat thick. The pea-shaped (sepals broadly triangular, upper shorter than lower) fragrant flowers are 1 to 2 inches long. Bright magenta or purple, they hang in long clusters at the tips of the stems. The seeds grow in flattened pods with constrictions between each seedpod; hairless, stalked; two to five seeds per pod.

Range: **western Alaska to northern Manitoba.**

Carole Williams of Whitehorse, Yukon, writes, "Its common name is wild sweet pea, reflecting its heady fragrance which is so pervasive in June that it defines summer for me at Kluane [National Park]. . . . For me the colour landscape at Kluane is dominated by the particular purple-pink of many of the flowers (*Saxifraga oppositifolia*, Lapland rosebay, moss campion, *Hedysarum*, fireweed). The combination of this colour with the turquoise blue of the sky and Kluane Lake strikes a colour chord with me of pure happiness."

Hedysarum mackenzii is the poisonous relative of *H. alpinum*, known as **Indian potato** or **bear root**, which has an edible root tasting a bit like carrot and well-known to the indigenous peoples of the North. For all we know, the root of *H. mackenzii* tastes the same, but few people have lived to describe it. In 1848, some of the men with Sir John Richardson's party, who were searching for the missing Franklin expedition, ate *H. mackenzii* and fell ill. More recently, the plant was blamed for the death of young Christopher McCandless, whose short life and early death in the Arctic were described in the book *Into the Wild* by Jon Krakauer. McCandless was a young man from Washington, D.C., who had an idealistic love of the wild. In 1992 he trekked into Alaska on the Stampede Trail and lived for some months by hunting and gathering. However, he is believed to have accidentally eaten parts of *H. mackenzii*, mistaking it for *H. alpinum*. He became weak and unable to fend for himself and died near Denali National Park in Alaska.[77]

Later, however, Edward Treadwell, a graduate student at the chemistry department of University of Alaska–Fairbanks, who had heard of the Richardson expedition and the McCandless stories, conducted tests on both *H. mackenzii* and *H. alpinum* to see if the former really is poisonous. McCandless was reported to have eaten the seeds of *H. mackenzii*; this assumption is drawn from his diary, in which he writes: "Extremely weak. Fault of pot[ato] seed. Much trouble just to stand up. Starving. Great jeopardy." Treadwell found no evidence of poison and concluded that McCandless simply died of starvation. Looking at the Richardson expedition's journals of 1850, he postulated that it was something other than wild sweet pea that killed the members of the party. They had also been eating lichens, rotten meat, fish entrails, leather, and warble fly dung. Nonetheless, even if it is not poisonous, there is no evidence that wild sweet pea was ever eaten by natives anywhere. Thomas Clausen, a chemist, argues that this is only because the roots of the wild sweet pea, while edible, are so much smaller than those of the Indian potato (*H. alpinum*) and so hard to harvest that it simply wasn't worth the effort.[78] The jury is still out.

Araliaceae

Wild Sarsaparilla

Aralia nudicaulis, American sarsaparilla, false sarsaparilla, small spikenard, rabbit's-root, wild licorice, shotbush, aralia, salsepareille, bamboo brier, oscar

Aralia nudicaulis

grows to 16 inches high from underground branching rhizomes which send up short stems bearing three longer stalks, each with five leaflets. Rhizomes often form large clonal colonies. The compound leaves have five leaflets, oval with pointed ends and toothed along the edges. The greenish-white flowers bloom in two to seven umbels, 1 to 2 inches across. The flowering stems are hidden underneath the leaves. The seeds are borne in small berries, greenish ripening to purple.

Range: all Canadian provinces and territories and in the midwestern and eastern United States.

Sarsaparilla always makes me think of gunslingers in old movies knocking back drinks in the saloon. Back in the 1950s, making homemade root beer was as popular as making your own wine is today. The sounds of bottles exploding in closets forbidden to children was a regular event and tasting this brew a treat grown-ups reserved for special occasions.

The foot-long dried-up-looking roots were the "roots" in root beer: in the wild the fleshy roots spread out as far as 6 feet. The scientific name *nudicaulis* comes from the Latin *nudus* meaning "bare" or "naked" and the Greek *kaulos* or "stem." The flower stem rises from the roots "naked"—without any leaves around it. These naked stems poke through the leaf litter in early spring before the leaves have time to unfurl. It takes only a month for the fruit to mature. The seeds are spread by birds and other animals, especially bears, who favor this as a spring food after a long winter's hibernation. Moose, white-tailed deer, as well as birds such as ruffed grouse and thrushes also feast on the drupes—the one-seeded fruit. The berries with the inner stone have no particular taste, but the roots are described as "aromatic," "balsamic," and "sweetish."

Native Americans have traditionally used sarsaparilla as food and medicine. They made teas and poultices from the roots to treat coughs and colds, bladder infections, arthritis, and rheumatism. As with many other plants it was considered a "blood purifier" and a tonic that imparted energy. These tonics were taken regularly, usually in the spring, to wake up a metabolism that had been going into a bit of a dormancy itself as less and less food was available in the last gasps of winter. The roots were considered emergency food for hunting trips or during times of war. Some groups also fermented the berries to make a wine or tonic. The Penobscot mixed the dried powdered roots with sweet flag (*Acorus calamus*) for a cough remedy[79] and the Wood Cree boiled up the roots to give to teething children.[80]

European reports on the uses of this plant started coming in by 1624. Father Theodat Gabriel Sagard wrote, "I inquired of them [the Huron] respecting the chief plants and roots which they use for curing their illnesses and among others they highly esteem the one called *oscar*, which does wonders in healing all kinds of wounds, ulcers and sores."[81]

Three hundred years later, Frank Speck observed similar practices among the Montagnais. "[Sarsaparilla] . . . roots steeped in case of weakness. The dark berries are made into a kind of wine by the Montagnais and used as a tonic. . . . Women cut up pieces of the root, tie them on a string and keep them in their tents until needed. The same is done by the Penobscot. The berries are put into cold water and allowed to ferment in making the wine referred to."[82] Ethnobotanist Huron Smith wrote in 1932: "The Flambeau Ojibwe recognise the root of this plant as strong medicine, but do not steep it to make a tea. The fresh root is pounded and applied as a poultice to bring a boil to a head or to cure a carbuncle. . . . This root is mixed with sweet flag root to make a tea to soak a gill net before setting it to catch fish during the night. Big George Skye, at Lac de Flambeau, was quite successful in catching them."[83]

Early European settlers also made a strong tea from the roots that was administered to people suffering from syphilis. The plant is related to ginseng, also considered a tonic. Charlotte Erichsen-Brown cites a recipe printed in the *Canadian Pharmaceutical Journal*, under the name "New Orleans Mead": "Take 8 ozs. each of the contused roots of sarsaparilla, licorice, cassia and ginger, 2 ozs. of cloves and 3 ozs. of coriander seeds. Boil for fifteen minutes in 8 gallons of water; let it stand until cold. . . . Then strain through flannel and add to it in the soda-fountain; syrup 12 pints [thick sugar syrup],

honey 4 pints, tincture of ginger 4 ozs. and solution of citric acid 4 ozs."[84] If that mixture didn't waken up a sluggish body, nothing would.

Dorothy Louise Molter became one of the most prolific makers of root beer. She was born in 1907, spent most of her life in the area on the border between Minnesota and Ontario, and became famous as the "Root Beer Lady." From 1948 to 1975, she operated the Isle of Pines Resort on Knife Lake, a popular spot with travelers. Dorothy made between 11,000 and 12,000 bottles of root beer a year. In the early 1970s, under the U.S. Wilderness Act, Dorothy's property was purchased by the U.S. government and she was ordered to leave. Her friends petitioned the government and she was granted lifetime tenancy in 1975. She stayed until her death in 1986.[85]

Traditionally, wild sarsaparilla was harvested by hand, harvesters being careful to take the primary root of a mature plant, leaving the secondary root to propagate. The roots were washed, dried, and stored in dark ventilated areas. Today few companies still make traditional root beers using sarsaparilla, most preferring the ease of synthetic flavorings, but there is nothing, absolutely nothing as good as the original, the real root beer.

Empetraceae

Crowberry

Empetrum nigrum, black crowberry, curlewberry, crakeberry

Empetrum nigrum

grows 6 to 12 inches high as a low-growing, creeping shrub. The leaves are needlelike, spreading, edges rolled under, deeply grooved underneath. Flowers solitary in leaf axils, inconspicuous,

purplish-crimson; plant monoecious or dioecious. Fruits blackish drupes, with six to nine white nutlets; edible but not palatable. Crowberry bushes grow in cold coniferous forests, rocky mountain slopes, and acid bogs and peatlands.

Range: Alaska, across northern Canada from Yukon to Newfoundland, along the Pacific coast as far as northern California, around the Great Lakes, and in New England.

Empetrum nigrum, crowberry

This shrub looks like a miniature fir tree. The purple flowers are almost invisible, and produce small black berries. Until they are touched by frost, the berries have little taste. After a hit of frost, they acquire a slightly acidic, refreshing taste. They can hang on to the branches of the plant even under the snow. The crowberry begins growing again in early spring, before the snow is completely melted. People have recorded seeing crowberry blossoms "emerging through a half-frozen crust of ice!"[86] The root system is largely destroyed by fire, so it thrives where there are few fires.[87]

The crowberry plays an important part of the fall and winter diet of more than 40 species of birds, in particular the grouse and the ptarmigan. The red-backed vole depends on crowberry seeds for sustenance throughout the fall, along with reindeer, caribou, and bear. The low-growing shrub provides cover for small birds and mammals.[88]

The common name may refer to the crow-black berries, or to the fact that crows eat them. *Empetrum* means "growing on rock" and *nigrum* means "black." In the writings of Samuel Hearne, circa 1795, we find an explanation for one of the Indian names given to the fruit: "It is also the favorite repast of the gray goose, which is why Indians called it 'Nishca-Minnick' or gray gooseberry. Juice makes pleasant bevorage [sic] although seeds detract from the fruit itself."[89]

The berries were used by some native peoples to slake their thirst when no water was available. The Inupiat Inuit ate the sweet acidic berries with fatty fish livers.[90] Crowberries are generally mixed with other berries to make jams, jellies, desserts, or drinks. From a nutritional standpoint, a crowberry has almost twice as much Vitamin C as a blueberry, and some medicinal uses. The Upper Tanaina cooked and ate the berries as a treatment for diarrhea.[91]

Judith Colwell tried eating crowberries on a trip to Alaska. "We discovered crowberries along the road and began to gather them up. Crowberries are small very dark berries which look like large black ball bearings perched on tiny thick leafed shrubs hugging the ground. Crowberry juice is good, but the skins are astringent and fibrous. For me, it was better to munch and dejuice the berries, after which spit out the skins."[92]

Salicaceae

Quaking Aspen

Populus tremuloides, trembling aspen, American aspen, aspen poplar, golden aspen, small-toothed aspen, poplar, smoothbark poplar, popple

Populus tremuloides

grows up to 50 or 60 feet high. The pale gray-green bark, which has dark horizontal scars, darkens and furrows with age. The leaves are round with a pointed tip, finely toothed, shiny green on top, and dull green underneath. The petioles are flattened, so the leaves flutter in the slightest breeze. The flowers are drooping catkins that appear before the leaves, and the seeds are borne in capsules 1 to 2 inches long. It has 2-inch catkins in very early spring, producing small narrow cones that split to release small cottonlike seeds. Aspens form stands by suckering from their extensive and shallow root system.

Range: throughout the Rockies, and in the North from Labrador west to Alaska, south to Virginia in the East and New Mexico in the West.

Quaking aspen was once known as "noisy leaf," a native name that has stuck because the leaf does

Populus tremuloides, quaking aspen

make slight sounds as it turns in the breeze. It is a species easy to identify in autumn, when the long pale trunk is bedecked with fluttering golden leaves. It's the one tree that can safely be described as the most ubiquitous in North America, and the dominant deciduous species in mountainous areas. It can grow anywhere except the Alpine and Arctic tundra. The true pioneer of the country, it will move into any space, no matter how unpromising, and is one of the few hardwoods that can thrive in dry climates. This was the savior of the prairies. As settlers ripped out the life-giving grasses to plant wheat, what soil was held down in the Dirty Thirties was anchored by this tree.

Stands of pure aspen grow where the tree has cloned itself. In Utah, one aspen tree has covered a 200-acre area with its roots. Scientists have nicknamed this grove *Pando*, or "I spread."[93] Although an individual aspen is short-lived—about 150 years in the West; 50 to 70 years in the Great Lakes region—it's estimated that a continuous clone might date back 1 million years.[94] The roots can remain underground, alive and well, just biding their time until sunlight hits the soil, and then resprout almost immediately. Male clones dominate in dry climates and in higher elevations, female clones in wet areas and those lower in elevation.

Aspens reproduce sexually as well. The fluffy seeds are produced in huge volumes and spread everywhere the wind takes them. A female tree can produce as many as 54 million seeds.[95]

The leaves quiver in the slightest breeze because the leaf petiole is flat and flexible. According to a Native American legend, "In ages past, the Great Spirit came down to earth to see what he had done. As he looked around, all the trees and bushes bowed down to him in reverence, except for the Quaking aspen, which was so proud of its young, beautiful beauty, it continued to stand erect: so therefore the Great Spirit decreed that forever after the leaves must tremble."[96]

The Lodger

Many a night I start,
Sudden awake,
Feeling my smothered heart
Flutter and quake;

Like an aspen at dead of noon,
When not a breath
Is stirring to trouble the boon
Valley. A wraith

Or a fetch, it must be, shivers
The soul of the tree
Till every leaf of it quivers.
And so with me.

Bliss Carman

There is also a legend to explain the brilliant fall foliage of the aspen: "When the Great Bear smelled the hunter's fire in fall, the ensuing fight splattered yellow cooking grease and red blood on the leaves of the aspen forest."[97]

Aspens are so slow to burn in a fire that they have been described by firefighters as the "asbestos tree." Once burned, they recover quickly, because the bark contains a living layer that can photosynthesize.

So many animals feast on quaking aspen it's surprising the tree survives in the numbers it does. It is an important browse for elk, moose, and deer. Rabbits, hares, mice, squirrels, pocket gophers, and porcupines munch away at the buds, twigs, and especially the bark, often girdling trees and consequently killing them. Rabbits, porcupines, and snowshoe hares also eat bark, and the winter buds can appease the appetites of grouse and quail.[98] Beavers can go through aspens like a chainsaw, consuming leaves, bark, twigs, and branches, which they also use to build their dams and lodges. A single beaver can eat up to four pounds of quaking aspen bark a day; it's estimated that as many as 200 quaking aspen stems are needed to support one beaver for a year. Birds such as the ruffed grouse depend on aspen stands for foraging, breeding, courting, and nesting sites and food. They eat the leaves, catkins, and buds (gobbling the last at the rate of 45 buds per minute).[99]

Aspen is related to the willow, and contains the same chemicals—salicin and populin—the relatives of aspirin. These substances are found in the bark, which Native Americans used in various way to make medicine. It was made into a tealike drink for fevers, menstrual pain, headaches, or other ailments, or a poultice to put on wounds. They also made a variety of medicines from it for sore eyes, earaches, rheumatism, coughs, and insect stings. The Chipewyan used the ash of green aspenwood to make soap for softening up moose hides and removing rust from traps.[100] The white powdery material which sometimes appears on the bark makes a natural sunscreen that can protect human skin from ultraviolet radiation.

Thoreau was familiar with the medicinal properties of aspen, and used the tree to test an Indian guide's knowledge: "Our Indian said that he was a doctor, and could tell me some medicinal use for every plant I could show him. I immediately tried him. He said that the inner bark of the aspen (*P. tremuloides*) was good for sore eyes; and so with various other plants, proving himself as good as his word. According to his account, he had acquired such knowledge in his youth from a wise old Indian with whom he associated, and he lamented that the present generation of Indians 'had lost a great deal.' "[101]

The quivering of the aspen leaves has inspired many poets, including Pauline Johnson, who wrote in "Autumn's Orchestra":

A sweet high treble threads its silvery
    song,
Voice of the restless aspen, fine and thin
It trills its pure soprano, light and long—
Like the vibrato of a mandolin.[102]

The wood is very weak but, because it doesn't splinter, is used for things like Popsicle sticks, chopsticks, tongue depressors, sauna benches, and playground equipment. It's also used in paper pulp and is grown commercially in the East for a variety of uses. When newly cut, it is too heavy to float in a river, so it must be sent to the mills by truck. The wood is straight-grained and soft. The soft wood makes everything from newsprint to fine book paper, but most of it is used for magazine paper. Part of almost any house may be constructed of this wood, since it's also used for fiberboard, particleboard, and chipboard. Because it had no smell, it is often used in food containers. At least one company shreds the wood into chips to make bedding for hamsters and other small pets.

Aspen wood chips have also been used as fuel. During the 1974 oil embargo, among suggestions for alternative energy sources was the conversion of aspen wood chips into burnable gases like alcohol, gasohol, and hydrogen to power steam boilers or generators.

Today research is being conducted on the aspen as an agent of "phytoremediation." Because of the strong and unique pumping action of the root system of the *Populus* species, it has the ability to control or remove "organic contaminants from groundwater or . . . leachate from landfills." According to one optimistic report, by 2005 aspen will be "cleaning up everything from lead to trichloroethylene."[103]

Elaeagnaceae

Soapberry

Shepherdia canadensis, foamberry, buffalo berry, russet buffalo berry, Canada buffalo berry, soopolallie

---

Shepherdia canadensis

grows 3 to 8 feet high. It is deciduous, and has bright green, opposite leaves covered with hairs that give it a silver-green appearance in summer. The paired branches have gray, scaly twigs; young twigs and buds are covered with reddish-brown scales. The small, yellowish or brownish flowers measure about ¼ inch wide, male and female on different plants. The fruits, elliptical, red or yellowish, and about ¼ inch long, look nearly transparent.

Range: **Alaska to Oregon, eastward to the Atlantic Coast.**

The juicy soapberries look enticing, but taste bitter and feel soapy. Because the berries are nasty to touch, during the harvest Native Americans put a mat under the shrub and hit the branches with a stick until the fruit fell onto the mat. The berries, rich in iron and vitamin C, were used in medicine and added to stews. The early gourmet enjoyed soapberry sauce or jelly with his buffalo meat. The berries were also pressed in cakes and, like so many other plants, stuck in a pipe and smoked. The peoples of Alaska and British Columbia beat them into a dessert known as "Indian ice cream." It's now made with sugar, but originally would have been sweetened using other berries. It is important not to get any oil near the berries, or they won't whip.[104] Here's the classic recipe: Mix 2 cups soapberries with ½ cup

Shepherdia canadensis, soapberry

water and beat until foamy. Slowly add ½ cup sugar, a little at a time, and whip until a stiff foam results. Serve at once.[105]

This is one of those famous native dishes traditionally eaten at feasts and family gatherings. Special dishes and spoons were used both to make it and to eat it. Indian ice cream is said to be an acquired taste, and some people find it somewhat like turpentine at first. When the whipped confection was being eaten in the traditional way, it had to be "swished in and out of the mouth to get the air out of it before being swallowed."[106] In large amounts, the saponin that causes the berries to be soapy can irritate the stomach and cause diarrhea, vomiting, or abdominal pain. In small amounts, however, it does no harm. (Beets, alfalfa seeds, and green tea also contain saponin, for example.)

The wood of the soapberry is never used for firewood, because it smells appalling when burned. The roots were used by the Sioux to make a tea that cured diarrhea. The Salish and Kootenai boiled branches and twigs to make an eyewash to treat eye infections. The Dena'ina used a similar infusion to treat cuts and swellings. Wet'suwet'en women chewed the berries to induce labor.[107]

Native peoples used soapberries as an important item of trade. Inland groups, for instance, would trade them with coastal groups for fish. Even today, the berries are exchanged as gifts between interior and coastal groups.[108]

Soapberry was also valuable as the source of a reddish dye, hence the name "russet buffalo berry." The dye is made by putting the berries in a nylon stocking and simmering them in a pot of water. Clean fabric, either wool or cotton, is added and simmered until the right color is achieved.

Sphagnaceae

Sphagnum Moss

Sphagnum spp., peat moss, bog moss

Sphagnum spp.

vary in color from green to yellow to red to brown. About 20 species can be found in the boreal forest, in bogs and peatlands. Each clump is a collection of tiny individual plants, growing upright, with bundles (known as fascicles) of scaly branches emerging from the stem. At the top, a cluster of these scaly branches forms a pompom called a capitulum.

Range: wetlands throughout North America.

Peat bogs act like filters, collecting water and accumulating carbon, and are home to hundreds of thousands of living creatures, from diatoms to worms. Sphagnum moss is a fascinating plant that grows from spores. The top part is alive, but as it dies it compresses into peat moss. Growth and decay go on at the same time. Above the water the moss is springy and full of color. Below the soil, it's dead and brown.

Sphagnum moss has been an economically important plant over the years. Not only does it absorb water like a sponge (it can hold up to 20 times its weight in water), but it also has bactericidal properties. Native Americans used the moss to line menstrual pads and babies' diapers—their babies rarely suffered from diaper rash. In the First World War, it was used for surgical dressings. Millions of dressings were made of sphagnum moss encased in muslin. Early settlers in North America used sphagnum to chink their log cabins and stuff their mattresses, and campers still use it to make bedding in the wild. The moss can be used on the floors of stables, because it absorbs liquid manure and neutralizes

its odor. A friend uses it in her kitty litter box and finds it excellent for the purpose. Sphagnol, a distillate of peat tar, is used to soothe and heal skin ailments such as eczema, scabies, and hemorrhoids.

Dead and decomposed sphagnum moss is the main component of peat, which is used in the horticultural industry as a growing medium for gardens and pot plants. This is called sphagnum peat moss—not to be confused with sphagnum moss, which is alive and is used in the florist industry for floral arrangements and wreaths. The floral moss should be handled with caution: it is believed to carry a substance which, if it gets into skin lesions, can cause infection.

Peat was once widely burned as fuel, although this was more common in northern Europe (especially Scotland) than in North America. Ireland almost destroyed all of its peat bogs for fuel during hard times.

Though Canada has a thriving peat moss industry, the largest peat producer in the world is Russia. But there is a serious environmental cost to this industry. Once harvested, a peat bog takes at least 20 years to regenerate, sometimes longer. Habitats for plants and animals disappear. John Eastman gives this frightening view of sphagnum moss:

According to one estimate, the world holds about 223 billion dry tons of sphagnum, about half of this in Russia. From the small Canadian province of New Brunswick alone, some six million bales (each 70 to 80 pounds dry weight) of sphagnum are exported annually. Almost half of U.S. boglands have now been drained for agricultural expansion; some 90 million acres remain. The largest surviving American bogs are located in Minnesota and Maine (sphagnum covers about 2.3 percent of the latter state). Most commercial horticultural peat comes from New Brunswick, Quebec, and Florida. . . .

Sphagnum bogs also produce methane gas; often a hollow rod inserted into a bog will nurture a flame. Small clouds of released methane sometimes spontaneously ignite, producing those mysterious, hovering "bog-lights" or "will-o-the-wisps" seen over a bog at night. Methane is one reason why many fires in drier bogs are notoriously difficult or impossible to extinguish once begun; some of them smolder for years.[109]

Peatlands are another habitat that we must look after or we will lose yet another valuable commercial resource. Next time you buy a bag of sphagnum peat moss, think about where it came from and what it once housed. There are now excellent, easily renewable substitutes for gardening, such as coconut-husk fiber.

The Prairie

The true prairie, compared with the traditional image of wheat fields stretching monotonously to the horizon, is a rare and dazzling sight, a sea of grasses interlaced with some of the most enchanting flowers on earth. It is a landscape of many colors and constantly shifting light, but most profoundly it is a place of haunting sounds—the busyness of insects, the rustling of grasses, where nothing breaks the onrush of the winds sweeping down from the Rockies.

Many writers, including Sharon Butala, have struggled to capture the feeling of being on the Great Plains:

True prairie is a dazzling tapestry of color and texture.

The landscape is so huge that our imaginations can't contain it or outstrip it, and the climate is concomitantly arbitrary and severe. It is geology stripped bare, leaving behind only a vast sky and land stretched out in long, sweeping lines that blend into the distant horizon with a line that is sometimes so clear and sharp it is surreal, and sometimes exists at the edge of metaphysics, oscillating in heat waves or, summer or winter, blending into mirages and the realm of dreams and visions which wavers just the other side of the horizon. The Great Plains are a land for visionaries, they induce visions, they are themselves visions, the line between fact and dream is so blurred. What other landscape around the world

281

produces the mystic psyche so powerfully? Sky and land, that is all, and grass, and what Nature leaves bare the human psyche fills.[1]

The prairies appear empty because most of the really serious business of life goes on underground. In the tropics, root systems are shallow and relatively small compared with the lush growth above ground. In a prairie there's the same dramatic growth rate but it's all below soil level—rather like an upside-down forest—with comparatively small growth above. Grasses have labyrinthine root systems delving down 15 feet, becoming thick enough to support the stems above ground but also to store the massive amount of nutrients necessary to feed all this underground growth. The deep roots, constantly being renewed, reach down into the subsoils to bring even more minerals toward the surface. When the grasses die back, they decay quickly, adding organic matter to the top 6 inches of soil. Prairie soil teems with life—creatures tunneling, nesting, procreating, digging, shifting, defecating. This constant churn of life and death results in some of the richest soil on earth. But it doesn't happen overnight: it took 300 years to establish the rich organic content of prairie soil.

The prairies have, as well, been repeatedly cleared by fire over the centuries. Without fire, huge amounts of plant litter would end up in a choking mass preventing low-growing plants from reaching sunlight. When fire sweeps through, the litter vanishes, leaving in its wake a layer of ash that not only absorbs energy from the sun and insulates the soil at night, but is also rich in phosphorus and potash to supply elemental needs.

When people first drifted onto the prairie 12,000 years ago or more, they brought fire with them. They discovered that periodic burning of the grasses promoted the numbers of plants and animals in the area and kept fresh grass growing to draw large game in for the hunt. Fire kept away

infestations of mosquitoes and blackflies and cleared the areas for settlement or growing medicinal plants. It was also a method of warfare (scorched-earth policies were not unknown). And sometimes people set fires for the hell of it. Lewis and Clark on their monumental cross-country trek in 1803–1805 described seeing Indians set afire a fir tree as the evening's entertainment; it went off like a Roman candle.[2]

Bison also shaped the land. From time to time, the great beasts threw themselves onto the ground, the impact often heard miles away, and rolled around, perhaps to combat ticks or to get rid of the shaggy winter coat that fell out in great clumps. Their actions disturbed the earth, loosening it so that rain could get into the soil. The evidence of their slow rambling movement as they passed by was huge droppings that fertilized the land and moved seeds from the plants they nibbled into new areas. The bison created an entire culture for the native peoples of the Plains.

The prairies, however, spooked the European newcomers, who came from a culture of trees. The immense grasslands were unlike anything they had experienced. When the Spanish conquistadors under Francisco Vásquez de Coronado rode north from Mexico as far as Kansas in 1541, the grasses reached as high as their horses' bellies and, in places, came right to their shoulders. What really scared them was the fact that they left no tracks. As soon as they passed, the grasses sprang right back behind them as though they had never been there. At one point, two of Coronado's soldiers got separated from their compatriots. The main column of men rode in circles for days searching for the pair but they were never seen again.

Three hundred years later came the farmers in big, wooden wagons loaded up with their possessions. These wagons became known as "prairie schooners" because the wind puffed out the great canvas tops, making them look like sails as they passed through the oceans of grass.

The settlers began what they felt was not just an assault on the recalcitrant grasses with their tenacious roots but a war. No one before had seen root systems such as these grasses possessed. The heavy iron plows pulled by oxen snapped like matchsticks. It took years to wrest from this virgin soil something resembling a farm.

Pa was breaking the prairie sod, with Pet and Patty [horses] hitched to the breaking-plow. The sod was a tough, thick mass of grass-roots. Pet and Patty slowly pulled with all their might and the sharp plow slowly turned over a long, unbroken strip of that sod.

The dead grass was so tall and thick that it held up the sod. Where Pa had plowed, he didn't have a plowed field. The long strips of grass-roots lay on top of grass, and grass stuck out between them.

But Pa and Pet and Patty kept on working. He said that sod potatoes and sod corn would grow this year, and next year the roots and the dead grasses would be rotted. In two or three years he would have nicely plowed fields. Pa liked the land because it was so rich, and there wasn't a tree or a stump or a rock in it.

Laura Ingalls Wilder,
Little House on the Prairie

In 1837, John Deere came up with the brilliant idea of a forged-metal plow. It was much lighter, sharper, and more efficient than any of its predecessors. The prairies finally had a conqueror. What no one realized then was that this precious sod, once plowed, would never return.

The settlers also failed to understand that droughts afflict the prairies roughly every 20 years. Droughts in the 1870s and 1890s, and from 1918 to 1924, discouraged many settlers. Those of the 1930s were exceptional in that they lasted so long, coming after years of overgrazing and plowing, which created a fine sandy loam easily scoured away by the wind. The deeply rooted grasses that had held the soil intact through the eons had been removed.

So began the first of the great modern eco-catastrophes. 1933 was drier than 1932. Early in 1934, planting seemed ridiculous; plows turned up a dust as dry and fine as cornstarch. In February, the wind began to blow and the dust began to fly. In mid-April, a giant dust cloud, black at the base and tan at the top, rose from the fields of eastern Colorado and western Kansas and began to move south. Inside the cloud, darkness was total, and remained for hours after the cloud passed. People in the cloud's path thought the end of the world was nigh, and went to churches to await it. The storm left dead birds and rabbits in its wake, and drifts of dust six feet deep against the sides of houses. On May 10, another dust storm came up, on a wind from the west. This one blew all the way to the Eastern Seaboard, and blocked out the sun in New York City for five hours. . . . Great Plains dust showered ships three hundred miles out in the Atlantic.[3]

Today, the plains are irrigated using water from underground aquifers, but they have been changed forever. The Great Plains cover 1,125,000 square miles of North America, including parts of 10 American states and three Canadian provinces. They are 3,000 miles long and 300 to 700 miles wide.

Tallgrass prairie, which once carpeted most of the Great Plains, contains a mix of sod-forming grasses and bunchgrasses including big bluestem, Indian grass, witchgrass, and Canada wild rye. Where there is no burning, trees and shrubs tend to move in. Hardly anything remains of the tallgrass prairie in North America—it's estimated at less than 1 percent of its former glory. Huge tracts of land are now given over to annual crops such as soybeans, wheat, and corn. Small patches of the tallgrass prairie are scattered from southern Manitoba south to Minnesota, Texas, west to Oklahoma, eastern Nebraska, and to the Eastern deciduous forest, with a scrap here and there in southern Ontario.

The **mixed-grass prairie** runs from central Alberta to Mexico through central Nebraska, central Kansas, and western Oklahoma, to the Canadian River, Texas. The composition of this prairie changes with the amount of rain. In wet years, the middle to tallgrasses find favor; in the dry years, the shortgrasses flourish. In the mixed prairie there is a canopy at about 4 feet high, with taller grasses including little bluestem, needle and thread, sideoats grama, and wheatgrass. Underneath is a second canopy just 1 foot off the ground. This area is characterized by blue grama, hairy grama, buffalo grass, and a profusion of wildflowers and legumes that enrich the soil with nitrogen.

Shortgrass prairie grows largely in the rain shadow of the Rocky Mountains. Most grasses are 8 to 19 inches high, and are drought-tolerant. These areas may receive as little as 10 inches of rain a year and are particularly vulnerable to overgrazing and therefore wind erosion.

Poaceae

Big Bluestem
Andropogon gerardii, turkeyfoot

Andropogon gerardii

grows from 3 feet on a slope up to 12 feet in rich bottomlands with plenty of moisture. Short rhizomes produce erect clumps of arching, flat, sheathed, linear, blue-green leaves which turn bronze in winter. The stems have red-purple racemes 4 inches tall. The flowers are 2- to 4-inch spikelets in pairs along the racemes. One is sessile and awned (this is the beard or bristle that carries the seed on the wind); the other is sterile and awnless. Blooms from June to September.

Range: **southern Canada from Quebec to Saskatchewan, south to Florida, west to Arizona.**

Big bluestem is the signature plant of the tallgrass prairie and the most elegant of all the grasses. Two hundred years ago, when the prairie was still in its prime, the giant grass covered thousands of square miles in the tall- and mixed-grass prairies. Today it is found in a few ragtag areas. To catch sight of it with the sun shining through the silky flower heads is breathtaking. Thousands of miles of the stuff must have been terrifying to a newcomer lost in it, with nothing but the sound of the beards rubbing together, the dripping dew, and the impossibility of seeing beyond.

The delicate feathery blooms, more properly called inflorescences, give the impression of giant turkey feet springing from the end of the long willowy stems, except that they glow with a rich ruby red in autumn. At the base of the stem is a sheath called a culm which opens to the sun, allowing photosynthesis to take place. The two or three racemes of flowers at the head of each stem are composed of tiny complex florets containing the sexual organs. The pollen originates in anthers on the protruding stamens of the flowers.

Like all grasses, big bluestem depends on the wind. Its pollen, caught by the ceaseless gusts, is whirled thousands of feet high in the air and as far as the winds will take it. Fertilization is a chance event: if the pollen happens to land on a stigma (the female part of the plant), it sends pollen into the ovary of the flower. If there is plenty of spring rain, seeds are produced. But the rain crucial to germination is uncertain even

Andropogon gerardii, big bluestem

in good times, and sexual reproduction consumes vast amounts of energy. No plant can risk this much vigor going to waste. In a clever bit of evolutionary strategy, instead of producing seeds each year, big bluestem concentrates on building up its root system, and reproducing itself in place. Sexual reproduction waits for the good times.

Thick rhizomes radiate from the parent plant. They send up shoots called tillers from every node, or joint. The tillers, which eventually become stalks, have another function: where the tillers emerge, more roots are formed. The plant also has reproductive stems called stolons that snake their way along the surface of the soil. Everywhere the stolons touch, new hairy rootlets develop. Any nutrients not needed aboveground

are immediately transferred back into the rhizome for storage.

A square yard of big bluestem contains 25 miles of roots, rootlets, and root hairs. A foot below the thin crust of soil, the roots are so thick nothing can penetrate them. They dive into the soil for a dozen, perhaps 15 feet. Though not the fastest-growing of all plants, big bluestem's roots can grow more quickly than those of a tree, which concentrates its dense growth in the upper foot of soil. An acre of big bluestem can manufacture 500 pounds of raw organic matter a year for the top 6 inches of the soil around it. Along with decaying roots and surface matter, this adds up to 900 pounds a year per acre.

The use of this plant was widespread among prairie peoples. The Chippewa mixed up the roots to make a tonic for stomach pain and as a diuretic. The Omaha boiled up the

blades to combat what Daniel Moerman, in his *Native American Ethnobotany*, calls a cure for "general debility and languor." For their lodges, the Omaha used poles to make a form, then covered the form with grass and earth. The Apache, Chiricahua, and Mescalero laid the moist grass on hot stones to prevent steam from escaping, and covered fruit with it to speed up ripening or drying. The little boys of many tribes used the stiff-jointed stems to make arrows for toy bows.[4]

Poaceae

Little Bluestem
Schizachyrium scoparium

Schizachyrium scoparium

grows 3 to 5 feet tall from blue-green basal shoots. It is erect with broad leaves. The blue-green leaves turn red in autumn with a silvery stalk. The inflorescences are a fluffy white or beige. Grows in dry sandy soil and spreads by both seeds and rhizomes.

Range: **New England west to Oklahoma and as far north as Saskatchewan.**

Little bluestem, named for its bluish-green basal shoots, is a dominant grass in both the short-grass prairie and in the mixed prairie. The leaves change color with every movement of the wind. Sometimes they appear to be light green; at other moments a deep sea green. In summer they turn reddish-brown, and in autumn a pure copper.

Little bluestem is considered a midsized grass, one that uses moisture more efficiently

Schizachyrium scoparium, little bluestem

than giants such as big bluestem. It does best on dry slopes, where it grows in bunches to 4 feet.

The smooth sheath has fine hairy margins. The culms, when in flower, have up to 20 hairy spikelets. These spikelets have twisted bristles known as awns that can attach themselves to the coat of an animal. Irritated by their prickly sharpness, the animal moves away, trying to get these little devils out of its hair. When it eventually gets rid of the awns, the seeds go with them. The possibility of germinating in a new area, especially one with favorable conditions, increases greatly once a seed has been moved far away from the mother plant.

Little bluestem is a major forage plant for large animals such as deer, antelope, and bison, but it can't take heavy grazing. The Comanche used the ashes from burning the stems to rub on syphilitic sores, and bundles of the stems were used as switches in the sweat lodge. The Lakota rubbed the grass until it became soft and used it as a furlike insulation in their moccasins during the winter.[5]

Poaceae

Sideoats Grama
Bouteloua curtipendula

Bouteloua curtipendula

grows 18 to 32 inches tall, with ribbonlike leaves that are clustered at the base, and a few leaves appearing alternately up the stem. The flowers are spikelets that droop to one side. It flowers from June to September. Found in dry plains and desert grasslands.

Range: **throughout the central and eastern United States, and in the Canadian prairie provinces.**

The flowering spikelets droop over to one side to give the sideoats grama the appearance of having come straight from a Japanese brush painting. One of the commonest grasses of the Great Plains, it has a high nutritional value, and feeds not only a huge number of birds and animals, but also butterfly larvae. If spring rains are abundant, it nourishes all the insects and animals in the neighborhood, many of which also depend on it for a summer food when other plants have disappeared or gone into dormancy. Bison have a great preference for it, as does the northern pocket gopher, which depends on it as an important food source.

Sideoats grama may look sparse in its growing habitat, but it actually has short rhizomes with fine fibrous roots spreading out for great distances. Some roots, called adventitious roots (they occur in unexpected places), may penetrate 5 or 6 feet into the soil, which helps prevent erosion. Because it is so drought-tolerant, it is now being used for erosion control.

The air was really as hot as the air in an
oven, and it smelled faintly like baking bread.
Pa said the smell came from all the grass seeds
parching in the heat.

Laura Ingalls Wilder,
Little House on the Prairie

The Plains people had many uses for sideoats grama. The Tewa bundled and dried the grass to make brooms and hairbrushes. Because it resembled a feathered lance, the Kiowa used it as decoration to be worn by those who had killed an enemy in battle with a lance. The western Apache and Zuni used the stems as combs, brushes, and broom material. The latter also tied brushes together to make a sieve to strain goat's milk. The Costanoan used the hollow stems of *Bouteloua* species as straws. It became an important food for the western Apache, who ground the seeds and mixed them with cornmeal

Bouteloua curtipendula, sideoats grama

and water to make a nutritious mush, and the White Mountain Apache ground the seeds to make a kind of bread. The Montana used the grass as a weather indicator: one fruit spike meant a mild winter, several fruit spikes meant a more severe winter.[6]

B. gracilis, a related species, is commonly known as **blue grama grass** or **prairie wool**. It grows to 12 to 16 inches in the northern part of its range; 23 to 30 inches in the south of its range on the shortgrass prairie. It forms a thick dense sod mat; has erect stalks with spikelets arranged like a comb maturing into plumes. Dung beetles found in cattle manure may carry off the seeds and bury them. Otherwise, they are spread by wind, water, and animals.

Poaceae

| **Buffalo Grass** |
| *Buchloe dactyloides* |

Buchloe dactyloides

grows an average of 5 inches high (4 to 12 inches in different parts of its range) and flowers in May to June. It's a mat of sod-forming grass in well-drained soil, reproducing by seeds and stolons. The narrow leaves are rolled when they are in bud and flatten out as they mature. Male and female plants are separate (males have two or three comblike seed heads; females are in separate colonies and produce seeds in small hard burs close to the ground). As a warm-season grass, it grows quickly in May and flowers from June to September, then turns brown in autumn.

Range: central Montana, east to Minnesota, south to eastern coastal Louisiana, west to Texas, New Mexico, and eastern Arizona.

Buffalo grass is a warm-season grass: one of the grasses that start their growth as soon as the soil warms up. It's found in the shortgrass prairie along with blue grama grass. It is the most important forage food in this and the mixed-

Buchloe dactyloides, buffalo grass

grass prairie, especially for animals as various in size as the bison and the black-tailed prairie dog. To protect itself from the hungry hordes, the leaves grow close to the ground, leaving the stalks for food. The nutritional value is so complete that bison eat it year-round.

The Belgian Jesuit Father de Smet, in South Dakota on one of his many travels over the Oregon Trail, wrote in 1868: "The plains are covered with a short but very nutritious grass called buffalo grass, which will some day serve to support and fatten numberless domestic herds."[7] Unfortunately he was wrong in his predictions, but right in his assumption that it's an important grass. Settlers used it in other ways: the very early sod houses were almost entirely constructed of buffalo grass.

Farmers talk about being able to hear their corn grow. But buffalo grass afforded the same experience for thousands of years before. It can spring upward by about 2¼ inches a day and, at the end of the first summer, its roots, finer than most other prairie grasses, form a dense, 2-foot-thick mass beneath the soil. As a plant of the wind, it counts on the constant prairie flow of air to move its seeds around. Once borne away, the seeds need a cold winter before they can break dormancy and germinate. The plant is so tough and resilient that seeds found in abandoned houses in Kansas germinated 25 years later.

The Latin name of buffalo grass comes from *bufalus* (antelope or buffalo—which is what the Europeans called the bison that lived on the plains) and *chloe* (grass). Presumably it got the name because it was one of the major sources of food for the bison which were the key to survival for those first peoples of the Great Plains. This creature was so perfectly adapted to the grasslands, so heavy with meat, with such a superb hairy coat that it was like a supermarket on the hoof. The herds never grazed any area clean, dotting themselves about the prairie as they grazed gently over its surface. Grass and buffalo evolved together, one adapting to the needs of the other with exquisite timing. The grasses developed tougher and tougher stems, and some of them secreted silica. The animals learned to nibble only at the tips of the grasses, where new growth was found, leaving the rest of the plant to recover and regrow. Buffalo grass actually improves with grazing and is now often used to control wind erosion; it is even occasionally used as a turf grass for golf courses, schoolyards, and picnic sites. Proponents of native grasses recommend it as a useful lawn grass, and some garden designers are now installing lawns of buffalo grass.

Poaceae

June Grass
Koeleria macrantha, prairie June grass

The bright, spikelike flowers of June grass are a fairly common sight in summer on the dry prairie. It is considered one of the superior range grasses for domestic livestock, but wildlife don't care much for it. It reseeds quickly in bare areas after a disturbance, making it invaluable in restoration projects. The shallow, fine root system radiates in every direction near the surface, but can also work itself at least 30 inches into the soil.

Koeleria macrantha grows 8 to 24 inches tall, singly or in bunches, with 10 to 30 stalks in a few square yards of soil. The hollow erect culms rise out of downy basal leaves. The pale green species has a spikelike head up to 5 inches long, which loosens as it gets older. It mixes with other grasses such as buffalo grass and little bluestem.

This grass was considered one of the most important food grasses before wheat was introduced. The Havasupai used the seed to make breads and cakes and as a general food which they stored in blankets or bags of skin in caves. The Isleta made a nourishing mush as well as an all-purpose flour from it.[8]

June grass had important ritual and magical functions. The Lakota wore the feathery flowers on top of their heads as a war charm. The Cheyenne used June grass in their sun dance ceremony because it was considered a stimulant. Tied to the dancer's head, it would prevent him from getting tired—important, since this ritual could last up to 10 days. The male dancers fasted, going without water, dancing until they dropped as they stared at an effigy of the sun. They also used June grass to alleviate the pain of cuts they inflicted on themselves during the ceremony.[9]

Range: **British Columbia to Maine and south to Minnesota, Texas, and New Mexico.**

Poaceae

Switch Grass
Panicum virgatum, tall panic grass

Panicum virgatum

grows 3 to 5 feet tall. It is a coarse warm-season grass with long, scaly rhizomes and fibrous roots. It has distinctive veined leaves. Sheaths are purplish-red. The panicles contain one fertile floret and one sterile floret. It is a sod-forming and bunch-forming perennial.

Range: **most of the United States east of the Rockies; from Saskatchewan east to Quebec.**

During the 1980s, the American government started to experiment with switch grass, to restore land degraded by farming, and as a forage crop for livestock. They chose this grass because it is a perennial, unlike corn which is an annual requiring massive amounts of energy and fertilizer to grow. Most livestock enjoy the sprouts of the young plants. Later in the season the grass can be harvested for hay. Apart from food for farm animals, it provides seeds and nesting cover for many birds and small animals. Switch grass also shows promise as a source of bioenergy (plant matter that can be used as a source of energy).

Unlike many native grasses, switch grass will grow just about anywhere, on dry upland soil or in boggy lowlands. In the far south of its range, it's tall—up to 11 feet; but in the northern part of its range it is shorter and has a finer stem. It

Panicum virgatum, switch grass

has deep roots—as much as 12 feet below the surface—which protect it during drought, and an efficient metabolism, and will grow in the most arid regions of the Midwest. Switch grass is also amazingly flexible: it can reproduce by seed and by rhizome, and lives on indefinitely once established.

The need to provide help for heavily cultivated lands has become more and more obvious as the fertility of the soil dwindles. Switch grass can be used to reseed these wasted lands and return their native grass cover. Annual grain crops are grown with the help of pesticides, which contribute to erosion and threaten water quality, but the reestablishment of native grasses corrects both problems.

Panicum species were used by the Creek, who made a warm infusion from the leaves to help alleviate diseases such as malaria. Other Native Americans bathed in an infusion of the plant for "gopher-tortoise sickness" (translated as dry throat and a noisy cough). The Cherokee insulated their moccasins with the stems.[10]

The grass seems to have only one drawback. According to anthropologist Melvin Gilmore: "On the buffalo hunt, in cutting up the meat the people were careful to avoid laying it on grass of this species . . . because the glumes of the spikelets would adhere to the meat and afterwards would stick in the throat of one eating it."[11]

Poaceae

Indian Grass
Sorghastrum nutans, wood grass

Sorghastrum nutans

grows 3 to 8 feet high, depending on the site. It has short rhizomes, small flat leaves, and blue-green branches (they look a bit like rabbit ears

where the leaf attaches to the stem) that turn golden in autumn. The copper-colored plumelike panicles are 4 to 10 inches long and appear in early autumn.

Range: Quebec and Maine, west to central Saskatchewan, south to Arizona, east to Florida. State grass of Oklahoma.

When the wind whips the seed heads of Indian grass into motion, the prairie looks like a green sea. Indian grass is found in deep moist soil, from rich bottomlands, along roadsides, or in the open woods.

This grass is fire-dependent. When burned in late spring, it becomes denser and stronger, encouraging many more flowering culms. Without burning, its capacity for reproduction is jeopardized and it declines. Rather than seeds—which would perish in fire—this plant reproduces by rhizomes.

During a drought, the leaf rolls up into a long tube which hangs on to precious water vapor.[12] Songbirds are attracted to its seeds, and many prairie birds depend on it to conceal their nests. Cattle graze on it all summer long and it is also forage for wildlife because it provides protein and vitamin A. More recently it has been used in overgrazed areas and is planted to reclaim strip-mined soil.[13]

Poaceae

Prairie Cord Grass
Spartina pectinata, ripgut, slough grass

The graceful arching leaves of *Spartina pectinata* turn a magnificent golden in autumn. It will grow up to 7 feet tall, with flower stalks up to

Sorghastrum nutans, Indian grass

8 feet high; the 10 to 20 spikes on each plant bloom from July to September, producing a flat paperlike seed with dangerously barbed awns.

On still nights Pa kept piles of damp grass burning all around the house and the stable. The damp grass made a smudge of smoke, to keep the mosquitoes away. But a good many mosquitoes came, anyway.

Laura Ingalls Wilder,
Little House on the Prairie

This warm-season grass forms an almost impenetrable sod as its roots spread out from the mother plant. It's also called slough grass, because it thrives in wet, boggy places. Prairie cord grass originally covered hundreds of square miles of bottomland, and as the wet prairie has all but disappeared, its range has become somewhat restricted, though it is still common as an individual plant in the Midwest. The stems can often reach higher than a horse. The shade it casts is so dense that little else can grow alongside, including its own seedlings. It's not generally useful as forage for livestock, because it grows mostly in swampy spots. However, it does act as a wonderful cover and shade for game and songbirds.

Cord grass's amazing growth pattern is reflected in the sharp points on the newly rising shoots. They are so tough they can push through a foot of sand or silt. The rhizomes, which grow from 5 to 10 feet a year, can reach 13 feet deep. The roots growing from the rhizomes make an open network in the first foot of the soil. Cord grass reproduces both sexually and vegetatively. It cannot, however, survive drought. During dry periods, plants such as big bluestem and Indian grass will move into prairie cord grass's territory as quickly as possible.[14]

The traditional moons of the Oglala Lakota roughly correspond to the months of the Roman calendar. These moons tell the story of an existence dependent on hunting and foraging:

Moon of Frost in the Tepee—January
Moon of the Dark Red Calf (Bison)—
    February
Moon of the Snowblind—March
Moon of the Red Grass Appearing—
    April
Moon When the Ponies Shed—May
Moon of Making Fat—June, when the
    growing power of the world is
    strongest
Moon When the Cherries Are Ripe (red
    cherries)—July
Moon of the Black Cherries—August
Moon When the Calves Grow Hair, or
    Moon of the Black Calf, or Moon
    When the Plums Are Scarlet—
    September
Moon of the Changing Season—October
Moon of the Falling Leaves—November
Moon of the Popping Trees—December

The Native Americans used cord grass for thatching their lodges, so it was logical for the pioneers to follow suit. They used it not only for thatching, but also for covering hay and corn, and burned it as fuel. Handling it could be a real problem: razor-sharp edges earned it the epithet "ripgut."

Early settlers on the prairies used the sod created by big bluestem and cord grass for their homes by slicing enormous pieces to make bricks 2 feet long, which they stacked to form walls. One old settler said that walls made of this sod dripped two days before the rain, and three days after the rain.[15]

Cord grass is used today to stabilize soil

where it has been seriously disturbed. It prevents erosion in flood areas and around dams.

Range: **Newfoundland to Quebec in the east, to Washington and Oregon in the west, and to North Carolina, Arkansas, Texas, and New Mexico in the south.**

Poaceae

Prairie Dropseed
Sporobolus heterolepsis

When the flowers of prairie dropseed come out in the summer, they smell like buttered popcorn (some people say like sunflower seeds). The soft green of summer turns to a rich rust in autumn. *Sporobolus heterolepsis* is a bunchgrass which grows 24 to 36 inches high in a fountain shape. The slender leaves are 20 inches long and less than ⅛ inch wide. It is a warm-season sod-forming grass that is also remarkably drought-resistant. It regenerates by seed and will grow in moraines, rocky outcrops, and as high as a mile above sea level in Colorado. This tremendous span in its range is possible because the seeds are widely dispersed, thanks to their awns, those pesky little bristles that will cling to the coat of a passing animal (or person). Awns ensure that seeds are carried far away from the mother plant, and actually help the seeds take hold in the soil; as they fall, the awns twist around, pushing the seed points into the ground so that birds can't peck them out. The grass forms a dense clump and develops slowly in spring, with a handsome flower stalk developing in late summer and blooming from August to October.

The Navajo and Ramah applied a cold infusion of this plant to sores or bruises on horses' legs. Many other tribes, such as the Apache,

Chiricahua, and Mescalero, threshed the seeds, winnowed them, and then ground the flour to make bread. The Hopi used the plant to make bread and pudding, while the Navajo made dumplings, rolls, griddle cakes, and tortillas, as well as mush or bread. Breads were important because they were easily portable and could be eaten as the family walked from camp to camp. Prairie dropseed was also an excellent forage food for animals.[16]

Though threatened with extinction in some areas, this grass has been rediscovered for its superb properties in roadside revegetation.

Range: **Scattered from Wyoming and Colorado east to Connecticut and Massachusetts and south to Texas; endangered in Ohio and North Carolina, possibly Kentucky.**

FLOWERS OF THE PRAIRIES

Prairie grasses have their brilliant flowering companions—the forbs. Forbs (the word for any herbaceous plant that is not a grass) animate the prairies from early spring to late autumn, making the land a never-ending tapestry of color. Their flowers rev up activity among insects. Forbs use many strategies to get pollen moving from male to female flower parts (from anther to ovary). The blossoms of most prairie forbs depend on bees, moths, or butterflies for pollination. Some plants rely on a particular insect: the soldier beetle specializes in goldenrod, for instance; other plants draw in hummingbirds, monarch butterflies, or bumblebees for pollination.

All flowers secrete nectar to attract specific pollinators. If a flower needs an insect with a long "nose," or proboscis, bumblebees, butterflies, and moths will do the job. Other plants appeal to insects such as honeybees, ants, wasps, and flies with short probosces. The insects drift busily from flower to flower to get their fill, covering themselves with pollen as they go. When the insect moves to another flower, the pollen brushes off on the stigma and immediately starts growing a little tube down through the style and into the ovary, where an egg awaits fertilization. A seed is born.

Many of the prairie forbs have hairy leaves or stalks to conserve precious plant fluids. The movement of the hair cools the plant and slows down transpiration. Anemones, geums, or vetches can also close off their stomata (pores on the underside of the leaves) so that respiration slows down in dry periods.

Some forbs grow as tall as grasses, so their pollinators can see them waving the nectary flag above the grasses. Others bloom early in the season before the canopy of grasses grows and shades them out. The first low-growing forbs of the season belong to the aster family (Asteraceae). Every few weeks, new ones come into bloom, shading out the earlier flowers. Always there is color.

Forbs and grasses have a symbiotic relationship. Grasses need nitrogen to survive. Many forbs have the nodules on their roots that fix nitrogen from the air and make it available in the soil. Other flowers attract the animals needed to keep the grasses healthy: from pollinating insects to the burrowing mammals that aerate the soil. The forbs also cover any exposed soil between bunches of grass to protect the soil from sun damage. In death they add to the fertility of the soil. At the same time, the grasses protect and support the forbs against the winds.

Asteraceae

Pale Purple Coneflower

Echinacea pallida, black sampson, black Susan, combflower, droop, hedgehog coneflower, Indian head, Kansas snakeroot, purple daisy, rattlesnake weed, red sunflower, scurvy root
E. purpurea, purple coneflower
E. angustifolia, purple coneflower

Echinacea pallida

grows to 3 to 4 feet and has a long taproot. A single flower grows at the end of each stem; the cone is an orange-brown color and the pale purple-pink rays droop down. The rays are longer and narrow and droop in a more pronounced way than those of other coneflowers.

Range: Minnesota south to northeast Texas, east to Illinois and Georgia.

Of all the prairie plants, the purple coneflowers are probably the best known. The early settlers embraced them because they are so similar to their European counterparts. There are nine species of *Echinacea* worldwide, all native to North America. Three of them in particular (*E. purpurea*, *E. pallida*, and *E. angustifolia*) have a reputation as medicinal plants to boost the immune system. Since there are subtle differences in the structure of the different species, each has proponents who see it as *the* plant for this purpose.

Anthropologist Melvin R. Gilmore wrote in 1914 that, among the Native Americans of the northern Great Plains:

Echinacea pallida, pale purple coneflower

This plant was universally used as an antidote for snakebite and other venomous bites and stings. . . . Echinacea seems to have been used as a remedy for more ailments than any other plant. It was employed in the smoke treatment for headache in persons and distemper in horses. It was used also as a remedy for toothache, a piece being kept on the painful tooth until there was relief, and for enlarged glands, as in the mumps.[17]

Dr. J.S. Leachman of Sharon, Oklahoma, writing in the October 1914 edition of *The Gleaner*, wrote: "Old settlers all believe firmly in the virtues of *Echinacea* root, and use it as an aid in nearly every sickness. If a cow or horse does not eat well, the people administer *Echinacea*, cut up and put in feed. I have noticed that puny stock treated in this manner soon begin to thrive."[18]

According to an old story, circus performers used to bathe their hands and arms in the juice of *Echinacea* so they could take a piece of meat from a boiling kettle with their bare hands without suffering pain, to the wonderment of onlookers.[19] Alas, there are no pictures of this, and no confirmation that it was anything beyond a good yarn.

In 1871, Dr. H.C.F. Meyer, a patent medicine salesman of Pawnee City, Nebraska, marketed a tincture of the root as Meyer's Blood Purifier, promising that it would cure everything from carbuncles, piles, and eczema (wet or dry), to gangrene, diphtheria, and hydrophobia. *Echinacea* was later discovered by physicians and, despite denunciations by the American Medical Association, it became one of the most widely used medicinal plants in America in the late nineteenth and early twentieth centuries.[20]

Neil Diboll, a man obsessed with prairie plants, who grew up with them and finally turned his passion into a business, supplying seed for prairie plants all over North America, says:

Pale purple coneflower was one of the most important medicinal plants of the Plains Indians. It was used as a treatment for stings, cuts, burns, and almost any skin wound. Today, it is one of many species of Echinacea *that are taken orally as a blood purifier and immune system enhancer. There is little question that this plant is "Big Medicine." The roots typically contain the highest concentrations of active ingredients.*

I conducted an experiment using the roots of this plant on a friend who'd sustained numerous chigger bites on her ankles while camping. I chewed up a root of Echinacea pallida *and applied it to one ankle as a poultice. I left the other ankle untreated. Within 24 hours the swelling and itching had greatly subsided in the treated ankle. The untreated ankle showed no improvement.*

Echinacea is not an antibiotic, as Diboll points out, but helps build up the immune system so that the body can produce its own antibiotics naturally. According to Steven Foster, a herbalist and author of many books on the subject:

[Echinacea] *simulates phagocytosis in the blood stream. Phagocytosis is the first defense component of the cellular immune system, a process that helps to prevent the invasion of foreign substances in the body. One important factor in immunostimulation is an increase in phagocytosis (by macrophages and granulocytes). Macrophages and granulocytes are cells in the blood that "ingest" invading pathogens or particles acting like janitors or guardians of the blood stream.*[21]

E. angustifolia, a related species, now has a reputation as having the most powerful healing effects. *E. angustifolia* grows 18 to 40 inches high, and blooms from June to July. The leaves pop up from the bottom of the stem. The central portion of the flower (the cone) has more than 100 brown disk florets surrounded by about 24 rays. The rose-purple flower heads droop gently at the ends of the flower stalks. It grows in dry upland prairies; on the Great Plains, east of the Rockies, in Texas, Montana, Saskatchewan, eastern Oklahoma, western Iowa, and western Minnesota. Because of its stellar reputation, it is being pulled out of the wild at such a rate that it will be eliminated altogether without a good deal of vigilance. In 1999 alone, 70,000 pounds of dried root were collected, much of which was shipped abroad. The same year, the U.S. government stopped issuing licenses for this purpose in fear that the herb might be wiped out.[22]

Today, *E. purpurea* is the best known of all the purple coneflowers and is grown as both a garden plant and a commercial crop. At first, *E. angustifolia* was the only species used in medicine. But in 1939, *E. purpurea* was used by Dr. Gerhard Madaus, a German manufacturer of *Echinacea* products. He thought he was buying seeds of *E. angustifolia*, but they turned out to be from *E. purpurea*. Since that time, most of the research done on *Echinacea* in Europe has been on *E. purpurea*.

Asteraceae

Yellow Prairie Coneflower

Ratibida columnifera, longheaded coneflower, prairie coneflower, yellow coneflower, Mexican hat, upright prairie coneflower

Ratibida columnifera

grows up to 6 feet high from a taproot and branches out from the base with stiff hairy stems with long grooves. The deeply divided alternate leaves are 2 to 4 inches long. The flowers are borne at ends of long stalks 2 to 12 inches high, with slightly reflexed florets and a gray or purple disk in the center.

Range: **southeastern British Columbia, to Manitoba and Michigan, south to Illinois and Louisiana, west through Texas, northern Mexico, and Arizona.**

With its drooping yellow rays and velvety brown cone head, *Ratibida columnifera* is common on dry prairie and along roadsides. The 2-inch-long knobby cones are actually hundreds of minuscule purple-brown flowers bunched together, surrounded by the familiar rays. This is where its common name, "Mexican hat," comes from.

According to Neil Diboll, yellow coneflower:

> *is prone to taking regular siestas. When the heat of the afternoon become oppressive, it simply wilts and shuts down. Never one to exert itself under less than optimal conditions, it waits until evening and regains turgor in its leaves. By morning, it looks fine and is ready for business. At least for the morning. By afternoon, if no rain comes, once again it's siesta time!*

The young plants supply forage for all sorts of wildlife, including white-tailed deer and pronghorn antelope. Even the exotic honeybees which have only been here for a few hundred years use this plant.

There are at least 700 seeds on every plant. Once a fire has swept through, with this many seeds available, it is one of the first plants to move back in to recolonize the area since there are bound to be some survivors. This may seem invasive, but it is part of a survival mechanism important to the health of the prairies.

Almost all parts of the yellow coneflower were useful. The Oglala brewed a tea from its

leaves and made a rich dye from the flowers. The Cheyenne boiled the leaves and stems to counteract snakebites and made a tea to relieve the itching of poison ivy.

Asteraceae

Joe-Pye Weed

Eupatorium maculatum, spotted Joe-Pye weed, smokeweed, gravel root, gravelweed, purple boneset, purple Joe-Pye weed, queen-of-the-meadow root, trumpetweed

Eupatorium maculatum

grows 2 to 6 feet high on purple-spotted stems. It has whorls of lance-shaped, serrated-edged leaves up to 10 inches long. Numerous flower heads are clustered in a flat-topped pink to light purple inflorescence 6 to 8 inches across. Found in wet meadows, streams, ditches, and bogs.

Range: Newfoundland to British Columbia, Maryland to Ohio, the Dakotas, and Utah.

In late summer, the big flat heads of Joe-Pye weed may be covered with the softly palpitating bodies of butterflies. They get their long tongues into the tubular disk flowers and hang on to this easy resting place sucking away on the nectary. In spite of the attraction Joe-Pye weed has for butterflies, it will self-pollinate. The flowers are so jammed together that the pollen-bearing stamens of one can touch the long pollen-catching stigma of a neighbor. Bumblebees which are native to the prairies will roost on the big pink blooms overnight, shaking themselves slowly awake as the sun warms them up.

Ratibida columnifera, yellow prairie coneflower

Eupatorium maculatum, Joe-Pye weed

Joe-Pye weed is a familiar sight in abandoned moist fields or the verges beside highways. Despite its reputation as a weed, it is a valuable wetland plant which stabilizes streambanks, and grows well on the wet prairies, in marshes, and in fens.

The pink flowers of *Eupatorium maculatum* were used by Native Americans to combat venereal disease in females, and the white flowers of a related species, *E. perfoliatum,* were used for males. Cherokee shamans used a section of the stem to blow or spray medicines over their patients. An infusion made from the roots was used for women suffering from postpartum pain and just about anything else dubbed as "women's problems." It was also a well-known cure for dropsy and other maladies, and was used to make a tonic revitalizing drooping spirits.[23] In his charming book on plant lore Jack Sanders writes:

The Potawatomi made poultices for burns from its leaves and considered the flower heads to be good luck charms, especially when gambling. The Ojibwe believed that washing a papoose up to the age of six in a solution made from the roots would strengthen the child. Children who were fretful and could not sleep were put in a bath to which a Joe-Pye weed decoction had been added, and they were supposed to have relaxed and gone to sleep.[24]

Joe Pye himself is supposed to have been an Indian herbalist from the Massachusetts Bay Colony who sold medicinal concoctions to settlers. Joe Pye, who came from a Maine Nation, was, legend has it, so good at plying his trade that he became connected with this particular plant, which he recommended for curing typhoid fever. Legend also has it that if you carry a few leaves of the plant with you, you will gain respect.

The botanical name comes from Mithridates Eupator (a.k.a. Mithridates the Great), a ruler of the Asia Minor kingdom of Pontus from 120 to 63 B.C. He was the first person to discover that this plant was an antidote to poison. To foil his many enemies, he decided to eat small amounts daily to build up his immunity to poison. When he was ultimately captured, he'd eaten so much of the antidote he couldn't poison himself to escape from a slow and painful death in prison. Unwilling to face this sentence, he had a slave stab him to death.

In the nineteenth century, Americans used *Eupatorium* to treat kidney and urinary tract dis-

eases. But not every species of *Eupatorium* had benign effects. One species, *E. urticifolium* (also known as *E. rugosum* or *Ageratina altissima*), **white snakeroot**, is poisonous in all its parts. It has similar lance-shaped, toothed leaves 4 inches long, but with a petiole (stalk) and white flowers in branched clusters. This plant is lethal, as Abraham Lincoln's mother, Nancy Hanks Lincoln, found out. She was only 35 in 1818 when she died in Indiana of what was then called "milk disease." This was one of the most dreaded diseases to hit frontier families. When cows were desperate for forage, they ate whatever the fields offered. White snakeroot was everywhere, since it is an extremely invasive plant. It made the cows lethargic and poisoned their milk. Anyone drinking the milk usually died within a few days.[25]

The source of milk disease was identified in 1914. After an outbreak of milk sickness near Beecher City, Illinois, a Dr. E.R. Brooks believed that white snakeroot was to blame. He fed the plant to healthy animals. They all developed the symptoms of lassitude, shaking, and weight loss and finally keeled over.[26]

Today, *Eupatorium* is considered a first-class garden plant and many different cultivars have become popular. But you can still see the old-fashioned wild Joe-Pye weed blossoming in ditches and fields, vibrating with butterflies in the late summer sun.

Asteraceae

Jerusalem Artichoke
Helianthus tuberosus, sunchoke, girasole

Helianthus tuberosus
grows up to 10 feet high and 3 feet wide with saw-toothed rough, hairy, lanceolate, pointed

leaves 12 inches long. It blooms in autumn with deep yellow ray florets and yellow disk florets. The roots have knobby white-, red-, or purple-skinned edible tubers.

Range: **southeastern Canada, southeastern United States. It has spread to other areas through cultivation.**

Helianthus tuberosus, Jerusalem artichoke

The Italian word for this plant is *girasole* (meaning "turning toward the sun") and somehow that name was corrupted fairly early on into "Jerusalem." Then, when the vegetable was brought to Europe in the seventeenth century, it reminded people of artichokes. Hence the common name.

Jerusalem artichokes have been eaten for thousands of years in North America, and constituted an important crop for those groups of people who farmed. Many Native Americans ate the roots raw; sometimes boiled or fried. In their journey across the continent, Meriwether Lewis and William Clark reported that they ate the tubers prepared for them by the Indians in North Dakota. The Dakota knew that eating too many roots caused flatulence. Indeed, the root contains inulin, which causes fermenting in the gut, and this in turn produces an explosive amount of gas. In some areas, Jerusalem artichokes were eaten strictly as starvation food.

In the 1600s, some English visitors to North America took the tubers back home and used them like potatoes. The plant wouldn't bloom in gloomy England as well as it does here, but nevertheless, it caught on in Europe.

Jerusalem artichokes have found a whole new audience in North America and are once again being grown commercially. The roots taste like early potatoes, but slightly sweeter. They can be eaten raw in salads or cooked, and are usually harvested after the first frost. The plant is being touted as a herbal medicine that might be useful in treating diabetes. However, this is highly speculative, and extracts from the plant should not be taken in combination with other drugs or used without the advice of a physician.

Even gardeners are looking on them with favor, in spite of the fact that they can get completely out of control and take over a garden. The seeds are much loved by goldfinches, chickadees, and hummingbirds. The nectar attracts several species of butterflies, including swallowtails and fritillaries.

Asteraceae

Prairie Blazing Star

Liatris pycnostachya, tall blazing star, Kansas gayfeather

Liatris pycnostachya

grows 2 to 4 or even 5 feet high. It has fluffy rose-colored flower heads that open from the top to the bottom on 18-inch spikes that appear in midsummer. The leaves are lance-shaped; multiple stems rise from a basal tuft with an almost grasslike appearance. The leaves at the top of the stem are small; the ones lower down are larger.

Range: North Dakota, east to Wisconsin, south to Kentucky and Louisiana, west to Texas and central Nebraska.

Standing dramatically erect in the midst of the prairies is the spectacular blazing star. This is the tallest of all the species. Others, such as *Liatris spicata,* **gayfeather**, have smoother stems and the leaves are tightly pressed together. But they have use beyond beauty. This is an indicator plant, which means that its presence can tell you a lot about the land it sits on. It will seed freely on land that was once savanna or open woodland.

The flowers at the top of the spike open first followed by the lower ones. All species of *Liatris* produce these spikes of unusual, downy-looking, pinky-purple flowers. *Pycnostachya* means "crowded spike" in Latin.

Liatris was once called "the pine of the prairie" because it had so many uses.[27] Native Americans from what is now the St. Louis area used it to cure gonorrhea. The Cherokee used the roots for backaches or pain in the limbs and

as a diuretic.[28] The plant was also edible—the Osage would store the starchy roots until they had turned to sugar before eating them.

In folklore it was said that the root was used against witchcraft and spells: The root is cut in two with one half carved into a cross. Both halves were worn for protection.[29]

Nurseryman Neil Diboll remembers seeing *Liatris* growing along a railroad track in a small prairie remnant in eastern Wisconsin more than two decades ago. This was long before it became ubiquitous in floral arrangements. "This plant became my first love of the prairie flora. As I gazed upon its exquisite beauty, I could only ask myself how this plant could have possibly been neglected by the august corps of American horticulturists for an entire century. I resolved at that very moment that I should learn how to grow it immediately." He has achieved his goal and now this stately plant has become one of the most sought-after garden plants and a star of prairie restoration projects.

Bob Stadyk, a nurseryman from Alberta, was equally enchanted by the plant. One day while walking with a friend outside Lethbridge, "I was amazed to see the southern prairies ablaze with *Liatris.* Looking at that scene made me see the prairie in a whole different way. Same with *Ratibida pinnata.*" The beauty of *Liatris* is among the reasons he too became a nurseryman with a specialty in native plants.[30]

Liatris pycnostachya, prairie blazing star

Asteraceae

Rudbeckia hirta

grows 12 to 20 inches tall. It can be an annual, a biennial, or a short-lived perennial. It has ribbed, hairy, lanceolate leaves up to 6 inches long. The flowers have orange-yellow rays with purple or dark brown central disks and grow at the end of long stalks. Usually found in elevations under 9,000 feet at the bases of hills, along coulee bottoms, or in the low prairie just shoreward of wetlands.

Range: throughout North America, particularly east of the Rockies.

State flower of Maryland.

The bright yellow rays of black-eyed Susan seem to hold the promise of sunshine. Looking into the face of this plant is full of rewards. The central cone, like many others in this family, is made up of hundreds of tiny seed-producing flowers and is surrounded by a halo of yellow petals.

In its first year, the black-eyed Susan is not much more than a rosette of fuzzy leaves storing up food. The following year the brilliant yellow of its rays breaks out, attracting insects by the dozens. The flowers contain an enormous supply of nectar and produce masses of pollen. Short-tongued insects brush the pollen around as they look for the nectar they know is there; then along come the bees, butterflies, and moths, the long-tongued specialists, which can get inside to the florets holding the nectar and take a good long sip.

The Chippewa used *Rudbeckia* for indigestion and burns. They called it *gizuswebigwai* ("it is scattering"), which might refer to its propensity either to spread seeds or to multiply quickly. Originally a plains and prairie dweller, *Rudbeckia* spread much more widely as newcomers cleared out the Eastern forest. Sometimes it was deliberately planted, sometimes it was spread by accident when it arrived in barrels of seeds from the prairies. It became either a beloved plant or a weed, depending on whether you were a gardener or a farmer. By the 1830s, it was running amok in Eastern fields and, it was thought, poisoning farmers' sheep and hogs (in fact, the plant is not poisonous to livestock).

Despite this reputation, in 1893 it was proposed as the national flower. In 1896, the gardeners of Maryland wanted it as their state flower and kept up pressure until 1918, in spite of farmers' protests, when it was officially designated. Black and gold were the colors of Lord Balti-

Rudbeckia hirta, black-eyed Susan

more, who gave his name to Maryland's capital city. Today they are the colors of the Baltimore Orioles baseball team (and of course the bird has the same colors). The state planted many acres of wildflowers with black-eyed Susans predominating. The idea is to use these forbs to provide new habitat for wildlife, reduce air pollution, and enhance the look of the roadsides.[31]

Rudbeckia was named by Linnaeus, who became famous when he invented an ordering of plants into groups according to the number and arrangement of flower parts. Having come up with the system, he also got to give the plants Latin names, many of them based on the names of the people who first described them. In this case, the name was a thank-you to Olaf Rudbeck the Younger (1660–1740), a professor at the University of Uppsala in Sweden, who had given Linnaeus a job. Full of gratitude, Linnaeus wrote: "So long as the Earth shall survive, and as each spring shall see it covered with flowers, the *Rudbeckia* will preserve your glorious name."[32]

The common name probably comes from the ballad written in the early eighteenth century by John Gay, of *Threepenny Opera* fame, which tells of black-eyed Susan, who went on board ship to say a touching good-bye to her "sweet William" (another flower name). The song begins:

> All in the Downs the fleet was moor'd,
> The streamers waving in the wind,
> When black-ey'd Susan came on board;
> "Oh, where shall I my true-love find!
> Tell me ye jovial sailors, tell me true,
> If my sweet William sails among the
>   crew."[33]

Most sailors knew the song, so when they saw a beautiful black-centered flower, the familiar and fanciful words must have sprung to mind.

Today *Rudbeckia* is used to rehabilitate strip-mined areas in the West. But in Hawaii it has become an unwelcome weed, because it will readily invade any disturbed site.

Asteraceae

Compass Plant
Silphium laciniatum, cut-leaved silphium, gum plant, polar plant, rosinweed, turpentine weed

Silphium laciniatum

grows to 3 to 7 feet and can often reach 10 feet. It is rough and bristly with huge hairy, deeply cut, toothed basal leaves up to 2 feet long. The 3-inch yellow rays bloom from summer to autumn. Taproots may go to 6 feet deep. Grows on moist to moderately dry soil.

Range: Ohio, Kansas, Missouri, Wisconsin, the Dakotas, south to Texas, east from Ontario to New York.

It may look a bit like a sunflower, but compass plant has quite a different personality. In his many years of breeding native plants for their seeds, Neil Diboll calls compass plant "the classic prairie plant." It is related to the sunflowers, and like its cousins, has nutritious seeds that birds love. Its long taproot and rough leaves slow down transpiration, making it an extremely tough, drought-resistant plant. "Individual plants have been documented to live for decades," Diboll says. "Some people believe that the compass plant can live as long as some trees."

It's called compass plant because the leaves position themselves so the edges rather than the surfaces of the leaves face the sun, making things a little cooler for themselves. This accounts for

Silphium laciniatum, compass plant

their north–south orientation. The other peculiarity of the compass plant is that the ray flowers around the rim of the flower head are female and the flowers in the center are male; therefore seeds are produced only around the edges of the flower heads.

Plants in this genus are also called "rosinweed" because they produce a gummy sap on the upper stem when they are in bloom. The sap was used by both Native Americans and settlers as a gum. William Bartram, during his travels in west Florida (1776), wrote:

> *The stem is usually seen bowing on one side or other, occasioned by the weight of the flowers, and many of them are broken, just under the panicle or spike, by their own weight, after storms and heavy rains, which often crack or split the stem, whence exudes a gummy or resinous substance, which the sun and air harden into semi-pellucid drops or tears of a pale amber colour. This resin possesses a very agreeable fragrance and bitterish taste, somewhat like frankincense or turpentine; it is chewed by the Indians and traders, to cleanse their teeth and mouth.*[34]

According to writer John Madson, "It has an odd, pine-resin taste that's pleasant enough, but it must be firmed up before it's chewed. A couple of summers ago, I tried some of this sap while it was still liquid. It's surely the stickiest stuff in all creation, and I literally had to clean it from my teeth with lighter fluid."[35]

Aldo Leopold, a legendary writer about prairie ecology, remembers finding a graveyard untouched since the 1840s, when it was established.

Heretofore unreachable by scythe or mower, this yard-square relic of original Wisconsin gives

BOTANICA NORTH AMERICA

birth, each July, to a man-high stalk of compass plant or cutleaf Silphium, *spangled with saucer-sized yellow blooms resembling sunflowers. It is the sole remnant of this plant along this highway, and perhaps the sole remnant in the western half of our county. What a thousand acres of* Silphium *looked like when they tickled the bellies of buffaloes is a question never again to be answered, and perhaps not even asked.*[36]

Leopold writes of trying to transplant a compass plant from the prairie to his farm. "It was like digging an oak sapling. After half an hour of hot grimy labor, the root was still engaging. Like a great vertical sweet-potato. As far as I know, that *Silphium* root went clear through to bedrock."[37]

The Dakota, Omaha, Pawnee, and others burned the dried root of compass plant during storms to act as a charm against lightning. A boiled-down extraction of the roots was also given to horses as a tonic, and used to induce vomiting.[38]

Asteraceae

Cup Plant
Silphium perfoliatum, Indian cup, Indian gum, pitcher plant, ragged cup, rosinweed

Silphium perfoliatum

has a horizontal rhizome, a smooth square stem, and grows to 7 feet high. The opposite leaves are fused at the base to form a cup-shaped disk. The

Silphium perfoliatum, cup plant

leaves are 8 to 14 inches long by up to 7 inches wide. The terminal flowers have 20 to 30 yellow rays and a central, darker-toned disk.

Range: **North Dakota east to Ontario, south to Oklahoma and Georgia.**

The cup plant has the perfect name: the leaves become cups that hold rainwater. Birds and butterflies will stop and drink from them.

The Indians living along the Missouri River smoked the roots to relieve a head cold or rheumatism or made vapor baths with it.[39] The Winnebago believed the plant had supernatural powers. The braves made a tonic from the rhizomes to purify themselves before hunting bison. For the Chippewa, an extract made from the roots eased back and chest pain and prevented excessive menstrual bleeding. The Iroquois put the ashes of the burned root on children's cheeks to prevent them from seeing ghosts. Similar treatment was supposed to prevent sickness caused by the dead.[40] Both settlers and Native Americans used the resinous sap of cup plants to keep their teeth clean.[41]

Asteraceae

Goldenrod
Solidago spp.

Solidago canadensis

reaches up to 5 feet tall on rhizomes and grows in clumps with multiple stems and a deep taproot down to 11 feet in good prairie soil. The leaves are lance-shaped, toothed, alternate, usually 2 to 4 inches long, crowded on the lower stem with fewer higher up. The yellow flower heads are small (1 inch), often up to 17 rays and flowers cluster in plumes. The tiny one-seeded nutlike fruits come

equipped with white bristles. It grows throughout North America and has five regional varieties.

Goldenrod, harbinger of autumn on the prairies, is taken for granted in North America. It's much too common to think of its strange and marvelous healing powers or its function within the ecosystem. It is one of the most important of all bee plants. In Europe, where only a few relatively poor species grow, the North American goldenrod is practically worshipped as a magnificent garden plant. Gardeners there see it as an exotic of great beauty.

Its poor reputation is reinforced by a major misconception. Since it blooms at the same time as ragweed (*Ambrosia* spp.), and is frequently found near it, goldenrod is often accused of being the cause of allergies. This is not true; goldenrod is pollinated by insects, and its pollen is too heavy to be borne by the wind. Ragweed pollen, on the other hand, can float up to the seventh story of a building and affect people with allergies for miles around.

Solidago species hybridize freely so that no one knows exactly how many different varieties there are. But their overwhelming beauty is apparent to anyone who takes the time to look at the plant carefully. Almost all the 100 or so species originate in North America and there is a bewildering array of tiny local variations in every species. A walk through any country property can fetch up at least seven or eight distinct varieties. They can grow on a single stalk or have many stems. Most species have an enlarged stem base called a simple caudex. The leaves are usually smooth on top and 2 to 4 inches long. Goldenrod tolerates such an enormous variety of soil conditions that it can be found almost anywhere, from junkyards to parking lots in cities, from abandoned fields to country farmyards. Any area not sprayed with herbicides will have a magnificent display of goldenrod, often mixed with local asters, making an eye-catching tapestry of yellows, pinks, and purples from September on. It flowers from top to bottom to compete for pollinating insects at the time of year when asters and sunflowers are also beckoning.

Goldenrod has an allelopathic effect on some other plants which can discourage gardeners. Allelopathy is a plant's ability to produce chemicals (often from its root system) to keep other plants from growing nearby. Allelopathic plants include black walnut, wormwood, and sunflowers. Goldenrod can prevent sugar maple seedlings from developing, for instance.[42] A few goldenrods are a good way to prevent those carpets of seedlings under the trees.

The Last Walk in Autumn

Along the river's summer walk
The withered tufts of asters nod;
And trembles on its arid stalk
The hoar plume of the golden-rod.

John Greenleaf Whittier

Solidago comes from the Latin *solidare*, "to strengthen or make whole." The plant was known by Europeans as a "vulnerary," a plant used to heal open wounds. The English herbalists knew about *S. virgaurea*, and there is one famous story told about a boy of 10 who, in 1788, was given an infusion of goldenrod leaves for several months. He passed gravel (as all kidney and other stones were called then) that weighed in at 1¼ ounces.[43]

The Chippewa word for goldenrod was *gizisomukiki*, or "sun medicine." The whole plant was dried and the roots boiled down in water to make an essence for treating colds and liver problems. A piece of the chewed plant laid against a tooth is an old treatment for

Solidago spp., goldenrod

toothache.[44] Root poultices were applied to boils and burns, and for rheumatism, neuralgia, and headache. A tea made from the flowers has been used for sore throats, snakebite, fever, kidney and bladder problems. It's also a laxative.

The Cherokee made goldenrod tea from the roots; this was applied to bruises and sores to alleviate pain. A more contemporary salve for stings consists of a cup of goldenrod flowers, chopped, then heated in a double boiler with two cups of shortening cooked for 45 minutes, and cooled into a salve.[45]

The flowers of goldenrod were plucked to make a yellow dye not only by natives but also by settlers. Ancient diviners believed the plant could point the way to underground sources of water, hidden springs, or even to troves of silver and gold.[46]

The Meskwaki had an unusual use for the plant. They would boil up goldenrod along with the bones of an animal which had died when a child was born. They would bathe the baby in this wash to ensure it would grow up with a sense of humor.[47] Native Americans also used it for a tea to apply to bruises and wounds. Early physicians recommended it as a diuretic, for gas pains, and to promote sweating.[48]

Though many people, even lovers of prairie plants, hate the goldenrod, it can't be all bad. *S. gigantea* is the state flower of Nebraska, and *S. altissima*, a 7-foot-tall giant, is the state flower of Kentucky. In 1895 goldenrod was actually considered as a possibility for the American national flower, but now the plant is ignored, not celebrated.

Tall varieties:

S. canadensis is described above.

S. rugosa, as the name implies, has wrinkled leaves. It is found in dry soils and roadsides and grows to 6 feet. It has sharply toothed leaves, and inflorescences 6 inches long with tiny flower heads holding six to nine rays and four to seven disk flowers.

S. altissima can reach 8 feet tall and is found from the Atlantic to Wyoming and Arizona. The Chippewa, who called it "squirrel's tail," used it to relieve cramps.

S. gigantea is much like *S. canadensis*, but has a white bloom on the stem and grows throughout North America.

Short varieties:

S. odora, known as **blue mountain tea** or **sweet goldenrod**, grows 20 to 40 inches high with narrow smooth-edged leaves 5 inches long. The tiny yellow flower heads curve slightly out-ward. When the leaves are crushed, they smell faintly of anise. This is one of the plants (*Monarda* was another) that the American colonists used after they dumped all their highly taxed tea into Boston Harbor. It came to be called Liberty Tea. Apparently it proved to be such a hit that it was eventually exported to China.[49] It was also used as an astringent, for ulcerations of the intestines, and its essence was used for infants against headaches, flatulence, and to induce vomiting. It is found throughout New England, south to Florida and southeast as far as Texas.

S. rigida, known as **stiff goldenrod** or **rigid goldenrod**, has thick flat leaves that are rough on both sides. The leaves at the base of the plant are very large and grow on stalks. The flower stalks spread out and make the top of the plant rounded. It was used to make a lotion that soothed bee stings. Since bees were so attracted to these plants, it's easy to consider what was at hand as an antidote. The blossoms were believed to counteract hemorrhaging, and were used as an astringent, an oil, and a diuretic. It is found in New England and southern Ontario, as far west as southern Alberta, and south as far as Texas, Louisiana, and Georgia.

Asteraceae

Ironweed
Vernonia fasciculata

Vernonia fasciculata

grows to 3 to 6 feet tall. The leaves are 4 to 8 inches long and 1 to 2 inches wide with simple lance-shaped, ovate, alternate leaves. The purple flowers grow in loose, flat-topped clusters. It is a stout, erect plant that blooms from July until

Vernonia fasciculata, ironweed

autumn. Grows in just about any soil, from wet to clay to peat.

 ***Range:* Minnesota, Ohio, and through the central regions of Canada**

A group called the Appalachian Women's Alliance holds an Ironweed Festival each year. In song, dance, and poetry, they work to celebrate women and to be as strong and as tough as this purple wildflower.

 Though any plant called a "weed" is usually considered a nuisance plant, weeds are actually very highly developed species. There is an old saying among gardeners: A weed is simply a plant in the wrong place. To botanists, however, a weed is a generally invasive species that takes root in disturbed areas and has one or more ways of ensuring its survival in adverse conditions.

 Ironweed is common in woods and prairies all over most of North America east of the Rockies. Find an abandoned pasture and you will find ironweed. Cattle dislike the taste and texture of the tough fibrous stems. It has little medicinal value, although according to herbalists a bitter tonic made from it is said to have been useful in combatting scrofula (swelling in the neck glands). The name commemorates William Vernon, a seventeenth-century plant collector from Maryland.

 A visitor to the native plants forum on the Garden Web noted, "I've heard ironweed is

called ironweed because of the strong stems which can*not* be picked. (What an odd plant that would rather be completely uprooted than let its stalk be severed!)"[50]

Apiaceae

Eryngium yuccifolium

grows 2 to 6 feet tall. It has stiff stalks with sword-shaped spiny leaves 1 to 4 inches wide and up to 3 feet long. The basal leaves are hooked to the stem and the leaves have prickly margins. The stiff, rounded flowers are whitish to pale blue and bloom on the ends of the stalk.

Range: **Minnesota, south to Iowa, New Jersey, Florida, and Texas.**

Although its foliage *looks* like yucca (hence the Latin name), this *Eryngium* has nothing to do with yuccas, which are desert plants. Its relations are really closer to Queen Anne's lace. The similarity to yucca lies in its pointy leaves. Rattlesnake master reaches a large size in moist areas, although it will adapt to drought and excessive sun. It's a magnet for butterflies and other insects.

The plant was well known to English doctors, who made a tonic with it right into the eighteenth and nineteenth centuries. If the settlers who moved to North America had been more observant, they would have noticed that the Indians around them made similar use of the plant.

There are two theories about the plant's common name. According to James Adair, an eighteenth-century Indian trader, he'd seen a Chickasaw shaman chew on the root of this

Eryngium yuccifolium, rattlesnake master

plant, blow it on his hands, and handle rattlesnakes without any danger. No one could confirm this feat, but the tale was widespread (no doubt by Adair himself), and the name stuck. The second theory was that it was an antidote to rattlesnake bites. However, the tonic most Native Americans made with the plant was an antidote to venereal disease and worms. The Meskwaki used the root for bladder trouble and other poisons. It is also thought that some natives used the gigantic leaves to make sandals—again a story without confirmation, though it makes a wonderful image.

In *American Indian Medicine*, Virgil Vogel tells us that this plant was discovered by Dr. Alexander Garden of Charleston (1685–1756, the man after whom the gardenia is named) when he was traveling through Cherokee country. He called it "button snakeroot" (though that name is usually used for *E. aquaticum*). In 1755 he sent seeds to Cadwallader Colden (1688–1776), the governor of New York before the Revolution, and an amateur botanist, who reported it as a "powerful attenuant & Diaphoretic." *E. yuccifolium* was registered in the official U.S. Pharmacopeia in 1820–1873 as a diaphoretic expectorant and emetic.[51]

Asclepiadaceae

Asclepias tuberosa

grows 5 to 30 inches tall. It has fuzzy, lance-shaped, alternate leaves (and is the only one of four different milkweed species to have alternate leaves). The yellow to orange flowers consist of umbels 2 inches long, each with five petals below the crown, or corona. It spreads by seed or creeping underground tubers and has a long tuberous root that makes it drought-tolerant. It grows in sandy loam to heavy clay, provided the soil is not too wet.

Range: Minnesota to Ontario to New England, south to Florida, west to Texas and Arizona.

The brilliant yellow and fiery orange of butterfly milkweed is often a surprise, because the plant lodges in the most unpromising places imaginable—abandoned pastures, highway rights-of-way, and other disturbed places—as well as on the prairies. It is irresistible to butterflies and according to Neil Diboll, it is one of the best all-around plants in a prairie garden. "Blooming in the heart of summer, it survives drought and hot weather because of its deep taproot, which can plunge over 10 feet deep into the soil. . . . Unlike other members of the milkweed family, which have sticky, milky sap, butterfly milkweed has clear sap."

It was named after Asklepios, the Greek god of healing. The *tuberosa* part of its name is for the tuberous roots. To some insects, the attraction of the flower is fatal. Inside each part of the corona there is a curved horn arching over the center of the flower. The pollen-bearing anthers adhere to the stigma and form a gynostegium, a specialized structure for the pollen. The pollen is squeezed into a waxy mass called a pollinium (orchids share this same feature). The unwary little honeybee or fly enters and is trapped in this mass. It is swooped down upon by yellow jackets lurking nearby and is devoured instantly. Butterfly milkweed is pollinated by wasps that have adapted to this flower structure and can escape safely.

At night, butterfly milkweed releases a fragrance to attract night-flying insects and moths. Hummingbirds and both monarch and viceroy butterflies also feed on its nectar, and the

Asclepias tuberosa, butterfly milkweed

monarch lays its eggs on the plant. The eggs become the caterpillars which eat the leaves. Since the plant is poisonous, any munching insect becomes poisonous to birds. Though birds won't die after eating a caterpillar, they do get sick and usually one such meal will keep them away from the plant and its attendant insects forever.

At one time, butterfly weed was called "pleurisy root" because it was used to treat bronchial and pulmonary afflictions. It was also applied to cuts, wounds, and bruises by many different Native Americans. The roots were boiled to make medicine for diarrhea, asthma, and other respiratory ailments. So many claims were made for the plant by early American medical writers that by the nineteenth century Charles Millspaugh wrote:

> *The pleurisy-root has received more attention as a medicine than any other species of this genus, having been regarded since the discovery of this country, as a subtonic, diaphoretic, alterative, expectorant, diuretic, laxative . . . astringent, anti-rheumatic, anti syphilitic and what not. . . . It has been recommended in low typhoid states, pneumonia, catarrh, bronchitis, pleurisy, dyspepsia, indigestion, dysentery . . . and obstinate eczema.*[52]

The seedpods become dark brown with age and split at the side to release soft, fluffy down. Birds have been known to use the down for nesting material. In hard times, this material was used to stuff beds and cushions. Botanist Diane Beresford-Kroeger says that the down can be spun into silk to make a nonallergenic yarn and dyed just the way silk is.[53]

Fabaceae

Leadplant
Amorpha canescens, wild tea, shoestring

Amorpha canescens

grows 2 to 4 feet tall. The small shrub has dense white hairs, odd-pinnate leaves 2 to 6 inches long with oval leaflets. Hundreds of minuscule blue to purple flowers with orange anthers are clustered at the end of each stem, making racemes 2 to 6 inches long. The pods are short and curved and almost ½ inch long. It has a long, deep taproot.

Range: **central Canada, south to Texas, east to Michigan and Kansas.**

The leadplant has much strange lore connected with it. At one time, people firmly believed that this plant's presence meant that there was lead ore in the soil. No one has proven this theory,

Amorpha canescens, leadplant

even with modern technology, but the common name persists. More likely it was given the name because of its grayish appearance. In the right light it looks almost ghostly because of the fine hairs that envelop the whole plant, including the unopened flowers. The botanical name *Amorpha* means "without shape or form" and *canescens* means "hairy."

According to anthropologist Melvin R. Gilmore, the Omaha-Ponca called this the "buffalo bellow plant" because it blooms during the rutting season of the bison. The Indians connected the plant and the sound of the bellowing bison. The long, tough, stringy roots probably gave it the alternative common name of "shoestring."[54]

Native Americans used the plant by smoking the stems or made a tea to treat pinworms, eczema, and rheumatism. Gilmore reports that "the small stems, broken in short pieces, were attached to the skin by moistening one end with the tongue. Then they were fired and allowed to burn down to the skin. An Oglala said the leaves were sometimes used to make a hot drink like tea, and sometimes for smoking material."[55]

This plant interested John Frémont in his expedition in 1842 through Nebraska to the Rocky Mountains. As he neared the Platte River, he noted, "The *amorpha* in full bloom was remarkable for its large and luxuriant purple clusters."[56] Frémont was the first to identify it as a nitrogen-fixing plant.

Fabaceae

Ground-Plum Milk Vetch

Astragalus crassicarpus, buffalo bean, buffalo pea

The exquisite, showy, deep blue flowers of the ground-plum milk vetch seem like a frippery compared to the serious business going on underground. The Latin name means, roughly speaking, "thick-fruited bean." But it's the stout, woody, branching taproot and the leguminous nodes underground that are important. The pods fix nitrogen, adding to the nutrient value of the soil. It grows on many different kinds of soil, from uplands to woody hillsides and roadsides.

Astragalus crassicarpus grows along the ground with stems 4 to 12 inches long; blue-purple racemes have 5 to 25 flowers each, with black hairs in the corolla, which is purple, pink, or, rarely, white. The hairy leaves are 1½ to 5 inches long, with 15 to 30 alternate pinnate leaflets arranged on a central axis. The stems are upright until the pods develop and bend to the ground in June. The inch-long fleshy pods resemble small plums.

The renowned English naturalist Thomas Nuttall (1786–1859) was the first European to pay attention to ground-plum milk vetch. In 1810, at the age of 22, he began to collect specimens in the wilderness, as most people referred to the far distant West. He traveled thousands of miles on foot and hitching rides with fur traders. Many of his trips were solitary, while on others he had a single companion. In 1834, he made his way to Oregon, becoming the first scientist to cross North America by land. He collected hundreds of plants, and gave his name to many of them. Eventually he became professor of natural science at Harvard and director of the botanical garden there.

The flowers of ground plum appear in spring. Songbirds, wild turkey, mice, and other small mammals use the seeds as forage. The pods were often eaten raw or boiled by Native Americans. The Dakota called this "the food of the buffalo."

Botanist Julie Hrapko says that David Melting Towel, a member of the Blood Nation, who did translations for the Alberta Provincial

Museum, told her that the name given to the plant by the Blood meant "the flower that bloomed when the bison returned each spring." "No wonder these bright yellow flowers were called buffalo beans," she says. "My Ukrainian grandmother would boil the bright yellow flowers and then boil eggs in the water to get yellow Easter eggs."[57]

Ken Parker, who owns the Sweetgrass Gardens Nursery on the Six Nations reserve near Brantford, Ontario, says that he has heard from his neighbors that the boiled roots of buffalo bean can be used to reduce fever.

Range: **Illinois south to Louisiana, west to the Rockies.**

Fabaceae

Blue False Indigo
Baptisia australis

Baptisia australis

grows 1 to 4 feet tall in clumps from rhizomes. It has bright blue-green stems with lupine- or pealike flowers in a bright blue. Flowers May through June. The cloverlike leaflets are 1½ inches long. After flowering it has dark seedpods. It grows along riverbanks and in woody thickets.

***Range:* Nebraska to New Hampshire, south to northern Georgia, west to Texas.**

I was born on the prairies and lived there for some years before moving to Labrador. It was not the wild and gorgeous place of infinite variety known a few hundred years ago. At that time prairie was wheat and there was no other possibility for this land. It is the wind even more than waves of wheat that I remember most. The wind soughing through the prairies in autumn is a haunting sound, but one of the most enchanting was that of seeds rattling in the pods of the *Baptisia*. As very small children, we used the stems loaded with pods for rattles.

There are two major regional varieties of blue false indigo. The western plant *B.a.* var. *minor* is smaller than the one found in the East. Both are known for their cobalt blue flowers and deep, widespread root systems, and both tolerate poor soil and dry sites.

Native Americans and settlers alike used *Baptisia* as an antiseptic for cuts and wounds.

Baptisia australis, blue false indigo

Many different nations, including the Mohegan of New England, steeped the root to make a medicine to stimulate bruised areas or areas of the body with poor circulation. Herbalists still use exactly the same technique. The Cree made a poultice of the powdered rhizomes to use on syphilitic sores; other groups a wash of boiled-down roots to stimulate drowsy, listless children.[58]

The early settlers made dye from it but generally with disappointing results. Real indigo (*Indigofera*) provides a brilliant blue dye, and the blue of this plant was a very poor substitute for the real thing. Once the stems are cut open, the sap usually turns slate-colored when exposed to the air.

In the early nineteenth century the U.S. Pharmacopeia, an independent organization founded in 1820 to create a drug compendium, ran trials on extracts of this plant as a remedy for typhoid fever. The Pharmacopeia researches found that "experimental doses and overdoses of the root tincture and powder resulted in symptoms similar to those of the onset of typhoid, and this led practitioners of homeopathy (a medical system based on the doctrine that 'like cures like') to hope for cures in actual cases of the disease."[59] Unfortunately, there is no scientific evidence that it worked.

Fabaceae

Purple Prairie Clover
Dalea purpurea

Dalea purpurea

has flower spikes composed of 100 or more tiny purple flowers forming themselves into cylinders ½ inch wide and up to 5 inches long. It grows 8 to 35 inches high with one to three branches per stem. The leaf is composed of five to seven tiny narrow leaflets set in pairs. The taproot on a mature plant plunges into the soil as deep as 6½ feet with branched lateral roots.

Range: southeastern Alberta, southern Saskatchewan and Manitoba, south to Texas and New Mexico, east to Illinois, Missouri, and Arkansas.

Purple prairie clover is found almost everywhere on the prairies. There are at least 150 species of clover alone, most of them in Mexico and the United States.

This plant was named in honor of English botanist Samuel Dale (1659–1739) and was first described by French botanist Étienne Ventenat (1757–1808). Native Americans ate it as a vegetable by collecting the fresh leaves and boiling them quickly. The bruised leaves were also steeped in water and then applied directly to wounds. *D. purpurea* was also known to remedy heart trouble; infusions of the root were taken for measles; the Navajo used it as a remedy for pneumonia, and the Pawnee made an infusion of roots to use as a prophylactic against disease.

Dalea purpurea, purple prairie clover

The roots are very tasty, and many people used them for sweetening, or chewed them as a gum. The Oglala used the leaves to make a tealike beverage. The Pawnee also fashioned the tough elastic stems into brooms to sweep out their lodges.[60]

Fabaceae

Lupine
Lupinus perennis

Lupinus perennis

grows from rhizomes and forms clumps 1 to 2 feet tall. The palmate leaves radiate around the stem. The complex indigo blue flowers have two lips which make it special to three rare butterflies. The pea-shaped flowers bloom in early May and turn into pods by July. The seeds and pods are toxic. It grows on the eastern prairies, in oak savannas, and in sandy soil.

***Range:* Along the Eastern Seaboard, west to Minnesota, south to Arkansas and Texas.**

Lupine can spread out on the poorest soil to look like a sea of blue in spring and early summer. It must have open spaces to survive, but does well even in pure sand or almost any disturbed area. It has the marvel of nitrogen-fixing bacteria in nodes on its roots that provide each plant with its own nutrients. However, lupines are threatened by too much shade and do not thrive under trees.

The lupine's seedpods have stiff hairs which make them dry out quickly. With a twist in the wind, the pods pop open and the seed scatters within a few feet of the plant. Pick a few pods and leave them in bright sunlight and you can watch this action up close. The seeds remain viable in the soil for up to three years, but to ensure survival the plant also reproduces by sending up roots from the rhizomes.

The Cherokee used a cold infusion from the plant as a wash to stop hemorrhages; the Menominee used it to fatten up a horse and make him spirited and full of fire. Lupine rubbed on the hands was said to give a person power to control horses.[61] It was never taken internally, since the seeds cause nausea, vomiting, and twitching.

Lupines are the food source for three endangered butterflies—the Karner blue, the Persius dusky wing, and the frosted elfin. The tiny Karner blue has a 1-inch wingspan. It can lay eggs only in very specific habitats (it must be one with 80 days of snow), and there should be a dense enough population of lupines to supply food for at least two generations of larvae as they crawl up the stems. If the lupine emerges late in spring or dies too early, the larvae will starve. Ants also benefit from the relationship between the Karner blue and the lupine. The caterpillar has a specialized gland that secretes a liquid that ants love to eat.

Adequate plant populations are crucial. A decrease in the number of plants is a disaster for the butterflies that depend upon it. This relationship is so fragile that if the population of plants is poorly concentrated, the butterflies may not survive. The lupine, in turn, depends on these butterflies for pollination. The fewer pollinators it has, the fewer plants reproduce.

Very simple things can upset the extremely delicate balances in nature. When a species becomes rare or extinct, the world doesn't just lose a pretty plant, it also loses a place for insects that grow on or alongside the plant. The loss of lupines means a loss of genetic diversity within the plant population leading to inbreeding, and ultimately to weaker plants with fewer and fewer valuable nectaries for insects.

Steps are being taken to help save some of

the Karner blue's habitat so this intricate web of life can be studied. The Wisconsin Department of Natural Resources has a program called Partners in Protection to protect the Karner blue's dwindling habitat; it involves homeowners, developers, and government departments. Among other measures, mowing in areas where the lupine grows is halted between April and September.

Fabaceae

Indian Breadroot

Psoralea esculenta syn. *Pediomelum esculentum*, plains apple, Cree potato, Cree turnip, Dakota tipsinna, Dakota turnip, Indian turnip, Missouri breadroot, *navet de prairie, pomme blanche, pomme de prairie*, prairie apple, prairie potato, scurf pea, tipsin, tipsinna

Psoralea esculenta syn. *Pediomelum esculentum*

grows 2 to 3 feet tall and rises from one to several hairy stems. It has a thick brown root. Three to 4 inches belowground, the root forms a tuberlike body up to 3 inches long and 1 inch wide. Each light green leaf is edged in white and is divided into five leaflets. About 20 to 30 white and blue or purple flowers are densely clustered in spikes about 2 to 4 inches long at the top of the plant. The pods are flat and have a long slender tip. It propagates by seed.

Range: on plains and grassy hills from the Northwest Territories to Texas.

Indian breadroot is one of the best known of all the survival foods of the prairies. It was recog-

Lupinus perennis, lupine

nized very early on by the itinerant botanists who covered the Western trails and sent back samples and notes to the East Coast and Europe. As a result, we have a good record of the dozen or so common names *Psoralea* picked up in its long history. The botanical name comes from the Greek *psoraleos* ("scabby"), referring to the little dots on the leaves. *Esculenta* means "edible" in Latin.

Because it grows on disturbed sites, any place animals roamed and grazed meant that a niche for breadroot popped open. In fact, along with the presence of bison, *Psoralea* is a prime indicator of a healthy prairie. It starts growing in May, with the flowers emerging in June to mature in July and August. When the plant dies, the stem and leaves break off and blow away in the wind.

The Plains peoples were observed early on digging out breadroot left in the wallows made by the huge bodies of the bison. University of Toronto agriculture professor Henry Hind, who chronicled the Red River expedition of 1857, wrote:

> *The lower prairie consisted of a sandy loam, in which the Indian turnip was very abundant. We soon came up with a group of squaws and children . . . who were gathering and drying this root. . . . Many bushels had been collected . . . and when we came to their tents they were employed in peeling the roots, cutting them into shreds and drying them in the sun. . . . The Crees consume this important vegetable in various ways; they eat it uncooked, or they boil it, or roast it in the embers, or dry it, and crush it to powder and make soup of it. Large quantities are stored in buffalo skin bags for winter use. A sort of pudding made of the flour of the root and the mesaskatomina [Saskatoon] berry is very palatable, and a favourite dish among the Plain Crees.[62]*

Psoralea esculenta, Indian breadroot

Various groups, from the Blackfoot in the West to the Osage in the East, ate *Psoralea*. They used sharp-pointed sticks which they carried with them to dig in the earth for the bulbous turnip-shaped roots. They had to harvest them in late spring when the roots are mature, because once the flowers and the stalk become brittle, they break in the wind and disappear, making it almost impossible to find the roots.

Psoralea was a boon to the people of the prairies who followed the bison south in the winter and north in the summer on their traditional treks. The plant has a higher carbohydrate content (70 percent) than potato; it can be eaten in just about any manner devised by humankind, including some unusual recipes.

For example, a fur trader named Edwin Denig (1812–1858) reported that when "sliced, dried and boiled with the dried paunch of the buffalo, or the peas extracted from the mice's nests cooked with dried beaver's tail . . . [it is] much admired and considered fit for soldiers, chiefs and distinguished visitors."[63]

Once the Europeans moved onto the prairies, they were often forced to barter the goods they had brought with them for the roots the Native Americans naturally had in their possession. There are stories of trappers surviving for days on nothing but this root.

The first reference to *Psoralea* in William Clark's journal of the monumental Lewis and Clark expedition was as "ground potato." He says it was also known as white apple, *pomme blanche*, prairie apple, and prairie turnip. Clark found it somewhat tasteless, but wrote "our epicures would admire this root very much, it would serve them in their ragouts and gravies instead of the truffles morella."[64]

German botanist Frederick Pursh (1774–1820) described it in 1814, using the specimen that Lewis had brought back (the specimen became part of the Lewis and Clark Herbarium collection in Philadelphia). "The present plant," wrote Pursh, "produces the famous Bread-root of the American Western Indians, on which they partly subsist in winter. They collect them in large quantities. . . . This root has been frequently found by travelers in the canoes of the Indians, but the plant which produces it has not been known until lately."[65]

Father Pierre Jean de Smet (1801–1873), in South Dakota, wrote in 1868, "Everywhere the *pomme blanche* is found in abundance. . . . When an Indian is pressed by hunger, he has only to dismount from his horse, and armed with a little pointed stick of hard wood . . . he will pull out roots enough in ten minutes to satisfy him for the moment."[66]

Annora Brown in *Old Man's Garden* points

out (and this is confirmed in the journals of the explorers) that it all depended on what else you had to eat as to whether or not you found it wanting in taste.[67]

The plant also had medicinal uses. A poultice of chewed roots was applied to sprains and fractures (Blackfoot). Chewed roots were blown into a baby's rectum to cure colic. Older children with bowel complaints got to chew on the roots for themselves, as did teething infants. An infusion of roots was taken for chest troubles, sore throats, and gastroenteritis.[68]

Lamiaceae

Bergamot
Monarda fistulosa, bee balm

Monarda fistulosa

grows 2 to 3 feet tall. It has coarse, toothed, oval or lance-shaped, opposite leaves 1½ to 4 inches long. It grows from slender creeping rhizomes to form clumps. The flowers are narrow-lipped tubes of pale pink or pale purple with purple-tinged bracts, borne in tight whorls 2 to 3 inches across. They bloom in summer. It is found in dry soil in fields, thickets, and roadsides.

Range: throughout southern Canada and the United States east of the Rockies.

The scent of bergamot has a note of citrus, which is why one of Charles I's gardeners dubbed it *bergamot* after the oranges (*Citrus bergamia*) grown in Bergamo, Italy. The botanical naming was left to Linnaeus, who honored the Spanish doctor Nicholas Monardes (1493–1588) after he wrote *Joyful Newes Out of the Newe Founde Worlde* in 1577. Monardes saw the economic value of the plants he discovered as he

traveled in the New World, including this herb. It was taken back to England in 1637, and made its way from the physic gardens there and so spread across Europe.

There are several species of *Monarda*, four of which were recognized by Native Americans, who had known its value for thousands of years. *M. fistulosa* has a much stronger mintlike flavor than *M. didyma*. *Fistulosa* means "tubular," and although the moplike little heads of monarda don't look tubular, they are in fact made up of many small tubes. The flowers are perfect nectaries for bees, butterflies, and hummingbirds. Like all plants in the mint family, the stems are square in cross section, which is one way to suss out what family a plant belongs to: feel the shape of the stem.

The Oswego Nation in what is now upper New York state used *M. didyma* to make a tea which became known as "Oswego tea" and settlers soon followed the practice of drying the leaves to make a half-decent cuppa. After the Boston Tea Party, colonists looked for alternatives to imported teas. Oswego tea became the norm in the colonies.

Monarda

Monarda or Oswego tea
First gladdened Indian and bee.
Then came the English, French and
    Dutch,
I wonder if they liked it much.
It paid no tax to King and Crown
And so our forbears gulped it down,
To patriotic zeal I bow,
Delighted not to drink it now.

Elisabeth Morss,
from *Herbs of a Rhyming Gardener*

The most important properties of *Monarda* are the oils in its leaves, including thymol. Thymol is an antiseptic applied to many different kinds of wounds. The Blackfoot of Alberta boiled the leaves and applied them to skin eruptions, and used an infusion of leaves to relieve inflamed eyes. The aromatic leaves were also boiled with meat in very much the same way we use thyme or rosemary to flavor meat today. The plant could also be boiled down to make a hair pomade important to grooming. A warm tea made with the plant was used to induce sweating, a cool tea to soothe a woman's reproductive system. The tea was also used to relieve flatulence and vomiting; the steam could be inhaled to relieve the congestion of a cold.

The Shakers, always observant, borrowed the habit of their Indian neighbors to make a tonic out of the leaves and flowers, and cultivated the plants in their gardens for the same reason.

Bergamots are not as important medicinally as they once were now that the essential oil thymol can be produced synthetically. The dried leaves are more likely to be used to scent closets or added to cosmetics and perfume than as a medicine. They make good dried flowers: just hang them upside down in a cool airy place and let them dry out. And it is still possible to take the dried leaves from a plant in the garden and make a refreshing tea.

Lorrie Otto, one of the leaders in the fight against DDT in the 1960s, is an avid grower of native plants and knows how much butterflies and bees love her *Monarda:*

Monarda fistulosa, bergamot

On one cool summer evening, as I was lifting clumps of blooming Monarda *out of a holding bed of sand, a passing neighbor said, "Lorrie! Do you know that you are just covered with bees?" My goodness! We counted 37 bumble bees clinging to the long sleeves and back of my bright yellow shirt. Like bantam chickens, they had gone to roost but unlike birds, their feet did not clamp and lock around branches. Those fuzzy little balls just tumbled loose and then their hooked appendages caught on to my sweatshirt. I've seen photographs of people coated with a swarm of honey bees but never a shirt sprinkled with bumble bees. We gently put each one back on the bergamot and watched as they slowly readjusted themselves. I love all of my native wildflowers, but this species may well be my favorite with hawk moth wings "puddering" in the heat of the day and bumble bees roosting in their bee balm in the cool of the evening.*[69]

Ranunculaceae

Prairie Crocus

Anemone patens, pasqueflower, ear of the earth, April fool, cat's-eye, gosling, hartshorn plant, headache plant, lion's beard, Mayflower, old man, prairie hen flower, prairie smoke, red calf flower, rock lily, stone lily, wild crocus, windflower

Anemone patens

grows 3 to 8 inches tall from a rosette of basal leaves. It has a thick woody taproot. The conspicuous flowers are petal-like sepals to 1½ inches wide in pale blue or mauve and very occasionally white. They bloom in April, before the leaves appear, on erect sessile (stalkless) stems. The leaves and stem are hairy. The flowers are solitary (one to a stem). The achene (a one-seeded fruit with thin walls) produces a light airy plume. It grows in open prairies, glacial moraines, burned-over ground, railway rights-of-way, and over-grazed pastures.

Range: Illinois, west to the Rockies, and north to the Arctic.

State flower of South Dakota, provincial flower of Manitoba.

It's early spring, so early that there is still snow on the ground. But in bare patches, all across the cold soil, the tiniest of budlike tips begin to appear. They have been waiting since last autumn to push upward and they are lonely presences on the austere prairie landscape. In

How the Prairie Anemone Got Its Fur Coat

To enter the world of chiefs, Wapee, son of a Chief was to spend four days and nights atop a lonely hill until a vision of the man he was to be came to him.

The first night, no visions appeared to him and he was downhearted. But with the dawn, the warming sun beamed upon a beautiful flower who opened her petals and nodded towards Wapee as if to welcome him. Wapee no longer felt alone.

When night came again, Wapee curled his body around his new friend to protect her from the icy night winds. Three times he did this and three times when the Morning Star rose, visions came to him foretelling of great things to come.

When Wapee rose to leave he said, "You have comforted and counseled me well these past three days and nights. What three wishes would you have me ask of the Great Spirit?"

"Pray that I may have the purple blue of the distant mountains in my petals, a small

Anemone patens, prairie crocus

golden sun to hold close to my heart on dull days, and a furry coat to face the cold winds in the spring."

The Great Spirit was so pleased with Wapee's thoughtfulness, he fulfilled his prayer.

Annora Brown, *Old Man's Garden*

any sun-warmed spot in April *Anemone patens* can be found braving the last fall of snow. This little plant was dubbed prairie crocus probably because it reminded settlers of the crocuses back home. The explorer Alexander Mackenzie (1764–1820), standing on the banks of the Peace River, described the tiny purple flower with its furry stem as "a yellow button encircled with six leaves of a light purple." It gave him his first sense of hope after a long and lonely winter.[70]

The prairie crocus comes out in the raw weather of early spring, when insects are scarce. It is the job of these enchanting flowers to attract as many insects as possible. The prairie crocus copes with the cold because it has a thick toughened stem base, called a caudex, hairy leaves, and a long taproot. These small plants need the furry protection they've developed. On cloudy cold days with the wind whirling about, the anemone closes in on itself. When the sun comes out, its blooms unfold to provide food and protection for any insects in the neighborhood.

The prairie crocus is a heliotropic plant—that is, it always faces the sun. This helps the pollen and seeds to develop very early in the season. The violet to white flowers are enclosed by sepals that reflect the sunlight back into the interior of the flower with its profusion of yellow stamens, so that sluggish insects can find some warmth in the center.

Benign and charming as this plant may seem, it is toxic to sheep, causing a serious impairment to the digestive system if grazed. Deer, elk, and ground squirrels have no problems with it, and goldfinches can eat the seed with impunity.

For humans, the plant can cause serious irritation if it is touched and then rubbed into the eyes. The Indians exploited this characteristic and gently kneaded the plant into boils to make them blister and pop open. They also made a poultice of the crushed leaves which they used to treat rheumatism and neuralgia.[71]

Melvin R. Gilmore wrote in 1914 that the Arikara had a spring ceremony that included finding a cedar tree (which was considered holy) and floating it down the Missouri River with a well-worn baby's moccasin and a prairie crocus on it. This was to let the villages downstream know the Arikara were alive and had survived another winter. The Lakota called the plant *hosi 'cehpa*, which means "child's navel."

Annora Brown, in *Old Man's Garden*, wrote: "The name of 'gosling' given the downy buds by prairie children is eminently suitable, but the Indian name is even better. The Indians, unhampered by Greek tradition and all the old world sentiment, had a perfect genius for choosing the most poetic and significant name for things about them. 'Ears of the Earth' they called these furry ears which, so soon after the snowdrifts melt, the prairie thrusts up to listen for the first faint rustle of summer."[72] The Blackfoot word *Napi*, or "old man," referred to the grayish seed heads which appear in early summer. More recently, a Yukon child coined the term "elephant's Q-tips" for the fuzzy pointed buds.

Rosaceae

Prairie Smoke
Geum triflorum, three-flowered avens, torch flower, long-plumed avens, lion's beard, old-man's-whiskers

Geum triflorum

grows 6 to 18 inches high from thick perennial black roots. The many dissected leaves rise from the base. They are pinnately divided, almost ferny, with small wedge-shaped leaves mixed with smaller leaflets. Each stem usually has three urn-shaped flowers, about ½ inch across, with elongated styles that become the feathery plumes that carry seeds away. The sepals are purplish-pink and the petals may be pink, yellow, or flesh-colored. Spreads by rhizomes as well as wind-dispersed seed. Grows in dry prairie soil.

Range: **Ontario, south to Illinois, and west to British Columbia, Washington, Oregon, and California.**

Prairie smoke comes into bloom after the prairie crocus on almost any dry prairie site. The bright evergreen leaves add a little color to the beige litter covering the prairies after the winter. It is pollinated by small bees that crawl inside to get at the nectar and pollen. The sepals make up the outer part of the dark red flowers, and the petals of yellowish-white are the inner part of the flower. Each plant has bunched-up flower stalks making a glorious display, especially when the seed heads form. The seed heads are the smoke of prairie smoke. The small feathery wisps act like sails for the seeds and carry them off on the winds.

Ken Parker, who lives on the Six Nations Reserve in Ontario, says that each reserve had its own name for the prairie plants. On his, they call *Geum triflorum* "Grandpa's whiskers" because of its distinctive wispy seed tufts.

The Blackfoot crushed the seedpods as a perfume or used them as a soothing eyewash. The roots were also boiled down to make a mix to relieve sore eyes. The Blood boiled the roots and used the mixture to soothe sore gums, women's painful nipples, or saddle sores on horses. The roots were dried and made into a tea to soothe a cough.

Rosaceae

Prairie Rose
Rosa acicularis, prickly rose

Rosa acicularis

grows up to 4 feet tall. A shrub with several main stems, many branches up to 4 feet high, covered

Rosa acicularis, prairie rose

with weak bristles; flowers are a deep rose red, 2 to 3 inches across; fruit is bright red.

Range: across Canada from British Columbia to Quebec, south to Montana, Wyoming, Colorado, the Dakotas, Minnesota, Wisconsin, and Michigan.

Provincial flower of Alberta.

Rosa acicularis is one of three rose species known as prairie rose; the others are *R. arkansana* and *R. woodsii*.

The peoples of the prairies seem to have had a great fondness for the prairie rose. The Blackfoot and Blood ate the fruits fresh or roasted after removing the seeds. Before trade beads became the norm, the hips, which were often made into necklaces, were used in the perpetual

Geum triflorum, prairie smoke

trading that went on all over the prairies. The roots could be boiled to make a tea to treat diarrhea, and a poultice of rose hips was used to treat boils. Rose hips were also mixed with grease and eaten with dried meat. Frozen in the snow, they became a special winter treat for children. There were even some arcane treatments such as sticking the thorn into a painful part of the skin and setting it on fire. After the thorn had burned away, a blister was left. Believe it or not, this treatment was intended to cure pain.[73]

The early European explorers of the prairies learned to eat roses, but not all of them were happy about it. Daniel Williams Harmon (1778–1845) wrote in 1820: "For some time after our arrival we subsisted on *Rose-buds!* which we gathered in the fields, but they are neither very nourishing nor palatable, yet they are much better than nothing at all, but where to procure anything better I knew not, for the Buffaloe at that time were a great distance out into the Plains."[74]

Harmon clearly failed to appreciate the

nutritional value of rose hips, which are crammed with vitamin C. The bitter seeds contain vitamin E. Some people such as Steve Brill and Evelyn Dean find them delightful: "The best rose hips are the largest. They taste like a combination of apricot and persimmon. . . . You can cook whole rose hips in fruit juice, and strain out the seeds with a food mill. . . . Rose hips are also great raw."[75]

Today, the prairie rose is the emblem of the province of Alberta. There is, however, some confusion over the species. The act naming the provincial flower of Alberta in 1935 specifies *R. acicularis*, even though the two other wild roses in the province are more common: *R. arkansana* (sometimes called the **cattle rose**), found in the Peace River country and from Edmonton south, and *R. woodsii*, which has the widest distribution. As Julie Hrapko, curator of botany at the Provincial Museum of Alberta, recalls, "The MLAs of Alberta decided to have a contest amongst schoolchildren to choose a provincial flower. They picked the wild rose. But the MLA in charge didn't know which of the wild roses they meant, so he turned to a friend at the University of Alberta for expert advice—only the professor was in the English Faculty. I've had more people arguing with me in all my years as curator of botany over which one is the provincial flower, but the act says *Rosa acicularis*."[76]

Scrophulariaceae

Culver's Root

Veronicastrum virginicum, Beaumont root, Bowman's root, black root

Veronicastrum virginicum

grows 4 to 6 feet high. The leaves are simple and whorled around the unbranched hairless stems.

The 6-inch-long toothed leaves are lance-shaped. The flowers are usually white, sometimes pale blue or pale pink. The terminal racemes look like little spikes. Once the terminal shoots have flowered, the side shoots will bloom. It is a plant of moist woodlands and meadows of the eastern prairies.

Range: from Texas north to Ontario and all across the eastern United States.

Culver's root is like a giant candelabrum rising above the magic feast a prairie presents. When it is in bloom, the spikes are jammed with flowers, each with five white petals forming a tube. They attract butterflies in great quantities. It's almost impossible to figure out how insects latch on, but any given plant in summer will have its slender stems bending in half under their weight. The seeds are eaten by birds. The roots and rhizomes have been used by people for millennia.

The botanical suffix *-astrum* means "partial resemblance," referring to its similarity to *Veronica* species. There is only one relative of this plant and that's in Asia, reminding us once again how close the continents were millions of years ago.

Veronicastrum has been known in the North American health system ever since a member of the Seneca Nation of the eastern United States introduced it to a physician. It acts on the membranes of the intestines and increases mucus, which washes out worms and other parasites. The Native Americans knew that if the root was eaten fresh, it was toxic, so they aged it for a year after digging it up in autumn. They used it in cleansing ceremonies to purify the blood and the soul. It was also used as a treatment for fainting and to remove the agony of kidney stones.

Settlers made a powerful laxative by steeping the root in hot water or milk. Cotton Mather (1663–1728), the Puritan leader who not only

Veronicastrum virginicum, Culver's root

EXOTIC ALIENS

When the first Europeans stepped on the shores of the East Coast, they brought European honeybees with them. These bees spread so wildly and so efficiently that by the early nineteenth century they had begun to elbow out the native North American bumblebees. The honeybee was a portent of the other exotic insects and diseases that would reach these shores all too soon.

These newcomers brought not only their possessions, but also their plants. Sometimes seeds arrived unwittingly in pant cuffs, on the bottoms of shoes, or clinging to a hat. When alien plants invade new territory, they don't just do well, they explode into a whole new niche uninterrupted by either native plants or insects. And most of them have no enemies in this new home.

Unfortunately, many of them are cool-season plants with the ability to emerge when the ground is still cold. This happens usually long before the native grasses, which need warmth to germinate. In this way, the exotics often out-competed the native species. Once the plow was introduced, these invaders spread even more quickly and efficiently.

Bromus tectorum (cheatgrass, downy chess) has become a huge problem in the prairies. An annual grass, it's much like crabgrass but even more mobile because it has prickly awns. It is a disaster for prairie animals because it's edible for only a short time. Once the prickly awns develop, it becomes a painful and often dangerous meal. When the cheatgrass disappears in winter, there is no other food for these animals and they starve to death.

Another prairie invader is Russian thistle (*Salsola pestifer*). Max Braithwaite, who taught in

wrote about but probably also took part in the Salem witch trials, felt confident in prescribing Culver's root for his daughter's tuberculosis. It was a well-known emetic and it must have been a last resort for the poor man. His daughter died shortly after. Today, Culver's root is still used by herbalists, to improve the skin tone by cleaning out the kidneys.

a one-room prairie schoolhouse in the 1930s, describes the problems it caused:

> *Russian thistle, the third plague of that plagued spring, was, in its own way, as mean as the other two [dust storms and grasshoppers]. . . . The whole field becomes a mat of prickly green stems and leaves that by autumn are big enough to foul binders and threshing machines. Then the big branching plants break off at the base and go tumbling across the summer fallow, scattering billions of spiral-shaped seeds that burrow into the ground for next year's crop.*
>
> *These round, barrel-sized plants piled up against barbed-wire fences until their sheer weight broke the posts. They clogged corrals and banked the sides of outbuildings. And they doubled field work.*[77]

Dozens of plants such as purple loosestrife, which has taken over whole areas of wetland, do so because they have no enemies here. A look at only part of the list of these exotics tells the story: garlic mustard *(Alliaria petiolata)*; Queen Anne's lace *(Anthriscus sylvestris)*; tall fescue *(Festuca elatior)*; nodding, bull, and Canada thistles *(Cirsium* and *Carduus* spp.); periwinkles *(Vinca major* and *V. minor)*; ground ivy *(Glechoma hederacea)*; Kobe, Korean, and Sericea les pedeza *(Lespedeza* spp.); red, white, Dutch white, and yellow sweet clover *(Melilotus* and *Trifolium* spp.); leafy spurge *(Euphorbia esula)*; kudzu *(Pueraria lobata)*; common chickweed *(Stellaria* spp.); wild teasel *(Dipsacus* spp.); dame's rocket *(Hesperis matronalis)*; and crown vetch *(Coronilla varia)*. What they cost the economy is staggering. Fifteen of them have been carefully documented to cost the American economy more than $600 million annually. The worst of a list of bad hats in the plant kingdom are *Melaleuca* trees, purple

loosestrife *(Lythrum salcaria)*, and the parasitical witchweed *(Striga asiatica)*.[78]

These plants elbow out the native insects that have evolved along with the plants they depend upon. They also make restoration of native flora incredibly difficult, because their seeds are everywhere—blown on the wind, carried by animals from one area to another, lurking about for years if not decades in the soil itself. Land they touch, even if left alone completely, does not revert to its original vegetation; rather, huge fields of these exotic aliens emerge.

To sort out what is an introduced plant and what is a native plant isn't always easy. Take Queen Anne's lace, for instance. Can we imagine what the landscape would look like without it? Or dame's rocket, which blooms everywhere with such grace? But what have they eliminated by their success? In most cases we can only surmise, and since it's a long and difficult process to restore landscapes to what we imagine is their original condition, this is a learning process that will go on for many decades. In the process, we must all become guardians of the land. There are native plant associations and societies in every part of the continent. They can provide lists of what is native and what is not, what the native plants will do, and what the aliens have done. It's our responsibility to educate ourselves about this. The fragility of native populations is obvious in the number of plants that disappear with and without our knowledge. What insects, microbes, and other living creatures have disappeared along with them we will never know.

What is more deeply satisfying: a field of huge diversity, rich with every color of the rainbow in a harmonious mix, or vigorous swaths of purple loosestrife? At first the latter enchants, and then it bores. This is not what nature intended and, looking at a monotonous landscape flung in front of us, we know instinctively that something's wrong. Our attachment to nature is called "biophilia": a deep-seated love of nature and our need for it. If we understand that it is part of our humanity to be part of nature and live within nature, we will want to know more and more about the natural systems we live in, and care about what role we can play in preserving them. We are, after all, only stewards on a planet that will last long after our species has disappeared.

The Desert

A first-time visitor to the Sonoran Desert is in for a shock. Where is the sandy dune-filled vista stretching in all directions so familiar from *Lawrence of Arabia*? The gently rolling hills of the Sonoran are covered with plants huddled together as far as the eye can see—hundreds of varieties dominated by the huge, ancient saguaro cactus dwarfing everything around it. The air is full of the murmur of insects, punctuated by the occasional call of the birds that live within the cactus.

Though the temperature might be 100°F, there is life everywhere in this desert. During

Never an empty place, the desert explodes with color.

seasonal droughts, ocotillo plants denuded of leaves are emblazoned with rich red flowers at the end of each prickly stalk. Palo verde trees, creosote, teddy-bear cholla, barrel cactus, and opuntia sprout from unpromising-looking soil.

As deserts go, the Sonoran Desert is wet: it receives 7 to 8 inches of rain a year. Below 3,000 feet there are few freezing lows and it rarely falls below 25°F. If it does, the saguaros suffer. The profusion of plants changes at every level, from sea level to mountains known locally as sky islands. The cacti and trees grow on the plains. At higher elevations yuccas, nolinas, agaves, and dasylirions grow.

It is a miracle that living things can survive

on the desert, given the heat and relentless sun. But cacti and succulents have evolved highly water-efficient forms of the process of photosynthesis crucial to all plant life. At night when it is cooler and there's more humidity, the stomata (porelike openings) on the surface of the plants open up to gather in carbon dioxide, which is then used for photosynthesis during the day. When it gets really dry, the systems idle. The stomata stay closed all the time, gas exchange halts, and water loss comes to a stop. Metabolism carries on in the moist tissues, albeit at a very low level.

Desert plants have a variety of survival strategies. The saying is that everything in the desert bites, sticks, or stings. Some plants have developed spines and toxic substances to protect themselves from the animals and birds that live in the desert. In addition, spines provide shade along the surface of the plants and collect water. Mesquite grows long roots to get at the water table; ocotillo goes dormant during dry periods. Some ephemeral plants, such as brittlebush, germinate only during rain and bloom with such ferocity that the desert seems on fire.

The Sonoran Desert, however, is not what it once was. As early as the 1890s, settlers were looking for ways to tame the Colorado River. After some disastrous attempts, they succeeded in erecting the Hoover Dam in 1936 in the name of preventing floods and providing hydroelectric power. The result, says Thomas Sheridan, ethnohistory curator, has been to turn rivers such as the Colorado, the Yaqui, and the Mayo into "ghosts of the past, victims of the twentieth century, carcasses of sand whose lifeblood has been diverted into cotton fields, copper mines, and vast, sprawling cities."[1]

Livestock have also dramatically altered the landscape. Between 1870 and 1890, the number of livestock grazing in Arizona territory increased from about 38,000 head of cattle to 1.5 million head and more than a million sheep.[2] This increase took its toll quickly. By 1901, the chief botanist of the Arizona Experiment Station in Tucson, D.A. Griffiths, sadly noted that these desert grasslands were the most "degraded" of all the grasslands in the western United States.[3] Wildfire suppression, copper mining, irrigation, and urbanization have also affected the area, and the rapidly increasing population is putting a strain on water resources.

An early 1990s study determined that of Arizona's 3,200 plant species, approximately 330 are nonnative, compared to 190 in 1942.[4] This shows an alarming continuation of a trend that began in the 1930s, when the South African Lehmann love grass (*Eragrostis lehmanniana*) was introduced in an attempt to reduce erosion. Today there are 400,000 acres of Lehmann love grass in Arizona. In the 1960s, more than a million acres of desert and subtropical thorn scrub were destroyed to plant nonnative buffel grass (*Pennisetum ciliare*), which was considered a superior forage crop.[5] Buffel grass has staked its claim with great success, perhaps because cattle seem to prefer the native grasses, leaving the buffel grass untouched, and able to triumph by the end of the growing season.[6]

The Sonoran Desert is not the only North American desert ecosystem. The Great Basin Desert, which lies west of the Sonoran Desert, is mainly mountainous, rising 4,000 to 10,000 feet. It contains alkali flats of dry lakes. The rainfall is 8 to 12 inches, evenly distributed throughout the year, including snow in winter. The Mojave Desert, a transitional area between the Great Basin and Sonoran Deserts, is mainly 2,000 to 4,000 feet with lower mountains than the Great Basin. It receives an average of 5 inches of rain a year. The Chihuahuan Desert lies entirely to the east of the Continental Divide. Half of it is more than 4,000 feet. Its plains and high mountains make it 10° to 20°F cooler on average than the Sonoran Desert, and frosts are common. It gets 8 to 12 inches of precipitation, mostly as summer rain, but there is some snow in winter. Part of it is in New Mexico and Texas, the main area in Mexico.

THE GREAT CACTI

Cactaceae

Saguaro

Carnegiea gigantea, sahuaro, giant cactus, pitahaya

Carnegiea gigantea
has tall, thick, columnar, spiny stems up to 24 inches in diameter with thick, 2-inch spines. Funnel-shaped white flowers grow in clusters near the end of the branches from May to June. Growth is very slow: ¼ inch the first year; 1 foot after 15 years; the arms develop at age 75 and how many there are depends on each plant's genetic makeup. Grows up to 50 feet high on desert slopes and flats.

Range: southeastern California to southern Arizona.

State flower of Arizona.

A lonely outpost? A vulnerable target? A symbol? A sign? The saguaro, the giant cactus, is all of these things and more. A forest of this cactus

Carnegiea gigantea, saguaro

has a quality all its own. Yet its chances of survival are slim. A mature plant can send out 50 million seeds in a lifetime, but few make it to germination. Even fewer, perhaps only three or four, become mature plants.

A saguaro is like a giant water container. The waxy, accordion-pleated exterior is so supple that, when there is plenty of moisture, it can expand to hold enough water to last for two years—as much as five or six tons of water. According to Ruth Kirk, "Once in a while the roots of the saguaro will draw up more water than the cells can hold, and a cactus will burst. Usually the ruptures heal without serious injury, although sometimes infection sets in."[7]

A story told by writer-photographer Bill Thomas illustrates how the saguaro functions:

I remember being caught in a flash flood some years go. . . . I was parked in the downpour near a desert wash not far from Tucson. The rain came down in sheets, and within forty-five minutes the desert became a lake. Everywhere I looked was water. It was no more than a foot deep, perhaps only inches, but the dry wash nearby became a roaring stream from which emerged snakes, lizards, and other creatures bent upon survival. The great saguaros which stood sparsely around me began to swell before my eyes. Their rib-accordion structures began to expand and I knew the water was being soaked up by the expansive root system. I returned the next day just out of curiosity; the saguaros were perhaps a third larger than they had been prior to the rain.[8]

The knife-sharp 2-inch spines, which are actually modified leaves, saturated with pectin and calcium carbonate that makes them stiff, don't seem to discourage the dozens of birds that nest in the saguaro as it ages. The spines not only protect the plant from predators, they also provide it with shade. Breezes blowing over the spines cool them down; the coolness is passed to the interior of the saguaro. The surface is covered with stomata which close down in the heat and open up in the cool of the evening.

The taproot is only 3 feet long—not much to hold up such a huge superstructure. Short but thick radial roots act as anchors, and surround rocks to make sure they are secure. Additional, much finer, radial roots stay within 4 inches of the surface, with tiny water-absorbing ones in the top half inch. They spread as wide as the plant is high, which could be up to 50 feet.

Flickers and gila woodpeckers build nests in the stems of the saguaro. It's such a desirable residence that birds compete for available space. The birds dig holes with their beaks; in a self-protective reaction, the saguaro develops a kind of scar tissue around the holes. Saguaros with several holes in them can remain quite healthy. The holes may be used by other birds, including the rare pygmy owl, once the original birds have finished with them. When the saguaro dies, these pockets of scar tissue remain, and are known as "boots." This also works in another way. When there is rain, water collects in the holes and is absorbed by the cactus.

Some saguaros develop "crests." This is a mutation that makes the cactus sprout dozens of small bulbous outgrowths from its top. When saguaros are about 50 to 60 years old, they blossom at the top of the stem and the arms. The flowers are a creamy white with a thicket of yellow stamens in the center. They begin to open in April or May and keep right on doing so until June. Each night a few flowers unfold after the sun is down and close the following day around noon. Each blossom lasts no more than 24 hours.

During the night, the flowers attract nectar-feeding bats. The reason for all this action, according to desert writer John Alcock, is that saguaros have energy-rich nectar, digestible pollen, and a huge amount of protein, twice as much as insect-pollinated cacti such as prickly pear or barrel cactus. The bats feast all night and then in the morning, bees, wasps, ants, and butterflies arrive to enjoy the remaining pollen and nectar.[9]

A word about the scientific name: In 1848, botanist Dr. George Engelmann first described it as *Cereus giganteus*; then it was changed at New York Botanical Garden to *Carnegiea gigantea* as a ploy to squeeze money out of Andrew Carnegie for further research into the desert. As a fund-raising ploy it worked—until Carnegie traveled west and found them growing all over the desert. It wasn't the rare plant he thought it was.

If pollination is successful, green fruit develop in late May or June. In the heat and dryness it continues to mature; even after three months without rain, it turns pale orange or red, swells, and splits. The fruit starts ripening in June and with a great thud plops to the ground. The egg-shaped fleshy fruit is up to 3½ inches long with brilliant red pulp inside a green casing. The bright red interior encloses a sticky mass of between 2,000 and 4,000 black seeds.

To germinate, the seeds must land in the shade of another plant, known as a nurse plant, usually palo verde or mesquite. These plants are crucial to its survival, protecting saguaro seeds with their shade as the seeds struggle to germinate. Even then, the seeds are vulnerable to everything around them, such as being inadvertently dragged out of the shade by an animal that gets too close.

The saguaro has always been held in high regard by the Papago Nation, also known as the Tohono O'odham Nation (*Papago* means "bean

eaters"). According to Gary Nabhan of the Arizona Desert Museum: "Saguaros are not seen as a 'separate' life form at all, not something of an 'other,' outside world. Papago classify saguaros as part of humankind; a saguaro cactus is 'that which is human and habitually stands on earth.'" Nabhan suggests that if saguaros are considered human, it's not because they have arms but "because no matter how much they tend to dominate a landscape, they are still vulnerable."[10] According to one legend, the saguaro was a child who searched for its mother, holding up its arms.

Saguaro was like corn to the earliest inhabitants of the Sonoran Desert. They carried balls of saguaro pulp and clay jars of syrup to exchange for grain, pottery, and abalone shell. The fruit, which tastes like raspberries, is rich in oils and sugar and is a source of B vitamins. When the fruit ripened, the people would stuff themselves for weeks in preparation for the lean times before the corn and beans matured. As with many other Native American groups over hundreds of years, their metabolisms adjusted to this pattern of feast or famine.

The saguaro was even used to make wine. According to legend, I'itoi, the creator, taught the people how to do this. Some of the fruit would be set aside in large ollas (earthenware pots) and placed in a specially built house. Outside, the people danced to help the fermentation along. When the wine was ready, everyone would party until they threw up or passed out. As anthropologist Ruth Underhill explains: "[Vomiting] is recognized as a ceremonial feature, and people say with pleasure, pointing out a man so affected: 'Look, he is throwing up the clouds.' The regular procedure during the twenty-four hours of the feasts is to drink, vomit, sleep, and drink again, until the result is a thorough purging."[11] This was not only a social event, it was also a ritual to encourage the rain, "the famous drinking ceremony when the rain

Coyote Scatter Saguaro Seed

At that time Turtle lived with his friend by the ocean shore. He had saguaro cactus and when it ripened, he would gather the fruit, dry it with the seeds in it and store it in his house. That's the seed that's scattered under the saguaro.

Then Turtle would pick them up and go to the ocean and throw them into the water so the seeds wouldn't grow. This way they were the only ones who ate the fruit.

So Coyote was sent to see if he could get some saguaro seed so the people could also plant it.

Coyote was thinking of what he could do to deceive Turtle when he met him. He went to the ocean and was wandering around on the shore when he saw him. Turtle was coming down from the mountain, so Coyote went to meet him.

When they met, Coyote said, "Where are you going?"

Turtle said, "I'm going to the water to swim."

Then Coyote said, "What's that in your hand?"

Turtle answered, "It's a strange thing and is very dangerous for people. I'm going now to put it under the water. That way it will never come out on the land."

Then Coyote said, "Oh, so it's some kind of seed. Let me see what it's like."

But Turtle said, "If you try to see it, it will make you sick."

"It won't really make me sick. Don't you know that I am also a medicine man? That's why nothing ever makes me sick."

So Turtle held out his hand, and just opened it a little bit.

Coyote said, "Wait, wait! I want to see it real close. It's not clear from here." When he

said this, he crept up on Turtle. "There! Now we'll see what it is." And just as Turtle was opening his hand, he hit it from below, and the seed was scattered wherever there are saguaro growing now.

When Coyote had done this he ran back, telling everyone as he ran, "Even though I did not get the seed, I scattered it everywhere on this land. Maybe when the Saguaro comes up and ripens, you will gather it and eat it."

This is why Coyote is good for something for people. Then they gave him a wife who was beautiful, and Coyote married her, and said, "From now on, I will not just wander around. Whoever wants to see me for any reason will go over there looking for me. I will be living in the east where they have already spoken for land."

So Coyote went to the east with his wife. And, because he was a survivor, and saw many things and suffered much and knew the earth everywhere from the beginning until now, he was a very wise person.

Saxton and Saxton,
Legends and Lore

gods of the four directions were invoked to wet the earth to satiety, even as the drinkers were saturated by the magic liquor."[12] The Papago still make wine from saguaros to celebrate their new year and show respect for the elders and tradition.

The Papago also used the saguaro's woody ribs to build walls, doors, ceilings, shelves, and furniture; the long sticks were used to form cradles, baskets, small animal traps, and birdcages. The spines were used as needles for tattooing. Kids played with saguaro dollhouses and saguaro noisemakers; adults depended on the cactus for items in games and wood to carve religious fetishes. Medicinally, pieces of cactus served as painkillers; saguaro gruel helped mother's milk flow; ribs made convenient splints.

Today the Tohono O'odham still make long poles from the ribs of fallen saguaro to get at the plum-sized, dark red fruit which grows on top; they hook it off and collect it in baskets. One fingernail is groomed and allowed to grow long specifically for separating the husk from the pulp. Husks placed on the ground fresh side up are a prayer for good summer rains. The pulp, eaten raw or mixed with other foods, is processed by cleaning, soaking, and boiling. The seeds are used in cakes or fed to chickens.

Saguaro, although it is so closely identified with the Sonoran Desert, is a relative newcomer. People have been here longer than saguaros. The saguaro evolved in the humid, tropical climate that characterized the area from Los Alamos to the Mogollon Rim of Arizona and New Mexico about 10,000 years ago. Over several thousand years, this area dried up. The water-loving saguaro adapted by storing rainwater to preserve itself during dry spells. Now the saguaro is so closely associated with the Arizona landscape that its blossom was adopted as the official territorial flower in 1901 and as the state flower in 1931.

Today saguaro grows only in southern Arizona and parts of Mexico; it used to grow in other parts of the United States, but has since died out. In the 1920s California had large forests of saguaros along the Colorado River. But as it became trendy to have a saguaro cactus in the garden or as part of a movie set, people would drive through the deserts looking for them. Movie people used them with great abandon in Westerns. When the movie was finished, the cacti were trashed. As for those that ended up in gardens, they were doomed to a slow and grisly death in the wet coastal plains to which they had been dragged.

Hollywood loved saguaros so much it seemed unable to film a desert scene without them. In the classic 1951 Jimmy Stewart Western *Broken Arrow*, the movie crew made imitation saguaro out of plaster of Paris and scattered them around the set, which was near Sedona, a place far too cold for the saguaro. Set decoration in those days did not concern itself with veracity.

Saguaros are still commonly used in landscaping in Arizona, even though only a small percentage survive the trauma of transplanting. A group of researchers at the University of Arizona is monitoring about 800 transplanted saguaros on a golf course in Tucson, and comparing their survival rate to naturally occurring saguaro near the golf course and in a wilderness area. After two years, 6 percent of the transplanted saguaro had died. The long-term outlook is worrying.

Saguaros are often kidnapped by "cactus rustlers" who dig them up and sell them to landscapers, despite stringent efforts by the armed "cactus cops" of the Arizona Department of Agriculture to halt this trade. Kidnapping saguaros is a felony in Arizona, and conviction may result in a fine of up to $150,000 and three years in prison. Many of the stolen cacti, which are protected by Arizona's Native Plant Protection Program, are sent to collectors overseas. Tourists who don't know any better and try to take home a living souvenir may get off with a misdemeanor charge, a fine of $500 per plant, and only 30 days in jail.

The cactus cops also have to ensure that new subdivisions in this rapidly growing state do not threaten native plants. The plants are carefully transplanted, tagged, and, when the development is complete, replanted, even if their chances for survival aren't all that good. These officials know who has the equipment to move the huge cacti around, and take a great interest in unfamiliar people transporting a cactus that hasn't been tagged. Even in national parks, where you'd expect people to make the effort not to touch, there is widespread theft. At Lake

Mead National Recreation Area in Arizona and Nevada, federal officers are implanting tags into the biggest, best, showiest, and most tempting plants. If they see cacti in a truck, they can check whether they are stolen.

Saddest of all, living saguaros are used as gun targets. On a website devoted to Arizona cacti, a correspondent in Carefree, Arizona, posted the following story about a saguaro living beside a water tank:

> *Wide-tire prints indicate a great sportsman drove a macho vehicle to within a few steps of the scene, then proceeded to fire 63 twelve-gauge shotgun rounds into the magnificent century-old saguaro. That's two and a half boxes of shotgun shells! Distinct piles of spent shotgun shells showed that this fiend had fired repeatedly from three different positions in order to sever and finally fell the beautiful cactus plant. Such acts make me ashamed of (some of) the human race.[13]*

In 1854, J. R. Bartlett reported that it was not uncommon to see saguaros in the vicinity of what is now Yuma, Arizona, with arrows projecting from them. *Plus ça change . . .*

At least once, however, the saguaro exacted its own revenge. In 1982, David Grundman, a 27-year-old Phoenix man, fired a shotgun at least twice at a 26-foot-tall cactus. The shots broke the base of the cactus, which then toppled and crushed him to death.[14]

For such a huge plant, the saguaro is actually quite fragile and its numbers are in decline. For 60 years there was no reseeding: between 1900 and 1960, rainfall dropped to less than the normal 7 inches in July and August. Cattle browsed on the palo verde and mesquite which provide shade for the saguaro seeds. Overgrazing also attracted ant colonies; the ants carried saguaro seeds underground: alas, too deep for germination.

Saguaros are also susceptible to lightning strikes, and this area gets plenty of lightning: 15,000 strikes were recorded in one day in the Tucson area.[15] Standing up as they do like giant telephone poles, saguaros act like lightning rods. Wind is another problem. Since the anchoring roots may be quite short and shallow, the top-heavy saguaros sway back and forth and sometimes blow right over. Prolonged frost (more than 24 hours) is death to the saguaro. Air pollution may also now be affecting saguaros as cities expand into the desert. Scientists believe that pollution damages the cactus's protective outer coating, debilitating the whole plant.

The plants and animals that help the saguaro reproduce are also in danger. The most important nurse tree for saguaro is mesquite, but this plant was whacked down all over the place during the era of wood-burning stoves and is now harvested for barbecue charcoal. Bats, which pollinate the plants, are often exterminated by humans. Perhaps other pollinators like cactus wren, white-winged doves, even honeybees, may take the place of the bats.

There has been a relatively long history of conservation efforts to make up for some of the massive losses. In 1933, President Herbert Hoover decided to preserve an area of the Sonoran Desert in Arizona—now known as Saguaro National Park. Before that time, almost nothing was known about the saguaro. Since then scientists have been studying it intensely.

In 1939, after an unprecedented cold snap in the Tucson area, hundreds of cacti in the park died from a strange disease. They toppled over and upon examination, were found to be pocked with brown rot. Park rangers tried to halt the epidemic by bulldozing and burning 320 acres of cactus forest. This tactic failed to contain the problem and simply destroyed hundreds more saguaros. What was affecting them, however, was not a disease but a shift in mean temperature and too many days of frost.

By 1991, the stand that had been huge in Hoover's time was reduced by half and it was feared that the rest would be gone by the year 2000. The saguaros are still there, although much reduced in numbers. In the mid-1960s, scientists from the University of Arizona set up 20 saguaro-monitoring plots containing 213 mature plants in the Rankin plot east of Tucson; 30 years later, 42 remained. But they found 119 new plants that had seeded since 1960.

Artist Bruce Law of Phoenix, who uses the saguaro skeleton in his wood sculptures, has found a way to propagate the saguaro from its arms. In collaboration with the Maricopa County Parks and Recreation Department, he has established a nursery that rescues the arms of otherwise fire-damaged cacti. Other Arizonans have created businesses based on saguaro fruit. Cathy Lambert runs a company called Desert Decadence, which makes and sells Saguaro Blossom Cactus Tea. The tea is made with the dried pulp of the saguaro fruit, blended with rose hips, rose leaves, and strawberries. Although large-scale harvesting of saguaro fruit is forbidden in Arizona, it is permitted to gather a few fruits from cacti growing on public land. These may be used in recipes such as saguaro salsa or saguaro cream.[16]

Cactaceae

Organ-Pipe Cactus
Stenocereus thurberi

Stenocereus thurberi
grows 8 to 20 feet high in great clusters on south-facing slopes. White flowers with white inner petals and reddish outer petals. Globe-shaped, very sweet, edible fruit with red pulp and black seeds.

Range: **mostly in Organ Pipe Cactus National Monument on the Mexico-Arizona border.**

The organ-pipe cactus makes an unforgettable profile in the desert because it *is* shaped like an extremely large set of organ pipes. The wood skeleton has no trunk and grows in huge, distinctive clusters up to 8 feet in diameter. The branches can be as thick as 8 inches in diameter with 12 to 19 ribs. The spines are almost black, but get grayer as they age, eventually developing

Stenocereus thurberi, organ-pipe cactus

black tips. The flowers begin in June and continue through the summer, blooming at sunset and closing by morning. The nectaries attract creatures of the night such as *Leptonycteris*, a migratory long-nosed bat, and hummingbirds, which feed it to their young. The sticky sweet red pulp holds jet black seeds.

In Mexico the very sweet fruits are cooked with prickly pear fruit to make a candy called *pitahaya dulce*. This candy is not exported. Mexicans also grow the cacti as living fences around their properties.

The name honors George Thurber (1821–1890), botanist for the Mexican boundary survey in the 1850s. Most grow in Organ Pipe Cactus National Monument, set aside by presidential proclamation in 1937.

Cactaceae

Senita
Pachycereus schottii, whisker cactus, old-man cactus, bearded cactus

Senita can be a long-lived cactus judging by one well-identified specimen in Baja, California. It was in photographs dated about 1905.[17] Other photographs more than 100 years old of the Grand Canyon show senita which are still there today.[18] Senita's most recent claim to fame is that it has been found to have a mutualistic relationship (one in which two different species both benefit) with the senita moth, *Upiga virescent*, which pollinates the flowers and uses the developing fruit as food for its larvae.[19]

Pachycereus schottii has a gray-green color and grows up to a stately 20 feet tall. It looks a bit like an organ-pipe cactus, although it has only five to seven ribs on each stem. Juvenile stems have clusters of 1-inch bristly spines. In contrast, the tips of mature stems are covered with long, hairlike, downward-facing gray spines, hence the names "whisker," "bearded," and "old-man cactus." Its greatest attraction is the pink flower, which opens at night; its oval edible fruits are just over 1 inch long with red pulp.

Senita is a relative of the saguaro and organ-pipe cactus and grows in a small area south of Organ-Pipe Cactus National Monument in Arizona. It is distinguished by two features. First, it produces up to five pink flowers from each areole, whereas other cacti generally produce only one. Second, the flowering stem produces bristles from the areoles for several years. The Seri people believe that the senita had great powers handed down by Icor, the Spirit of Vegetation. Senita was consulted for aid in placing curses resulting in an enemy's sickness and death.[20]

According to Park Nobel in *Remarkable Agaves and Cacti*, the plant also contains natural steroids which "act as hormones for humans and animals."[21]

Range: **Sonoran Desert, Baja California, Organ Pipe Cactus National Monument, and Tohono O'odham Reservation.**

Cactaceae

Hedgehog Cactus
Echinocereus spp., including *E. engelmannii, E. triglochidiatus,* and *E. coccineus*

There are so many species in this cold-hardy genus that hedgehog cacti have their own online magazines (one in English, another in German) and several fan clubs. They are hugely popular with collectors who seek out almost any variant and plant it with loving care. To the untrained eye, these cacti may all look alike, but don't tell that to a cactophile. Each plant has its own personality, shape, and form.

There are about 45 *Echinocereus* species; most of them grow in the Sonoran Desert. *Echinus* means "spiny" and they have similar common names (hedgehog this, hedgehog that) so it's very confusing for the nonexpert. The most common is *E. engelmannii*, **strawberry hedgehog**, but many other species are also called strawberry hedgehog. They all have certain things in common: a low columnar, ribbed trunk with clusters which form a clump of 5 to 15 stems. All are completely covered with whitish, pink, or brown spines that are straight or only slightly curved but never hooked (that sets them off from many other species). They grow in rocky or gravelly soil.

To bloom, this plant must have a cold period (but not prolonged cold). In fact, in cool winters, it goes dormant. After a brief cold snap, glorious blooms sprout from the tips of each arm. The flower buds form inside the stems and the tips come bursting through the skin rather than the areoles, as is usual on most other cacti. They leave behind a small scar. The funnel-shaped flowers (all the better to sip from) may be red, orange, yellow, pink, purple, or white. Green stamens and yellow anthers seem stuffed into the interior.

When the temperature gets above 85°F, the hedgehog loses water very slowly, at a fraction of the rate of a nonadapted species. Consequently it is found all over the arid Southwest. It was used by Native Americans as a staple—if they could get to it before the animals did. Rodents, desert tortoises, and birds all fed on the fruits.

The striking flowers may attract insects for pollination, but they also bring in the cactus thieves. It's too easy to rip this one out of the desert and smuggle it to a waiting buyer because of its shallow roots. People can collect cacti legally by getting a permit (as is required in Arizona, for instance) that allows them to collect on private or public land. But this doesn't stop the rustling. John Alcock, the poetic writer of the Sonoran Desert, says: "The damage done by cactus rustlers cannot be repaired; they reduce the diversity of the desert and rob it of its color for a lifetime, leaving the desert impoverished for all."[22]

Range: **central Mexico to as far north as the Dakotas.**

Cactaceae

Night-Blooming Cereus
Peniocereus greggii, queen of the night, *reina de la noche*

Peniocereus greggii

has thin spiny stems, up to 4 feet tall, and grows from a tuber. The nocturnal white flowers are 2 to 3 inches in diameter, have long floral tubes, and flower in May and early June. The orange-red edible fruit is about 3 inches long, has red pulp and black seeds.

Range: **southern California, southern Arizona, west to Texas, and south into Baja California and central Mexico.**

This eccentric plant puts most of its bulk below ground in a huge tuber like a giant turnip. The tubers usually weigh 5 to 15 pounds, but some specimens have been found weighing 80 pounds. What emerges from the tuber looks exactly like a bunch of dead sticks. These gray stems are the branches with small spines along the ribs. Night-blooming cereus clusters near ironwood, creosote, and palo verde bushes. It sprouts new stems from the tuber if they are nibbled at by pack rats and cactus borers.

At night, the hauntingly beautiful, fragrant white flower opens up its many petals. It is said that Native Americans would go sniffing through the desert at night to find this almost mystical plant. The flower dies after its single night of glory, but there are three to five flushes of flowers in May and June. It is pollinated by hawk moths. When the bright red fruit develops, humans seldom get a chance to sample it, since the birds usually get there first.

The roots have been used by Native Americans as a remedy for diabetes. They contain a substance called B-si-tosterol, which may reduce cholesterol levels. The roots also have the properties of a heart stimulant, and were used to treat syphilis and gonorrhea. Modern herbal companies still offer products from the night-blooming cereus, but their numbers are dwindling because of overharvesting.

According to the experts at the Sonoran Desert Museum, if pesticides are used in quantity near any natural areas, the hawk moth populations crash and the flowers of this plant will not fruit. This is, they say, "an example of chemical habitat fragmentation; the habitat appears to be intact, but some of its ecological processes have been destroyed or degraded."[23]

Cactaceae

Fishhook-Barrel Cactus

Ferocactus wislizenii, candy-barrel cactus, Southwestern barrel cactus, Arizona barrel cactus, biznaga, bisnaga, viznaga, visnaga

Ferocactus wislizenii

has cup-shaped flowers about 2 inches in diameter that last for only four days during late summer and early autumn. It usually grows 2 to 4 feet tall, but the occasional one soars to 10 feet high. The flowers come later than other barrel cacti (August and September). The yellow, pineapple-shaped fruit hangs on the plant for up to a year or until some animal comes along and eats it.

Range: southwest United States to northwest Mexico.

Ferocactus means "fierce cactus." Indeed, the tough spines look lethal. The scientific name honors Frederick Adolphus Wislizenus, a European naturalist who traveled through the desert in the mid–nineteenth century collecting plants. It was named by George Engelmann (1809–1884) of the Missouri Botanic Garden, in 1848. He was a German-born physician and botanist who spent several years tootling about the area happily identifying and naming plants. Three plant genera and dozens of species bear his name.

The common name is bang on: it does have the stumpy look of a barrel. The red, orange, or yellow flowers grow right on the top of the plant, giving it a rakish appearance. But backlit by sunlight, the flowers look like silk. The evil-looking spines are red with a surface layer of gray and some have an obvious fishhook that looks as if it would tear leather. This cactus doesn't stretch out until it's at least a foot in diameter and each one's shape gives it individuality. Many desert dwellers who had access to any form of water used the hooked central spines to hold bait for fishing.

A barrel cactus can survive for six years without water, or even soil. Daniel T. MacDougal, a plant physiologist who came to the desert in 1906, made this discovery when he "excavated a barrel cactus, stuck it in an out-of-the-way corner of his laboratory, and weighed it periodically to see how much water it had lost by transpiration. When he ended the experiment after six years, the weight had been reduced by a third, from eighty-one to fifty-three pounds."[24]

Ferocactus wislizenii, fishhook-barrel cactus

Cactaceae

The barrel cactus may live 50 to 130 years along desert washes, gravelly slopes, and beneath desert canyon walls of all hot deserts.

The peoples of the desert sometimes squeezed water out of these plants and drank it, but only in dire emergencies. The Seri could survive on the flat-tasting juice for up to a month, but it probably caused diarrhea. The Tohono O'odham ate the fruit as emergency food all year. The Seri cooked the flowers for food, made gruel from ground seeds, and used hollowed-out plants as honey containers. The spines have been used as fishhooks and even early phonograph needles. Today, some people make cactus candy by boiling the succulent tissue and adding sugar and flavoring.[25]

Compass-Barrel Cactus
Ferocactus cylindraceus, California barrel cactus

Ferocactus cylindraceus

grows to 10 feet in ideal conditions. It is usually single-stemmed and has red spines. The central spines are flat with curved tips; radial spines emerge from the areoles and completely obscure the body. Flowers at the stem tips are usually yellow but sometimes orange to red in March and April.

Range: Sonoran and Mojave Deserts.

The Seri name for this cactus is "thinks it's a saguaro," and it does look like a saguaro when it is young. It's also called the compass cactus because the apex tilts toward the southwest, into the sun. The afternoon sun slows growth on the southwest side of the plant and the cells on the shaded side expand, making the plant lean over toward the sun.

Closely spaced yellow or red spines spiral up the plant and this heavy spiny covering provides shade and lowers the temperature on the surface. The wood ribs make up a skeleton, which appears as a series of ridges on the outside of the cactus. These ribs expand easily when the spongy tissue between them stores up water. Ants crawl over the surface, even when the flowers are not blooming, and provide some kind of symbiotic protection for the cactus.

The spongy pulp is unpalatable but, in a pinch, can be used as a source of water. The desert peoples sliced the top off the cactus, scooped out the pulp, and squeezed it to get the water. Bighorn sheep also eat the tissue during seasonal droughts. Neil Luebke issues this warning: "Some contain water that is drinkable while others are reported to contain juice that is slimy and nauseating, and best drunk only as a last resort."[26] Don't try this yourself; in fact, don't ever rely on any cactus as a source of water. You're better off with a large water bottle when walking in the desert.

THE CHOLLAS

There are 20 species of cholla (pronounced "choya") in the *Cylindropuntia* genus, which is in the Opuntioideae subfamily of the cactus family. It is a subtle genus in which new species are being discovered all the time. All members of this genus have a papery sheath covering the spines, distinguishing them from other cacti. Like most cacti, they have tubercles—little warty growths sticking out of the stems from which the spines spring. They also have glochids: tiny, sharp, brittle, barbed spines, easily detached. Rudimentary leaves grow on the new joints. The seeds have a covering—called an aril—and are pale green, whereas most other cacti have black seeds. They usually have orange or acid yellow flowers. Chollas, which have cylindrical stem segments, grow 1 to 12 feet high.

Cactaceae

Jumping Cholla
Cylindropuntia fulgida, chain-fruit cholla

Jumping cholla no doubt got one of its common names because the joints come off so easily they seem to leap out and cling to the passerby. Ann Zwinger writes:

My hand doesn't even brush one, not even close enough to feel the spine penetrate, and as my hand swings forward with my stride, I am suddenly the unhappy possessor of a joint of cactus, the size of a small baking potato and as heavy, implanted on the back of my hand with, as William Gabb, traveling with Browne, remarked, "a pertinacity worthy of a better cause."[27]

Cylindropuntia fulgida, jumping cholla, has irregular jointed branches and is the largest of the cholla species, growing to 12 feet tall and up to 6 feet wide at its most gargantuan, though most of them are smaller than that. Delicate inconspicuous white or pink blossoms project from the dangling fruits of previous years, adding to the colorful quality of this cactus. The fruits are retained by the plant, with new fruit being added to the fruit from the previous season. This gives the plant its alternative name "chain-fruit cholla." One year's fruits become the next year's reddish blossoms, which in turn give rise to a new link in the chain. The spineless, green, pear-shaped fruits are 1½ inches long. Eventually portions of chain fruit fall and give rise vegetatively to descendants, though joints that break off can form new plants as well.

In a drought, deer and bighorn sheep rely on this plant to provide moisture in their diets. Now, with widespread ranching in some desert areas, cattle have developed a taste for the plant as well.

Range: **Sonoran Desert in both Arizona and Mexico.**

Cactaceae

Teddy-Bear Cholla
Cylindropuntia bigelovii

Cylindropuntia bigelovii

grows to 9 feet high. A covering of silvery spines hides the surface of this cactus, protecting it from sun and animals. The inconspicuous flowers are green or yellow, sometimes streaked with lavender, 1½ inches wide. They appear near the ends of joints in March or April. The stems are cylindrical with 1-inch-long, yellow, egg-shaped knobby fruit.

Range: **southeastern California to western Arizona, south to northwestern Mexico.**

With the sunshine backlighting it, the teddy-bear cholla looks charming enough to embrace. There is something endearing about this furry-looking plant which is probably why it acquired such an affectionate common name. John Alcock waxes eloquent about the plant:

The teddy-bear cholla seem aflame as May sunlight scatters through the thousands of long creamy white spines that cover the cactus. From a thin black trunk the limbs radiate out in top-heavy profusion, rather like a Kachina doll with an elaborately spiny headdress. The limbs are segmented, prone to fracture at the constrictions; spiny grenades lie everywhere on the ground beneath the plants.[28]

Bits and pieces snap off, putting it in the category of those chollas with a reputation for "jumping" which is how the plant reproduces itself—hitching a ride with any passing stranger, human or animal. Wherever it is eventually plunked down, it roots. This kind of asexual reproduction produces clones; identical offspring sprout up around each teddy-bear cactus. It is not a rapid process, however, since it requires years of growth to produce the tissues in each joint. Excreted wax conserves moisture in the limb after it falls to the ground, where it takes root and starts again. A good thing, too, since few insects visit the insipid little flowers.[29]

The spines that protect the plant from grazing animals also protect those creatures with enough audacity to nest in the nearly impenetrable spiky mass. Cactus wrens, curve-billed thrashers, and mourning doves flit in and out of the plant to their nests without suffering

FOLLOWING PAGES: *Cylindropuntia bigelovii,* teddy-bear cholla

any harm. Wood rats (pack rats) nibble at the cholla and use the joints to cover their dens. They have developed the ability to snip off a patch of spines without getting impaled, then eat the edible portions and use the remaining spines in their dens for protection. Wood rats never need to drink water because eating cholla provides all the moisture they require. Alas, the babies may not be so adept. Tiny wood rats sometimes become impaled on the very spines their mothers have drawn into the den for safety. Likewise, baby birds whose parents have built nests in the cholla sometimes get entangled. Life in the desert is full of danger.[30]

Wood rats are particularly defenseless against coyotes, which is one of the reasons they make their nests from piled-up cholla joints. But the spines don't always protect them from rattlesnakes, which find wood rat a tasty feast; once they've consumed their prey, they take over the den. Other animals that munch away on this cholla are jackrabbits and javelina (desert pigs), both of which can digest the oxalic acid produced by the plant, at least in small quantities.

The Havasupai recognized the spines of these cacti as being unusually difficult to remove because of the thin sheath that covers the spine and remains in the flesh after the spine itself has been removed. Getting caught by this teddy bear is a disagreeable experience because even trying to shake them off just gets the spines more deeply embedded. One early traveler in the desert wrote, "The plant is the horror of man and beast. Our mules are as fearful of it as ourselves."[31] Leave them alone, and if you do get spiked, cut the spine out with scissors or pull it out with pliers.

The Havasupai called it *Kwatha'vuwaka* and used it in pottery making. Descriptions of the cactus they used indicate that it was either a short, stocky cholla or some form of an elongated barrel cactus. One anthropologist said it was about 1 foot tall, as big as her forearm, and had red flowers. Others said it was small, like a barrel cactus. The Havasupai burned off the spines and peeled away the "skin." The sticky interior was then rubbed over the molded, but unfired, pot to provide a protective glaze. Some people, however, broke the cactus open, cut the inside into small pieces, put these into water, and worked them with the fingers to create a liquid that was rubbed all over the pot.

The cactus has even found a thoroughly modern use: according to a 1977 edition of *National Geographic*, the computer company Honeywell used to use the spines to remove tiny specks of solder from delicate computer parts.[32]

The common name of **buckhorn cholla**, *C. acanthocarpa*, refers to the fact that the shape of the plant is a little like deer antlers. Buckhorn cholla is very difficult to differentiate from **staghorn cholla**, *C. versicolor*, except when they are in bloom. The fruit of the buckhorn cholla is covered with barbed spines and hangs on the plant for several months. Staghorn cholla is usually spineless and grows with teddy-bear cholla. The Tohono O'odham and other Native Americans love the cholla buds on buckhorn, staghorn, and pencil chollas.

Golden or **silver cholla**, *C. echinocarpa*, is a low shrub with lots of spiny branches growing 2 to 4 feet tall. The spines occur in clusters of 3 to 10, each one up to 1 inch long, but not barbed. The spines are yellow when young, grayish in age (this may be the source of the two different common names), and fall off eventually in extreme old age. The flowers, which appear in May, have yellow-green petals with red edges, up to 2½ inches in diameter. The fruit is spiny and green, turning brown over time, and edible (once the spines have been knocked off). Range: Arizona, Utah, Nevada, California.

Tree cholla, *C. imbricata*, also called **walking-stick cholla, cardensia,** or **coyonostle,** is a spiny, leafless tree that grows 3 to 7 feet high. Reddish-purple flowers bloom near the ends of the branches between May and July, followed by egg-shaped yellow fruit 1 to 2 inches long. Range: southern Colorado, Kansas, Arizona, New Mexico, Texas, and northern Mexico.

Christmas cholla, *C. leptocaulis*, is also called **desert Christmas cactus, holy cross cactus, tsejo, tasajillo, pencil-joint cholla, diamond cactus, darning-needle cactus,** and **turkey pear**. It is the skinniest of the chollas and has long smooth joints with yellow to bronze flowers. The bright red fruits, about the size of a grape, appear during the Christmas season and stay most of the winter. But this is not the familiar houseplant called **Christmas cactus** (*Schlumbergera*), which is an altogether different member of the cactus family. It usually grows to a height of 3 feet (among desert trees such as creosote it can reach up to 6 feet). Hummingbirds, honeybees, and one of the cactus bees come to it, but so far no one knows why the flowers open for only three hours a day. According to Daniel Moerman, the edible fruit has a "narcotic effect,"[33] and according to Park Nobel, it can be "slightly hallucinogenic."[34]

Cactaceae

Prickly-Pear Cactus

Opuntia spp., including *O. phaeacantha, O. polyacantha, O. englemannii, O. macrocentra, O. humifusa*

Opuntia spp.

are sprawling plants up to 5 feet high and as much as 6 feet in diameter. Grows near mesquite and palo verde; horses and cattle will browse on it when the plant is young. No trunk, but distinctive oval-shaped joints branching in all directions. It has yellow flowers; pale green pads with many fine bristles; fruit of most types edible, very prickly; found in dry sandy soil.

Range: **southern Saskatchewan to the southwestern and south-central United States.**

There are dozens of types of prickly pear in the Southwest, characterized by the way their jointed structure is expressed in paddle-shaped branches, usually covered with spines (only a few species are spineless). Sorting out the species is a tough job because of the presence of hybrids, nonnative imports, and general disagreement among botanists on which scientific name corresponds to which common name.

Some ethnobotanists, such as Park Nobel, speculate that Christopher Columbus may have returned to Spain with *Opuntia indica*.[35] Certainly those who came after him were attracted to the plant and carried specimens back to Europe. One look at the *jardins exotiques* of the Côte d'Azur and you can see that this was a collector's prize.

Prickly pears can survive in very poor soils because of their efficient root systems that rapidly soak up rainwater during storms and store it. The cup-shaped flowers are saffron yellow, violet, or red, filled with golden stamens that respond to any bee hovering around. They produce a rich nectar and attract many insects. Edward Abbey writes: "I've done my best to annoy them, poking and prodding with a stem of grass, but a bee in a cactus bloom will not be provoked; it stays until the flower wilts. Until closing time."[36]

Since the plant isn't produced from seed, it doesn't matter whether bees fertilize the flowers or whether the seed is broadcast. *Opuntia* species grow in a zigzag pattern, developing a new plant from an older one, year after year. Flowers may be produced from the areole of the fruit from the year before. In some plants the chains have been found to have more than

Opuntia spp., prickly-pear cactus

20 years' worth of growth hanging down from the older joints.

The prickly pear produces a sweet fruit (known in Spanish as *tuna*) that ripens in August and September. Native Americans removed the fruit with tongs made of coyote willow, *Salix exigua*. To remove fine spines, the fruits were rolled on the ground and brushed with a handful of Mormon tea or any other weed that was handy. Finally, they were rubbed with a rag and placed in a basket to return to camp. Prickly pears were eaten fresh or split in half, the seeds removed, and dried in the sun. The dried fruit might be pounded into a cake for storage or eaten on its own. If food was scarce, the stems were roasted, too. Prickly-pear fruit was one of the few sweet foods eaten before sugar was introduced. Unlike mescal, this fruit was not used in season.

During his venture into what is now the United States, a soldier on the expedition of Spanish explorer Francisco de Coronado (1510–1554) noted in 1540: "In a province named Vacapan there were large quantities of prickly pears, of which the natives made large amounts of preserves. They brought much of this preserve as a present, and when the people of the army ate of it they all became drowsy with headaches and fever, so that the Indians could have done great harm to them if they had wished. This illness lasted intermittently for twenty-four hours."[37]

The cladodes, or young stem segments (also called joints, pads, or nopales), are eaten as a green vegetable. Sliced or diced they are called *nopalitos* and they taste rather like green beans. Mexicans and Mexican Americans eat them and also feed them to dairy cows. They impart a special flavor to milk and butter, which many Mexicans prize.

Luther Burbank (1849–1926), a notable plant breeder who developed more than 800 strains of new plants (it was his potato that saved the Irish potato industry after blight wiped out their own), had high hopes for *Opuntia*. Between 1901 and 1915, he promoted the fruits of prickly pears as human food and the cladodes as cattle fodder. He overflowed with enthusiasm for the prickly pear. In the July 19, 1911, *Los Angeles Examiner*, he wrote that the cactus "promises to be of as great or even greater value to the human race than the discovery of steam."[38]

Although this turned out to be an overstatement, there is still a good market for prickly-pear fruit and products made from it. The fruits are high in sugar, most of which is fructose, rather than glucose or sucrose, and therefore better tolerated by people with diabetes mellitus. They are also high in vitamin C. In 1992, 10,000 tons were imported to the United States from Mexico.

The explorer Peter Fidler (1769–1822), who kept a journal in 1792–1793, mentions finding *Opuntia* on the banks of the Red Deer River in what is now Alberta:

The men are also busily employed making arrows—of the Sasvutten wood, which is very hard . . . there is great plenty of it here along the river under the Bank. I found on this side near the top of the Bank—a particular kind of Grass—very full of sharp Prickles of 2 Inches long & the thickness of a Pin—which grew upon a wrinkled round flatt knob of 1½ Inch diameter, the outer Skin has much the appearance of a Cucumber. . . . They are very bad to walk amongst—running immediately thro the shoes usually worn here of dressd leather—the Indians say that far to the Southward, about & beyond the Mississury river, the ground is almost covered with them for a great distance & so very large— & form large bushes that at a Distance they have often been taken for Buffalo laid down—when on their war excursions—in these parts—the only method they have to fortify their feet against these formidable & very bad things, they make shoes of the raw hide of the Buffalo—which the prickles are not strong enough to penetrate thro![39]

Melvin R. Gilmore records the "cactus game" played by small boys of the Dakota Nation.

They gathered on the prairie and the swiftest boy was chosen as "it." He stabbed a small cactus with a stick which he held up for the others to shoot with bows and arrows. Once the target was hit, the holder of the cactus took off after the shooter and tried to hit him with the cactus. Then he returned to the goal while other boys tried to hit him. It went on indefinitely and you wonder how many of the children got hit or hurt. This certainly would have sharpened their running and shooting skills.[40] The Havasupai also used the spines for tattooing by pricking the design into the skin with cactus spines moistened and dipped into ground charcoal. It took a few weeks to heal. Boys would let girls tattoo them.

The mucilaginous juice of the stems was used to fix color painted on hides by rubbing freshly peeled stem over the painted surface. Mexicans also mix the sticky sap with whitewash and mortar to make adobe buildings more durable. This ancient technique of strengthening adobe is currently being used in the restoration of the San Xavier Mission just outside Tucson, Arizona. Prickly-pear sap might also prove to be valuable in stabilizing dirt footpaths and erosion-prone slopes.[41]

The juice makes a good wound dressing. The pads contain a sap with healing properties similar to aloe vera. The pads may also be ground up to produce a laxative. According to Luther Burbank, the sap was also an effective mosquito repellent.[42]

The cochineal insect (*Dactylopius*) feeds on the pads of the prickly pear. When crushed, the female insect releases a substance called carminic acid, which produces a rich red dye. In the sixteenth century, the export of cochineal from Mexico was second only in importance to the export of silver. Cochineal was used in cosmetics and also to dye cloth. The Aztec and Maya used the dye extensively in their ritual robes. When the Spaniards realized the commercial potential of the dye, they monopolized its production from the sixteenth to the nineteenth centuries. Robes worn by European royalty, the red coats of the British army, and the tunics of the Royal Canadian Mounted Police were once dyed with cochineal. Cochineal was eventually replaced by cheaper synthetic dyes, but once it was found out that the synthetic dyes contained potential carcinogenic properties, some dyers once more turned to cochineal for its scarlet color.

Prickly pears have spread to many other areas of the world. In 1992, 150,000 acres of land were cultivated worldwide in prickly pears. You see them all around the Mediterranean, especially in railway cuttings. The Moors took prickly-pear cacti with them to North Africa, when they were forced out of Spain in 1610. Others took them to Israel. They have become a nuisance species in Australia, where they spread rapidly, invading the grasslands, and now huge efforts are needed to control their spread. The Australians are not the only ones who rue the introduction of this plant. *O. stricta* was collected and taken to South Africa, where it almost instantly became invasive. It is also found in Yemen, Eritrea, Ethiopia, and Somalia, where it is causing so many problems that drastic measures have had to be taken to get rid of it.

Even in North America, ranchers considered the prickly pear a weed to be exterminated. The spines can be harmful to livestock, causing ulceration and bacterial infections in the mouths of animals grazing on them—a syndrome known as "pearmouth." The indigestible seeds of the prickly pear can also cause intestinal problems for sheep and goats. Janice Bowers, in *The Mountains Next Door*, writes of the interaction between ranching and prickly pears:

The abundant prickly pears on the gentle slopes [of Saguaro National Monument], so thick you cannot walk a straight line in many places, are a legacy of cattle ranching. Cows wouldn't eat them, so the cacti

multiplied beyond any reasonable or attractive level. They became weeds, out of place and out of control. . . . Now, fifteen years after cattle were kicked off the monument for good, the vegetation shows little signs of recovery. To the uneducated eye, the landscape looks ordinary enough, but to someone who knows, it's a war zone after the peace treaty has been signed.[43]

Unfortunately, it is very difficult to get rid of prickly pears. Over the years, the plant has evolved a variety of ways to survive in dry, harsh climates and ensure reproduction, and even the ranchers' most determined efforts to exterminate them may be futile. The seeds can remain viable in the ground for years and may pop back up even after all the standing prickly pears have been removed. A year or so later, the conditions may be just right for germination, and a whole new crop will sprout. Prickly pears also spread by livestock, because the seeds pass through the digestive systems of animals quite unharmed, and may fall on fallow ground wherever the animals excrete them.

The reason for the name **beavertail cactus**, *O. basilaris*, is obvious at first glance: the pads are shaped like beavers' tails. The branches are wrinkly when dried and about ¼ to ⅜ inch thick after a rain shower. Though the beavertail is only 12 inches tall, it will grow in clumps 6 feet in diameter. New pads grow from the old, but it doesn't get to be more than two pads tall. The spineless pads have hairy glochids that can cause a lot of pain or intense itching from the sneaky little barbs. Large cherry-red flowers are 2 inches in diameter with purple filaments. Dun-colored spineless fruit contains seeds which get very large. There are various species of beavertail cactus—one, *O. basilaris* var. *ramosa*, forms a treelike plant, but only 20 inches tall.

Beavertail cacti grow on desert slopes or flats in southeastern California, western Arizona, southern Utah, and northern Mexico. The flowers open between February and June, depending on the elevation—the higher up they grow, the later they appear.

The beavertail is a popular ornamental plant because it takes root quickly. Though it doesn't have the lethal spines of other *Opuntia* species, the hairy little barbs can still be extremely painful. The *Sonoran Desert Natural History* gives the following warning:

> *Removing hundreds or thousands of them after falling into such a plant is an exhausting and tormenting task. Some people shave them off at skin level, which somewhat reduces the irritation, even though this leaves the tips beneath the skin. A better remedy is to gently draw very sticky tape across the afflicted skin. Another effective treatment is to cover the area with a layer of white glue, then peel it off after it dries.[44]*

Better yet, try not to fall into a cactus of any sort.

Cactaceae

Peyote
Lophophora williamsii, mescal buttons

Lophophora williamsii

is a small, spineless cactus with fat gray stems growing singly or in clumps 1 to 3 inches high. Pink, many-petaled flowers appear on top throughout the summer.

Range: southern Texas and northern Mexico.

This low, gray, spineless cactus looks rather like a large mushroom, and certainly has some of the properties of "magic mushrooms." Chewing dried peyote buttons, which contain mescaline, produces hallucinations. This practice was once part of the religious rituals of certain Native American peoples.

Melvin R. Gilmore, in his research on the uses of plants by native peoples, discovered that the cult of peyote (misnamed mescal, which is a liquor derived from certain species of agave) was brought north to the Omaha in the winter of 1906–1907 by a member returning from a visit to the Oto in Oklahoma. The Winnebago had already been introduced to it and eventually half the tribe came to use peyote, although people were much against it in the beginning. "The peyote plant and its cult appeal strongly to the Indians' sense of the mysterious and occult. The religious exercises connected with it are attended by many ceremonies and much symbolism. The average Indian, with his psychic inheritance and his physical and psychic environment, naturally attributes to the peyote most wonderful mystic powers."[45]

Gilmore (1868–1940) was a Nebraskan botanist who spent most of his life studying the Native American use of plants. He felt the Europeans who first came to North America were indifferent to the Indians' use of local plants. The newcomers wanted the comfort of their own, familiar plants. In his observations about the Native American use of peyote, he figured if Europeans could accept a deity in a human body, why not one in a plant body? He connected the tradition of eating peyote to the bread and wine in Christian ceremonies. He described supplicants sitting around a fire, the leader holding a staff decorated with feathers; the chanting, the rhythm of the water drum, the people gazing into the fire or sitting with bowed heads. He pointed out that it's mainly the hypnotic effect of the ceremony which brings on the visions, though peyote can affect the optic nerve. People described wonderful visions of spirits.

The Havasupai used "mescal mats" (roasted peyote buttons), which were convenient for transport and trade. In early spring, groups of families gathered peyote. The collection of the buttons was primarily the task of the women. The men gathered firewood and prepared a roasting pit in sandy soil (never in gravel, which allowed too much steam to escape). The pit was usually 4 to 6 feet in diameter and 4 feet deep, filled with brush placed on top of a thick layer of stones. The fire had to be ignited by someone born in summer. If a person born in winter lit the fire, it was believed the peyote would not light properly. After a couple of hours the men took poles and tamped down the hot rocks; any gaps were filled with gravel. Each family was allotted a pie-shaped section. The roasting pit was covered with more grass or a gunnysack and topped with dirt. After 48 hours of cooking, the peyote buttons were piled up, cut into thin slices, and dried in the sun.[46]

The Havasupai also used peyote fibers as a hairbrush. They took great care over the styling of their long hair, which was drawn back and tied at the nape of the neck and kept in place with grease from bighorn sheep.

THE LEGUMES

The Sonoran Desert is home to 53 species of legumes, out of more than 16,000 species worldwide. The plants of the legume family are excellent arid-country plants. They have nitrogen-fixing bacteria in the nodules of their roots. When the legumes decompose, the nitrogen is released to enrich the soil. Wildlife also benefit from their great big seeds (plenty of protein there). Legumes are among the nurse plants for other desert species that grow near them.

Fabaceae

Yellow Palo Verde

Cercidium microphyllum syn. *Parkinsonia microphylla,* foothill palo verde, little-leaf palo verde

Cercidium microphyllum syn. *Parkinsonia microphylla*

is a many-branched large shrub or small tree opening out toward the top. It has a multi-stemmed photosynthetic trunk; the tiny short-lived leaves are blue-green. It grows to 15 to 30 feet (the latter height is quite rare) in deep soil; the yellowish-green branches are upright without spines; the pods contain two to four bean-size pods in a thin shell. The flowers bloom intensely for two weeks in a pale yellow.

Range: throughout central and southwestern Arizona, southeastern California, and in Baja California.

State tree of Arizona.

The palo verde is a nurse plant par excellence—particularly for the saguaro cactus. Without the shade of this elegant tree, the saguaro seeds wouldn't stand a chance of germinating in the killer desert heat. The plant also hosts birds such as black-tailed gnatcatchers and Gambel's quail that roost in it. All kinds of animals, from bighorn sheep and mule deer to jackrabbits and wood rats, like to graze on the palo verde.

Although the palo verde looks like a tree, it's actually a legume: its roots contain bacteria capable of capturing nitrogen from the atmosphere and fixing it so that the plant can use it. But the most extraordinary thing about the palo verde is the fact that it is leafless for much of the year. Photosynthesis, which normally goes on in the leaves, is carried out by the bark on the trunk and branches. The trunk is short and squat with branches sprouting out and upward about 8 inches from the ground. The green pinnate compound leaves are about 1 inch long. They appear in spring, but as the days get hotter and drier, they turn brown and fall off in preparation for flowering. The fallen leaves collect in crevices in the desert and provide tiny pockets of decayed matter that permit other plants to grow, perhaps even another palo verde.

Palo verde means "green stick," and refers to the photosynthetic trunk. The cream-colored or pale yellow flowers appear in 1-inch-long clusters that eventually turn into long cylindrical pods containing the seeds. Like the saguaro, the palo verde produces a large number of offspring, very few of which ever see maturity, let alone old age. Germination requires a minor miracle. First, the tough seeds have to be crushed. Then, there must be a heavy rainfall to ensure the right amount of moisture to establish the roots. Then the seeds and seedlings must avoid being eaten by jackrabbits and desert rodents such as pocket mice and antelope ground squirrels. And finally, any seeds left by the animals must escape the clutches of bruchid beetles, which are particularly fond of them. The beetles lay their eggs on the seedpods. When a larva hatches, it burrows into the seedpod and munches away as it grows, leaving the palo verde pretty much seedless.

Although very few palo verde seeds survive beyond the seedling state, once it gets going, the tree is a hardy survivor. It grows so slowly that maturity takes about 100 years, and since palo verdes do not have growth rings, it's hard to tell their exact age. In the desert around Tucson in 1911, Forrest Shreve, a desert botanist at the Carnegie Desert Laboratory, found palo verde trees that he estimated were 400 years old.

Nature writer John Alcock says: "Like so many desert plants, palo verdes are advertisements for adaptation. On a mere six inches of

Cercidium microphylla, yellow palo verde

rain a year, they grow slowly but steadily to small tree size. . . . During a prolonged dry spell, a palo verde may even permit some of its limbs to die, dropping the . . . branches to the ground where they will no longer demand scarce nutrients and water of the tree."[47]

Yellow palo verde seeds were ground up for food by the Pima, and the Tohono O'odham ate the green seeds or pods, which taste like green peas. The Seri ate fresh green seeds, and made the mature seeds into a porridge. To find them easily, they scoured pack rats' nests for mature pods. Today, the plant is used in landscaping.

Fabaceae

Blue Palo Verde
Cercidium floridum syn. *Parkinsonia florida*

The deep root system of the blue palo verde makes it less vulnerable to drought than many other plants, because it can reach down to deeply buried sources of groundwater. It is also an important nurse plant for the saguaro cactus. This is a chancy relationship for the palo verde, since the cactus may crowd out and kill off its nurse plant.

The yellow flowers that smother the tree usually appear in spring, but may appear as late as August in some areas.

The heavy, soft, close-grained wood can be used as fuel. The Pima used to carve the wood to make spoons and the Pima and Papago ground the seeds to make a type of porridge. Fast-growing and short-lived, blue palo verde usually lasts about 30 years.

Range: **southwestern United States.**

Prosopis spp., mesquite

Fabaceae

MESQUITE
Prosopis spp.

Prosopis spp.
remain bare all winter, and leafs out in March or April, followed first by the yellow, tassel-shaped flowers and later by the beans, which are full of sugar and protein. Growth may be slow, but in older groves, or bosques, some have a diameter of 5 feet. The sheer weight sends them sprawling. The bosques support entire communities of plant and animal life.

The name *mesquite* comes from the Aztec word for the plant: *mizquitl*. There are 44 species of *Prosopis* worldwide, of which three are native to North America: **velvet mesquite** (*P. velutina*),

honey mesquite (*P. glandulosa*), and **screwbean mesquite** (*P. pubescens*).

This thorny plant was considered the key to the settling of the West in the mid–nineteenth century. It was as useful to mankind as any plant: for food, firewood, and lumber. Probably the most common shrub in the Southwest, the mesquite, like the palo verde, is not a tree but a legume.

The taproots can be larger than the trunk and are capable of marathon journeys in search of water. Raymond Cowles, author of *Desert Journey*, writes:

I saw a large mesquite perched on a stilt-like framework of roots. One root extended almost horizontally about 180 feet away from the tree and then turned sharply downward in search of the water table. In the 180 feet of horizontal growth, the root diminished in size to about a fourth of its original dimension. If the downward turning portion of the root is assumed to have penetrated an additional 100 feet or so, it appears that the total length of a root in pursuit of water may have extended much beyond 250 feet . . . a phenomenal length, but in its way it matches the other accomplishments of this remarkable tree.[48]

Because mesquite can fix nitrogen in the soil, plants that can't survive elsewhere have a future growing alongside them. They are also a host to mistletoe, a notorious parasite that lives off the moisture and nutrients of the host plant. In turn, the mistletoe attracts a bird called the phainopepla (of the silky flycatcher family) which eats the mistletoe berries. Other animals cluster around the mesquite for shelter, moisture, and food. The coyote is particularly fond of mesquite beans, and the flowers attract honeybees. Even the dead branches support communities of wood-boring beetles and termites.

People, too, have long been attracted to mesquite groves. Archeological evidence shows that the Native Americans of the Southwest set up camps in the groves, which provided them with shade, firewood, construction materials, and food. Because it attracted animals, the groves offered a steady supply of meat as well. As Cowles put it, "Where the mesquite exists, it is as completely used as the proverbial pig of the Chicago packing plants, of which all was used but the squeal. But the mesquite tree does the pig one better: even its shade is used."[49]

Hohokam and, later, Pima gathered the beans and pounded them into a meal called *pinole* that could be stored for up to two years. Mesquite sap was used to make glue or dye, and the leaves were made into infusions to treat eye problems, headaches, and stomachaches. The Apache were able to travel long distances because they ate powdered mesquite meal mixed with water, which kept them going for hours (a habit which baffled their white adversaries). The Havasupai also processed and stored the beans by pounding them into a powder. One layer of powder was sprinkled over mesquite seeds as the base layer of a cake. Layers of powder and water were built up until the cake was 2 to 3 inches thick. It was either eaten as is or soaked in water. The inedible seeds forming the base were spat out with great relish.

The Havasupai also used the wood to make cradle boards. A maternal grandmother or great-grandmother prepared an oval frame of hardwood, such as mesquite. A branch was bent into an oval while still green; a row of crosspieces lashed together (arrowweed or Apache plum); then smoothed over with a layer of pinyon pitch; a basketlike bumper of catclaw or some other light twig was fixed over the child's head in case the little one fell over or a blanket was placed over the face.[50]

One of the major uses of mesquite now is as fuel for fires and barbecues. It ignites quickly and produces a very intense heat (mesquite has the highest BTU output of all fuel woods). It

Story of Coyote's Devil's Claw

One day Coyote was walking all over the desert, trying to find something to eat. He couldn't find anything, and he was too lazy to grow anything himself. So he walked and walked until he found what looked like a bone in the sand.

He tasted it. It had no taste. It was too dry. So he sat down, thinking. Then he started to jump up and down, yelling, "I think this bone wants to tell me that I will find something to eat around here."

So he ran around. All he saw was desert, no food. Then he came to a wide wash. He tried to jump across, but he landed in the middle of it, on top of a little green plant half-buried in the sand.

"This looks like it would be good to eat," Coyote said, and he gobbled down the whole plant—root, bony fruit, seeds, and all.

Glad that he didn't have to work to get his food, he decided to lie down and sleep. But after a while, he woke up with a big pain in his stomach. He got so sick that he had to get the plant out of his insides. He buried it in the sand and hoped he would not see it again. "They don't like me and I don't like them."

But each year when the rain comes, those plants come up again. The floods carry the bony fruit and bury more of them in the sand where they can grow. Pretty soon, Coyote sees those plants he doesn't like all around. When the Desert People learn that they make him sick, they decide to say it is his plant. His devil's claw. Ban I'hug-ga.

Papago elder, as told to Gary Paul Nabhan, in *The Desert Smells Like Rain*

also produces a pleasant-smelling smoke and is sold as wood chips, sticks, nuggets, and charcoal briquets throughout the United States. In Texas alone, about 15,000 tons of mesquite chips and chunks are sold as barbecue fuel. The popularity of mesquite has changed the look of the landscape. All kinds of fences from the old days were made out of mesquite. Once the trend for barbecue wood took off, many of these old reminders of an earlier way of life were dismantled and sold for firewood.

Mesquite wood is, however, an industry subject to environmental abuses. In some areas, mesquite is clear-cut for fuel wood, denuding large areas that once relied on the plant to support communities of plants and animals. The production of mesquite charcoal produces such high levels of air pollution that the industry cannot pass U.S. Environmental Agency standards for air quality, so the charcoal is produced in Mexico and imported into the United States. An alternative exists: the pods produce the same characteristic smell, and could be burned with other woods to create a similar effect.

Burning is a poor use of mesquite. As fuel, it is worth only about half its value as lumber, flooring, furniture, or fine woodworking. The National Hardwood Lumber Association is working with a group called Los Amigos des Mesquite to promote the use of the wood in construction and for flooring. Another company, called the San Pedro Mesquite Company, offers wood for everything from door frames to musical instruments. As they point out, mesquite wood is harder than oak, takes a good finish, and is subject to less shrinkage than other hardwoods.

Fabaceae

Velvet Mesquite
Prosopis velutina

This plant has been a major source of food for many Native Americans. The pods are very sweet and the seeds are 35 percent protein, higher than soybeans. The Tohono O'odham use mesquite meal in their cooking today and hope to make it a commercial crop.

It's called "velvet mesquite" because the leaves and pods are covered in fine, velvety hairs. As the tallest of the mesquites, it produces long lengths of wood and has often been used to make fence posts, corral walls, and lumber for buildings. Both wild and domestic animals feed on the sweet, nutritious seedpods. Honeybees love it, making it one of the most important honey plants in Arizona. As a desert ornamental shade tree it is non pareil.

Though it can grow up to 40 feet, it has a short trunk, long slender branches, small thorns, and hairy compound leaves. Light yellow, fragrant flowers appear in spring and summer with 4- to 8-inch pods covered in fine hairs. They have a sweet pulp and several bean-like seeds within.

It grows naturally in central and southern Arizona, southwestern New Mexico, and northern Mexico, and has been introduced in California and Texas. Alas, it is dying out in some areas where land has been cleared for agriculture or urban development and because of the lowering of the water table in areas where underground water is being pumped out.

Range: **central and southern Arizona, southwestern New Mexico, northern Mexico, California, Texas.**

Fabaceae

Honey Mesquite
Prosopis glandulosa, common mesquite, prairie mesquite, glandular mesquite

Perhaps the commonest of the native species of mesquite, the honey mesquite was originally confined to the Southwest, but its range has increased with the grazing of cattle. The seeds evolved to pass through the gut of animals, and sprout in the resulting manure pile. This has allowed it to invade former grassland and prairie areas. A variant called **Western honey mesquite** (*P. glandulosa* var. *torreyana*) has its range west of the Pecos River in Texas, while honey mesquite grows east of the Pecos. Honey mesquite has longer, more widely spaced, less hairy leaves than the velvet mesquite, and 3-inch thorns along the branches. Clusters of light yellow flowers bloom in spring and summer. The largest grow along riverbanks and get up to 40 feet high, but in drier areas they are confined to 10 to 15 feet high at the most. Honey mesquites often have deep taproots that grow to 40 feet long to reach groundwater. If groundwater is not available, the roots may grow sideways in their search for water, and move up to 60 feet away from the base of the plant.

If undisturbed, the honey mesquite usually develops a tree shape with a single trunk. If it is frozen, burned, trampled, or chewed by animals when it is young, the plant develops multiple stems and a bushy shape. Honey mesquites up to 200 years old have been cataloged, but most mature plants are under 100.

Honey mesquite wood is probably the most widely used fuel wood for barbecues, although it is also popular for woodworking, since it is strong, hard, has attractive grain patterns, and tends not to warp. The wood is orangey or reddish. Few trees, however, grow large enough to produce commercial lengths of wood. The inner bark was once used for baskets and the making of coarse fibers. Gum from the stem was used for making gumdrops, mucilage for mending pottery, and black dye.[51]

Range: **Texas, west through New Mexico and Arizona, and north to Oklahoma and Kansas.**

Fabaceae

Screwbean Mesquite
Prosopis pubescens

Screwbean mesquite has tightly coiled pods, unlike the straight or slightly curved pods of the other varieties. The pod is not as flavorful as that of honey or velvet mesquite. The leaves are smaller than those of other mesquites and are covered with fine gray hairs. The pale yellow flowers appear in tight clusters, late in spring or early summer. It grows to a height of about 20 feet, with a short trunk and slender branches.

The screwbean mesquite can be found in river valleys throughout the Southwest. Wild animals eat the pods, and the wood is used for fuel or lumber.

Range: Texas to California, northern Mexico, Utah, and Nevada.

Fabaceae

Desert Ironwood
Olneya tesota, ironwood, tesota, palo de hierro, pallo fierro

Olneya tesota
grows to 45 feet. It is a spiny, evergreen tree with pinnate blue-green, hairy leaves, gray, smooth bark, and purplish flowers appearing in late spring.
Range: southern Arizona, southeastern California, and northwestern Mexico.

The endurance of ironwood is legendary. The trees can survive for up to 600 years and its wood may last 1,500 years. It is one of the largest and longest-lived of all desert plants and is found only in the Sonoran Desert. The slow-growing ironwood needs a dry climate to survive, and is threatened by overwatering and irrigation. In areas that are being turned over to agriculture, ironwood is dying out. According to Jack Hill, a biologist at the University of Arizona, "Ironwood indicates that this is a frost-free area, so when the citrus growers came along, they knew this and would take out the ironwood and put in citrus groves. Now they are taking out the citrus groves and putting in condominiums."[52]

John Alcock writes:

In years when the proper conditions apply, these trees [ironwood], like paloverde, metamorphose from a rather ordinary green shrubby plant to a glorious burst of color produced by thousands of flowers. The exact nature of the color changes with every shift of viewing angle and every alteration in the position of the sun. A luminescent purple, a subdued pink, a flat blue gray—all are possible in the chameleon-like tree. Female digger bees forage urgently at the flowers, as if they know that the ironwood flowers are their last brief chance to harvest. Soon the desert floor beneath the ironwoods will be obscured by the ghostly gray fragments of myriad flowers, fragments that will dry and disappear.[53]

Ironwood creates an oasis in the desert. It provides shade from the sun for the plants growing beneath by lowering the temperature about 15°F. It is a nurse plant for up to 165 species including organ-pipe cactus, night-blooming cereus, ocotillo, saguaro, and cholla. Once a tree dies, it decomposes so slowly that it's been called nonbiodegradable. This explains why it persists for hundreds of years if left alone. It makes a

Olneya tesota, desert ironwood

great fire, burning long and hot—another reason it's being wiped out in some areas.

The scientific name honors Rhode Island botanist Stephen Olney, *tesota* comes from the Spanish *tieso*, meaning "stiff." The common name describes the hardness of the dark brown wood, which is so dense that it won't float. The name is also given to other extremely dense types of wood; this species, however, is the second heaviest wood native to North America (the heaviest is leadwood, *Krugiodendron ferreum*, which grows in Florida).

The Seri believed that looking after the ironwood helped sustain their life on the desert. There is a legend that a giant chewed ironwood seeds and blew the pulp over the sea to calm it so he could harpoon fish. This small nation living on the coast of Sonora in Mexico now uses ironwood to carve figurines for tourists. José Astorga was a Seri who began by carving fish and animals out of local ironwood in the 1960s, using fairly crude tools, when the local fishery was suffering from competition with the main Mexican fishing industry. Soon his friends and neighbors started carving as a way of making a living. Each Seri sculpture takes a week or more to make and may sell for about $20. Understandably, the price is dropping with the number of Seri carvers. Now many Mexicans have entered the field and are producing ironwood sculptures for tourists, using power tools. Meanwhile, ironwood is becoming scarce, and the carvers are turning to different woods.

Today, the living wood is protected by law in Arizona. Though good-quality deadwood is difficult to find—what can be scavenged is hollow, cracked, and gnarled—only deadwood less than 100 years old may be gathered by woodworkers. Now that the wood has become increasingly scarce, the largest and most wondrous of their carvings are becoming more valuable.

The desert tempered muscles and will and spirit; and the sea blessed with its bounty, for it was the Sea Turtle god who created this world, raising the Seris' land from the water with his great back. Poetry poured from the mind and hands of these seagoing desert people who lived closely and fondly with their dual environment. Women painted their faces using a clam shell of water as a mirror and the chewed tip of a mesquite twig as a brush. One of their motifs formed a series of wavy lines in blue: the legs of an octopus. Another, painted in red, symbolized the roots, pods, and seeds of ironwood trees, a thoroughly desert species.

[They] speared porpoises and sea lions and sea turtles from jutting rock ledges. Their lances were carved from the hard taproot of ironwood trees and the ropes they used to haul in the prey were twisted from mesquite roots. . . . They sang to the music of a harp made from the flower stalk of an agave strung with gut from seals.

Ruth Kirk, *Desert: The American Southwest*

One reason for ironwood's scarcity is that it is often gathered, either inadvertently or carelessly, by those who harvest mesquite to make charcoal or barbecue chips. According to Gary Nabhan, "Ironwood has been a by-catch of mesquite harvests in much the way dolphins are a by-catch of tuna harvests."[54] There is, however, little incentive for harvesters to stop this practice, because ironwood burns very well and makes excellent charcoal. But given its slow rate of growth, harvesting it for charcoal seems foolish. In 1991, a joint task force from Mexico and the United States formed a group called the Ironwood Alliance to try to solve the problem of the plant's depletion.

Fabaceae

Catclaw Acacia

Acacia greggii, catclaw, cat's-claw acacia, Gregg acacia, devil's-claw, devil's-claw acacia, paradise flower, long-flowered catclaw, Texas mimosa, *una de gato*

Acacia greggii

is a spiny shrub that can grow to the height and shape of a tree. The hooked thorns grow along the branches and twigs. The double compound leaves are hairy. Light yellow flowers appear in clusters in spring, followed by thin, flat, brown seedpods containing round, flat, beanlike seeds.

***Range:* southeastern California, southern Arizona, southern Texas, northern Mexico.**

A spiky, thorny legume that grows as tall as 30 feet, the catclaw is considered a pest species by ranchers, who do their best to exterminate it. The stout, clawlike thorns can do serious damage to clothing or unprotected flesh, and even animals tend to give these shrubs a wide berth.

The plant was named for explorer and naturalist Josiah Gregg (1806–1850), who had the misfortune to die at the age of 44 after enduring a wet winter trapped in a forest of giant redwoods, where he had been leading a prospecting party.

For Native Americans, the catclaw acacia was an important source of fibers for basketry. The twigs were stripped of leaves and thorns by running the fingers, protected by a cloth, over them, sorted, and the butt ends trimmed off to a length of 2 and 3 feet. Short twigs were then mashed between two stones. Long twigs were split in three beginning at the tip, using the teeth and hands. The strips, collected green in July or August, were stored and soaked, then scraped to a uniform thickness with a knife.[55] Fruits were collected ripe, spread on a blanket, beaten with a stick to free the seeds from the pods, roasted in a parching tray, ground on a grinding stone, and made into bread to be stored in a skin sack or blanket bag.

Birds, ground squirrels, and wood rats feast on the seeds, and mule deer on the fruit. Mistletoe often establishes itself in the acacia.

Fouquieriaceae

Ocotillo

Fouquieria splendens, coachwhip cactus, fishing-rod cactus, Jacob's staff, vine cactus, devil's paintbrush, candlewood, slimwood, flaming sword, monkey tail, wolf's candles

Fouquieria splendens

is a shrublike plant with spiny stems that can grow in an upside-down cone shape up to 20 feet high, with as many as 75 branches.

***Range:* throughout the Southwest, from western Texas to southeastern California to northern Mexico.**

The name ocotillo (pronounced "o-co-TEE-yo") is a diminutive of an Aztec word meaning "pine tree"—*ocote* or *ocotl*. It has a variety of common names ("more aliases than a check artist," according to Reg Manning[56]). It is sometimes called the "coachwhip" or "fishing-rod cactus," because it looks like bundles of spreading whips or fishing rods united at the base and bearing a bunch of flame-colored flowers at the tips.

Ocotillo has adapted well to its surroundings. When there's enough rain, it will grow leaves. When there's too little water, the leaves wilt and drop. The cycle can be repeated several times a year. The sight of a dead-looking many-pronged shrub with bright red, tubular

Acacia greggii, catclaw acacia

Fouquieria splendens, ocotillo

flowers is one of the desert's most dramatic spring sights. Even better, there are usually hummingbirds darting around the carmine flowers.

Desert peoples used to bathe in water containing the crushed roots of the ocotillo to relieve fatigue. The flowers and roots were also placed over fresh wounds to stanch bleeding. The Papago used the thorns to pierce the ears of both men and women. European women took a hint from the Indians who used the thorny spines as roofs of their adobe buildings. To keep out dust and insects, pioneer women tacked up cloth strips underneath the rafters of their ocotillo roofs.[57]

Today, the ocotillo is often used to make living fences for livestock. Jack Hill of the University of Arizona says: "I've seen Mexicans hedge in their chickens, duck and geese from coyotes with ocotillo, and when they didn't have enough they'd use creosote in between. They would lay it down or stuff it in there. The spiny form also keeps out livestock and rodents."[58] This ancient tradition is now being revived all over the area around Tucson. Ocotillo is sold in native plant nurseries and used widely in xeriscape gardens.

Bignoniaceae

Desert Willow
Chilopsis linearis, flowering willow, willow-leaved catalpa, desert catalpa, catalpa willow, false willow, bow willow, mimbre, *flor de mimbre,* jano

This is a perplexing plant. It looks like a willow, but it isn't; nor is it a legume, like most of the other treelike species in the Sonoran Desert. It's actually a member of the trumpet-creeper family, Bignoniaceae, related to the catalpa, famous for its long drooping seedpods.

Like real willows, desert willows cluster near watercourses, including dried-up streambeds, where water is available underground all year. It keeps its leaves throughout the summer and into the early fall. The flowers attract hummingbirds and honeybees. It has a slender trunk and branches and narrow leaves, and may grow 10 to 30 feet high. The twigs are sometimes hairy or sticky. The showy orchidlike flowers are white, pink, or purple and appear in late spring or early summer. The fruit looks like a cigar: a dark brown cylinder 4 to 8 inches long and ¼ inch in diameter.

Native Americans used the desert willow branches for weaving baskets and to make bows for hunting. The leaves were also used to make tea or a hot poultice for chest problems. The leaves and bark are antiseptic and can be used to make a wash for skin problems, ringworm, or yeast infections.

In the 1930s, the Civilian Conservation Corps planted desert willows in shelter belts in the Southwest. Today, it is used as an ornamental in gardens and along roads.

Range: **southwestern Arizona, southern California.**

Burseraceae

Elephant Tree
Bursera microphylla, copal, torote

Bursera microphylla

is usually 6 to 8 feet tall, but can grow up to 18 feet high. It has an immensely stout trunk with crooked, tapering branches, papery pale bark, and very small

leaves. **White flowers appear in early summer, followed by small red fruits that mature in fall.**

Range: **southwestern Arizona, southern California.**

The elephant tree is an example of arboreal obesity. It has a huge trunk, all the better to store water, and disproportionately small leaves, about 1 inch long.

When the papery bark is peeled off, the first inner layer is green, and the one after that is red. The wood smells somewhat like turpentine and, when pierced, exudes an aromatic resin called copal, which was used by Native Americans as a healing ointment for various ailments, including skin problems. It has also been used as a dye, a tanning agent, an adhesive, and a base for varnish.

The elephant tree is a member of the torchwood family (Burseraceae) of aromatic, resinous plants that produce, among other things, frankincense and myrrh. Elephant tree is also the name

Bursera microphylla, elephant tree

given to a completely different species, *Pachycormus discolor*, which grows in Baja California, and has a similarly swollen trunk and light bark.

THE AGAVES

The waxy coating of the agave leaf comes off like a powder on my fingers. The coating, formed by a deposit of hydrocarbons, protects the large leaf surface by reflecting up to 75 percent of both ultraviolet and near-infrared heat radiation, and lowers the chance of damage when the leaves are dehydrated. It also slows down evaporation, conserving both the water and minerals within.

Ann Zwinger, *The Mysterious Lands*

Agaves have been used by people for at least 9,000 years and in the twelfth century it is estimated that at least 100,000 were being cultivated in the Sonoran Desert alone. *Agave* comes from the Greek word for "noble." In Spanish, agaves are known as *mescal*. One Native American nation had so many uses for agaves that they became known as the Mescalero Apache.

Most species have edible fruit, fibrous leaves that can be used to make ropes, sandals, mats, and baskets, and roots full of saponin that can be soaked and pounded to provide soap. The young flower buds were roasted and eaten and the sap is still fermented to make an alcoholic drink. Mescal pits, which are rock-lined pits used to roast young flower stalks, are often found in archeological sites in the Southwest.

Because agaves contain a high concentration of sugar and starch in their leaves and sap, they are used to make liquors such as pulque, tequila, and mescal. Pulque has been created by fermentation for thousands of years. The Aztec used pulque in their religious ceremonies, and worshipped a goddess of pulque, called Mayahuel.

The Mexicans still make pulque with an alcohol content between 3 and 4 percent, both commercially and at home, and drink it like beer.

Tequila and mescal are distilled, a process introduced to North America in the sixteenth century from Europe. Currently, commercial production is centered in the state of Oaxaca, Mexico, about 250 miles south of Mexico City.

Agavaceae

Century Plant
Agave parryi, Parry's century plant

Agave parryi
is a low bush, about 2 feet high with gray-green, pointed leaves with long spines at the ends. The flower stalk may be 10 to 15 feet high and 4 inches in diameter; the flowers are yellow or orange and bloom between June and August.
Range: western Arizona, southern Florida.

The name "century plant," is something of an exaggeration. Once, people thought the plant took 100 years to produce a flower. In fact, small species flower when they are three or four years old; larger species take 40 to 50 years to bloom. Many flower just once, then die—this is a monocarpic plant. All the plant's energy is devoted to producing its one gigantic flower stalk, then, spent with exhaustion, it dies. Parry's agave blooms when it is 20 years old or more.

The flowers attract the bats which come to it at night to sip the heady nectar. Flitting about in a feeding frenzy, they get the pollen stuck to their foreheads and they fertilize every flower they visit. The sheer size of the plants ensures the seeds are sent out into the desert as far as the breezes can take them.

It was named for Charles Christopher Parry (1823–1890), a physician born in England who grew up in upstate New York. Like most physicians who made their own medicines in those days, he had to study botany and found he preferred it to medicine. He was the student of the famous botanist, John Torrey. Parry served as botanist from 1849 to 1852 on the survey of the United States–Mexico border, and made several further journeys to the Southwest, Mexico, and California. He furnished museums all over the world with seeds, branches, and cones he found in Colorado (where he described 80 species that had previously been unknown), Mexico, and the American Southwest.

The plant had a few uses: the stems were carved into spoons and the leaves could be made into a brush for the hair or for cleaning grinding stones. To make the brush, the dried matter of a dead and rotten leaf was knocked free from the fibers, which were then bent in two. The upper end of this brush was wrapped with a cord, and the bent portion was covered with buckskin or cloth. The loose fibers were cut to the right length and hardened by burning the ends.

Agave parryi, Parry's century plant

Agavaceae

Desert Agave
Agave deserti

It takes the desert agave at least 20 years to flower. Its rosettes form large circular colonies in the desert. Some of these rings are up to 20 feet in diameter and may be more than 1,000 years old. Normally, it is a relatively small plant, 1 to 2 feet tall and 18 inches wide. In the dry climate of Arizona it sends out a few suckers and, in California, where it may receive more rainfall, it might produce more.[59] In most of the dry areas it grows in, however, all its energy is required to produce the 13-foot-tall flower stalk. The pale yellow flowers, which have long stamens and cluster along the flower stalk, attract hummingbirds which are not in the least put off by the teeth along the edges of the greeny-gray leaves.

Some of these agaves have bright red buds. Massive amounts of sugars, other organic compounds, and about 40 pounds of water must be moved from the leaves to the flower stalk. No wonder it collapses immediately after flowering. As naturalist Park Nobel says: "If the flowering stalk is cut off when relatively small, water and photosynthetic reserves are not transferred from the leaves, and *A. deserti* . . . can continue to live for many years."[60]

The flowers and flower stalks have long been used as a food by Native Americans. Roasted, pounded stalks and leaves were made into cakes and sun-dried. The flowers were parboiled to release the bitterness, then eaten. The hearts were dried and stored indefinitely. These days it's still being turned into traditional beverages. In Mexico, they make pulque or a kind of beer from the fermented mash. The leaf fibers were also once used to make brushes, ropes, or baskets and the thorns were used as needles in basket-making or as tools for tattooing.

A. deserti is native to rocky or gravelly soils in the Lower Colorado River Valley subdivision of the Sonoran Desert. This is the most frequently encountered agave on the northeastern boundary of the Sonoran Desert north and east of both Phoenix and Tucson.

Range: **extends barely into the Arizona Upland and Mojave Desert.**

Agavaceae

Shindagger
Agave lechuguilla, lechuguilla

Shindagger is poisonous to cattle, goats, and sheep. But pocket mice and mule deer eat the new shoots, which aren't as toxic as the mature plants. The common name refers to the daggerlike leaves, which in the early days could cut the legs of passing horses, and even impale a rider who had the misfortune to fall off. Today they may pierce the tires of off-road vehicles. Not only are these leaves edged with spines, they also fold together at the tip, creating a dagger several inches long.

The sharply pointed leaves of *Agave lechuguilla* grow up to 20 inches long from a central base. The flower stalk may grow to 11 feet with yellow or purplish flowers in small clusters. Shindagger is monocarpic, which means that it flowers only once near the end of the plant's life cycle, which might last 3 to 20 years. The seeds are shaken by animals and wind and usually fall near the plant. It reproduces primarily through offsets from its extensive system of rhizomes.

Oddly enough, the scientific name *lechuguilla* comes from the Spanish diminutive of *lechuga*, which means "lettuce." Native Ameri-

cans ate the flower stalk, stems, and unfolded leaves either roasted or boiled. Hummingbirds and bats consume the pollen; various birds consume the syrupy sweet nectar, a good source of water and energy for these animals as well as for people.

The shindagger's central bud is a source of hard fibers, which have long been used to make twine, rope, and brushes. Recent research has suggested that its fibers are comparable to glass fibers and could be used in construction materials. The leaves also contain smilagenin, a steroid precursor (a biochemical substance from which more stable products are formed), and are being considered for use in drug manufacturing.

Range: **New Mexico and Texas.**

THE YUCCAS

The yuccas are very special in more ways than one. For instance, the only creatures that can pollinate the flowers are moths of the genus *Pronuba*. Their partnership begins with the moth laying its eggs in the ovary of the flower. As the larvae mature, they feed on the yucca's seeds, but leave some in the pod to be dispersed by the wind for next year's crop. The moth in its turn moves the flower's pollen from the male (anther) to the female (pistil) parts of the flower.

The flowers and buds are edible when raw, but steaming improves the flavor. The new shoots, which look a bit like asparagus, were peeled and then cooked or baked. The raw fruit is also edible, but the flavor improves with baking, boiling, or sun-drying. Native Americans worked the fruit into a paste which was formed into cakes and dried in the sun for later use, or fermented to make an alcoholic beverage. Seeds can be ground and used as pastry flour. The roots of the yucca are full of saponin, which is a soaplike substance. A vegetable soap prepared from yuccas, particularly *Yucca baccata* (**banana yucca**) and *Y. glauca* (**soapweed**) is called amole soap. The yucca suds were used by the Tewa and the Hopi as part of their important ceremonies. In their *Ethnobotany of the Tewa Indians*, Robbins, Harrington, and Freire-Marreco tell us:

The Tewa of Hano, like the Hopi, accompany all ceremonies of adoptions and name-giving by washing with yucca suds. Thus, when an infant is named before sunrise on the twentieth day after birth, its head is washed by the paternal grandmother and each member of the father's clan who gives an additional name smears the child's head with suds. The bride is bathed by the bridegroom's mother at the beginning of her bridal visit to the bridegroom's house, and at the end of the visit, when she is about to return to her own clanhouse, women of the bridegroom's clan wash her hair before sunrise and give her a new name.[61]

Native Southwestern peoples twisted yucca fibers to make bowstrings, fishnets, brushes, and fabric (sometimes combined with cotton or animal fur), and plaited strips of yucca into sandals. The Anasazi sandal makers even incorporated designs into their intricate weaving.[62] Dried leaves were collected and boiled with pinyon gum and cooled. The resulting hard mixture was pounded to a powder, then mixed with water to plug holes in baskets and make them watertight. A coat of red paint, then of pinyon gum followed.

Today, new uses are being found for yuccas. For example, Peter Spoecker in California makes yuccas into didgeridoos, a wind instrument up to seven feet long, invented by the Aboriginals in Australia, now popular in North America and Europe.

Agavaceae

Joshua Tree

Yucca brevifolia, yucca tree, tree yucca, yucca cactus, yucca palm, praying plant

Yucca brevifolia

grows 15 to 40 feet tall in a tree shape with a thick fibrous trunk and short branches. Trunk and branches are covered with dead, dry leaves pointing downward. Spiny, sharp evergreen leaves grow at the ends of the branches. Blooms once or twice a year, producing cream-colored waxy blossoms with a musty smell that open at night. Pollinated by the yucca moth. Fruit is elliptical, 2½ to 4 inches long, green to brown, with many flat seeds.

Range: Mojave Desert, Joshua Tree National Park in southern California, and Joshua Forest Parkway in western Arizona.

Of the stories about how this tree got its name, the most charming is quoted by writer Bill Thomas: "The early pioneers happened upon it when wagon trains first rolled into California from the east. Because of its many upstretched branches, or arms, which they thought resembled the biblical Joshua waving them onward, they christened it the Joshua tree. And while botanists and foresters insist that it's not a true tree, this is still its most commonly accepted name."[63]

This is one of the oldest known desert plants and is restricted to a few areas because it needs cold winters and grows at 3,000 feet above sea level. It was more widely spread in prehistoric times, when the climate was more humid. We are told it can thrive for 800 years, but individual plants have branches of widely different ages and

Yucca brevifolia, Joshua tree

it's hard to tell the real age of any given plant. Tests show these trees grow about 3 inches in the first 10 years and then slow down to about 1½ inches a year from then on[64]; where it's too cold, or the soil is rocky or acid, stunted ones 150 years old are only 5 feet high and never blossom.

Like other plants in the Agavaceae family, the Joshua tree relies on the small white yucca moth, *Pronuba* species, for pollination in the years when it happens to flower, which is not every year. The yucca moth lays its eggs in the pollinated flowers; these in turn produce the seeds which feed the larvae hatched from the eggs. The yucca moth eggs need a particular temperature and amount of rainfall to hatch.

A veritable menagerie of creatures depend on the Joshua tree to store up food for the future: 25 species of birds use it for nest sites; deer, squirrels, and birds eat the blossoms and fruit; insect larvae, termites, spiders, beetles, and ants are all drawn to it for food and shelter.

My first encounter with a Joshua tree was the same as that experienced by many other people— it was hiking over a far ridge with a pack on its back and waved at me in greeting. I waved back. And wondered why it didn't respond.

Ann Zwinger, *The Mysterious Lands*

The Native Americans of the Southwest used the trunk and branches (which are fibrous rather than woody) in construction, made the fibers into baskets and sandals, and ground or roasted the seeds for food. The roots were used to make a red dye. In the late nineteenth century, an English company based in San Francisco tried using Joshua trees to make paper. In 1897, several editions of the *London Telegraph* were printed on the resulting paper, but the process proved to be too expensive and was eventually abandoned. Around the same time, another English entrepreneur tried to create a type of beer from the roots, but it was undrinkable.

Joshua Tree National Park in southern California was created in the 1920s when Minerva Hamilton Holt and other conservationists became concerned about people ripping out cacti and succulents in the California desert to use them in gardens. She lobbied to preserve the desert landscape and its plants and wildlife and to educate people about the value of the desert. She proposed a desert federal park. In 1936, President Franklin Delano Roosevelt proclaimed 825,000 acres as a national monument. Because of pressure by miners and others who wished to extract resources inside the monument, the area was reduced by 250,000 acres in 1950. On October 31, 1994, as a result of the California Desert Protection Act, Joshua Tree (along with Death Valley) was upgraded from national monument to national park status.

Agavaceae

Soap-Tree Yucca

Yucca elata, Our Lord's candle, palmella, palmilla, Spanish bayonet, datil, amole

Yucca elata

is a shrub or small tree that grows 10 to 17 feet high, with a single gray trunklike stem or multiple stems, no branches. Long, thin, yellow-green, leathery leaves ending in sharp spines. White bell-shaped flowers; brown fruit 1½ to 3 inches long, maturing in early summer, filled with small black seeds.

Range: **western Texas, New Mexico, and southern Arizona.**

State flower of New Mexico.

Yucca elata, soap-tree yucca

This is what could be called a pointy plant. Each stiff blade has a needlelike tip with outward-reaching, thick, spiky leaves. From the center of this formidable mass comes a slender stalk, as elegant as a spear of asparagus. Eventually, creamy white flowers bloom when the stalk reaches full height. The heady perfume drifts out on the evening breeze.

The leaves are not as stiff as those of other yuccas, and animals from wood rats to cattle graze on them. The soap tree used to be harvested to provide emergency food for livestock in the early twentieth century. The roots of the soap tree (and other yuccas) produce a detergent used in the manufacture of shampoo and soap.

Yuccas were once a part of Native American footraces. Children were trained from an early age to be as swift as the wind. As they got older, longer and longer races took place regularly. Before an important race, the participants washed their hair in yucca root as part of the purification of the body, and then washed again after the race.

The leaf fibers are used to make baskets, mats, and ropes. The buds and flowers are edible and rich in vitamin C. Today, soap-tree yuccas are used in landscaping as ornamentals or hedges.

Agavaceae

Torrey's Yucca
Yucca torreyi, Spanish dagger, old shag

Yucca torreyi has a basal clump of daggerlike leaves emerging from the central part of the brown, scaly trunk. Above the crown of leaves, the trunk emerges, covered with dead leaves. Torrey's yucca grows to a height of about 13 to

18 feet and may be up to 1 foot in diameter. Clusters of white, bell-shaped flowers appear in spring, followed by fruits that are either banana- or egg-shaped, brown or black, with many small black seeds.

As with the other yuccas, the Native Americans of the Southwest used the fibers for baskets, mats, and rope. Sometimes the red roots were woven into a basket or mat to make a pattern. The fruit could be roasted and stored indefinitely, or it could be pounded into a pulp, drained, and drunk. Young Indian boys were given yucca tea before they set off on the standard 5- to 10-mile daily run to collect pine, fir, and scrub oak boughs. The boys first threw up (the tea is an emetic) and then they would get on the road.

It was named by Charles Christopher Parry for his friend Dr. John Torrey (1796–1873), one of the leading botanists of his day. Torrey was coauthor of *A Flora of North America* and wrote *A Flora of the State of New York*. An interesting historical footnote: Torrey first learned about botanical classification from Amos Eaton, botanist and inmate at the prison where Torrey's father worked. (Eaton was incarcerated for a few years on trumped-up charges relating to land speculation. He went on to become the first president of Rensselaer.)

Range: **southwestern Texas, southern New Mexico, northern Mexico.**

Agavaceae

Chaparral Yucca
Hesperoyucca whipplei, Our Lord's candle, Quixote yucca

This is yet another yucca known as Our Lord's candle, along with the soap-tree yucca. Spanish Californians call the plant *quióte*. This is probably nothing to do with Don Quixote, but is more likely a corruption of the Aztec word for the agave, *quiotl*. The scientific name was given by Charles Christopher Parry in honor of Lieutenant (later General) Amiel Weeks Whipple (1818–1863), a surveyor of the desert Southwest during the 1840s and 1850s who was killed fighting on the Union side during the Civil War.

Hesperoyucca whipplei is a low shrub with narrow spiky leaves densely packed in a pompom shape. Each leaf is 1 to 3 feet long and ends in a sharp spine. It flowers between late February and early June and the individual flowers last two to seven weeks. The flowers are thickly clustered on the stalk, rising high above the shrub, which is quite variable and divides up into five subspecies: *caespitosa, intermedia, parishii, percursa,* and *whipplei*. In most subspecies, the white, creamy, or purple-tinged flowers emerge on a long stalk from the shrub. In the tallest subspecies, *parishii,* the flower stalk may be 21 feet long. The flowers emerge when the plant is 9 to 10 years old. *Caespitosa, intermedia,* and *percursa* live for many years and flower, but the *parishii* and *whipplei,* after flowering, are reduced to a dried, fibrous base that remains for several years.

Native Americans made flour from the seeds and used fibers from the leaves to make ropes and baskets. They also used to roast and eat the flower stalk, which tasted a little like baked apple. Like those of other yuccas, the roots can be soaked and pounded to produce soap. The dead flower stalks are light and full of pith. *The Western Flower Guide* of 1917 by Charles Francis Saunders notes that curio manufacturers used to cut up the dried stalks and make them into pincushions for tourists.

Range: **western Arizona, southern Florida.**

Agavaceae

Banana Yucca

Yucca baccata, datil, blue yucca, amole,
Spanish bayonet, Spanish dagger, wild date

The fruit of the banana yucca is regarded by the Zuni as a great luxury. Before they obtained wagons, it was gathered and carried in blankets on their backs, and later on the backs of burros. The fruit, which is called *tsu'piyane* ("long oval"), after being pared, is eaten raw or boiled. When the boiled fruit becomes cold, the skin is loosened with a knife and pulled off. The fruit can be eaten at this point, or made into a conserve.[65] Other Native Americans baked the sweet fruit and gathered and ate the flowers.

Yucca baccata grows in bushes up to 5 feet high, with long, daggerlike leaves radiating from a central point on the ground. It flowers between April and July. The sweet-smelling white, cream, or purple-tinged flowers, with their six petal-like segments, emerge in the center of the leaves. The fruit is a long cylinder, rounded at the ends, fleshy and green.

Fermented drinks were made from the sap, which was also used as a color-fixing agent for pottery paint. The roots were used to make soap, and fiber from the leaves for ropes and baskets. Banana yucca is used today in landscaping, and is excellent for disturbed sites, since it resists trampling.

Range: throughout southeastern California, southern Nevada, southwest Colorado, much of Arizona and New Mexico, western Texas, and northern Mexico.

Agavaceae

Soapweed Yucca

Yucca glauca, soapwell, bear grass, Great
Plains yucca, small soapweed

This is not just a desert plant; it also grows throughout much of the Great Plains. *Yucca glauca* is a perennial, deep-rooted, long-lived shrub that can grow up to 6 feet tall. It has light bluish-green, stiff leaves, sharply pointed at the ends. Greenish-white flowers on a tall flower stalk appear in June or July. The fruit is oblong and develops into a woody capsule containing many black, winged seeds.

The roots were soaked and pounded to release saponin for soap, and the fibers of the leaves used to make baskets, sandals, mats, and ropes. During World War I, soapweed yucca was used to produce about 80 million pounds of fiber for burlap products. In World War II, it was also used to make paper for the United States Navy.

Range: **Texas and Arizona, north to Montana and the Dakotas and into Canada, portions of Kansas, Nebraska, Iowa, and Missouri.**

Nolinaceae

Sotol

Dasylirion wheeleri, spoonflower, spoonleaf,
desert spoon

Dasylirion wheeleri

is a shrub of narrow leaves emerging from the top of the trunk, which may be underground or rise up to 3 feet aboveground. The leaves are strap-like affairs 3 feet long with soft, almost ragged ends, which turn ecru as they mature and have edges with sharp prickles. The bulbous trunk becomes more obvious as the plant matures. It is dioecious, with male and female flowers on separate plants. Greenish-white (female) and yellow (male) flowers emerge on a flower stalk between May and July. The yellow male flowers have protruding stamens; the female flowers are more like lilies. The seeds are winged capsules.

Range: Arizona, New Mexico, Texas, and Mexico.

Dasylirion is Greek for "thick or tufted lily," a name that refers to the thick stem. Sometimes the plant is called "spoonleaf" or "desert spoon," because the leaves are concave, especially at the base.

This is a transition plant from the Chihuahuan to the Sonoran Desert. It starts out looking a bit like an agave, but changes in maturity to look more like a tree. It grows on the grassy lands of the desert at altitudes between 3,000 and 6,000 feet, mostly in rocky hillsides. Its partners are the ocotillo, creosote bush, and *Agave palmeri*. Sotol grows in arid conditions and may be killed by wet, slushy winters. It is estimated that sotol lives to be 150 years old, because it has few enemies and animals avoid the sharp leaves.

The leaves of *Dasylirion* have been used for thatching and fiber, as well as the production of an alcoholic beverage—not surprisingly known as sotol. Native Americans used the fiber for making baskets, mats, and roof thatch. The leaf hearts were an important source of food (apparently they taste like raw cabbage). The long stalk was sometimes used to make spears for hunting, tipped with a point from a harder wood, such as mahogany. Today, a company in Texas makes lightweight walking sticks and hiking sticks from the dried flower stalks.

Zygophyllaceae

Creosote
Larrea tridentata, creosote bush, greasewood

Larrea tridentata

is an evergreen shrub usually about 4 feet high, but up to 12 feet in areas with a good water supply. The small yellow-green leaves have a pungent smell, especially when wet. Yellow flowers appear throughout the spring and summer; some plants have flowers all year long.

Range: throughout the southwestern United States.

This is an invasive plant: but the invasion from South America occurred between 40,000 and 10,000 years ago, probably because birds left the seeds in their droppings as they migrated north. The creosote bush may be one of the oldest living species on earth.[66]

Creosote has a complicated set of survival adaptations to the desert environment. The tiny leaves close their stomata during the day to avoid water loss and then open them at night to absorb moisture. Creosote has not only deep taproots, but also roots that radiate out widely, up to twice the diameter of the canopy, so it can pull in both surface- and groundwater from the soil. During a drought, it sheds leaves to conserve water in the stems and then goes into a deep dormancy. After a good rainfall, creosote takes only a few weeks to return to normal growth. Like pine, it produces a sticky resin, which appears as a bright sheen on the surface of the leaves. The resin protects the plants by slowing down water loss. (The resin accounts for the common name greasewood, but this plant should not be confused with other Southwestern genera known as greasewood: *Sarcobatus* and *Atriplex*.)

Creosote is not just a survivor; individual plants continue to grow for hundreds of years. Some species have been documented at more than 600 years old. Nothing seems to bother it much. In 1962, at Yucca Flat, Nevada, a huge thermonuclear test blast was set off; creosote was the only perennial plant around before the explosion. Ten years later, ecologist Janet Beatley recorded that 20 of the original 21 plants had resprouted.[67] In sites contaminated with heavy metals, creosote is able to absorb and recover metal ions from polluted water by using its phyto-filtration system.[68] Maybe creosote and cockroaches will be the last living things on earth.

The smelly, unpleasant-tasting compounds in creosote resin repel most animals and insects. Sonorans call it hediondilla, "little stinker."[69] However, a few creatures can tolerate these compounds. One is a grasshopper called *Astroma*, which actually resembles the plant it feeds on: the female looks like old stems; the young male looks like a leaf spray.

Larrea tridentata, creosote

As darkness washed up against itself, a spirit grew inside it: Earth Maker. Earth Maker took from his breast the soil stuck to it, and he began to flatten this soil like a tortilla in the palm of his hand. He shaped this mound of earth, and from it, the first thing grew: the greasewood. From its branches, the first animal came. It was a tiny, scaly insect that could use the resin of greasewood to produce its own cover of lac. Earth Maker gathered this gumlike lac. He began to sing. Pounding out various shapes while singing, he formed the mountains. They hardened like shellac, making a hard crust for the earth. The space which brushed against their edges became the sky.

Gary Paul Nabhan, *Gathering the Desert*

As ecologist Jack Hill has pointed out, most smelly plants have a reputation as cure-alls.[70] It has some historical basis.

Ignaz Pfefferkorn, a Jesuit missionary and

botanist (1725–1795) noted that although creosote "gives off an odor which is almost unendurable to a somewhat sensitive nose . . . it is really a very powerful remedy . . . for worms in children as well as in adults."[71] His contemporary Juan Nentvig (1713–1768) believed it to be effective against syphilis and, as Gary Paul Nabhan writes, "an ointment made from creosote branches fried in tallow was efficacious when massaged onto gnarled rheumatic limbs . . . [but] if the masseur washes his hands after applying the ointment, his will become gnarled."[72]

Other botanists, including David Barrow and Katherine Saubel, have documented the plant's use as a remedy for more than a dozen afflictions, from colds and cancer to dandruff, distemper, and postnasal drip.[73]

Ruth Underhill said that the creosote was the Papago's universal remedy. They called it *shegai*. "It was used for stiff limbs and sores and poisonous bites. Men, after running all day barefoot, would make a fire, and when it had burned out, heap creosote branches on the ashes and hold their aching feet in the smoke. Women after childbirth or with menstrual cramps would lie on a bed of such heated branches."[74]

The Papago still grow it for medicinal uses. It's found in health food stores packaged as chaparral tea (nobody would buy it under the name creosote). Even today, says Angela Powers of Tucson, the old people "use it for all their ills. They make a tea with it and gargle to get rid of a sore throat. They have these remedies and they add a little alcohol and put the small leaves in it for several days. You open the refrigerator doors and you can see these bottles of painkillers. They swear by it."[75]

The creosote bush is the habitat of an insect called *Tachardiella larreae*, which produces lac deposits on stems. When heated, then cooled, lac becomes like a sealing wax. Lac was once an important element in native trade. The Havasu-

pai used to obtain it in trade from the Mojave and they in turn traded it to the Hopi. A chunk of lac, 2 to 3 inches in diameter, could be traded for a blanket. They used lac to attach arrowheads to the shafts.[76] Creosote lac was also used to seal the pottery jars full of food and seed that many peoples buried with their dead.[77] In more modern times, lac has been used to mend everything from ceramic pottery to cracked engine blocks.[78]

Creosote has hairy seeds designed to be spread by the wind, but the plant can also reproduce by cloning and form colonies. At Old Woman Spring in California, there is an oval ring of creosote with a radius of 24 feet, with a bare spot of earth and deadwood in the middle. It is actually all one plant, a genetically identical clone. It's known as King Clone. Given the rate of expansion and the hole in the middle, F. C. Vasek estimates that it is 9,400 years old and suggests that it began to grow when junipers still dominated the valley. Now the junipers have retreated up the slope. Gary Paul Nabhan says: "Whatever King Clone's exact age may be, it is older than the most ancient bristlecone pine known to humankind."[79]

Chenopodiaceae

Four-Wing Saltbush
Atriplex canescens, hoary saltbush, wing scale, shad scale, chamiza, cenizo, buckwheat shrub, busy atriplex, salt sage, wafer sage

Atriplex canescens
is a shrub, 4 to 6 feet high, sometimes more, with narrow, scaly gray-green leaves 1 to 2 inches long, and whitish stems. It produces separate male and female flowers and when in bloom is covered in

yellow spikes. The light tan-colored fruit has four wings, each about ½ inch long, surrounding each seed. Occurs usually in sandy, desert areas, but also on slopes up to 7,000 feet.

Range: throughout western North America.

Atriplex canescens is a plant that has always stayed put; it has not moved from its range, which extends from southern Canada down to northern Mexico. Since it has few spines, animals, including big ones such as elk and little ones such as jackrabbits, love to browse on its small leaves.

This is a dioecious plant: the female has masses of seeds on it; and the male has no seeds at all. All plants are affected by the type of soil they grow in, and the saltbush is no exception. On neutral soil it matures quickly and has a life-span of 10 or so years. But on dry alkaline soil, it lasts much longer. It can grow as high in elevation as 1,600 feet on shale, but only at lower elevations if it's on igneous rocks. In the northern parts of its range it survives because the salts from the soil accumulate in the leaf tissue of the plants and act as a kind of antifreeze.

The leaves have healing and soothing properties, and Native Americans used them to make poultices to put on ant bites and rashes. The leaves could also be worked up into a lather for washing hair. The wood is hard enough to make arrowheads.

Today, cattlemen see it as a fail-safe plant: when all other plants fail, this one will be there to provide fodder. It is often planted along roads and highways in the Southwest as a low-maintenance erosion-control plant.

Ephedraceae

Mormon Tea

Ephedra spp., including *E. nevadensis, E. viridis,* and *E. torreyana,* joint pine, tepote, teamsters tea, Mexican tea, Brigham tea, shrubby horsetail, canutillo (Spanish for "little tube"), canutilla ("little reed"), tepopote ("little straw"), popotillo, American ephedra

---

Ephedra

is the only genus in the joint-fir family of plants. Its jointed stems grow to about 4 feet and it has tiny, scalelike leaves. It blooms in spring, when it's covered with tiny, pale yellow, flowerlike cones (there are male and female plants). It prefers dry hillsides and grasslands.

Atriplex canescens, four-wing saltbush

Range: **arid regions throughout the southwest-ern United States.**

Mormon tea's vivid green stands out from the brown and gray of the desert. Because the leaves are so tiny—more like scales than leaves—the branches appear leafless. The stems of *Ephedra* contain caffeine and ephedrine and an extract from the plant is used in asthma and cold remedies. There is a good deal of controversy about the effectiveness of this remedy. Many people swear by it, and it's long been used in Chinese medicine (the Chinese call it *ma huang*[80]). Native Americans have used the plant for hundreds of years, whipping up a soapy lather and rubbing it over the body to treat an itch or a rash caused by chicken pox or measles. The Havasupai used it in this manner as well as for shampooing hair.

The Havasupai also liked to make a hot drink with the plant. They collected the upper portion of *Ephedra*, put it into a pot with cold water, and brought it to a boil. The tea was allowed to boil for just two minutes; if boiled longer, it would become bitter. Some say it tastes best when made during the fall months. This drink became a well-known substitute for coffee among early European settlers in the West.

Animals browse on the plants, and today it is grown as fodder for livestock. It may be cultivated as an ornamental plant, especially *E. viridis* (**green ephedra** or Mormon tea), a broomlike plant with jointed stems and small yellow conelets that smother the bush in summer.

The Navajo used green ephedra to make a light tan dye. They also used the tops of plants boiled up into a cough medicine.

E. nevadensis was also called **Indian tea, Mormon tea, Mexican tea,** or **tutut** by the Cahuilla, who used it fresh or made a tea from its dry twigs; they ground the seeds into a meal and made it into a nutritious mush. Other people

Ephedra spp., Mormon tea

To make Mormon tea:

Take 10 grams of *Ephedra* and boil it for 10 minutes in 2 cups of water; pour through a coffee filter. It's also supposed to reduce the appetite and speed up metabolism. Like any herb, it has detractors and a whole Web site of its own. Take it only on the advice of an expert.

used it to make a flour for breads or cakes. But tea was its most famous use.

It came to be called Brigham tea or Mormon tea, the story goes, because Mormons were not allowed to drink coffee. It packs a certain punch and is considered a stimulant. The settlers caught on to it quickly as a good pick-me-up.

Asteraceae

Brittlebush
Encelia farinosa

Encelia farinosa

is a relative of the sunflower that grows up to 5 feet tall on dry slopes and washes. It is a round, silvery-gray, leafy bush with bright yellow daisy-like flowers that bloom some time between March and June (depending on the winter rains). The hairy leaves are 2 to 3 inches long and oval.

Range: **southeastern California, across southern Nevada, to southwestern Utah and western Arizona. A similar species,** *E. californica,* **grows near the coast in southern California, but has only one head on each stalk.**

John Alcock calls the brittlebush "a surrealistic pincushion" because of its thin yellow stalks which rise a foot above the leafy mass, each one opening into a luminous flower. The stems of brittlebush exude a fragrant resin much favored as chewing gum by Native Americans.

It responds quickly to winter rains and produces seeds like mad. The seeds are borne huge distances by the winds. If there's no rain, it simply closes down reproduction until good times return. To protect itself against intense sunlight, the white outer leaves are covered with white hairs. In areas with plenty of rain, it grows bigger, less hairy leaves designed to collect as much sunlight as possible. The blooms are up to 3 inches wide, with a central yellow disk (they become brown in the southern part of its range) and yellow ray florets. The seed-like fruit is without hairs or scales at the top. The plant has a woody base and an umbrella shape of leaves with flowering heads clustered on long branchlets.[81]

Brittlebush exudes toxins in the soil around it to discourage competition by other plants. However, it attracts aphids which in turn attract ladybird beetles and their larvae to feed on them. It is food for the meloid or blister beetle, *Lytta magister*. Moth caterpillars cut off the petals, hook them onto the hair of their backs, and hide underneath.[82]

In spite of its shallow rooting system, brittlebush grows to 5 feet but is generally short-lived. At one time the resin was used for incense in the churches of Baja California. Priests also sent the gum off to Europe to earn money for their missions. The Seri of Sonora, Mexico, used the twigs as a remedy for toothache. The Tohono O'odham still use it. Because brittlebush transplants easily and requires little maintenance, it is used today in Arizona to prevent erosion near highways.

Asteraceae

White Bur Sage
Ambrosia dumosa, burrobush, burroweed

Ragweed is a scourge to most people and this plant is one of its relatives. But it is far more useful in every way: it is the nurse plant for creosote and isn't bothered by the toxins the latter sends into the soil. White bur sage has brittle branches, silvery leaves, and inconspicuous flowers that appear

Encelia farinosa, brittlebush

between March and May. The fruits are like little burs, hence the name. The parasitic plant, sand food *(Pholisma sonorae)*, may grow on the roots of this pale, woody, 2-foot shrub.

Animals graze on white bur sage, especially when conditions are dry and there is not much else around to eat. It is often planted along highways in California for erosion control and landscaping, because it requires little water and no maintenance. Naturalist Ruth Kirk writes: "The bur sage . . . is a species that by itself produces a great deal of organic debris. New shoots are continually sent out from a subsurface crown. These root and produce offspring shoots of their own, thereby perpetuating the individual shrub into virtual immortality—forever the same bush yet constantly new."[83]

Range: **southern Nevada, southwestern Utah, Mojave Desert, southern California, and Sonoran Desert.**

Simmondsiaceae

Jojoba
Simmondsia chinensis, goat nut, coffeebush, quinine plant

Simmondsia chinensis

is a wide-spreading shrub with small, blue-green, leathery, oval leaves; may grow to 6 feet tall, 8 feet in moist sites. Between March and May, it produces separate male and female bushes: the male flowers are yellow and hang down in clusters, the females appear singly, without petals. After pollination, the female flowers give way to fruit capsules containing large acornlike seeds up to 1 inch long. The seeds can remain viable in the soil for years before sprouting. It can live 100 to 200 years.

Range: **Sonoran Desert regions of southern California, Arizona, and northern Mexico.**

Now familiar as an everyday ingredient in so many things we use, jojoba (pronounced ho-HO-ba) was taken from the wild and domesticated only within the past 100 years. This is a rare phenomenon in the plant world. Though there are 200,000 plants worldwide, we seem to have done all our domesticating a long time ago and never taken a step further. Jojoba is an exception.

Long known as a therapeutic plant, the seeds contain a unique chemical compound of long-chain fatty acids and fatty alcohols. Native Americans rubbed the oil from the seeds on sores to speed healing. They also ground them up to produce a coffeelike drink, or dried and ground them to make a food with the consistency of peanut butter. It has also been a folk remedy for cancer as well as for colds, headaches, obesity, poison ivy, and warts. The oil was used by people in Mexico as a hair restorer. The Seri applied jojoba to head sores and aching eyes, and drank jojoba-ade for colds and to facilitate childbirth.

Entrepreneur Gary Tremper has a company devoted to this plant and its history. He supplies oil for commercial uses and has another explanation for the plant's name:

The credit for naming jojoba has been assigned to several individuals from Hernan Cortez the despoiling conquistador of the Aztec culture, to a Jesuit priest Francisco Clavijera in Baja California. It is assumed that the native cultures had been using the seeds for centuries as a food, the oil that was taken from the seeds as a cosmetic and on skin afflictions and wounds as a medicine. The story goes that an unknown observer asked the Native Americans the name of the oil they were rubbing on their bodies and hair and they wrote the name "jojoba." And so it came to pass that a native plant of North America began its exodus out of obscurity into the modern world of today.[84]

Jojoba oil's enormous potential was first noticed in 1933, when a paint company began to use it in its paint mixtures. During the Second World War, when other sources of oil were in short supply, jojoba oil became an additive in motor oil, transmission oil, and differential gear oil because it could withstand high temperatures; "even machine guns were lubricated with jojoba as it wouldn't gum up as petroleum would."[85]

Simmondsia chinensis, jojoba

The jojoba industry really took off in 1972. In 1969, the sperm whale was put on the endangered species list, and sperm whale oil, widely used as a lubricant in many industries, was banned. Trials had shown that jojoba oil might be the ideal substitute for sperm whale oil. Researchers from federal agencies and the University of California, working with the residents of the San Carlos Apache Reservation, began to collect and process the seeds. By 1982, the seeds were being commercially grown and harvested, making jojoba the first wild plant to be domesticated in the United States in more than a century.

Today, jojoba oil is used in cosmetics, pharmaceuticals, waxes, candles, varnishes, rubber adhesives, linoleum, detergents, disinfectants, and emulsifiers. It is so popular that in some areas, widespread planting of jojoba is driving out native vegetation. Cultivated, irrigated trees are usually much bigger than those growing in the wild. The United States currently produces about 880 tons a year, and this number is increasing each year. There is even a possibility that jojoba may have a terrific future as an edible low-calorie nonsaturated oil.

Liliaceae

Sego Lily
Calochortus nuttallii, mariposa lily

Calochortus nuttallii

has white, three-petaled flowers 1 to 2 inches wide that appear between May and July from an unbranched stem 6 to 18 inches tall. The petals may have a red or purple tinge.

Range: southwestern and western United States as far north as eastern Montana and western North Dakota.

State flower of Utah.

In 1903, the sego lily was on the Utah state flag designed by the Daughters of the American Revolution. Today it is the state flower of Utah. Dazzled when they came upon the flower, the early Spanish explorers described it as being like butterflies and called it *mariposa.*

Native Americans used to eat the walnut-sized bulbs of these plants and taught Mormon settlers to do the same in winter, when food was scarce. In the 1840s, when Utah suffered a plague of crickets, these bulbs were among the few things the Mormons had to eat during a long period of food rationing.

Poaceae

Big Galleta
Hilaria rigida syn. *Pleuraphis rigida*

Big galleta may look like a shrub, but it is actually a type of grass. After a rainfall, it is green and attractive to cattle, horses, and other animals. It is sometimes planted on dunes to keep the sand from blowing. It is, however, losing ground to creosote and other bushes, which invade areas where the grass has been eaten away by livestock. A group in California (the Mohveg Project) is trying to reseed desert areas with big galleta and other desert grasses to restore the natural ecosystems, which were held together by grasses. It is coarse, tough, drought-tolerant grass growing in 3-foot-tall clumps on sand dunes and rocky slopes.

Range: **throughout the Southwest.**

Lennoaceae

Sand Food
Pholisma sonorae

Pholisma sonorae
is a flowering parasite that grows in sand dunes. It has 6-foot succulent, underground stems. It sends up a 5-inch flower head.

FOLLOWING PAGE: *Calochortus nuttallii*, sego lily

Range: **California, Arizona, and northern Mexico.**

This is one weird plant. It lacks chlorophyll, does not photosynthesize, and lives almost entirely underground. The thick, scaly stem extends as deep as 6 feet into the dunes in search of the roots of shrubs such as bur sage or creosote. It locks on to them and sucks up nutrients. In the spring, the flower head, which looks like a gray mushroom, pokes up above the surface of the sand to flower and find pollinators. The 5-inch-diameter flower head is actually dozens of tiny lavender-colored flowers all clumped together. Flies, beetles, and butterflies visit, and after pollination, the plant puts forth a dry capsule containing 12 to 20 seeds arranged in a circle.

The Papago roasted and ate the stems, which tasted like sweet potatoes. Writer Ruth Kirk describes it this way:

Sand Papagos could locate the plant at any time of year, although to the eyes of modern men no hint is apparent except during the brief flowering period in May. Some inexplicable clue evidently guided the Indians, much as Bushmen in the Kalahari Desert are able to rely on bi *and* ga *tubers totally indiscernible to outsiders; and, as in the Bushmen's case, the fleshy stalks of sand food substituted for water, carrying men through days when death would otherwise have been certain.*[86]

Gary Paul Nabhan spoke to one of the O'odham people about sand food, who told him, "Because it doesn't stick up above the ground like other plants, I had to learn to see where the little dried-up ones from the year before broke the surface. That's where I would dig."

California

To most of the world, California defines North America. It has been depicted and reflected in movies, plays, television, and song for over a century—the promised land, the golden Eden filled with sweet fruit and endless sunshine. California has spoiled us. We know the problems of its sprawling cities and lush fields that feed a continent, but we love the movies and enjoy eating out-of-season vegetables, without giving a thought to what has been done to the native soil and the surrounding landscape.

California is not just blessed with good

California has more native species than any other region of North America.

weather and green fields, it is the state richest in flora as well. In fact, it possesses more varieties of native plants than all of Canada. According to the California Native Plant Society, there are about 6,300 native flowering plants, conifers, and ferns. Yet in spite of this seeming plenty, the statistics are frightening: some say that native grasslands are down to 1 percent of their original extent (some researchers argue that none of them are left, since they were replaced by Mediterranean annuals 150 years ago) and vernal pools have almost entirely disappeared.[1]

California's varied topography and climate support hundreds of endemic species—plants that have evolved to grow under very specific conditions, usually in small, restricted areas. Only certain areas in Chile, around the Mediterranean, Australia, and South Africa have anything resembling the weather of California and, therefore, some of the same plants. Since California's economy took off in the 1950s, many of these plants, already rare, are becoming scarce and their long-term survival is endangered.

Today, it is America's most populous state—but that's nothing new. When Columbus made it to the eastern shores, there were more Native Americans living in California than in the rest of the continent. It was home to between 300,000 and 2 million people in 120 ethnic groups, speaking up to 200 distinct languages.

They were healthy people who ate fish, birds, animals, and a huge variety of seeds, roots, shoots, and acorns. Entire civilizations were built around acorns, which were easy to store and high in oil, though not an easy resource to process.

European settlers faced landscape and culture shock wherever they went in North America, and nothing could have prepared them for California. They didn't know what to make of this landscape of dwarf trees, unusual grasses, and wildflowers that flourished in carpets of brief glory. So they cleared whole areas, replacing native plants with their own, cutting down trees, and shoving the Indians aside. They piped and channeled water in from elsewhere. Any bit of flat space was sown with exotic alien crops (mustard, crabgrass, and thistle trailed in their wake). They introduced grazing animals, which stripped the landscape and drove off local wildlife.

Once the land was irrigated, European plants thrived as escapees into the wilderness. The native species did not. They had evolved methods of surviving long hot summers with little water. They could make chlorophyll in their wood as well as their foliage. They could fold up their leaves or even drop them to conserve moisture, but, most important, they held down the soil, thin and poor as it was.

The changed environment became dangerous. By the 1880s, according to Stephen J. Pyne, this was one of the most fire-prone environments in the world. Early on in settlement history, burning, which had formerly occurred as natural, low-intensity burns, began to be controlled. It led in the twentieth century to what Pyne calls "the mercurial mixture of suburb and brush, neither able to drive the other out."[2] Hot fire leveled the chaparral, and when the land was cleared, hotter fires ensued, leaving more land cleared, ready to be developed. Anything could set off a fire, he says, from little boys with matches to power lines arcing during a Santa Ana wind. This wind sucks all the moisture out of the air (like the mistral in France), setting up the perfect conditions for fire. Nathanael West ends his novel *The Day of the Locust* with an apocalyptic fire. He was, no doubt, thinking of the chaparral fires that often consumed entire hillsides.

In the late 1800s, Theodore Payne, an Englishman, predicted California's native wildflower population would disappear if something wasn't done to protect it. He had fallen in love with the wildflowers of California at a time when most people ignored what was growing around them. He opened a nursery and sold the seeds of native plants, mostly to Europe (his 1906 catalog listed 56 varieties). By the end of his life in 1963, he had introduced 400 to 500 species of Western plants into cultivation, turning what most people regarded as weeds into superstars. Many of them he saved as development was carpeting the hills around Los Angeles. Today, the Theodore Payne Foundation is devoted to preserving California flora, as are a number of other foundations and strong native plant associations. Cherishing the incredible diversity of flora, they see the plants of the chaparral as individuals, and the land's stark and compelling beauty as something to be treasured.

This section is devoted to California's scrub,

grasslands, and chaparral and their distinct plant communities. Its rain forest, mountain, desert, and marsh plants are described in other sections. The plants here often become dormant during the long, dry, hot summers and autumns, then grow profusely in the short intense winters, when they are deluged with rain. Fire is as important and life-giving to this land as the rain, rejuvenating the plants, prompting seeds to germinate, and restoring nutrients to the soil for the next generation.

Papaveraceae

California Poppy
Eschscholzia californica, flameflower

Eschscholzia californica

is an annual or short-lived perennial flowering plant with a heavy taproot, which grows up to 2 feet tall. Purple when new, the stems turn green and bend over as the plant grows. The gray-green leaves are finely segmented. Showy yellow or orange flowers, 2 inches across, appear at the end of the stems in early spring, unfurling from fused sepals that resemble a dunce cap. Wild varieties have four petals, but cultivated varieties may be different. The fruit is a thin, ribbed capsule, 3½ inches long. As the capsule dries, it twists open with a pop, dispersing the tiny black or brown seeds.

Range: grassy and open places.
State flower of California.

Eschscholzia californica, California poppy

The most potent image of California is one of brilliant poppies blooming on its hills, tucked next to rocks, and in ditches. It is the state's official flower, protected by a state law with a stiff fine for picking or mutilating. The law is intended to keep this native poppy species from being wiped out by road construction and land development. (The ban doesn't apply to homegrown poppies cultivated from seed.) The golden blooms have inspired effusive praise ("the gladsome beauty of this peerless flower," wrote Emory E. Smith in 1902[3]), poetic outpourings ("whole hill-slopes streaked along the streamlines with nearly stemless, pale-gold cruciferae"[4]), and songs ("Golden Poppies Goodnight"[5]).

The first botanist to record this glorious poppy was Archibald Menzies, the surgeon and naturalist on George Vancouver's expedition (1792–1793). He brought poppy seeds to Kew, but they produced fragile plants that died without reproducing. But 25 years later, Adelbert von Chamisso, naturalist and member of the Prussian Academy of Sciences, rediscovered the poppy and cultivated it successfully. The scientific name honors von Chamisso's friend Johann Friedrich Eschscholz, the physician and naturalist on the Russian expedition led by Otto von Kotzebue to California. In Spanish, it is called *la amapola* (flameflower), *copa de oro* (cup of gold), or sometimes *dormidera*, which means "the drowsy one" because the flowers close at dusk.

According to Jake Brouwer:

In the early days of Pasadena and Altadena, the hills and fields below the foot of Echo Mountain were covered with poppies much to the delight of all who laid eyes on them. The Spanish whilst sailing up and down the newly discovered coast, looked inland and saw the flame of orange red poppies spread upon the hillsides. This was the "land of fire." They said the altar-cloth of San Pascal was spread upon the hills and with devotional spirit the explorers withdrew from their ships to worship on the shore.[6]

The color was so bright these early navigators could spot the flowers from 25 miles away and set their course by them.

Unlike other flowers, the poppy's sepals (the leafy structures outside the petals) are joined together, forming its distinctive "elf cap." When this separates and the petals unfurl, the sepals usually fall off, leaving a rim called the torus, a distinguishing feature of the California poppy.

The four fan-shaped petals unroll on warm, sunny days. At night or on chilly, rainy days, the petals recoil and almost disappear. The flowers are pollinated by bees and butterflies.

Photographer John O'Neill describes it thus: "An unabashed sun-worshiper, the California poppy thinks nothing of closing up shop during a cold, overcast day or lingering north wind . . . but will continue to bloom until it has exhausted the last bit of available moisture and its flowers have shrunk to the size of buttercups."[7]

The California poppy colors many Native American legends.

[On the journey to the afterlife] the soul comes to a place where there are two gigantic qaq *(ravens) perched on each side of the trail, and who each peck out an eye as the soul goes by. But there are many poppies growing there in the ravine and the soul quickly picks two of these and inserts them in each eyesocket and so is able to see again immediately. When the soul finally gets to* Šimilaqša, *it is given eyes made of blue abalone.[8]*

The California poppy contains many of the poppy family's analgesic compounds without the narcotic opiates. The Yuki relieved a toothache by placing the fresh poppy root into the cavity of the tooth. The root extract was used as a wash or liniment for headache or open sores, and taken internally to induce vomiting, cure stomachache, or prevent tuberculosis. The Kashaya Pomo applied mashed seedpods to a nursing mother's breast to dry up milk when it was time to wean a child. Many native peoples ate the leaves as greens.[9] Spanish Californians boiled the leaves with olive oil and added perfume to make a hairdressing lotion. Today, California poppy is used as a nonaddictive herbal sedative, and as an alternative medication to reduce "overexcitability" in children. (This echoes, in an extreme form, the traditional advice of putting one or two flowers under the bed to get a child to sleep.)[10]

We looked forward to the coming of the poppies, did Bess and I, looked forward as only creatures of the city may look who have been long denied. . . . The first poppies came, orange-yellow and golden in the standing grain, and we went about gleefully, as though drunken with their wine, and told each other that the poppies were there. We laughed at unexpected moments, in the midst of silences, and at times grew ashamed and stole forth secretly to gaze upon our treasury. But when the great wave of poppy-flame finally spilled itself down the field, we shouted aloud, and danced, and clapped our hands, freely and frankly mad.

And then came the Goths. . . . At the far end I saw a little girl and a little boy, their arms filled with yellow spoil. Ah, I thought, an unwonted benevolence burgeoning, what a delight to me is their delight! It is sweet that children should pick poppies in my field. . . . I did not glance out of the window again until the operation was completed. And then I was bewildered. Surely this was not my poppy field. No—and yes, for there were the tall pines clustered austerely together on one side, the magnolia tree burdened with bloom and the Japanese quinces splashing the driveway hedge with blood. Yes, it was the field, but no wave of poppy-flame spilled down it, nor did the great golden fellows nod in the wheat beneath my window. I rushed into a jacket and out of the house. In the far distance were disappearing two huge balls of color, orange and yellow, for all the world like perambulating poppies of cyclopean breed.

Jack London, from
"The Golden Poppy," in
Revolution and Other Essays

Gold miners setting sail for new fields in Chile, New Zealand, and Australia used sand from the bluffs at San Francisco as ballast. The sand contained poppy seeds, and since then, those countries in the southern hemisphere have had a new, but not unwelcome, golden weed.[11]

April 6 is official California Poppy Day. Potter and artist J. R. Dunster waxes enthusiastic on the celebration:

It is hard to describe how much this little flower means to Californians. For a lot of us, it is a big deal. We . . . announce on the news and newspapers when the poppies are in bloom. We flock out in droves to see hillsides saturated orange with these glorious wildflowers. We absolutely love our state flower![12]

Pinaceae

Pinyon Pine

Pinus edulis, piñon pine, Arizona pine, Arizona tall pine, Colorado pine, Colorado pinyon, foxtail pine, nut pine, two-leaved pinyon, two-needled pinyon

Pinus edulis

is a small tree that grows 10 to 50 feet tall, with a rounded crown and red-brown bark with shallow furrows. The dark green needles, 1 to 2 inches long, come in bundles of two. Oval, yellow-brown seed cones sit erect on the branches, 2 to 3 inches high. The large, oily seeds are edible. It is a long-lived, slow-growing, monoecious species that can survive for up to 1,000 years, but most mature trees in California are 250 to 400 years old. Significant cone crops start when the tree is 75 to 100 years old.

Range: **Utah, Colorado, Arizona, Wyoming, Texas, New Mexico, California.**

Pinyon pine is the source of the pine nuts beloved by pesto makers. The word *edulis* means "edible," referring to the nuts, also known as pinyon nuts, piñones, or just pinyons. In the wild, deer, black bears, squirrels, and wood rats relish the seeds, along with birds such as wild turkeys and pinyon jays, which have a special relationship with this tree. A single jay can hold up to 56 pinyon seeds in its expandable esophagus. The birds cache the seeds just under the leaf litter, an ideal spot for germination. So close is the relationship between pine and pinyon jay that the tree produces different-colored seed coats, which help the birds distinguish between edible and aborted seeds, and orients the cones to make the seeds more visible to jays.[13]

Heavy crops of cones are produced every four to seven years and each cone bears an average of 12 to 20 seeds. Once dispersed, the seeds remain viable up to three years—if they aren't gobbled up first. But it regenerates entirely by seed, so the pine has to get those seeds out there and dispersed into favorable ground.

People's fondness for pine nuts goes back well before the days of pesto sauce, to prehistoric times. In central Nevada, the seed coats of pinyon pines have been found in the remains of human shelters at least 6,000 years old. Some anthropologists believe that the pinyon pine helped Native Americans of the Southwest evolve into an agricultural society. The nuts were an early staple food high in fat and as protein-rich as prime steak (and a good deal less damaging to the arteries), and may have been used in trade for corn, beans, and squash from other tribes. An isolated stand of *P. edulis* in Owl Canyon, Colorado, lies about 93 miles north of the nearest pinyon pines; it may have been planted with pine nuts carried along an ancient trading route.

Harvesting pine nuts was a family affair. American conservationist John Muir wrote about harvesting of the pinyon nut by the Mono Lake Paiute Indians of California in about 1870:

When the crop is ripe, the Indians make ready the long beating-poles; bags, baskets, mats, and sacks are collected; . . . all are mounted on ponies and start in great glee to the nut-lands. . . . Arriving at some well-known central point where grass and water are found, the squaws with baskets, the men with poles ascend the ridges to the laden trees, followed by the children. Then the beating begins right merrily, the burs fly in every direction, rolling down the slopes, lodging here and there against rocks and sage-bushes, chased and gathered by the women and children with fine natural gladness.[14]

The Native Americans of the Great Basin region collected cones with hooked sticks and either stored them or opened them over a fire for immediate use. The people of the Southwestern Pueblo cultures took to temporary camps in the mountains in autumn when the nuts began to fall. They either picked the nuts off the ground or shook the trees to release the nuts onto blankets spread out below. A really efficient nut gatherer could collect up to 20 pounds of nuts in a single day. After hauling the harvest home in wagons, the harvesters roasted the nuts on a griddle and stored them in earthenware jars for the winter.[15]

The nuts were often ground for flour. The Navajo mashed them into a butterlike paste to spread on hot corn cakes. The Chiracahua and Mescalero Apache mixed them with yucca fruit for a pudding, or ground and then rolled them into balls. The Mescalero Apache mixed ground pine nuts with datil fruit, mescal, mesquite beans, or sotol, while the Western Apache mixed them with corn flour for mush. The nuts were not the only food from the tree. The immature female cones were roasted until they turned soft and syrupy in the center. The inner bark, which sweetens with cooking, was cut into strips and prepared like spaghetti.

Pinyon is medicinal as well as a source of food. The Havasupai, Hopi, Isleta, Western Keres, Navajo, and Tewa put the gum or pitch on sores and cuts. The Mescalero Apache inhaled the needle smoke to relieve cold symptoms, and the White Mountain Apache chewed the leaves as a treatment for venereal disease. The inner bark made an expectorant tea, and an infusion of the leaves was used as an emetic. Before the invention of the razor, heated pitch was slapped on the face and pulled off to remove facial hair.

The wood was also used to make houses, corrals, and furniture such as cradles. The Ramah Navajo used the resin in pottery and basketry, and the Hopi, Jemez, and Tewa added it to red and black dyes. The black dye, made by combining the gum with sumac leaves, twigs, and native yellow ocher, was used to color leather and buckskin. Many Native American groups used the resin and pitch to waterproof woven water jugs or baskets.[16]

Today, the fragrant wood is a popular firewood because it burns so well and doesn't throw off sparks. It is also occasionally processed into charcoal. Pinyons are now the thirteenth most important Christmas tree species grown commercially in the United States. Woodworkers have been known to make small items out of pinyonwood and, because it is tough and resists breakage, a few sawmills use it to produce railroad ties and shoring timbers for mines.

But the big business lies in the nuts: the annual crop of pine nuts ranges from 1 million to 3 million pounds, second in value only to pecans among uncultivated nuts in the United States. Every autumn, residents of the areas in which pinyons grow, especially the Navajo and the Spanish Americans, harvest quantities for the local and gourmet markets. The nuts are sold raw or roasted, and sometimes shelled.

Pinaceae

Monterey Pine

Pinus radiata, insignis pine, insular pine, insular two-leaved pine, Monterey small-coned pine, radiata pine, remarkable cone pine, remarkable pine, spreading-cone pine

Pinus radiata grows up to 100 feet tall. It has a symmetrical form when young, but develops an irregular rounded crown as it ages. The thick, ridged, and furrowed mature bark is dark reddish-brown and the branches are slender and dark orange. Its leaves are slender, shiny, and green, up to 6 inches long; they grow in groups of three, and remain on the tree for up to three years. Male cones are yellow-brown, $\frac{1}{2}$ inch long, in clusters on the tips of branches; female cones are up to 5 inches long, unevenly conical, with little or no stalk, and point downward. The cones may remain on the tree for several years. The winged, paper-thin seeds are dark and pimpled.

Range: rare in the wild, but one of the most commonly planted trees in the world; native to the central coast of California, around San Mateo, Santa Cruz, Monterey, and San Luis Obispo Counties.

The Monterey pine stands near Carmel have been so well photographed that, for many people, they symbolize this part of the California coast, as they have for hundreds of years. "In 1602, the Spanish government commissioned

Pinus radiata, Monterey pine

Sebastian Viscaino to map the coastline; he traveled as far north as the Mendocino coast. In his journal, [he referred to the] 'great pine trees, smooth and straight, suitable for the masts and yards of ships' that he saw while anchored in Monterey Bay."[17]

Pinus radiata was noted in 1830 at Monterey by Irish botanist Thomas Coulter (1793–1843), after whom another pine native to California was named (*P. coulteri*). *Radiata* refers to the strong markings on the cone scales. The cones are serotinous, which means that they remain closed until opened by heat, such as the heat of a forest fire or, to a lesser degree, hot weather. The abundant seeds are then discharged to regenerate the burned forest.

Unfortunately, the three remaining native stands of this pine in California are under threat of extinction from pitch canker (*Fusarium subglutans pini*), a fungal disease plaguing the southeastern United States, introduced into California in the 1980s. When trees begin to sicken, they attract bark beetles, which provide a pathway for infection. In some stands, 80 to 90 percent of the trees are infected. The Sierra Club voted unanimously to support listing the Monterey pine as "threatened" in July 1998. It is hoped that countries such as New Zealand, where the tree has been widely planted, are spared from pitch canker. University of California scientists are working with seedlings of trees showing no symptoms to try to develop a fungus-resistant variety.

The wood of Monterey pine is light, soft, brittle, and coarse-grained. Because so many trees are diseased, it is not used commercially in the United States, except occasionally for rough lumber or fuel; in other countries, though, it is a highly valued wood. Most Monterey pines in the

United States are planted as ornamentals in parks and urban areas. Because it grows so quickly, it's planted as a windbreak or on slopes to control erosion, or on barren areas where blowing soils are a problem, and is sometimes used for Christmas trees.

A Native American story gives an unusual meaning to the phrase "the eye of the needle." In this legend, Bat and Pine Squirrel have fought and Pine Squirrel has plastered pitch over Bat's eyes to blind him.

Bat felt his way down the country. Night came on and it got cold. He camped and went to sleep, thinking about what to do. Along about daylight, he started singing and singing and singing. Bat lay there, looking up. North Wind came and blew the pine needles on the pine trees. A pine needle swept right down and poked Bat in the eye. That is what Bat wanted it to do. Bat could see daylight then. He kept on a singing, and another pine needle came down and hit the other eye. He looked around. "That's good." He quit singing. He had eyes. He could see all right then. Bat camped around the lava flats, and he lives there yet. . . . And Bat's eye? Why, it's tiny, just the size of a pine needle.[18]

Taxodiaceae

California Redwood
Sequoia sempervirens, coastal redwood, palo colorado

Sequoia sempervirens

is a tall, fast-growing, monoecious, evergreen conifer that usually reaches about 300 feet tall. The needles are ½ to 1 inch long, a dark yellow-green, lined up in two rows; at the tips of new twigs, the leaves are scalelike. The branches droop downward. The cones have open scales. Life expectancy is about 1,000 to 1,500 years.

Range: in the coastal forest from California to Oregon.

The redwood, one of the oldest trees on earth, was scattered throughout the northern hemisphere about 60 million years ago. Now the redwoods are confined to a narrow coastal belt 5 to 35 miles wide and 450 miles long, including the Avenue of the Giants along Highway 101 north of San Francisco.[19] When Don Gaspar de Portolà found them in 1769, there were almost 2 million acres of virgin forest along the coast.[20] Once sawmills were built, beginning in the mid-nineteenth century, about 95 percent of the timber was culled over the decades that followed.[21] Now the redwood's range has shrunk to about 100,000 acres.[22]

Redwoods that evolved in the Pacific climate came to depend on summer fogs to compensate for the lack of rain. The fog condenses on the leaves and water runs down the trunk, dropping off nutrients as it goes. It has shallow roots spreading out 50 feet in every direction, tending to interlock with the roots of other redwoods. Trees in the middle of the forest stand firm. The ones at the edges may fall over if the soil is saturated and the wind is right. This is where Native Americans would gather up the wood.

Redwood is able to tolerate periods of flooding and the burial of its root system by sediment by sending up vertical roots. Then it sends out a horizontal root system below the surface of the deposit. Eventually, a new root system develops just below the ground close to the trunk. With each new pile of sediment, a

FOLLOWING PAGE: *Sequoia sempervirens,* California redwood

root system is formed. As a new one forms, the older one dies, leaving only one root system active at a time.

Among its many great mysteries is why the great redwood is so incredibly durable. Insects and microbes seem to make little headway in breaking down its tough wood. The red in its wood contains chemicals that are either revolting or toxic to most other living creatures. These tannins and other phenolic compounds are highest in the heartwood of the trunk. Padre Junípero Serra, who started several missions in California, asked to be buried in a redwood coffin at Carmel in 1784. The roof of the mission collapsed in 1852, and when restoration work was started in 1882, the coffin was found to be in perfect condition.

The Native Americans who lived along the coast generally stuck to using fallen timbers, rather than trying to fell these unwieldy giants and split the wood. They made shelters from debris picked up at the edges of the forest, and used the wood for sweat baths and occasionally canoes. There are a few recorded instances of people burning down large redwoods to get at the supply of acorns that birds and squirrels had cached in the upper parts of the trees, but mainly the tree was too intimidating to mischief-makers.

The first Europeans to see the redwood were the Spanish explorers who came to California in the eighteenth century. The Spaniards called it *palo colorado*, or red tree. Father Juan Crespi, Franciscan missionary and diarist of the Don Gaspar de Portolà expedition (first by land up the Pacific Coast) in 1769, recorded it in the vicinity of the Pajaro River in Monterey County.

Over plains and low hills, [it is] well forested with very high trees of a red color, not known to us. They have a very different leaf from cedars, and although the wood resembles cedar somewhat in color, it is very different and has not the same
odor; moreover, the wood of the trees that we found is very brittle. In this region, there is great abundance of these trees and because none of the expedition recognizes them, they are named redwood from their color.[23]

In the sixteenth and seventeenth centuries, Juan Bautista de Anza, the Spanish explorer, started from Monterey to journey through the San Francisco peninsula. He made a complete survey of the Bay region and first reported the East Bay redwoods. He and his colleagues used two of the tallest trees as a navigational point and they were marked as such on early maps. When someone suggested chopping down these great space-grabbing trees, every sailor in the area rose up in protest.

The first collection of specimens was made by Thadeys Haenke, a German botanist with the Alejandre Malaspena expedition of 1791, which landed at Monterey Bay. He sent back seeds collected from the Santa Cruz area but didn't bother naming it. In Spain there is still one redwood growing from seed collected on this expedition. In the end, it was the intrepid Archibald Menzies who sent back a specimen, in 1794, which became the basis of the description by Aylmer Bourke Lambert in 1824. Lambert thought it looked like a member of the Taxodium family so he called it *Taxodium sempervirens* (because it is evergreen, not just because it can live for thousands of years). In 1847, botanist Steven Endlicher called the coast redwood *Sequoia sempervirens*.

It has been suggested that *Sequoia* was named after a Cherokee known as Sequoya or Sikwayi who developed the alphabet for a written Cherokee language. This led to the founding of the first Indian newspaper, *The Cherokee Phoenix*, and to written laws. The tree is said to have been named to commemorate his death in 1843. Another theory is that the great American botanist Asa Gray invented the name from the Latin word *sequi* or *sequor*, meaning "sequence" or "following"—an allusion to the fact that redwoods were remnants or followers of numerous fossil ancestors.

The Spaniards prized the wood and used it to make rafters, beams, and door and window frames. They banned the practice of burning. When the Russians moved in from Alaska, expanding their fur trade to include the otters living and breeding on the Pacific Coast, they felled numerous redwoods. The Mexicans who succeeded the Spaniards in 1822 relaxed the Spanish rules against burning. But their destruction of the forest was nothing compared to the inroads of the Yankees who arrived in 1834 and immediately started cutting down the biggest and best trees. The first large-scale cutting in the redwood region started in Monterey County.

At first, the trees were cut one at a time, by two men with a large saw. By 1852, however, technology had improved so much that 50,000 board feet a day were being produced from steam mills. Lumbermen in those days selected the best trees rather than clear-cutting, the method of choice only a few decades later. Never before had men dealt with 300-foot trees weighing 300 tons each. A single log might weigh up to 60 tons. They had to build mills at the mouths of rivers and streams to accommodate them. It sometimes took a week to fell a large tree with their seven-foot saws and double-bladed axes. If the tree fell the wrong way, it shattered, so there was a huge amount of waste as gigantic logs were lost in rivers and streams.

After 30 years of logging, the 12-foot crosscut saw was developed. But the ultimate blow to the forest was the steam-donkey engine. A system of ropes and cables moved logs through the woods, hauling big trains of logs. New machinery was brought in that could lift giant logs up into the air, swing them down the mountainsides to the landing area near the donkey, then load them onto railroad cars and send them to mills. Not only were selected trees taken out, everything around them was knocked down in the process. The debris was allowed to dry on the ground for a year and then burned. It left the forest vulnerable to wildfires, which meant seedlings didn't stand a chance. The lands were then sold to ranchers.

Though the damage to these great old forests was appalling, the wood spurred the building boom in California and was used for railroad ties that led to even more development. It was also used to make casks, since redwood imparts no flavoring to liquid stored in it. Early California vintners constructed enormous tanks of redwood which absorbed liquid and swelled to create a very tight seal for long-term storage of aging wine. Today many of these old tanks, still in pristine condition, are used for water storage.

By 1900 Mendocino county was largely burned over—mostly from accidental fires spreading out of control. Fire control became aggressive after the turn of the century in redwood country, and laws were enacted to control burning. The logging, however, continued apace. Replanting programs were introduced, but the second-growth forest could not replace what had been lost.

The battle for the forests continues. "Environmental activist Julia 'Butterfly' Hill lived on a 200-foot-tall ancient redwood tree named Luna from December 19, 1997 to December 18, 1999, above the community of Stafford, California, about 230 miles north of San Francisco. 'Here I can be the voice and face of this tree, and for the whole forest that can't speak for itself,' Julia said."[24] In spite of her courage and the enormous amount of information now coming from the canopies of these old trees, these remnants are still under threat because of encroaching development.

Taxodiaceae

Sequoiadendron giganteum

is an evergreen conifer that can grow up to 300 feet tall (some larger specimens have been documented) and may be as much as 100 feet in circumference at the base. The small blue-green leaves on mature trees cover the twigs like tiny overlapping scales. The bark is reddish and deeply seamed, and the cones are oval and reddish-brown with thick scales. As the tree ages, it expands in girth more than height.

Range: western slopes of the Sierra Nevada.

Sequoiadendron giganteum, giant sequoia

Standing within a grove of the giant sequoia is a breathtaking experience. The wind has a different sound in here. The tops of the trees seem so very distant that it makes a viewer dizzy looking upward. The giant sequoia is the largest of all living things. Once upon a time sequoias covered huge expanses in Europe, Greenland, and North America. It became extinct elsewhere about 2 million years ago. Today it is restricted to 75 groves covering 35,607 acres along the western Sierra Nevada in California, about 4,500 to 7,500 feet above sea level. Each grove is named. They survive only because some stands are so remote they are seldom visited.

Natural historian Donald Culross Peattie notes: "The lowest [bough] is still so high above the ground that it would stretch out over the top of a twelve-story office building."[25] The largest specimen is known as General Sherman and grows in Sequoia National Park. It's estimated to contain 30,000 board feet. It is 274.9 feet high, 102 feet in circumference at the ground; the height of the first branch is at 130 feet; the diameter at 60 feet above the base is 17.5 feet; the diameter of the largest branch is

6.8 feet. It's almost impossible to grasp the scale of this tree and even less possible to photograph it in a way that captures the effect of seeing it in person.

What keeps the giant sequoia humming along century after century is the remarkable sap, which contains tannic acid. The sap protects it from fire, and has the healing properties that help it recover after a fire. Stories abound of living trees with holes created by fire that were so big a man on horseback could ride through them. John Muir, who wandered around the California groves in the 1880s, found fallen logs that showed no signs of decay after 10,000 years.

A young tree has a smooth gray bark, soft blue-green leaves, and a conical shape from its tip to the lower boughs of its skirt. After 100 years, the strong central trunk drops its lower boughs and the foliage becomes a more metallic green. The red bark indicates that it is coming into maturity. As the centuries wear on, the ridges become more marked and deeper in color.

The seeds are so tiny it takes 3,000 of them to make an ounce, and there are up to 300 seeds to a cone, which is only 3 to 3½ inches long. When it flowers, millions of male and female cones (the tree is monoecious) are released and fall to the ground. Not many have even the ghost of a chance of getting to germinate—they are much too tasty. Birds and animals eat most of them. Seeds germinate where they fall, but need soil with a high mineral content and plenty of sunshine. Any seeds that manage to sprout are usually nibbled by deer. Few survive the first year to grow into trees.

The tree can also reproduce asexually, since root crowns send up shoots. In the 1800s the loggers left very high tree stumps, and shoots appeared on the stumps. In some place the shoots have been cut off again and again, yet they continue to come back. Because of its thick bark,

Sequoiadendron is a fire-tolerant plant, which starts resprouting from the base after a fire.

The tree is seldom attacked by insects or pathogens. However, the gray squirrel may harm it by girdling the trunk. Deer browse the bark and if bears get hungry enough, they'll rip at it to get at the cambium.

Sequoiadendron first came to the attention of nonnative North Americans in the 1850s when a hunter chased a wounded grizzly into a forest in the Sierra Nevada and was astonished to find himself surrounded by enormous trees much bigger than most buildings. Pretty soon, the trees were being touted as a tourist draw, and parties of sightseers were brought by railroad or steamboat from San Francisco to a nearby town and from there by coaches to the Calaveras Grove in the Sierra Nevada. Some people were particularly impressed by the idea that many of the trees had been standing at the time of Jesus Christ. Others were simply awed by their sheer size. Photographers and painters flocked to the area. One engraving from this period shows a group of people dancing a cotillion on the sawn-off stump of an enormous redwood. Even today, we treat these trees like museum pieces, to be admired but having nothing much to do with the real world.

The first botanical description of the tree was written by English botanist John Lindley. Lindley named it *Wellingtonia gigantea* in honor of the Duke of Wellington and his triumph over Napoleon at Waterloo. It was added to the *Sequoia* genus in 1847 and became known as the Sierra redwood, *Sequoia giganteum*. In America some patriotic botanists tried to change the name some years later to *S. washingtonia*. Eventually it was accorded its own genus and became *Sequoiadendron giganteum*.

The story of the immense trees spread far and wide. Sequoia seeds sold like hot cakes in England and Europe. Every city, it seemed, needed an avenue of sequoias. Over there it was still known as *Wellingtonia* and Tennyson's son wrote: "The great event of the year 1864 was the visit of Garibaldi to the Tennysons, an incident of which was the planting of a *Wellingtonia* by the great Italian and ceremonies connected with it."[26]

In America, the tree eventually came to be seen less as an impressive ornamental than as a supply of wood. It was not much used for structural timber, since the trees tended to shatter when they were felled. Most giant sequoias were milled into fence posts, grape stakes, and shingles. In the late nineteenth century, however, a movement began to preserve the ancient trees. In 1864 President Abraham Lincoln signed legislation that established the Mariposa Grove as the first preserve for the giant sequoia. In 1890 the National Park of Sequoia was created in California (along with Yosemite and General Grant Parks).

John Muir took President Theodore Roosevelt to visit the forests of the Sierra Nevada in 1903. Roosevelt, who was duly impressed, said, "I feel most emphatically that we should not turn into shingles a tree which was old when the first Egyptian conqueror penetrated to the valley of the Euphrates."[27] In 1908, Roosevelt paid tribute to Muir by designating Muir Woods, a redwood forest north of San Francisco, a national monument.

Today, redwoods are cultivated as unique pieces of American heritage. In April 2000, Giant Sequoia National Monument was established when President Bill Clinton announced federal protection of the hitherto unprotected half of the giant sequoia ecosystem. This action is intended to preserve the trees, watersheds, and lands around the largest remaining groves of the giants. The monument includes about 400,000 acres of publicly owned national forest lands, which were added to the 402,000 acres of the existing Sequoia National Park.

Song of the Redwood-Tree

A California song!
A prophecy and indirection—a thought
   impalpable, to breathe, as air;
A chorus of dryads, fading, departing—or
   hamadryads departing;
A murmuring, fateful, giant voice, out of
   the earth and sky,
Voice of a mighty dying tree in the
   Redwood forest dense.

Farewell, my brethren,
Farewell, O earth and sky—farewell, ye
   neighboring waters;
My time has ended, my term has come.

Along the northern coast,
Just back from the rock-bound shore, and
   the caves,
In the saline air from the sea, in the
   Mendocino country,

With the surge for bass and
   accompaniment low and hoarse,
With crackling blows of axes, sounding
   musically, driven by strong arms,
Riven deep by the sharp tongues of the
   axes—there in the Redwood forest
   dense,
I heard the mighty tree its death-chant
   chanting.

The choppers heard not—the camp
   shanties echoed not;
The quick-ear'd teamsters, and chain and
   jack-screw men, heard not,
As the wood-spirits came from their
   haunts of a thousand years, to join the
   refrain;
But in my soul I plainly heard.

Walt Whitman

THE OAKS

Oak trees meant life not only to California's Native Americans but also to a host of birds and animals. Historian Malcolm Margolin says time itself was measured by oaks:

The acorn harvest marked the beginning of the new year. Winter was spoken of as so many months (moons) after the acorn harvest [summer as so many moons before the next acorn harvest]. The rhythms of the oak trees marked the passage of the year and defined the rhythms of life.[28]

At the height of its acorn-processing culture, California had roughly double the population density of the Great Plains. The acorn was everything to these people. It was central to their rites; it was central to their diets. The relationship held mutual benefits. By knocking the acorns off the trees with long poles, the native peoples knocked down dead and diseased wood, thereby stimulating new growth and discouraging harmful burrowing insects. Regular burning kept meadows surrounding the oaks clear of vegetation that might harbor disease and pests. The surface fires burned only the undergrowth, and protected the trees from hot, harmful crown fires.

A woman of the Karok Nation was recorded in 1933 describing the process her ancestors practiced:

Our kind of people never used the plow. . . . All they used to do was burn the brush at various places so that some good things will grow up. . . . And sometimes they burn where the tan oak trees are lest it be brushy when they pick up the acorns. . . . Some kinds of trees are better when it is burned off. They come up better ones again.

Fire suppression in recent times has allowed shrubs and conifers to encroach and overwhelm the oak savanna, eliminating valuable grasses and wild herbaceous plants that rely on fire to create a suitable seedbed.

Today, the oaks are threatened by urban sprawl, changes in habitat, a rise in acorn-eating animals through lack of predators, and worst of all—disease. Several species are being ravaged by a pathogen called *Phytophthora ramorum*. First seen in **tanoak** (*Lithocarpus densiflorus*) in 1995, it has spread to thousands of coast live oaks, black oaks, and Shreve oaks, resulting in what is called sudden oak death. A number of environmental organizations have sprung up to monitor the destruction, create awareness of the problem, and lobby for protection for all the native oaks—such a vital part not only of history but also of living ecosystems.[29]

Fagaceae

Coast Live Oak
Quercus agrifolia, encina live oak, California live oak

Quercus agrifolia

is a broad-topped tree that grows up to 80 feet tall. The bark may be whitish and smooth on an old tree, but is often dark brown to black, tinged with red. The large lower branches may rest on the ground. The leaves are usually wide and convex, 4 inches long, with a dark green, shiny upper surface, usually hairless and paler beneath, and smooth or finely toothed edges. The chestnut brown acorns are narrowly cone-shaped, slender, and pointed, with bowl-shaped cups that enclose up to one-third of their length.

Range: native from Sonoma County south to Baja California.

The coast live oak was once one of the dominant trees along the shores of early California. Here was another area where the Indians used acorns as the staple of their diet. According to a historian of the oak, Bruce Pavlik, at least 12 major groups of Indians used this tree. Father Junípero Serra said his first Mass in California in 1770 under a large coast live oak at Monterey Bay.[30] Cities such as Sherman Oaks and Thousand Oaks get their names from this species, as does the town of Encino (from *encina*, the Spanish word for this tree).

The coast live oak has a dense, helmet-shaped, broad canopy. It grows so quickly when it's young that it can reach 8 feet in two years and 40 feet with a spread of 90 feet by the age of 25. The thick bark makes it fire-tolerant; and in its native sites, which may get as little as 15 inches of rain a year, it is completely drought-tolerant. Coast live oaks can live for 250 years or more, but most existing stands are 40 to 110 years old.

The Cahuilla, Luiseño, and Kashaya Pomo ground the acorns into flour to make cakes, bread, soup, or mush. It was once a cash crop for the Cahuilla, who exchanged acorn meal for pinyon nuts, mesquite beans, and palm tree fruit.[31]

Traditionally, oak groves were owned by various nations, and trees within a grove were owned by specific families. As in the harvesting of pine nuts, acorn harvesting was a family affair and temporary camps were set up near the groves. The men and boys climbed the trees and shook the acorns down to be gathered in baskets by the women and children. Sometimes they used long sticks to knock the acorns down. An average tree yielded up to 200 pounds of acorns,

and occasionally a bumper crop of 500 pounds. The women pounded the acorns into a meal, using a pestle and either a portable mortar or a bedrock mortar (a depression chipped out of a large rock on site).

Most of the crop was dried in the sun or over a low fire, then stored for later use. Storage structures varied. The northern Miwok built raised basketlike structures from willow boughs. They lined them with the leaves and branches of wormwood (*Artemisia* spp.), which contains a natural insect repellent, to deter acorn worms. The bottom was about three feet off the ground to keep rodents and squirrels out. The top was covered with cedar bark (also an insect repellent) and could hold up to 50 bushels of acorns. Some Native Americans stored acorns in baskets inside their homes, caves, or in rock crevices. Others buried them near cold springs (unearthed more than 30 years later, the acorns in some of these caches were still preserved).[32]

The importance of acorns and oaks was reflected in the spiritual life of the native peoples, and acorns figured in many legends.

The man who made this world left a basket filled with acorn soup. He said, "Always leave a little soup in the bottom of the basket. And it will fill itself again." Coyote came along one day and gulped down most of the soup. He peered about— nobody was watching. Then he drank all the acorn soup, and ran off. No more soup filled the basket. The man who made the world wanted to make it easy for people to gather acorns. He said, "I'll make big acorns, and put them low on oaks, where even a young person can reach them." "No," said Coyote. "That's not right. Make small acorns and scatter them in the trees. That way men can work hard knocking down the acorns with long, heavy poles. Then they show their young wives how strong they are." The big man never argues. So Coyote won. People always had to work hard to get acorn soup after that.[33]

Early settlers harvested coast live oak primarily for fuel. The wood was not considered suitable for making furniture because of its strange growth patterns and frequent branching near the base, but it burned well and was often used for fuel on ships.

Today, coast live oaks are in a state of decline. Beset by urbanization or weakened by the pathogen *Phytophthora ramorum*, they are susceptible to attack by *Hypoxylon* spp. fungus, western oak-bark beetles, and oak-ambrosia beetles, which bore through the bark and the wood. The larvae feed on the fungus that enters the tree through these insect tunnels.

Another enemy is the **mistletoe**, *Phoradendron villosum*. This parasitic plant sends its roots into the oak's branches and steals nutrients and water. It produces small white berries in the fall. These are devoured by birds, which then excrete the seeds in their droppings. If the droppings land on a tree limb, the seeds can sprout and infect the new tree.[34]

Some forestry experts believe that the oaks were originally weakened by the severe drought that occurred from 1987 to 1993, followed by two very wet years (caused by El Niño). Others say that the stands logged by early settlers grew back from stump sprouts, and these roots are now old and declining. Moreover, since fire has been successfully prevented in some areas for more than 50 years, competition for space is very high. Another possible factor is the live oaks' habit of holding on to their acorns longer than other California oaks. They don't drop them until autumn, when few other acorns are available, making them a rare treat for acorn-loving animals. Oak seedlings are also a favorite food of deer and other wildlife.

Fagaceae

Canyon Live Oak

Quercus chrysolepis, golden-cup oak, blue oak, California oak, canyon oak, drooping oak, golden-leaf oak, Georgia oak, hickory oak, iron oak, laurel oak, maul oak, mountain oak, pin oak, rock oak, Spanish live oak, Spanish oak, Valparaiso oak

Canyon live oak grows both as a tree and a dense shrub. The tree is large, rugged, and evergreen; the shrub is usually short (under 15 feet) and broad. Naturalist John Muir described this tree:

A tough, rugged mountaineer of a tree, growing bravely and attaining noble dimensions on the roughest earthquake taluses in deep cañons and yosemite valleys. The trunk is usually short, dividing near the ground into great, wide-spreading limbs, and these again into a multitude of slender sprays, many of them cord-like and drooping to the ground. . . . The top of the tree where there is plenty of space is broad and bossy, with a dense covering of shining leaves, making delightful canopies, the complicated system of gray, interlacing, arching branches as seen from beneath being exceedingly rich and picturesque.[35]

The bark is slate gray and smooth, with small gray-brown scales on older trees. The leaves are about 2 inches long, leathery, oblong to oval, with a pointed tip, sometimes with toothed edges, dark green above, golden-hairy maturing to dull, smooth, and grayish below. Male flowers are borne in catkins and female flowers are solitary or appear in short spikes with the leaves. The acorns mature in two years; they are about 1 inch long, with a saucer-shaped, thick-walled cup.

Although the Mendocino believed the acorns were poisonous, they were eaten by Cahuilla, Diegueño, Karok, Kawaiisu, Luiseño, Pomo, Shasa, Tubatulabal, and Wintoon. The most common use was as flour, but the Wintoon also ate them roasted. The Karok buried the acorns for one to four years to kill the bugs and worms before eating them.[36] All acorns are bitter to some degree because of their tannic acid content and a time-consuming leaching process had to take place before they were edible:

First the acorn meal was mixed with water to form a paste. Then it was spread depending on technique either into a sandy basin or into baskets. Warm or cold water was poured through the meal several times until the tannic acid was removed. Two grades of acorn meal were prepared: a fine grade for mush and a coarse grade for bread. Preparing a few days supply for a family could take up to seven hours. It is no wonder that the Cahuilla Indians believed that an angry god had put a curse on the acorns to make them bitter. . . . John Muir . . . used to carry several loaves of acorn bread with him on his long journeys through the Sierra Nevada. He was very fond of this sweet and nutritious bread. A woman's status often depended on the quality of acorn mush and bread she prepared.[37]

The traditional way to make acorn bread was to mix the acorn flour with enough water to make a soft dough. The dough was patted flat on a rock and both sides were cooked on the ashes of a campfire until crisp.[38] For the Cahuilla, acorn meal was a cash crop, and the acorns were used for trap bait, jewelry, and toys. The Diegueño soaked acorn caps in water containing iron to make a black dye.[39]

The Kawaiisu used canyon live oak wood for house construction,[40] and early settlers com-

monly employed it to make furniture and farm implements, as well as for shipbuilding and fuel. The common name "maul oak" came from its use as a splitting maul. Today, it is sometimes made into wall paneling.

Canyon live oaks make attractive street trees in many California cities, and are often planted to control erosion on slopes.

Range: **Oregon, Arizona, California.**

Fagaceae

Blue Oak
Quercus douglasii, California blue oak, iron oak, mountain white oak, mountain oak

The blue oak, named for its sensuous blue-green foliage, has a rounded crown and many crooked branches, giving it a very distinctive look. It can withstand the harsh, arid conditions of California's Central Valley better than other oaks, although it drops its leaves when things get too dry. The extensive root system reaches deep into the ground for water, allowing the tree to spend as little energy as possible on lateral roots. When surface water becomes available, it switches its resources over to speedily developing lateral roots.

Quercus douglasii grows to heights of 60 feet with bark shallowly checked by small thin scales. The blue-green, finely hairy leaves are 1 to 4 inches long, and may have shallow lobes. Flowers which appear in April and May are monoecious: the males are borne in drooping yellow-green catkins; the females are small, often solitary, and appear in the axils of leaves on the current year's twigs. Oval acorns with pointed tips mature in one year.

The acorns are an important food for deer,

farm livestock, small animals such as western gray squirrels, and several kinds of birds, particularly the acorn woodpecker. This close relationship, explained in one Native American tale, is a flood story recounted by Maria Solares to John Peabody Harrington before 1922:

Maqutikok, Spotted Woodpecker, was the only one saved in the flood. He was Sun's nephew. Maria doesn't know why the flood came or how it started, but it kept raining and the water kept rising higher and higher until even the mountains were covered. All the people drowned except Maqutikok, who found refuge on top of a tree that was the tallest in the world. The water kept rising until it touched his feet, and the bird cried out, "Help me, Uncle, I am drowning, pity me!" Sun's two daughters heard him, and told their father that his kinsman was calling for help. "He is stiff from cold and hunger," they said. Sun held his firebrand down low and the water began to subside. Maqutikok was warmed by the heat. Then Sun tossed him two acorns. They fell in the water near the tree and Maqutikok picked them up and swallowed them. Then Sun threw two more acorns down and the bird ate them and was content. That is why he likes acorns so much— they are still his food. After the water was gone, only Maqutikok remained.[41]

The blue oak harbors a large number of insects: one study recorded 38 species of insects in 21 families. Developing acorns may be attacked by the filbert weevil (*Curculio uniformis*) or the filbert worm (*Melissopus latiferreanus*). Larvae of the filbert weevil are short, fat, glistening white worms, which mine the acorn, destroying its contents. Not just the nuts suffer—all parts of the tree are affected. Leaf skeletonizers do exactly what their name says, along with a host of other sucking and chewing insects. Borers infest the roots, trunk, and limbs. While many of the insects are found in low numbers, epidemics do occur, and the damage is severe.

The tree also suffers from bumps called galls which may be found on all parts of blue oak. These galls are caused by more than 40 species of cynipid wasps, including the jumping oak-gall wasp (*Neuoterus saltatorius*), which stings the underside of the leaves of blue oak and lays its eggs inside the leaf. The larvae emerge in July and August and form a tiny gall less than 1/16 inch in diameter. In mid-August, the galls fall to the ground. Sometimes there are so many larvae on the ground that the area around the tree seems to writhe in agony.[42]

Blue oaks provided native peoples with varied resources: the Kawaiisu boiled the inner bark and drank the brew to relieve arthritis. They also wove baskets out of its flexible shoots. Strong, hard, and heavy, the wood was used as posts in the houses of the Maidu of Chico Rancheria. The Yana and Achomawi crafted cooking utensils from it, and the Sierra Miwok charred forked oak sticks and appended them as fake antlers on deer masks for hunting. The Gabrielino and Sierra Miwok baited their traps and snares with acorns to catch pigeons and other birds.

Blue oak wood has little or no commercial value these days, because of the crookedness of the branches, and is mainly used for fence posts and fuel. Always decorative, it has been used to make wall hangings and table centerpieces—large branches hollowed out by heart rot are sawn into sections, cleaned, coated with resin and hardener, and filled with dried seed stalks.

The blue oak population is now declining. Oaks and grasses are strongly interdependent, and researchers speculate that the deep root systems of native perennial bunchgrasses help oaks by taking water to a different depth or making water more easily available to oaks. But introduced annual grasses such as **wild oats** (*Avena fatua* and *A. barbata*) have taken over from the bunchgrasses, depleting water and producing chemical inhibitors that affect acorn germination. In addition, grazing livestock damage oak seedlings much more than browsing deer do. Reducing the livestock population in certain areas and restoring native grasses could reverse this pattern and ensure blue oak's survival. Studies are also being conducted to see whether various mycorrhizae (beneficial soil-borne fungi that attach themselves to roots) can make a difference. Greater mycorrhizal diversity may help the oaks to survive as different mycorrhizae are able to exploit water and nutrients and then transfer them to the oaks.[43]

Range: **California.**

Fagaceae

Nuttall's Scrub Oak
Quercus dumosa, California scrub oak, scrub oak

Quercus dumosa

is an evergreen shrub that grows up to 9 feet tall. The twigs are slender, sparsely hairy at first, later becoming smooth and reddish-brown. The leaves are oblong, elliptical, or round, with toothed edges; the upper surface is slightly shiny, the lower a dull, pale green. The oval acorns taper toward the tip and sit in bowl-shaped cups. Found on dry slopes and chaparral.

   Range: **California.**

Thicket-forming and full of small, stiff, spiny leaves, the scrub oak is the main deciduous component of chaparral. In fact, the old Spanish name for this shrub, *el chaparro,* is the origin of the words *chaparral* and *chaps.* You definitely want the latter to protect your legs against the former. Francis Fultz, who wrote about the chaparral landscape, noted:

> *Although thoroughly hated by the cattle men,* Quercus dumosa *is one of the most useful and beneficent of the [chaparral]. Its roots search out every rock crevice and strike as deeply into the earth as the formation will allow, thus not only holding the soil in place, but hiding its own life away from the fires which so often sweep over the face of the land. Its branches may be consumed and its trunk burned down to the merest stub, yet the roots immediately send up a new crop of sprouts, and within two or three years have furnished the land with a fresh forest cover. . . . No shrub of all the chaparral makes a greater yearly foliage contribution to the leaf mold that goes into the making of the soil.[44]*

The tree had many uses. The branches were crafted into baskets and the framework for cradles by the Diegueño, who also boiled up galls for an eyewash. The Luiseño treated sores and wounds with its gallnuts.[45] The Cahuilla used the acorns for trap bait and necklaces, and strung them to make a musical instrument that made a noise when swung against the teeth. With their high tannic acid content, the scrub oak acorns were not a primary food choice of the native peoples. The Cahuilla, Diegueño, Kawaiisu, and Luiseño ground the acorns into flour for mush, cakes, and bread, but often it was merely an additive to other kinds of acorn meal in times of shortage.

Not every creature finds the scrub oak unpalatable: its leaves are relished by the larvae of the gold hunter's hairstreak butterfly (*Satyrium auretorum*).

The landscape is subject to fire, and many of its species are adapted to frequent fires as it is described here by Linda Thomas:

Fire in the chaparral is an amazing sight. The Santa Ana winds pass over desert and arrive in the foothills of southern California in hot, bone-dry, ten to forty mile-per-hour gusts that lower the relative humidity to three percent. The condition is perfect for fire that can rush up a canyon like a locomotive. After a particularly wet spring, chaparral shrubs can grow so densely that with the heat of summer and the moisture-sapping Santa Ana winds, they are kindling for the fire that devours them in whirlwinds of flame.[47]

Nevertheless, the chaparral possesses its own peculiar charms. In 1878, Helen Hunt Jackson wrote: "From each ascent we gain we can see only hills. All the fertile beauty is gone. But the greasewood is in full white flower, and looks like a heath; and the ground is gay with low flowers."[48]

The word "chaparral" comes from the Basque word *chabarra*, which refers to the scrub oak common in the Pyrenees. It was corrupted to *chaparro*, meaning "brushlands." (Every language seems to have its own word for this kind of landscape. In France there is the *garrigue*, in Spain the *maquia*, in Chile *matarral*, in South Africa *macchin fynbosch*, and in South Australia *mallee-scrub*, *mulga scrub*, and *brialow*.)

Chaparral is found in three distinct brushlands: the coniferous forest of northern California and the middle slopes of the Sierra; the woodland-grass slopes of the Sierra, northern and central coast ranges; and the brush fields in the mountains of southern California.[46]

THE CHAPARRAL

Ericaceae

Pointleaf Manzanita

Arctostaphylos pungens, Mexican manzanita, manzanilla, kunth, common buttonbush

Arctostaphylos pungens

is an evergreen shrub or small tree up to 12 feet tall, with smooth, red-brown bark and leathery, cone-shaped leaves. It produces white or pinkish urn-shaped flowers in bunches, followed by small edible red berries. The manzanita tends to expand in concentric rings. Plants that have survived (that is, not been subject to fire) for 50 years can grow to a diameter of 20 feet. The older center eventually dies and a ring of vigorous new plants grows around it. If there is a fire, the manzanita quickly germinates by seed.

Range: **California.**

Manzanita is Spanish for "little apple." The small red or brownish fleshy berries are sweet, and provide forage for deer and other animals; they can even be made into jelly. The berries were eaten by the Cahuilla and Yavapai, either fresh or sun-dried. They could be stored, ground into flour, and made into mush, squeezed for juice, or, as the Northern Maidu found, made into "manzanita cider."

Jack London once wrote: "The blossoms of the wine-wooded manzanita filled the air with springtime odors, while the leaves, wise with experience, were already beginning their vertical twist against the coming aridity of summer."[49]

Wise or not, the Cahuilla and Ramah Navajo smoked the leaves with tobacco. And the Cahuilla used a leaf infusion to treat diarrhea and poison oak rash.[50]

Because of its hard, dense wood, manzanita is sometimes (mistakenly) called "ironwood," though that name more properly refers to *Olneya tesota*, the ironwood of the Sonoran Desert (see page 364). Still, manzanita comes a close second. The Cahuilla commonly used the wood for house construction, pipes, tools, and as firewood.

Fire is very important to the manzanita life cycle. Horticulturists and native-plant preservationists often urge suppression of fire, not realizing its role in growth and renewal. As Maureen Gilmer writes:

Without fire the native shrubs in these mountains have grown old and decadent. . . . We see stands of manzanita . . . that would historically have been culled early in the successional process by landscape fire. Today they are abnormally aged and dying out . . . leaving behind a forest of skeletal branches. The dynamics of this problem . . . became painfully apparent to me . . . [after] an enormous fire ran through this community two seasons ago leaving behind a 5,000-acre tract of burned-out wildland. That first spring, a ridge once cloaked in manzanita and studded with a dog-hair stand of ponderosa pine trees was still black with ash and skeletal trees. Nothing had sprouted over winter despite copious rains. [Then,] to my amazement the hill burst into a mass of brilliant orange, producing a forest of

Arctostaphylos pungens, pointleaf manzanita

tiger lilies Lillium humboldtii . . . *six feet tall bearing dozens of flowers per stem. These bulbs had lain quietly for many years under a cloak of brush waiting patiently for a fire to clean house and give them a clear view of the sun.*[51]

Lauraceae

California Laurel

Umbellularia californica, California bay, pepperwood

Umbellularia californica

is an evergreen tree or shrub found in dry, open habitats that can grow up to 80 feet tall. Its leaves are simple, dark green, thick, leathery, and aromatic. The small yellowish flowers have six sepals and appear in stalked umbels. The fruit is an oval drupe, 1 inch long, yellowish-green maturing to purplish; when it is dried, it looks like an olive. Found in canyons, valleys, and chaparral.

Range: Oregon, California.

Violent storms often bend the California laurel into a near-horizontal shape causing contorted trunks to sprout upright branches. It's not a sociable tree. The leaves produce water-soluble compounds that drop to the ground when it rains. These compounds slow root growth in conifers, thereby eliminating the competition.

California laurel was such an important medicinal plant for the West Coast Native Americans, it was the modern-day equivalent of aspirin. Many nations (the Cahuilla, Mendocino, Miwok, and others) used it to relieve headaches, by drinking a decoction of the plant or making a poultice that was applied directly to the forehead. For the Yuki, inhaling crushed leaves seemed to do the

trick. Karok women drank a laurel infusion to relieve the traumatic pain of afterbirth, while the Pomo and Kashaya Pomo slapped on poultices of heated leaves to relieve rheumatism and neuralgia.

The fruits were eaten fresh or cooked or dried (which makes them look and taste a little like nuts). Seeds could be ground into flour or roasted and shelled as a crunchy snack. The distinctive smell also was important: the Costanoan hung leaves in bunches to freshen the air; the Karok placed leaves on the fire during the brush dance to drive away evil spirits; and the Yurok burned leaves to banish bad luck. Before a Kashaya Pomo man went hunting, he rubbed leaves on his body to disguise his smell. The leaves were also burned to keep fleas away.

The wood of California laurel, or myrtle as it is referred to in Oregon, is characterized by unique grain patterns. Minerals drawn from the soil color the wood. Any stress, including rain or the lack of it during its growth cycle, causes patterns such as burl and fiddleback to appear.[52]

A piece of California laurel played a role in the historic meeting of the Central Pacific and Union Pacific railroads on May 10, 1869, as P. J. Gladnick reports:

At about noontime the trains were close enough that the last tie could be laid down. This tie was made of California laurel and had a silver plate in the middle engraved with the date and the names of the officials of the two companies. [After the final four gold and silver spikes had been driven in] the two engines moved up and touched each other and the event was christened by pouring champagne on the last rail to much cheering from the crowd of workers.

The laurel tie and one of the gold spikes had been donated by David Hewes, a great supporter of a transcontinental railway. Unfortunately, the ceremonial tie was a casualty of the San Francisco earthquake and fire in 1906.[53]

The wood is still valuable but not cultivated right now though it may be one day if the decline of other hardwood species continues. However, California laurel leaves can be used in stews as a substitute for real bay leaves (hence the name "California bay"). Retailers have caught on, as Michael Ellis laments:

The upstart California bay is rapidly supplanting the Old World bay, at least in the West. At the corner market a small jar of 20 "premium" California-bay leaves costs over three bucks. That's 17 cents a leaf! It is indicative of our detachment from the environment that we feel more comfortable using leaves purchased in a bottle than those picked from the trees growing in our backyard.

We have some magnificent bay trees. With wide arching branches they make the perfect climbing tree and an ideal fantasy site for a Swiss Family Robinson treehouse. They are one of the few plants that flower in the winter, reminding us of the mild tropical climate in which they once dwelled.[54]

Anacardiaceae

Laurel Sumac
Malosma laurina, taco plant

Malosma laurina

(sometimes known as *Rhus laurina*) is a shrub or small tree, growing 15 to 20 feet tall. The simple, leathery, evergreen leaves are elliptical or lance-shaped with a pointed tip, about 4 inches long, folded along the midrib. The flowers may be bisexual or unisexual. The petals are usually white. The blooms appear in a dense branched **panicle 6 inches long. The fruit is small and fleshy. Found on dry coastal slopes and chaparral.** *Range:* **southwest California.**

In the middle of the day, the leaves on the highest part of the laurel sumac partly fold up to protect themselves from the hot sun (giving rise to the nickname "taco plant"); in lower shadier parts of the plant, they remain open. Since this plant won't tolerate cold, the pioneer farmers of southern California used it as a guide to planting citrus orchards. A severe cold snap in 1978 killed off many of the laurel sumacs in southern California, but since then they've made a dramatic comeback.

Heavy grazing is devastating to most plants, but not this one—the more it's chewed on, the better it grows. Stands of laurel sumac on the offshore islands of California have expanded since livestock were introduced there, even though it is not a particularly favored food of sheep or goats.

The evocative scent of the laurel sumac is distinctive and hard to forget. The flowers are much favored by honeybees. When the little flower bunches dry out, they resemble miniature trees—prized by model railroad enthusiasts who dye them green to make them into forests for their Lilliputian landscapes. The small, fleshy fruits are seldom harvested now, but at one time were dried and made into flour by the Chumash people. They also used the root bark to make a tea for treating dysentery.

A distinctive member of the chaparral community, the laurel sumac actually encourages fire: aromatic compounds in the leaves increase a fire's intensity, and the mature plant produces a deep, nicely combustible litter. Like other chaparral shrubs, it stores nutrients in its roots and lignotuber (a swelling at the base of the stem at soil level or just below containing dormant buds which spring to life if the top growth is destroyed). After

FOLLOWING PAGES: *Malosma laurina*, laurel sumac

a fire, these reserves help the plant resprout quickly; it's one of the fastest-regenerating species. The coats of its seeds are cracked open by fire, and that is their cue to start growing. Areas that have been devastated by fire recover quickly with laurel sumac. One writer describes laurel sumac after a fire:

Naked laurel sumac hold their blackened branches stiffly, bristling in cindered shock. Charcoal stains the earth like some abandoned hearth of invading warriors. The tale they tell is an old one, a secret one. One of breathing fire to scour the hills, one of mixing sweat and blood and entrails to make gold. To make sense out of a land that shape-shifts and tilts in its sleep.[55]

Rosaceae

Toyon

Heteromeles arbutifolia syn. *Photinia arbutifolia*, Christmasberry, California holly

Heteromeles arbutifolia syn. *Photinia arbutifolia*

is a large evergreen shrub or small tree that grows up to 16 feet high. The thick, leathery leaves are oblong to elliptical, up to 4 inches long, hairy, dark green and glossy above, paler beneath. Small flowers with five creamy white petals and 10 stamens appear in panicles 2 to 4 inches long in June. The bright red (occasionally yellow) fruit is pulpy and mealy with three to six small brown seeds inside and appears in November. It is found in chaparral, oak woodland, and mixed evergreen forest.

Range: **California.**

This is the plant that gave Hollywood its name. In the early days, its dark green leaves and red berries made it into a West Coast faux holly for homesick English settlers. Horace Henderson Wilcox, who bought the land near Los Angeles, named his subdivision Hollywood because the plant grew abundantly in the hills there.

The toyon grows in open woodlands, canyons, rocky slopes, sand, and typical chaparral regions. Depending on its habitat, it varies in size and shape, growing close to the ground and dense in exposed places, or open and treelike in wooded areas. Unlike other chaparral species, it tolerates shade well and needs very long—ideally a century—fire-free periods for seedlings to become established.

Viable seed is produced at about three years of age and seeds will germinate within 10 to 40 days if they land on an appropriate site. However, seedling survival rates are low due to fungus disease and browsing animals.

The toyon produces masses of fruit in the fall and early winter. These are prized by many chaparral mammals and birds, which depend on them for sustenance when other fruit is scarce. The berries could be eaten raw (although they were rather sour) or tossed in baskets with hot pebbles or wood coals to cook them. They were also used to make a drink similar to cider, enjoyed by Native Americans and early settlers alike. Spanish settlers reportedly made a pudding by baking them with sugar in a slow oven.

Toyon leaves were used in an infusion by Olhonean Costanoan women to suppress or regularize menstruation. A leaf extract was a panacea for various aches and pains, including stomachache, for the Mendocino Indians, while the Diegueño infused the bark as well as leaves to make a wash for infected wounds.[56]

The Native Americans found that toyon leaves and stems yielded a golden brown dye, the leaves and berries, a dark olive green or black dye.[57] Channel Island fishermen once used toyon bark to tan their fishnets.

In the early 1900s, toyon boughs and berries became so popular as Christmas decorations that bushes in the wild were in danger of disappearing. One of those raising the alarm was a Mrs. Bertha Rice, who, with her son Roland, published a small book in 1920 called *Popular Studies of California Wild Flowers*. The last section of the book was devoted to the plight of the toyon and the thoughtless manner in which it was being gathered. During the 1920s a law was passed making it illegal to pick or dig wild plants without permission, a law that remains in effect today.[58]

That they flourish today in wild profusion is confirmed by Arnel Guanlao, who wrote:

The toyon bushes were still heavily burdened with masses of bright, scarlet berries, hanging from their branches like strings of glass beads. . . . The robins gathered around the toyon bushes in enormous flocks that filled the damp air with their soft voices. Completely ignoring the rain, they constantly alighted on the toyon's branches, greedily plucking its berries. Then, with a berry or two firmly clenched in their bills, they flew off to a branch or dense patch of underbrush, where they could consume their prize undisturbed. With the whispery fall of rain, the trees looming like gray-green shadows all around me, and the sound of the robins' cries drifting through the air, I almost felt as if I were hiking through a tropical cloud forest. The feeling was eerie, haunting, and beautiful all at once.[59]

One of the first Europeans to notice this species was Archibald Menzies in the 1790s. He sent some seeds to England, where it soon became a popular garden plant the English called "maybush" or "mayblossom" because of its resemblance to the hawthorn. Victoria Bease-

ley of Boston grew up in the Channel Islands, United Kingdom, and fondly recalls the maybush or maytree, an exotic transplant from distant California:

In the days before ubiquitous yellow buses transported children to school, we walked in small groups for two miles along country lanes, enjoying nature's bounty the way small children do, picking and tasting. Even though we were given full English breakfasts and carried substantial lunches, we were always hungry. One treat we enjoyed were the leaves and berries of the maybush, which we called "bread and cheese." The bright red, pea-sized, mealy berries, which grew in clusters of four, resembled miniature Gouda cheeses, with their waxy red outsides and creamy insides; the leaves were the "bread" completing the snack. . . .

Years sped by. Soon the "bread and cheese" eating children were old enough to go to the cinema to watch Hollywood films. Little did we realize that the glamorous place name was derived from the transplanted snack of our early childhood years.

Today, toyon is widely grown as an ornamental in home gardens, where it attracts and sustains many native birds. The U.S. Forest Service plants it to control erosion and restore streambanks.

Rosaceae

Chamise
Adenostoma fasciculatum, greasewood

The most characteristic and widely distributed chaparral species in California, chamise predominates on the hottest, driest sites. *Adeno-*

stoma comes from the Greek *aden* meaning "gland" and *stoma* "mouth," referring to the glands at the mouth of the sepals. *Fasciculatum* refers to the arrangement of the leaves in bundles called fascicles. The common name "chamise" comes from a Spanish word meaning "half-burned wood."

Chamise, a many-branching shrubby plant, reaches 8 feet in height. The evergreen, leathery, and needlelike leaves grow in clusters. The greenish-white flowers, which appear in May and June, have five petals and grow in dense terminal clusters. Each flower produces a single-seeded fruit or achene.

The chamise, the manzanita—in fact, most chaparral plants—are full of solvent extractives that burn intensely and ignite easily. Their leaves are glossy with oils and resins that seal in moisture during hot dry periods and serve the dual purpose of responding explosively to flame. . . . [The chaparral] burns as if it were soaked with gasoline. Chaparral plants typically have multiple stems emerging from a single root crown, and this contributes not only to the density of the thickets but, ultimately, to the surface area of combustible material. . . . Hundreds of acres can be burned clean in minutes.[60]

Chamise produces two types of seed, one that germinates under normal conditions and one that needs the heat of a fire to germinate. The latter seed type is called a "refractory" or "resistant" seed. The presence of burned organic material from plants and trees is a chemical cue for dormant chamise seeds to germinate. Another adaptation of the chamise is its ability to sprout after a fire from underground woody plant structures called lignotubers that can penetrate fractured rock to depths of 10 to 12 feet. The tubers have vegetative buds several inches

below the soil surface which remain dormant until the winter rains following a fire.

Tough and hard, the chamise wood was fashioned into tools by the Chumash, and arrow shafts by the Luiseño. The Cahuilla built fences and houses using chamise branches, and made the leaves into a tea to help with childbirth, menstrual complications, skin infections, or syphilis. Sometimes even sick cows were treated with it.[61]

Chamise can live for a long time; some specimens have been dated at 100 to 200 years old. It can form almost pure stands (called chamisal) that make tough going for anyone trying to get through. These dense stands of chamise often conceal open spaces where animals can find shelter.[62]

Range: coastal hills of California, Baja California to northern California west of the deserts.

Asteraceae

California Sagebrush

Artemisia californica, coast sagebrush, coastal sage, sagebrush, old-man sage, romerillo, spreading California sagebrush, California sagewort

Artemisia californica

is a shallow-rooted shrub growing up to 5 feet tall, often branching out so that it is wider than it is high. The silvery gray-green leaves are 1½ inches long, finely divided into threadlike segments and clustered in bunches. Terpenes in the leaves give them a pungent fragrance. The stems are smooth and grayish. The tiny greenish-yellow flowers, which appear in late summer through February, are borne in dense panicles at the twig

ends. The fruit is a small achene. Found in coastal scrub, chaparral, and dry foothills.

Range: California.

Sagebrush suggests the novels of Zane Grey, and the children who once imagined riding a horse (Trigger, Silver, or Champion) through the sagebrush after the bad guys. It was part of the landscape of the loner, the outsider, and of the hero with his white hat who always emerged from the sagebrush to fulfill the promise of a new land.

The silvery quality of this plant seems a perfect medium for reflecting the glittery sun. It is a shrubby, medium-sized plant, dominating the coastal sage scrub community. It is not a true sage (that's *Salvia*), but belongs in the aster/sunflower family. There are two varieties of California sagebrush: var. *californica* (mainland) and var. *insularis*, which grows on the islands of San Clemente, Santa Barbara, and San Nicholas. The island variety has wider leaf segments than the mainland variety.

Many species of birds make their homes among the branches of *Artemisia californica*, and the California gnatcatcher (*Polioptila californica*) depends on insects that live on this shrub. Sagebrush belongs to an ecosystem known as coastal scrub—also called "soft chaparral" because it combines shrubs with a soft-stemmed herbaceous understory. The coastal scrub shrubs shelter and protect dozens of creatures, including the orange-throated whiptail lizard, a threatened species in California. As coastal scrub disappears, so do the creatures that live there.

Today, coastal sage scrub occupies only about 2 percent of the state's total area and about a quarter of the chaparral area. Fortunately, California sagebrush colonizes new areas easily and sprouts rapidly in the first year after a fire. Animals browse on it—in fact, the feral goats and sheep on Santa Catalina and Santa Cruz Islands have nibbled whole stands of sage-

Artemisia californica, California sagebrush

brush to death; but if the animals are kept away, the sagebrush usually bounces right back. Like chamise, California sagebrush is very oily and can reproduce both by sprouting and producing seeds immediately after a fire. The seeds are so light they can float on the wind for great distances.

Southern areas of coastal sage scrub receive

as little as 9 inches of rainfall annually. But they do get fog, and chaparral plants make the most of it. They collect drops of moisture on the fine threadlike leaves. Once laden, the leaves arch over, letting the drops slide to earth.

Like most *Artemisias*, the leaves of California sagebrush are full of scented oils that fill the air with an intoxicating smell if brushed. Settling in the soil below, the oils inhibit the growth of other plants, including their own seedlings, so the area around is usually bare.

Because of its scent and oils, the sagebrush was important in Native American rituals. It was soaked in water and used for ceremonial cleansing, or burned to produce smoke in religious ceremonies. The Luiseño believed that its smoke absolved them of breaches of social observance which would otherwise bring bad luck. The Cahuilla chewed and smoked the leaves with tobacco, and the Chumash made fire sticks and arrow foreshafts out of sagebrush branches.

Sagebrush's healing properties were widely credited. While the Costanoan made the leaves into poultices to relieve toothache, rheumatism, wounds, and asthma,[63] the Spanish settlers considered California sagebrush an all-round panacea. They made sagebrush tea to treat bronchial problems and sagebrush wash for wounds and swellings. Fresh leaves were mashed and put on sores. Standing in the smoke of a sagebrush fire was supposed to eliminate the smell from a skunk. Miners put sprigs of it into their beds to repel fleas, and even tossed a few leaves into pinole (a meal of ground corn and mesquite beans) for flavoring.

When Sarah Royce made her westward journey with her husband and son in 1849, they became lost in the desert with perilously little food or water. A burning sage bush inspired her with the strength to reach the meadows where they were able to replenish their supplies.

Just in the heat of noon-day we came to where the sage bushes were nearer together; and a fire, left by campers or Indians, had spread for some distance, leaving beds of ashes, and occasionally charred skeletons of bushes to make the scene more dreary. Smoke was still sluggishly curling up here and there, but no fire was visible; when suddenly just before me a bright flame sprang up at the foot of a small bush, ran rapidly up it, leaped from one little branch to another till all, for a few seconds were ablaze together; then went out, leaving nothing but a little smouldering trunk.[64]

Asteraceae

Coyote Brush
Baccharis pilularis, coyote bush, chaparral broom

Baccharis pilularis

is an evergreen, mat-forming shrub. In exposed coastal areas, the mats are about 6 inches high, but in protected areas the shrub can grow up to 6 feet tall. The leaves are oval, coarsely toothed at the ends, ½ inch long, thick, and resinous. They grow closely together, making for a very dense shrub. Male and female flowers appear on different plants. The male flowers are short, flattish, and tan-colored, while the female flowers are long, white, and glistening, and expand into cottony, creamy white seed heads with dandelionlike seeds.

Range: **California.**

Because of its dense root crowns, coyote brush, also called coyote bush, regenerates quickly from disturbances such as fire, flooding, or road and trail clearing. It succeeds in almost any soil, from heavy clay to pure sand, as long as it gets plenty of sun. The leaves become sticky and strong-smelling on hot summer days. Many animals dislike the odor and avoid browsing on the coyote brush. The evaporation of the oils from the surface of the leaves also helps to cool the plant. Just like those of the sagebrush, these oils may inhibit the germination of seeds from other plants and so reduce competition for space when the coyote brush is colonizing an area.

Baccharis comes from the Roman god Bacchus, a god of vegetation and wine, and *pilularis* means "pill-shaped," referring to the small, round, flat flower heads. The Spanish name is *rama china*, meaning "curly branch."

The coyote brush was used much like other chaparral plants. The Coast Miwok heated the leaves and applied them to swellings, while the Chumash, whose tribes inhabited the Santa Barbara area, used it in a wash to soothe poison oak rash, which is just as miserable as a poison ivy rash. An infusion of the plant was considered a general remedy. The branches and wood were used as construction material and arrow shafts.

Coyote brush has never been an endangered species, but as Jane Strong writes, it "clothes the dunes and bluffs, canyon bottoms, trail sides and road edges of the coastal sage community with a normally unnoticed green cover. Overlooked, taken for granted, ignored, tolerated, coyote bush seems to be everywhere, yet seldom seen."[65] It's a little like Wile E. Coyote himself.

In spite of, or perhaps because of, its durability, coyote brush was adopted by early groups involved with the preservation of native plants. It was recognized as a part of a coastal scrub community or a successional to mixed evergreen woodlands, not an environmental disaster, but a component in a complex riddle. As Judith Lowry of the group Friends of the Coyote Bush writes,

Baccharis pilularis, coyote brush

We chose coyote bush as an example of an important native species. . . . Its role is reminiscent of the role of mesquite in the Southwest, where old-grove mesquite bosques are an increasingly rare critical habitat, but where mesquite is also an invader of grasslands devoid of livestock or fire. Like coyote bush, it is both these things, and many more besides. Coyote bush functions differently according to soil, aspect, slope, distance from the coast, and land-use history. . . . Walking by the coyote bush islands, you hear a satisfying rustling, which could be quail, lizards, garter snakes, deer, or wren-tits.

Lowry adds that, at the Larner Seeds Demonstration Garden in Bolinas, California, coyote brush forms the framework of the garden, bringing unity to the whole. It is now being used in gardens, and even the native bunchgrasses grow better around it.[66]

All aspects of the coyote brush are valuable. The wide-spreading, fibrous roots make for excellent erosion control, the branches provide food and nesting sites for migratory birds, and the nectar sustains hundreds of insect species in fall when little else is blooming.

Rhamnaceae

California Coffeeberry
Rhamnus californica, California buckthorn

Rhamnus californica

is a long-lived, evergreen shrub or small tree, with oblong to elliptical alternate leaves 1 to 2 inches long, shiny and smooth on the upper sides. In dry sites, the leaves are small and thick; on moister sites, larger and thinner. The incon-spicuous bisexual green flowers are borne in clusters in the leaf axils. They have five pointy petals and are star-shaped. The juicy berrylike fruit (actually drupes) are green, black, or red, and usually contain two seeds. Found in coastal sage scrub, chaparral, woodlands, and sandy and rocky areas.

***Range:* California, southwest Oregon.**

The seeds of the coffeeberry bear only a superficial resemblance to coffee beans, and they have never been used as such. This is a close relative of cascara, so it's more of a laxative than a refreshing drink. However, wildlife and birds, which aid in the plant's seed dispersal, eat the berries with no apparent ill effect.

The Cahuilla steeped the berries in water to make a laxative tonic. The Atsugewi and the Cahuilla made a bark tea to treat rheumatic pains, while the Costanoan and Maidu used an extract from the leaves to soothe poison oak rash.

Native Americans of the northern Sierras hunted the huge, fearsome, but now-extinct California grizzly by turning themselves into "grizzlymen." The disguise, which involved donning the head of a dead grizzly, incorporated the roots of the coffeeberry as an important element. The writer known as Coyote Man once interviewed a fellow called Leland, who remembered hearing about grizzlymen from his grandfather, Oregon City Charlie. "They rigged the grizzly skull, before they put it inside the hide, so the mouth would open and close and the ears move. They also dried coffeeberry roots and sharpened them until they cut like a knife. These, they attached with rawhide to their forearms—the sharpened ends sticking out beyond their elbows. This is the way they turned into grizzly."[67]

Today, the shrub is grown as an ornamental, and having it in one's garden provides excellent opportunities for bird-watching. Jeff Caldwell writes: "[A while ago] I was gathering ripe coffeeberry fruits and was astonished at what a good look I was able to get of the western bluebirds which continued to feed on them, or retreated but a very short distance, eyeing me as I eyed them! Yellow-rumped warblers were with them."[68]

A representative from the Las Pilitas Nursery specializing in native California plants believed for many years that coffeeberry was one of their most deer-proof plants.

We had it on sites throughout California and the deer didn't bother it until the 6th year of drought and they decided it was edible. What I can't figure out is how they got together state wide and changed their minds all at once. On some sites the bushes had been untouched for 20 years and in one week they were defoliated. The same week in Cambria, Paso Robles, Monterey, and the lower Sierras? Maybe they're smarter than we thought.[69]

Recently, coffeeberry played a different role when it was used together with coyote brush, rosemary, bamboo, rockroses, and other plants to create a "Garden of Healing" around the Temple Beth Israel in Redding, California. The garden was erected as an alternative to barriers such as a chain-link fence in an effort to protect this synagogue after three synagogues in Sacramento had been burned.[70]

Some determined souls insist that coffeeberry can indeed be used as a coffee substitute. Christopher Nyerges offers the following advice:

Collect coffeeberry fruits in the fall when they will have turned nearly black. Remove all the seeds; this is most easily done by simply rubbing the fruits between your hands in a dish pan, then washing away the pulp (the pulp can then be put into your compost pile). When you're down to just

seed, let them dry, then roast them until brown. Grind the roasted seed and percolate as you would ordinary coffee. Though a bit on the weak side, the flavor and aroma of coffeeberry is very much like traditional coffee and, with some honey and cream, it really can pass for the old familiar—but without the caffeine.[71]

Anacardiaceae

Sugar-Bush
Rhus ovata, sugar sumac

Rhus ovata

is an evergreen shrub or small tree with stout, reddish twigs, growing 5 to 20 feet tall. It has rough, very scaly, gray-brown bark. The shiny, light green, oval leaves are thick and leathery, $1\frac{1}{2}$ to 3 inches long, folded along the midrib. Rose-colored buds open to showy white or pinkish five-petaled flowers in dense clusters at the ends of the twigs. The small fruits are reddish drupes, covered in short, sticky hairs, each containing one seed. Very drought-tolerant, it grows on dry slopes, mostly chaparral, away from the coast.

Range: Santa Barbara County to lower California.

Rhus ovata gets its common name because it exudes a sweet, waxy substance coating its fruit that tastes like sugar. The fruit of the sugar-bush can be eaten raw or cooked, or soaked for up to an hour in water to make a thirst-quenching beverage. The leaves can also be boiled for a tea.

The Native Americans used the sugar-bush for food and medicine. The Cahuilla and Yavapai ate the fresh berries, or ground them into a flour for mush. They also used the fruit sap as a sweet-

Rhus ovata, sugar-bush

Sterculiaceae

ener. The Cahuilla made an infusion of leaves for colds, coughs, and chest pain, and a similar tonic was administered to pregnant Diegueño women just before childbirth for easy delivery.[72]

The leaves have finely pointed tips which curl under, so any drops of moisture captured from fog run down and are shunted to the base of the plant. Rich in tannin, the fallen leaves can be collected and used to make a brown dye. A tallowlike oil extracted from the seeds may be made into candles, which burn very brightly but give off a pungent smoke.

Like other chaparral species, the sugar-bush is adapted to fire. Its seedlings appear in burned areas and it resprouts quickly. It is used for erosion control because it takes hold quickly and firmly on dry slopes. It is also an important source of food and shelter for wildlife.

Fremontodendron californicum syn. *Fremontia californica*

is an open, spreading shrub that grows 12 feet high, sometimes more, with deeply fissured bark and short lateral branchlets. The dull green leaves are round or oval, usually three-lobed, 1 inch or so long. The waxy flowers, which appear between April and July, have a clear yellow, flat calyx and intense yellow sepals with reddish margins. The egg-shaped seed capsule is hairy, with silky hairs on the seeds. Found in the chaparral, oak/pine woodlands, and on rocky ridges.

Range: **California, Arizona.**

The largely tropical family to which *Fremonto-dendron* belongs includes such plants as cola (a source of caffeine) and *Theobroma* (cocoa). Flannelbush is a dominant shrub in desert chaparral communities, which burn less frequently and intensely than other chaparral. The shrub may be used for erosion control at elevations of 500 to 6,000 feet. It tolerates smog and drought—the ideal freeway plant—though it does best in well-watered, well-drained situations.

The bark makes good rope, as the Sierra Miwok, Western Mono, Tubatulabal, and Owens Valley Paiute peoples found (they also made carrying straps from the braided fibers). The Western Mono used young split branches to tie together stirring sticks and to make cone-shaped storage bins for acorns and manzanita berries. Kawaiisu babies slept in cradles made from flannelbush wood.

The plant gets its common name, flannelbush, from the bristly hairs found on the leaves and fruit. These bristles can irritate the skin—flannelbush gardeners would be wise to make use of gloves and long sleeves.

The Latin name comes from John Charles Frémont (1813–1890), a U.S. military officer who led several scientific and mapping expeditions to the West in the 1840s. Intensely interested in all the natural sciences, he collected dozens of plant specimens in California, many of which also bear tribute to him in their species name (*fremontii*). A controversial man regarded by some as a braggart, by others as a courageous explorer, Frémont was pivotal in the acquisition of California from Mexico in 1847, later becoming a senator and presidential candidate. His accounts of his travels were hugely popular, in large part thanks to his wife, Jessie Benton Frémont. A talented writer, she not only helped her husband turn his expedition notes into highly readable and popular reports, but also wrote her own account of these experiences (including *A Year of American Travel and Far West Sketches*).

When Frémont lost his fortune in an effort to build a railroad to the Pacific, it was Jessie's writing that helped support them. The genus name should honor her as much as it does him.

Anacardiaceae

Poison Oak
Toxicodendron diversilobum

Toxicodendron diversilobum

can be a shrub or a vine. The compound leaves have three leaflets with wavy or lobed edges, and are 1 to 3 inches long. They are bright green and shiny on the upper surface and paler underneath. The yellow-green flowers appear in the leaf axils, and give way to small whitish seeds. Grows in canyons, on slopes, chaparral, and oak woodland.

Range: throughout the West Coast, from California to British Columbia.

This is the true poison oak; the so-called "poison oak" found in eastern North America is a variant of **poison ivy** (*T. radicans*). It is, however, every bit as unpleasant as poison ivy. The plant contains urushiol, an oil that causes contact dermatitis in humans. If exposed to poison oak toxins, the best immediate measures are to take a shower without soap, and clean off the skin with rubbing alcohol, according to an expert on poison oak, dermatologist Dr. William Epstein, of the University of California at San Francisco.

It isn't called *Toxicodendron* ("poison tree") for nothing. Poison oak and poison ivy account for 10 percent of lost time in the U.S. Forest Service. Firefighters are especially affected, because the smoke of burning poison oak produces serious inflammation of mucous membranes. Like

Toxicodendron diversilobum, poison oak

from plant poisons. The leaves come in various shapes (hence the appellation *diversilobum*), and were used fresh in poultices by the Wailaki against rattlesnake bites. Somehow, the Karok managed to chew the leaves like tobacco, and the Costanoan and Mendocino made baskets from the stems. The Diegueño treated tiny sores in the eyes with a decoction from the roots, and the Yuki used plant juice on warts. The Pomo used the ashes from the plant for black dye (though one hopes they made themselves scarce while it was burning).

Most Californians have at least one poison oak story and this is Peter Hacker's:

As a young teenager I spent my summers in Northern California where the poison oak was as prevalent as Bermuda grass and as tall as the trees. I remember one summer I had it so bad that it covered my face and neck, making it appear that I had eaten a poison oak salad in a fit of ravenous insanity. Despite knowing what the plant looked like, I could never avoid it and could always count on a harsh outbreak every summer. Improper treatment often made it worse, like the year my Dad applied the very greasy Neosporin ointment to my rash. I welcomed death after the rash spread over my entire body.[73]

other chaparral plants, it grows back quickly and vigorously after a fire, so the poor firefighters can't even comfort themselves that fire eliminates this pest. The plant has extensive underground rhizomes which sprout again as soon as the coast is clear.

Though we generally curse it, Native Americans had an amazing variety of uses for the plant, especially in protecting themselves from other poisons. According to Daniel Moerman, the Mahuna drank an infusion of dried roots and the Tolowa ate the early spring buds for immunity

Folk remedies also abound. Some people swear by aloe vera, others try more high-tech remedies such as steroids. One suggestion for removing the plant is to keep goats, which can apparently eat the stuff with no ill effect. But whatever you do, don't burn it!

Rhamnaceae

Ceanothus

Ceanothus cuneatus, buckbrush, wedgeleaf ceanothus

Ceanothus cuneatus

is an evergreen shrub with stout, ridged branches that grows up to 8 feet tall. The dull gray-green leaves are narrow at the base, flaring out toward the ends, ¾ inch long. The flowers are white to pale lavender and appear in loose clusters in early spring. The light brown fruit capsules containing small shiny seeds are less than ¼ inch in diameter, and each has three little horns.

Range: central Oregon to southern California.

Ceanothus comes from the Greek *kean othos,* "corn thistle." The genus *Ceanothus* is often called wild lilac or California lilac because many of the 40-odd species often have fragrant blue, lavender, or white flowers. The most widespread species is *C. cuneatus,* which forms impenetrable thickets on dry rocky mountain slopes from the

Ceanothus cuneatus, ceanothus

Columbia River in Oregon to lower California. Its flowers aren't as showy as some other species, and the common name, "buckbrush," comes from the fact that its twigs and leaves are an important food for deer, particularly during the winter months. (New sprouts and seedlings are also a deer delicacy.)

Although not a legume, buckbrush has nitrogen-fixing root nodules, which allow it to survive in nitrogen-poor soils. The flowers also contain substances called saponins which give them the unusual property of lathering when rubbed in water. Native Americans used the flowers to make soaps and shampoos. (In some areas, wild lilac flowers were used as part of the wedding ceremony to wash the beloved's hair, an act of tenderness and a symbol of commitment.)[74] The flowers and leaves were also boiled for teas and tonics.

The wood was equally valued: by the Owens Valley Paiute and Miwok for digging sticks, by the Miwok for baskets, and by the Kawaiisu for arrows. Ear piercing was common among the Achomawi, who used needles made of ceanothus. And the Yuki and Mendocino fashioned fish dams out of its stout angular branches.[75]

The hard-coated seeds of *C. cuneatus* are dispersed when the capsule holding them explodes, hurling some of them away from the parent plant. Fire stimulates germination because the heat increases the seed coat's permeability to water, so they sprout like crazy. The seeds are also relished by many insects and small birds such as quail, as well as by mice, squirrels, and the mariposa brush rabbit. The buckbrush is the preferred home of the chaparral mouse, and a common dwelling or hiding place for his mousy relations, including the California pocket mouse, house mouse, and deer mouse.

Bees value the buckbrush's flowers in early spring, finding nourishment when few other blooms are available. Modern hikers have also noted its allure: "On the way up the trail, we were often surrounded by blooming ceanothus bush (here, I believe it is *Ceanothus cuneatus*). Chris thought ceanothus smelled like tortillas. I thought it smelled like honey. We compromised and said it smelled like sopapillas."[76]

Many species of ceanothus, including buckbrush (which has been cultivated since 1848), make wonderful ornamentals, especially as plants to attract beneficial insects. One vineyard, for example, uses ceanothus, toyon, and coffeeberry to draw ladybird beetles, lacewings, and predatory flies, which attack pests such as leafhoppers and aphids on the vines. This is the kind of gardening using native plants which is being promoted by native plant societies across the continent.

Fabaceae

Silverleaf Lupine
Lupinus albifrons var. *collinus*, whiteleaf bush lupine, silver lupine

Lupinus albifrons var. *collinus* is a deciduous shrub that is usually about 1 foot high. The stems are not woody and tend to grow parallel to the ground. The compound leaves, which are whitish-gray, have six to nine leaflets clustered near the base. The flowers, blue to red-purple to lavender, grow in racemes up to 1 foot long, but not densely clustered, in whorls up the flower stalk. Each small flower is similar to a small pea flower. Found in sand and rocky environments.

Range: California.

Lupines can be hardscrabble plants and those native to California will grow everywhere except in salt marshes. These resourceful plants thrive in rocky rubble and on sand dunes, where their root system may reach a depth of 20 feet. They grow where other plants fail because they are legumes and as such can fix nitrogen via bacteria in nodules on their roots. California has 70 species of lupines, at least 17 of which are native to the Santa Lucia Mountains.

All lupines produce seeds in pods that look much like peapods, but lupine pods and seeds are poisonous. In spite of this, the leaves were eaten as greens by the Luiseño, and the Yuki roasted young plants on stones over a hot fire. The Owens Valley Paiute used the plant to treat bladder trouble.[77]

The silverleaf lupine, along with two other lupine species, *L. variicolor* and *L. formosus*, is an important host plant for the mission blue butterfly, which searches for just the right flower on which to lay its eggs. Once the eggs are laid, ants come in to guard the larvae, in exchange for the mature larvae's honeydewlike carbohydrate secretions. A grassland species, the mission blue butterfly was declared endangered on June 1, 1976, by the U.S. Fish and Wildlife Service. Unfortunately, the coastal scrub in which the lupines thrive is also endangered. On San Bruno Mountain, 2,000 acres of critical habitat is reserved for the butterfly, managed by the local county department of parks and recreation.

Texas bluebonnet, *L. texensis*, closely resembles another Texas lupine, *L. subcarnosus*. Unable to play favorites, Texas declared them both (along with four other lupine species) the state flower. *L. texensis*, the more familiar one, has deep blue flowers with white centers and grows on the dry prairies of the south.

Scrophulariaceae

Monkey Flower
Mimulus spp.

Mimulus spp.

include about 80 species indigenous to the West Coast. They are annual or perennial herbs or low shrubs, with woody stems and sticky, opposite leaves, often toothed, and oval, oblong, or lance-shaped. The flowers grow from the leaf axils with five sepals fused together into a tube. The petals that emerge from the tube may be purple, red, violet, yellow, or buff, often with a lower lip and four stamens. The fruit is usually oval and fragile, with tiny yellowish to dark brown seeds. Found in moist, sandy areas, often near streams, or in meadows, rocky foothills, canyon slopes, and chaparral borders.

Range: British Columbia to California, Utah, Arizona.

The delicate mimulus, with tube-shaped petals in glorious shades of purple, red, violet, yellow, or buff, is actually a woody plant. The genus is called *Mimulus*, a diminutive form of the Latin word *mimus*, "actor," because in some species the flower looks like a grinning face. The common name also alludes to the facelike flower.

The sticky substance exuded by the leaves prevents water loss from transpiration on hot, summer days and also helps protect the plant from being munched on by insects. This did not, however, deter many West Coast Native Americans from eating them, along with the green shoots. The roots from a variety of *Mimulus* species were also used by the Pomo peoples to make a healing eyewash. People

Mimulus spp., monkey flower

lived in smoky dwellings with such poor ventilation it often led to early blindness; consequently eyewashes were very important for their general health. The Karok used a plant infusion of *M. cardinalis* (**scarlet monkey flower**) as a wash for newborns. The Kayenta Navajo treated the hiccups with *M. eastwoodiae* (**Eastwood monkey flower**), and the Kawaiisu put a decoction of *M. guttatus* (**common yellow monkey flower**) in a steam bath for chest and back soreness.[78]

Most of the early plant collectors in the region found examples of the genus, including Meriwether Lewis, Georg Heinrich von Langsdorff, John Charles Frémont, John Milton Bigelow, Albert Kellogg, Edward Palmer, and David Douglas, all of whom have species named for them. Archibald Menzies collected *M. guttatus* in California in the 1790s.

Different species may appear in very restricted areas. *M. cupriphilus* grows only in two small copper mining areas in Calaveras County; *M. marmoratus* grows atop Table Mountain in Calaveras County; *M. glaucescens* (**shield-bracted monkey flower**) is restricted to a small area of Butte and Tehama Counties; *M. laciniatus* (**cut-leaved monkey flower**) is restricted to the high Sierra Nevada. (There is some dispute about the species; some scientists think that most or all of these are just variants of *M. guttatus*.)[79]

The distinctively patterned *M. tricolor* (**tricolor monkey flower**) prefers to live in vernal pools—natural depressions which are waterlogged from winter to spring and typically parched over summer. These miniature ecosystems are home to many native plants and provide important resting sites for migrating birds. Like many habitats, California's vernal pools are dwindling, threatened by industry and development.

The tricolor monkey flower made a surprise reappearance in Oregon in 1999, when some amateur botanists found it in a former ryegrass field near Corvallis. The flower, once common in the Willamette Valley before plowing and irrigation changed the site, hadn't been seen since 1991. It's thought that some seeds may have survived in their nutlike capsules, and germinated when a flood left muddy puddles. It now looks as if these plants don't have to be reintroduced to areas where they were once wiped out. Some ecologists, including Peter Chesson of the University of California at Davis, think that the seed bank in the soil is so rich that it's often more efficient to create favorable conditions that encourage germination of these old seeds in the restoration of ecosystems.[80] Create the favorable conditions, and they will germinate.

Because *Mimulus* flowers come in a wide range of colors, from yellow to orange to red, they attract different pollinators, from hummingbirds to insects. *M. aurantiacus* **(bush monkey flower)** is a favorite of the common checkerspot butterfly. Female checkerspots looking to lay their eggs determine if they have the right plant by scratching the surface of the leaves and "tasting" the plant with their feet. Their caterpillars eat the leaves, which contain toxins. These in turn make the caterpillar and adult butterfly taste nasty to birds, which leave them alone.

The plants hybridize easily and many hybrids that have been cultivated often change qualities along the way. For example, *M. moschatus* **(musk flower)** had a strong smell until about 1900, when all cultivated plants lost their smell, probably because of crossbreeding with a single

odorless wild variety. The smelly version has simply disappeared.[81]

This trait of easy hybridization made monkey flowers the plant of choice in a study which revealed that evolution need not move in countless tiny steps, as long-standing theory has held, but can proceed by leaps and bounds. As Carol Kaesuk Yoon explained in a *New York Times* article: "The subjects of the new study were two species of monkey flower found in the western United States. Lewis's monkey flower has pale pink flowers designed to attract bumblebees. The flowers have a landing platform with yellow stripes that guide the bee to the pollen and a rich nectar deposit deep within the flower. The cardinal monkey flower, on the other hand, has red blossoms that appeal to hummingbirds. The flowers offer a more diluted nectar, and daub pollen on the birds' heads as they poke their beaks into the nectar well."[82]

Hybridization doesn't happen in the wild because different pollinating mechanisms keep the species separate. In the greenhouse, the scientists created two generations of hybrids, then analyzed the DNA to determine where the information that determined the color, size, and shape of the flower was located. Most of these characteristics seemed to be controlled by a single gene, which made the difference between a Lewis's monkey flower and a cardinal monkey flower. "The biologists say they believe that with such powerful genes at their disposal, new species have the potential to appear nearly full blown in relatively few evolutionary steps."[83]

Delving further into the sex life of the *Mimulus*, one discovers that the stigma in bee-pollinated species is two-lobed and sensitive: when touched, the lobes close together, reducing the chance of its picking up its own pollen when the bee backs out. If it has received pollen, it remains closed. If not, it slowly opens again, a feature that makes hand pollination very easy.

Iridaceae

Douglas Iris
Iris douglasiana, coast iris

Iris douglasiana

grows from a creeping tuberlike rhizome, about ⅓ inch in diameter, with fibrous shallow roots. The long sword-shaped leaves are ⅖ to ⅘ inch wide, with prominent veins. Reaching 3 feet in height, the leaves are yellowish to deep green with reddish bases. Stems are flattened and occasionally branched. Two or three cream, lavender, or deep reddish-purple flowers appear on each stem March to May. The fruit capsules are 1½ to 2½ inches long, and contain small black, finely wrinkled seeds.

Range: western California.

Iris is Greek for "rainbow," and with its great variation in colors and patterns, the Douglas iris fits the name more aptly than most. The flower parts come in threes: three spreading or drooping petal-like sepals called falls, three erect petals called standards, and three petal-like structures that cover the stamens. Colors range from royal purple to palest cream, with patches of yellow or white, or both, bearing dark or light veins. Douglas iris is one of two coastal species "named and described by William Herbert (1778–1841) . . . an English politician and later a churchman, dean of Manchester, and an authority on Amaryllidaceae (the umbellate lilies)." Both specimens were probably collected near Monterey in the early 1830s by David Douglas.[85] The second species, *I. longipetala*, is more robust than *I. douglasiana* and likes to have wetter feet.

The edges of the long, swordlike leaves of Douglas iris yield a remarkably fine, strong fiber that was once patiently worked by the Pomo and Mendocino into nets and ropes. David Rogers explains:

As the margin of the leaf is alone used, the work of making anything of it was exceedingly laborious. The silky strands were separated from the leaf and thoroughly cleaned from other tissues by means of a sharp-edged oblong piece of abalone shell, which was fastened to the thumb and used to scrape the fibre. Frank Youree informed me that it took nearly six weeks to make a rope 12 feet long. The rope, which was exceedingly strong and very pliable, was especially valuable in making snares to catch deer; and on this account it was known as "deer rope."[86]

The Pomo also used the leaves as a covering when acorn meal was being processed. The meal was placed in shallow pits and covered with iris leaves, then water was poured over the meal to leach out the bitter tannins. The Monache and Southern Yokuts pounded the iris seeds into flour. Medicinally, the Yana chewed the roots to relieve a cough, and the Modoc deemed a root decoction to be just the thing for sore eyes.

I. douglasiana grows naturally along the coast; it is common on bluffs and treeless grassy hillsides. It also may grow inland in open woodlands, where it has crossed with other local iris species more adapted to shade. Some long-established natural hybrid populations have been given their own names, such as **Thompson's iris** on the California-Oregon border, and the **Marin iris** in the coastal range just north of the Golden Gate.

But open grasslands are the Douglas iris's favored habitat and it can form large, extremely dense colonies. In fact, it is even regarded as a weed in areas where its crowded stands overwhelm meadow forbs and other valuable wildlife forage plants. This happens when livestock heavily trample and overgraze a field: many plants can't withstand this treatment and so weaken and fade, but Douglas iris, despite trampling, rapidly regenerates from rootstocks, and lives to rule the domain.

Iris douglasiana, Douglas iris

Montane

A journey to the North American Continental Divide is like a trip to the top of the world. There are larger mountains in the world, but surely few places are wilder and more beautiful, or better at restoring a sense of awe to jaded city dwellers. To ascend from the lowest foothills to the top of the divide is like traveling across the continent from south to north. The higher one goes, the colder and wetter it gets. Every 300 feet upward is like driving 170 miles north; each type of plant community exists in miniature, from desert to alpine tundra.

The montane is a land of extremes: the

Wild and beautiful, the montane is a land of extremes.

hottest, the coldest, the wettest, the driest. For anyone who has made the trip through the mountains, the fluctuations of wind and temperature can be torturous: boiling one minute and freezing the next. The terrain is equally varied, with gullies, glaciers, Alpine lakes and meadows, bogs and fens, as well as bone-dry prairies, and the plant communities belonging to each one. This complex mosaic of plants is further varied according to whether it's on the west (wet) side of a mountain, or in the drier rain shadow on the east.

The mountain ranges of the West are the youngest in North America, still flexing their muscles with quakes and the occasional huge volcanic eruption. By comparison, in eastern

North America, the mountains are the oldest on earth, softened by millions more years of erosion and compression from the great ice fields.

In the mountains, summer is brief, cool, and dry. Winters are cold and wet, with 70 to 100 inches of precipitation a year, most of it in the form of snow. Nevertheless, the Alpine meadows, often saturated with water for long periods of time, will produce wildflowers of incomparable beauty. They bloom swiftly and showily to attract insects and hummingbirds. These meadows are transition zones that may eventually become forest, if the land dries out and the pioneer species of lodgepole pine move in.

Lodgepole pine, which came along about 10 million years ago, has turned into one tough plant, able to recover from almost any kind of calamity. After a fire, it regenerates with astonishing vigor. The litter of its needles is so acidic that little grows beneath the trees. When a lodgepole pine dies off, Douglas fir, sub-Alpine fir, or Englemann spruce move in. Higher up, the bristlecone pines cling to rocky ridges. Past the timberline, the trees are stunted and dwarfed by the harsh conditions. This is called the Krummholz zone and it looks like a forest of bonsai. Above the Krummholz, conifers cannot grow. Small plants hug the ground and hide behind rocks to keep out of the ferocious, desiccating winds.

The montane region also includes the foothills forests, crammed with lodgepole pine, trembling aspen, and a shrub-dominated understory. In the southern part of the Rockies in Canada, the foothills give way to prairie, and in the United States they roll into scrubland.

The amount of self-protection in the mountain environment is seen in the evergreens. The trees of the montane forest up to the treeline have downswept branches that shed snow and allow layering (roots form if the branches touch the ground for any length of time). The spruce, firs, and pines of this forest can photosynthesize at temperatures lower than other trees. They tend to shut down in midwinter, reducing transpiration when everything, including their roots, is frozen, and because their needles are covered by a cuticle, evaporation is reduced as well.

Fire was suppressed during the first part of the twentieth century, thereby changing the habitat for the animals that evolved here: the woodland caribou, mule deer, bighorn sheep, and badgers. As roads were built to allow travel or logging, exotic plants invaded. Here you see outbursts of *Euphorbia esula* (**leafy spurge**), *Linaria genistifolia* (**Dalmatian** or **broomleaf toadflax**), and *Potentilla recta* (**sulphur cinquefoil**), along with *Lythrum salicaria* (**purple loosestrife**). They may present us with vivid colors, but they elbow native plants completely out of the way, decreasing the biodiversity and leaving us all bereft.

Asteraceae

Arnica

Arnica spp., mountain arnica, mountain tobacco, mountain snuff, leopard's-bane

Arnica spp.

include more than 30 species of herbaceous perennial, varying from 2 inches to 2 feet high, the majority of them native to North America. Most have few branches, deep roots, with basal leaves in a cluster and opposite, simple stem leaves. The yellow to orange compound flowers appear at the top of the upright stems.

Range: native to the plains and Pacific Coast regions, northward to Alaska; also cultivated throughout North America.

The fuzzy exterior of arnica may be endearing and the pretty yellow daisylike flower head cer-

Arnica spp., arnica

tainly has charm, but this is a poisonous plant. It's thought that the word *arnica* comes from the Greek *arnakis*, meaning "lamb's coat," and refers to the feltlike sepals covered in soft hairs that surround the flower. It might also be a corruption of *ptarmica*, meaning "causing sneezing." The appearance of the arnica influences each arnica species' name, such as **heart-leaved arnica** (*A. cordifolia*), **hairy arnica** (*A. mollis*), or **long-leaved arnica** (*A. longifolia*). Arnicas are often confused with *Doronicum* (also called **leopard's-bane**) and *Inula*, both popular as cultivated garden plants and somewhat similar in appearance.

A. cordifolia is found at elevations ranging from 3,500 feet to 10,000 feet. In Montana's Madison range, it forms the primary ground cover in sub-Alpine fir forests, and in northern Utah, it is a staple food in summer for mule deer and elk. Its populations ebb and flow: a major disturbance such as logging prompts mass flowering a year or two later, and a subsequent population explosion. Not content with casting its seed to the wind, it also sprouts from rhizomes.

Both the flowers and roots of arnica are poisonous, and the sap may cause minor skin irritation, but they have traditionally been used to make poultices to soothe bruises, muscle aches, or sprains. According to Daniel Moerman, the roots of *A. cordifolia* and *A. latifolia* were used by the Okanagan-Colville of British Columbia and Washington state to make a love potion. The recipe also included a robin's heart and tongue, and ocher paint. These elements were mixed together, dried, and powdered. The smitten one went into the water, faced east, recited certain words, including the name of the object of desire, and marked his face with the arnica mixture.

Today, arnica (derived largely from the flowers and roots of *A. montana*) is a popular ingredient in herbal and homeopathic treatments, primarily for external use. Often called the "black and blue remedy," it is used in creams and compresses for bruises, sprains, and muscle or joint pain. Flowers were gathered carefully; it takes up to six pounds of fresh flowers to get a single pound of dried powder. The dried flowers are steeped in alcohol to make a tincture. (In Europe, where the flowers are often wild-collected, *A. montana* is now a protected species in several countries.)

Arnica is a favorite of rock gardeners, though providing the ideal growing conditions is tricky and they are prone to summer rot in lowland areas. But all rock gardeners love to have a plant that's tough to grow.

Boraginaceae

Bluebells

Mertensia spp., Alpine bluebells, mountain bluebells

Mertensia spp.

include several native mountain dwellers. They are herbaceous perennials ranging from 8 inches above the timberline to 36 inches in very moist sites. Leaves are green to blue-green (often turning subdued red-purple in fall), mostly lance-shaped, and hairy. The dainty tubular flowers are usually blue, borne in terminal panicles in spring and summer. Found in Alpine and sub-Alpine meadows and woods, often along mountain streams.

Range: **Hudson Bay to Alaska, through the Rockies, South Dakota, central Oregon, southern Idaho, Montana, south to Colorado, northern New Mexico, and California.**

The sight of brilliant bluebells and magenta monkey flowers blooming in splendid profusion prompted Ann Zwinger to write, "I think the Elysian fields must be like this—redolent with blowing flowers beside quiet waters."[1]

Named after F. K. Mertens, a German botanist working in the early 1800s, *Mertensia* belongs to the borage family and comprises about 50 species, many of them native to North America. (These bluebells are not the English bluebells, which are *Hyacinthoides nonscripta*, members of the lily family, or the bluebells of Scotland, *Campanula rotundifolia*, members of the bellflower family.) Among those at home in the mountains is the aptly named *Mertensia alpina* (**Alpine bluebells**), which at 8 inches is one of the shorter species. *M. ciliata* (**mountain bluebells**) grows about 2 feet high, with large long-throated flowers. Along mountain streams it can form extensive colonies with thousands of blooms, and the distinctive blue-green color of its leaves stands out even at a distance.

M. paniculata (**tall bluebells**, **tall lungwort**, **chiming bells,** or **languid lady**) grows up to 3 feet high, with nodding flowers that change color. The young blossoms contain anthocyanin, which gives them a pinkish tinge. Once the flower develops, the sap increases in alkalinity, masking the red pigment and making the flowers appear blue. This strategy brings certain insects that are more attracted to blue than red at just the right time to pollinate the flowers.[2]

Another bluebell strategy was explained by a naturalist to a group of kindergarteners on a field trip to a park near the Colorado foothills. "I showed the children chiming bells and pointed out their pink buds, which opened into sky blue flowers. When the flower is pollinated and begins seed making, the flower tube drops off as a unit so the bee does not waste his time looking for pollen."[3]

OPPOSITE PAGE: *Mertensia* spp., bluebells

Some Native Americans used infusions of mountain bluebells to increase the milk flow of nursing mothers, and to relieve the itching caused by smallpox and measles.[4]

Crassulaceae

Roseroot

Sedum integrifolium, also known as *Rhodiola integrifolia*, *Sedum rosea* var. *integrifolium*, or *Tolmachevia integrifolia*

Sedum integrifolium

is a perennial, succulent, herbaceous plant with clustered stems growing 6 to 10 inches high from a rhizome. The tiny, alternate, fleshy, egg-shaped leaves crowd along the stem. They are green but often covered with a bluish-white powder. The flowers, each with five fleshy petals, appear in clusters at the ends of the stems, The plant is dioecious; the male flowers are yellow or white, the females dark red to purple. The seeds grow in red or purple clustered capsules with pointed tips. Found on rocky, open slopes in sub-Alpine to Alpine environments.

Range: Alaska to Newfoundland; farther south on higher elevations into Alberta, Idaho, Montana, Colorado, Nevada; near the Great Lakes, and in Northeastern states.

Sedum integrifolium was named roseroot for the pleasant fragrance of its fleshy roots.

The succulent leaves are edible and thirst-quenching in an emergency. Traditionally, Native Americans gathered the leaves, shoots,

Sedum integrifolium, roseroot

and rhizomes, all high in vitamins A and C, and ate them either raw or cooked. People from eastern Alaska and western Yukon made a tea out of the red plant tops. To treat mouth sores, they'd chew the roots and spit them into the mouth of the sufferer.[5] The Thompson used the plant as a poultice or decoction in treating hemorrhoids.[6] Researchers are currently investigating roseroot's effectiveness in regulating metabolism.[7]

A subspecies, *S. integrifolium* subsp. *leedyi*, or **Leedy's roseroot**, is classified as a threatened species—only four populations exist, two in Minnesota and two in New York. It is now being suggested that Leedy's roseroot is a glacial relict—one of the more romantic terms applied to native plants. After the last great period of glaciation this was a widespread plant, but as things warmed up, more and more of its habitat has been lost and we find it now in isolation—a relict.[8]

A close relative, **lance-leaved stonecrop,** *S. lanceolatum,* is found from southern Alaska and Yukon south to California and east to Alberta, and in South Dakota, Nebraska, Colorado, and New Mexico. It has a similar form, though its reddish or purplish leaves are, as the name implies, more lance-shaped. Like roseroot, they can also ease the thirst of a weary traveler, with one caveat: the Okanagan-Colville considered them a laxative. Women brewed the leaves into a tea for drinking after childbirth to "clean out the womb."[9] Lance-leaved stonecrops were the equivalent of Sensen for the Gitxsan and Wet'suwet'en Nations, who often ate the leaves to freshen their breath following a meal.[10]

Cupressaceae

Juniper

is a much-branched sprawling evergreen shrub around 5 feet tall, though it may sometimes grow into a small tree 15 to 40 feet high, and at other times form a prostrate ground cover, all with reddish-brown, scaly bark. Small needlelike leaves are gray- to blue-green with a white band, and tightly crowded, three to a whorl, on the twigs. The male and female cones (each about ½ inch long) appear on different plants, in clusters at the tips of the branches. The female matures into a berrylike seed cone, reddish at first, turning to dark blue as it ripens. Its elongated triangular seeds require a long germination period to grow in dry open sites and in forests.

Range: circumboreal; almost circumpolar except for gap in Bering Sea region; widespread in North America beyond treeline; extends southward through New England to the Carolinas and westward through northeastern Illinois, Indiana, northern Ohio, Minnesota, and Nebraska, to the western mountains of Washington, California, Arizona, and New Mexico.

Juniper is considered the most widely distributed tree in the world. In Europe it was known as the "bastard plant" because it was used to induce abortions.[11]

Naturalist Constantine Rafinesque wrote of the common juniper, "The berries have a strong, pungent, aromatic smell and taste, somewhat sweet and bitter, containing an essential oil, tannin, and a sweet mucilage. The leaves and wood contain some of the oil also, in which resides the active properties."[12] The volatile oil is terpinen-4-ol, which increases kidney action. The Gitxsan knew of this effect, since they boiled the twigs and berries to make a tea for people suffering from kidney disease, as well as for those with tuberculosis, colds, heart trouble, and respiratory problems.

The Gitxsan of the Pacific Northwest also attributed supernatural powers to the common juniper, and valued it as a fumigant, deodorizer, and cleanser. They burned or boiled the boughs, which created a pungent smoke or smell, to purify a house and protect its inhabitants from infection and harmful spirits. Sometimes juniper combined with other plants was used to make "smudge" mixtures for cultural ceremonies. The Navajo burned juniper to make good-luck smoke for hunters. The Blackfoot of Canada used it to treat lung and venereal diseases.[13]

Kathleen Keeler of Nebraska recalls hearing that long ago superstitious people believed juniper could protect them from witches. "Hang a branch over the door. The witch has to count all the needles correctly to come in (so if you use a big branch, you should be safe)." Even today, people have been known to plant a juniper near the door in an attempt to ward off accidents and theft in their home.

Some wild foragers still advocate limited use of juniper berries as a spice in cooking, and Henriette Kress suggests putting whole juniper twigs in with your beef, then "tell your guests that it's moose . . . the taste really changes that much."[14] Juniper extract, which can be fatal in even fairly small amounts, was once used as a meat preservative and an ingredient in gin, particularly in "bathtub gin" during Prohibition. The highly inconsistent results inspired humorist Will

Rogers to remark that Prohibition was "better than no alcohol at all."[15]

Cupressaceae

Rocky Mountain Juniper

Juniperus scopulorum, mountain red cedar, weeping juniper, Rocky Mountain red cedar, Colorado red cedar

Rocky Mountain juniper is less drought-resistant than other Western junipers, but it can live longer—to 300 years and more—finding a perch on cliffsides and in stony ground (*scopulorum* means "of rocky cliffs or crags"). This slow-growing erect shrub or small tree, usually with a conical crown, grows up to 40 feet high. The bark is reddish-gray or brown, furrowed, and scaly. Blue-green, scalelike leaves lie flat on the branch. The male cones are brown; females are berrylike, green at first; blue with white waxy bloom when ripe; each contains two seeds.

Every two to five years, a juniper tree bears an especially heavy crop of "berries." They attract small mammals, and birds such as Townsend's solitaires, Bohemian waxwings, and wild turkeys. Dozens of junipers were planted years ago in Garden of the Gods park in Colorado Springs. As a local put it, "In a good berry year the solitaires act like this place is Miami Beach or something. If only the solitaires would shut up for a while, then you could hear the waxwings."[16]

Many peoples also ate juniper berries raw or cooked in the late summer and fall. The Straits Salish, Okanagan-Colville, and Thompson of the Pacific Northwest used Rocky Mountain juniper to combat evil spirits associated with illness and death. Hilda Austin, a Thompson interviewed in 1981–1982, said:

If you know there's sickness coming, the old people break the branches [and] put them on a hot stove—burns, smokes—smells strong all over the house . . . it keeps the sickness out. Also when there's a lot of mosquitoes, you do that—all the mosquitoes "run away." Also if everyone has a cold or is sick, you do that. It keeps the air fresh. When a person has been sick and then dies in the house, you do that—fill the house with smoke—like an air freshener.[17]

In addition to the smoke, a juniper concentrate was used as a disinfectant for scrubbing floors, walls, and furniture when there had been a death in the house.

The wood of Rocky Mountain juniper is durable, rot-resistant, and clean-burning. It has long been used as firewood and for fence posts (it's said there are century-old posts still in use today) and, more recently, for paneling, furniture, pulp, and handmade items. Sculptor Clark Schreibeis explains why it's his favorite wood:

Juniper is a beautiful wood with hues of red, brown, purple and with occasional streaks of yellow running through the grain. Being very aromatic, it is also prized for the lining of cedar chests. Due to these qualities, plus the tightness and character of the grain, it makes an excellent carving wood for either hand or power tools. . . . No two trees are alike and many hours are spent looking for just the right tree to get the design I intend to carve.[18]

The Jardine Juniper in Logan Canyon, Utah, is one of the oldest living things on earth—a core sample taken in the 1950s determined that it was 1,500 years old. Discovered in 1923 by a botany student, Maurice Linford, it was named in honor of William J. Jardine, who was secretary of agriculture under Herbert Hoover. Growing from a stony cliff, the gnarled and ancient tree is truly clinging to life.

In 1996, another old Rocky Mountain juniper was saved. At 180, it was just a youngster compared to the Jardine Juniper, but it was one of the largest and oldest junipers on the Fort Lewis College Campus in Durango, Colorado. It had to be moved so the sidewalk could be widened. First, crews excavated around the tree's fibrous root system; a week later, arborists wrapped the giant root ball in canvas and welded a metal basket around it as a support. Then a 120-ton construction crane lifted the 22-ton tree from the ground and safely moved it to a spot 25 feet away. It cost about $5,000 to move the tree, which arborist David Temple figured was worth about $12,000. But it went beyond the monetary value. "More people see it in one day than just about any other tree in town," he said.

One aspect of close encounters with a Rocky Mountain juniper that home gardeners should take into account is its recognizable scent, which has been described as similar to that of a cat litter box. Don't drive it home in a car with the windows up.

Range: **throughout the Rocky Mountains, from British Columbia south to New Mexico; also the Dakotas, Utah, Nevada, Arizona, and Texas.**

Ericaceae

Grouseberry
Vaccinium scoparium, whortleberry, dwarf blueberry, dwarf bilberry

Vaccinium scoparium
is a deciduous rhizomatous dwarf shrub about 1 foot high that grows from a shallow root system

in thin soil. The numerous slender branches are greenish or yellowish-green, smooth or slightly hairy. Its small leaves are lance-shaped to somewhat oval, smooth, light green, $1/2$ inch long, and veiny on the lower surfaces. The tiny pink urn-shaped flowers give way to bright red edible berries. Found in areas of heavy snowpack, sometimes above the timberline.

Range: British Columbia south to California, east to Alberta, Montana, and South Dakota, throughout the Rocky Mountains, south to Colorado, Utah, and northern New Mexico.

Grouseberry forms light green mats on the floor of the coniferous forest. The foliage is high in carotene, the berries are full of vitamin C, and both have high energy content. Animals from elk, deer, and moose to skunks, and birds such as ptarmigans and, of course, grouse browse on this shrub and eat its fruit; bears relish it and even like to sleep on mats of grouseberry in the woods. Unfortunately grazing sheep have reduced populations of the once-common grouseberry, not just by browsing, but also constant trampling (the plant can withstand the occasional bear siesta, but not a lot of sheep feet).

The sweet berries were also eaten fresh or dried like raisins by the Native Americans of the Pacific Northwest.[19] Those who have enough patience to collect the berries say that they can be used to make jam and wine. As Doug Wilson of Crowsnest Pass says, "While their modest size proves to be a discouragement to most berry pickers, there can be no denying that their taste matches or exceeds the huckleberry."[20]

With shallow roots and rhizomes, grouseberry is vulnerable to fire and clear-cutting, and can take up to five years to recover from a major disturbance.

Liliaceae

Glacier Lily

Erythronium grandiflorum, snow lily, yellow glacier lily, dog's-tooth violet, adder's tongue, fawn lily, trout lily, avalanche lily

Erythronium grandiflorum

grows 5 to 15 inches high from an elongate, deep-seated corm. The two leathery basal leaves are elliptical, up to 10 inches long. The nodding yellow flowers have sepals that curl back to reveal the stamens and style, making the flowers look like turbans. Each plant produces one to five flowering stems. It blooms from March until August according to latitude and elevation. The flowers give way to brown papery seed capsules, 1 to 2 inches long. Found in moist, fertile soils along streambanks, in shaded woods, and in sub-Alpine meadows, following the melting snow line from the valleys up to sub-Alpine.

Range: western North America from southern British Columbia to northern California, east to Alberta, Wyoming, and Colorado.

Erythronium grandiflorum is the showiest of all mountain wildflowers. Its nodding yellow flowers look like little turbans. Glacier lily is aptly named because it tolerates the worst winter weather. It can squeeze in between rocks for shelter and the tough little flowers generate enough heat to work their way through the last patches of snow to fill mountain meadows with clear yellow blooms.

They are mostly pollinated by bumblebees, among the first seekers of pollen, and occasionally by hummingbirds. Like their woodland cousins, *E. americanum,* glacier lilies are spring

Erythronium grandiflorum, glacier lily

ephemerals, which grow, bloom, and fruit in 10 short weeks. Then the plants disappear altogether and remain dormant until the following year's snow melt.

Sometimes when the snow lingers, it can turn blooming seasons topsy-turvy. In September 1999, Mary Ann Spahr visited Chinook Pass in Mt. Rainer National Park and was astounded to see glacier lilies in full flower beside 6-foot snowbanks on the north-facing slopes, just a few steps away from a display of fall-blooming asters on a south-facing slope.[21]

Once upon a time, this was an important root vegetable, along with camas (*Camassia*) and rice root (*Fritillaria*). Because the glacier lily grows best in open meadows, the Native Americans burned off hillsides to preserve this habitat. The corms lie deep in the soil, up to 7 inches below the surface. Even after a destructive fire they are safe and sprout again the following spring. Ethnobotanist Nancy Turner marvels at the way in which these people enhanced the stocks of glacier lilies (and the other roots they ate). They would turn over a whole section of turf, she says, select the largest roots, and leave the small ones behind to grow. They scattered the seeds, replanted the propagules attached to the plant, and then left the area alone for three or four years before returning. In this way the glacier lily and other plants were able to be very successful plants even in what were often tough terrains.

The glacier lily's edible corms, reportedly as big as one's fist and described as "white clear tubes," were enjoyed by the Okanagan-Colville, Shuswap, Thompson, and Montana. The corms contain the carbohydrate inulin, which converts to the sweeter, more edible fructose when it's cooked. They were usually boiled and eaten immediately or dried in sacks, to be rehydrated later and put into soups, stews, and puddings.[22] The Thompson traded strings of dried corms and used them as wagers in gambling. They could be medicinal as well. The Montana applied

a poultice of roots to boils and the Okanagan-Colville found them effective for bad colds.[23]

Animals, especially bears, are also partial to the lily corms. Modern hikers look for glacier lily "dig sites" to determine whether there are bears in the area. Brian and Dee Keating were part of a group who safely observed such a site, and saw how difficult they are to dig up.

Just after supper . . . Alex noticed a bear out the window! Sure enough, a grizzly was plowing up the earth, eating glacier lily bulbs . . . She would dig her front paws into the earth, move her back legs forward, then using the weight of her whole body she would tug heavily. With her bum momentum in full force, she'd tug up a mound full of glacier lily bulbs that she would neatly nip off and consume.[24]

A rare white variety of glacier lily (*E.g.* var. *candidum*) occurs in a few scattered populations in northeastern Washington, northern Idaho, and northwestern Montana. It deserves protection, especially since some of its habitat in Montana's Glacier National Park has been destroyed by road maintenance work.

Pinaceae

Sub-Alpine Fir
Abies bifolia, Rocky Mountain Alpine fir, corkbark fir

The ramrod-stiff form of the sub-Alpine fir can grow from 9,000 feet above sea level right up to the timberline. Along with Engelmann spruce, it's one of the most prominent trees of the sub-Alpine forests, and one of the toughest. It grows in areas with only two frost-free months and flourishes in glacial deposits or on igneous bedrock. In spite of the harsh conditions, it's a long-lived tree. One sub-Alpine fir in Wyoming's Medicine Bow Mountains is more than 490 years old.

Abies comes from the Latin *abeo*, meaning "to rise," and most of the fir species grow straight up from the ground. A sub-Alpine fir can reach 65 feet high, but in the montane environment it may be stunted or twisted. Its thin, smooth, gray bark becomes furrowed with age. The branches are stiff and straight. The flattened bluish-green needles, which have a camphorlike scent, are 1 to 2 inches long with tiny white stomata. They cluster thickly on the twigs and tend to bend upward. Purplish male and female cones grow on the same tree; the small male cones hang down from the branches and the larger female cones sit upright on the upper branches. They produce winged brown seeds.

A. lasiocarpa is also known as sub-Alpine fir, and *A. bifolia* was once thought to be part of the *A. lasiocarpa* species, because the two are so similar, but there are some minor chemical differences and the layer under the bark is tan-colored in *A. bifolia* and red in *A. lasiocarpa*. *A. lasiocarpa* usually exists in small stands at high elevations in the coastal ranges from Yukon and British Columbia south to California. One of the tallest specimens, at 170 feet, overlooks the Icicle Creek Trail in the Alpine Lakes Wilderness of Washington.

The name *lasiocarpa* means "hairy-fruited" and refers to the cones. The seeds are eaten by songbirds and mammals; grouse munch on the leaves, and the bark is browsed by deer, elk, bighorn sheep, and moose. Though reproduction is generally by seed, when the lowest branches are weighted to the ground with snow, they sometimes take root and form new shoots, a process called layering.

White fir, *A. concolor*, is another related species which is found in the Rocky Mountains from Idaho and Wyoming south to New Mex-

ico. It has longer, flatter needles curving upward from two rows, one on either side of the twig. *Concolor* means "of uniform color," and refers to the needle surfaces. It grows in warmer and drier conditions than the other native firs. The wood

In Their Time

(for Nicholas Pearson)
I like to be there—
late spring at the far reaches of treeline—
when the mountain hemlock and
    subalpine fir
first break out of the deep snowpack:
soft-sliding blanket that had laid them
bough and stem to the slope
while the weight of winter moved past.
It's the warmth of the life in these small
    trees
slowly melts through the frozen grip,
and on a day of sun-loosened crust,
a break-through-to-your-knees day full
    of juncos
and the skittery tracks of marmot,
they will upturn like a drawn bow
and with a sudden springing burst of snow
right themselves once more into
    treehood.
I like to think of one winter
when each of them, thickened with the
    years
of snowmelt and wind,
find that singular strength to hold straight
through the deepening snows;
to have turned the great bows of their
    trunks
into the slope, and held there;
lifting, finally
out of the slow dance of the years
as all things lift in their time.

Tim McNulty,
In Blue Mountain Dusk

is used as lumber for pulpwood, furniture, boxes, crates, and, because it has no scent, was once used to make tubs for butter.

All the firs were widely used for medicinal purposes by the Native Americans. Among the Blackfoot, a poultice of needles was applied to relieve chest colds or fevers. Burning the needles in a "smudge" was thought to soothe a fever and heal sick horses, while an infusion of the needles was used to treat tuberculosis. The outer and inner bark were reduced to a syrup by the Gitxsan as a tonic and a treatment for colds, coughs, and flu.

The firs also had cosmetic uses: powdered bark was used as a deodorant by the Okanagan-Colville, while the Flathead mixed dried, pounded needles with deer grease for a hair tonic. And firs figured in the Blackfoot dessert menu: they ate the cones, pulverized and mixed with back fat and marrow, as candy.[25]

Range: **Rocky Mountains from Yukon to New Mexico.**

Pinaceae

Grand Fir
Abies grandis

The name says it all: this is one grand tree. A magnificent specimen in Washington state's Olympic National Park is 231 feet tall with a circumference of 20 feet 8 inches. That is a true giant, since the grand fir generally grows to 150 feet. It also grows quickly—up to 10 inches a year—with a narrow, pointed crown and slightly curved branches. The grayish-brown to reddish-brown bark is thin and smooth, with pockets of resin. As it ages, the bark acquires fissures and ridges. The flexible evergreen needles are 1 to 2 inches long, arranged in two bushy rows, flat and shiny green above, two silvery white bands below. The pale yellow male cones are up to 1 inch long, and the yellow-green female cones are slightly larger, narrow, and cylindrical. Both grow at the top of the tree. The paired winged brown seeds don't develop until the tree is at least 20 years old. Grand fir grows best in valleys and on mountain slopes in cool, humid climates.

Not surprisingly, it shelters dozens of animals, insects, and birds such as the marbled murrelet and the northern spotted owl. Not only is it a bed-and-breakfast for animals, the grand fir was once like a corner store for Native Americans. The Thompson ate the inner bark, used the branch tips to make a tea, and bathed their newborn babies in a bark solution to make them strong. The bark was used by the Straits Salish to make brown and pink dyes for their basketry; the Ditidaht Nation brought fir boughs inside as an air freshener; the Nitinaht and Shuswap ate the gum as candy; and the pitch mixed with oil was used as both a deodorant and an antibalding treatment.[26]

The arboreal store also included a pharmacy: Kwakwaka'wakw people boiled the bark (sometimes mixed with stinging nettles) to bathe wounds or to drink as a general tonic. Fir gum was applied directly to sore eyes, while the young shoots and bark were boiled down for treating stomach trouble. Other solutions were used for gonorrhea and as an eyewash.[27]

Like related firs, the trunks of grand fir have swellings that, when pinched, release a squirt of fragrant resin. The Salish of Vancouver Island collected pitch from grand fir blisters, rubbed it into wooden implements, and scorched it to provide a varnished finish.

Unfortunately, grand fir is susceptible to a number of pathogens, and the mortality rate is highest in stands where they dominate. Massive harvesting of pine and larch, which began in the early twentieth century, resulted in an abundance of grand fir, but today many of the trees are diseased or dead. Grand fir does not exude heavy pitch to heal its wounds, and the wood lacks decay inhibitors, leaving it open to rotting fungi such as annosus (*Fomes annosus*) and Indian paint fungus (*Echinodontium tinctorium*). It is a slow death. Insects also take their toll, but the most damaging are the Western spruce budworm, Douglas fir tussock moth, Western balsam bark beetle, and fir engraver beetle. Even invaded by fungi or infested with insects, however, grand firs can still reach 250 or 300 years of age.

Despite its afflictions and the fact that it's lighter and weaker than most firs, grand fir is a valuable timber species. Many coastal groups, including Straits Salish and Ditidaht, shaped, steamed, and carved fir knots into fishhooks, while the Okanagan people made canoes from grand fir bark. Today, grand fir wood is valued for light-duty uses because of its compatibility with adhesives and its tendency not to shrink. It is also used for pulp, and grown for Christmas trees.

Range: **western British Columbia, Washington, Oregon, northern California, eastern Cascades, Idaho.**

Pinaceae

Bristlecone Fir

Abies bracteata, Santa Lucia fir, silver fir, leafy-coned silver fir, fringe-cone fir, bracted fir

This is the rarest of all fir species—it is an *Abies* subgenus unto itself (*Pseudotorreya*). It has bracts around the cone scales, which end in long, narrow, leaflike structures and large, spindle-shaped, resinless, winter twig buds. The top of the tree is narrow and sharply pointed, similar to the tops of sub-Alpine and Arctic fir and spruce species. Bristlecone fir grows up to 100 feet tall, with a dense branching pattern, drooping branches, and a tall, narrow crown. The mature bark is smooth or slightly fissured, broken into scales. The flat, stiff needles are 1 to 2 inches long, and the mature 3- to 4-inch seed cones produce winged seeds to be dispersed by the wind.

David Douglas and Thomas Coulter were the first botanists to collect specimens of this tree, in 1831, probably near Cone Peak at Mission San Antonio. In the late 1800s, Albert Kellogg of the California Academy of Sciences traveled to the Santa Lucia Mountains, and was sufficiently enraptured to write:

> *This exceedingly elegant steeple-shaped Fringe Cone Fir is of the most extraordinary aspiring beauty, . . . clad in a light green dress of silvery sheened foliage nearly or quite to the feet, it gives them the most exquisitely feminine expression it is possible to conceive. Besides the modest plumy-fringed cones, evanishing up in the blue amid a kind of gossamery webby haze . . . the foliage is gemmed with golden drops of gum, that glitter in the sunlight like radiant beaded jewels, thus sparkling all over, from crown to foot.*[28]

The missionaries at Mission San Antonio de Padua in California called the tree *incensio*, because they used its resin to make incense. In the mid-1800s, New England whalers cut trees along the Big Sur coast to replace masts smashed up by storms.

By 1898, botanists were aware of the exceptionally small distribution of *A. bracteata*. Charles Sargent noted that the largest groves contained no more than 200 trees, and expressed relief that "this beautiful tree . . . [has] found a foothold in the Old World," since he thought it was destined for extinction in America.[29] Stanford University in California boasts a Santa Lucia fir planted in 1898 in what was known as Encina Garden that is now treasured as one of the heritage trees on campus.

Fossil evidence in western Nevada suggests that 13 million years ago, during the Miocene epoch, *A. bracteata* enjoyed a much wider distribution throughout western North America. Presumably, during the ice ages of the Pleistocene epoch the bristlecone fir retreated to the Santa Lucia Mountains, where it is found today.[30]

Despite its vulnerability to fire, and to a small wasp which attacks its seeds, the Santa Lucia fir's future seems assured, as most of its stands lie in protected areas. The California Native Plant Society has it on its watch list because its distribution is so limited.

Range: **Santa Lucia Mountains in Monterey County, California.**

Pinaceae

Engelmann Spruce
Picea engelmannii

P. engelmannii

is a long-lived conifer that may survive for 400 years; some specimens are 600 years old. Cone production begins at 15 to 40 years. When it's not Krummholz, it can grow to about 150 feet, sometimes taller, with a narrow, conical crown, gray to reddish-brown bark; the branches spread out horizontally or droop slightly. Pointed blue-green needles are about 1 inch long, with a squarish cross section, and have a skunky smell when crushed. The seed cones are violet or deep purple, ripening to beige or brown, up to 3 inches long, with flexible diamond-shaped or elliptical scales; the seeds are blackish and long-winged.

Range: **throughout the Rockies, from Alberta and British Columbia south to Arizona and New Mexico.**

This sub-Alpine inhabitant was named after George Engelmann (1809–1884), a German-born physician and botanist from St. Louis, who was an authority on conifers. Normally tall proud trees, Engelmann spruces become progressively more stunted as elevations increase. At the tree-line in northern Idaho, they rarely exceed 65 feet; higher than that they form Krummholz. As Robert Devine explains, "'*Krummholz*' is German for 'crooked wood.' The trees of the *krummholz*, many of them hundreds of years old, are the same species that grow lower down to heights of 150 feet, but up here about 11,000 feet the fierce climate doesn't brook such loftiness. *Krummholz* trees prostrate themselves as if begging to be spared from the cold and ferocious wind."[31]

The spruce was another all-round resource for the Native Americans. The Thompson made baskets and canoes from its bark (the inner side of the bark forming the outside of the canoe). They also used a concentrate of the needles and gum for cancer; if this treatment didn't work, they said, nothing would. The prickliness of the needles was thought to offer protection against evil. Babies, warriors, hunters, and young men reaching puberty were washed in a strong elixir made of the bark or the branches. The Okanagan-Colville used an infusion of bark for tuberculosis and respiratory problems, and still make spruce tea from the branches.[32]

The wood, which is white, odorless, straight-grained, and easy to work, was used by the Hoh and the Quileute for shakes, clapboards, and framing timbers, an activity which still goes on today. It is also fashioned into specialty items such as food containers and aircraft parts. Because of its resonant qualities as well as its singular beauty, it is much sought after for violins and pianos.

One major pest of Engelmann spruce is the spruce beetle. Unabated logging leaves lots of downed trees in its wake and this provides the beetle with plenty of food. This is probably why there are now many serious outbreaks.

Pinaceae

Lodgepole Pine
Pinus contorta

Pinus contorta

grows up to 80 feet tall, with a narrow, conical crown, has scaly light brown bark with needles growing in groups of two, from yellow-green to dark green, and slightly twisted. The cones are tightly closed and prickly, sealed shut with resin. Fire is needed to melt the resin and release the seeds. The trees thrive in high mountain areas, in disturbed or marginal sites, in pure stands or mixed with other conifers. Those that grow in bogs near the sea are called shore pines; in these areas the trees are usually stunted.[33]

Range: from the foothills to the montane in southeast Alaska, central Yukon south on the Pacific Coast, from the southern Rocky Mountains into southern Colorado, in the Black Hills of South Dakota, and in California south through the Sierra Nevadas.

Pinus contorta, lodgepole pine

This tree was given the name "lodgepole pine" by explorers Lewis and Clark, who noticed that the peoples of the Great Plains went into the mountains and cut down the pine tree to use as support for their lodges or tepees. This is a use we still have for this tree—as poles and logs for log cabins.

Lodgepole pine is one of the most important trees of the montane. It is known as a "fire tree" because the cones are serotinous, meaning they need fire for them to burst open the film of resins holding the scales shut. But it must be a hot forest fire, not a smoldering ground fire. Therefore, the tree evolved two kinds of cones: serotinous and nonserotinous. The latter don't

need a hot fire and fall to the ground or scatter with the wind. The higher the elevation the tree grows, the more likely it is to have serotinous cones and have a huge seed production in these cones, which can store seeds for 10 years. Once they are released by a hot fire, there are enough seeds to cover the ground like a blanket. When this happens, they become the pioneer tree of a charred area. In stands of mature trees, however, the seedlings can't survive in the deep shade of the lodgepole's own canopy and are replaced by Engelmann spruce high up in the mountains and by Douglas fir lower down.

As well as gratifying chipmunks and many songbirds, lodgepole pine seeds top the pine squirrel's menu and are practically its only food all winter long. Porcupines enjoy chewing on the wood, and have made a conspicuous dent in the Cascades of Oregon, where it was found that one animal chews on about 100 lodgepoles in a single winter.

The Thompson liked to eat the cambium layer within the trunk, which they dried in strips. They made the pitch into an ointment for blemishes and to rub on babies. Rheumatic and other pains were treated with a poultice made with boiled gum from the cones and bark mixed with deer fat.[34] The Haisla and Hanaksiala applied smoldering twigs to arthritic joints, then used the pigment from burned twigs to tattoo their skin. The Nitinaht mixed pitch with melted deer tallow and used it as a skin cosmetic.[35]

Today, lodgepole pine is an important timber tree, used for lumber, plywood, posts, railway ties, mine timbers, and paper. Many stands of the tree are infected by the parasitic lodgepole-pine dwarf mistletoe (*Arceuthobium americanum*). Unfortunately, the most effective treatment of infected stands so far has been clear-cutting.

Pinaceae

Ponderosa Pine

Pinus ponderosa, Western yellow pine, yellow pine, pino real, pinabete, bull pine, blackjack pine, Western red pine, Western pitch pine, Sierra brown-bark pine, Western longleaf pine, ponderosa white pine

Pinus ponderosa

usually grows 150 to 180 feet tall with a narrow, straight trunk and a conical crown. The orange-brown bark grows into deeply grooved, jigsawlike plates, each one covered with small scales. The dark yellow-green needles are 4 to 10 inches long and grow in groups of two or three. The cones are prickly, 3 to 6 inches long, reddish-brown. They release winged seeds after fertilization.

Range: throughout the Rockies from Alberta and British Columbia south to Arizona and New Mexico.

State tree of Montana.

Scottish botanist David Douglas spotted this tall pine in the 1820s, near what is now Spokane, Washington. He called it *ponderosa*, meaning "weighty," because the specimens he saw were massive and about 400 years old; a mature ponderosa can grow to a diameter of 10 feet and a height of more than 200 feet. The sheer size made European settlers in the Northwest call it the "bull pine." Lumbermen called it "blackjack pine" because of the color of the bark, which is very dark when the tree is young. The town of Flagstaff, Arizona, got its name from a particularly tall ponderosa pine used for a flagstaff.

This pine is one of the stalwarts of the Sierran montane forest. As Rob Curtis describes it:

Just above the oak woodlands and chaparral of the foothills, Ponderosa Pines form open, parklike forests with grassy floors and scattered shrubs. Black Oaks and Douglas-firs may or may not be present, along with various trees that are more commonly found at lower elevations. Water is obviously a problem here. In spring the grasses are green and tough yellow daisies are conspicuous among the annuals sprinkled among the pines. By midsummer the grasses are straw and the flowers have gone to seed. On cool mornings, the forests are lively with the sounds of birds, the scurry of chipmunks and lizards, and the hum and flutter of insects. But as afternoon approaches, life retreats to shady corners and burrows. The air is

*pungent with the resinous fragrance of pine nee-
dles, which quaver ever so slightly in the ruffled
afternoon air.*[36]

Pinus ponderosa, ponderosa pine

He could have added that, on a hot day, the
bark smells like vanilla, and the young twigs
have a slight orange scent. And that its seeds
must have bare, moist, mineral soils in bright
sunlight to germinate. And that a young seedling
can put down a 2-foot taproot, but a mature root
can plunge 6 feet in soil or up to 40 feet in fractured bedrock and it must have space around to
prevent overcrowding.[37]

Before European settlement, fires regularly
swept through pine forests, killing young
seedlings but sparing mature trees, which were
protected by their thick bark. Some of these fires
originated in lightning strikes; others were set
by Native Americans. Fires shaped the landscape, helping to create the majestic parklike
expanses of trees and grasses that astonished the
early explorers. With increased livestock grazing
and fire suppression policies, modern forests are
denser, which means when a fire occurs, it is
more destructive. The change in the environment has also made the ponderosa more susceptible to disease and insect infestations such as the
mountain pine beetle.

Ponderosa pine bark and seeds provided
Native Americans with food, and the resin was
widely used to make ointments and salves to treat
sores on human or animal skin, or as a hairdressing. The Cheyenne used the roots to make a blue
dye, and the Paiute used its pitch as glue in arrow
making and to protect pictures painted on rocks.
For the Navajo and Ramah, it had a special role
in baby cribs. They would choose a young tree in
a secluded area, cover it with corn pollen, and
take a solid piece from the east side. As they
made the cradle, they said special prayers, then
rubbed the finished product with red ocher and
tallow to protect it from evil spirits.[38]

One distinctive inhabitant of the ponderosa
pines on the Colorado Plateau is the tassel-eared
Abert's squirrel. This acrobatic little guy (he can
jump 40 feet or more to the ground unharmed)
relies on the tree for nesting, shelter, and food
(eating both seeds and inner bark).[39]

John Muir, the West Coast naturalist who
founded the Sierra Club, reveled in ponderosa
pines.

*Climbing these grand trees, especially when they
are waving and singing in worship in wind-
storms, is a glorious experience. Ascending from
the lowest branch to the topmost is like stepping
up stairs through a blaze of white light, every
needle thrilling and shining as if with religious
ecstasy.*[40]

It still thrills hikers and backpackers like James Ratzloff: "I love being in a ponderosa pine forest in a windstorm, hearing the sound the wind makes in the long needles of pines."[41]

Portulacaceae

Bitterroot

Lewisia rediviva, konah, mountain rose, *racine amère*, redhead Louisa, resurrection plant, rockrose, sand rose, spatlum, wild portulaca

---

Lewisia rediviva

is a low-growing plant, 1 to 3 inches high, that springs from a fleshy taproot. The basal leaves are long and narrow, succulent when they first appear in spring, dying back in summer. The flowers appear just above the ground. One to two inches in diameter, they have 12 to 18 white to pink petals with yellow or orange centers.

Range: throughout the Rockies, from Alberta and British Columbia south to Arizona.

State flower of Montana.

Lewisia is named for Meriwether Lewis of Lewis and Clark fame. He collected the plant on their monumental march across the country for President Thomas Jefferson to map the unknown territories. He brought the roots back to botanist Frederick Pursh, who established this new genus and named it *Lewisia*. The *rediviva* ("revivable") comes from the fact that an apparently dead root can still put forth shoots. Lewis and Clark came across bitterroot by accident on August 22, 1805. A small band of Indians attempted to steal a rifle; in the aftermath of the scuffle, they escaped, leaving their baggage behind. Captain Lewis said:

[It contained] a couple bags woven with the fingers of the bark of the silkgrass containing about a bushel of dried serviceberries, some chokecherry cakes, and about a bushel of roots of three different kinds dried and prepared for use which were folded in as many parchment hides of buffalo . . . one species of the root was . . . much mutilated, but appeared to be fibrous. The parts were brittle, hard, and of the size of a small quill, cylindric, and white as snow throughout, except for some small parts of the hard black rind which they had not separated in the preparation.[42]

This was of course the bitterroot. Lewis and Clark were given the plant to eat, and reported that it had a "bitter taste, which was nauseous to us, though the Indians seemed to relish it." The following year, they collected specimens of the plant in Montana and caused quite a sensation with demonstrations of the plant's ability to revive when it appeared to be completely dead and dried out.

L. rediviva continued to amaze people 60 years later, when a specimen collected in British Columbia in 1860, which became the model for an illustration, was miraculously revived to enjoy life at London's Kew Gardens. W. J. Hooker, then director of Kew, remembered how the plant "was immersed in boiling water on account of its well-known tenacity of life. More than a year and a half after, it notwithstanding showed symptoms of vitality, and produced its beautiful flowers in great profusion in May of the present year [1863]."[43]

Bitterroot was an important plant for the Native Americans of the West. The Flathead made an infusion of the roots to soothe heart pain, and the Nez Perce took the infusion to purify the blood; nursing mothers of both groups took it to stimulate the flow of milk. For some, it was a valuable trade commodity; in 1844, according to explorer Carl Geyer, "a sack . . . commands generally the price of a good horse."[44] The roots were sometimes eaten raw, but more commonly boiled and dried, steamed, fried up with fish or meat, or cooked in soups and stews.

The Flathead have a story as to how the plant was created.

Long ago, . . . in what we now call the Bitterroot Valley, Flathead Indians were experiencing a famine. One old woman had no meat or fish to feed her sons. . . . Believing that her sons were slowly starving to death, she went down to the river early one morning to weep alone and sing a death song. The sun, rising above the eastern mountains, heard the woman singing . . . [and] sent a guardian spirit in the form of a red bird to comfort her with food and beauty.

"A new plant will be formed," said the bird, "from your sorrowful tears which have fallen into the soil. Its flower will have the rose of my wing feathers and the white of your hair. It will have leaves close to the ground. Your people will eat the roots of this plant. Though it will be bitter from your sorrow, it will be good for them. When they see these flowers they will say, 'Here is the silver of our mother's hair upon the ground and the rose from the wings of the spirit bird. Our mother's tears of bitterness have given us food.' "[45]

The Kootenai say they learned about bitterroot when one of their men married a Southern woman, who brought camas and bitterroot with her. She didn't like the North and returned south, but on the way back she threw away the bitterroot at Cranbrook and Tobacco Plains because she preferred the camas. For this reason, bitterroot is found in those areas, although camas is not.[46]

Both the Flathead and the Kootenai performed a "first roots ceremony" before the harvest. The Kootenai ceremony was led by an

Lewisia rediviva, bitterroot

elderly woman who had lived her life free of misfortune.

When they arrived at the first bitterroot, . . . the leader stuck her digging stick in the ground at the base of the selected plant, and the other women followed. All then knelt down and prayed for protection against harm. The leader uprooted the plant, . . . and thus formally opened the bitterroot-digging season. They gave the first root to the chief, and thereafter the women were allowed to dig their own. . . . The next day, . . . the Kootenai held a feast, cooking bitterroot in a broth of boiled blue grouse and deer which the men had hunted the day before. The chief then gave a prayer of thanksgiving, after which the food was distributed among the people.[47]

It was well known among other groups that the Kootenai and the Flathead would be absorbed with this ritual every year; on several occasions, Blackfoot raiders took the opportunity to plunder the Kootenai and Flathead camps at this time.[48]

Although a sackful of bitterroot will not purchase a horse nowadays, its discovery still occasions excitement for botanists. Ethnobotanist Nancy Turner remembers:

My very early experiences with plants were in Medula, Montana. From the time I was two or three I roamed around the hills with my sister. We knew all of the local plants around there though we didn't know the names for them. We weren't eating anything but chokeberries and sarviceberries (saskatoon berries in B.C.). This kindled an interest in plants. I remember vividly our neighbors coming along and saying they were going to look for bitterroot and I was four and didn't know what it was. But there was the Bitterroot River flowing through Medula and this notion of somebody going on an expedition to find bitterroot intrigued me. It was very ephemeral in its blooming and then the whole plant dries up. I never saw one until I was on my honeymoon. But the idea always stuck with me as a romantic goal to search for . . . because it's a small plant and so beautiful.[49]

Primulaceae

Shooting Star
Dodecatheon spp., American cowslip

Dodecatheon spp.
have leaves growing in a basal rosette, and showy white, pink, or purple pendant flowers emerging on a flower stalk that may be up to 20 inches tall. The petals are upswept, which gives them the look of a shooting star, and the stamens protrude. Stems are somewhat reddish.

Range: Rocky Mountains from Alaska to Mexico.

Dodecatheon species enliven meadows, bogs, marshes, and streambanks throughout the Rockies. *Dodecatheon* comes from the Greek *dodeka*, meaning "12," and *theos*, "gods"; Pliny coined this word referring to the top 12 gods, so this little flower has some mighty protectors.

Gene Bush of Depauw, Indiana, has a story about the common name of shooting star: "According to myth shooting stars streak across the evening sky and fall to earth creating a flower with each impact on earth. A living replica of the journey is created by individual flowers shaped like rockets or comets from the debris."[50]

Like their celestial namesakes, shooting stars' appearance is relatively brief and they are dormant during the summer. Native Americans took advantage of their fleeting emergence. *D. jeffreyi* (**tall mountain shooting star** or **Sierra shooting star**) was used by Thompson women as a good-luck charm and to make a love potion. Beadwork designs were based on the deep pink flowers of *D. pulchellum* (**pretty shooting star** or **few-flowered shooting star**). The Okanagan-Colville smeared mashed flowers on arrows to turn them red, while the Pomo and Kashaya hung the flowers of *D. hendersonii* on cradles to make babies sleepy.[51]

The blooms are small, but can appear in the thousands, as shown in an entry in the 1874 San Diego city directory: "As we rise higher, the American Cowslip or Pride of Ohio or shooting stars . . . abound in favored spots, sometimes covering many square rods, and hanging their pink and white banners with a singular abandon."[52]

To a child's eye, these little plants are magical. When Opal Whiteley, of Cottage Grove,

Dodecatheon spp., shooting star

Oregon, was seven (ca. 1904), she began to keep a diary inspired by her surroundings, and the animals and plants she saw each day:

March 17th—Shooting stars are in blossom. We children counted one hundred and three on the way from school—and left them blooming there—those quaint, purple-pink flowers, with their nice little noses. Other names have they beside Shooting Star—Bird Bills, Prairie Pointers, Crow's Bills, and American Cowslip. They belong to the Primrose family. Have you watched the Bumblebees come to the blossoms?[53]

The shuttlecock-shaped flowers with their upraised petals and projecting cone-shaped anthers are "buzz-pollinated"—bees latch on to the cone and, by vibrating their thoracic muscles, shake the pollen from small pores at the anther tips. The pollen settles on their bodies ready to be taken to the next plant. After pollination, the flowers turn skyward.

In 1979, a new species was described, *D. austrofrigidum*, or **frigid shooting star**. Because the plant clings to nearly inaccessible nettle-lined ridges and rocky riversides, studying it "involves a precarious balancing act and more than one researcher has slipped into waist-deep frigid water."[54] So far, it has been found in seven locations in western Oregon and Washington.

Pyrolaceae

One-Sided Wintergreen

Pyrola secunda or *Orthilia secunda*, yevering bells, side bells

One-sided wintergreen is not the plant that wintergreen oil comes from (that's *Gaultheria procumbens*). This plant got the name simply because it is green all winter. *Pyrola* is a diminutive of *Pyrus* (pear), from the resemblance of its leaves to those of the pear. The common name "yevering bells" was coined from its bell-shaped flowers hung one above the other.

Pyrola secunda is found from the montane to the sub-Alpine. It is a perennial, evergreen herb that grows from rhizomes in montane and sub-Alpine woods, usually in dry acid soils under spruce and firs. It can form mats as the rhizomes branch over the forest floor. The basal leaves are oval or round and finely toothed. A flower stalk up to 8 inches high bears white or pale green nodding flowers all lined up on one side of the stalk. A related species, known as **single delight**

(*P. uniflora* or *Moneses uniflora*) bears one waxy white nodding flower.

The Native Americans of the West used wintergreen for a wide range of ailments, from eye inflammations, sore throats, skin problems, and indigestion, to hemorrhage, menorrhagia, gonorrhea, and cancer. Its medicinal powers reached London's John Gerard, whose "herball historie" of plants published in 1633 included the following entry:

Pyrola *is a most singular wound hearbe, either given inwardly, or applied outwardly: the leaves whereof stamped and strained, and the juice made into an unguent, or healing salve, with waxe, oile, and turpentine, doth cure wounds, ulcers, and fistulas. . . . The decoction hereof made with wine, is commended to close up and heale wounds of the entrailes, and inward parts: it is also good for ulcers of the kidneys, especially made with water, and the roots of Comfrey added thereto.*[55]

Range: **throughout the Rockies from Alaska to New Mexico.**

Delphinium spp., larkspur

Ranunculaceae

Delphinium spp.

are perennial herbs that grow from fibrous or tuberous roots. The leaves are stalked and lobed or finely divided, basal in some species, opposite in others. The flowers are blue or purple, with five sepals, the upper one forming a spur that gives the plant its common name. Found in open woods, meadows, and on dry grassy slopes in the mountains.

Range: throughout western regions of Canada and the United States; some species found in the East.

Larkspur is a relation of one of the most favored of all garden flowers, the delphinium, but it grows nowhere near as tall as the garden variety. Called larkspur because the flower looks like the spur on a lark's foot, it has also been called "lark's heel" and "lark's claw." The spur contains the nectar, which long-tongued butterflies and moths or the more aggressive bees are able to reach.

Although delphiniums grow everywhere, they are most prevalent in the West. Mountain species include *D. bicolor* (**low larkspur, montane larkspur, showy larkspur**), which has bright indigo flowers 12 inches tall; and the **tall larkspur**, *D. barbeyi*, which grows in high meadows and along streams, and *D. × occidentale* (**dunce-cap larkspur**), which reaches heights of 6 feet.

Ranchers call larkspur "staggerweed" or "locoweed" because it is poisonous to cattle (sheep, horses, elk, and deer, however, can ingest greater quantities before they are affected). The toxin is a mixture of alkaloids, including ajacine and delphinine, which blocks communication between nerves and muscles. It is equally toxic to humans.

These poisonous properties have led to its use as an insecticide, particularly for dealing with body lice and scabies. The plant has other uses. The seeds were ground and steeped in rubbing alcohol or vinegar or made into a soap. The Hopi also extracted a dye that they used to turn cornmeal blue, and the Thompson used the dark blue flowers as a paint and a clothing dye.[56]

Rosaceae

Dryas octopetala

is an evergreen, perennial semishrub. The woody stems trail on the ground and sometimes form mats. The thick, leathery leaves grow alternately on the stem, and in clusters at the base of the flower stalk. They are oblong to oval, with scalloped edges, the upper surface dark green and wrinkled, and the lower surface slightly hairy. In late spring, the white, saucer-shaped flowers 1 inch across appear individually on leafless stems 2 to 6 inches high, with 8 to 10 egg-shaped petals around numerous yellow stamens. The seeds have feathery bristles that form a fluffy head on the flower stalk.

Range: Alaska, south to northern Cascade Mountains, east to northeastern Oregon, Rocky Mountains from Alberta south to northwestern United States.

Territorial flower of the Northwest Territories.

Dryas octopetala, white mountain avens

These low-growing, tough-rooted plants hug the ground to protect themselves from cold and wind. The crowded leathery leaves are shiny on top and a fuzzy white beneath; being evergreen, they are ready to photosynthesize whenever there is enough sunlight. The shallow bowl shape of the flowers focuses the sun's warmth on the stigma at the center, to promote fertilization. In late spring they appear as if by magic on the leafless stems to be as obvious as possible to passing insects—an invitation to come in and do a little fertilizing. Bright yellow stamens act like little flags of welcome. Once the pistils have matured, they turn into delightful fluffy seed heads easily blown away on the wind.

Dryas comes from the Latin *dryas* or *dryad*, the wood nymphs associated with the oak tree—the *Dryas* leaves are a little like tiny oak leaves.

The plant is unique in that a major climatic period was named after it, the Younger Dryas. This was the last major period of glaciation in North America, about 13,000 years ago, and got its name because of the associated increase in sediment cores of pollen of the cold-resistant *D. octopetala*. Like Leedy's roseroot, *Dryas* is a glacial relict, or outcast. As the glaciers retreated, the plant became isolated from other populations and survived only in areas where conditions were similar to the Arctic zone. It grows in the Arctic and all over Iceland, and in the mountains of Scandinavia, Russia, Europe, Korea, and Japan.

Scrophulariaceae

Indian Paintbrush
Castilleja spp.

Castilleja spp.

include about 200 species of annual, biennial, and perennial herbs that are hemiparasitic; that is, they can photosynthesize but require root contact with another plant. Indian paintbrushes have long, hairy, unbranched stems and alternate leaves. The upper leaves or bracts are brightly colored, red, yellow, purple, pink, orange, even white, and showier than the flowers. The fruit is a two-celled capsule.

Range: Different species grow in eastern and western North America, in mountainous and prairie habitats.

State flower of Wyoming *(C. linariaefolia).*

Indian paintbrush and hummingbirds go together perfectly—they are mutually reliant, and may have evolved together. Hummingbirds love the paintbrush's nectar, and are equally attracted by the crimson petal-like bracts.[57]

The genus was named for Domingo Castilleja, an early Spanish botanist. There are about 250 species in the West alone, but they are difficult to classify, prompting a prominent botanist to write that the genus "is one that makes botanists wish they had embraced some

Castilleja spp., Indian paintbrush

easy branch of science such as theoretical physics."[58]

C. occidentalis (**Western Indian paintbrush**) is often found above the timberline; its bracts are usually yellow, but may be red or purple. *C. gracillima* (**slender paintbrush**) has pale yellow bracts, narrowly clustered up the stem in early summer. *C. rustica* (**rustic paintbrush**) has furry, deeply parted, spidery leaves and is found on sagebrush-covered slopes. *C. linariaefolia* or *C. linariifolia* (**Wyoming paintbrush**) is brilliant red and found mostly in the West, and *C. coccinea* (**scarlet paintbrush**), also red, grows in the eastern United States.

Henry David Thoreau noted the eastern variety in Massachusetts, writing on May 8, 1853:

At the foot of Annursnack, rising from the Jesse Hosmer meadow, was surprised by the brilliant pale scarlet flowers of the painted-cup (Castilleja coccinea) just coming into bloom. . . . Methinks this is the most high-colored and brilliant flower yet, not excepting the columbine. . . . It is all the more interesting for being a painted leaf and not a petal, and its spidery leaves, pinnatifid with linear division, increases its strangeness.[59]

The plant doesn't just look like a paintbrush. *C. linariaefolia* was actually used by the Tewa for decorating pottery and wood. Among the Wet'suwet'en, the **giant red paintbrush** (*C. miniata*) was considered sacred and children were forbidden to pick it. (Its name comes from red lead, "minium," used to paint illuminated manuscripts in the Middle Ages.) The Gitxsan called it *ihlee'em ts'ak* ("bleeding nose") because they used the plant medicinally for bleeding, stiff lungs, as a purgative, and against sore backs, kidney trouble, and sore eyes.

Others used the plant medicinally as well: The Costanoan used a reduction of *C. affinis* as a wash on infected sores; the Quileute used an infusion of *C. angustifolia* to regulate menstruation, while the Rama Navajo used the roots of *C. integra* boiled down into a syrup to clean the blood after injury. *C. linariaefolia* is contraceptive, and was used by Hopi, Rama Navajo, Shoshoni, and Tewa for this purpose.[60]

In *The Legend of the Indian Paintbrush*, a contemporary book for children, Tomie de Paola retells an old Texas legend about how the paintbrush first bloomed and how a young native boy dreams of creating a painting that will capture the sunset's beauty. What does he find all around him? Indian paintbrush, of course.

Viscaceae

Dwarf Mistletoe
Arceuthobium spp.

Arceuthobium spp.

are perennial parasitic flowering plants that grow on conifers, particularly the lodgepole pine. They are fleshy plants with tiny, scalelike, yellowish-green leaves and segmented or whorled branches. The yellowish-green flowers grow in clusters. The male flowers have three or four petal-like structures; the female flowers are inconspicuous. After fertilization, the plant produces sticky bluish berries that dry out and explode to release the single seed within.

Range: throughout the Rocky Mountains from British Columbia to New Mexico.

Mistletoe may draw kisses in the parlor at Christmastime, but in the wild it can suck the life right out of its host. This is a parasite and, unlike many other native parasites, it can be a killer. It latches on to conifers—the lodgepole pine, for instance—causing loss of vigor and growth until the tree dies. Each species is associated with a principal host, but can also attack secondary hosts. The mistletoe survives as long as the host does. Its leafless shoots are anchored into the tree and are designed to remove both nutrients and water.

In the autumn, the fruit explodes, scattering the seeds in all directions. The seeds, which can travel at speeds up to 50 miles an hour, adhere to foliage, twigs, or branches with a substance called viscin. When it rains, the viscin becomes slippery and the seeds attached to needles slither onto the twigs. Germination can occur as early as February. The developing plant penetrates the host bark with threadlike filaments, growing along and around the inner bark. It heads toward the cambium, extending itself by a system of filaments called the sinker, which embed themselves farther and farther into the wood as the tree grows. Aerial shoots do not appear until the third year after infestation. The shoots are less than 2 inches long in larch and Douglas fir mistletoes, up to 5 inches long in hemlock and lodgepole pine mistletoes.

In the second year, flowers come out after the shoots appear. The plant is pollinated by the wind or by insects (ants or bees) that like the nectar, which is up to 95 percent sugar. The pollen sticks to hairs on the ant's body and is brushed off as the ant moves from flower to flower.[61]

Native Americans made some powerful remedies from this parasite for treating stomachache, lung hemorrhage, tuberculosis, and emaciation.[62]

Each species is named for its principal host, including **hemlock dwarf mistletoe** (*A. tsugense*), **lodgepole-pine dwarf mistletoe** (*A. americanum*), **larch dwarf mistletoe** (*A. laricis*), and **Douglas-fir dwarf mistletoe** (*A. douglasii*). It is found both in the eastern and western hemispheres, and of the 34 species in the New World, most are in western North America.

Arceuthobium spp., dwarf mistletoe

The Tall Trees:

THE PACIFIC NORTHWEST

The forest that gives us a true intimation of both history and mortality is the old-growth forest. It takes more than a lifetime (95 to 140 years) for its trees just to bulk out, and another 500 years for its systems to be firmly enough in place to support the millions of creatures that work within its finely honed web of life. Once established, it rumbles on like a sleeping giant. As Sallie Tisdale put it: "Left alone, it will pulse its own slow pulse, exhale its own slow breath, forever. There is little on this earth more immortal."[1] But the forest has not been left alone and it is not immortal. About 150 years

An old-growth forest is centuries in the making.

ago, the forest of the Pacific Northwest still held many trees that were 1,000 years old. Today, we have only a few relics of a once-vast population.

In an old-growth forest the canopy provides shade but lets some sun filter in, creating a soft green light. A magnificent tangle of salal, vine maple, ferns, and thick, soft moss carpets the floor of the forest and the leaning snags. Trees and rocks are adorned with a tapestry of lichens. At first everything appears to be chaotic in this junglelike setting, but a subtle logic informs every creature's place on living and dead plants, including gigantic snags (standing dead trees) and nurse logs (fallen dead trees).

Every part of the forest contributes to the

whole, each part depending on the others, so that the forest is itself a type of living, breathing organism. This network is so intricate that if a few snags are taken out, 10 percent of the wildlife (with the exception of birds), disappear; if nurse logs are carted off, 29 percent of wildlife species vanish.[2]

The canopy is another mysterious ecosystem, a frontier barely understood. Red tree voles live only in the tops of Douglas firs—they never see the earth, yet flourish in this unique environment high overhead. Certain mosses build soil in the bright clean air, creating a world completely separate from their mossy relatives hundreds of feet below. In places, the canopy is so thick that a heavy snowfall never makes it to the ground. Fog and rain condense into drops along the needles of the trees, which absorb enough nutrients to feed the plants below once they fall to the ground. This is a wet forest. Water seeps through it in the form of dew, mist, fog, sleet, snow, and, of course, rain: at least 50 inches a year, and in some places as much as 200 inches. It is a region of moderate temperatures and little frost.

The Pacific Northwest has everything: there's a succession of ridges with the continent's highest mountain ranges running parallel to the sea. There is a 400-mile-wide strip of forest in some valleys and basins; there are thousands of nooks and crannies where unusual and interesting plants escaped from the last period of glaciation; there are islands that rose out of the sea after the ice released its heavy grip. There are highlands and lowlands. There is a forest of trees tall enough to be compared with cathedrals. No one would seriously consider knocking down a thousand-year-old building, but these trees are more vulnerable than buildings. These are unique monuments in the natural world and will never be seen again.

The cathedral analogy does not tell the story of just what a lively place this forest is—hardly one of serene repose. It's a cacophony of birds singing, animals snuffling in the underbrush,

and cannonlike reports as giant trees do a little self-pruning—a place in perpetual motion. Litter rains down: needles and leaves fall from trees, dander and dung drop from animals who make their living among the branches. Each individual tree is so complex, so layered with thousands, perhaps millions, of invertebrates living in it that each is an ecosystem unto itself.

In this place belonging neither to sea nor to land I came upon an old man dressed in nothing but a brief shirt. He was sawing the limbs from a fallen tree. The swish of the sea tried to drown out the purr of his saw. The purr of the saw tried to sneak back into the forest, but the forest threw it out again into the sea. Sea and forest were always at this game of toss with noises.

Emily Carr, *Klee Wyck* (1941)

The forest is intensely alive. As Jonathan Raban puts it, "Here, more than anywhere else I know, the tenure of civilization appears unexpectedly provisional and insecure." He has no difficulty imagining it without humans. "If some disaster hit the people here—if Boeing and Microsoft both went bust—it wouldn't be long, a few months at most, before the advance troops of vine and bramble took over the highways and strip-malls, closely followed by an occupying army of young Douglas firs. In less than half a human lifetime the place would be back to wilderness again. Bears and cougars . . . would make their dens in the urban ruins."[3] It wouldn't be wilderness, but it would be wild with plants and animals once more.

The keystone animal in this forest is the bear. Bears fish in streams, taking the salmon and throwing the uneaten parts into the forest, where animals and insects depend on this detritus for survival. This web of life, in which species depend on one another, can be broken and when the web is broken, in most cases it can't be fixed. (See the introduction for more on this.)

Clear-cutting was the norm in these forests all through the twentieth century. Today it is still a threat. Logging goes on right to the edges of protected areas and often into them. Fifty percent of 500-year-old trees go directly into pulp. Logging roads are routinely run through our parks, changing everything around them. No amount of protest seems to be able to protect these incredible places as long as they mean a profit to some, jobs to others. It is a huge conundrum and, it is hoped, not too late for a movement to encourage sustainable logging: just taking certain trees, leaving the forest as intact as possible. Clear-cutting breaks every fragile connection. Replanting in this complex system is so chancy it's almost useless. Who can wait hundreds of years for the ecosystems of the trees to reestablish themselves? But there are many people who see the value of these forests, who see that to destroy them is to destroy an area the world depends on not only for oxygen (it produces more than the Amazonian rain forest), but for the succor of wilderness as well. We need this forest, because we may need it for our own survival.

Pinaceae

Douglas Fir
Pseudotsuga menziesii syn. *Douglasii taxifolia*

Pseudotsuga menziesii syn. *Douglasii taxifolia*

grows 125 to 200 feet tall. When mature, it has a symmetrically tapered, massive trunk with no branches for the first 40 to 60 feet. In youth, the bark is a smooth, soft gray, but older trees have thick, deeply ridged, reddish-brown bark up to 1 foot thick. The 4-inch cones hang down and have three-lobed bracts that extend beyond the scales. The dark green flat or bottle-brush needles are less prickly than spruce. The seeds disperse easily with or without fire and will germinate in mineral soil. The wood has a distinctive orange color.

Range: British Columbia to California.

This tree, among the very tallest on earth, was first reported in English botanical circles in 1795 by Archibald Menzies. Menzies was a busybody of a botanist. One story goes that, during an expedition to Chile in 1791, he pocketed the kernels of the nuts being served for dessert one night during a gala dinner thrown by the viceroy. Menzies germinated the kernels on the ship during the return trip to England and planted the seedlings at Kew Gardens, where they grew into the previously unreported monkey puzzle tree (*Araucaria araucana*).

In the same zealous spirit, on a slightly later voyage, Menzies signed on as surgeon-naturalist (a common posting in those days) with Captain George Vancouver's expedition to chart the Pacific Northwest. He clipped off a twig from one of the huge trees that dominated the area. These monsters soared to 200 feet or more, and remarkably they grew in soil so impoverished few other plants seemed able to survive. Menzies returned, twig in hand, to England in 1795 and gave his find the name *Pinus taxifolia*. More than half a century later, in 1867, the tree acquired its botanical name, *Pseudotsuga menziesii*, in honor of the man who snapped off the twig.

The genus name is misleading. It means "false hemlock," although this giant is commonly known as a fir. Some consider it a spruce, even a red spruce. The tree, in fact, is neither a false hemlock nor a fir nor a spruce of any kind. It's a pine and has the typical downward-hanging pinecones. Nevertheless it was and always will be known as the Douglas fir after yet another botanist, David Douglas.

Douglas had a lot in common with Menzies and was even more fanatical. On his major trip to the Pacific Northwest in 1824 aboard the *William and Ann*, Douglas lugged 30 quires (a quire is 24 sheets) of special paper weighing 102 pounds for packing his plants. As he traveled through the wilderness, he tromped about on foot. His horse carried the paper. He often slept under his upturned canoe with his precious specimens under oilskins, though his own body was soaked through. All that mattered to him was the Collection. He brought back so many conifers to England he once remarked to the famous botanist Sir Joseph D. Hooker, "You will begin to think that I manufacture Pines at my pleasure." He scrupulously measured the trees he inspected and recorded their dimensions. The largest was 227 feet high, 48 feet in circumference.[4]

Douglas collected the cones and seeds and sent them to Britain, where they were successfully propagated and survive to this day. He knew the tree would be a good timber tree, and recommended using its resin as well. His description for the Royal Horticultural Society contains the following information:

The trees which are interspersed in groups or standing solitary in dry upland, thin, gravelly soils or on rocky situations, are thickly clad to the very ground with widespreading pendent branches, and from the gigantic size which they attain in such places and from the compact habit uniformly preserved they form one of the most striking and truly graceful objects in Nature. Those on the other hand which are in dense gloomy forests, two-thirds of which are composed of this species, are more than usually straight, the trunks being destitute of branches to the height of 100 to 140 feet, being in many places so close together that they naturally prune themselves.[5]

Pseudotsuga menziesii, Douglas fir

Douglas was looking at 1,000-year-old trees. For anyone who hasn't seen an old-growth Douglas fir, it's almost impossible to imagine the scale. Think, however, of a 14-story building and add branches the size of old city trees. There was nothing else like it on earth.

It took 30 million years for conifers to evolve to this level of sophistication. The efficient evergreen needles are able to photosynthesize beyond the usual growing season; they survive winter drought and have a remarkable ability to extract nutrients from the soil. Male and female cones are produced on one plant. The male cone falls off the tree after releasing its pollen; the female cone stays on, getting tougher and tougher, before shedding its seeds. It tends to be longer-lived than other trees, dominating large areas, and in some places Douglas firs outnumber hardwoods by a thousand to one.

Over the course of their long lives, the Douglas firs grow taller and taller, reaching toward the sunlight they need. In Seymour Valley near Vancouver, British Columbia, a tree was felled in 1895 that was reportedly 417 feet high with a base 25 feet in diameter. There were no limbs for the first 300 feet.

P. menziesii needs fire to survive. If a jagged edge of lightning struck one of the gigantic trees, or if the native peoples set fire to clear part of the forest, there might be a raging inferno for miles around, yet the Douglas fir, which holds tons of water in its system, could remain unscathed. There are old trees with scorch marks up and down the trunk still alive 500 years after a fire burned itself out. The needles may turn to flame on its lower branches, but the cones close themselves off to save the seeds, which do particularly well in fire-ravaged soil. Best of all, the fire bumps off the competition, leaving behind space and light for the Douglas fir seedlings to grow.

The Native Americans all along the Pacific Northwest depended in every way on the Douglas fir. Since it was difficult, if not impossible, to chop down one of these monsters with an ax made of stone and shell, they relied on fallen limbs for most of their needs. The list of uses for the wood reads like a camping and fishing catalog: tepee poles, drying scaffolds, smoking racks, spear shafts, gaff-hook poles, spoons, dip-net poles, fish rakes, harpoon barbs, and salmon weirs. It was considered an excellent fuel. Torches could be made from the heartwood, and the rotting wood used to cook or smoke hides. The peoples of the interior used the fragrant boughs as flooring and seating or to make mats and beds of boughs piled six layers high.

The Comox, who had a particular taste for highly aromatic food, used Douglas fir to prepare dogfish. First, they lined a pit with powdered rotten Douglas fir wood, then they stuffed a dogfish with the same rotten wood, covered it, and cooked it. The Salish made halibut and cod hooks from fir knots (they also used Western hemlock) by steaming, then placing them in a section of bull kelp overnight to give them the right curvature. The hooks were dried out and rubbed with tallow to make them waterproof. The Okanagan ice fishermen blackened their Douglas fir spears in fire to make them invisible to fish. The Kwakwaka'wakw even made coffins of firwood.[6]

The Secwepemc tied Douglas fir branches with loose knots and suspended them as targets for bow-and-arrow practice. To toughen up their leg muscles, they tied bundles of fir around their feet and ran through the water—roughly like running through surf on snowshoes. They also discovered that boiling up the needles created an attractive light yellow or brown dye which they used for baskets and clothes. Many groups used the pitch to seal the joints of implements and to caulk canoes and water vessels. The Stl'atl'imx mixed the pitch with soil and sand to make soles for fishskin moccasins.[7]

The peoples of the West Coast, like those of the Great Plains, were accustomed to using fire to modify the landscape. Though normally the

fires were well managed, things sometimes went awry. An Alsea Indian named William Smith recalled a fire in 1849 which was one of the worst he'd ever seen and one of the greatest conflagrations to hit the Willamette area. This massive fire consumed 500,000 acres of forest:

We were coming back from Suislaw when long ago, the world was in flames. Then it seemed to be getting dark all over. . . . Although the sun stood high, nevertheless it threatened to get dark . . . what on earth was nature going to do? The fire was falling all around us. Wherever it would drop, another fire would start there. Everybody was staying near the ocean on the beach. The fire was flying around just like the birds. All the hills were on fire. Even the hills that were near the sea were burning as soon as the fire arrived at the sea. . . . For probably ten days it was dark all over. . . . Even the trees [that] lay in the water caught fire.[8]

The pioneers who moved into the area were aghast at the annual burning. It fit into no pattern they recognized from their own forest-management practices. Though many of them learned to follow the local people's example in clearing land, in the end they got fed up with the constant smoke and haze. The high rate of respiratory illness among an otherwise healthy people had not escaped their notice—this was before European diseases laid the Native American people low.

In the name of environmental conservation, but more likely because of the demands of the lumber barons, burning was halted by the late nineteenth century. The Douglas fir suffered once the fire-burned mineralized soils were no longer available to help new seedlings sprout.

At the same time, the use of European woodcutting technology meant that the forest giants were no longer safe from human predation. Mind you, even with a metal saw, cutting down a Douglas fir was a huge job. Audrey Grescoe recounts the story of Clem Bradbury, who lived in the nineteenth century. He spent hours and hours chopping into the sapwood of a particularly large specimen. Pitch ran out of the tree all night long, he reported. It took another six hours of work to take down the rest.[9]

When the donkey engine was invented, lumber companies were not only able to bring out the giants, but they dragged just about everything else that went with them as well. Clear-cutting became a norm. One old-time forester calls it the rape-and-run form of lumbering. They'd pick out the best and knock down the rest. Profit was almost the only thing the forest represented to these lumbermen.

Douglas fir was one of the greatest commercial lumber trees of the Pacific Coast. It was unusually strong and even-grained, and didn't warp. "Douglas fir two-by-fours form the hidden framework of millions of American homes," writes Jon Luoma, "its strong veneers the layers of plywood that sheath them. One 1990 estimate suggested that about one fifth of the total volume of timber harvested in all of the United States comes from this single species of tree, which grows commercially solely in the Northwest."[10]

Now a new generation of entrepreneurs has created an industry out of reclaimed Douglas fir wood. Desmond Mayne of Environmental Timber Recovery Co. Ltd., reclaims about 20,000 to 30,000 board feet of Douglas fir a year from British Columbia's lakes. Timber logged in the remote forests of B.C. was traditionally rolled into rivers and lakes and then hauled by water to sawmills. About 6 percent of the logs sink to the bottom. Over the years thousands of logs have sunk, and now lie waiting to be brought to the surface.

Mayne built his own equipment to raise the waterlogged timber, some of it as much as 7 feet in diameter and weighing 22 tons. Fortunately, Douglas fir doesn't warp as it dries. He sends the wood to a sawmill in Madeira Park, where it is custom-cut for clients willing to pay a premium for what Desmond Mayne calls "the best Douglas fir God made." And the government gets a second round of stumpage fees for logs that were felled 100 years ago. Understandably, the same practice is also being encouraged by governments in Ontario and Michigan, who see it as a way of making another buck out of the wood.

The value of this wood is also being seen by salvagers such as Alain Dubreuil, who began his business in Vancouver by rescuing doors, windows, moldings, fireplaces, and other architectural elements from houses that were being demolished. More recently he began to salvage Douglas fir timbers from Second World War buildings such as hangars and shipyards. Steel was at a premium during the war and more than 100 bases were built in Canada with Douglas fir timbers cut in British Columbia and sent to Montreal to be dressed and drilled for prefabrication.

Most of Dubreuil's clients are in North America, but he also has clients in Chile. Chile once exported saltpeter to the United States, where it was used to make gunpowder. The ships unloaded their cargo and then filled up with Douglas fir logs from Oregon as ballast on the way back. The logs were used to build houses in Santiago, where they were called "pino Oregano." Now that some of the houses are aging and in need of restoration, Chile is buying up Douglas fir from Canada.

Reclaimed Douglas fir is also used by certain artists, such as Brent Comber. He started making picture frames with refound wood and telling the stories he'd heard about the source of the wood and the community it was used in. Gradually, he built himself a business. Today he turns these old logs into stunning furniture. "Douglas fir is such an honest wood," he says. "You can trust it. The grain is reliable and I love the color because it is soft and warm, rich. It's so much a part of our heritage as well."

Pinaceae

Sitka Spruce

Picea sitchensis, coast spruce, silver spruce, tideland spruce, Menzies spruce, yellow spruce

---

Picea sitchensis

is a conical tree with spreading branches that grows to 160 feet or more. Purplish-gray or brown bark forms coarse plates. Needles are four-sided, sharply pointed, dark green above, and white beneath. Cones 2 to 4 inches long have distinctive wavy-edged scales, producing winged seeds. Sitka spruce naturally hybridizes with other spruce such as the white spruce, *Picea glanca,* and they form *P × lutzii,* Lutz spruce.

**   *Range:* Alaska, south to northern California. State tree of Alaska.**

This is the only spruce in the foggy coastal forests. It can live for 500 to 750 years, possibly more, and it's one of the most useful and versatile of the evergreens. A rapid grower, it can soar to 200 feet in 100 years. It retains moisture from sea spray, along with other minerals such as phosphorus, calcium, and magnesium. Sitka spruce grows mainly below 1,500 feet altitude, although dwarf spruce have been seen as high as 3,900 feet on rocky outcrops in Alaska.

The Sitka spruce was named after the Russian-American Company's first permanent settlement in North America on Sitka Island. Although probably nobody knew it at the time, it was from this area that the spruce made its launch into new territory from Asia millions of years ago. Spruce began its conquest of North America by traveling over the thousand-mile-wide band of land known as Beringia that once connected Asia and North America. As the glaciers retreated, the temperatures rose and spruce seed was carried by the Arctic winds blowing from Siberia. Spruce flourished in the powdery soil left behind by the grinding of glaciers against rock. The trees grew quickly in the glacial soil, with its high mineral content and low organic content. Today, because of global climate change, the spruce are retreating north again.

Of the approximately 40 species of spruce around the world, Sitka spruce is found only in the Pacific Northwest. After surviving the wild temperature changes between glacial periods and extreme winds, the Sitka spruce has become a fog-dependent, fire-hating tree. It grows best in wet mild areas where summers are not too hot and winters are not too cold, drawing moisture from dense fogs that surge inland on a regular basis. Water particles adhere to the needles, accumulating until they form huge drops. The weight carries them through the tree, absorbing moisture and nutrients on the downward journey. Fog also lowers the rate of evaporation during a drought, adding the equivalent of 5 inches of rain in a dry summer.[11] Fog and tree are partners, and they in turn support the teeming world of moisture-dependent plants thriving in the coastal forest.

The tree's tolerance for salt developed into a *need* for salty water, allowing trees a life clinging to impossible places and exposed sites where salt spray can reach them. The Sitka spruce, however, has a thin bark, which makes it intolerant of fire. If it catches fire, the cambium layer is scorched and the whole tree dies. Yet, if the wind is right, the seeds drift into a burn site and start putting down roots almost immediately.

Blowdowns are one of the dramatic elements of the coastal forest. Sitka spruce have shallow

Picea sitchensis, Sitka spruce

roots and sway crazily in heavy winds, even though they are packed tightly into the forest. When a wind strikes at just the right angle, a creaking old giant can suddenly topple. The sheer force of the weight, and the massive branches stretching out on every side, can crush even 100-foot-tall hemlocks growing in its shadow. Smaller trees—maples, mahonias, and rhododendrons—are smashed under the fallen spruce. The voles that lived in the tree are thrown aside or squashed beneath its weight. Birds, salamanders, toads, and the frogs that live off this tree scramble for safety.

The almighty crash is followed by an eerie silence. After a while, the ambrosia beetle moves in on what is now not a corpse but a nurse log. The beetle takes two days to bore a tunnel into

the thick bark, but it eventually makes a nest for fungi and bacteria. Mites, termites, and carpenter ants follow and settle in different parts of the tree, munching, burrowing, making nests. Fungi break up the cellulose and lignin of the sapwood. Spiders skitter up and down the log, pursued by the salamanders, which feed off the other animals munching their way through the log generation after generation. Minute creatures barge in by the thousands and can take up to two days to bore into the thicker bark of other trees. The thick bark slowly turns into a soft emerald surface. Mosses and lichens grow along the composting spine and hold the seeds dropped by birds, animals, and other trees or borne on the wind.

As Deanna Kawatski, who lived in a tiny cabin in the mountains of northern British Columbia for 13 years, notes, "About one-third of a tree's useful life occurs after it is dead. The place between the bottom of a fallen tree and the soil is one of the richest areas of the forest in terms of nutrient exchange. They also provide habitat for the little mammals that spread the spores of the vital mycorrhizal fungus. This fungus is essential to the survival of all conifers. In fact, a large, decaying fallen tree has twice the number of living cells found in a live tree."[12]

At the same time, sunlight pierces the forest where the tree once stood, allowing the seeds of new Sitkas or other plants to germinate. Sitka seedlings often form a line on either side of the nurse log. As they mature, their roots straddle the old trunk and eventually obscure the base from which they draw their own lives. If it is particularly waterlogged, a massive nurse tree might take up to 400 years to decompose completely.

If a Sitka dies but does not fall down, it becomes a snag and develops its own community of insects, animals, and birds, such as woodpeckers and bats. Snags are apartment buildings for birds, refuges for animals, and a support system for millions of invertebrates. They have enough nutrients not only to feed a complex population but to hold water for them all.

Humans developed a dependency on Sitka spruce as well. The Kwakwaka'wakw used the light wood to make digging sticks, herring rakes, arrows, bark peelers, and even armor to protect themselves against animals or for battling with other groups. The Haida and Gitxsan used the wood and bark for fuel. The Nisga'a and Haida obtained extra-wide boards for the houses of important people by cutting curved pieces from the circumference of the largest trees, then steaming them and weighting them with stones until they became flat. The Dena'ina used Sitka spruce to make sleds, shovels, tongs, drums, dishes, mauls, bows, dugout canoes, fish traps, and many other items.[13]

No part of the tree was wasted. The Kwakwaka'wakw skillfully exploited spruce branches for rope. Some people used the boughs to make camping shelters and occasionally for roofing if cedar bark was unavailable. The Dena'ina boiled up the bark to dye their fishnets. Spruce gum cemented the joints of harpoons and spears. The Kwakwaka'wakw greased their canoes with spruce-scented tallow to mask human odors when they went off to hunt porpoises. The roots were made into baskets, nets, and rope for tying together house planks, binding fishhooks and harpoons, and for fishlines. The Haida and Tlingit of Alaska were exceptionally skilled in creating exquisite spruce-root baskets. They could twine the roots so tightly that the baskets were watertight. They also made distinctive spruce-root hats.[14]

Today, Sitka spruce is an important commercial forest species. It is light but strong with a texture that makes it ideal for trim, paneling, siding, and furniture. But the wood is most highly prized for musical instruments because of its superb resonating qualities, and has been used for organ pipes, the sounding boards of pianos, and for violins.

Because of its quick growth, Sitka spruce is close-ringed. It produces straight boards that are tough yet light—a very special quality. It can be used for ladders, bleachers for sports arenas (they don't splinter with use), and racing shells. Natural historian Donald Culross Peattie writes: "The seat of such a shell of Sitka Spruce, for instance, will carry a 200-pound man, yet it weighs only 1¼ pounds . . . pound for pound it is stronger than steel."[15]

In 1947, billionaire Howard Hughes made his very own gigantic airplane out of Sitka spruce wood and called it the *Spruce Goose*. He admired the wood's lightness and ability to hold its shape. Alas, the airplane was airborne for only 70 seconds. But the Second World War was fought with sprucewood. More than 340 million board feet were shipped to Britain to make feisty little Mosquitoes, the heroic planes of the Royal Air Force. Today, Sitka spruce produces high-grade wood pulp for newsprint and plywood. It's also still used in the construction of turbine blades for wind-driven electrical generators and for all kinds of sailboat masts.

Like other Pacific Coast giants, the spruce comes in several sizes. There is big and even bigger. In the Carmanah Valley of British Columbia there are 239 Sitka spruce more than 230 feet tall; the Carmanah Giant is 315 feet high. Nevertheless, lumber companies are still allowed to clear-cut in these areas. Only a few small relict populations were saved in the 1990s by environmentalists who waged perpetual war against the loggers. But nothing could save one 1,000-year-old giant from a developer who wanted the space in Tofino on Vancouver Island. In the winter of 2000, the tree was felled and mushed up into pulp, despite the fact that it had stood longer than the cathedrals of Europe.

Sitka spruce is the state tree of Alaska. Ketchikan, Alaska, one of the small tourist settlements on the Inside Passage, sits within an area of 2.3 million acres of spruce, hemlock, and cedar forest including many pure stands of Sitka spruce. However, the 17-million-acre Tongass National Forest (Misty Fjords National Monument) and the 4-million-acre Chugach National Forest have been the source of most of the Sitka logged in the state.

Tsuga heterophylla, Western hemlock

Pinaceae

> ### Western Hemlock
> *Tsuga heterophylla*, Alaska pine, Pacific hemlock, lowland hemlock

Tsuga heterophylla
grows to 225 feet with a graceful droopy top and downward-sloping branches. The shiny green needles are white beneath; small reddish-brown cones hang down and produce winged seeds dotted with resin. It is found in areas of high rainfall below 2,000 feet.

Range: **east side of the Cascade Mountains to northern Idaho.**

State tree of Washington.

This is the largest kind of hemlock in the world. It dominates many old-growth forests of the Northwest, but with its thin bark it is vulnerable to fires, and has such a shallow root system it may blow down in high winds. It has one of the most distinctive characteristics of all coastal trees: a limp top, the leader drooping over to one side looking for all the world like the "tassel of a tall toque."[16] Over time, the lower branches of Western hemlock die and drop off. As the tree grows, it prunes itself and develops straight, unbranched trunks up to three-quarters of its overall height. In the eyes of loggers, this natural tendency is one of its most desirable characteristics.

The European who gave the tree its scientific name was Stephen L. Endlicher, a botanist and linguist, in 1847. He called the tree *Tsuga*, which in Japanese means "yew-leaved," because of the tree's short, flat needles. (Endlicher had studied Oriental languages, which is presumably why he chose a Japanese name.) *Heterophylla* means a plant with more than one type of leaf.

The Western hemlock is found alongside Douglas fir, Western red cedar, and Sitka spruce. The shade-tolerant hemlock grows well in their dense shadows and casts a dense shade of its own. It starts good cone production at about 25 to 30 years old. A stand of 100-year-old hemlocks can disperse 8 million seeds per acre each year. The small, light seeds can travel more than half a mile in a strong breeze to land in moss, decay, litter, rotten stumps, or mineral-dense soil—anywhere there is enough moisture, they will germinate. Eventually the trees form such dense stands that an acre of 100-year-old trees on a good site can yield more timber than an acre of less-crowded, larger Douglas fir.

The West Coast peoples used the Western hemlock to make spoons, roasting spits, dip-net poles, combs, spear shafts, mallets, digging sticks, and elderberry-picking hooks. They also made curved fishhooks from the knots formed by the trunk ends of limbs. Sometimes they attached barbs made of bone or iron. According to Willy Matthews of Massett, British Columbia, the ends of the strongly curved hook were held apart with a small stick. When a fish was caught, the stick sprang free and floated to the surface, indicating the fisherman's success. The Haida carved large feast dishes out of bent hemlock trunks, and also made fishing weirs, wedges, octopus spears, net anchors, children's bows, and ridgepoles from hemlock. They sometimes spliced the roots onto bull-kelp fishing lines to strengthen them.[17]

The bark, which has a high tannin content, was used as a tanning agent, pigment, and cleaning solution. The Saanich and other Coast Salish pounded and boiled the bark in water to make a reddish dye. Young Saanich women rubbed the dye on their faces as a cosmetic. The Kwakwaka'wakw steeped the bark in urine to make a black dye. The Nuxalk soaked it in water to color their fishnets brown and rubbed the liquid on traps to remove rust and give them a clean scent.[18]

The boughs were used along the coast to collect herring spawn. From March to June, people lowered bundles of boughs into the ocean near river estuaries. Later, they pulled up the boughs, scraped off the spawn, and ate it fresh or dried. The boughs made wonderfully scented bedding material. The Squamish used them as "rags" to wipe the slime off fish.[19]

To blaze trails as they made their way through the dark forest, the Kwakwaka'wakw hunters snapped hemlock branches to reveal the white underside, which showed up well even in deeply shaded areas.

For special celebrations, the dancers of coastal nations wore skirts, headdresses, and headbands of hemlock boughs. It was also a special plant used in traditional rites of purification. When young girls menstruated for the first time, they lived in huts made of hemlock boughs for the first four days. Many peoples used the boughs as scrubbers for ritual bathing and purification. The Nlaka'pamux had a special name for Western hemlock; their word means "scrubber plant."[20]

When European settlers moved into the area, they did not find immediate uses for hemlock, because the wood tended to warp easily. But once modern drying methods were invented, it was not only considered useful but appreciated for its beauty. Houses all over Japan and the Eastern states are paneled in hemlock. Then, in the 1860s, Benjamin C. Tilghman, a Philadelphia chemist, invented a digester that turned hemlock into pulp for making paper and other products. (Tilghman, a busy man, was also known as Brigadier General Tilghman, commander of the 3rd United States Colored Troops in the Civil War, and was credited with inventing the process of sandblasting.) In the early twentieth century hemlock became very important in manufacturing cellophane, rayon, and plastics. But once researchers found a way to bleach pulp, *Tsuga* lost its glamor as the palest pulpwood.

In 1946, an article in an Oregon newspaper pointed out that Washington state did not have a state tree. Washington newspapers chose the Western red cedar, but the Portland *Oregonian* suggested the Western hemlock. State Representative George Adams of Mason County introduced a bill to the Washington legislature to make Western hemlock the state tree.

The hemlock, he said, would become "the backbone of this state's forest industry." The legislature agreed and signed the bill in 1947.

Cupressaceae

Western Red Cedar
Thuja plicata, giant arborvitae, canoe cedar

Thuja plicata

grows slowly to 130 feet with branches that curve upward at the tips; the twigs grow in flattened sprays. It has stringy gray or brown bark. The small, light brown cones are borne in clusters. It is monoecious (separate male and female flowers, but they grow on the same plant). The ½-inch female cone stays on the tree; the male falls immediately after releasing pollen.

Range: west side of the Cascade Mountains.

This was the tree of life (*arbor vitae*) to the European newcomers.[21] They called all cedars that because the cedar saved Jacques Cartier's men from death by scurvy—after the local people showed them how to use it. The leaves and bark contain vitamin C.

Thuja is from the Greek word *thuia*, a common name for an aromatic tree; *plicata* means "pleated." Red cedar is the most common tree on the west side of the Cascade Mountains, yet each tree develops its own individual patterns as it ages. Arborists swear they have so much individuality they can tell one tree from another.

The number of cones produced varies from year to year with the amount of rain, the temperature, and the quality of the soil. Most evergreens drop their needles every three years, but *Thuja* species lose the oldest ones each year. This process speeds up when a conifer is stressed either by too much or too little rain. Western red cedar is particularly prone to this process, which is known as "cedar flagging." Because of the high oil content of its wood, the cedar is also susceptible to fire and grows best in damp or swampy areas that offer some protection.

The beginnings of red cedar are unknown because it's almost impossible to distinguish juniper or yellow cedar from red cedar in the fossil record. But something like it grew 6,000 to 10,000 years ago. The climate changed; the cedar thrived and developed to massive proportions. The Native Americans who were first into the area were familiar with trees 200 feet tall and up to 60 feet in circumference.

The culture of the West Coast native groups reflected the size and strength of the trees around them. Francis La Flesche noted in 1911: "In [the Western peoples'] creation myths, the cedar is associated with the advent of the human race; other myths connect this tree with the thunder. The thunder birds were said to live 'in a forest of cedars.'"[22] The ethnologist Melvin R. Gilmore wrote in 1914: "Because the cedar tree is sacred to the mythical thunderbird, his nest being 'in the cedar of the western mountains,' cedar boughs were put on tipi poles to ward off lightning, 'as white men put up lightning rods,' my informant said."[23]

Monumental woodworks were created because there were monumental trees to inspire the builders. From 3,000 to 3,500 years ago, the people of the West Coast constructed amazing wooden buildings from 30 to 50 feet wide by 50 to 200 feet long. And red cedar was the linchpin, so to speak.[24] They worked out how to remove the inner bark as well as the outer bark of the cedar. They could split fine planks from a tree without felling it or otherwise damaging the whole tree. The harvesting was done with such care that the trees recovered, even if it took a couple of centuries to do so. These are called culturally modified trees and they are easy to spot in an old-growth forest because of their slightly wonky shapes.

The Haida used red cedar in every aspect of their lives, from canoes, to houses, to mats and baskets. They lived inside cedar houses, fished with implements made of cedar, went to sea in cedar canoes, dressed in ceremonial outfits made of cedar, cured illnesses with cedar, and were buried in cedar coffins. Their lives were filled with a rapturous scent. Emily Carr describes a native village:

On the point at either end of the bay crouched a huddle of houses—large, squat houses made of thick, hand-hewn cedar planks, pegged and slotted together. They had flat, square fronts. The side walls were made of driftwood. Bark and shakes, weighted with stones against the wind,

were used for roofs. Every house stood separated from the next.

Wind roared through narrow spaces between.[25]

Cutting down and preparing the wood for use in a house or canoe was a cooperative effort among the Haida. After first blessing the trees, a work party cut around the base with adzes and chisels. Sometimes the workers burned the trunk at the bottom until the tree toppled. The Nuxalk, for instance, wrapped a Rocky Mountain maple branch around the trunk, secured it, and set it on fire. It might smolder for many hours, but gradually it burned its way through the soft cedar until the tree toppled over.[26]

Fallen logs were much easier to deal with, and to make transporting one of these behemoths easier the people started hollowing out the old logs right where they had been found. Once hollowed out, they were light enough for a group of men to drag them back to the village or camp. When planks were split from standing trees, the men usually worked in a group at this difficult and time-consuming work. They selected a tree, then hammered a series of wedges made of yew or antler along the grain until a plank was freed from the edge of the tree. It was smoothed with stones or shells. Once a family had gone through the difficult task of wrestling a smooth plank from a tree, it was considered not only a valuable possession to be taken everywhere they moved, but also something of a bankable asset. It could always be traded for food and other goods in difficult times.

Ethnobotanist Nancy Turner, who has spent decades collecting material about how the coastal peoples developed their technology, describes the arduous task of making a canoe. Canoe making was an honored profession, one that took years of

Thuja plicata, Western red cedar

training before an artisan would even attempt a large one. The canoe maker passed on his artistry from one generation to the next.[27]

The whole family helped hollow out the interior. Sometimes it would be carved out; at other times it was carefully burned down until the plugs became visible. The canoe was dragged to a beach, the hull filled with water, and hot rocks put inside to make the water boil. As the wood became flexible, the workers opened the top with long stakes to make it wider than the original tree trunk.

They sewed thwarts into place and sometimes added separate prow and stern pieces to make the craft more seaworthy. Finally, they polished the hull with dogfish skin or horsetail and decorated it. These astonishing craft were produced in many sizes: from ones small enough for a single person to medium-sized craft that could hold a family of six to those large enough to carry a band of 60, with three masts and sails and a mainstay sail. When Lewis and Clark arrived at the coast, Lewis wrote:

> *I have seen the natives near the coast riding waves in these canoes with safety and apparently without concern where I should have thought it impossible for any vessel of the same size to live a minute. . . . They were waxed, painted and ornamented with curious images at bow and stern. Everything about their work was astonishing. A person would suppose that the forming of a large canoe with an instrument like this was the work of several years . . . but these people make them in a few weeks.*[28]

The coming of iron tools in the eighteenth century speeded up the pace at which these canoes could be made. The English and the Yankees who saw them being made were dazzled. It's said that they took away models that influenced their own designs for the clippers

Woman's Prayer When Gathering Roots from a Young Cedar Tree

Look at me, friend!
I have come to ask
for your dress,
for you have come
to take pity on us;
for there is nothing
for which you cannot be used,
because it is your way
that there is nothing
for which we cannot use you,
for you are really willing
to give us your dress.
I have come to beg you,
for this, long life-maker,
for I am going to make
a basket for lily roots out of you.
          Traditional Kwakiutl prayer

that dominated the seas by the late eighteenth century.

The Haida were not the only boat builders. The Nez Perce also hollowed out logs with fire, then smoothed and trimmed them to make canoes. The Southern Kwakiutl steamed the wood to make it more malleable for canoe building. Canoe building among the diverse First Nations all along the West Coast using the old methods is still practiced in some communities today.

The people of the Pacific Coast also used Western red cedar to create baskets of such extraordinary quality that the civilization of the West Coast was known as the Basket Culture. In spring and early fall, they cut the bark into strips and split the roots to make coiled, watertight baskets in which they boiled their food. The magnificent split-cedar-root baskets produced by the Salish people were decorated by using bitter cherry bark (dyed red or black) or reed canary grass stalks with designs that incorporated plant and animal motifs. The style with the flat bottom and flared sides with rounded corners is now found in museums around the world.

Cedar boxes, also called bentwood boxes, were made in many sizes and shapes by groups of the central and northern coast, including Nuu-chah-nulth, Nuxalk, Haida, Kwakwaka'wakw, Tsimshian, and Tlingit. Bentwood boxes "were used as storage containers, watertight ones for holding hot rocks and water for cooking, and the highly decorated ones as symbols of wealth. They range in size from small (measured in inches) to massive (large enough to provide seating)."[29] Cedar boxes were widely traded, and many people still own them today. The art of making them is now being revived by a few modern native craftsmen.

In his monumental *Native American Ethnobotany*, Daniel Moerman lists 368 uses of this plant by Native Americans.[30] Western red cedar was used as a drug, a food, and a dye, as well as a source of wood for construction, tools, and objects. Many different groups used the leaves and bark to make poultices to soothe skin eruptions, or brewed up an infusion of leaves or boughs, or a decoction of the plant's tips and roots for a cough and cold remedy.

The red cedar also had magical uses. The Pawnee burned cedar twigs to prevent bad dreams. The Oweekeno also used the wood and fiber from the tree to make shamanistic soul catchers in ritual healing. The Thompson believed that sleeping under *Thuja plicata* would cause vivid dreams. The inner bark was dyed and then tied to Haisla pets to protect them from the dog eater (a bad spirit who stole domestic animals). Among the Thompson, the strange formation of the root called a "doll formation" was considered to bring good luck to anyone who found it. The scent of the wood was useful, too. The Lummi chewed the tips of branches to avoid nausea while burying a corpse.

We don't normally think of evergreen trees as a source of food, but the peoples of the Pacific Northwest had thousands of years to experiment with all the plants they found around them. When they were facing starvation in a bad year, they boiled up their fish traps (made of red cedar, naturally) and produced a thin but nutritious soup from the fish-flavored sticks. The Southern Kwakiutl used the pitch as chewing gum. The Montana Indian and Coast Salish ate the spring cambium fresh or dried.

Cedar fits into the stories told around the fire. The Hesquiat, who called cedar nuhtume, *have a story that tells of the time Deer stole fire.*

This version has Deer taking fire from the wolves; another says Deer stole it from Chief Red-winged Blackbird. In both versions, Deer attached the soft, shredded cedar bark to his elbows, knees and horns. He allowed this dry cedar bark to catch fire when he visited the fire's owner. Deer escaped by jumping out through the roof. Since, however, he had to jump into water while fleeing, the only fire that continued to burn was that in the cedar bark on his horns. Thus, Deer did bring back fire but suffered burnt knees and elbows.

Daniel Moerman,
Native American Ethnobotany

People even wore cedar. The capes, shaman collars, and hats of the Bella Coola are one of the most unusual uses of the tree. The Haisla and Hanaksiala used the bark to make hats, aprons, and rain capes, or wove strips of bark into hip-length leggings to wear through deep snow. Finely shredded bark was used as padding for infants' cradles, sanitary pads, and towels, or to make waterproof hats and capes. Diapers, bandages, blankets, ropes, brushes, lamp wicks, sponges, and tarpaulins—you name it, it could be made from red cedar.

When Lewis and Clark saw the Clatsop Indians, they admired their conical hats made from cedar bark and bear grass (which the Clat-sop got by trading with other nations upriver). The weaving was so intricate and tight that it was completely rainproof. Lewis and Clark immediately ordered two made-to-measure hats for themselves.

But the most impressive and unforgettable objects made with cedar were the totem poles. The very earliest ones, created before more sophisticated tools were used, were relatively small. A carver went from village to village depicting the stories and myths that were the traditions of the coastal peoples and, over time, the poles became larger and more elaborate. The arrival of the Europeans, however, was the beginning of the end for this traditional craft. By 1916 only 125 poles remained, left when they'd been abandoned as the Haida people forsook their carving and their totems to adopt the Europeans' religion.

Emily Carr described some abandoned totem poles in an ancient burial ground beside a lonely village:

Skedans Beach was wide. Sea-drift was scattered over it. Behind the logs the ground sloped up a little to the old village site. It was smothered now under a green tangle, just one grey roof still squatted there among the bushes, and a battered row of totem poles circled the bay; many of them were mortuary poles, high with square fronts on top. The fronts were carved with totem designs of birds and beasts. The tops of the poles behind these carved fronts were hollowed out and the coffins stood, each in its hole on its end, the square front hiding it. Some of the old mortuary poles were broken and you saw skulls peeping out through the cracks. . . . I went out to sketch the poles.

They were in a long straggling row the entire length of the bay and pointed this way and that; but no matter how drunken their tilt, the Haida poles never lost their dignity. They looked sadder, perhaps, when they bowed forward and

more stern when they tipped back. They were bleached to a pinkish silver colour and cracked by the sun, but nothing could make them mean or poor, because the Indians had put strong thought into them and had believed sincerely in what they were trying to express.[31]

Today it is possible to see totem poles, looking much as they might have looked hundreds of years ago, in Saxman Totem Park in Ketchikan, Alaska. With the dripping rain forest all around, emerging into the light of the glade filled with buildings fronted by their totems is a relief and an amazement. The richness of the dyes, the scale of the work, is breathtaking. Though these are all copies, they have been done by modern master-carvers who practice their art using the same techniques and dyes as their ancestors, and who used all their skill and knowledge of a tradition that must be at least 5,000 years old. It is both inspiring and saddening—inspiring that the gift still exists, saddening to think of what has been lost.

Furniture designer Brent Comber talks about watching Haida carvers use their tools with such dexterity that they can look at a design or have it in their heads and simply start carving a huge piece of wood straightaway with an adze almost without looking at the wood before them. The design unfolds to perfection—a vast difference from the mold makers in Taiwan churning out "totem poles" for the tourists.

Today red cedar is one of the most important lumber products of the West Coast. Cedar shakes are valuable in a wet climate because of their ability to resist rot. Cedar also makes a magnificent laminated wood. The sapwood, which is almost pure white, contrasts with the dark reddish-brown to yellowish heart. Once it has weathered, it turns a silvery gray. When dried, it is easy to ship. One of its attractive properties in home building is its insulating quality when used for shingles, roof boards, subflooring, siding, and sheathing; it tends to keep the house cool in summer and warm in winter. Some of the most famous architects of the twentieth century, such as Arthur Erickson, favored cedar in their designs for this reason. And for shingles it is unmatched because of its fine straight grain and lack of knots. Naturalist Donald C. Peattie says: "On San Juan Island in Puget Sound, the roof of a house built in 1856 was covered with Red Cedar shingles which were still intact in 1916."[32] Handmade shakes are a rare find these days. But some people still practice the craft, using material salvaged from logging sites.

Decks and fences all over North America are made with cedar. And it is still used in cooking, in a method known as "planking." The red cedar imparts a smoky taste to foods that are grilled or baked in the oven on planks. Planked salmon needs a short piece of red cedar heated in a 400°F oven then brushed with walnut oil. A 16-ounce salmon fillet placed on it, sprinkled with sprigs of rosemary and thyme, and drizzled with 2 ounces of maple syrup should be cooked in the oven for 20 to 25 minutes.

Cupressaceae

Yellow Cedar

Chamaecyparis nootkatensis syn. *Cupressus nootkatensis*, Nootka false cypress, Alaska yellow cedar, yellow cypress

Chamaecyparis nootkatensis syn. *Cupressus nootkatensis*

grows slowly from 40 to 120 feet high (in British Columbia, some specimens reach 150 feet). The leaves are yellowish-green and scaly, with shredding gray or brown bark and aromatic, finely tex-

Chamaecyparis nootkatensis, yellow cedar

tured wood. It has tiny cones. The seeds are produced at irregular intervals. It requires moist conditions, but will tolerate lean soils in full sun. It tends to accept more shade than most native conifers.

Range: south-central Alaska and along coastal areas south to Washington and northern California, with a few inland populations in southeast British Columbia and central Oregon.

Nootkatensis means "of Nootka Sound," on Vancouver Island. This tree with its slightly drooping heavy branches seems to symbolize the heavy weight of rain and snow so characteristic of the Pacific Northwest. Yellow cedar is long-lived: specimens up to 3,500 years old have been found. Even a snag (an upright dead tree) can stand for more than 100 years. It grows from sea level to the treeline in moist conditions and endures more shade than most conifers. In full sun, however, it tolerates lean soil.

Natural historian Donald Culross Peattie writes poetically about the trouble it has seeding: "Slipping from the cones in the autumn gales, [the seeds] glide far on their little wings, but probably the number that struggle through to success is not great."[33]

Yellow cedar is a slow-growing tree that may take 200 years before it reaches 15 to 20 inches in diameter, the point at which it can be used for timber. The finely textured wood is highly aromatic when crushed.

Most of the coastal natives used this dramatic tree with its straight-grained wood for their carvings. The wood was so highly esteemed that it was traded to the peoples in the interior. All along Vancouver Island, the Salish used it to carve paddles and in more recent times to make knitting needles.[34]

The inner bark has much of the same fibrous quality as the red cedar, but is finer, softer, and lighter. The people pulled it off the tree in long strands and then split and dried it to use in their weaving, especially capes and blankets.[35] Today's uses are equally varied. It can be found in window frames, doors, boat and greenhouse construction, utility poles, and cabinets, and is used by artists for carving. There is an attempt to raise this tree on plantations. And it is now being planted in areas where there are frequent avalanches because of its good recovery rate. The wood is shipped to Japan where it is used as a substitute for the rare **hinoki cypress** (*Chamaecyparis obtusa*). It was hybridized with *Cupressus macrocarpa* to form *Cupressocyparis × leylandii* 100 years ago and planted as cedar hedges throughout the United Kingdom.

Around the turn of the twentieth century, the yellow cedar started to die back. No one is quite sure why, because there are no apparent pathogens affecting it. A warming trend began in Alaska in the 1880s and it has speeded up more recently. This might mean that the fine root systems are more susceptible to frost damage.

In Oregon it may occur with the rare **Port Orford cedar**, *Chamaecyparis lawsoniana* (a cedar that is endangered by an introduced disease and overexploitation in the past). In Kelsey Bay, Vancouver Island, there is one yellow cedar with a

---

Spring on the River

Oh! Restlessly whirls the river;
The rivulets run and the cataract drones;
The spiders are flitting over the stones;
Summer winds float and the cedar
  moans;
And the eddies gleam and quiver.
O Sun, shine hot, shine long and abide
In the glory and power of thy summer
  tide
On the swift longing face of the river.

Archibald Lampman

diameter of 14 feet and a height of 200 feet. In Olympic Park, Washington, there's a giant 12 feet in diameter and 124 feet tall. In the Sechelt Peninsula of British Columbia is one that's 1,834 years old and another 1,636 years old. Mt. Rainier National Park has a pure old-growth stand of these trees.

Taxaceae

Pacific Yew
Taxus brevifolia, Western yew

Taxus brevifolia

grows to 75 feet tall in areas with well-drained soil. It is an evergreen shrub or small tree which is classed as a conifer but produces only a tiny short-lived pollen cone. The pointed dark green needles have two dull yellowish-green bands. The seeds are borne singly and are surrounded by a scarlet fleshy cup called an aril.

Range: the Alaskan islands along the slopes of British Columbia, the Queen Charlotte Islands and into the Rockies of Montana and Idaho, south to Oregon and the Redwood Belt of California.

This distinctive small evergreen needs sun and well-drained soil, unlike the fir or hemlocks it resembles. It is identified mainly by its flat, prickly-tipped needles and its peeling bark, which reveals the purply tones below. In some parts of Oregon it is the dominant tree and forms dense, almost junglelike forests. Like the yellow cedar, it does not germinate easily, producing a berry-like red fruit called an aril with gooey sweet pulp. Birds gulp down the fruits as fast as possible and may fly great distances before excreting the seeds. Although the long-distance travel doesn't harm the seeds, it doesn't seem to help them produce vast quantities of seedlings either.

This plant represented a challenge in Indian culture. Any Southern Kwakiutl who could bend a yew tree from crown to butt was considered the local strong man. Branches were used as scrubbers by young Nitinaht boys in training rituals for manhood. And in death, the smoke from the branches was used by the Thompson to purify the bereaved.

Pacific yew wood is dark, hard, strong, and decay-resistant. It was used for making bows, arrows, harpoons, war paddles and war clubs, wedges for splitting logs, digging sticks for roots and clams, and spoons and dishes. Ethnologist Erna Gunther adds that the "Makah made trinket boxes of yew which are square, burnt out of one piece, and have lids." The Cowlitz also used it for the frame of their drums.[36]

Although yew is poisonous, from very early times the tree was recognized in Indian culture as containing health-giving substances. An infusion of crushed leaves was used as a wash to make a person perspire and improve general health. It was thought to be particularly useful for babies and old people. Some groups made a poultice of ground-up leaves and applied them to wounds. In more recent times, the Okanagan have used wood scrapings from the yew mixed with Vaseline to make a sunburn ointment.

The brightly colored arils, the only nonpoisonous parts of the tree, were eaten by the Haida, but in very small quantities because it was believed that too many made a woman sterile. The Saanich and many Salish groups used to smoke yew needles. They dried and pulverized the needles, which they used in place of tobacco, often mixed with the ubiquitous ground covering kinnikinnick (*Arctostaphylos uva-ursi*). This blend was very strong, and potentially harmful, but was smoked mainly by older men who were considered past their prime. The wood was used for making pipes by the Klallam, Samish, Swinomish, and Snohomish.

Taxol, a chemical substance known as a diterpenoid, which comes from the bark of the Pacific Coast yew, is being tested as a potential cure for a variety of cancers. Most modern taxol is synthesized from related yew species grown for this purpose, but at one time, Pacific yews were being stripped of their bark for this compound. Because of this rampant overharvesting, many Pacific yew stands are now protected and may not be cut for commercial purposes.

Aceraceae

Vine Maple
Acer circinatum

Acer circinatum

is a small tree or shrub up to 25 feet high. It grows in the deep shade of the Douglas fir, and any branches touching the ground will root. The new stems are green to red, becoming greenish-gray with age. Rich red flowers bloom in April, and it is one of the few West Coast plants with spectacular autumn color. As an understory shrub it tolerates shade, but grows into a distinctive tree in full sun.

Range: British Columbia, Washington, Oregon.

The early explorers had no great love for this tree. It took a machete to get through it in the forest. David Douglas records somewhat glumly that it was "called by the voyageurs, *Bois de Diable* (wood of the devil) from the obstruction it gives them in passing through the woods."[37] In his journal he describes his first experience with the plant: "After smoking with a few straggling

Acer circinatum, vine maple

Indians belonging to the Umpqua tribe, we resumed our route on the banks of the small stream; track mountainous and rugged, thickly covered with wood in many places; and in some parts, where *Acer circinnatum* forms the underwood, a small hatchet or large knife like a hedge-bill is indispensably necessary."[38]

The vine maple can grow in the shade of a Douglas fir and become a large shrub. But if exposed to the sun, it grows to become a small tree, looking a bit like a Japanese maple with its distinctive palmate leaves with toothed margins.

Even though it is a hardwood, vine maple has limited modern commercial use since it's likely to warp over time. Native Americans, however, found many innovative uses for this prolific plant. The Quinault tribe of Washington took vine maple charcoal and mixed it with oil to make a black paint. They also made fish traps and large, loosely woven carrying baskets out of the splints from this maple. It was known by many tribes as the "basket tree."[39]

Other tribes used it for spoons, salmon tongs, snowshoes, armorlike vests, and frames for their baby baskets. The Cowichan, now famous for their heavy-duty waterproof sweaters, used the wood to make knitting needles. The Squamish and Katzie sometimes made bows from the long straight branches of vine maple.[40]

Today, vine maple is a popular landscape plant in contemporary gardens. It can even be bonsaied since it seems to accept any form of pruning, no matter how extreme. It is considered a formidable hedge, and is used by some gardeners as a substitute for Japanese maples.

Aceraceae

Big-Leaf Maple
Acer macrophyllum, Oregon maple

Acer macrophyllum

has a distinctive form with a squat main trunk and several enormous lateral branches that may soar upward. The branches have an astounding 50-foot lateral spread and may grow up to 100 feet high. The huge leaves, each one up to 12 inches wide, have three to five deep lobes like a giant hand. The fall color is mainly yellow. Creamy-yellow flowers in midspring are borne in large, vertical, chainlike clusters before the leaves appear.

Range: **British Columbia, Washington, Oregon, northern California.**

Big-leaf maple has one of the largest leaves found in North American native plants. The huge leaves, which can be up to 1 foot across, are an adaptation that allows the tree to make the most of the light in dim, dark forests. Though an imposing tree, it has a squat main trunk with enormous lateral branches, giving it a rather odd appearance.

Pollinated by insects, the seeds need rich, moist soil for germination, though few make it to germination. Most are eaten by squirrels and mice and, if they do germinate, they are destined to be consumed by slugs.

Big-leaf maple is one of the most distinctive of the Pacific Northwest natives. Epiphytes—plants that need a host and seem to live on pure air—drape themselves over the plant, giving it a prematurely aged appearance. Epiphytes thrive because the big-leaf maple is one of the most moisture-retentive plants in the forest, and the amount of calcium stored in its bark is particularly enriching. The tree is also host to the licorice fern, *Polypodium glycyrrhiza*. These hangers-on supply nutrients to the host by absorbing them from the air and rain. This symbiosis between the epiphytes and the big-leaf maples is one reason these trees survive even in thin soil leached of all minerals. Epiphytes also protect the host tree during a drought, because anything under these plants tends to stay moist.

Big-leaf maple trees started a whole new branch of botany: studying tree canopies for their plants and animals. One of the first people to get up into the canopy and study the life was Professor Nalini Nadkarni, who in the 1980s was a graduate student at the University of Washington. She used mountain-climbing gear to ascend into the canopy of the Hoh rain forest.

She was particularly interested in the big-leaf maple, which is the most prominent host for epiphytes. When she got to the top, she made an extraordinary discovery. As Ruth Kirk writes:

She peeled back mosses, selaginella, ferns, liverworts, and lichens padding branches and trunks—and found maple roots! Whole networks, from tender new root tips to thick woody strands, laced the branches and reached into the epiphytic mats. Individual roots measured as much as twenty feet long. Some angled upward. A few reached from the canopy into the ground. Expanding her rope web and hauling self and equipment up and down, Nadkarni mapped and measured the surface structure and branch sizes and positions of several [big-leaf] maples. She also checked mature alders, cottonwoods, and vine maples. All had roots reaching from their trunks into their mats of epiphytes.[41]

Apparently, the trees were growing roots to draw moisture and nutrients from the thick layer of epiphytes at the top of the tree.

Big-leaf maple was highly esteemed in Native American culture. Lives were spent in a perpetually smoky atmosphere—fires indoors, burning outdoors—and much of the population suffered regular eye infections or were racked with a persistent cough. The wood of this shrub is extremely hard and produces a hot, almost smokeless flame, so it was a valuable fuel, especially good for a small lodge in the middle of winter.

Even though the wood warped over time, it was used in a variety of ways. The Coast Salish peoples called big-leaf maple the "paddle tree" and carved dishes, spoons, fish lures, hairpins, combs, cedar-bark shredders, and adze handles from its wood. The Squamish, Swinomish, Chehalis, and Quinault of Washington used the decayed wood to smoke fish. Even the inner bark was used to make rope. Many groups spread the leaves under and over food in steaming pits to impart an agreeable flavor to meat while it cooked, or used them to line berry baskets. The Squamish used the leaves to wipe the slime off freshly caught fish. The Comox created leaf-lined pits in which salmon roe were buried to make "stink eggs" (fermented salmon eggs), a great delicacy for some Native Americans, but a dish that tended to make European visitors gag.[42]

Big-leaf maple caught the attention of the early botanical explorers of the West Coast. Archibald Menzies, surgeon-botanist of the Vancouver expedition, was the first British scientist to collect specimens in 1790; Lewis and Clark were the first Americans to do so. David Douglas sent back its seeds to England, where it was quickly introduced into cultivation.

The patterns in the wood attracted the pioneers. The bird's-eye was used for gunstocks. Veneer formed from maplewood was used in knife handles that were much esteemed by both Indians and Hudson's Bay employees. Today it is cherished by makers of musical instruments. Any symphony orchestra worthy of the name contains stringed instruments with the distinctive curly grain running at right angles to the growth rings. Audrey Grescoe tells the story of a Vancouver Islander who bought a solid piece of big-leaf maple wood for $100. He cut it into smaller pieces, sold it piece by piece over the next four years and eventually amassed $40,000 from the original piece. She also knows a man who has taken cuttings from a big-leaf maple with the desired pattern and has planted them with a goal in mind: "It's a 75-year turnaround and a gamble . . . but he's planning to leave them for his grandchildren."[43]

According to Alan Courtright of Poulsbo, Washington, the sought-after wood used for making musical instruments comes from trees that grow on slopes and move enough to cause the trunks to bend. This causes "fiddleback" and "quilted" grain over time. This type of wood is so worthwhile that some unscrupulous people cut large, older maples on land that doesn't belong to them. Wood rustling like this is probably going to be an increasing problem as these trees become more valuable.

Acer macrophyllum was invasive in pioneer gardens (it is shunned now) and considered a major impediment in the work of hauling wood out of the forest. When loggers go about clear-cutting, this is one of the first trees to sprout back to life, just as it revives quickly after a fire. It is not, in the lumber parlance, a merchantable tree. Loggers usually use bulldozers to dig it out so that the much more valued Douglas firs can grow unimpeded. Today the wood is used primarily as a furniture veneer. It's not used in house construction, since it might rot the wallpaper off the kitchen wall. But it is possible to make maple syrup from this species. It takes about 35 to 50 gallons of sap to produce a gallon of maple syrup, depending on the sap's sugar content.

Betulaceae

Red Alder

Alnus rubra, Oregon alder, Western alder, Pacific Coast alder

Alnus rubra

is a deciduous tree 80 to 100 feet tall with a straight trunk. The oval leaves are doubly toothed along the edge, and the smooth bark is gray splotched with white. Male flowers are borne on 6-inch catkins which form over the fall and winter, then bloom in spring. Female flowers turn into conelike fruits bearing winged nutlets. The color of the trunk depends on what lichens live on it. Grows up mountains to elevations of 1,000 feet in rocky soil, but is also found in stream bottoms and at the edge of the sea. It's a short-lived tree—50 to 80 years.

Range: **British Columbia, Washington, Oregon.**

The red alder appeared some time after the last great ice melt (the Laurentide ice sheet 16,000 to 12,500 years ago), along with Sitka spruce, and became a true pioneer species—able to thrive in disturbed soil. It is also the companion plant of the Douglas fir and helped it to become a great tree.

Alders generated the conditions for conifers to flourish all up and down the coast. It has small nodules on its roots that convert nitrogen in the air into a soluble form so that plants can absorb it from the soil. It is also host to many bacteria that help keep conifers healthy by fighting invasive and harmful pathogens. Any spruce rooted in soil already enriched by the alder grew faster, but its own growth eventually put the alder into the shade. After the spruce came the hemlock, then the Douglas fir. The presence of alder improved the location for the seedlings of the bigger trees: the conifer's foliage would be a darker green, its needles bigger, and it would even produce harder wood.

Red alder has a straight trunk covered with light gray bark. Underneath, there is a rich red glow. In April, the tree is covered with a delicate foam of catkins. Scorned as being weedy and invasive, this practical plant has only recently received the appreciation it deserves.

In British Columbia, the native peoples used alder—red, as well as **mountain alder** (*A. tenuifolia*) and **green alder** (*A. viridis* subsp. *crispa*)—extensively. The trees were used as a dye and, by varying the preparation, people could produce colors from the wood and bark ranging from almost black to dark brown to russet to bright orange-red. Alder dyes were used on bas-

ket materials, cedar bark, ropes, fishnets, wooden articles (such as canoes, masks, rattles, and totem poles), mountain goat wool, feathers, porcupine quills, human hair, and buckskin. Bark was even used as a pigment for face and body tattooing. The simplest method was to boil the bark or wood, or both, in a small amount of water, then steep the material to be dyed in the solution. Many people used a reddish-brown color for their nets and lines to make them invisible to fish.[44]

Several coastal groups have traditional narratives in which heroes feign bleeding at the mouth by chewing pieces of red alder bark and letting the saliva ooze from the lips, fooling the enemy into thinking they are dead.

Well-seasoned alderwood was often used to smoke salmon and cook deer meat because it has a relatively low pitch content and doesn't add an unpleasant flavor to the food. The smooth texture of the wood was also ideal for carving spoons, serving platters, rattles, masks, adze handles, headdresses, arrow points, pendants, canoe bailers, and paddles.[45]

During the early years of the lumbering industry, red alder was considered a pest because lumberjacks thought it competed with Douglas fir, a far more valuable tree. They pulled it out with donkey engines and used the wood for fuel. David Tirrell Hellyer, however, extols its virtues: "Not too many decades ago alder was considered by most foresters as a weed tree—now we know better. Dry alder burns brightly in the fireplace with little sparking and only a small residual of fine white ash, while when still green it is an idea fuel for smoking and barbecuing meat—especially fish. Much of our alder, cut while still undried into dimensional lumber, is shipped to furniture factories in California and elsewhere and manufactured into light inexpensive furniture, sometimes labeled 'maple.' Others who

Alnus rubra, red alder

have carved alder say it is easy to work and has a fine grain and texture."[46]

With the help of modern drying processes that prevent warping, this strong wood has now become valuable for posts and beams, and as a veneer. Although it starts out a lovely red-brown, it is stained to make an imitation walnut or mahogany. It is also used extensively for land reclamation in areas where erosion control is needed after disastrous clear-cutting, something of increasing importance as we see the long-term effects of this destructive kind of lumbering. It restores some of the health to the soil, helping other trees to thrive.

Ericaceae

Arbutus

Arbutus menziesii, madrone, Pacific madrone, strawberry tree

Arbutus menziesii

grows 20 to 80 feet tall and can live 200 to 250 years. It's a broad-leaved evergreen with 6-inch leathery green leaves (the new ones push out the old ones in midsummer), glossy on top and grayish underneath. The bark is reddish-brown and the berrylike fruits are orange-red. Found in well-drained soils in coastal areas.

Range: southern British Columbia to southern California, west of the Cascades.

One of the most romantic of all the West Coast trees, the madrone got its name from Father Juan Crespi in 1769 while he was looking for the lost bay of Monterey. He noted "many madronos, though with smaller fruit than the Spanish." *Madrono* is Spanish for "strawberry

most superficial observer."[48] Both Crespi and Menzies recognized the kinship of Pacific madrone to its European relative, *A. unedo*, also known as strawberry tree. In 1827, David Douglas sent seeds back to England and Europe, where the madrone became a treasured evergreen.

Birds are strongly attracted to the bright red berries that follow the drooping clusters of white spring flowers. The fruit is variously described as bland or sour. Native Americans ate the fruits raw, boiled, or dried.

The most eye-catching part of the tree, however, is its satiny exfoliating bark. It goes from soft orange to dull chartreuse and ages to a reddish-brown. Cameron Young, in *The Forests of British Columbia*, writes:

The arbutus bark . . . is a deep reddish brown colour and peels off every spring in paper-thin strips to reveal a shiny green underneath. Throughout the summer, the green slowly turns to brown, while the tree sheds about half its leaves. The following summer it drops the other half. Sweeping crinkled, brown arbutus leaves from the sun deck is almost a daily chore of any coastline cottager.[49]

The bark is a source of tannin, used both medicinally as an astringent to treat cuts and a gargle for sore throats, and as a preservative for wood or rope.

The wood has a pinkish tint, and often dark swirling patterns in the grain. However, it is heavy and hard, tends to be brittle, and since it may crack when drying, is used mainly for woodworking by skillful carpenters, hobbyists, or furniture makers. But Wallace Hansen winkled out other even more interesting uses:

Mexican caballeros were known to make their spurs from this wood because of the hardness and

Arbutus menziesii, arbutus

tree"; presumably Crespi thought the berries looked like strawberries.[47]

The beautiful tree transfixed the European botanists who first saw it. Archibald Menzies, the intrepid surgeon-naturalist with Captain Vancouver, noticed it in the Olympic Peninsula in 1792, hence the plant's Latin name. He referred to his discovery as the Oriental strawberry tree, writing that "its peculiar smooth bark of a reddish brown colour will at all times attract the notice of the

the beautiful color. . . . A curious natural deer repellant is produced during the time when the leaves are young and attractive to our forest friends. A liquid containing reduced sugars is secreted just at the tip of the leaf bud which attracts ants and flies. The presence of these insects renders those succulent leaves unpalatable to deer.[50]

Ericaceae

Salal
Gaultheria shallon

Gaultheria shallon
usually grows 2 to 3 feet tall, but in some sites it may grow as tall as 10 feet. It is an erect to creeping evergreen shrub with 4-inch oval leaves and bell-like pink flowers that are followed by purplish berries ¼ to ½ inch in diameter. Found in coastal sites with well-drained, dampish soil.

***Range:* Alaska to northern California.**

Salal moves about on underground stems colonizing just about any space in every direction it can go. It's one of the seminal understory shrubs of the West Coast forests, especially in damp soil alongside its regular companions—cedar, spruce, and hemlock.

When David Douglas landed at Fort Vancouver in April 1825, he wrote:

Constant heavy rain, cold thermometer 47 deg. Saturday the 9th . . . I went on shore on Cape Disappointment as the ship could not proceed up the river on consequence of heavy rains and thick fogs. On stepping on the shore Gaultheria shallon *was the first plant I took in my hands. So*

pleased was I that I could scarcely see anything but it. Mr. Menzies correctly observes that it grows under thick pine-forests in great luxuriance and would make a valuable addition to our gardens. It grows most luxuriantly in the margins of woods, particularly near the ocean.[51]

Douglas describes a celebratory meal held in his honor by one of the Hudson's Bay traders:

Mackay made us some fine steaks, and roasted a shoulder of the doe for breakfast, with an infusion of Mentha borealis *sweetened with a small portion of sugar. The meal laid on the clean mossy foliage of* Gaultheria shallon *in lieu of a plate and our tea in a large wooden dish hewn out of the solid, and*

Gaultheria shallon, salal

supping it with spoons made from the horns of the mountain sheep or Mouton Gris of the voyageurs.[52]

The spicy berries were also something to be enjoyed. The native peoples of the Northwest gathered the fruit to use over and under camas bulbs when they were cooking them in their steaming pits. The berries were also dipped in oolichan grease as a treat and sometimes mixed with currants and elderberries as a kind of fruit salad. They were used to sweeten other food and thicken salmon eggs.[53] Douglas even tried pickling the salal berries in alcohol and sending them back to Britain, but unfortunately on the return trip they were stolen "by some evil disposed person . . . for the sake of the spirits they were in."[54]

John Browne Jr., a plant maven from Vashon Island, Washington, feels that the potential of salal berries has yet to be fully enjoyed. "[While making wine] I pressed my first salal 'juice' (it's actually more like syrup) around 1973; and discovered that it had twice the sugar content of ripe blackberries. . . . Because of its low acid content, however, it is legally 'unavailable' to commercial wine makers."

Nowadays salal is popular with florists from Hong Kong to Europe, as well as everywhere in North America where it doesn't grow naturally. Rick Ross of Courtenay, British Columbia, is one of the wholesalers supplying florists with salal. He hires pickers and pays them enough to keep them going from August to the end of March. New shoots appear between May and mid-July, when picking is called to a halt. Plenty of retired people, ex-loggers, aging hippies, new immigrants, and followers of alternative lifestyles are happy to take the harvesting on as a way of making a living in the forest.

His pickers stockpile salal in February to tide them over in the nonpicking periods from May to August. He stores it in his wholesale warehouses in coolers for up to 14 weeks; florists can store it in their own coolers for another 6 to 8 weeks. He has found that once an area has been logged over, salal takes about 30 years to really flourish once again. However, studies have shown that salal's roots engulf the roots of large forest trees and siphon off nutrients. In another study, he says, fertilizer put down to help out new tree seedlings was lapped up by salal and all the little trees died.

Ericaceae

Red Huckleberry
Vaccinium parvifolium, red whortleberry

Vaccinium parvifolium
is an erect deciduous shrub with sharply angled branches that grows 3 to 12 feet tall in the understory of the forest. It's a relative of the blueberry. The urn-shaped yellow, pink, or red flowers bloom in May, and give way to an edible red berry.

Range: west side of the Cascades from southern British Columbia to Oregon and central California.

This is a relative of the blueberry and one of the most attractive understory shrubs of the Pacific Northwest forest. The seeds are sometimes dropped through a canopy opening by a bird, and if they fall on a decaying stump and germinate, the seedlings will probably thrive.

Archibald Menzies was a fan of the berries and much preferred them to cranberries. Like the latter, they have a pleasantly sour taste and make an excellent jelly. To collect the berries Indians clubbed a branch, letting the berries fall into baskets held underneath. It was an important food and a major source of vitamin C for the Nuxalk Indians of the Bella Coola islands. They ate the berries fresh, preserved in grease or oil, or smoke-dried

Vaccinium parvifolium, red huckleberry

Ericaceae

(they turned out a bit like raisins). They could also be mashed into cakes and dried for winter eating.

Red huckleberry is considered an extremely important browse plant for animals such as grouse, bear, deer, elk, and goats, which like to nibble off the tender green shoots, keeping the plants well-pruned. Biologists are now finding that it's extremely important to the well-being of grizzly bears. The seeds are distributed through their feces wherever these animals travel. It's also important as one of the plants that survives in the scarred landscape after logging operations have ceased.

Pacific Rhododendron
Rhododendron macrophyllum, California rosebay

Rhododendron macrophyllum
grows up to 12 feet high, with grayish-brown, smooth bark. The simple, alternate leaves are 3 to 6 inches long, shiny dark green on the upper side and paler underneath. The showy bell-shaped flowers are a pale purple, sometimes almost white, 1 to 1½ inches wide, borne in loose clusters. The fruits are pale brown, ½-inch-long capsules in five parts.

Following Pages: *Rhododendron macrophyllum,* Pacific rhododendron

Range: **coastal and montane areas, from southern British Columbia to northern California. State flower of Washington.**

In the deep green of a conifer forest the elegant, shiny, evergreen leaves of this rhododendron gleam. Out in the open they are compact, but in the deep forest they seem to relax and spread out. In either condition they have pale purple flowers. This is one of more than 1,000 species of rhododendron that grow around the world.

Rhododendrons are so easy to grow in the coastal forest that in the Kitsop peninsula of Washington, an entire week is devoted to rhododendrons of all species. The Rhododendron Festival is in full swing when the spring blooms are at their height.

Pacific rhododendron was chosen as the state flower of Washington in 1892 by the women of Washington. Even before they had the right to vote in an election, these women voted for their favorite flower in booths set up all over the state. The idea was to have it announced at the World's Fair in Chicago in 1893.

Because of the popularity of rhododendrons, they were rooted up willy-nilly and sold to gardeners. Dr. Arthur Kruckeberg remembers in the 1950s "seeing trucks parked along Seattle highways advertising their forlorn wares—rather sad-looking, spindly specimens of *R. macrophyllum* that had been collected in the wild, and balled and burlapped for the carriage trade."[55] Most people, it is hoped, know better now. Unless they are about to be destroyed by bulldozers, native plants should never be culled from the wild, but should be grown from seeds.

Rhododendrons contain a toxin called grayanotoxin. This substance affects honey produced from the nectar of rhododendrons, causing a syndrome known as rhododendron poisoning, "mad honey" intoxication, or grayanotoxin poisoning. It is rarely fatal and generally lasts for no more than 24 hours. Its effects are usually dizziness, weakness, excessive perspiration, and vomiting.

Cornaceae

Pacific Dogwood

Cornus nuttallii, Western flowering dogwood, mountain dogwood

Cornus nuttallii
is a deciduous tree with several stems that can grow 20 to 60 feet tall. The 4-inch-long leaves cluster at the tips of branchlets. Blooms appear in spring, and what look like petals are actually bracts (or false leaves) that come in a creamy white turning pale pink as they get older. The flowers produce bright red berries, and the foliage turns yellow to red in autumn.

Range: **west of the Cascade Mountains, from southern British Columbia to California.**

The queen of all the native hardwoods was named after English plant collector Thomas Nuttall. He collected it in 1833 and took seeds with him when he returned to England in 1841.

Dogwood was the main bow-making material for the Lower Stl'atl'imx in the Pemberton Valley, who also used it for arrows and combs. Other tribes used it to make handles for their implements as well as bows and arrows. The Skagit, Klallam, and Green River groups in Washington made gambling disks from dogwood branches.[56] Gambling was a favorite pastime with many of the First Nations. They were quick with games and quite happily whiled away a good deal of time when they gathered in large groups to trade goods or celebrate important occasions.

Several native groups also used the bark as a tanning agent and a deep brown dye. They

Cornus nuttallii, Pacific dogwood

Cornus, the ancient name, from cornu, *a horn, referred to the hardness of the wood, a European species having long been used for skewers by butchers and for daggers and other sharp implements, whence the colloquial names in some English provinces,* skewerwood *and* dagwood, *the latter coming from the Old English* dagge, *a dagger or sharp, pointed object.* Cornus *and dagwood are then apparently closely related in meaning and only by an erroneous etymology did dogwood become established as the English name of* Cornus. [58]

Modern uses of the heavy hardwood include thread spindles, mallet heads, golf club heads, and piano keys.[59]

Berberidaceae

Oregon Grape
Mahonia aquifolium, tall Oregon grape

Mahonia aquifolium

grows 3 to 10 feet tall. In early spring, tiny new leaves appear from the center of the shrub. It has small, intense yellow flowers in spring, followed by blue berries in terminal clusters. It thrives in open woods and is usually found in clearings.

Range: British Columbia, south to northern California, west of the Cascades.

State flower of Oregon.

mixed it with fir bark (*Abies grandis*) to make a black dye for color in baskets made of bitter cherry bark.[57]

The tree was pictured for the very first time in James Audubon's *Birds of North America* (1827–1838) as the branch on which a band-tailed pigeon is perched. Many years later, noted Harvard botanist Merritt L. Fernald solved the mystery of the common name, dogwood. He wrote:

The sharp edges of the Oregon grape's evergreen foliage probably gave it the name *aquifolium* (*aquila* for eagle—think sharp claws; *folium* for leaf). Birds feast on the berries, which appear in terminal clusters arranged so a bird can strip the shrub clean in short order.

David Douglas collected *Mahonia aquifolium* and sent seeds back to England. In Europe and all over North America it has become a very important garden plant.

The inner bark of the stems and roots has a bright yellow pigment. The native peoples of the Northwest shredded sticks and roots and boiled them in a small amount of water to produce a dye for basket materials. The material to be dyed (often bear grass for basketry) was steeped in the solution. Since the 1970s, the Upper Skagit have been using Oregon grape to dye rags for braided rugs.[60] The Okanagan dyed porcupine quills and also made a thick, concentrated paint by boiling the dye solution until the water evaporated, then mixing the remaining powder with ocher paint or cottonwood bud resin. They often intensified the color by boiling wolf lichen (*Letharia vulpina*) with the Oregon grape. Some Vancouver Island Salish groups used an extract of Oregon grape roots as a detergent lotion for washing the hands.[61] Oregon grape jelly has long been a Northwest delicacy.

M. nervosa, **dull Oregon grape,** is a more prostrate form that suckers all over the place. It has crooked stems with yellow wood, and leathery pinnate leaves with a hollylike appearance. The yellow flowers come in terminal racemes, and the fruit is blue to black berries. It grows about 24 inches tall. Natives used branches from it along with wild roses to adorn graves, and placed them around the walls of the house of a person who'd recently died. The sharp edges of the foliage fended off the spirit of the dead person. Like *M. aquifolium*, it provided a dye. It is now used to make a treatment for skin problems because it has antifungal and antibacterial properties; it is considered particularly effective in the treatment of psoriasis.

Mahonia aquifolium, Oregon grape

Berberidaceae

Inside-Out Flower
Vancouveria hexandra

Vancouveria hexandra

grows to 12 inches tall with open white flowers. The double compound leaves grow in three groups of three on threadlike stalks. The leaves are shaped like duck's feet. It grows in shady open spaces and spreads by underground rhizomes.

Range: from western Washington to coastal northern California.

This plant was named in honor of Captain George Vancouver, the eighteenth-century explorer. He was a fat, bluff fellow and considered a tormentor by his crew members, but was nevertheless a faithful chart maker of this complicated coast. The plant named after him is quite unlike the man himself: from the delicate compound leaves (almost like maidenhair fern) to the pretty tiny flowers. This is an understory shade-tolerant ground cover which is an important part of the forest.

It was named by Jacques Édouard Morren of Liège and Joseph Decaisne, director of Le Jardin des Plantes, Paris. The common name, "inside-out flower," refers to the fact that the white sepals and petals are swept backward to reveal the pistil and six stamens (hence *hexandra*).

Vancouveria is becoming increasingly popular with people who cultivate native plants. Sydney Eddison, writing in *Fine Gardening*, notes: "Without knowing anything about it, I once bought a tiny pot of *Vancouveria hexandra* from a North American Rock Garden Society plant sale. I have since come to delight in this graceful cousin of *Epimedium* as a lacy edging for large-leaved hostas. The white flowers of *Vancouveria* are so minute that they really don't count, though if you examine them carefully, their shape is intriguing."[62]

Caprifoliaceae

Red Elderberry
Sambucus pubens syn. *S. racemosa* subsp. *pubens*, scarlet elder, stinking elder

---

Sambucus pubens syn. *S. racemosa* subsp. *pubens*

grows up to 20 feet tall. It has rough bark and large, opposite, compound, lance-shaped leaves. The creamy-white flowers are pollinated by insects. Berries hang down in huge, bright red, pyramidal clusters. Found near streams and banks in mountains and lowlands.

Range: west of the Cascades. Blue elderberry, *S. caerulea*, is more common on the east side.

The genus name *Sambucus* comes from the Greek *sambuca*, a stringed instrument, that was said to be made from elderwood. The species name *pubens* means "downy," referring to the young stems and leaves.

The creamy-white flowers give way to fleshy red drupes, usually containing three to five seeds each. The seeds are dispersed by the birds and animals that eat the fruit, especially bears, which eat huge quantities of the ripe berries and leave their steaming piles of ordure all through the forest, fertilizing as well as dropping seed. Birds can wipe a tree clean in a matter of days and spread the seed as far away as they can fly. The seed coat is so hard it usually lies dormant in the ground for about two years before germination. Red elderberry also regenerates from rhizomes or root crowns after a fire. It prefers moist, rich, rocky soils and is an excellent soil stabilizer, aiding in erosion control of moist sites.

Sambucus pubens, red elderberry

The raw berries are toxic to people because they contain glycosides. Indeed, the whole plant, including the roots, stems, and foliage, is poisonous. Even so, Native Americans found many uses for it. Elderberry stems were easy to hollow out and kids used them like peashooters with bits of kelp as missiles. The hollow stems were also used for whistles, drinking straws, and blowguns. The Salish made pipe stems and the Nuxalk pipe bowls from red elderberry, and the Flathead of Montana and the Okanagan of Washington made small flutes. The Haida used red elderberry pith to fasten flint tips onto arrow shafts. The Kwakwaka'wakw used pieces of stem as bases for feather shuttlecocks used in a game.[63]

Elderberries can be eaten cooked. The fruit and stems are boiled for sauce, then strained to make a jelly. In the old days they were cooked with oolichan grease, mixed with huckleberries, and made into cakes. The bark and roots were often boiled up to make a drink to cause vomiting—long believed one of the best ways of purifying the body of a sick person.

Rosaceae

Salmonberry
Rubus spectabilis

Rubus spectabilis

is a thicket-forming deciduous plant growing 2 to 10 feet tall with erect and curved biennial stems. Irregularly toothed leaflets appear in groups of three. The flowers are large (1½ inches across) and showy, with pink to reddish-purple petals and many purplish stamens. The fruit, which looks like an elongated raspberry, is yellow-orange to dark red, edible with a mild taste, but seedy. Grows in moist soil in lowland forests, clearings, and along streams.

Range: Alaska, south to northern California, along the coast.

One of the first botanists of the colonial era was Frederick T. Pursh. He wrote an early American flora, which means he collected all the information then available on the local plants and published it in book form in England, where the taste for North American plants was almost insatiable. Salmonberry was one of those he named.

One of the many transient glories of spring, salmonberry in bloom can light up a whole section of the forest. In autumn, the underbrush is crammed with fruit, and bears can be seen munching away on the ripe berries. Deer, elk,

and even the more recently introduced sheep browse on it as well. Rufous hummingbirds and many insects are attracted to the nectar in its showy pink to red flowers.

Like other members of the *Rubus* genus, salmonberry is reproductively versatile; that is, it can reproduce vegetatively or sexually. It sprouts prolifically from almost every part—the stump, root crown, stem base, rootstock, and, of course, from the dense networks of rhizomes, especially after a fire or a disturbance such as bulldozing. Vegetative regeneration helps keep colonies alive in areas where the understory is heavily shaded. Because it is a pioneer species, it moves rapidly into any open space after clear-cutting.

Salmonberry produces only about 30 berries per bush each year, but each berry contains many seeds. The seeds are covered with a tough, impermeable coat, and germination may be slow unless something happens to crack the seeds. The seeds also need very precise conditions to

Rubus spectabilis, salmonberry

germinate: 90 days of warm weather (68° to 86°F) followed by 90 days of cold weather (36° to 41°F). The berries make an excellent jelly, but are considered too seedy for jam. Salmonberry was an important source of vitamin C for Native Americans, who used the fresh fruit in autumn but also turned the berries into a mush and dried them into cakes. In spring they were keen on eating the early sprouts. Salmonberry was recommended to combat ailments connected with eating too much salmon.

Salmonberry has been cultivated since 1827. Because of its deep roots, it proves to be a really good plant to hold back soil erosion on steep slopes.

Rosaceae

Ocean Spray
Holodiscus discolor, creambush, ironwood

Holodiscus discolor, ocean spray

Holodiscus discolor

is a tall, multistemmed shrub that grows 10 to 12 feet tall. In midsummer it is covered with large clusters of tiny creamy-white flowers. The stems are long and straight, covered with leaves up to 3½ inches long. The leaves may be either oval or triangular, and are deeply lobed and doubly toothed. The seeds are produced in tiny dry capsules.

Range: **British Columbia, south to California, east to northeastern Oregon, northern Idaho, eastern and western Montana, and in the Cascade Mountains.**

Holodiscus is Greek for "whole disk" and refers to the unlobed disk to which the petals are attached. *Discolor* means "of two colors," and refers to its leaves, which are dark green on top and light underneath. The names "ocean spray" and "creambush" refer to the foamy clusters of tiny cream flowers, which bloom in June and July.

This is a common and often abundant understory shrub in forests dominated by Douglas fir. It makes dense shrubby stands even in areas that have been burned over and survives in dry, rocky, well-drained areas. As well, it is a good fire-resistant plant, since heat scarifies the seeds and helps them get on with the process of germination. A related species, *H. dumosus* **(desert ocean spray)**, grows in the volcanic ash soils of south central Oregon.

Ethnobotanist Richard Hebda says it was known as ironwood to most of the southern native peoples of British Columbia. "Ocean-spray wood has many uses. Branches and stems were once hardened over a fire and shaped into digging sticks, spear shafts, arrows and bows. Other uses included teepee pins, fish clubs, baby cradle hoops, armor, salmon barbecuing sticks and knitting needles. Sa'anich peoples steeped the dry flower heads in hot water to make medicine against diarrhea."[64]

Today, it's a great favorite in coastal gardens, with elegant arching stems from which the new young shoots arise, gradually replacing the parental arches as they die. Nevertheless, one gardener cautions that although it has a "sweet and agreeable" scent from a distance of 20 to 30 feet, "up close it smells like old newspapers!"[65]

Rhamnaceae

Cascara
Rhamnus purshiana

Rhamnus purshiana

is a small, multistemmed deciduous tree that grows 20 to 35 feet tall, depending on its situation. It is highly tolerant of shade, so grows in the Pacific Northwest forest, but also on the floodplain of the Willamette Valley in Oregon. It has 3- to 5-inch-long, broadly oval, glossy green leaves, embossed with bold veins.

Range: west of the Cascade Mountains.

This small deciduous tree was named for the botanist and writer of an early American flora, Frederick T. Pursh. A well-known laxative, cascara, comes from the bright yellow inner bark of this tree, which people have been using as medicine for thousands for years. It contains

hydroxymethylanthraquinones, which cause peristalsis of the large intestine but not in the small intestine.

The bark is harvested in spring when the sap runs and the bark can be peeled easily. It is dried like a herb in the shade. The Kootenai and Flathead of western Montana had strict rules about how to prepare cascara. The bark had to be stripped downward to be effective. If it was stripped upward, it would be an emetic and cause vomiting. Small animals that ate the plant and in turn were eaten by people were said to retain the purgative qualities of the plant.

John Browne Jr. of Vashon Island explains the harvesting of cascara:

Dry it like an herb, in the shade, with good air circulation around. If you tie bundles of bark together (with twine, hay wire, etc.) you can hang them from conifers with low limbs (six to ten feet from the ground), and they should dry nicely. If you want to dry it in the oven, that's also a possibility, but use the smaller pieces from the limbs or side trunks, which are the easiest to prepare, anyway. The main trunk bark needs to dry out about a year, in the air; and herb users prefer the smaller stuff. If the tree snapped off instead of being uprooted (it is kind of brittle wood), and you want it to resprout, cut the stump off at a bit of a slant. Don't peel the trunk below the cut. It should resprout in the spring, and can be pruned back to 2 or 3 trunks in a couple of years, giving you an additional harvest in the process.[66]

Every year about 5 million pounds of dried cascara bark are processed and shipped from the West Coast by drug companies.

Liliaceae

Camas

Camassia quamash, common camas, Indian hyacinth

Camassia quamash

grows to 2 feet on long slender stalks from basal leaves, which grow from an edible bulb. It is a member of the lily family. The grasslike leaves cluster at the base of the plant, and the blue, purple, or (rarely) white starlike flowers grow in a dense erect cluster on the top of a leafless stem. It has small black seeds held in a pale brown capsule. Found in open spaces, sub-Alpine and damp prairie meadows, grassy bluffs, and rocky outcrops.

Range: British Columbia south to California, inland as far as Alberta in the north and Wyoming and Utah in the south.

There is nothing about the brilliant blue, star-like blooms of *Camassia* to distinguish it as one of the West Coast's most important foods, almost a culture-creating plant. *Camassia* was known as Indian hyacinth and became one of the most famous of all Pacific Northwest plants. The big edible bulb does look a bit like a hyacinth bulb, but the flowers are singular and spectacular with a brief but brilliant life in May. The dense blue or purple clusters tower over the 2-foot slender stalks emerging from the basal leaves. In the forest, once a large enough space has been made in the canopy, camas moves right in. As the forest changes, camas emerges wherever there is a patch of sunlight.

It was an important source of food, but it was also used for both commerce and gambling (though some native peoples made no distinction between the two). Many tribes used the bulbs strictly as a cash crop, including the Kootenai, Nez Perce, Salish, Halkomelem, and Shoshoni. Hundreds of years before tulipomania raged in Holland, the tribes of the coastal and interior of the Pacific Northwest had their own lively trade in camas. Coastal peoples traveled inland with their treasures (boards, cedar baskets, camas) to special areas where several tribes gathered to trade and take part in games on which serious betting would take place.

Camas was a staple in the diets of all the Western peoples. Recent lab analysis has shown what a good choice this was. It is rich in protein and a good source of fiber, calcium, phosphorus, and iron. Probably no other food was more widely traded. The natives knew how to use the bulbs while preserving what they needed for the future.

The history of camas indicates the highly developed nature of agriculture along the coast. Camas fields were considered tribal property and jealously guarded. Once an area with camas was discovered, the same woman would go back to it year after year. Beds were often divided up so that each family had its own plot, which was cleared each season, sometimes by a small controlled burn. Between May and August, the women and children collected the bulbs. They lifted out the soil in small sections, and took out only the bigger bulbs to ensure a good crop the following year. Then they replaced the soil and cleaned up the beds. Families filled four to five sacks at a time. The highlight of the harvest invariably included the traditional feast and the exchanging of gifts.

When cooked, blue camas bulbs are soft, brownish, and sugary. They were often used to sweeten other foods in the days before sugar was available. It took slow cooking for at least three days for the chemical—in this case inulin, a complex sugar—to turn into fructose (a sweet digestible sugar). In what is now Washington state, people mashed up bulbs and pressed them

Camassia quamash, camas

until they had the consistency of cheese. This goo was used to thicken salmon stews. Other groups dried them in the sun after cooking, and stored them against the long winter months of privation. Camas bulbs could also be roasted. The women dug pits, covered the bottom with rocks, and made a fire. This was covered with sheets of thoroughly soaked woven cedar. Then the bulbs were laid on top to roast.

To steam them, they dug pits 3 to 6 feet across and almost 3 feet deep. A fire was lit in the bottom and allowed to burn until the rocks lining the pit were red-hot. After removing the ashes, the cooks leveled the bottom of the pit, adding seaweed, blackberry and salal branches, fern fronds, or grand fir boughs. Then they added the camas bulbs, as much as 125 pounds at a time. Sometimes, they mixed them with red alder or arbutus bark to give the bulbs a reddish color. The pit was covered with more branches, then soil or sand and old mats or sacking. Water was poured in through a hole made with a stick, and the bulbs were allowed to steam for a day and a half.

Native peoples were very careful not to confuse this plant with the **death camas**, *Zigadenus* spp., which looks similar in size and shape. The flower, however, is creamy with a smaller, tighter cluster.[67]

Early explorers compared the bulb to onions. What they didn't know was that eating the plant had unpleasant side effects. David Douglas writes of enjoying a good feed of camas with his Indian friends and then staggering out of the hut clutching at his throat from the smell of uncontrolled passing of gas.

Lewis and Clark made their monumental forced march over the Continental Divide on their way to the Pacific Northwest, on blistered feet in tattered boots, lugging hundreds of pounds of equipment. Lewis met up with Clark in September 1805, in a Nez Perce village: "I found Capt Lewis & the party encamped," he wrote, "much fatigued, & hungery, much rejoiced to find

something to eate of which They appeared to partake plentifully." Clark's experience was that too many roots had made his hunters violently ill, so "I cautioned them of the Consequences of eateing too much &c." The captains called these two villages in the area "Quawmash flats" or the "camas flats" (near Weippe, Idaho), where "Indian women gathered great quantities of camas roots which they made into a kind of bread or cake."[68]

In his journal, Patrick Gass, one of the members of the expedition, writes: "The provisions which we got consisted of roots which they call cams, and which resemble onions in shape, but are of a sweet taste. This bread is manufactured by steaming, pounding, and baking the roots on a kiln they have for the purpose."[69]

In Idaho, the evidence of how widespread the camas was during the time of settlement is shown in place-names: Five waterways are called Camas Creek; there's also Camas Slough; a grassland area called Camas Meadows and another called Camas Flats; Camas Butte; Camas County; and Camas National Wildlife Refuge. On the Little Camas Prairie, Little Camas Dam stops up Little Camas Creek to form Little Camas Reservoir.

Today, camas is still eaten by the Native Americans. Mostly, however, we know *Camassia* as a wonderful garden plant standing erect and elegant in the late spring border.

Orchidaceae

Fairy-Slipper
Calypso bulbosa, Venus's-slipper

Calypso bulbosa
rises from a bulbous corm. The pleated leaf is oval, comes out in late summer, and survives winter. The slipper-shaped flower has purple-pink

sepals and a whitish lower lip streaked with purple. The leaf withers after flowering.

Range: throughout North America except for the Deep South and the driest areas of the Great Plains.

Calypso bulbosa is a monotypic (there is only one genus in the species) terrestrial orchid.

Calypso was named for the Greek sea nymph, Kalypso, which means "I hide." She was the beautiful nymph who waylaid Ulysses on his way home and refused to give him up until Zeus forced her to. "On the tenth black night the gods brought me near the island, Ogygia, where Calypso of the fair braids dwells, dread goddess. She took me in and kindly befriended me, nourished me, and said she would make me immortal and ageless all my days."[70]

The spring flower is pink to purple and occasionally white. It is pollinated by bumblebees, which are fooled by the yellow bristles on the lip of the flower into thinking it's a nectary, and as they bumble around they spread the pollen.

It's a rare plant which few have seen, and in most places in the southern part of its range it is in decline. Given its preference for cool soils, some feel its disappearance in the South is an early indicator of global warming.

Bulbosa means "having bulbs." Actually, the fairy-slipper has a pseudobulb. Native peoples chewed on the bulbs or sucked on the flowers as an anticonvulsive for mild epilepsy. Craig Childs says:

Calypso bulbosa, fairy-slipper

The dip and roundness of its corm and its hood have the grace of Roman archways. Haida women once ate these buttery corms in hopes of enhancing their breasts. Considering the evolution of all plants and animals it is probably one of the most architecturally advanced pieces of reproduction ever generated. It is here alone, its lip fac- *ing the log as if the remainder of the world is of no concern.*[71]

In 1864, John Muir was in Meaford, Ontario. When he found this delicate plant, he was deeply moved:

It seemed the most spiritual of all the flower peo-
ple I had ever met. I sat down beside it and fairly
cried for joy. It seems wonderful that so frail and
lowly a plant has such power over human hearts.
This Calypso *meeting happened some forty five*
years ago, and it is more memorable and impres-
sive than any of my meetings with human beings
excepting, perhaps, Emerson and one or two
others. [72]

Fumariaceae

Western Bleeding Heart

Dicentra formosa, Pacific bleeding heart,
lady's-locket, Our-Lady-in-a-boat

Dicentra formosa, Western bleeding heart

Dicentra formosa

grows from a rhizome, from 6 to 20 inches high.
The leaves spring from the base of the plant and can
reach right up to the top of the stem. Numerous
ferny leaves almost obscure the heart-shaped pink
flowers that bloom from spring to late summer.

Range: British Columbia to northern California.

Formosa means "beautiful," and this flower is
certainly a beauty. The name comes from the
Greek: *dis* ("two"), *kentron* ("spur"). The shape
of the pink flower is spurred like Dutchman's-
breeches. There is also a white-flowering form.
They both have delicate, lacy, blue-green
foliage, which makes a wonderful show until it
goes a mushy yellow and fades in fall. The common name of "bleeding heart" comes from a
Chinese legend which says that the blossom of
the plant resembles a heart with a drop of blood.

Gardeners know that *Dicentra* can cause dermatitis, so are a bit wary of working too close to
the plant. It grows in shade in rich, moist soil
and blooms from April to late summer west of
the Cascades right to the coast. *D. formosa* subsp.
oregona has bluish foliage and cream or pink
flowers and is found in southwestern Oregon
and California. Like the trilliums, much of
Dicentra propagation is performed by ants which
transport the seeds because they contain an oil
that the ants love to eat.

Western bleeding heart was used as a drug
by the Skagit, who made a decoction of pounded
roots and administered it as a worm medicine.
They crushed leaves and put them in the water
used for washing hair to make hair grow better,
and chewed the raw roots to relieve toothaches.

Araceae

Skunk Cabbage

Lysichiton americanus syn. *L. americanum*,
swamp cabbage, swamp lantern, yellow arum

Lysichiton americanus syn. *L. americanum*

is a stemless, hardy herb that grows from a rhizome and produces large yellow "flowers" that are actually showy modified leaves called spathes wrapped around a club (the spadix) containing tiny male and female flowerlets. It has 2- to 4-foot-long leaves that form a rosette once the flowering is over. It flowers in March. Found in marshy areas.

Range: Alaska to California.

The astonishing look of the skunk cabbage is of a huge yellow flower rising out of the bog with no stem in view. The interior of this "flower," which is really a bract (a leaf structure), is opaque and when backlit by the sun it reveals a texture close to silk. The large leaves form a rosette around the flower stalk once flowering is over. (*Lysichiton* means "loose tunic.") It attracts carrion flies, beetles, and bees as pollinators.

Though the aroid family has about 3,000 species in it, this genus has only two. Its relative, **Asian skunk cabbage,** *L. camtschatcensis,* probably evolved separately when Asia and North America separated. Aroids are distinguished by the spathe (a large, colorful, modified leaf) surrounding the spadix (a long cylindrical spike of innocuous flowers). Some are found in the wild in California, and all but the skunk cabbage are weeds or escaped ornamentals.

Skunk cabbage grows along the edges of streams and in marshy areas. It's the harbinger of

spring in many areas and makes its appearance in March. The smell of skunk cabbage has been described variously as fetid, odoriferous, putrid, vile, and revolting. Here is Emily Carr's description: "In the swampy places and ditches of Greenville skunk cabbages grew—gold and brimming with rank smell—hypocrites of loveliness peeping from the lush green of their great leaves. The smell of them was sickening."[73]

Native Americans all up and down the coast knew this plant and used it well. In some places in the interior it was known as "Indian wax paper." Those great big, water-repellent, rather funky leaves were inedible, but they could do duty as cooking and serving surfaces without imparting any noticeable odor to the food. They were used as plates, pots, lids, and liners for cooking in the local steaming pit. The storage pits for fermented salmon eggs were always lined with skunk cabbage leaves, and people wrapped salmon in them for cooking. Berry boxes were lined with the leaves to keep the boxes from leaking.[74]

According to ethnobotanist Nancy Turner,

The Nuxalk, Ditidaht and other coastal people made an ingenious temporary drinking cup and water dipper by folding a large skunk cabbage leaf in half from top to bottom, bending the two layers to form a U-shaped trough and pulling the edges back to the lower end, holding them together with the stem as a handle. They also constructed makeshift berry containers by folding and pinning the leaf edges together with sticks. Some people even used the larger leaves for sun shades on hot summer days. And the children, probably even today, use them as spears.[75]

Given the odor of the plant, it may seem an unlikely prospect as a food, but people may have

FOLLOWING PAGES: *Lysichiton americanus* syn. *L. americanum*, skunk cabbage

seen bears eat them in spring and experimented with them (an early example of animal testing). The Skokomish steamed and ate young leaves and the Tolowa and Yurok ate the root centers after boiling them eight times. The Cowlitz cooked the blossoms overnight but ate no more than two or three at a time, otherwise they became sick. The Hoh placed the leaves over roasting camas, wild onion, or garlic for flavoring.

Almost all coastal groups used the leaves to dry berries on. They removed the fleshy midribs and sometimes heated the leaves to make them more pliable. Overlapping leaves were placed on the ground or a wooden rack. Rectangular wooden frames were placed around them and the berries, cooked to a jamlike consistency, were poured into the frames. Once these berry cakes were dry, the leaves could be peeled away and the berry cakes could be stored for winter feasts.[76]

Skunk cabbage, like all aroids, contains calcium oxalate crystals (raphides) in their cells. These small crystals, if not destroyed by processing, can not only cause irritation but also the possibility of a serious injury to the mouth and throat when eaten. (*Dieffenbachia*, also known as "dumb cane," has the same effect.) Nevertheless, Native Americans used skunk cabbage medicinally in many ways. Poultices made from the roots of skunk cabbage were used for sores and swellings by the Klallam and Kwakiutl. The Cowlitz made a poultice of heated blossoms and applied it to the body for rheumatism. Plasters made out of a compound that included the roots of skunk cabbage were used to treat lung hemorrhages, rheumatism, blood poisoning, and boils. From headaches to pains in the knees, the skunk cabbage had a use. Shamans knew of the plant's toxicity and used it pulverized in minute amounts. The Thompson put skunk cabbage leaves under their pillows to induce "power dreams." Smoke from the root was inhaled to banish persistent bad dreams.

Araliaceae

Devil's Club

Oplopanax horridus, Alaskan ginseng, Pacific ginseng, wild armored Alaskan ginseng

Oplopanax horridus

has 1-foot-wide maplelike leaves and bright red berries (drupes) which come out as the foliage yellows in autumn. The spiny stems grow 6 to 10 feet high. Found in moist, shady woods.

Range: **Alaska to Oregon, east to Idaho and Montana, with disjunct populations in Michigan and Ontario.**

The devil's club has many names, including *heshkeghka'a*, which means "prickle big big" to the Dena'ina of Alaska. The genus name *Oplopanax* incorporates the Greek word for tool or weapon; *horridus* means "prickly." It's a deciduous understory plant that turns into a jungly mass when it grows with alders in the forest. The spiny stems grow 6 to 10 feet tall, ending in giant leaves. Deer and elk love to browse on the young shoots. Botanists have little information on how devil's club regenerates, though it appears to reproduce vegetatively, possibly by layering, or rhizomes. Peter Stickney believes it may be a "root crown sprouter." Whatever the exact method, seedling growth is very slow.[77]

The Dena'ina used it as a medicine rather than a food. The Outer Inlet people boiled the stems and branches and drank the tea to lower a fever. Many Indians brewed tea from the bitter bark or stripped off the thorns to eat the green bark as a tonic.

The devil's club was also part of occult rituals. In their book on the trees and shrubs of Alaska, authors Leslie Viereck and Elbert Little tell us:

"Years ago the stalks were used by Indians for beating suspected witches to obtain confessions. Even today old people will nail the devil's club stalk over their door or window to protect the house from witches, evil influences, and bad luck."[78]

The Haida used devil's club to fish for octopus. They perched outside the octopus's den and hooked the animal on the lethal thorns as it emerged. Some tribes peeled the sticks and cut them into small pieces as fish lures. The Manhousaht Nuu-chah-nulth and Ditidaht carved them to resemble small fish.[79]

The Straits Salish, Squamish, Ditidaht, Haisla, and others mixed devil's club charcoal with bear grease to make a black face paint for ceremonial occasions. Some people inserted it under the skin to produce a bluish-colored tattoo. Erna Gunther in 1945 reported that the Lummi were still using this charcoal for face paint, but that they were mixing it with Vaseline instead of grease. This use continued into the 1990s in some areas.

Oplopanax horridus, devil's club

The Dena'ina of Alaska also used the charcoal mixed with water to make a black dye.[80]

Native Americans classified plants according to a system of partners. Here is a story about their plant classifications and how they classified devil's club:

After the earth was created, all the trees came to be related in pairs. For example, cottonwood and birch got together, and so on, until spruce was left. This was because it had sores all over it. But instead of being left to be useless, it was blessed by being made the most valuable plant to the human race. It was told to spruce, "Whatever is on your skin will be useful. The scabs on your skin will be medicine." So the scabs which are the pitch became a medicine. Cottonwood, on the other hand, which is the prettiest tree, is the most worthless. All of the other plants got partners too. For example, the rose and the devil's club went together [because both had thorns]. The animals, too, were separated into pairs, and also the birds.[81]

Unfortunately, the reputation of this plant as one with healing powers has meant that it is being hauled out of the forest indiscriminately, without thought for its scarcity.

David Tirrell Hellyer writes in his memoirs as a physician and naturalist on the West Coast: "In 1980 I saw the first devil's club by the edge of the bog, its huge maple like leaves rearing upward to catch the filtered light beneath the cedar trees. Later in the spring, bunches of white flowers, then red berries, give an exotic appearance to this armed and dangerous plant, which only the elks seem to be able to browse with impunity."[82]

Polypodiaceae

Sword Fern
Polystichum munitum, Christmas fern

Polystichum munitum, an evergreen fern, grows to 5 feet tall. The cinnamon-colored fiddleheads turn into 100 arching, dark green, almost leatherlike fronds. Each has simple, toothed leaflets. It is ubiquitous in hemlock-cedar forests of the Pacific Northwest and grows almost anywhere, from shady woodlands to sunny areas, from wet to dry soil.

***Range:* British Columbia to southern California along the coast.**

These plants are a remnant of the once-tropical climate of North America and they are still to be found in the tropics.

The light, wind-borne spores are produced by the millions from early to midsummer. These spores can remain viable for up to four years, provided they have enough moisture and temperatures between 59° and 86°F for germination.

Initially the germinating spores produce gametophytes (tiny plants without root systems that do not bear spores). The gametophyte which contains the male and female sexual reproductive organs also sustains the egg until it becomes a spore-bearing plant. Unlike other spore-bearing species, the sword fern needs to be fertilized from a different plant, thus ensuring variety and adaptability.[83]

Polystichum means "many rowed" (referring to the lines of spore clusters) in Greek; *munitum* means "armed." No doubt *munitum* and the common name, "sword fern," refer to the enlarged basal lobe found at the base of each leaflet, which looks like the hilt of a sword.

The native peoples of the coast and the interior gathered the long, stiff fronds for lining steaming pits. The fronds were also useful in making racks for drying berries or wiping off freshly caught fish, and were also woven into crude rugs, used to cover the floor of summer houses, or made into a kind of sleeping mat. Chewed and swallowed raw, the young leaves eased sore throats.[84]

P. munitum is a relatively long-lived fern and today is used widely in gardens, sometimes lasting for generations. Florists promote the sword fern as a Yuletide decoration, hence the common name Christmas fern (a name that is also applied to *P. acrostichoides,* an eastern North American native fern that is still green in late December).

Polystichum munitum, sword fern

The Tundra

The tundra has been called barren. It may look empty, even devoid of interest at first glance, but in fact it is neither. Millions of creatures from insects to caribou make it their summer home, as do plants in the thousands. John Muir's description of Kotzebue Sound in his book *The Cruise of the Corwin* in 1881 gives an idea of the Far North's riches in the middle of September:

> *The tundra glowed in the mellow sunshine with the colors of the ripe foliage of vaccinium, empetrum,*

The vast reaches of the tundra cover a fifth of the earth's surface.

arctostaphylos, and dwarf birch; red, purple, and yellow, in pure bright tones, while the berries, hardly less beautiful, were scattered everywhere as if they had been . . . broadcast with a lavish hand, the whole blending harmoniously with the neutral tints of the furred beds of lichens and mosses on which the bright leaves and berries were painted.[1]

Occupying about a fifth of the earth's surface, tundra is the land of tempestuous winds, winter snow, and permafrost—the eternally frozen ground whose icy grasp extends as deep as 3,000 feet. There are no trees at all. What trees could endure the cold, the winds, and the permanently frozen underpinning? Not surprisingly, *tundra* comes from the Finnish word *tunturia* meaning "treeless plain." It's also as dry as a desert—less than 5 inches of precipitation falls each year. A thin layer of soil, perhaps a few feet deep, covers the ground, and what there is has little organic content. Decomposition, which in more southern areas helps build up layers of humus, doesn't occur as much here because of the poor drainage and permafrost. But this is the active layer where plants make the most of the short, 6- to 10-week growing season and long days of midnight sun. They get the job of fertilization done as quickly as possible and then head straight into dormancy.

Most plants are ground-hugging, able to photosynthesize at low temperatures (which, in winter, can average −30°F) and low light levels. Rosettes on some plants protect the growth bud as the plant dies down to the ground. Woody plants tend to lie prostrate out of the worst of the wind. In spring, reproduction is normally by buds and division, because plants cannot always count on insects for pollination.

In the High Arctic, plants thrive in islands of growth: lichens and mosses on rock surfaces, and some forbs or green plants growing around the edges or in protected spots. Farther south, in the Middle Arctic, or Arctic Coastal Plain, freeze-thaw cycles heave up pebbles and rock fragments, forming a mosaic pattern when seen from the air. Mosses and sedges grow in waterlogged areas, while forbs and heaths favor drier spots. The most southerly area has woody shrubs such as willow and birch, with alders and willows along the waterways and spruce and fir on south-facing slopes.

For many millennia—perhaps 24,000 years—people traveled back and forth across the Beringian land bridge connecting Siberia and Alaska, where the Bering Strait is now located. Plants in Siberia resemble those of Alaska and Yukon, and we know they too moved eastward, if not westward. We also know there were tundralike conditions south of the ice, because the remains of both tundra flora and fauna (woolly mammoths, lemmings, caribou, and musk-oxen) have been found frozen in the permafrost.

Gradually, the glaciers retreated. The rocks left behind by the walls of ice were pounded furiously by the wind; sand and dust swooped around from place to place creating dunes. Wind was crucial to plant migration. Their seeds are blown hither and yon and, when they land on fertile ground, they germinate. This happened at the edges of the melting ice: as water retreated from lakes, the lake beds became the perfect places for blown seeds to germinate.

Animals adapted in turn. Caribou eat practically anything, including the less toxic sedges and herbs, as well as lichens and shrubs such as willow. Songbirds eat up the insects, which develop defenses against plant toxins. Waterfowl, ptarmigan, and small mammals consume the tender belowground materials and follow receding snow beds to harvest the crop of new shoots. Bears, sandhill cranes—and people—harvest berries.

Now global warming is affecting the Arctic ice. It's hard to tell this when you see gigantic icebergs or watch a glacier calving and listen to the cracking of the ice breaking off. But Tim Flannery says that everything about North American ecological history suggests that the continent responds to the least bit of climate change with a more strenuous response than any other continent on earth.[2]

SHRUBS

Betulaceae

Dwarf Birch

Betula glandulosa, bog birch, glandular birch, resin birch, swamp birch, ground birch, dwarf birch, mountain birch, marsh birch, Arctic dwarf birch, shrub birch, scrub birch, buckbrush

This is one of the smallest birches; in a sheltered moist site it can reach 5 feet, but in dry, cold, exposed spots it may creep along the ground, growing to a mere 8 inches high. Plants that can be a normal shrub size in one place, and miniature in another, are typical of the Arctic and tundra regions. The farther north the dwarf birch grows, the more likely it will be to clone itself, since there's only a small chance seeds will fall on the ideal site. The birch hugs the ground and forms a mat of roots and laterals running under the moss layer which pop up farther away. Although it looks like any other birch in its overall appearance, it is not to be confused with *Betula pumila* var. *glandulifera*, also known as **swamp birch**, a taller deciduous shrub many miles south along the treeline. *B. glandulosa* is similar to *B. nana*, also called **dwarf birch** (in fact, some taxonomists lump the two together), but *B. nana* is even smaller and has no resinous glands.

B. glandulosa has smooth, reddish-brown bark and oval, leathery, toothed leaves about 1 inch long. It is a monoecious plant with male and female catkins appearing on the same plant in spring. The males hang down and fall off after shedding their pollen; the females look a bit like very small cones and sit upright on the branches. In late summer, the wind carries off the winged seed produced by the mature catkins. In autumn, the birches are a flame of gold.

Birch twigs with their yellow resin glands are extremely flammable, and tundra fires that rejuvenate the cottongrass tussocks eliminate dwarf birch.[3]

The peoples of the tundra found many uses for dwarf or bog birch, and considered it lucky. The Dena'ina of Alaska sometimes strewed branches of the bog birch behind them on a trail to keep away bad luck. When carrying meat, they placed birch branches on their backs to prevent the dripping blood from staining

their clothes.[4] The Chipewyan used birch twigs to treat bleeding cuts by chewing the twigs and then applying them to a wound. Bog birch leaves when boiled up produced a yellow dye, and the branches a light brown dye. Bundles of twigs were used to make brooms. Merritt L. Fernald noticed the Inuit on the shores of Hudson Bay near Depot Island using *B. glandulosa* "as a matting between their bedding and the snow."[5]

Range: **interior of Alaska and British Columbia, across northern Canada to Labrador; may be found farther south in wetland areas, from northern California to Maine.**

Salicaceae

Arctic Willow
Salix arctica, creeping willow

This captivating shrub is one of the rare woody plants of the High Arctic, although it is abundant in the Low Arctic. *Salix arctica* creeps through both moist, sheltered sites and dry, exposed ones. Its stems may be upright or may lie flat on the ground, and the hairy, leathery leaves are oval, 1 to 3 inches long, with male and female catkins appearing on separate plants. The male catkins have transparent hairs that stick out; the female catkins are more compact. Both secrete a nectar which attracts insects.

E. C. Pielou, in her book on the Arctic, explains the roles of male and female catkins:

Although it lacks colorful petals, its catkins (both sexes) are attractive to pollinating insects (including bumblebees) because they secrete plenty of nectar. . . . The catkins warm up in the sun, possibly

Salix arctica, Arctic willow

slightly. It is conjectured that this may result from adaptations that enable female plants to crowd out the males competing with them for space at the better sites; the females need good soil, as they have to nourish ripening seeds.[6]

John Burroughs, writing about the Harriman expedition in 1899, said:

Gustavus Peninsula seems to be a recent deposit of the glaciers and our experts thought it not much over a century old. The botanists here found a good illustration of the successive steps nature takes in foresting or reforesting the land—how she creeps before she walks. The first shrub is a small creeping willow that look like a kind of "pusley." Then comes a larger willow, less creeping, then two or more other species that become quite large upright bushes; then follow the alders and with them various herbaceous plants and grasses, till finally the spruce comes in and takes possession of the land.[7]

Willow bark contains salicin, a precursor to salicylic acid, the substance used to make aspirin. Native peoples in Alaska chewed on willow twigs when they suffered from mouth sores. A cough syrup is made from willow buds, and anyone stung by a bee grabbed a handful of willow leaves, chewed them, and spat them onto the sting to help soothe the pain.

The Dena'ina fulfill many of their daily needs with various parts of the willow: The inner bark is removed when sap is running and used to make string; ice scoops are made from willow cambium; young, long stems make fish hangers, fish weirs, animal snares, and twine for lashing wood together. Those pliable stems can also be waved in the air and the high-pitched sound is said to scare away wolves. When the

because their "fur" of clear transparent hairs functions like greenhouse windows. In sunshine, female catkins warm to 4° or 5° above ambient temperatures. . . . Studies in Devon Island show that male and female plants do best in different habitats. In moist, fertile, sheltered sites, females greatly outnumber males; in dry, exposed habitats, males outnumber females, though only

mosquitoes are biting during the short summer months, a screen made of young branches may be woven to protect a baby's face. They were used for thatching, boiled to make a yellow dye, and just for pure pleasure the young leaves were eaten and the sap sucked from the stem.[8]

Range: **Alaska, across northern Canada to Labrador, northern United States, Idaho, Montana, Oregon, and Washington.**

Ericaceae

Lowbush Cranberry

Vaccinium vitis-idaea var. *minus*, mountain cranberry, lingonberry, lingenberry, mountainberry, Alpine cranberry, rock cranberry, cowberry, partridgeberry, foxberry

Vaccinium vitis-idaea **var.** *minus*
is a creeping evergreen shrub that seldom grows more than 6 inches high. It forms large, dense colonies or mats from branching rhizomes. The leathery oval leaves are about ¾ inch long, dark green and shiny on top, pale green with black glandular dots underneath. The white or pale pink bell-shaped flowers bloom between May and July on stalks at the ends of twigs, and may appear singly or in clusters. It is self-pollinating or pollinated by bees. The edible berries are red and juicy. It grows in boreal pine forests and tundra regions, and can be found in most habitats: bogs, barrens, and dry woods.

Range: **Alaska, all Canadian provinces and territories, and the New England states.**

Cranberries and lingonberries are often compared with each other, since both have a sharp

taste when raw and are more palatable when cooked. Long before anyone even speculated about a cranberry industry, a lingonberry industry thrived in Canada. By the late 1700s, the Hudson's Bay Company recorded considerable quantities of lingonberries being sent back to England.[9] Today, there's a move to revive the lingonberry industry.

Lingonberries have long been used in pies, jams, and jellies, but can also be squeezed to make juice or even wine. The juice is full of vitamin C, helps with digestion, and can ease heartburn or an upset stomach. The species

Vaccinium vitis-idaea var. *minus*, lowbush cranberry

Vaccinium vitis-idaea var. *minus* grows taller (1 foot or more), while the European form of this plant, *V. vitis-idaea* var. *majus*, blooms twice a season and is the one usually found in nurseries.

These plants were known by different names in different parts of the country. In Canada, they were most often known as "lowbush cranberry" or "mountain cranberry," but Nova Scotians call them "foxberries," Newfoundlanders call them "partridgeberries," and Swedes and Finns know them as "lingonberries." The Haida of British Columbia call the berries "dog-salmon eggs."

The berries were immensely useful because they could be stored in a cool place for great lengths of time. There was one drawback: they also come out at the same time as every bug known in the North, especially blackflies. But the berries can hang on to the branches well into winter and the Inuit used to pick the berries after the first frost and throughout the winter. In the spring, children pick any of the remaining berries as a special treat. These days the berries are picked and stored frozen in *siglauqs*, or ice cellars, dug right into the permafrost. These are the iceboxes of the North.

Ericaceae

Bearberry

Arctostaphylos uva-ursi, red bearberry, kinnikinnick, sagakomi, jackashepuck, mountain box, mountain tobacco, mountain crawberry, barren myrtle, Yukon holly, universe vine, mealberry, sandberry, hog cranberry, chipmunk's-apples

Arctostaphylos uva-ursi

grows to 4 inches as a creeping evergreen shrub with hairy bark and flexible branchlets that even-tually become smooth. Wherever the branches touch the soil, they put down roots. Plants can also form mats on rocky or sandy soil. The leathery, smooth, spatula-shaped leaves are dark green on top and lighter on the underside. They are small—¼ to ½ inch long—with white urn-shaped flowers tinted pink at the end of twigs in late spring. Bright red, mealy berries appear in late summer.

Range: Alaska and across northern Canada to Labrador, in mountainous areas as far south as California.

Bearberry got its name because bears like to eat the berries. The botanical name is redundant: *arcto* and *ursi* both mean "bear," and *staphylos* and *uva* both mean "grape." The first word is Greek, the second Latin. It is one of the circumpolar plants with a major reputation for relieving kidney irritations. Even as far back as Roman times people knew this plant and used it.

The seed is usually moved by animals, and unless there's been a fire or disturbance of the duff (the organic litter on the forest floor), there won't be enough light for the seeds to germinate. In that case they reproduce and rejuvenate from nodules, actually latent buds on the stems. They are usually successful because wherever they touch the ground, they develop roots (layering); there are also dormant buds on the stem base which allow the plant to survive fire. In open places, bearberry forms a spreading mat, but in shadier spots it grows upright. It prefers open sites and tends to languish if there is too much shade.

Bearberry is found in mixed forests because it can grow in coarse soil low in nutrients. In British Columbia, it is an indicator that drainage is so swift there is little water in the soil. It is an important food for animals, with year-round appeal. The berries spoil so slowly that what's left hang on frozen all winter, when elk polish them off. In spring moose browse on the new leaves. Black and grizzly bears eat the ripened fruit in autumn. In spite of this constant use by animals it is one of the few edible berries of the North (full of vitamins A and C) and highly prized. When people were out hunting, the frozen berries were considered a prime emergency food. The fruit is almost tasteless or very sour straight off the plant, but once cooked it tastes a bit like cranberries.

Bearberries can be confused with lingonberries, which have a much better flavor. Lingonberries have black dots on the bottom of the leaf, bearberries do not. Indigenous peoples from the Mi'kmaq of the East Coast to the Salish of British Columbia used bearberries in cooking, often with meat or fish. The Inupiaq and Dena'ina of Alaska stored them in oil or bear fat so they would be available during the winter. These berries were notorious for causing constipation, so the oil alleviated that somewhat. One Dena'ina elder from Upper Inlet remembers that several people who became stranded while traveling were saved from starvation when they found a cache of bearberries mixed with lard.[10]

Many different groups also smoked this plant, which was known as kinnikinnick (an Algonquian word for a plant that can be smoked). This is a bit confusing because many other plants and some mixtures were called by the same name. It is still smoked by many First Nations people. Sometimes, it is mixed with tobacco to make a stronger smoke. Prince Maximilian, visiting Fort McKenzie in 1833, describes the smoking ritual:

When the Blackfeet smoke, they put a piece of dried earth or a round mass made of the filaments and pods of certain water plants on the ground, to rest the pipe on. Their tobacco con-

sists of the small, roundish, dried leaves of the sakakomi plant. When you visit an Indian in his tent, the pipe is immediately taken up and passed round in the company, each person handing it to his left-hand neighbour. The master of the tent often blows the smoke towards the sun and the earth; every one takes some puffs and hands it on.[11]

Hudson's Bay clerks carried sacks of this herb and mixed it with tobacco. The name "sagakomi" comes from *sac-à-commis* (literally "clerk's sac"). There were other mixes as well combined with bunchberry and salal.[12]

The leaves are filled with antiseptic and anti-inflammatory compounds; it's not surprising it made people feel pretty good. Brewed into a tea it was served as a diuretic which turns the urine bright green. Today this tea is widely touted in various forms as a help in urinary tract disease, and the list of help for internal woes goes on: it alleviates the symptoms of bladder stones, gallstones, and cystitis as well. Because of the antiseptic quality, washes made from bearberry leaves could be used to treat the skin; once it was powdered, it was applied to sores by Native Americans. Scandinavians use it to tan leather commercially and as a dye for subtle tones of gray or blue-green. This is not a plant to be taken casually and is never recommended for pregnant women in any form.

Bearberry is now being used in rejuvenation projects all across the North. In urban areas, bearberry may be planted as ground cover along roads or near parking lots. It grows well on disturbed ground, doesn't need much water in summer, and is salt-tolerant.

FOLLOWING PAGES: *Arctostaphylos uva-ursi,* bearberry

SEDGES, GRASSES, AND GRASSLIKE PLANTS

Brassicaceae

Scurvy Grass
Cochlearia officinalis, spoonwort

Scurvy was a threat to just about every sailor who set out for the New World. It causes swelling of the stomach and tongue, and bleeding gums. The legs become wobbly, making walking or work just about impossible. Without treatment, the victim usually dies. Scurvy grass contains vitamin C, which helped prevent the disease, and nineteenth-century European sailors ate it, or put it in beer and drank it.

In spite of the name, it doesn't look like a grass. "Spoonwort" is more apt since the fleshy leaves are shaped like small spoons (*cochlear* is Greek for "spoon"). It is a succulent, spreading from a central point on the ground, and grows up to 12 inches high. It is normally a biennial, though sometimes perennial, with arched or creeping stems and globular to elliptical fruit up to ¼ inch long. Tiny white flowers appear in midsummer, and later give way to little oval seedpods. It grows near water and has a reputation of smelling like celery; though I've never tried any, it's supposed to taste like horseradish.[13]

In 1616, William Baffin recorded that when his ship's company were suffering from scurvy, "going ashore on a little island, we found great abundance of the herb called scurvy grass which we boiled in beer, and drank thereof, using it also in salads with sorrel and orpin, which here grow in abundance. By means of these, and the blessings of God, all our men were in perfect health within the space of eight or nine days."[14]

The Northern indigenous peoples also ate the leaves of scurvy grass and it is still eaten in salads, although the taste takes a bit of getting used to according to Michael McRae, who went foraging with Janice Schofield: "I took a nibble. The flavor was similar to that of arugula, though a bit stronger. It had a bitter undertone that I couldn't place. I took another bite, then a third. All at once the back of my tongue was assaulted by an explosion of astringency—a hot, acrid flavor that suggested rubbing alcohol. . . . 'This is fabulous stuff, Jan, really delicious,' I gasped, turning to spit out a wad of green pulp."[15]

Wild-food enthusiasts, on the other hand, are made of sterner stuff. "Scurvy-grass leaves, like watercress, make a good salad. They are also delicious in sandwiches with a meat, cheese or egg filling, or alone with bread and butter."[16] Try it, if you happen to be in the Arctic and hungry for greens. But don't say you weren't warned.

Range: **up the Pacific Coast from Vancouver Island, along the Arctic and Atlantic Coasts to Labrador, Newfoundland, and along the Gulf of St. Lawrence.**

Brassicaceae

Whitlow Grass
Draba spp.

A whitlow is a sore under or around a finger- or toenail, and presumably this plant was used as a longtime remedy. Throughout the tundra, the brave little whitlow grasses are the perfect example of plants' surviving bitter cold and a very short growing season. There are so many species, they are hard to tell apart (unless you are a rock garden collector of the genus—they can

Draba spp., whitlow grass

spot a difference at 50 feet). In spite of the fact that they grow in exposed sites, beset by heavy winds, they still manage to produce seeds. The tight little buns hug the ground where the air is warmest and where cross-pollinating insects are likely to hang out. The flower buds open when they are still close to the rosette at ground level. As the days of summer lengthen, so does the flower stalk, raising the seed off the ground, to be picked up by a breeze or animals passing by.[17]

Whitlow grasses stay small, forming mats in simple rosettes about 4 inches high. The hairy oval leaves protect the plant from being buffeted about by winds. The yellow or white flowers with four petals are inconspicuous enough to demand a magnifying glass to examine their interior charms. They spread across the Arctic in July and August, then give way to seedpods, each containing up to 20 seeds. *Draba fladnizensis*, **Arctic draba**, grows on Alpine tundra on dry, open slopes above the timberline; the flowers are white.

Aldo Leopold wrote in 1949:

Draba *asks, and gets, but scant allowance of warmth and comfort; it subsists on the leavings of unwanted time and space. Botany books give it two or three lines, but never a plate or portrait. Sand too poor and sun too weak for bigger, better blooms are good enough for* Draba. *After all it is no spring flower, but only a postscript to a hope . . . just a small creature that does a small job quickly and well.*[18]

Range: **Alaska and across northern Canada; also found in the Rocky Mountains as far south as Colorado and Utah.**

Cyperaceae

Cotton Grass

Eriophorum spp., cotton sedge, Alaska cotton, swamp cotton, mouse nuts

---

Eriophorum spp.

range from 6 inches to 2½ feet high, with thin stems and equally slender, stiff, grasslike leaves. The flowers are round balls of white fluff up to 1 inch wide; some species, such as *E. angustifolium*, have several flowers on a stalk; others have only one.

Range: Alaska and across northern Canada to Labrador and Newfoundland; *E. angustifolium* grows in boggy areas in the West from British Columbia south to New Mexico, and in New England.

Cotton grasses are really sedges, not grasses. They cover thousands of acres of the Arctic with dense, firmly rooted tussocks, forming distinct areas known as "tussock tundra." Indeed, they may be the single most abundant plant in the Low Arctic.

From a distance, the tussock tundra looks smooth and flat. Up close, however, each clump is as large as a football and connected firmly to the ground. In between there's usually wet moss.[19] This is prime mosquito-breeding ground. Here's how Robert Marshall describes trying to walk through this landscape:

From Nolan we headed west for Pasco Pass. . . . There we had our first taste of arctic sedge tussocks. These curses are tufts mostly of cotton grass, which gradually build up out of the swamp, the younger plants growing out of the dead remains of the earlier ones. As they grow larger, they also grow wider so that they are much bigger on top

Eriophorum spp., cotton grass

than below, becoming more or less mushroom-shaped. They get to be eighteen inches high, some even higher. They are very top heavy, and when you step on them they are almost certain to bend over and pitch you off into the swamp. When you try to walk in the swamp you have to step over these high humps, and sometimes they grow so close together your foot catches in between. Three quarters of a mile of this seemed like five. . . . As we continued along North Fork [of the Koyukuk River in northern Alaska], we hit three miles of the worst sedge tussocks we had yet encountered. It was mount up and fall off and stumble and sink into mud above your shoe tops and drag the horses and, in general, wear yourself out.[20]

Cotton grass is the only type of tundra vegetation to allow the spread of wildfire. It is both dense and peaty; therefore fires can smolder for weeks with the burning usually in the outer layers of the hump. This removes dead stems and leaves pretty effectively and the whole tussock is rejuvenated as more nutrients become available and competition disappears.[21]

The lower stems and roots are edible, and mice collect the little bulb-shaped root, giving it the name "mouse nuts." "These little bulbs are good to eat, but hard to gather, so people let the mice do the work. [They] stamp around on the ground until they find a soft spot which is the mouse cache. Then they dig it open and get the mouse nuts. When you take mouse nuts, leave a piece of blubber to thank the mice for their work. To preserve mouse nuts, store in seal oil. To eat, pour boiling water over the mouse roots to clean the black hairs off them. They can be eaten in oil, or boiled. Either way they taste good."[22]

In the past, people tried to spin the cotton from the flowers to make cloth, and though the fibers were too brittle and tended to shred, they could be used as candlewicks.[23]

WILDFLOWERS

Asteraceae

Arctic Coltsfoot
Petasites frigidus, son-before-father, sweet coltsfoot, butterbur, coughwort, British tobacco, flower velure, pestilencewort, owl's-blanket, wolverine's-foot

Petasites frigidus
grows 12 to 18 inches tall. It is a perennial with underground rhizomes forming large colonies many yards square in bogs, meadows, and other moist locations. The triangular, lobed, hairy leaves are green on top and white on the bottom. They grow from the rhizomes on thick hollow stalks with reddish scales. The creamy-white flowers are clusters of 30 or more florets, with both male and female flowers appearing in spring usually on separate heads, before leaves appear. It has a celerylike smell.

Range: Alaska and across northern Canada, as far south as Queen Charlotte Islands.

Coltsfoot is sometimes called "son-before-father," because the flowers appear before the leaves, giving the plant a strange, vulnerable appearance as it rises out of the earth. The flowers provide some of the first available nectar for bees in spring. Once the flowers have died off, leaves appear and the plant quickly grows to 1 foot or more high during the long Northern days. The felty texture of the leaves might discourage some people from eating them, but the young leaves are recommended for salads by foraging enthusiasts. Coltsfoot syrups and teas are available from health food stores. Because it's a circumpo-

Petasites frigidus, Arctic coltsfoot

lar plant, various species were known in Europe, a fact that led to the string of fascinating common names.

Janice Schofield tells us how, when she had once a hacking cough while living alone in the bush and knew coltsfoot was recommended to soothe coughs, "I headed for the moist spruce woods by the river," she says "and collected coltsfoot to make a syrup. After two small doses, the coughing ceased and I was able to enjoy some regenerative sleep."[24]

The Dena'ina of inland Alaska call this plant "owl's-blanket," presumably because of the

shape of the leaves. Some Dena'ina soaked the roots in water and made a medicine for treating sore throats, chest ailments, and stomach ulcers. The native peoples of Washington state used coltsfoot leaves to cover berries as they cooked them in steaming pits,[25] and the Alaskan peoples used the leaves to cover stored food. Coltsfoot leaves may also be dried and smoked like tobacco. Once upon a time, smoking coltsfoot was recommended as a treatment for asthma.

The leaves are boiled to produce a yellow-green dye for wool. Apart from being eaten raw, the young leaves are steamed or sautéed. An infusion of the dried leaves was used as a cold remedy by the Inupiat. Native Alaskans and Inuit used the seed heads as mattress stuffing, along with goose and duck feathers. The leaves were even used in place of chewing tobacco or as snuff.[26]

Boraginaceae

Alpine Forget-Me-Not
Myosotis alpestris

Myosotis alpestris

grows 6 to 15 inches tall from fibrous roots. It has narrow, lance-shaped, fuzzy, gray-green alternate leaves, unlike the garden variety forget-me-not, with its smoother, brighter green leaves. Tiny blue flowers with yellow centers cluster at the top of the stem. Each flower produces four seeds, separately packaged like little nuts. They can be found in Alpine meadows and along streambanks.

Range: Alaska, Yukon, Washington, Idaho, and Colorado.

State flower of Alaska.

Myosotis alpestris, Alpine forget-me-not

Silene acaulis, moss campion

The clear blue flowers of the Alpine forget-me-not can bloom at different times depending on the slightest difference in their exposure. *Myosotis* on south-facing slopes blossom in late May, while those in northern snow beds may wait until August. There are four members of this family and all have blue flowers, though occasionally, rarely, they may come out white or pink.

Myosotis means "mouse ear" and refers to the shape of the leaves. According to tradition, wearing the forget-me-not flower meant never to be forgotten by one's lover. The plant has an astringent quality, and a lotion made from it was used to soothe eye problems. It was also ground into powder and applied externally to wounds and nosebleeds.[27]

John Muir tells this story:

While wandering about the banks of these gold-besprinkled streams, I was so fortunate as to meet an interesting French-Canadian, an old coureur de bois, *who after a few minutes conversation invited me to accompany him to his gold mine on the head of Defot Creek. . . . Before entering it he threw down his burden and made haste to show me his favorite flower, a blue forget-me-not, a specimen of which he found within a few rods of the cabin, and proudly handed it to me with the finest respect, and telling its many charms and lifelong associations showed in every endearing look and touch and gesture that the tender little plant of the mountain wilderness was truly his best-loved darling.*[28]

Caryophyllaceae

Moss Campion
Silene acaulis, cushion pink

Silene acaulis
appears in cushions of tiny, close-packed, dark green leaves up to 2 feet in diameter. The bright

pink flowers lie flat against the cushion. It can be found in sandy or rocky soil in Arctic and Alpine environments, up to 8,000 feet above sea level.

Range: **Alaska and across northern Canada.**

Arctic plants make a fascinating study for field botanists. Consider moss campion, which tolerates freezing temperatures as low as −22°F and puts up with bouts of freezing all during the summer. Though temperatures all about them may fall, they maintain their internal temperatures at a constant level—they know how to keep themselves warm. This is partly due to large root areas and a very long taproot.

It is not really a moss, rather it's a mossy-looking cushion. *Acaulis* means "stalkless." When the moss campion blooms, it smothers its own needlelike leaves and looks like a bright pink sofa cushion. Butterflies and moths flock to the brilliant plant.

Although most tundra plants have been used at some time or another for emergency foods, moss campion isn't edible; it contains saponins. Because of its extraordinary beauty and connection with the North, however, it caught the eye of the U.S. Postal Service and ended up on a 29-cent stamp issued in 1992.

Brassicaceae

Parrya
Parrya nudicaulis, P. arctica, Parry's wallflower

Plant names are sometimes linked with an explorer, not because of his interest in botany but to mark a historic event. *Parrya*, a showy little plant, was named for Sir William Edward Parry (1790–1855), who led three expeditions to the Arctic in search of the Northwest Passage. In her book on Arctic explorers, Jeannette Mirsky writes:

> *When he was given the command [of the Arctic expedition of 1818], he was not thirty, though he had already been in the Navy fifteen years. He was more than a good sailor and an accomplished officer with a special interest in hydrography and nautical astronomy: he was a masterly navigator. His was the perfect combination of fearlessness and caution; he knew when to take risks and when to play it safe. He was an athlete whose hobby was to play the violin.*[29]

Certainly sounds like a man who deserved to be commemorated with a plant name.

Parrya nudicaulis grows in the western Arctic and Alpine regions; *P. arctica* grows in the east. The forms are variable, but both kinds have fragrant purple or white flowers. The 3-inch-long, dark green basal leaves are smooth or coarsely toothed on 8-inch stems. The seeds are encased in long thin seedpods.

The young leaves were eaten raw or stored in seal oil by the coastal Inuit. They also sometimes added the root to fish and meat stews as flavoring. Apparently it's another plant that tastes rather like horseradish.[30]

Range: **Alaska and across northern Canada, also in the Rocky Mountains.**

Cassiope tetragona, white heather

Ericaceae

White Heather

Cassiope tetragona, Arctic bell heather, Lapland cassiope

Cassiope tetragona

is a low, ground-covering plant, seldom more than 1 foot high, with dark evergreen overlapping leaves along the stem, giving the plant the look of braids. The small flowers are white and bell-shaped.

Range: Alaska and across northern Canada; also in Alpine environments in the Rocky Mountains.

The tiny bells of *Cassiope tetragona* resemble the flowers of lily of the valley. White heather grows in the High Arctic, in areas protected in winter by deep banks of snow.

White heather was used by native Alaskans as insulation for houses, along with mosses and lichens; the Inuktituk used it for tinder and fuel, green or frozen.[31]

Dr. Greg Henry of the Department of Geography, University of British Columbia, has studied the growth of *C. tetragona* as a way of determining historical climate data and gaining insight into the regional variations of climate in the North.[32] Other researchers compared living heather to subfossil heather dating from the fifteenth and sixteenth centuries, which has been found in Ellesmere Island, to learn about the onset of glaciation after about 1610.

Liliaceae

Wild Chives

Allium schoenoprasum var. *sibiricum*, oniongrass

Allium schoenoprasum var. *sibiricum* **grows as tall as 20 inches. The leaves are thin green hollow tubes that smell like onions, and the blooms are pink or purple, with many tiny flowers clustered at the top of the slender stem. The seed capsule has three lobes with black seeds inside.**

Range: **Alaska and across northern Canada to Labrador and Newfoundland, southwest as far as Oregon, and in northern New England.**

Many of us have kept a pot of chives, just to perk up the winter blahs, get a march on spring, and eat something fresh and green. Well, *A. schoenoprasum* var. *sibiricum* is the wild version of good old common domestic chives, and looks (and smells) much the same. Even the round flower clusters are similar. Wild chives grow from bulbs and can be found in open areas such as Alpine meadows, mountain slopes, and lake shores in boreal and tundra regions.

The Dena'ina of Alaska used both bulbs and leaves for food: the leaves for flavoring stews, soups, or alone as greens; the bulbs for flavoring, chopped and used like garlic. Leaves are usually gathered while tender, in early summer, then are preserved by drying, freezing, or packing in rock salt. To preserve them, the leaves and bulbs are chopped and placed in layers alternating with rock salt, which is washed off when they are eaten.[33]

The Dene people of Canada's north and the Inupiaq of Alaska ate this plant in exactly the same way with similar storage methods. The plant is so rich in iron, it was given to people suffering from anemia. The Chipewyan of northern Saskatchewan used to boil the bulbs with trout or other fish. The Wood Cree of east-central Saskatchewan ate the fresh leaves and used them as flavoring for fish.[34]

Wild chives can be used like cultivated chives in cooking, but be careful if you collect them. The mountain death camas (*Zygadenus elegans*) can be mistaken for wild chives by the unwary. The poisonous lily doesn't have the distinctive onionlike smell of chives and the flowers are very different (they are yellow or white). The mountain death camas contains a poison called zygadenine, which causes vomiting, leading usually to coma and death. The bulb is not the most dangerous part—animals and children have died after eating the leaves.

Liliaceae

Chocolate Lily

Fritillaria camschatcensis, black lily, skunk lily, rice lily, Indian rice, rice root, Northern rice root, Eskimo potato, black sarana, Kamchatka mission bells

Fritillaria camschatcensis **grows up to 18 inches tall from an underground bulb, with many overlapping scales. The bulb produces tiny ricelike bulblets for reproduction. The leaves sprout in two or three whorls spaced up the stem. The flowers are light brown to dark brown, each with six petals. Two to six flowers appear at the top of the stem. The lily grows in wetlands and meadows.**

Range: **Alaska south to northern California.**

Fritillaria camschatcensis, chocolate lily

rather unpleasant smell. The peoples of Alaska pounded the bulbs to produce a ricelike flour.

The Alaska Dena'ina use the bulbs of chocolate lily for food by breaking them apart, soaking them in one or more changes of water. Once soaked, they are boiled and eaten, often with lard or oil. They are put in soups or stews and are eaten raw with fish eggs. They are sometimes preserved for winter or preserved in oil, the latter being more common. Anthropologist Cornelius Osgood reported the preparation to be 12 hours of soaking, one hour of boiling, and mixing with any kind of oil before anyone would eat them.[36]

Camschatcensis refers to Kamchatka, a peninsula on the Bering Sea, where David Nelson, botanist on the *Resolution,* saw local people eating the bulbs in 1778.

Papaveraceae

Arctic Poppy
Papaver radicatum, P. lapponicum

Fritillaria species are related to both onions and garlic; they have the same scent, only a good deal stronger. Where this plant differs is in its stunning, almost black bloom with six petals. The bulbs with their bulblets resemble tight clusters of rice and contain plenty of starch. "*Fritillaria* bulbs have been an important food in Indian life in the Northwest," Janice Schofield says. "Traditionally, bulbs were harvested in autumn by women while the men were out hunting geese."[35] The bulbs can be eaten raw, but they are mainly boiled, dried, or preserved in oil. They taste bitter if they have not been thoroughly soaked, and the whole plant has a strong,

Papaver radicatum
grows up to 6 inches tall. The flowers are bright yellow or white. There are two subspecies: subsp. *radicatum* has short, crowded, blue-green, lance-shaped leaves on elegant stems, large sulfur yellow flowers, and an oval capsule; subsp. *polare* has stems covered in densely brown or black hairs, pale yellow, white, or pinkish-yellow petals, and prefers damper conditions.

**   *Range:* circumpolar, Alaska, Yukon.**
**   Territorial flower of Nunavut.**

Arctic poppies fluttering in the breeze, scattered across the Arctic meadows, is a heart-stopping sight. It is not just the lavish quality nature bestows on remote places, it's the fact that some-

Papaver radicatum, Arctic poppy

thing so delicate does so well in this harsh environment. The flower is heliotropic—turning its face to follow the sun. This warms up the interior of the bloom and makes its golden heart a cozy place for insects to bask.

The subtle differences between species is based on the shape of the seed capsule, the number of rays on the bloom, and the shape of the leaves.

E. C. Pielou, in *A Naturalist's Guide to the Arctic*, notes that in sunny areas, the poppies tend to be yellow, because yellow flowers absorb sunlight better, but in areas where it is cloudy most of the time, the poppies are white, since the plants cannot spare the photosynthetic energy necessary to make the yellow pigment.[37] If a flower has streaks or blotches of green, this doesn't mean it's a separate species, it's just been bruised inside the tight calyx. As soon as the snow melts, the poppy spills its seeds and germinates in the remaining moisture.

The indigenous peoples of inland Alaska say that people with power can talk to a certain flower and make the weather change; Priscilla Russell Kari suggests this poppy may be that flower.[38]

Polygonaceae

Mountain Sorrel
Oxyria digyna, sour grass

Mountain sorrel, which is common throughout the Arctic, can be found in a variety of environments, from sheltered wetlands, where it grows up to 12 inches high, to exposed rocky outcrops, where it may appear as a tiny plant only 1 inch high. The kidney-shaped basal leaves are edible. They start out dark green and turn red or purple as they age. The leaves are clustered at the base of longish stalks. The flowers grow in dark red or reddish-green spikes, with four sepals per flower, producing bright red fruit with winged seeds.

Mountain sorrel is one of the most important plants of the Arctic, for both humans and animals. The leaves are eaten by caribou, geese, musk-oxen, and the rhizomes by Arctic hares and lemmings. The Inuit boil the leaves with those of other edible plants, allow it to ferment in sealskin, store it frozen underground, then mix it with seal oil and serve it as a dressing for meat.[39] In some places the tradition was to cook it in sugar.

E. C. Pielou says:

The plant is rather like ordinary sorrel in form; its flowers are bright red and its kidney-shaped leaves turn purple in fall. They're more tasty and refreshing than those of ordinary sorrel; the trouble is that in the High Arctic the plants are so dwarfed and the leaves so tiny, and the plants so widely scattered, it's difficult to pick even a handful. The Inuit as a whole haven't much use for plants: they relied on caribou, muskoxen, seals and the occasional whale for food, fuel (seal oil), plus bones for implements and weapons. But it deserves a mention because it saved the early Franklin expeditions to Cape Turnagain from certain starvation.[40]

A. E. Porsild in his book about emergency foods of the Arctic says: "The succulent, juicy leaves and stems are edible. When raw they are somewhat acid but very refreshing, when cooked their flavour and appearance resemble spinach.

A very pleasant dish, resembling stewed rhubarb, may be prepared from the sweetened juice thickened with a small quantity of flour."[41] That sharp taste comes from oxalic acid, which can be poisonous in large quantities. However, the leaves contain large amounts of vitamins A, B, and C.

Sorrel is a wonderful herb and Turner and Szczawinski suggest making sorrel soup by cooking the sorrel, puréeing it, and adding sour cream.[42]

Range: **throughout the Arctic and boreal regions; also in Nova Scotia and northern New Hampshire.**

Rosaceae

Arctic Dryad
Dryas integrifolia, mountain avens, white dryad, entire-leaved avens

Dryas integrifolia

is a mat-forming shrublet that grows from a long taproot. The small leaves are shaped like arrowheads about 1 inch long, a shiny dark green on the upper side, white and woolly on the underside. The flowers appear in early summer on 4-inch stalks looking like wild roses, with 7 to 10 creamy-white petals. After it flowers, the seed heads appear: at first the plumes attached to the seeds are twisted in a spiral, then they uncurl slowly.

Range: **Alaska, across northern Canada to Labrador, and in the Canadian Rockies.**

The Arctic dryad, *Dryas integrifolia*, is found throughout the tundra regions on gravelly or

Dryas integrifolia, Arctic dryad

all winter to be ready to photosynthesize at the first glimmer of sun in the earliest part of spring.

The simple roselike flowers (as in wild roses, rather than cultivated garden hybrids) are charming, and give way to distinctive seed heads twisted into a tight spiral. As they dry, says E. C. Pielou, "They spread wide like a feather duster."[43] Other writers such as D. B. O. Savile have said the saucer-shaped flowers have another special quality, similar to the Arctic poppy: like a miniature satellite dish, the flower focuses the rays of the sun toward the center where the ovaries are, and turns to track the sun through the sky. This works so well that insects bask in the concentrated warmth generated in the center of the flower.[44] Pollination may be accomplished by syrphid flies, but the plant can also self-pollinate or spread by its roots. One genetic individual (clone) can be several hundred years old.

Pielou says calling this plant "mountain avens" just confuses it with *Geum*, which is more properly called avens and is a different genus. There are three species of dryads, creeping prostrate subshrubs with evergreen leaves. **Yellow dryad** (*D. drummondii*) has pale yellow flowers, *D. octopetala* has white flowers and scalloped leaves, while *D. integrifolia* has smooth-edged leaves.

Before the eastern Northwest Territories were incorporated as Nunavut, it was recommended in 1957 by A. E. Porsild, the chief botanist of Canada's National Museum and author of the *Illustrated Flora of the Canadian Arctic*, that Arctic dryad be the official territorial flower. Common to the western Arctic, as he pointed out, when it flowers it gives the whole landscape a creamy color.

Like *Cassiope tetragona*, *D. integrifolia* has been studied by climatologists to develop historical climate data. The flowers and seed plumes have also been used in the past to produce a green dye.

recently disturbed sites. Its long taproots help anchor gravel or soil. The plants hug the ground to keep out of the way of the fierce, drying winds. As a wind whips up, the woolly leaves curl inward to retain moisture. The leaves stay green

Rosaceae

Arctic Raspberry

Rubus acaulis syn. *R. arcticus* subsp. *acaulis*, dwarf nagoonberry, stemless raspberry

Canada is not exactly associated with raspberries, but it has 25 indigenous species of *Rubus*, most of them with edible berries.

Arctic raspberry, R. acaulis

is a low shrub up to 1 foot high, with three-part alternate leaves like those of garden raspberries, but without the familiar prickles. The plant grows from rhizomes, close to the ground, and bears bright pink flowers with narrow petals in early summer, followed by small, delicious raspberries in late summer. It can be found in wet woods, meadows, and peatland areas.

Range: Alaska, Yukon, northern British Columbia, across to Labrador and Newfoundland.

Wild raspberries have a sweetness of their own. I remember picking them as a child in Labrador. We used to sell our harvest for jam making, which, along with blueberry picking, was the August pastime. It was a hot sweaty job and we earned 25 cents a quart in those days. As I was making my way around one day, very intent on what I was doing, I came across a bear doing pretty much the same sort of thing. I dropped my berry bucket and ran. I was lucky there were other people close by who started to bang their tins to scare him off. Certainly I learned one lesson: don't go berry picking alone, although I'm sure the bear was just as terrified of me, completely decked out as I was in a huge hat with bug netting, gloves, and boots.

The Inupiat name for the berries is *aqpiks*.

The Dena'ina call them "nagoonberries." The native peoples of the North used to mix them with oil and sugar and eat them after meals. In winter they might be mixed with crowberries and dock leaves and stored in wooden kegs in cold cellars.[45] Lieutenant L. A. Zagorskin (1808–1890), a nineteenth-century Russian explorer in Alaska, noted: "The natives for winter supply themselves with the cranberries, cloudberry and blueberries; the *Rubus arcticus* has a very pleasant taste and is picked up only by the small children; the red and black currants are left for the bears only."[46]

The species is closely related to salmonberry and **cloudberry** (*R. chamaemorus*), which are also well-known fruits in the North. They were stored fresh in buried baskets by the Slave, Cree, and Chipewyan; large quantities were formerly picked at Lake Athabasca and Ile à la Crosse.[47]

Saxifragaceae

Northern Grass-of-Parnassus

Parnassia palustris, bog star

Parnassia palustris

grows from a thick rhizome and has leathery, 1-inch-long, rounded leaves. They grow from the base of each stem, which can be up to 1 foot high with one leaf (bract) growing halfway up the stalk. Each stem bears a single bisexual flower, white, green-veined, with five petals, about 1 inch across. It has a hint of fragrance. The flowers appear in mid- to late summer, followed by crammed seed capsules about ½ inch long. *P. palustris* creates small colonies and can be found in wet meadows, ditches, and bogs. A smaller relative, *P. kotzebuei*, is also found in the Arctic.

Range: Alaska, northern and eastern Canada.

Parnassia palustris, Northern grass-of-Parnassus

Saxifragaceae

Parnassia species are not grasses and have no connection to the mountain in Greece called Parnassus. The flowers are designed to attract flies. There are little green stripes along the white petals which steer pollinating insects into the central nectary, much like the beckoning of neon lights or the lights along a landing strip.[48] The plant also has false nectaries: the long stamens have what looks like a drop of nectar glistening at their tips. It's strictly a come-on, but it gets the insects into the right place and they go on to discover the real nectaries close to the base of the filaments.[49]

Saxifrage
Saxifraga spp.

Saxifraga means "rock breaker," named supposedly for this plant's ability to treat kidney stones, according to the doctrine of signatures, and because these plants can be found on exposed rocks, in crevices on cliffs, and in gravelly areas with strong taproots and many side roots securing them to rocky surfaces.[50] Many different species are found in the Arctic. The shape and size of the plant varies considerably, but generally they all have flowers with five petals and 10 stamens surrounding a two-part seed capsule shaped like two little horns.

S. oppositifolia (**purple saxifrage**) forms a cushion and has tiny overlapping gray-green leaves looking much like scales. Soon after the snow melts in spring, the purple flowers appear on short stalks all over the cushion. This is one of the earliest flowers to appear in the Arctic spring.

In *S. caespitosa* (**tufted saxifrage**), the creamy-white flowers appear on short stems above three-lobed leaves growing in tufts. *S. bronchialis* (**spotted saxifrage**) has white petals with orange and red spots, and *S. flagellaris* (**spider plant**) has yellow flowers and spreads by runners, like a strawberry plant.

Saxifrage leaves are edible and contain vitamin C. Janice Schofield recommends adding the chopped leaves to omelettes, stews, quiches, and casseroles. The Inuit are said to eat the blossoms of purple saxifrage. Gold and green dyes are still obtained from saxifrage.

The beauty of saxifrage appealed to Henry David Thoreau, who wrote in 1853: "The saxifrage is beginning to be abundant, elevating its flowers somewhat, pure trustful white amid its pretty notched and reddish cup of leaves . . . a response from earth to the increased light of the year."[51]

Range: **Alaska and throughout the Canadian Arctic, across northern Quebec, Labrador, and Newfoundland; also in the Rocky Mountains as far south as Wyoming.**

Pedicularis lanata, woolly lousewort

Scrophulariaceae

Woolly Lousewort

Pedicularis lanata, Arctic lousewort,
bumblebee plant, Indian warrior, fernweed

The inimitable Janice Schofield says: "The first lousewort I encountered was huddled in its woolly overcoat on Barrow's tundra, while I crouched in a warm parka marveling at our similar adaptations to the environment . . . this herb . . . sports a fuzzy layer of insulation until more clement weather arrives. As the sun beckons, spikes of rose-pink blossoms burst through the gray wool and brighten the landscape."[52]

All louseworts have one thing in common: they are partial parasites, and they have suckerlike attachments on their roots. They can fasten these attachments on to the roots of neighboring plants such as the dwarf birch, from which they draw moisture and nutrients. This characteristic makes louseworts almost impossible to transplant.

Lousewort grows 6 to 10 inches high from a yellow taproot, usually in dry, stony areas. The fernlike divided leaves, made up of many serrated leaflets, are in a cluster at the base of the stem. The woolly look comes from tiny white hairs covering the plant (hence the name *lanata*, meaning "woolly"). The hairs serve to keep the plant warm and help conserve moisture. The bright pink flowers emerge in a tight cluster from their woolly enclave in summer. Each small flower has three petals forming a lower platform and darker pink petals forming an upper hood. The flowers are designed to attract bees, who land on the platform and brush against the hood.

The leaves, stems, and roots are edible, and the taproot is a little like a carrot. The Inupiaq pick the flower tops in June when they are still small and closed, place them in a barrel, cover them with water, and ferment them like sauerkraut. Inuit children sometimes suck the nectar from the flowers and their elders dig out the roots as vegetables.[53] The tops are a browse for caribou.

Even though they are edible, some writers counsel against picking them. The plant has sedative properties, and Kuhnlein and Turner point out that louseworts, among the few wild vegetables to be found in the Far North, are vulnerable to harvesting. Some are rare or endangered, so they should not be used unless it's a dire emergency.

Range: **Alaska, northern Canada to northern Quebec, northern Rocky Mountains.**

MOSSES AND CLUBMOSSES

Moss and clubmoss often get muddled in people's minds because there has been a long tradition of calling one by the other's name, which indicates how widely they were used and how many subtle resemblances they have to one another. But each is special in its own way and has evolved methods of dealing with a large number of harsh environments. All these plants are worth looking at with a magnifying glass since much of their true beauty is hidden.

Mosses, the true mosses, have large colorless hollow cells which allow them to absorb water and hold it just like a sponge. They are a primitive plant with no roots, stems, or leaves.

Clubmosses look like mosses but are more closely related to ferns and horsetails than to the true mosses. They are actually primitive, miniature evergreens. More than 300 million years ago they were common and some grew 100 feet tall. They have roots, stems, and leaves but no flowers or seed. Reproduction takes place by spores or runners. There are two Arctic genera:

the *Lycopodium* has little conelike growths, and *Huperzia* has no cones.

Lichens may look dull but they are among some of the most fascinating of all flora. They are a combination of algae and fungi living together in symbiosis. The fungus anchors the alga to a surface, absorbs water and nutrients, and protects the alga from light. The alga contains the chlorophyll necessary for photosynthesis.

Lycopodiaceae

Mountain Clubmoss
Huperzia selago, shining clubmoss

Clubmosses are found in rocky areas of the tundra, and farther south in coniferous woods. Shining clubmoss is one of the oldest plants on earth. It has been around for about 300 million years. *Huperzia selago* (*selago* means "shining") grows right into the High Arctic. It has stiff little yellow-green branches standing about 2 to 3 inches high, with densely packed, tiny leaves. Indentations on the stem show the annual growth. Concealed spore cases are borne on the upper leaves, and when they are ripe they turn yellow or light brown.

These plants were never viewed as food, even in starvation situations, but the medicinal value was always apparent. The Dena'ina call the clubmosses "mouse's-tail" or "mouse's-tailbone." They placed the plants on the head as a remedy for headaches and boiled them to make a medicinal eyewash. The Nitinaht used the same method as a purgative, emetic, and gastrointestinal aid.[54] The absorbent moss was used to stanch a nosebleed—just inhale enough to stanch the blood; for probably the same reason the Potawatomi and Blackfeet dusted wounds with moss powder. The Montagnais blended clubmoss with hemlock bark to make a tea to reduce fever.[55]

Today, clubmoss is used in cures for eczema. They are also part of much we buy as Christmas greenery. Because the spores are rich in oil and are highly flammable, the first photographic cameras used spores of another clubmoss (*Lycopodium*) in their flash powder.[56]

Range: **Alaska, Yukon, British Columbia, south as far as Oregon.**

Bryaceae

Red Moss
Bryum cryophyllum

"A truly beautiful moss is Red Moss, *Bryum cryophyllum*, which forms great ruby-red cushions of an astonishing brightness. Sometimes it is found among cushions of apple-green mosses, making a dazzling mosaic. And sometimes it grows submerged in water, as do many other Arctic mosses. In late summer when Arctic flowers are over, vivid cushions of red moss may be the only splashes of color in the High Arctic tundra," says E. C. Pielou.[57]

Mosses took a different evolutionary path from flowering plants. The cells in moss contain only one set of chromosomes—all other flowering plants have two—and they have no vascular system whereby food and water can travel through internal structures in the plant. There are 18,000 species of moss, categorized as mosses, liverworts, and hornworts, all of which reproduce by spores. Water is crucial for the reproduction of mosses and they tend to grow well in wet places. They produce soft needlelike leaves attached to stalks but have growth spurts only when enough water is available to move through the moss's waxy cuticle. In spite of the need for water, they can withstand dry periods by simply going into dormancy. They can also be found on trunks of trees, in forest canopies, and on rock surfaces. Their importance in the Arctic ecosystem is to protect the permafrost by insulating it from drying out or thawing.

Here is a legend of the Passamaquoddy:

Long ago, before there were human beings, Koluskap and his brother Malsom lived together on the island of Oktomkuk. They were giants and both of them had great power. Because of his nature, Koluskap always tried to do things which would make life better for others. And because of his nature, Malsom always did things which made life difficult for everyone. When Koluskap made the rivers, he made them so that one bank flowed downstream and the other flowed upstream. That way, the people yet to come would find it easy to travel. But Malsom threw stones into the rivers, twisted their courses, and made them all flow downstream. Next Koluskap made all kinds of flowering and fruiting plants. He made beds of moss which would be soft places for people to sleep. But Malsom followed behind him and made plants which had thorns and plants which were poisonous, and he made the moss so moist that the people would grow cold if they tried to use it as a bed.[58]

Range: **throughout the Arctic.**

LICHENS

Lichens, another symbiotic coalition of fungi and algae, are among the most important plants of the Arctic, the tundra, and the boreal forest. Not only are they among the pioneer plants that move in after glaciation, fire, or any other dis-ruptive force of nature, they also provide food for many animals. As with so many pioneer plants, they make way for more advanced species to follow in their wake. They produce enzymes which, over time, make fissures in rock, and after a few thousands years, create soil. Then new plants get a chance to move in. The origin of the name "lichen" is the Greek word *leikhein*, meaning "to lick" or "to lick up," because of the way lichens "lick" their way across rocks and trees. They were given the name in 1700 by French scientist Joseph de Tournefort.[59]

Cladoniaceae

Reindeer Lichen

Cladina (formerly *Cladonia*) *rangiferina*, gray reindeer lichen, reindeer moss, caribou lichen, caribou moss

This is the best known and most easily identified of the Arctic lichens. It's often mistakenly called reindeer or caribou "moss" because it looks similar to moss. But it is a lichen, and one which grows faster than just about any other lichen—that is, so slowly even a small clump could be 100 years old. Its body is called the thallus. The podetium is a stalklike protuberance from the thallus. At first, the podetium grows steadily; after some years of growth it enters a 100-year phase of maintenance. Some parts of the plant grow while other parts of the plant die off at the same rate, after which the whole plant starts to degenerate. But each year it adds a new branch giving it the ability to reach a great age. It is sometimes possible to estimate the age of reindeer lichen fairly accurately, as E. C. Pielou points out, although older branches sometimes decompose, making calculation difficult.

Cladina rangiferina has a bushy, branching thallus, or body, which may be gray, greenish-gray, or yellowish-gray. They form clumps or mats made up of a large number of podetia. These lichens cover large areas of the Low Arctic in open areas, sub-Arctic forests, and bogs.

This lichen got its common name because reindeer and caribou like to graze on it. It's a circumpolar plant and in northern Europe it was once collected to feed livestock. The Inland Dena'ina used reindeer lichen for emergency food by smashing the dry plant and then boiling it or soaking it in hot water until it's soft. They ate it plain or mixed with berries, fish eggs, or fat. Sometimes they boiled it with caribou blood.[60] But it was strictly a hard-times food and usually given to dogs. The lichen may also be used to make a drink to help soothe tuberculosis and bleeding.

Between 1753 and 1761, when Peter Kalm was traveling through North America, he made northern forays and reported:

The reindeer moss grows plentifully in the woods around Quebec. M. Gauthier, and several other gentlemen, told me that the French, on their long journeys through the woods, on account of their fur trade with the Indians, sometimes boil this moss and drink the decoction for want of better food, when their provisions are at an end, and they say it is very nutritive. Several Frenchmen who have been in the Terra Labrador, where there are many reindeer (which the French and Indians here call cariboux), related that all the land there is in most places covered with this reindeer moss, so that the ground looks white as snow.[61]

Northern Inuit regard the partially digested reindeer lichen from the stomachs of newly killed musk-oxen and caribou as traditional winter survival food. This makes sense, since the cellulose which a human can't process is already on its way to breaking down in the animal's stomach. In Alaska they separate the lichen from the grass in the stomach and mix the lichen with oil. The word *teniyash* (which means "increase") is sung while the mixture is stirred to encourage it to rise and become light.[62]

Janice Schofield tells us:

[Lichens are not] a pop-in-your-mouth or toss-in-a-salad edible. They must be leached in a clean stream overnight, or boiled in several changes of water (preferably with baking soda added each time) to remove the acids which can cause intestinal irritation. Lichens, after leaching or parboiling, can be added to soups and stews as a thickener, boiled with fruits into a jelly, dried as a flour extender or substitute, simmered as a vegetable with wild game or fish, or cooked into a pudding or custard. In Iceland and Scandinavia, lichens are commercially harvested for a lichen powder that forms the basis of soups and desserts.[63]

The lichens' ability to absorb moisture from the air also means they can just as easily absorb contaminants, which are increasingly being found in air samples of the North. Radioactive materials from atmospheric nuclear testing concentrate in lichen tissue, which is then eaten by caribou. The Chernobyl disaster affected lichens in the Arctic and the caribou who ate them. More recently, core samples in ice have shown traces of aerosol deodorants which have drifted from the south, alerting us once again that all the systems on the planet are connected somehow.

Range: **Alaska and across northern Canada, in mountain areas as far south as California.**

Cladoniaceae

Cauliflower Lichen

Cladina stellaris, star reindeer lichen,
Northern reindeer lichen

This pale yellow-green, bushy, many-branched lichen forms clumps the size and color of cauliflower heads. At the top of each clump, four or five branches form a whorl around a central hole. It is pale yellow to almost white in open sites, darker gray in thick woods. It grows on rocks, soil, and occasionally on old wood and stumps.

Clumps of this lichen are soaked in glycerin, dyed, and used to simulate trees and shrubs in the miniature landscapes of model railway builders or scale models for architects. Most of the lichen used in this way comes from Scandinavia, but the same lichen grows in North American Arctic regions too. With the growth of model railroad hobbyists, this demand will no doubt increase. Anyone planning to muscle in on the business, however, should be warned to wear gloves when picking this lichen. It contains usnic acid, which can cause a form of contact dermatitis in some people. The symptoms are itchiness, scaliness, or reddened skin.

Range: **Alaska, throughout Canada, and the northern and eastern United States.**

Rhizocarpaceae

Map Lichen

Rhizocarpon geographicum

One of several species known as "map lichen," this is common in the Arctic and looks like an irregular black patch forming a crust on rocks. The hypothallus (bottom layer) is black, and the upper thallus looks like yellow polka dots on the black background. The irregular outline of the plant gives it its common name. It grows best on acid rocks such as granite, sandstone, or quartzite.

Scientists and archeologists use crust lichens to date Arctic sites. It grows very slowly in the Arctic, perhaps only about 1/1,000 to 1/50 inch a year, and may take hundreds of years to form a patch the size of a human hand. The technique of using lichens to estimate dates is called lichenometry and it is used in dating moraines left after retreating glaciers, and therefore in determining when the Ice Age ended.

There is some controversy about just how old some of the map lichens might be. There are estimates that they might be 4,500 years old,[64] while other scientists put them at more than 9,000 years old. Either way they are among the oldest living things on earth.[65] But as E. C. Pielou points out, the dating of lichens is highly speculative because it assumes growth remains constant over the centuries.

Range: **throughout the Arctic.**

Teloschistaceae

Jewel Lichen

Xanthoria elegans

This bright orange lichen grows on bird droppings on rocks and cliffs. The orange is of such neon brilliance biologists and bird watchers use it to locate the nests of eagles, falcons, and cliff-dwellings seabirds. It sticks close to the rock, with flat leaflike branches radiating from a central point. The thallus is orange to reddish-orange and the lobes are irregularly branched. Some lobes form rosettes. It usually grows on rocks rich in carbonate such as limestone, although it can grow on any kind of rock covered with bird droppings, on old bones, even on old wood.

Xanthoria elegans is also used to date sites of interest to scientists and archeologists. It grows very slowly: Jewel lichen formed on the gravestones of members of the third Franklin expedition who died in 1846 on Beechy Island, southwest of Devon Island. It is still less than 2 inches in diameter.

Range: **throughout the Arctic.**

Umbilicariaceae

Rock Tripe
Umbilicaria spp., tripe de roche

Umbilicaria spp.
have a black, dull thallus, shaped like a saucer with a raised outer edge attached to the underlying rock in its center.

Range: **throughout the Arctic.**

In the lichen literature, every time rock tripe comes up, so do epithets as to its disgusting qualities. There are several different species of rock tripe, all black and unappetizing-looking. When the thallus is dry, it is brittle and grayish to deep brown; when wet, it is limp, rubbery, blackish to dark green.[66]

This ugly growth has nevertheless saved many travelers in the Far North. The early Franklin expedition (1819–1822) was rescued from starvation by rock tripe.

Tripe-de-roche became their only food—an unpalatable, noxious weed that produced in many

an enfeebling diarrhea. For days they plodded on, twenty starving men, abandoning more and more of their equipment, desperately burning the canoes for a little warmth. . . . [Several men died.] More would have if help had not arrived. Three Indians sent by [George] Back [who had gone on ahead] came to their aid and with infinite tenderness fed and cared for the white men, whose emaciated bodies, sunken eyes and sepulchral voices told a pitiable story that needs no translating. . . . Had Back not chanced on those Indians, he must surely have perished, for his strength was fast going and his food consisted of a pair of leather trousers, a gun-cover, an old shoe, and a handful of tripe-de-roche![67]

It was, of course, a well-known survival food long before the Franklin expedition. A Jesuit named Rale, writing to his brother, mentions rock tripe being eaten by the Iroquois. He reports it was made into a "black and disagreeable porridge."[68] Claude-Jean Allouez, in *Journey to Lake Superior* (1665), reports: "We were forced to accustom ourselves to eat a certain moss growing upon the rocks. It is a sort of shell-shaped leaf which is always covered with caterpillars and spiders; and which, on being boiled, furnishes an insipid soup, black and viscous, that rather serves to ward off death than to impart life."[69]

At least one explorer, however, appeared to enjoy the stuff.

There is a black, hard, crumply moss, that grows on the rocks and large stones in those parts [the Barren Grounds], which is of infinite service to the natives, as it sometimes furnishes them with a temporary subsistence, when no animal food can be procured. This moss, when boiled, turns to a gummy consistence, and is more clammy in the mouth than sago; it may, by adding either moss or water, be made to almost any consistence. It is so palatable, that all who taste it generally grow fond of it. It is remarkably good and pleasing when used to thicken any kind of broth, but it is generally most esteemed when boiled in fish-liquor.[70]

This is most unusual. Most people consider rock tripe revolting. E. C. Pielou calls it "disgusting stuff."

Native Americans certainly used rock tripe as survival food. The Inuit considered it an emergency food, but one never to be eaten over a long period of time, because it was eventually harmful. It needs to be boiled to reduce the level of acid, which causes stomach upset. The acid, however, is considered effective against tapeworms, and the Chipewyan once chewed it or made a syrup from it to expel worms. The Wood Cree boiled it with fish broth, and gave it to the sick because it did not upset the stomach.[71] In Scandinavia rock tripe was used to produce a powder to form the base of easily digested soups and jellies.[72] In Scotland, where it also grows, rock tripe was used to make a purple dye.

Usneaceae

Beard Lichen
Usnea spp., old-man's-beard

Usnea spp.

are gray-green lichens growing in hairlike clumps up to 2 feet long, hanging down from tree branches.

Range: eastern Arctic and boreal regions, and in the southern ranges of the Rocky Mountains.

Usnea spp., beard lichen

The beard lichen gets its name from its wispy flowing strands and does resemble an ancient beard. This fascinating lichen grows around the world and has been used for thousands of years. There are notations about it in the annals of ancient Egypt, Greece, and China as well as Rome. People in all of these places used it medicinally because it is a natural antibiotic, though by the eighteenth century it was believed that simply holding beard lichen in the hand could stop bleeding.

Some species possess a white inner cord which provides tensile strength, and may be a source of sugarlike substances. White-corded species *(Usnea hirta)* are believed to be of medicinal use. They are mainly small, and found on Douglas fir trees.[73] Some species reproduce when a piece of the lichen thallus breaks off and is carried elsewhere by animals brushing against it.

Beard lichen is important winter food for animals because it is not covered up by snow. *Usnea* is edible if cooked, and is used as a leavening agent or a thickener for stews. The Inland Dena'ina of Alaska make an emergency food or camp food by boiling them up and eating them with fish, berries, or grease. The Dena'ina say they taste different depending on the tree they grow on; some are said to have a sweet taste.[74] The Dena'ina call it *ch'vala andazi*, which means "spruce hair."

Because beard lichen is dry and brittle, it makes good tinder for starting fires. The Haida used this and black tree lichen *(Bryoria fremontii)* to strain hot pitch to remove impurities before it was used as a medicine. It proved a comfortable bedding when camping. The Haisla and other coastal peoples made pigments from certain black and yellow lichens growing on rocks and trees. They mixed the powdered lichen with salmon eggs to make a paint used on spoons, bowls, and totem poles. Some lichens were used to wipe the slime off fish and for protecting food in earth ovens. They had their use in ritual as well: the Secwepemc and Nuxalk used them as false whiskers and artificial hair for decorating dance masks and for children's masquerades.

The Wood Cree of Saskatchewan used

Usnea to stop nosebleeds, by stuffing it into the sufferer's nostrils.[75] The Hanaksiala used it as a mattress in seasonal camps; Nitinaht used it for baby diapers and as sanitary napkins. Makah used any *Usnea* for yellow dye.[76]

Scientific experimentation beginning in 1944 discovered that *Usnea*, which contains usnic acid, is antifungal, antibacterial, and antiparasitic. It can even boost the immune system and may be used in treatment for chronic lung infections.[77] *Usnea* tincture is used to treat bacterial infections affecting the throat and lungs, as well as certain skin problems, from diaper rash to athlete's foot.

FUNGI

Fungi seem to be insignificant in the world of flora, but they are crucial to the health of the forest. One could not exist without the other. As the climate changed over millions of years, so did trees. In order to survive, mycorrhizae must develop. These are fungi that attach themselves to tree roots. They absorb nutrients and contribute to enriching the soil, protect the tree from toxins and the invasion of pests, and perhaps even more important, they help fix nitrogen for the next generation of plants. We are now just beginning to understand that there is a succession in fungi, just as there is in the whole forest. This means that trees need shorter-lived yeasts and molds to use up sugars at first, then fungi that specialize in far more complex carbohydrates. Once a tree dies, it becomes even more valuable as other fungi, insects, and animals move in to live off its bounty while aiding its decomposition.

Polyporaceae

Bracket Fungus
Polyporus oregonensis

Polyporus oregonensis
grows on living or dead tree trunks. The mycelium penetrates the wood and makes it crumbly. It may grow up to 1 foot across, with a hard upper surface of brown or gray and a white lower surface.
Range: throughout North America.

There is a lot of confusion about different fungi. Until recently, ethnographers and contemporary aboriginal people didn't always distinguish the different species of bracket fungi other than by specifying their host tree (birch fungus, pine fungus, and so on). Recent scientific studies using DNA have now identified many different varieties.

A number of First Nations peoples used the corky inner tissue of some species for "slow matches." When ignited, the tissue smolders for hours. A fire can be kindled simply by blowing on the smoldering fungus. Aboriginal peoples of the Pacific Northwest used clamshells, cedar bark, or rolls of birch bark to contain the smoldering fungus for transport. The Nlaka'pamux used a type of fungus from cottonwood trees to tan buckskin, or any bracket fungus if the cottonwood was unavailable. Annie York describes the process:

> *They chopped the fungus and mixed it with deer brains and water. Then they soaked the hide, already stripped of hair, in this mixture for a few hours. Adding a small amount of fish oil or, in modern times, butter, they left the skin to soak for four days. On the fifth day, the tanners removed the hide, laced it to a frame and worked it with the hands—first one side, then the other—until it was dry. Then they repeated the entire process. After the second soaking, they smudged the hide over a fire made with bracket fungus and rotten wood, covered with earth to prevent the hide from burning. When it had been smoked on both sides, they soaked the hide again in the fungus/brains/water mixture, and again stretched and worked it with the hands. By this time the hide was soft and white, suitable for making clothing or bags, although a perfectionist might subject it to the entire process of smoking, soaking and stretching one more time before using it.*[78]

In her wonderful book on the plant technology of Native Americans, Nancy Turner writes:

> *The Sechelt tanned hides by burying them for several weeks with a small, scalloplike polypore, Turkey Tails [Trametes versicolor]. . . . Some people also burned tree fungi as a smudge against mosquitoes. The Haida used the felt-like mycelium of a fungus growing on rotten Sitka Spruce—they rubbed the mycelium into a soft paste to make caulking for canoes and oil boxes. The Squamish used the corky inner tissue of various bracket fungi for scrubbing their hands. Nuxalk people painted faces on large specimens of bracket fungi, attached miniature bodies of cedar bark to them, and used them as dance symbols in a special "fungus dance" of the Kusiut ceremonials. . . . The Stolo, Haisla and many other peoples used bracket fungi as targets in a variety of spear-throwing and ball games. Many coastal peoples attributed protective properties to these fungi, believing that they could deflect evil thoughts and that they caused echoes in the woods.*[79]

Tinderwood polypore has been employed commercially in manufacturing German tinder or sticks for lighting cigars and pipes and for touching off fireworks.

The Three Sisters:

AGRICULTURE

Many kinds of living have been made from the great variety of ecologies in North America. There were hunters in the very far north, hunter-gatherers on the prairies, hunter-gatherer-fishers on the coasts, and farmers in parts of the continent suited to agriculture. Complex societies arose from each of these.

Farming spread rapidly over the continent and thrived long before the Spaniards and other Europeans arrived. Eight thousand years ago, nuts and acorns were a significant part of local diets and the trees they came from were treated much like farms. About 4,500 years ago there were vast plantings of the native sunflower, and the Three Sisters—corn, squash, and beans—

had been introduced into the diets of the peoples living across North America. They harvested berries and honey, living well off the land. They moved from the fields they cultivated in summer to autumn hunting grounds, which were usually located near a ready supply of wood. When the soil was exhausted after five or six years, they would move to another place.[1]

How people came to modify plants for their own use is open to much speculation, but some things are obvious: plants that were bigger and juicier and better-tasting would be the plants from which seeds were collected and replanted. Different crops were suited to each plant community, so farms varied from place to place. This

560

was a sharp contrast with the Europe of the fif-teenth century, where there were uniform annual crops in almost every country.[2] Native American peoples also used burning to clear land for hunt-ing and farming and to get rid of pests. Agricul-ture was well established by the time Hernando de Soto and his expedition came to these shores in 1539. He and his men marched through vast fields of corn, beans, and squash and other veg-etables which "were spread out as far as the eye could see across two leagues of plain."[3]

What the newcomers saw before them, how-ever, was not the tidy rows of a typical European farm. Plants grew helter-skelter over and around each other, including beans, corn, squash, sun-flowers, pumpkins, and gourds. By this time Native Americans had developed a highly sophis-ticated and very well-balanced diet. In the South-west, agaves were grown as a crop; oak groves were considered personal property; camas fields in the Pacific Northwest were tended by the same family year after year. A crop was not imposed on land where it might be difficult to grow. This would have been too time-consuming and too difficult in what was already a demanding life.

Many of the cultivated plants were entirely unknown to the Europeans and they had great difficulty seeing them as crops harvested in the traditional sense. Apart from the messy-looking fields, the Europeans were shocked to see that women tended the crops, tilling them with their clamshell hoes while the men went off to hunt. There was no need to domesticate animals, since they were all around, and so no need to fence in pastures, or to invent the wheel or the plow. The light tilling of the land meant that erosion was kept under control. Temporary villages might be established, but they would be abandoned when the hunting season came along. If animals were scarce, the Indians burned an area, creating an edge to attract animals back to new crops of berries and perennials. By doing so they had fresh fruit and honey for themselves.

A partial list of the plants domesticated by Native Americans shows how much we are still in their debt: corn, squash, beans were modified and improved. There were tomatoes, both white and sweet potatoes, peppers, grapes, nuts of all kinds including peanuts, strawberries, cassava, tobacco, and cotton; they knew how to make maple syrup. They developed techniques of dry-ing and preserving food. Today we still benefit from all these fruits and vegetables pioneered by the Indians of North America.

Poaceae

Corn
Zea mays

Zea mays,
an annual herbaceous seed producer, grows from 2 feet to nearly 20 feet high and has many sub-species. It is monoecious (both sexes are on the same plant in different places): the male organ is the tassel and the female, the ears. The leaves grow alternately on the stem. The fibrous roots penetrate deeply into the earth and when tassel-ing begins, they will send up aerial roots which anchor the soil for extra support. Pollination is by both gravity and wind. The plant requires from 70 to 140 frost-free days to reach maturity depending on the variety and will grow in sandy loam to clay loam, peat, or muck.

Range: originally Mexico, Central and South America; currently worldwide.

Corn is beyond food: it is religious, the basis of a culture, the beginning of history in North America. All North Americans are corn people, from the first peoples who learned how to nur-

ture corn 1,000 years ago, to the first Pilgrims who were kept alive during the fearful winter of 1621 by a cache of Indian corn, to the modern-day city dweller enjoying the ritual of breakfast cornflakes.

When the corn is ripening in midsummer, each stalk has from 8 to 20 nodes; the places between the nodes—the internodes—are where the growth takes place. Each node has a leaf and as the leaf slips out of its sheath it makes such a noise farmers swear they can hear it grow.[4]

Corn grew wild in South America more than 8,000 years ago, found its way north and was cultivated to become not only the mainstay of the food chain in cultures as sophisticated as the Aztec and Inca but also central to their religions. It was the kind of gold they treasured, not the cold metal later explorers lusted after. The Hopi called corn "milk for mankind," and the image of the "corn mother"[5] providing life and sustenance pervades the creation stories and legends of many tribes.

Some people think corn developed from a grassy relative named **teosinte,** *Zea mexicana.* In his search for the "primal ear" in 1960, anthropologist Paul Christoph Mangelsdorf found corncobs dating from between 500 and 7,000 years ago in caves in Oaxaca, Mexico:

As I stood . . . in the San Marcos Cave looking down on the alluvial terraces . . . I could visualize the prehistoric cave-dweller, primarily a hunter of small game and a food-gatherer, bringing back small ears of wild corn to his shelter, picking off the small, flinty kernels, much too hard for him to chew even with his excellent teeth, and exposing them one by one to the glowing coals of his fire. After absorbing heat for several minutes, the kernels would have exploded, transforming the stony little grains into tender, tasty morsels.[6]

Early corn was nothing like the fat, kernel-laden torpedoes we munch on today. It was more likely pencil-thin and about 1 inch long. It was probably regarded as a novelty, but as more peoples turned to agriculture to supplement hunting as a source of food, it became a staple. The cultivation of corn contributed in no small way to the complex civilization of the ancient Maya. Their knowledge inexorably flowed out among semi-nomadic tribes who lived on corn, beans, squash, pumpkins, and melons. By 700 B.C., maize was significant in southern Arizona,[7] and a staple of the Anasazi in Colorado by 100 B.C. The appearance of maize in the East is more of a mystery than its introduction to the Southwest. Maybe it went downstream to the Mississippi, then upstream to the Eastern woodlands.

Scientists have found that corn eaters have more of the isotope carbon 13 in their bones than eaters of most other plant foods (fewer breaks and fractures). They also have worse teeth—the result of carbohydrates being converted into sugars in the mouth.[8] They can analyze human remains to determine which peoples were corn eaters. By the time of European contact, hundreds of maize strains grew everywhere in North America, except on salt deserts, rock peaks, and in the Far North. The Hopi grew it on mesas. The Onondaga planted grain among the bodies of girdled trees. The Huron let their grain ferment into a black mass and ate it.

All parts of corn were used. Stalks became medicine containers; the shredded corn husks filled pillows and mattresses or were woven into mats or baskets; the silk was used for medicine; the cobs were fuel for smoking meat and hides, or made into pipes.[9]

The name "maize" comes from the Arawak *mahiz.* The first written record of corn was by Columbus, who found it in Haiti in November 1492.[10] In his diaries from 1604–1608, Samuel de Champlain reports on the Indian corn-growing practices in New England (the earliest such report among explorers to New England):

We saw their Indian corn, which they raise in gardens. Planting three or four kernels in one place, they then heap up about it a quantity of earth with shells of the signoc [horse shoe crab] before mentioned. Then three feet distant they plant as much more, and thus in succession. . . . The Indian corn which we saw at that time was about two feet high, some of it as high as three.[11]

Geneticist Walton Galinat says, "The American Indians were not simply the first corn breeders. They created corn in the first place."[12] They selected seeds to preserve their color and type. Each color had a sacred meaning as well as a food meaning. The Native Americans isolated different strains in different plots, bred new strains resistant to drought, and found new ways of irrigating their crops. For instance, the Hohokam ("the vanished ones") in the Southwest farmed the desert, where temperatures could rise as high as 119°F with less than 7 inches of rain a year. They invented a system of canals exploiting the flash floods raging temporarily out of the mountains. Phoenix is built over the ruins of 1,750 miles of these Hohokam canals.

Indians developed early-, middle-, and late-maturing strains for different uses and saved the perfect ears for seeds. Anthropologist George Will describes varieties which would ripen in 60 to 70 days. People in northern nations (present-day Minnesota and Canada) grew corn varieties that needed growing seasons of only 90 to 100 days, rather than the 120 to 160 days of other varieties.[13] These varieties became vital to European settlers who were unable to grow corn in the West until native Western strains became available to them after 1909. Between 1909 and

Now the corn, as we believe, has an enemy—the sun who tries to burn the corn. But at night, when the sun has gone down, the corn has magic power. It is the corn that brings the night moistures—the early morning mist and fog, and the dew—as you can see yourself in the morning from the water dripping from the corn leaves. Thus the corn grows and keeps on until it is ripe. The sun may scorch the corn and try hard to dry it up, but the corn takes care of itself, bringing the moistures that make the corn, and also the beans, sunflowers, squashes, and tobacco grow. The corn possesses all this magic power.
Gilbert Wilson, *Agriculture of the Hidatsa Indians*

1924, Montana's corn acreage increased from 10,000 acres to 420,000 acres.[14]

Though most have been lost, early varieties, such as a "flinty chapalote" and "reventalador popcorn," are still found among Indian farmers in Mexico, along with a "floury eight-rowed sixty-day corn."[15]

The Dutch traveler Isaack de Rasieres, writing in the 1620s, described how the Agawam of New England planted corn in "heaps like molehills," adding that corn "is a grain to which much labour must be given, with weeding and earthing-up, or it does not thrive." All crops except tobacco were the work of women among the Agawam. A woman on her own could raise 25 to 60 bushels of corn from one to two acres, enough for half the needs of a family of five. In combination with other harvested crops, women contributed as much as three-quarters of a family's food.[16]

The Agawam Indians did not add compost or manure to their fields; they moved on to new fields when the soil in the old ones was exhausted. Early settlers in Plymouth, however, fertilized with fish, and for 40 days after corn planting, all the dogs in the colony hopped around with one forepaw tied to the neck so they couldn't dig up the fish.[17] Added to the practical knowledge of corn growing was a measure of tradition and superstition. For example, when corn approached maturity and blackbirds attacked the corn, the men of the Omaha tribe chewed up some grains of corn and spat them around the field to prevent further damage. The Omaha did not eat the first mature ears for fear it would lead to birds' devouring the rest of the crop. A white leaf in a cornfield was seen as a portent of a bumper crop, and a good buffalo hunt, too. If a murderer passed near a cornfield, it could blight the entire crop.[18]

Buffalo Bird Woman was a Hidatsa born in 1837 who recounted her life to anthropologist Gilbert L. Wilson. Her book was published in 1917 and it is one of the most lyrical and elegiac accounts of a life that was fading even as she aged. Her knowledge is stunning to the uninitiated. "We Indians understood perfectly the need of keeping the strains pure. . . . We Indians knew that corn can travel." Here is Buffalo Bird Woman's description of growing corn:

The hills were in rows, and about four feet or a little less apart. They were rather irregularly placed the first year. It was easy to make a hill in the ashes where a brush heap had been fired, or in soil that was free of roots and stumps; but there were many stumps left over from the previous summer's clearing. . . . Corn planting began the second month after sunflower-seed was planted, that is in May; and it lasted about a month. . . . We knew it was time to begin planting corn . . . [when] the wild gooseberry bushes were almost in full leaf. . . . A small vessel, usually a wooden bowl, at my feet held the seed corn. . . . I planted about six to eight grains in a hill. Then with my

hands I raked the earth over the planted grains until the seed lay about the length of my fingers under the soil. Finally I patted the hill firm with my palms. . . . The corn hills I planted well apart, because later, on hilling up, I would need room to draw earth from all directions over the roots to protect them from the sun, that they might not dry out. . . . If the corn hills were so close together that the plants when they grew up, touched each other, we called them "smell-each-other"; and we knew that the ears they bore would not be plump nor large. . . .

The first corn was ready to be eaten green early in the harvest moon, when the blossoms of the prairie goldenrod are all in full, bright yellow; or about the end of the first week in August.

[To roast green corn she would make a fire of cottonwood.] I laid the fresh ear on the coals with the husk removed. As the corn roasted . . . the green kernels would pop sometimes with quite a sharp sound. If this popping noise was very loud, we would laugh and say to the one roasting the ear, "Ah we see you have stolen that ear from some other family's garden!"[19]

In a Hopi creation story, humans emerged from a world beneath the earth. The mockingbird divided them into tribes, then allowed each to take whatever variety of corn it liked. Each variety had particular qualities: For example, yellow brought prosperity and enjoyment, but also a short life; blue brought a life of hardship and hard work, but peace and longevity. All the tribes, beginning with the Navajo who took the yellow, grabbed a variety. At the end, only the Hopi remained, and they were left with the short blue corn, which gave them a hard but long life.[20]

In a Pima and Papago legend, corn is anthropomorphized into a young man from the East who comes west in search of a beautiful young

Subspecies of Corn

tunicata: **pod corn,** has each kernel enclosed in husk, as well as whole ear.

amylacea: **flour** or **soft corn,** used for cornmeal and hominy.

indurata: **flint,** has hard kernels which store well and are insect-resistant.

indentata: **dent,** the mature kernels are dented, hard outside and floury inside, used for elote, tamales, tortillas, corn beer, and animal feed.

everta: **popcorn,** a subset of flint corns.

ceratina: **waxy corn,** contains only one type of starch (amylopectin); all other types contain two (amylopectin and amylose).

wife. He shows the women how to cook corn. On his travels to find the young woman, he sings this song:

> Over there beneath the sunrise,
> The corner of the earth is my garden.
> In it flowery songs go forth in every
>   direction.
> Go along, corn, little corn.
> Over there beneath the sunset a woman
>   sits,
> Speaking bravely.
> I'll laugh at her and no doubt marry her.
> Go along, little corn.[21]

Corn is a metaphor for the life of the Hopi people and appears in some form at every Hopi ritual. A newborn baby is wrapped in a blanket with one or two ears of corn, which is called its mother. Before a wedding, the bride and her female friends and relatives grind vast amounts of cornmeal. Four days before the wedding, the bride stays at the house of the groom's family and grinds corn there. On her wedding day, the bride's face is dusted with cornmeal. The dead receive corn to take them on their journey into the spiritual world.[22]

Corn, beans, and squash were the fabled Three Sisters, common to dozens of tribes ranging from woodlands to near desert. Corn provides support for the beans, which in turn add nitrogen to the soil, while the lush squash foliage sprawls at their feet, protecting the earth and holding in moisture. Corn complements beans by providing the protein zein, while beans provide two essential amino acids lacking in corn: lysine and tryptophan.[23]

These staple foods were used so creatively that Iroquois cooks knew up to 40 methods for maize alone: "A squaw in one western tribe detailed to an inquiring anthropologist a hundred and fifty recipes of various kinds without exhausting her mental cookbooks of maize dishes."[24]

"Paper bread," used for expeditions, was made by boiling cornmeal and water, then spreading it on hot stones. The Iroquois, Navajo, and Zuni made "leaf cakes" by grinding green corn kernels into a paste, boiling them in corn leaves, then baking them. The Zuni made a hominy (a word derived from the Algonquin *rockahominy*) by boiling the kernels, then cooking them with meat and beans. The Chippewa roasted corn in the husk, and ate it boiled and sweetened with maple sugar.[25] Corn figured in beverages, too, from the Zuni's "corn wine," made by fermenting corn sprouts in water, to the "corn coffee" of the Iroquois, which was made from parched corn and boiling water, often sweetened with maple sugar.[26]

Grinding the corn often became a social occasion. Juanita Tiger Kavena describes how a group of girls would grind to the rhythm of a song sung by an uncle or grandfather. A small window was often located next to the grinding stones and provided a safe way for a boy to court a girl while she was working away. If the girl was interested in the young man, they would talk through the window. But if she wanted to discourage him, she would fling cornmeal at him instead.[27]

> Down to the cornfield let us go;
> Pulling up corn has been our trade,
> Ever since Adam and Eve was made.
>
> Jimmy crack corn, I don't care,
> Jimmy crack corn, I don't care,
> Jimmy crack corn, I don't care,
> Old Master's gone away.
>
> American folk song

Early settlers adapted native recipes. For instance, samp or new samp was corn beaten and boiled and eaten hot or cold with butter or milk (the Indians ate it without). Corn pone was made by pounding mature corn fine, sifting it, adding water or bear oil (and occasionally berries in season) to make a dough, then covering it with leaves or patting it into little inch-thick cakes and baking. Good but heavy.[28]

Corn balls were also popular: a nineteenth-century traveler called Henry the Younger first had them in 1806. He described them as "about the size of a hen's egg, made of pears, dried meat, and parched corn, beaten together in a mortar . . . Boiled for a short time . . . we found them most wholesome."[29] Piki (blue corn bread) is still eaten by the Hopi; made from blue corn flour, the batter is smeared onto a fire-heated stone (just like the old paper bread), flipped, and then rolled and folded.[30]

In a "cornhusking bee" or "husking bee," the terms used by early settlers since at least 1693, neighbors got together to help husk corn. A young man who found an ear of red corn got a kiss from the girl of his choice.[31]

The term "corncracker" was someone who cracked corn to make grits or cornmeal, eventually shortened to "cracker." After the Civil War, the word came to mean a poor white person, especially one from Georgia. Now regarded as a racial epithet, its use is a violation under the Florida Hate Crimes Act.[32]

Today, corn fuels millions of breakfast eaters, and corn roasts remain the traditional focus of gatherings from church fund-raisers to family reunions. We continue to make use of all parts of this amazing plant, employing it in more than 600 products, including paper, glue, textiles, and plastics. There are even extraordinary corn varieties: corn fanatic Walton Galinat has bred corn 2 feet long, corn with square ears (as airplane food, he jokes, because they won't roll off the plate), and red-white-and-blue "Old Glory" corn for events like the U.S. Bicentennial, with blue-dotted kernels that look like stars.[33]

Today one of our main concerns about corn is how much it should be genetically modified, and whether we can ever find our old-fashioned tasty corn varieties again. This has made a bonanza for organic farmers who have saved heritage corn and market it all across North America at a premium.

Fabaceae

Beans
Phaseolus spp.

Phaseolus spp.

are all native to the Americas, including common bean, *Phaseolus vulgaris;* tepary bean, *P. acutifolius;* lima bean and butter bean, *P. lunatus;* and scarlet runner bean, *P. coccineus.* Other beans include black, pinto, kidney, string, navy, and cranberry. All are annuals (though some wild beans are perennials).

There are three growth forms: climbing, bushy, and intermediate. Climbing varieties may reach 12 feet or more. Nonclimbing varieties usually grow to 8 to 24 inches. The three-part leaves grow alternately on the stem. Beans prefer light, airy soil with good drainage and are heat- and drought-tolerant. The somewhat showy flowers are self-fertile; elongated pods contain from three to more than a dozen seeds (beans), usually kidney-shaped.

Range: originally Peru to Mexico; currently worldwide.

The second of the Three Sisters, beans, is as well rooted in our history as corn, and the bean family is second only to the grasses in its importance to humans. Beans, along with maize, squash, chili, amaranth, and avocado, were an established part of the basic Mexican diet by 2000 B.C.[34] The Hohokam Indians (c. 300 B.C.–c. A.D. 350) first grew the desert-adapted **tepary bean** (*Phaseolus acutifolius*) in the floodplains of central Arizona. Tepary beans, still grown by the Hopi and other nations, are similar to navy or soup beans, but sweeter. The dried seeds contain 60 percent carbohydrate and 22 percent protein.[35]

Beans are nitrogen fixers (*rhizobia* bacteria, which live in nodules on the roots, convert nitrogen in the air into soluble nitrates usable by other plants through the soil) and contain complex sugars which combine with bacteria in the gut of humans to cause flatulence.

When explorers moved about the continent, people such as Jacques Cartier and John Josselyn noted the huge diversity of beans. Josselyn mentions "white, black, red, yellow, blue and spotted."[36] The Iroquois, who divided beans into "bread beans" (for corn bread) and "soup beans," were growing about 60 varieties in New York in 1908.

Buffalo Bird Woman says beans were planted immediately after squash in hills (mounds) about 2 feet apart[37] (the original derivation of the age-old expression "not worth a hill of beans"). Beans were often planted between rows of corn, and were threshed after a few days of drying on uprooted vines.[38]

Lowly as beans may be, their importance was acknowledged in ritual, legend, and song. The Seneca said the "cornstalk bean" grew from the Earth Mother's grave.[39] The Hopi did (and still do) perform a bean dance[40] and the Onondaga tell this story about corn and beans to account for their habit of planting them together:

A young man (corn) lives on a hill, but is lonely. One day a beautiful maiden (a pumpkin), dressed in a flowing robe covered with yellow bells comes along and tells him, "I have come to marry you." But he replies, "No, you are not the one, you wander too much from home and run over the ground so fast that I cannot keep you by my side." The rejected pumpkin maiden leaves sadly. But one morning, a tall slender maiden appears, covered with flowers and dangling leaves. After one look, the man is attracted to her. The two embraced, and to this day corn and beans grow intertwined in Indian cornfields.[41]

The Korosta Katsina Song of the Southwest Hopi includes the lovely lines:

Blue butterflies,
Over the blossoming virgin beans,
with pollen-painted faces
Chase one another in brilliant streams.[42]

Beans were boiled, baked, mashed, ground, dried, stewed, and parched. New World recipes combined beans with maple syrup and bear fat;

Old World recipes added sugar, molasses, ham, and bacon. Buffalo Bird Woman made a dish called Four Vegetables Mixed: "I put a clay pot with water on the fire. Into the pot I threw one double-handful of beans. This was a fixed quantity; I put in just one double-handful whether the family to be served was large or small, for a larger quantity of beans in this dish was apt to make gas on one's stomach." Slices of dried squash were cooked and mashed and added to the pot, along with four or five double-handfuls of pounded parched sunflower seed and pounded parched corn. "The whole was boiled for a few minutes more, and was ready for serving."[43]

In New England lumber camps, beans and cornmeal were cooked with stew meat or soup bones, a relatively high-fat dish. Sometimes camp cooks prepared bean porridge in quantity and set it outdoors to freeze. Bags of cooked beans were carried in the backpacks of the men who searched for gold in the great rush of '49. In either case the first utensil the cook reached for was an ax. Paul Bunyan and his mythical lumberjacks ate prodigious amounts of beans and made prodigious noises about it. You can understand the inspiration Mel Brooks had for the famous farting scene in *Blazing Saddles*. It was definitely a part of the Wild West.

Probably the most famous dish is Boston baked beans, consisting of navy beans *(P. vulgaris)* cooked slowly with molasses and salt pork. Some historians say the early settlers learned the technique from the Indians, while others argue that it was introduced by seafarers who had voyaged to North Africa and Spain where baked beans was a traditional Sabbath dish among resident Jews. Today, recipes for baked beans are associated with Boston (often called Bean Town), where in colonial days the Puritan women used to bake beans on Saturday so that they would not have to cook on the Sabbath day. Beans were served at Saturday dinner and again as leftovers on Sunday.[44]

More than 100 varieties of beans, including dry (kidney, navy, and pinto) or fresh (string, snap, or wax), are still cultivated in America, mostly in Michigan, North Dakota, Nebraska, and Colorado. American farms produce 1.5 million tons of beans every year.[45] From beanbag chairs to Beanie Babies, from baked beans to bean salads, we still love our beans.

Cucurbitaceae

Squash
Cucurbita spp., winter squash, ornamental gourd, pumpkin

Cucurbita spp.

include *C. pepo* (winter squash, ornamental gourd, pumpkin); *C. pepo* var. *melopepo* (summer squash); *C. moschata* (winter squash, pumpkin); *C. mixta* (winter squash, pumpkin); *C. maxima* (winter squash, pumpkin). Botanical differences are clearly not reflected in the common names. The species are all annuals, climbing or trailing with tendrils. The leaves are large, simple, alternate, and often prickly, and the fruits can be very large (one pumpkin was recorded as weighing almost 400 pounds). The plants are monoecious, with bright yellow male and female flowers borne singly in leaf axils. They are usually pollinated by bees. The roots are extensive, but shallow. Grows in fertile, well-drained soil.

Range: originally southern temperate North America and northern temperate South America; currently North and South America, with the winter types produced mainly in the north and the summer types mainly in the south.

The third sister, squash, is so called from New England tribes' name for it: *askutasquash*.[46] One

Hopi tribe is called the Patuñ, which means "the squash people."

Plants known as gourds actually belong to two different genera: *Cucurbita* and *Lagenaria*. The bottle gourd or white-flowered gourd, *L. siceraria*, probably originated in south central Africa. The seeds are thought to have drifted via ocean currents to South America. Their use predates the invention of pottery almost everywhere, from a few hundred to a thousand years.[47] Gourds were grown about 5000 B.C. in Illinois,[48] and made into utensils, fishnet floats, birdhouses, and rattles (filled with seeds or stones). In the 1530s, Alvar Núñez Cabeza de Vaca came upon an Indian village somewhere in Texas and was greeted by a party of Indians "shouting frightfully, and slapping their thighs. They carried perforated gourds . . . which are ceremonial objects of great importance."[49]

Blossoms were also a great delicacy. Buffalo Bird Woman describes drying them by opening the flowers and spreading them out on patches of a kind of grass called "antelope hair."[50] When the squash was dried, Buffalo Bird Woman says, "we strung the slices upon strings of twisted grass, each seven Indian fathoms long; an Indian fathom is the distance between a woman's two hands outstretched on either side."[51] The seeds, which contain oil and protein, were eaten toasted, or ground into pastes and thickeners for sauces.[52] The Chippewa ate their pumpkins and squashes fresh, or cut and dried them over a fire; dried, they could be stored for up to two years. Dried squash and pumpkin were boiled, alone or with game, and seasoned with maple sugar.[53] Early settlers used pumpkin to make a sweet custard.

Pumpkin also had its medicinal uses: the Maya used the sap to soothe burns and the Zuni used both seeds and blossoms to make a salve against cactus-needle injuries. The Menominee made a mortar of pulverized seeds and water as a way of "facilitating the passage of urine." The

Catawba chewed pumpkin seeds for kidney disease. Plains Indians attributed medicinal powers to the roots of the wild gourd, *C. foetidissima*, but believed it was dangerous to dig them up. Colonists used pumpkin seeds as a remedy against intestinal parasites, a practice eventually adopted by the medical profession.[54]

As a fruitful, adaptable staple, squash figures in myth and legend: some say that squash sprang from the Earth Mother's navel, tobacco from her head, and beans from her feet.[55] According to Pima legend, Corn Man and Tobacco Woman have a child, Squash Baby, whose neck is broken (the crookneck squash) when some children come into the house and play with it. After it is broken, Squash Baby sinks under the ground, then arises at Superstition Mountains and becomes a saguaro cactus.[56]

Buffalo Bird Woman explains that squash seeds were sprouted in bundles of grass, sage, and buck brush with their leaves folded inside pieces of tanned buffalo robe. At squash planting, "two or three women worked together. One woman went ahead and with her hoe loosened up the ground for a space of about fifteen inches in diameter. . . . Following her came another woman who planted the sprouted seeds. Four seeds were planted in each hill, in two pairs. . . . If we planted them in level ground the rains would beat down the soil, and it would pack hard and get somewhat crusted, so that the sprouts could not break through, but if we planted the sprouts on the side of the hill, the water from the rains would flow over them and keep the soil soft."[57]

"We picked a good many squashes in a season. One year my mother fetched in seventy baskets from our field. I have known families to bring in as many as eighty, or even a hundred baskets, in a season."[58]

The picture we get of Indian women in their long days of cultivating along with all their other duties toward their families was one of a lively companionship among the women as they worked cooperatively on their land. There was none of the isolation that must have been felt by the pioneer women living alone on farms far from other people in the community. They came together on occasions (church, quilting bees, and barn raisings), but it must have been a difficult life for what were originally city people.

Today, zucchini and yellow squash are the most commonly grown varieties of summer (soft-shelled) squash these days. Acorn, hubbard, and butternut squash top the winter (hard-shelled) squash list. Pumpkins are still favorites for pies in Thanksgiving season and Halloween jack-o'-lanterns, while their seeds are roasted and salted as snack food.

Asteraceae

Sunflower
Helianthus spp.

Helianthus spp.

are hardy herbaceous perennials or annuals found wild in North America that produce large yellow flowers in late summer and autumn.

Helianthus annuus, **common sunflower: annual; height 3 to 10 feet; the flowers consist of sterile yellow ray flowers and brownish, fruit-producing disk flowers; long, ovate, hairy leaves; flower heads 3 to 16 inches in diameter.**

Range: **originally western North America; currently Quebec, south to Georgia, west to Oklahoma, north to North Dakota.**

The state flower of Kansas.

The most easily recognizable of all the prairie flowers, *Helianthus* gets its name from the Greek *helios* (sun) and *anthos* (flower) because it is heliotropic, that is, it always has its face to the sun. *Helianthus* originally grew in the rich soil of the prairies. Now it has spread to roadside ditches and abandoned lands. Either because of its habit of lifting its face to the sun or simply because of its golden rays, this plant represented the sun god in some Indian legends.[59]

The earliest evidence we have of sunflowers is from 3000 B.C. in Arizona and New Mexico. In western North America, they grew plentifully in the wild, and were not cultivated as they were in Mexico and around the Great Lakes. Both wild and cultivated types were widely used. The seeds were used for flour for cakes, mush, and bread; mixed with beans, corn, and squash; or eaten as a snack.[60]

Buffalo Bird Woman describes the "sunflower seed ball" carried by warriors. Seeds were ground to meal, then pressed into a ball. The expressed oil holds the ball together. She says, "It was amazing what effect nibbling at the sunflower-seed ball had. If the warrior was weary, he began to feel fresh again; if sleepy, he grew wakeful."[61]

Sunflower oil, extracted by bruising the seed, boiling it, and then skimming off the oil, was used for cooking and also for oiling the body and dressing the hair.[62] The Hopi found that soaking the purple seeds of some varieties yielded a gorgeous purple dye.[63] A bright yellow dye was also obtained from the flower heads, and a fiber from the stalks. As part of the Dakota's pharmacopoeia, an infusion from sunflower heads was taken against chest pains. The Pawnee pounded the seeds with certain roots to be taken by pregnant women as protection against illness.

Lewis and Clark described finding sunflowers blooming in great abundance:

The Indians of the Missouri and more especially those who do not cultivate maize, make great use of the seeds of this plant both for bread and for thickening their soups. They first parch them and then pound them between two stones until they are reduced to a fine meal. Sometimes they add a portion of water and drink it thus diluted; at other times they add a sufficient proportion of marrow grease to reduce it to the consistency of common dough and eat it in this manner. The last composition we preferred to all the rest.[64]

According to Buffalo Bird Woman, sunflower seed was planted as soon as the soil could be worked. "Our native name for the lunar month that corresponds most nearly to April, is *Mapi'-o'c-mi'di*; or Sunflower-planting-moon. . . . Usually we planted sunflowers only around the edges of a field. The hills were placed eight or nine paces apart, for we never sowed sunflowers thickly. We thought a field surrounded thus by a sparse-sown row of sunflowers had a handsome appearance."[65]

Today commercial sunflower cultivation occupies some 3.5 million acres in the United States, mostly in the Dakotas, Minnesota, Kansas (where it is the state flower), and Nebraska. Most sunflower seeds are sold as bird-seed or crushed to make oil; about one-seventh of the crop is roasted and salted for direct human consumption. The plants are a treat for the farmer—about 18,000 plants can be grown on an acre of land.

A new market for sunflower seeds is in the ballpark, where many baseball players have switched from chewing tobacco to munching on sunflower seeds—in the shell. The process of shelling and eating them is part of the appeal. There's a competitive aspect, too. "Extracting the kernel from the shell 'takes tooth-tongue coordination,' notes Shawn Boskie, a California Angels pitcher. 'No player'd be caught dead opening a seed with his fingers.' There are,

indeed, legends of the seed-eating art. Among them is Reggie Jackson, the former Oakland A's and New York Yankees outfielder. 'He could eat 'em and spit the shells like a machine gun,' says Bell [one of his suppliers] with awe."[66]

On a more evocative note, there's a valley below a Hopi mesa called "Place of Many Sunflowers," where the unusually large *H. anomalus* grows. This flower is used for Hopi ceremonies in October and November. Hopi maidens gather wild sunflower petals, dry and grind them into a yellow powder, and apply the powder to their faces. It looks like gold.[67]

H. giganteus, **giant sunflower:** "giant" refers to the height of the plant, rather than the size of flowers; flowers with yellow disk and ray flowers; flower heads 3 inches in diameter; long, rough, lanceolate leaves, height 3 to 12 feet; grows in swamps, wet thickets, meadows.

H. maximiliani, **Maximilian's sunflower:** grows 8 to 10 feet tall; flowers with yellow disk and ray flowers; heads 3 inches wide; leaves long, narrow, rough; original habitat prairies, but escaped east from cultivation. Often used on prairie restoration projects. Named after Prussian naturalist and explorer Maximilian Alexander Philipp, Prinz von Wied-Neuwied (1782–1867). Range: originally bluegrass prairie of Saskatchewan, Manitoba, Minnesota, south to Missouri, Oklahoma, Texas; currently from original range east.

H. strumosus, **Woodland sunflower:** yellow ray and disk flowers; heads 2 to 3 inches wide; leaves long, ovate to lanceolate, rough; height 3 to 7 feet; grows in woods, thickets, clearings.

Solanaceae

Tobacco
Nicotiana spp.

Nicotiana tabacum,
grown in the South, resulted from natural hybridization of the wild species *N. sylvestris* and *N. tomentosiformis*. A frost-sensitive perennial, it is now grown commercially as an annual. The plant has a single erect stem with terminal inflorescence; the leaves are oval or lance-shaped. The flowers, which are about 2 inches long, can be pink, white, or red. *N. rustica*, which grows as far north as the Great Lakes, has yellow to greenish-yellow flowers. Both species have extensive, fibrous, shallow roots. Tobacco grows in a wide range of soils. The seeds are tiny—there are 4.5 million per pound.

Range: worldwide.

The name *tobacco* comes from Native American names *tabaga* and *tabaca*, although these originally referred to the pipes used for smoking rather than tobacco itself. Ancient pipes were connected to the nose rather than the mouth.[68] The earliest confirmed record of tobacco use is an Aztec pictograph from the third century A.D., found in Central America.

Tobacco was part of a communal smoking ritual among the Pima, whose shamans also used it to produce rain. Better known is the smoking of the peace pipe: ceremonial pipes were up to 4 or 5 feet long. The ceremony began by blowing smoke up toward the sun and downward to the earth. Then puffs were blown in the direction of the four points of the compass.[69]

According to John D. Hunter in his 1823 tract *Manners and Customs of Several Indian Tribes Located West of the Mississippi,* Indians west of the Mississippi used tobacco medicinally as wet and dry poultices on swellings, abscesses, and ulcers.[70] The Chickahominy, the Mohegan, and the Malecite blew tobacco smoke into the ear for earache,[71] while the Louisiana Choctaw blew smoke on snakebites.[72] The Creek made a tobacco tea against stomach cramps,[73] and the Catawba made a decoction for sick horses.[74]

Buffalo Bird Woman says young men were taught that smoking would injure their lungs, making them poor runners. "But when a man got to be about sixty years of age, we thought it right for him to smoke as much as he liked. His war days and hunting days were over."[75] The tobacco beds had their own mores as well. Anyone whose hair fell out was accused of stealing tobacco.

The blossoms were picked and the green parts were kept and dried for sacred objects. The rest of the plant was harvested just before frost. The leaves were dried and stored in a cache pit, and the seeds were carefully saved for planting. A good deal of tobacco was sold to the Sioux; a bunch 6 or 7 inches in diameter would fetch one tanned hide.[76]

The Europeans' discovery of tobacco came in 1492, when one of Columbus's landing parties explored Cuba. There, they found "men always with a firebrand in their hands, and certain herbs to take their smokes, which are some dry herbs put in a certain leaf. These are lit at one end and the other end they chew or suck, and take it in with their breath the smoke. These were claimed to drive away all the weariness, and were called tobago." The Europeans learned about snuff on Columbus's second voyage in 1494.[77]

In 1535, Jacques Cartier discovered tobacco among the Huron on Georgian Bay. "This [plant] they prize very much and men only use it, in the following manner. They dry it in the sun and carry it on their necks in a small animal skin, with a pipe of stone or wood. They never go anywhere without these things. We have tried this smoke; after taking some into our mouth it seemed like pepper, it was so hot."[78]

Tobacco was introduced to the Old World in the mid-sixteenth century initially for its supposed virtues as a panacea, first in Spain and Portugal, from where it spread across Europe, Asia, and Africa. Early European settlers used snuff to cure allergies and to induce labor through sneezing,[79] and applied wet tobacco to bee stings.[80] They also used it as an insecticide (nicotine is highly poisonous to insects).[81] The first tobacco to reach England was probably harvested in Virginia, where John Rolfe introduced tobacco as a crop as early as 1612. By 1619, tobacco had become a leading export of Virginia, where it was later used as a basis of currency. They traded future crops for the needed goods to survive in the new colony.

Most Native Americans used it for medicinal or ritual use, but the European settlers took to it as though they were genetically addicted. James I of England thought the habit was disgusting and tried to suppress the use and export of tobacco from the New World. But the cultivation of tobacco persisted, along with the habit, "throughout the Connecticut River valley in the late 1700s. Local shops first made plug and twist tobacco, but, in the 1830s, an unusually fine-textured and broad-leafed plant, especially well suited for the outer wrappers of cigars, was introduced into the area. Until the first quarter of the 20th century, when cigarette consumption increased substantially, this area was a center of cigar-wrapper production. Most recently, the long-leaf market supplied the wrappers for Cuban-grown fillers. But the combination of the loss of trade with Cuba and the decrease in the use of all tobacco products has left only a few growers still cultivating cigar-wrapper tobacco along the Connecticut River."[82]

In 1997, the United States produced nearly 1.7 billion pounds of tobacco (about one-tenth of world production), of which about 30 percent was exported.

Poaceae

Wild Rice
Zizania aquatica

Zizania is a type of grass unrelated to true rice (*Oryza*). It is a wind-pollinated, cross-fertilized annual, that grows in freshwater at the edges of lakes and streams, but not in stagnant ponds. If left alone, the seeds fall into the water, where they will germinate. The water must be shallow (about 2 to 3 feet, usually) so that the sunlight can get through and warm up the seeds on the bottom. When the seed germinates in the spring, a tiny hair root anchors it to the lake bottom and the stalk starts to grow upward. When the stalk reaches the water surface, it forms a float leaf, or banner leaf, which floats on the surface of the water at right angles to the stalk. While it is growing, the water levels must remain constant. If they rise too high, the weak stalks get ripped right out. If water levels fall, the stalk may collapse in on itself. If the water becomes turbulent or if there are high winds, the young plants may be swept away or blown over. This is a tricky crop requiring constant vigilance on the part of the farmer.

The seeds are borne on stems known as tillers. Two weeks after fertilization the green seeds appear, and after four weeks the seeds ripen and turn purple-black. The ripe grains drop off (or "shatter") at the slightest touch or when the wind blows. Birds love the grains, and water birds often nest in wild rice stands.

Wild rice was an important food for the peoples of the Great Lakes region. The grains contain more protein and carbohydrate than white rice, and are high in vitamin B, iron, and calcium. The indigenous peoples used to call the grains *manomin*, which means "good berry." The name of the Menominee of the Great Lakes region means "wild rice eaters." The French explorers thought it looked like oats, and called it *folle avoine*. It was the English explorers and settlers who called it wild rice even though it isn't a rice at all.

To harvest the grains, indigenous peoples used to float their canoes among the plants, bend the branches gently over the canoes, and tap them with poles called "knockers," then gather the grains that fell into their canoes. Two people working all day could usually gather a canoe full, or about 300 pounds, of rice. Whatever fell back into the water formed the next year's harvest, if the ducks didn't get it first.

The rice was then dried in the sun or over a slow fire in a metal tub, to loosen the husks. It was milled by pounding it in a wooden mortar to winnow away the husks. The grains were used in various ways. A mush made from wild rice was one of the few foods other than mother's milk that children under the age of five were fed. Another way of using it was "tassimanonny," made with wild rice, corn, and fish boiled together. For certain native groups, including the Ojibwe, wild rice was also a sacred food, important in religious rituals.

"Elder Maude Kegg (Naawakamigookwe), of Mille Lacs Ojibwe Tribe, was born around 1904, and raised at Portage Lake . . . in Minnesota. She recalled helping her uncle make a bootaagan, the mortar for pounding parched rice: ' . . . He cut a log, then sawed it straight. Then he pointed one end and carved some wooden pieces, pointing them so they'd fit well and make the bootaagan round. When he was through carving them, he dug a pit and put grass in it . . . Then he put a willow strip bent into a circle. He pressed the grass down. Then he fitted the boards together in it again. I held them as I watched him. . . . [Then] he tapped in the round piece of log. It looked just like a pail. . . .

Whiteman say to the Redman, "Is this
the Promised Land?"
"Groundnuts and wild rice and turkey in
the hand!"
Whiteman say to the Redman, "Just look
what you have got."
"Wild rice and wild thyme and turkey in
the pot."
Whiteman say to the Redman, "I think I
envy you."
"Wild rice and artichokes and
groundnuts in the stew."
Redman say to the Whiteman, "Do you
really have to push?"
"Redman and greener land, and turkey to
the bush."
Redman say to the Whiteman, "Are you
really having fun?"
"Nuts and bolts and wild, wild oats, and
the turkey on the run."
Blackman say to the Whiteman, "Just
look what you have done."
"Played your hand on the Redman's land,
and the turkey's on the run."[84]

J. A. Duke

That was where they pounded or trampled the rice. When . . . they were done picking rice, they took it apart and stored away the parts.' "[83]

Today, wild rice is considered a gourmet food. Many First Nations communities in Manitoba, Ontario, Minnesota, and Wisconsin still harvest and sell wild rice to food companies. In places like Minnesota, the traditional method of harvesting is still used; elsewhere the rice may be harvested mechanically. Strains that resist "shattering" are also being developed. Today, native growers are facing competition from commercial wild rice operations in northern California. The wild rice cultivators also have to do battle against redwing blackbirds, which given half a chance will devour the entire crop when the kernels are young. Metallic streamers keep most of them off.

Range: **from Lake Winnipeg in Manitoba south to the Gulf of Mexico, east to the Atlantic coast, and particularly around the Great Lakes.**

There is no part of any ecosystem that can be overlooked. Each plant, each animal, each microbe has a role to play in keeping the system alive and healthy. It is extremely complex; we know so little that masses of new information about the minute creatures of this earth are being newly published each year. It's impossible to keep up with all of them. But we all have a responsibility to think of ourselves as stewards of the planet, not just lucky exploiters of a great and glorious system. To be responsible, we must learn as much as we can and never stop learning about plants. They will always reveal something new if we just take the time to look carefully.

Glossary

A

Achene: A small, dry, one-seeded fruit.

Adventitious roots: Roots that arise from unexpected places.

Aerial root: A root emerging above the soil level.

Allelopathy: A plant's ability to produce chemicals that inhibit the growth of other plants.

Alpine: Mountain terrain that rises above the height and cold tolerance of trees.

Alternate: The staggered arrangement of plant parts, usually referring to single leaves arrayed alternately on a stem. Compare **opposite**.

Anaerobic: Without oxygen.

Annual: A plant that completes its life cycle, from germination to seed dispersal and death, within one year.

Anther: The terminal, pollen-containing part of a **stamen**. The female counterpart which receives the pollen is the **stigma**, on the **pistil**.

Anthocyanin: The pigment that gives leaves their red color.

Anthracnose: A fungal disease causing dieback of twigs and dead spots on leaves and fruit.

Apex: The tip or growing point of a stem.

Apical: Relating to the apex or tip of a plant organ.

Arctic: Term used to describe unforested regions north of the treeline.

Areole: A raised or depressed spot on the stem of a cactus, usually having spines.

Aril: A fleshy, brightly colored appendage to a seed.

Ascending: Curving branches that rise from the base, or branches of a flower head extending up at a 40- to 70-degree angle.

Awn: A bristlelike appendage, especially on grass seeds.

Axil: The angle formed between a stem and leaf stalk.

Axis: The central support or main stem of a plant.

B

Ballooning: The way caterpillars dangling from their silky threads are carried by the wind from tree to tree.

Barrens: Areas with low soil fertility able to sustain only low scrub with few trees. Treed areas that burn more often than every 5 to 10 years become barrens.

Basal: Near or forming part of the base of a plant or plant part.

Bast: The woody fibers of a tree's inner bark. Also called **phloem**.

Bearded: Having long **awns**, as in some grasses, or stiff hairs, as on the petals of some irises.

Berry: A fleshy, multiseeded fruit developed from a single ovary.

Biennial: A plant with a two-year life cycle, germinating and developing in the first year, flowering, fruiting, and dying the year after.

Biomass: The amount of living matter in a given habitat expressed either as the weight of the organisms per unit area or as the volume of organisms per unit volume of habitat.

Biome: A well-defined biotic climax community, characterized by the dominant plants and prevailing climate, as in tundra, desert, or coniferous forest.

Bisexual: A perfect flower that has both male and female sexual parts.

Blade: The broad part of a leaf.

Bloom: A whitish coating on fruit, stems, leaves, or flowers.

Bog: A wetland where water is trapped and stagnant, with high acid content. Compare **fen, marsh, swamp**.

Bole: The trunk of a tree.

Boreal forest: A cold-temperate forest dominated by evergreens such as spruce and fir. Called the taiga in Europe and Asia.

Bract: A modified leaf arising below a flower or inflorescence; usually tiny or scalelike, but sometimes large and brightly colored, as in Pacific dogwood.

Bunchgrass: Any grass that grows in tufts.

Buzz pollination: Intense buzzing by a bee in a flower, creating vibrations that force the release of pollen.

C

Callus: (1) A hard projection or swelling at the base of bracts enclosing the flowers in grasses. (2) Scar tissue formed in response to a wound.

Calyx: A collective term for a flower's sepals.

Cambium: The layer of growth tissue in woody plants, found between the inner bark and the **sapwood,** immediately beneath the bark.

Capitulum: A composite flower head; also a pompomlike cluster of branches, as in sphagnum moss.

Capsule: A dry fruit produced from a compound **pistil** that splits partly open at maturity.

Carnivorous plant: One adapted to capture insects and derive nutrition from them.

Catkins: Spikelike, often drooping, compact clusters of tiny flowers, either male or female, as in willow and birch trees.

Caudex: The thick, tough, sometimes woody base of a herbaceous stem.

Chaparral: An area densely populated by mostly small-leaved shrubs, with hot dry summers and cool moist winters, as in the foothills of California.

Chlorophyll: A green pigment found in plants that absorbs energy from the sun, enabling the process of photosynthesis.

Ciliate: Fringed with hairs.

Cladode: A flattened stem that functions as a leaf, as in cactus pads.

Clasping: Describes a stalkless leaf whose base partly surrounds the stem.

Cleistogamous: Having flowers that self-pollinate in the bud without opening.

Climax forest: The community of plants in the final, most stable stage of **succession,** growing and remaining dominant in an area.

Collar: The swollen base of a tree branch where it joins the trunk; the junction of the leaf sheath and blade in grasses.

Coma (plural **comae**): Tufts of silky hairs on certain seeds that act as parachutes.

Compound leaf: A leaf made up of two or more leaflets.

Compressed: Flattened in one plane, as in a **spikelet.**

Conifer: A cone-bearing tree or shrub, usually evergreen with needlelike leaves.

Cool-season grass: A grass that starts growth when the soil is still cold. Compare **warm-season grass.**

Corm: A bulblike underground stem structure with a fibrous covering, usually round with a flattened top.

Corolla: A collective term for the petals of a flower.

Corona: A crown or cup-like structure in a flower formed by either fused **stamens,** as in butterfly weed, or undifferentiated petals or **sepals,** as in narcissus.

Corymb: A flat-topped flower cluster in which the outer flowers open first.

Culm: An aboveground grass or sedge stem.

Cuticle: An outer waxy layer of cells, as on the stalks of mosses.

Cyme: A broad, flattened flower cluster in which the outer flowers open last.

D

Deciduous plants: Plants having parts that are shed at the end of their normal function, as in leaves of trees and shrubs in autumn.

Decoction: The process of boiling plant parts in water to extract their essences; also the resulting extract.

Decumbent: Lying horizontal or inclining with an upward-curving tip.

Dentate: Toothed along the edges, as in leaves.

Dioecious: Having unisexual flowers with each sex confined to separate plants of the same species. Compare **monoecious, polygamodioecious.**

Disk: The central area in the flower heads of the Asteraceae family bearing many tiny florets and usually surrounded by **ray flowers.**

Doctrine of signatures: A concept originating in the sixteenth century that a plant's form or color indicated its medicinal purpose, for example, hepatica's liver-shaped leaves suggested it could be used to treat the liver.

Drooping: Bending or downwardly arching.

Drupe: A fleshy fruit containing a single stony seed, such as cherries.

Duff: The ground-level mix of decaying leaves, twigs, needles, animal droppings, and insects that forms a rich mulch and keeps the forest soil cool and moist. Compare **litter.**

E

Ecosystem: A community of interacting plants and animals in an area functioning with their environment as an ecological unit.

Ecotone: A transition area between two distinct plant communities, containing species from

both and sometimes unique species.
Compare **edge effect**.

Edge effect: The often lively impact of two different types of habitat meeting, such as forest and meadow, or stream and prairie. Compare **ecotone**.

Elaiosome: A fatty, protein-rich blob attached to a seed, luring ants which carry it off, thereby dispersing the seed.

Emergent plant: One whose leaves and stems grow vertically above the water level, while its roots remain submerged.

Endemic: Plants confined to a particular location, usually a relatively small area, and frequently found nowhere else.

Ephemeral: A plant such as *Erythronium* that grows, flowers, and goes dormant in a short period of time in spring.

Epiphyte: A plant dependent on another plant only for support, not nourishment. Compare **hemiparasite, parasite**.

Exotic: A term referring to nonnative plants. Compare **introduced**.

F

Fen: A wetland where water seeps very slowly through decaying vegetation. Compare **bog, marsh, swamp**.

Fascicle: An arrangement of leaves in bundles.

Filament: The thin lower portion of a **stamen** that holds the **anther**.

Flower: A plant's reproductive structure, often brightly colored, containing male and/or female parts, and usually including petals and sepals.

Fog drip: Condensed water droplets that run down hanging, pointed leaves directly to the plant's root zone.

Forb: A herbaceous plant that is not a grass.

Foredune: The coastal area where beach meets dunes.

Frond: The leaflike part of a fern, seaweed, or lichen. Compare **thallus, hypothallus**.

Fugitive species: Opportunistic species adapted to colonize freshly disturbed habitats.

G

Gametophyte: A tiny haploid (containing a single set of chromosomes) plant which produces gametes (egg and sperm).

Glacial till: Sand, gravel, and stones deposited by a glacier.

Glochids: Tiny, sharp, brittle, barbed spines, easily detached.

Glume: One of a pair of bracts found at the base of a grass **spikelet**.

Grass bald: An open meadowlike site in an otherwise heavily forested area.

Grass stage: Referring to the young stemless seedlings resembling grass produced by some pines, a stage that lasts from two to many years, until conditions are favorable for growth.

Gynostegium: A specialized crown of united stamens, as in milkweeds.

H

Halophyte: A plant that can tolerate salt water.

Heartwood: The older, harder, and often darker-colored interior wood of a tree.

Heliotropic: Turning to face the sun at all times, as sunflowers.

Hemicryptophyte: A plant whose stems die back, leaving a remnant shoot system, with buds hidden by snow or litter.

Hemiparasite: A plant that conducts photosynthesis but also derives some nourishment from a host, such as *Castilleja* spp. Compare **parasite, epiphyte**.

Herb: Any nonwoody, vascular plant.

Herbaceous: Possessing little woody tissue; also describes plants that die back to the ground in winter, sending out new shoots in spring.

Hydrophyte: A plant adapted to growing in water.

Hypothallus: A marginal outgrowth from the **thallus,** in lichens. Compare **frond.**

I

Indicator plant: A plant whose appearance signals something else, as the blooming of dogwood indicates it is time to plant corn.

Inflorescence: A collection of flowers arranged in various ways on a stem, such as the spikes of grasses.

Internode: The part of a stem or culm between two successive nodes.

Intertidal beach: A seashore area that is covered at high tide and exposed at low tide.

Introduced: A term referring to plants that originated outside North America and were introduced by natural forces (wind, animals) and/or people either accidentally or on purpose. Compare **exotic.**

K

Knees: Knobby, above-water protuberances from the roots of cypress (*Taxodium*).

Krummholz: A distinctive zone of trees stunted and prostrated by severe conditions at the treeline.

L

Labellum: A pouchlike petal or lip, particularly in orchids.

Layering: The formation of roots where branches or stems touch the ground for a while.

Leaflet: A subdivision of a compound leaf.

Lenticel: A raised loose mass of cells on a stem forming an opening that allows the exchange of gases. Compare **stoma.**

Lignotuber: A woody swelling at the base of a tree or shrub at or just below soil level containing dormant buds which spring to life if the top growth is destroyed.

Litter: An accumulation of decaying organic material, animal dung, insects, and small animals on the ground; leaf litter in a forest makes an ideal growing medium for young plants. Compare **duff.**

Lobed: Having shallowly divided margins, as in leaves that are cleft into rounded segments but not separate leaflets.

M

Marsh: A wetland through which water floods intermittently and flows steadily. Compare **bog, fen, swamp.**

Mast: The fruit or seed of trees or shrubs. *Hard mast* refers to the nuts and acorns from oak, beech, hickory, witch hazel, black walnut, etc., pine seeds, and samaras from maple, ash, and elm. *Soft mast* is fleshy fruits such as black cherry, ironwood, and mountain ash. Mast trees are important food sources for wildlife. A season of heavy fruit production is called a *mast year*.

Monocarpic: Plants that flower and fruit once, then die, such as agaves.

Monoecious: Having male and female flowers growing separately on the same tree. Compare **dioecious, polygamodioecious.**

Monotypic: A taxonomic division that contains only one member, such as a family containing a single genus, or a genus containing a single species.

Montane: Mountain regions, especially forested slopes in subalpine zones.

Mutualistic relationship: An association between two different species in which both benefit.

Mycelium: The mass of branching, threadlike filaments called hyphae which form the vegetative part of a fungus, as in bracket fungus.

Mycorrhizae: Beneficial soil-borne fungi that attach themselves to roots.

Myrmecochorous: A plant that attracts ants to disperse its seeds.

N

Native: Indigenous, original to an area; specifically in this text, plants for which records (pollen, etc.) exist indicating that they grew in North America before 1500.

Nectary: A nectar-secreting gland in a flower or plant.

Nitrogen-fixing: The process by which bacteria in nodules on plant roots take nitrogen from the air and convert it into a form usable by the plants.

Nodding: Bent to one side.

Node: The place on a stem to which a leaf or branch is attached.

Nodule: A small bump or swelling.

Nurse log: A fallen tree that, as it rots down, provides a fertile site for seedlings. Compare **snag**.

Nurse plant: A plant that provides shade and protection for seedlings of other plants.

Nutlet: One of several tiny nuts, as found in members of the family Boraginaceae; the stone or pit of certain fruits such as cherry.

O

Opposite: The evenly paired arrangement of two plant parts, usually referring to leaves arrayed on either side of a stem. Compare **alternate**.

Ovary: The basal part of the **pistil** that encloses the young undeveloped seed and becomes a fruit.

P

Palmate: Radiating from a common point, as in veins on a maple leaf.

Panicle: A highly branched flower head, common in grasses and shrubs.

Parasite: A plant that derives nutrients directly from a host, usually to the latter's detriment. Compare **epiphyte**, **hemiparasite**.

Pedicel: The stalk of a **flower**, fruit, or **spikelet**.

Perennial: A plant that dies down in fall and returns to bloom and fruit the next year and for several growing seasons.

Perfect flower: One with both male and female sexual parts in one bloom. Also known as **bisexual** or hermaphroditic.

Perigynium: The vaselike sac, beaked at its apex, enclosing the ovary in sedges (*Carex* spp.).

Phloem: The food-conducting tissue of a plant; the inner bark of a tree or shrub, also called **bast**.

Phytochemicals: Plant substances such as flavonoids and carotenoids considered beneficial to human health.

Pinnate: A **compound leaf** with leaflets arranged along a central axis, like a feather.

Pioneer species: The first plants to colonize a new habitat in early stages of **succession**.

Pistil: A flower's female reproductive organ composed of an **ovary**, **style**, and **stigma**.

Pneumatophores: Vertical, pencil-like root extensions in aquatic plants such as mangroves which aid respiration.

Podetium: A stalklike outgrowth from the **thallus** of some lichens.

Podophyllotoxin: A glucoside derived from plants such as mayapple used to treat certain tumors.

Pollinium: A fused mass of pollen grains, as in milkweeds and orchids.

Polygamodioecious: Having **bisexual** and male flowers on some plants, and bisexual and female flowers on others. Compare **dioecious, monoecious**.

Polypore: A member of the genus *Polyporus*, mostly shelf-shaped wild mushrooms called bracket fungi that grow on wood.

Primary dune: Dune closest to the sea.

Pseudobulb: A thick, bulblike part of the stem in orchids, where food and water is stored.

R

Raceme: A flower head with an elongated stem and flowers on short stalks.

Rachilla: The axis of a **spikelet**.

Rachis: The main stalk bearing leaflets, an inflorescence, or the pinnae of a fern.

Raphides: Needlelike crystals of calcium oxalate found in all members of the arum family (Araceae).

Ray flowers: Small flowers resembling petals surrounding the central **disk** in flowers of the aster family (Asteraceae).

Reflexed: Bent backward.

Refractory seed: One requiring **scarification**, or the heat of a fire to germinate.

Relict species: A group of plants, remnants of a widespread population, now surviving in localized areas (called *refugia*).

Resin: A viscid, often aromatic secretion by many conifers that won't dissolve in water and hardens on contact with the air.

Rhizomatous: Having rhizomes.

Rhizome: A horizontal underground stem, usually with roots and shoots at the nodes.

Rhododendron hells: A term used by mountaineers to describe impenetrable thickets of native rhododendrons.

Rosin: A translucent resin derived from pines, used on violin bows and in products such as varnishes, inks, and adhesives.

S

Salt marsh: An area regularly flooded by salt water and inhabited by plants adapted to saline conditions.

Samara: A dry, one-seeded, winged fruit dispersed by the wind; also called a *key*.

Saponin: Any plant glucoside with a soapy action when mixed with water.

Saprophyte: An organism that obtains nutrients from decaying organic matter.

Sapwood: The newer and usually lighter-colored outer layer of wood in a tree.

Savanna: A transition zone between forest and grassland, dominated by grasses with intermittent canopies of trees.

Scape: A leafless flower stalk growing from the ground.

Scarification: The breaking or softening of a hard, impermeable seed coat, allowing water to enter and germination to start.

Sepal: One of the green, leaflike outer flower parts arranged outside the petals. When the petals unfurl, the sepals fall off leaving a rim called a *torus*.

Seral: Describing the stages of **succession**, and the communities of plants that successively appear during the process.

Serotinous: Late to bloom or develop; specifically, describing cones that remain closed until opened by extreme heat, a fire-survival mechanism in pines among other plants.

Sessile: Having no or almost no stalk.

Sheath: The lower part of a grass leaf that wraps around a stem.

Simple: Not compound, as in a leaf.

Sinker: The system of rootlike filaments that parasitic plants such as mistletoe send through the tissues of the host plant.

Slough: A term for a shallow lake system, **swamp**, or **marsh**.

Snag: A standing dead tree, a valuable habitat for many organisms. Compare **nurse log**.

Spadix: A succulent spike of minute flowers often enclosed in a **spathe**.

Spathe: A sheathlike modified leaf cradling the **spadix**, as in *Arisaema* spp.

Spikelet: One of the small spikes making up the inflorescence of a grass.

Spine: A modified leaf, which can be hairlike,

needlelike, barbed, or curved, as in cactus species.

Spore: The reproductive organ in nonflowering plants such as ferns and mosses.

Spur: A small tubular projection of the petal or sepal of a flower, as in *Aquilegia* and *Dicentra* spp.; also a short, woody side shoot on a tree branch.

Stamen: The male, pollen-bearing part of a flower, usually composed of **filament** and **anther**.

Stigma: The female, pollen-receiving end of a flower's **pistil**. The male counterpart that contains the pollen is the **anther**, on the **stamen**.

Stolon: An aboveground stem that produces roots and shoots at the tip and nodes.

Stoma (plural **stomata**): One of many tiny openings in the leaf surface that allow exchange of gases. Compare **lenticel**.

Strobilus (plural **strobili**): The usually cone-shaped male and female reproductive organs of conifers.

Strophiole: (Also called **caruncle**.) An outgrowth of the seed coat; may contain substances attractive to ants which carry off the seeds, aiding in dispersal. Compare **elaiosome**.

Style: The elongated part of the female **pistil**, carrying the **stigma**.

Sub-Alpine: Referring to mountainous zones just below the timberline.

Submontane: Referring to the steppe zone in mountainous areas that occurs below montane regions.

Subshrub: A shrublike plant with a partly woody stem that is mostly **herbaceous**.

Succession: The development of a plant community from initial simple stages to the complex and stable **climax** stage.

Swamp: A wetland where water flows through very slowly and which is dominated by trees or shrubs. Compare **bog**, **fen**, **marsh**.

T

Tannin (syn. **tannic acid**): Astringent phenolic substances found in many plants, possibly as a form of protection; extracted for use in tanning, dyeing, and pharmaceuticals.

Taproot: The primary, swollen, deep root from which the plant's root system extends.

Tepal: Formed when sepals and petals are undifferentiated, as in magnolia.

Thallus: The vegetative body of a lichen. Compare **hypothallus**, **frond**.

Thorn: A stiff, woody, modified stem with a sharp point.

Tidal wrack: Vegetative debris pushed up on a beach by waves.

Tiller: A shoot arising from a node on the rhizomes of grasses.

Transpiration: The evaporation of water from a plant, mostly through the leaves' **stomata**.

Tuber: A swollen root or underground stem that stores food for the plant.

Tubercles: Small warty growths on cactus stems from which spines spring.

Tundra: A treeless, grassy plain in the Arctic dominated by sedges, rushes, mosses, lichens, and small shrubs, growing over permanently frozen subsoil.

U

Umbel: A flat-topped clustered flower head in which the tiny flower stalks arise from the top of the stem like the stays of an umbrella.

Understory: Trees and shrubs in the layer immediately below the forest canopy.

V

Vernal pool: A depression intermittently covered by shallow water from winter to spring, and usually completely dry for most of summer and fall.

Warm-season grass: A grass that doesn't start growing until the soil warms up. Compare **cool-season grass**.

Xeriscape: A water-efficient landscape using drought-tolerant plants.

Xylem: The tissue that carries water and minerals upward in a plant.

Notes

Introduction

1. From the transcript of an interview for *The Sacred Balance* television series, by David Suzuki and Amanda McConnell, broadcast on CBC and PBS, 2002/2003.
2. William Cronon, *Changes in the Land: Indians, Colonists, and the Ecology of New England* (New York: Farrar Straus Giroux, 1983), pp. 20–21. This is a magnificent book.
3. Edward O. Wilson, *The Future of Life* (New York: Alfred A. Knopf, 2002), p. xvi.

Region 1: The Eastern Forests

1. Giovanni da Verrazzano, 1524, in Susan Tarrow, trans., "Translation of the Cellère Codex," *The Voyages of Giovanni da Verrazzano, 1524–1528*, ed. Lawrence C. Wroth (New Haven, Conn.: Yale University Press, 1970), pp. 133–143, this reference to p. 139, http://bc. barnard.columbia.edu/~/gordis/earlyAC/document/verrazan.htm.
2. William Wood, *New England's Prospect* (1634, republished, Boston: John Wilson and Son, 1865), p. 17.
3. William Cronon, *Changes in the Land: Indians, Colonists, and the Ecology of New England* (New York: Farrar Straus Giroux, 1983), pp. 142–144.
4. "National Register of Big Trees," American Forests, www.americanforests.org/resources/bigtrees.
5. Donald Culross Peattie, *A Natural History of Trees of Eastern and Central North America* (Boston: Houghton Mifflin, 1991), p. 4.
6. Cronon, *Changes in the Land*, pp. 109, 110.
7. François André Michaux, *The North American Sylva; or, A Description of the Forest Trees of the United States, Canada, and Nova Scotia . . .*, 3 vols., trans. A. L. Hillhouse, (Paris: C. d'Hautel, 1819).

8. Arthur Plotnik, *The Urban Tree Book: An Uncommon Field Guide for City and Town* (New York: Three Rivers Press, 2000), p. 315.

9. Craille Maguire Gillies, "Could, Should, Wood," *Globe and Mail*, March 24, 2001, p. R23.

10. Sheila Connor, *New England Natives* (Cambridge, Mass: Harvard University Press, 1994), p. 81.

11. Ibid., p. 86.

12. Rebecca Rupp, *Red Oaks and Black Birches: The Science and Lore of Trees* (Pownal, Vt.: Storey Communications, 1990), pp. 148–149.

13. Ibid., p. 150.

14. Peattie, *Trees of Eastern and Central North America*, p. 238.

15. Rupp, *Red Oaks and Black Birches*, pp. 148–149.

16. "Symbolic Concepts: Symbolic Memorial," Oklahoma City National Memorial, 2001, www.oklahomacitynationalmemorial.org/symbolic/concepts.html.

17. Nancy Wick, "Gardener Works to Preserve History of Campus Trees," *University Week*, March 3, 2001, University of Washington, http://depts.washington.edu/uweek/archives/2001.03.MAR_29/_article6.html.

18. Melvin Randolph Gilmore, *Uses of Plants by the Indians of the Missouri River Region* (Washington, D.C.: U.S. Government Printing Office, 1919), p. 63.

19. Peter Kalm, *Travels in North America* (1770), in Charlotte Erichsen-Brown, *Medicinal and Other Uses of North American Plants: A Historical Survey with Special Reference to the Eastern Indian Tribes* (New York: Dover, 1979, 1989), p. 31.

20. Raymond L. Taylor, *Plants of Colonial Days* (Mineola, N.Y.: Dover Publications, 1996), p. 71.

21. Brian "Fox" Ellis, "The Cottonwood," *Science and Children*, vol. 38, no. 4 (January 2001): 43–44.

22. Jennifer Morrissey, "Awed by Cottonwoods," *Green Line*, vol. 11, no. 1 (spring 1999), http://coloradoriparian.org/GreenLine/V11-1/Cottonwoods.html.

23. Mark Twain, *Adventures of Huckleberry Finn* (Oxford: Oxford University Press, 1999), p. 59.

24. Jon D. Johnson and Gorden Ekuan, "Hybrid Poplar Research Program," Washington State University, http://www.puyallup.wsu.edu/poplar/.

25. Lou Sebesta, "Balmville Tree," NYS Department of Environmental Conservation, reprinted from *New York State Conservationist*, April 1999, www.dec.state.ny.us/website/dpae/cons/balmvilletree.html.

26. Rupp, *Red Oaks and Black Birches*, pp. 27–28.

27. A. W. Brooks, "*Castanea dentata*," in Connor, *New England Natives*, p. 190.

28. Rupp, *Red Oaks and Black Birches*, p. 30.

29. Connor, *New England Natives*, p. 186.

30. Henry David Thoreau, "17 October 1860," *The Journal of Henry D. Thoreau*, ed. Bradford Torrey and Francis H. Allen (New York: Dover, 1962), vol. XIV, p. 137 [vol. #2, p. 1698].

31. "Research and Restoration," The American Chestnut Foundation, 2000, http://chestnut.acf.org/r_r.htm.

32. Rupp, *Red Oaks and Black Birches*, p. 41.

33. Ibid.

34. John James Audubon, *The Birds of America* (New York: Dover, 1967), pp. 27, 28.

35. "Wildman" Steve Brill and Evelyn Dean, *Identifying and Harvesting Edible and Medicinal Plants in Wild (and Not So Wild) Places* (New York: Hearst Books, 1994), p. 157.

36. Daniel Moerman, *Native American Ethnobotany* (Portland, Ore.: Timber Press, 1998), p. 141.

37. John Eastman, *The Book of Forest and Thicket: Trees, Shrubs, and Wildflowers of Eastern North America* (Mechanicsburg, Penn.: Stackpole Books, 1992), p. 108.

38. Peattie, *Trees of Eastern and Central North America*, p. 138.

39. Ibid. p. 130.

40. Moerman, *Native American Ethnobotany*, p. 140.

41. Daniel Drake, *Pioneer Life in Kentucky*, in Peattie, *Trees of Eastern and Central North America*, p. 131.

42. Benjamin Pressley, "Conquering the Darkness: Primitive Lighting Methods," *Bulletin of Primitive Technology*, no. 12 (fall 1996), http://www.hollowtop.com/spt_html/lighting.htm.

43. Peattie, *Trees of Eastern and Central North America*, p. 149.

44. "Pecans, So Good for You," National Pecan Shellers Association, 2002, http://www.ilovepecans.org/.

45. Guy Ames and Steve Driver, *Sustainable Pecan Production*, Appropriate Technology Transfer for Rural Areas, www.attra.org/attra-pub/pecan.html#pecanculture.

46. Henry Kock, interview, June 2001. Henry Kock, an interpretive horticulturist at the University of Guelph (Ontario) Arboretum, has written and spoken extensively on the trees of the Carolinian forest.

47. Moerman, *Native American Ethnobotany*, p. 465.
48. Peattie, *Trees of Eastern and Central North America*, p. 196.
49. Ibid., p. 197.
50. "Quiet Giant, the Wye Oak," Maryland Department of Natural Resources, Forest Service, 2002, http://www.dnr.state.md.us/forests/trees/giant.html.
51. Interview with Kevin Jackson.
52. ACC Public Information Office, "The Tree That Owns Itself," ACC Online, Unified Government of Athens–Clarke County, 1999–2002, www.athensclarkecounty.com/tour/tour10.htm.
53. Eastman, *Forest and Thicket*, p. 129.
54. Ibid., p. 132.
55. Frances Densmore, *Indian Use of Wild Plants, for Food, Medicine, and Charms*, reprint of *Uses of Plants by the Chippewa Indians*, from the 44th annual report of the U.S. Bureau of American Ethnology, 1926–1927 (Washington, D.C.: U.S. Government Printing Office, 1928; reprinted, Ohsweken, Ont.: Iroqrafts Ltd., 1993), p. 308.
56. Ibid., p. 312.
57. Christien LeClercq, *New Relations of Gaspesia with the Customs and Religion of the Gaspesian Indians*, in Erichsen-Brown, *Medicinal and Other Uses*, p. 79.
58. Susanna Moodie, *Roughing It in the Bush* (London: Virago Press, 1986), pp. 421–422.
59. Rupp, *Red Oaks and Black Birches*, p. 46.
60. Eastman, *Forest and Thicket*, p. 128.
61. R. B. Hillman, "Red Maple—*Acer rubrum*," September 18, 2001, web.vet.cornell.edu/CVM/HANDOUTS/plants/MAPLE.html.
62. Kalm, *Travels in North America*, p. 80.
63. Charles Sprague Sargent, in Connor, *New England Natives*, p. 196.
64. Moerman, *Native American Ethnobotany*, p. 40.
65. Rupp, *Red Oaks and Black Birches*, p. 81.
66. Connor, *New England Natives*, p. 115.
67. Moerman, *Native American Ethnobotany*, pp. 122, 125.
68. Huron Smith, *Ethnobotany of the Potawatomi*, in Erichsen-Brown, *Medicinal and Other Uses*, p. 46.
69. Connor, *New England Natives*, p. 242.
70. Euell Gibbons, *Stalking the Wild Asparagus* (New York: David McKay, 1962), p. 34.
71. Rupp, *Red Oaks and Black Birches*, p. 80.
72. Moerman, *Native American Ethnobotany*, p. 254.
73. C. F. Saunders, *Western Wild Flowers and Their Stories* (New York: Doubleday Doran and Co., 1933), p. 149.
74. Eastman, *Forest and Thicket*, p. 194.
75. *Oxford Companion to Gardens*, ed. Sir Geoffrey Jellicoe, et al. (Oxford: Oxford University Press, 1986), p. 560.
76. Campbell Hardy, *Forest Life in Acadie: Sketches of Sport and Natural History in the Lower Provinces of the Canadian Dominion* (London: Chapman and Hall, 1869), p. 32.
77. Samuel Strickland, *Twenty-Seven Years in Canada West*, in Mary Alice Downie and Mary Hamilton, *"And Some Brought Flowers": Plants in a New World* (Toronto: University of Toronto Press, 1980), "Hemlock."
78. Catherine Parr Traill, *Canadian Settler's Guide* (Toronto: McClelland and Stewart, 1969; first published 1885), p. 138.
79. Eastman, *Forest and Thicket*, p. 15.
80. Connor, *New England Natives*, p. 28.
81. Michael A. Dirr, *A Manual of Woody Landscape Plants: Their Identification, Ornamental Characteristics, Culture, Propagation, and Uses*, 3rd ed. (Champaign, Ill.: Stipes 1983), p. 624.
82. Taylor, *Plants of Colonial Days*, p. 57.
83. Connor, *New England Natives*, p. 105.
84. Rupp, *Red Oaks and Black Birches*, pp. 36–38.
85. Elizabeth Lawrence, *Gardening for Love: The Marketing Bulletins* (Durham, N.C.: Duke University Press, 1987), p. 148.
86. Eastman, *Forest and Thicket*, p. 72.
87. William Bartram, *Travels and Other Writings: Travels through North and South Carolina, Georgia, East and West Florida, Travels in Georgia and Florida, 1773–74—A Report to Dr. John Fothergill, Miscellaneous Writings* (New York: Library of America, 1996), pp. 321–322.
88. Peter Landry, "Blupete's Wildflowers of Nova Scotia," June 2001, http://www.blupete.com/Nature/Wildflowers/Wild.htm#Bunchberry.
89. Peattie, *Trees of Eastern and Central North America*, p. 546.
90. Frances Milton Trollope, *Domestic Manners of the Americans*, 4th ed. (London: Whittaker, Treacher, 1832), vol. 2, pp. 60–61.
91. Paul Le Jeune, *Jesuit Relations*, in Erichsen-Brown, *Medicinal and Other Uses*, p. 104.
92. Ibid., pp. 104–105.
93. Tim Flannery, *The Eternal Frontier: An Ecological History of North America and Its People* (New York: Atlantic Monthly Press, 2001), p. 305.

94. Cronon, *Changes in the Land*, p. 21.

95. Virgil J. Vogel, *American Indian Medicine* (Norman: University of Oklahoma Press, 1970), p. 39.

96. Rupp, *Red Oaks and Black Birches*, pp. 164–165.

97. Densmore, *Indian Use of Wild Plants*, p. 340.

98. Jacques Cartier, *First Voyage to Canada*, in Erichsen-Brown, *Medicinal and Other Uses*, p. 163.

99. Catherine Parr Traill, *The Backwoods of Canada: Being Letters from the Wife of an Emigrant Officer* (Toronto: McClelland and Stewart, 1966, 1971; first published 1846), p. 61.

100. Tom Isern, "Plains Folk: Plum Full," *News for North Dakotans*, September 24, 1998, Agriculture Communication, North Dakota State University, http://www.ext.nodak.edu/extnews/newsrelease/1998/092498/20plains.htm.

101. Trollope, *Domestic Manners of the Americans*, vol. 1, pp. 269–271.

102. Peattie, *Trees of Eastern and Central North America*, p. 519.

103. Frances Theodora Parsons [Mrs. William Starr Dana], *How to Know the Wild Flowers: A Guide to the Names, Haunts, and Habits of Our Common Wild Flowers* (New York: Dover, 1963), pp. 36–37.

104. Toby Musgrave, Will Musgrave, and Chris Gardner, *The Plant Hunters: Two Hundred Years of Adventure and Discovery Around the World* (New York: Sterling, 1998), p. 99.

105. Richard M. Steele, "Plant Hunting: Privilege Pleasure Posterity," *The Rosebay Newsletter*, vol. 27 (spring 2000), http://rosebay.org/chapterweb/rosebay/plant_hunting_in_nova_scotia.htm.

106. Ibid.

107. Jacob Bigelow, *American Medical Botany, Being a Collection of the Native Medicinal Plants of the United States*, in Connor, *New England Natives*, p. 158.

108. Dan McFeeley, "Added Contributions to Bee Tidbits," *Some Tidbits about Bees and Honey from Prehistory to Present*, April 6, 1999, http://www.chebucto.ns.ca/~ag151/addendum.html.

109. John Muir, *My First Summer in the Sierra*, in David Brower, ed., *Gentle Wilderness: The Sierra Nevada* (New York: Sierra Club and Ballantine Books, 1968), p. 21.

110. Thoreau, "June 4, 1853," *Journal*, vol. V, p. 221 [vol. 1, p. 583].

111. Roger Williams, *Key into the Language of America*, in Timothy Coffey, *The History and Folklore of North American Wildflowers* (Boston: Houghton Mifflin, 1993), p. 96.

112. Connor, *New England Natives*, p. 35.

113. Whitesbog Preservation Trust, "Discover Historic Whitesbog Village," Burlington County Library, www.whitesbog.org.

114. Hardy, *Forest Life in Acadie*, p. 200.

115. Samuel Hearne, *A Journey from Prince of Wales's Fort in Hudson's Bay, to the Northern Ocean: Undertaken by Order of the Hudson's Bay Company for the Discovery of Copper Mines, a North West Passage, &c. in the Years 1769, 1770, 1771, and 1772* (London: A. Strahan and T. Cadell, 1795; republished, Edmonton: M. G. Hurtig, 1971), pp. 449–450.

116. "The History of Cranberry Production," Cape Cod Cranberry Growers Association, 2001, www.cranberries.org/cchistory.html.

117. Brill and Dean, *Identifying and Harvesting*, pp. 180–181.

118. Ibid., p. 184.

119. Thoreau, "August 8, 1858," *Journal*, vol. XI, p. 84 [vol. 2, p. 1335].

120. Dirr, *Manual of Woody Landscape Plants*, p. 302.

121. Eastman, *Forest and Thicket*, pp. 116–117.

122. Coffey, *North American Wildflowers*, p. 288.

123. Parsons [Dana], *How to Know the Wild Flowers*, p. 269.

124. Kalm, *Travels in North America*, p. 236.

125. Catherine Parr Traill, *Canadian Wild Flowers* (Montreal: John Lovell, 1868; facsimile ed., Toronto: Coles, 1972), pp. 9–11.

126. Huron Smith, *Ethnobotany of the Meskwaki*, in Erichsen-Brown, *Medicinal and Other Uses*, p. 237.

127. Gibbons, *Stalking the Wild Asparagus*, pp. 284–285.

128. Barbara Hall, "Poison I Eye-Eye-Eye-Eye VEE," *Weeds and Wild Things*, Suite 101.com, February 23, 1998, http://www.suite101.com/article.cfm/weeds_and_wild_things/5847.

129. Anna Jameson, *Georgian Bay Lake Huron*, in *Littleflower's the Medicine of North American Plants*, Little Flower Publications, www.geocities.com/RodeoDrive/Mall/4992/poison_ivy.html.

130. Jane Vansittart, ed., *Lifelines: The Stacey Letters, 1836–1858* (London: Peter Davies, 1976), letter from George Stacey, October 17, 1847, p. 75.

131. Katherine Carter Ewel, "Effects of Fire and Wastewater on Understory Vegetation in

Cypress Domes," in *Cypress Swamps*, ed. Howard T. Odum and Katherine Carter Ewel (Gainesville: University Press of Florida, 1985), pp. 119–126.

132. Samuel de Champlain, *Journal of Champlain for the Years 1615–1618*, in Erichsen-Brown, *Medicinal and Other Uses*, p. 365.

133. Brill and Dean, *Identifying and Harvesting*, p. 214.

134. Jerry Doll, "Hemp Dogbane Biology and Management," University of Wisconsin, Department of Agronomy Publication, Agronomy Advice, Fd. Crops. 33.0, ipcm.wisc.edu/uw_weeds/articles/hempdog.htm.

135. Neltje Blanchan, *Nature's Garden*, in Jack Sanders, *Hedgemaids and Fairy Candles: The Lives and Lore of North American Wildflowers* (Camden, Maine: Ragged Mountain Press, 1993), p. 61.

136. Robert Hendrickson, *Ladybugs, Tiger Lilies, and Wallflowers: A Gardener's Book of Words* (Upper Saddle River, N.J.: Prentice Hall, 1993), p. 79.

137. Krista Foss, "Study Finds Ginseng Reduces Blood Sugar," *Globe and Mail*, April 10, 2000, p. A5.

138. Lawrence, *Gardening for Love*, p. 143.

139. Eastman, *Forest and Thicket*, pp. 86–87.

140. Lewis Gannett, *Cream Hill: Discoveries of a Week-end Countryman*, in Mary Durant, *Who Named the Daisy? Who Named the Rose? A Roving Dictionary of North American Wildflowers*, 1st paperback ed. (New York: Dodd-Mead, 1976), p. 79.

141. Moerman, *Native American Ethnobotany*, pp. 105–106.

142. Diamond Jenness, *The Ojibwa Indians of Parry Island, Their Social and Religious Life*, in Erichsen-Brown, *Medicinal and Other Uses*, p. 323.

143. Traill, *Backwoods of Canada*, p. 107.

144. Frances Densmore, *Uses of Plants by the Chippewa Indians*, in Erichsen-Brown, *Medicinal and Other Uses*, p. 385.

145. C. S. Rafinesque, *Medical Flora or Manual of Medical Botany of the United States*, in Erichsen-Brown, *Medicinal and Other Uses*, p. 385.

146. Barbara Hall, "A Jewel of a Weed," *Weeds and Wild Things*, Suite 101.com, August 25, 1998, http://www.suite101.com/article.cfm/weeds_and_wild_things/10044.

147. Brill and Dean, *Identifying and Harvesting*, p. 73.

148. Euell Gibbons, *Stalking the Healthful Herbs* (New York: David McKay, 1966), p. 209.

149. Brill and Dean, *Identifying and Harvesting*, p. 73.

150. Mathew Tekulsky, *The Hummingbird Garden: Turning Your Garden, Window Box, or Backyard into a Beautiful Home for Hummingbirds* (New York: Crown, 1990), p. 23.

151. Rafinesque, *Medical Flora*, in Erichsen-Brown, *Medicinal and Other Uses*, p. 356.

152. J. Auguste Mockle, *Contributions à l'étude des plantes médicinales du Canada*, in Erichsen-Brown, *Medicinal and Other Uses*, p. 357.

153. Catherine Parr Traill, *Female Emigrant's Guide, and Hints on Canadian Housekeeping* (Toronto: Maclear, 1854), p. 91. One of the best books on the early immigrant experience is Charlotte Gray's *Sisters in the Wilderness: The Lives of Susanna Moodie and Catherine Parr Traill* (Toronto: Viking, 1999). It gives an intimate and heart-rending portrait of the difficulties of the settler, unlike any other I have read.

154. John Bartram and William Bartram, *John and William Bartram's America: Selections from the Writings of the Philadelphia Naturalists*, ed. Helen Gere Cruickshank (New York: Devin-Adair, 1957), pp. 362–364.

155. Smith, *Ethnobotany of the Meskwaki*, in Erichsen-Brown, *Medicinal and Other Uses*, p. 244.

156. Oliver Medsger, *Edible Wild Plants* (New York: Macmillan, 1939), pp. 193–194.

157. Doug Elliott, *Wild Roots: A Forager's Guide to the Edible and Medicinal Roots, Tubers, Corms, and Rhizomes of North America* (Rochester, Vt.: Healing Arts Press, 1995), p. 73.

158. Thoreau, "October 12, 1858," *Journal*, Vol. XI, p. 208 [Vol. II, p. 1366].

159. Smith, *Ethnobotany of the Potawatomi*, in Erichsen-Brown, *Medicinal and Other Uses*, p. 307.

160. Gibbons, *Stalking the Healthful Herbs*, pp. 92–94.

161. Traill, *Canadian Settler's Guide*, p. 228.

162. Connor, *New England Natives*, p. 58.

163. Hardy, *Forest Life in Acadie*, p. 1.

164. Henry David Thoreau, *Walden*, ed. Stephen Allen Fender (Oxford: Oxford University Press, 1997), "House-Warming," p. 215.

165. Harvey Wickes Felter and John Uri Lloyd, "*Corydalis.—Turkey-Corn*," *King's American Dispensatory*, Henriette's Herbal Homepage, 2000–2002, http://www.ibiblio.org/herbmed/eclectic/kings/dicentra-cana.html.

166. John D. Gunn, *New Domestic Physician or Home Book of Health*, in Erichsen-Brown, *Medicinal and Other Uses*, p. 368.

167. Huron Smith, *Ethnobotany of the Menomini*, in Erichsen-Brown, *Medicinal and Other Uses*, p. 368.

168. Ibid., p. 366.

169. Brill and Dean, *Identifying and Harvesting*, p. 267.

170. Traill, *Backwoods of Canada*, p. 88.

171. Moerman, *Native American Ethnobotany*, p. 337.

172. Paige Peters, "Fruits, Roots and Shoots of the Northwoods," *Northbound*, vol. 11, no. 1 (spring 2001), p. 3, Trees for Tomorrow, http://www.treesfortomorrow.com/download/WILDEDIB.pdf.

173. Durant, *Who Named the Daisy?*, p. 183.

174. Elliott, *Wild Roots*, p. 27.

175. Moerman, *Native American Ethnobotany*, p. 422.

176. Smith, *Ethnobotany of the Menomini*, in Erichsen-Brown, *Medicinal and Other Uses*, p. 341.

177. John Gerard, *The Herball or General History of Plants Gathered by John Gerarde of London, Master in Chirurgerie*, in Erichsen-Brown, *Medicinal and Other Uses*, p. 340.

178. Brill and Dean, *Identifying and Harvesting*, p. 263.

179. Recipe adapted from Gerina Dunwich, *Wicca Candle Magick*, www.crystalforest1.homestead.com/oilrecipes.html.

180. Frederick W. Case Jr. and Roberta B. Case, *Trilliums* (Portland, Ore.: Timber Press, 1997), p. 26.

181. Traill, *Canadian Wild Flowers*, p. 31.

182. Elisabeth Beaubien, "Western Trillium," *Plantwatch*, University of Alberta, Devonian Botanic Garden, 1997–2002, www.devonian.ualberta.ca/pwatch/westtr.htm#NAMES.

183. Annora Brown, *Old Man's Garden* (Toronto: J. M. Dent and Sons, 1954), p. 144.

184. Parsons [Dana], *How to Know the Wild Flowers*, pp. 71, 73.

185. Mary Chiltoskey and Paul B. Hamel, *Cherokee Plants: Their Uses—A 400 Year History*, in "Indian Pipes: *Monotropa Uniflora*," *Wild Flowers and Plants of North Carolina*, NCNatural, 2000, http://ncnatural.com/wildflwr/indnpipe.html.

186. Ana Nez Heatherley, *Healing Plants: A Medicinal Guide to Native North American Plants and Herbs* (Toronto: HarperCollins, 1998), p. 84.

187. Brown, *Old Man's Garden*, p. 54.

188. Durant, *Who Named the Daisy?*, p. 101.

189. Parsons [Dana], *How to Know the Wild Flowers*, p. 253.

190. Traill, *Backwoods of Canada*, pp. 89–90.

191. Moerman, *Native American Ethnobotany*, p. 191.

192. John Smith, *Description of Virginia and Proceedings of the Colony*, in Erichsen-Brown, *Medicinal and Other Uses*, p. 333; Huron Smith, *Ethnobotany of the Potawatomi*, in Erichsen-Brown, *Medicinal and Other Uses*, p. 334.

193. Bernard Assiniwi, *Survival in the Bush* (Toronto: Copp Clark, 1972), p. 99.

194. Alice Morse Earle, *Old Time Gardens, Newly Set Forth* (New York: Macmillan, 1901), p. 457.

195. H. W. Youngken, "The Drugs of the North American Indian," in Erichsen-Brown, *Medicinal and Other Uses*, p. 320.

196. Durant, *Who Named the Daisy?*, pp. 21–22.

197. Coffey, *North American Wildflowers*, pp. 30–31.

198. Moerman, *Native American Ethnobotany*, pp. 515–516.

199. Angela Gillaspie, "Poke Salet: The Versatile Veggie," *Momma's Southern Stuff*, August 1998, http://home.talkcity.com/KudzuKorner/gnats-themomma/sostuf.html.

200. Traill, *Canadian Wild Flowers*, pp. 84–86.

201. Thoreau, "October 12, 1858," *Journal*, vol. XI, p. 207 [vol. 2, p. 1366].

202. See Hoffman, Millspaugh, Wood and Ruddock, Densmore, and Smith in Erichsen-Brown, *Medicinal and Other Uses*, p. 349.

203. Moerman, *Native American Ethnobotany*, p. 80.

204. Ralph Waldo Emerson, "Musketaquid," *Poems* (Boston: Houghton Mifflin, 1911), p. 144.

205. John Burroughs, *Signs and Seasons* (Cambridge, Mass.: Riverside Press, 1886), p. 188.

206. James Mooney, *The Sacred Formulas of the Cherokees*, and Huron Smith, *Ethnobotany of the Meskwaki*, in Erichsen-Brown, *Medicinal and Other Uses*, p. 306.

207. James Mooney, "Folklore of the Carolina Mountains," in Coffey, *North American Wildflowers*, p. 18.

208. Theodat Gabriel Sagard, *Le gran voyage du pays des Hurons, situé en l'Amerique vers la mer douce, es dernier confines de la Nouvelle France, dite Canada*, in Erichsen-Brown, *Medicinal and Other Uses*, p. 444.

209. Moerman, *Native American Ethnobotany*, p. 582.

210. Sanders, *Hedgemaids and Fairy Candles*, p. 35.

211. Densmore, *Use of Plants by the Chippewa*, p. 134.

212. Dirr, *Manual of Woody Landscape Plants*, p. 483.

213. Plotnik, *Urban Tree Book*, p. 136.

214. Dirr, *Manual of Woody Landscape Plants*, p. 600.

215. William Kennedy, *Texas: The Rise, Progress, and Prospects of the Republic of Texas* (London: Hastings, 1841).

216. George Wilkins Kendall, *Narrative of the Texan Santa Fé Expedition . . .* (New York: Harper and Bros., 1844), p. 139.

217. Peattie, *Trees of Eastern and Central North America*, pp. 221–222.

218. Moerman, *Native American Ethnobotany*, p. 460.

219. Ibid, p. 463.

220. Taylor, *Plants of Colonial Days*, p. 56.

221. W. P. Armstrong, "Major Types of Chemical Compounds in Plants and Animals," *Wayne's Word*, WOLFFIA, Inc., http://waynesword. palomar.edu/chemid1.htm.

222. Harriet Beecher Stowe, *Palmetto Leaves* (Gainsville: University Press of Florida, 1968), p. 45.

223. William Bartram, *Travels of William Bartram*, ed. Mark van Doren (New York: Dover, 1928), p. 91.

224. Shona Ellis et al., "*Diospyros virginiana*," *An Ethnobotany of the UBC Arboretum*, March 30, 1995, web adaptation August 1998, http:// botany.ubc.ca/arboretum/UBC030.HTM.

225. Lawrence, *Gardening for Love*, pp. 147–148.

226. Bartram, *Travels of William Bartram*, p. 41.

227. "Bill May Make Saw Palmetto Florida Crop," *Lubbock Online.com, Lubbock-Avalanche Journal*, May 6, 1997, http://www.lubbockonline.com/ news/040797/billmay.htm.

228. Family tradition has it that Mr. Tillman's great-grandfather, James Tillman, was born in Beaufort, South Carolina, in 1776, so Ms. Hicks wondered if these ties could account for the similarity between these legends. Dennis Adams, Information Services Coordinator, Beaufort County (South Carolina) Public Library, http://bcgov.org/bftlib/spanish.htm.

229. Adams, *Spanish Moss*; Raymond J. Martinez, *The Story of Spanish Moss and Its Relatives*, Home Publications, 1959, Communities Online, Inc., 1997, www.communityonline.com/local/culture/ spanishmoss/spanishmoss5.htm and ~spanish moss6.htm.

Region 2: Swamps and Wetlands

1. Henry David Thoreau, *The Major Essays of Henry David Thoreau*, ed. Richard Dillman (New York: Whitston, 2001), "Walking," p. 177.

2. William Bartram, *Travels of William Bartram*, ed. Mark van Doren (New York: Dover, 1928), p. 94.

3. Ann Vileisis, *Discovering the Unknown Landscape: A History of America's Wetlands* (Washington: Island Press, 1997), p. 118.

4. Ibid., pp. 121–122.

5. Sheila Connor, *New England Natives* (Cambridge, Mass.: Harvard University Press, 1994), pp. 192, 193.

6. Stephen A. Spongberg, *A Reunion of Trees: The Discovery of Exotic Plants and Their Introduction into North American and European Landscapes* (Cambridge, Mass.: Harvard University Press, 1990), p. 13.

7. Arthur Plotnik, *The Urban Tree Book: An Uncommon Field Guide for City and Town* (New York: Three Rivers Press, 2000), p. 61.

8. Rebecca Rupp, *Red Oaks and Black Birches: The Science and Lore of Trees* (Pownal, Vt.: Storey Communications, 1990), pp. 67–68.

9. Ibid., p. 68.

10. Thomas Jefferson, in Raymond L. Taylor, *Plants of Colonial Days* (Mineola, N.Y.: Dover, 1996), p. 84.

11. Rupp, *Red Oaks and Black Birches*, p. 74.

12. Euell Gibbons, *Stalking the Wild Asparagus* (New York: David McKay, 1962), pp. 118–119.

13. Plotnik, *Urban Tree Book*, p. 67.

14. Michael A. Dirr, *A Manual of Woody Landscape Plants*, 3rd ed. (Champaign, Ill.: Stipes, 1983), p. 634.

15. J. S. Pitcher and J. S. McKnight, "Black Willow—*Salix nigra*," *Trees of Western North Carolina*, reprinted online with permission of the USDA Forest Service, wildwnc.org/trees/ Salix_nigra.html.

16. John Gerard, *The Herball or General History of Plants Gathered by John Gerarde of London, Master in Chirurgerie*, in Erichsen-Brown, *Medicinal and Other Uses*, p. 91.

17. Bonnie Gale, "American Willow Growers Network," *English Basketry Willows*, October 3, 2000, http://www.msu.edu/user/shermanh/ galeb/awgndesc.htm.

18. Rupp, *Red Oaks and Black Birches*, p. 191.

19. Ibid.

20. Heather Sanft, "How to Plant a Living Lattice Fence," *East Coast Gardener*, vol. 3, no. 2 (March 1999), http://www.klis.com/fundy/ecg/ fence3–2.htm.

21. "How Pussy Willows Got Their Name," http://www.mooncrystal.com/~gypsy/cattrivia/pwillow.html.

22. Huron H. Smith, *Ethnobotany of the Potawatomi*, in Erichsen-Brown, *Medicinal and Other Uses*, p. 94.

23. Donald Culross Peattie, *A Natural History of Trees of Eastern and Central North America*, (Boston: Houghton Mufflin, 1991), p. 310.

24. David Williams and Catherine Hart, "American Sweetgum—*Liquidambar styraciflua*," Illinois Plant, Landscape and Nursery Technology, University of Illinois Extension, http://www.extension.uiuc.edu/IPLANT/plant_select/arboretum_trees/Sweetgum.htm.

25. Francisco Hernandez, "*De Xochiocotzo Quahuitl, seu Arbore Liquidambari Indici*," in Leslie Day, "The City Naturalist—Sweet Gum Tree," NY West Side Site: The 79th Street Boat Basin Flora and Fauna Society, 1996, www.nysite.com/nature/flora/sweetgum.htm.

26. Peattie, *Trees of Eastern and Central North America*, pp. 310–311.

27. Daniel Moerman, *Native American Ethnobotany* (Portland, Ore.: Timber Press, 1998), p. 170.

28. Gibbons, *Stalking the Wild Asparagus*, p. 162.

29. Dr. William Desmond, *Pomona*, in "Asimina," Tripple Brook Farm, www.tripplebrookfarm.com/iplants/Asimina.html.

30. Ibid.

31. Meriwether Lewis and William Clark, 1806, in Timothy Coffey, *The History and Folklore of North American Wildflowers* (Boston: Houghton Mifflin, 1993), pp. 284–285.

32. Peter Kalm, *Travels in North America*, in Erichsen-Brown, *Medicinal and Other Uses*, p. 236.

33. Janet Lyons and Sandra Jordan, *Walking the Wetlands: A Hiker's Guide to Common Plants and Animals of Marshes, Bogs, and Swamps* (New York: John Wiley and Sons, 1989), p. 29.

34. R. J. Marles, *The Ethnobotany of the Chipewyan of Northern Saskatchewan*, in Harriet V. Kuhnlein and Nancy J. Turner, *Traditional Plant Foods of Canadian Indigenous Peoples: Nutrition, Botany, and Use* (Philadelphia: Gordon and Breach Science, 1991), p. 76.

35. James A. Duke, *Handbook of Edible Weeds* (Boca Raton, Fla.: CRC Press, 1992), p. 178.

36. Melana Hiatt, "Bulrush (*Scirpus* species)," Edible Wild Kitchen, www.ediblewild.com/bill.html.

37. Peter Kalm, *The America of 1750: Peter Kalm's Travels in North America*, in Coffey, *North American Wildflowers*, p. 93.

38. Kalm, *Travels in North America*, in Erichsen-Brown, *Medicinal and Other Uses*, p. 228.

39. William Bartram, *Travels of William Bartram*, in Erichsen-Brown, *Medicinal and Other Uses*, p. 228–229.

40. Virgil J. Vogel, *American Indian Medicine*, (Norman, Okla.: University of Oklahoma Press, 1970), p. 284.

41. Frank G. Speck, *Medicine Practices of the Northeastern Algonquians*, in Erichsen-Brown, *Medicinal and Other Uses*, p. 231.

42. H. W. Youngken, "The Drugs of the North American Indian," in Erichsen-Brown, *Medicinal and Other Uses*, p. 231.

43. William A. Emboden Jr., *Narcotic Plants, Hallucinogens, Stimulants, Inebriants, and Hypnotics, Their Origins and Uses*, in Erichsen-Brown, *Medicinal and Other Uses*, p. 233.

44. Gibbons, *Stalking the Wild Asparagus*, p. 52.

45. Doug Elliott, *Wild Roots: A Forager's Guide to the Edible and Medicinal Tubers, Corns, and Rhizomes of North America* (Pownal, Vt.: Healing Arts Press, 1995), p. 104.

46. Lyons and Jordan, *Walking the Wetlands*, p. 55.

47. John Josselyn, *New-England's Rarities Discovered*, in Vogel, *American Indian Medicine*, p. 42.

48. John Josselyn, *New-England's Rarities*, in Erichsen-Brown, *Medicinal and Other Uses*, p. 248.

49. Kalm, *Travels in North America*, p. 249.

50. Harvey Wickes Felter and John Uri Lloyd, "*Veratrum Viride*," *King's American Dispensatory*, Henriette's Herbal Homepage, 2000–2002, http://www.ibiblio.org/herbmed/eclectic/kings/veratrum-viri.html.

51. Huron H. Smith, *Ethnobotany of the Menomini*, in Erichsen-Brown, *Medicinal and Other Uses*, p. 210.

52. Elliott, *Wild Roots*, p. 108.

53. www.wa.gov/ecology/wq/plants/native/nuphar.html.

54. Coffey, *North American Wildflowers*, pp. 57–58.

55. Rev. John Clayton, "Letter from Virginia to Dr. Grew in answer to several quaerys sent to him by that learned gentleman," in Erichsen-Brown, *Medicinal and Other Uses*, p. 219.

56. Erichsen-Brown, *Medicinal and Other Uses*, pp. 302–303.

57. Jonathan Carver, *Travels through the Interior Part of North America in the years 1766, 67, and 68*, in Erichsen-Brown, *Medicinal and Other Uses*, p. 302.

58. Charles A. Peck, a letter from him in Albany, N.Y., to I. U. Lloyd and C. G. Lloyd, in Coffey, *North American Wildflowers*, pp. 15–16.

59. Vogel, *American Indian Medicine*, p. 311.

60. Kuhnlein and Turner, *Traditional Plant Foods*, p. 234.

61. John Eastman, *The Book of Swamp and Bog: Trees, Shrubs, and Wildflowers of Eastern Freshwater Wetlands* (Mechanicsburg, Penn.: Stackpole Books, 1995), p. 197.

62. Janice J. Schofield, *Discovering Wild Plants: Alaska, Western Canada, the Northwest* (Anchorage: Alaska Northwest Books, 1973), p. 7.

63. Adam F. Szczawinski and Nancy J. Turner, *Wild Green Vegetables of Canada* (Ottawa: National Museum of Natural Sciences, 1980), p. 157.

64. P. H. Gosse, *The Canadian Naturalist: The Natural History of Lower Canada*, in Erichsen-Brown, *Medicinal and Other Uses*, p. 213.

65. Elliott, *Wild Roots*, p. 100.

66. Clayton, "Letter from Virginia to Dr. Grew," in Erichsen-Brown, *Medicinal and Other Uses*, p. 247.

67. www.shorejournal.com/9612/joc120a.html.

68. Frederick Burkhardt et al., eds., *The Correspondence of Charles Darwin*, vol. 8, *1860* (Cambridge: Cambridge University Press, 1993), p. 491, a letter to Charles Lyell, on November 26, 1860.

69. Henry D. Thoreau, *The Journal of Henry D. Thoreau*, ed. Bradford Torrey and Francis H. Allen (New York: Dover, 1962), "June 11, 1852," vol. IV, p. 91 [vol 1., p. 423].

70. Michel Sarrazin, *La flore du Canada*, in Erichsen-Brown, *Medicinal and Other Uses*, p. 203.

71. Freeman, Linda. *The Smithsonian Institution*, Mount Shasta Companion, 2001, www.siskiyous.edu/shasta/art/smith.

72. Rachel Carson, *The Sea Around Us* (New York: Oxford University Press, 1961), p. 119.

73. Walt Whitman, *Leaves of Grass* (Toronto: Musson Book Co., 1900), p. 413.

74. Stephen Facciola, *Cornucopia: A Source Book of Edible Plants* (Vista, Calif.: Kampong Publications, 1990), p. 160.

75. Jay Humphreys, "Building Better Sea Oats," *Fathom Magazine*, spring 1996, Florida Sea Grant College Program, 1998–2000, www.flseagrant.org/science/library/fathom_magazine/volume-8_issue-1/sea_oats.htm.

76. Henry D. Thoreau, *Cape Cod*, ed. Joseph J. Moldenhauer (Princeton, N.J.: Princeton University Press, 1988), p. 87.

77. "Wildman" Steve Brill and Evelyn Dean, *Identifying and Harvesting Edible and Medicinal Plants in Wild (and Not So Wild) Places* (New York: Hearst Books, 1994), pp. 256–257.

78. Christopher Nyerges, *Guide to Wild Foods and Useful Plants* (Chicago: Chicago Review Press, 1999), p. 152.

79. Szczawinski and Turner, *Wild Green Vegetables of Canada*, p. 97.

80. Alix Kates Shulman, *Drinking the Rain* (New York: Farrar Straus Giroux, 1995), pp. 45–46.

81. William H. Amos and Stephen H. Amos, *Audubon Guide to Atlantic and Gulf Coasts* (New York: Alfred A. Knopf, 1985), p. 548.

82. Thoreau, *Cape Cod*, p. 18.

83. Priscilla Russell Kari, *Tanaina Plantlore: Dena'ina K'et'una—An Ethnobotany of the Dena'ina Indians of Southcentral Alaska*, 2nd ed. (Anchorage: National Park Service Alaska Region, 1987), p. 97.

84. Edward W. Nelson, "The Eskimo About Bering Strait," in Wendell H. Oswalt, *Alaskan Eskimos* (Scranton, Penn.: Chandler, 1967), pp. 211–212.

85. Captain Thomas James, *The Strange and Dangerous Voyage of Captain Thomas James*, and Marc Lescarbot, *Nova Francia, a Description of Acadia*, in Erichsen-Brown, *Medicinal and Other Uses*, p. 265.

86. Thoreau, *Cape Cod*, p. 70.

87. Andris Petersons, "The Beach Pea of Salmon Cove," *Gazette*, vol. 32, no. 21 (July 13, 2000), Memorial University of Newfoundland, www.mun.ca/univrel/gazette/1999–2000/July.13/research.html.

88. Lisa Bernstein, "Time Capsule 2000," Cape Publishing, 1998–2002, www.capemay.com/lisa.htm.

89. John T. Walbran, *British Columbia Coast Names, Their Origin and History* (Ottawa: Government Printing Bureau, 1909, reprinted, Vancouver: Library's Press, 1971), pp. 240–241.

90. Gabriele Helmig, "Aboriginal Place Names in British Columbia," *Dreamspeaker*, Indian and

Northern Affairs Canada, spring 2000, www.
ainc-inac.gc.ca/nr/nwltr/drm/s2000/apn-e.html.

91. Arline Zatz, *New Jersey's Special Places*,
Countryman Press, 1998, adapted in "Hidden
New Jersey: Island Beach, Part 2," *Destinations*,
GORP, Countryman Press, http://www.gorp.
com/gorp/publishers/countryman/hik_njs1.htm.

92. The poem appeared in Edna St. Vincent Millay,
Second April (New York: Mitchell Kennerley,
1921), p. 20. Clive Driver, "Ending at the West
End," *Provincetown Banner*, December 5, 1996,
www.provincetownbanner.com/history/12/5/
1996/1.

93. "Watch the Weeds," Fishtalk Radio, 2000–2001,
www.fishtalkradio.net/watch_the_weeds.htm.

94. Curtis J. Badger, *Salt Tide: Cycles and Currents of
Life along the Coast* (Harrisburg, Penn.:
Stackpole Books, 1993), p. 29.

95. George Purchis, "Growing with the Tide,"
Tidal Tales, newsletter of the Hayward
Shoreline Interpretive Center, Hayward, Calif.,
Naturalist's Library, May/June 1998, www.hard.
dst.ca.us/hayshore/library/marsh5_98.htm.

96. Ibid.

97. Donald R. Strong and Mei-Yin Wu, "American
Insects Feed on English Cordgrass," *Biocentral
News and Information*, vol. 20, no. 2 (June 1999),
http://pest.cabweb.org/Journals/BNI/Bni20-2/
gennews.htm.

98. Carina K. Anttila et al., "Greater Male Fitness
of a Rare Invader (*Spartina alterniflora*, Poaceae)
Threatens a Common Native (*Spartina foliosa*)
with Hybridization," *American Journal of Botany*,
vol. 85, no. 11 (November 1998): 1600.

99. Jack Rudloe, *The Wilderness Coast: Adventures of
a Gulf Coast Naturalist* (New York: E. P. Dutton,
1988), pp. 6–7, 5–6.

100. Mead A. Allison, "Can the Loss of Texas Coastal
Wetlands Be Halted . . . or Reversed?" *Quarterdeck*,
vol. 5, no. 2 (summer 1997), Texas A&M
University Department of Oceanography, www.
ocean.tamu.edu/Quarterdeck/AD5.2/allison.html.

101. "Building a Defensive Line to Tackle
Phragmites," *University of Delaware Sea Grant
Reporter*, vol. 19, no. 1 (2000), Graduate College
of Marine Studies, University of Delaware,
http://www.ocean.udel.edu/publications/
Newsletter/reporter/special00/00.html#3.

102. Bessie M. Reid, "Vernacular Names of Texas
Plants," in Coffey, *North American Wildflowers*,
p. 249.

103. Mary Parker Buckles, *Margins: A Naturalist
Meets Long Island Sound* (New York: North
Point Press, 1997), p. 27.

104. Brill and Dean, *Identifying and Harvesting*, p. 81.

105. Nyerges, *Wild Foods and Useful Plants*, p. 74.

106. Mary Ann Spahr, Rainier Audubon Society, in
Schofield, *Discovering Wild Plants*, p. 276.

107. Sarah Orne Jewett, "Marsh Rosemary," *Atlantic
Monthly*, vol. 57, no. 343 (May 1886): 601.

108. C. Colston Burrel, "Environmental Concerns
of Harvesting Wild Materials," vol. 2 of a
report commissioned by the Saskatchewan
Dried Flowers Association, AgriCarta, 1999,
www.aginfonet.com/aglibrary/content/
sk_driedflower/environmental_concerns.html.

109. "Neighborhood Pride, Childhood Memories
Make the Town," *SouthCoast Today*, June 30,
2000, The Standard-Times, 2001, www.s-t.
com/daily/06-00/06-30-00/a061o030.htm.

110. Martha Blume, "Getting to Know the New
Birds on the Bay," *Bay Weekly Online*, March
1–7, 2001, www.bayweekly.com/year01/
issue9_9/lead9_9.html.

111. Doug Elliott, *Wild Roots*, p. 107.

112. Sean Manley and Robert Manley, *Beaches: Their
Lives, Legends, and Lore* (Philadelphia: Chilton,
1968), pp. 77–78.

113. Thoreau, *Cape Cod*, pp. 45, 52.

114. Nyerges, *Wild Foods and Useful Plants*, p. 157.

115. www.scituateharboronline.com/IrishMoss.html.

116. Salome MacLeod, *Memories of Beach Point and
Cape Bear* (Charlottetown, P.E.I.: Author),
p. 22.

117. Cheryl Daigle and Jim Dow, "Life Within the
Rockweed," *Quoddy Tides*, September 22, 2000,
www.cobscook.org/ART9-rock2.htm.

118. Schofield, *Discovering Wild Plants*, p. 260.

Region 3: Florida

1. Carl Hiaasen, *Tourist Season* (New York:
Putnam's Sons, 1986), p. 68.

2. "Wildlife," *Light Tackle, Snook, Redfish, Trout,
Tarpon*, Flying Fish & Company, 2001, www.
flyingfishnco.com/wildlife.htm.

3. Marjory Stoneman Douglas, *The Everglades:
River of Grass* (New York: Rinehart and
Company, 1947), pp. 13–14.

4. James A. Duke and K. K. Wain, *Medicinal Plants
of the World*, computer index with more than
85,000 entries, 3 vols., 1981; James A. Duke,
Handbook of Energy Crops (unpublished), in

Purdue University Center for New Crops and Plant Products, "*Rhizophora mangle* L.," New Crop: The New Crop Resource Online Program, 1983, www.hort.purdue.edu/newcrop/duke_energy/Rhizophora_mangle.html.

5. Julia F. Morton, "Can the Red Mangrove Provide Food, Feed, and Fertilizer?" *Economic Botany* (1968): 113–123. Duke, *Handbook of Energy Crops.*

6. "Everglades Plant Diversity," Andrews University, www.biol.andrews.edu/everglades/plant/plant.htm (website expired as of March 2002).

7. L. H. Burkill, *A Dictionary of Economic Products of the Malay Peninsula* (London: Crown Agents for the Colonies, 1935), p. 1898, vol. 2.

8. Marjory Stoneman Douglas, "The Mayor of Flamingo," from *River in Flood and Other Florida Stories* (Gainesville: University of Florida Press, 1998), p. 118.

9. Harriet Beecher Stowe, *Palmetto Leaves* (Gainesville: University of Florida Press, 1968, facsimile of 1873 ed.), pp. 252–253.

10. Sidney Lanier, *Florida: Its Scenery, Climate, and History* (Gainesville: University of Florida Press, 1973, facsimile of 1875 ed.), p. 85.

11. Jonathan Dickinson, *Jonathan Dickinson's Journal* . . . , ed. Evangeline Walker Andrews and Charles MacLean Andrews (New Haven, Conn.: Yale University Press, 1945), pp. 11–12.

12. Tina Bucuvalas, Peggy A. Bulger, and Stetson Kennedy, *South Florida Folklife* (Jackson: University of Mississippi Press, 1994), pp. 23–25.

13. Ibid., pp. 214–215.

14. Michel Oesterreicher, *Pioneer Family: Life on Florida's 20th Century Frontier* (Tuscaloosa: University of Alabama Press, 1996), pp. 14–16.

15. Marjory Stoneman Douglas, "Pineland," from *Nine Florida Stories* (Jacksonville: University of North Florida Press, 1990), pp. 4–5.

16. Daniel Moerman, *Native American Ethnobotany* (Portland, Ore.: Timber Press, 1998), p. 408.

17. Bucuvalas, Bulger, and Kennedy, *South Florida Folklife*, p. 58.

18. George M. Barbour, *Florida for Tourists, Invalids, and Settlers* . . . (Gainesville: University of Florida Press, 1964), p. 90.

19. Jeff Wasielewski, "Everybody, Gumbo Limbo Mon!" *Dade Florida Native Plant Society Newsletter*, February 2001, http://www.fnps.org/dade/pastnewslets/newslet102.html.

20. "Some Other Examples of Wood Bearing Use, from Our Readers," Woodex Bearing Company, www.woodex-meco.com/history 1.htm.

21. Moerman, *Native American Ethnobotany*, p. 233.

22. Tom Broome, "The Coonties of Florida," The Cycad Jungle, cycadjungle.8m.com/cycadjungle/The%20Coontie%20of%20Florida.htm.

23. Moerman, *Native American Ethnobotany*, pp. 419, 425.

24. Susan Orlean, *The Orchid Thief* (New York: Random House, 1998), pp. 48–51.

25. Ruben P. Sauleda, "The Orchids of Florida," The Everglades, www.members.tripod.com/~rsauleda/.

26. *Orchid Safari*, www.geocities.com/~marylois/index.html.

27. Douglas, *Everglades*, pp. 10–13.

28. Moerman, *Native American Ethnobotany*, p. 166.

Region 4: The Boreal Forest

1. Derek Johnson, Linda Kershaw, Andy McKinnon, and Jim Pojar, *Plants of the Western Boreal Forest and Aspen Parkland* (Edmonton, Alta.: Lone Pine Publishing, 1995), p. 11.

2. Senate Subcommittee on the Boreal Forest, *Competing Realities: Boreal Forest at Risk*, June 1999, http://www.parl.gc.ca/36/1/parlbus/commbus/senate/com-e/bore-e/rep-e/rep09jun99part2-e.htm.

3. Ibid.

4. Stephen J. Pyne, *Fire in America: A Cultural History of Wildland and Rural Fire* (Seattle: University of Washington Press, 1997), p. 499.

5. FEIS Online, "*Pinus banksiana*." www.fs.fed.us/database/feis/plants/tree/pinban/botanical_and_ecological_characteristics.html.

6. Daniel Moerman, *Native American Ethnobotany* (Portland, Ore.: Timber Press, 1998), p. 404.

7. Huron Smith, *Ethnobotany of the Ojibwe*, in Erichsen-Brown, *Medicinal and Other Uses*, p. 8.

8. Donald Culross Peattie, *A Natural History of Trees of Eastern and Central North America* (Boston: Houghton Mifflin, 1991), p. 18.

9. FEIS Online, "*Larix laricina*," www.fs.fed.us/database/feis/plants/larlar/botanical_and_ecological_characteristics.html.

10. Henry Wadsworth Longfellow, *The Works of Henry Wadsworth Longfellow* (Boston: Houghton Mifflin, 1886), vol. II, p. 164.

11. Johnson et al., *Western Boreal*, p. 29.

12. John Josselyn, *New England Rarities*, in Erichsen-Brown, *Medicinal and Other Uses*, p. 23.

13. A. R. M. Lower, *The North American Assault on the Canadian Forest: A History of the Lumber Trade between Canada and the United States* (Toronto: Ryerson Press, 1938), p. 25.

14. Peattie, *Trees of Eastern and Central North America*, p. 35.

15. FEIS Online, "*Betula papyrifera*," www.fs.fed.us/ database/feis/plants/tree/betpap/fire_ecology. html.

16. Arthur Plotnik, *The Urban Tree Book: An Uncommon Field Guide for City and Town* (New York: Three Rivers Press, 2000), p. 170.

17. Rebecca Rupp, *Red Oaks and Black Birches: The Science and Lore of Trees* (Pownal, Vt.: Storey Communications, 1990), pp. 77–78.

18. Samuel de Champlain, *Des sauvages, ou Voyages de Samuel Champlain de Brouage fait en la France Nouvelle, l'an mil six cens trois*, in Erichsen-Brown, *Medicinal and Other Uses*, p. 38.

19. George Heriot, *Travels Through the Canadas*, in Mary Alice Downie and Mary Hamilton, *"And Some Brought Flowers": Plants in a New World* (Toronto: University of Toronto Press, 1980), "Birch."

20. Johnson et al., *Western Boreal*, p. 36.

21. Major [Samuel] Strickland, *Twenty-Seven Years in Canada West: Or the Experience of an Early Settler*; ed. Agnes Strickland (Edmonton, Alta.: M. G. Hurtig, 1970), vol. II, p. 53.

22. Harriet V. Kuhnlein and Nancy J. Turner, *Traditional Plant Foods of Canadian Indigenous Peoples: Nutrition, Botany, and Use* (Philadelphia: Gordon and Breach Science, 1991), p. 138.

23. "Native Conifers of North America, *Picea glauca*," *Nearctica*, Nearctica.com, 2000, http:// www.nearctica.com/trees/conifer/picea/ Pglauca.htm.

24. FEIS Online, "*Picea glauca*," www.fs.fed.us/ database/feis/plants/tree/picgla/ botanical_and_ecological_characteristics.html.

25. Jacques Cartier, *Second Voyage to Canada*, in Erichsen-Brown, *Medicinal and Other Uses*, pp. 8–9.

26. Thomas Anburey, *Travels through the Interior Parts of America, 1771–1781*, in Downie and Hamilton, *"And Some Brought Flowers,"* "Spruce."

27. Christopher Middleton, wintering over in Churchill on Hudson Bay from 1743–1744, in Erichsen-Brown, *Medicinal and Other Uses*, p. 11.

28. Elizabeth Goudie, *Woman of Labrador* (Toronto: Peter Martin, 1973), p. 11.

29. Janice J. Schofield, *Discovering Wild Plants: Alaska, Western Canada, the Northwest.* (Anchorage: Alaska Northwest Books, 1989), pp. 71–72.

30. Erichsen-Brown, *Medicinal and Other Uses*, pp. 7–14; Johnson et al., *Western Boreal*, p. 24.

31. Jacques Cartier, *First Voyage to Canada*, in Erichsen-Brown, *Medicinal and Other Uses*, p. 8.

32. Johnson et al., *Western Boreal*, p. 28.

33. Longfellow, *Works*, vol. II, p. 164.

34. Shan Walshe, *Plants of Quetico and the Ontario Shield* (Toronto: University of Toronto Press, 1980), p. 76.

35. E. E. Gaertner, "Breadstuff from Fir (Abies balsamea)," *Economic Botany*, vol. 26, no. 1 (1970): 69–72.

36. Marc Lescarbot, *Nova Francia, a Description of Acadia 1606*, in Erichsen-Brown, *Medicinal and Other Uses*, pp. 18–19.

37. Baron Louis-Armand de Lom d'Arce de Lahontan, *New Voyages to North-America*, in Downie and Hamilton, *"And Some Brought Flowers,"* "Cherry."

38. Johnson et al., *Western Boreal*, p. 57.

39. Moerman, *Native American Ethnobotany*, pp. 442–443.

40. Earl J. S. Rook, "*Prunus pensylvanica*," *Flora, Fauna, Earth, and Sky . . . The Natural History of the Northwoods*, 1998, www.rook.org/earl/bwca/ nature/trees/prunuspen.html.

41. Johnson et al., *Western Boreal*, p. 149.

42. Earl J. S. Rook, "*Epilobium angustifolium*," *Flora, Fauna, Earth, and Sky*, www.rook.org/earl/ bwca/nature/herbs/epilobiuman.html.

43. www.articwildlife.org/fws/firewdra.html.beverly +skinner.

44. Timothy Coffey, *The History and Folklore of North American Wildflowers* (Boston: Houghton Mifflin, 1993), p. 135.

45. Patricia Russell Kari, *Tanaina Plantlore: Dena'ina K'et'una—An Ethnobotany of the Dena'ina Indians of Southcentral Alaska*, 2nd ed. (Anchorage: National Park Service Alaska Region, 1987), p. 125.

46. Johnson et al., *Western Boreal*, p. 149.

47. Personal e-mail.

48. *Canadian Pharmaceutical Journal*, in Erichsen-Brown, *Medicinal and Other Uses*, pp. 195.

49. Flora Beardy and Robert Coutts, eds., *Voices from Hudson Bay: Cree Stories from York Factory*

(Montreal and Kingston: McGill–Queen's University Press, 1996), p. 65.

50. Johnson et al., *Western Boreal*, p. 70.

51. Moerman, *Native American Ethnobotany*, p. 299–300.

52. Schofield, *Discovering Wild Plants*, p. 21.

53. Henry Ellis, *A Voyage to Hudson's-Bay, by the 'Dobbs Galley and California' in the Years 1746 and 1747*, in Downie and Hamilton, "*And Some Brought Flowers*," "Labrador Tea."

54. Nancy J. Turner and Adam Szczawinski, *Wild Coffee and Tea Substitutes of Canada* (Ottawa: National Museums of Canada, 1978), p. 57.

55. Kuhnlein and Turner, *Traditional Plant Foods*, p. 171.

56. *Journal of Senate of Canada*, 1887, in Erichsen-Brown, *Medicinal and Other Uses*, p. 195.

57. Mina Hubbard, *A Woman's Way through Unknown Labrador* (New York: McClure, 1908, reprinted, St. John's, Nfld.: Breakwater Books, 1981), pp. 84–85.

58. Monique Reed, "Sedge Meadow," *Lick Creek Park*, Herbarium, Department of Biology, Texas A&M University, 1996, www.csdl.tamu. edu/FLORA/LCP/LCP6.HTML.

59. Henry D. Thoreau, "March 25, 1859," *The Journal of Henry D. Thoreau*, ed. Bradford Torrey and Francis H. Allen (New York: Dover, 1962), vol. XII, p. 81 [vol. 2, p. 1452].

60. Kari, *Tanaina Plantlore*, p. 105.

61. Moerman, *Native American Ethnobotany*, p. 138.

62. John Eastman, *The Book of Swamp and Bog: Trees, Shrubs, and Wildflowers of Eastern North America* (Mechanicsburg, Penn.: Stackpole Books, 1995), p. 165.

63. Dave Leshuk, "John Muir's Wisconsin Days," *John Muir Exhibit*, Sierra Club, reprinted from *Wisconsin Natural Resources*, vol. 12, no. 3 (May/June 1988), http://www.sierraclub.org/ john_muir_exhibit/life/ muir_wisconsin_dave_leshuk.html.

64. Whit Gibbons, "Hey Kids, Put Down the Video Games and Play Outside," Online Athens, July 23, 2000, www.onlineathens.com/ stories/072400/opi_0724000003.shtml.

65. Kem Luther, "The Company of Weeds," *Saturday Night Magazine*, July/August 1996, www.sheridanc.on.ca/~kem/weeds/weeds.htm.

66. Moerman, *Native American Ethnobotany*, pp. 213–217.

67. Carla Allen, "Horsetail," *The Vine*, June 5,

1998, reprinted from *The Yarmouth Vanguard*, www.klis.com/scove/060598.htm.

68. Charles Millspaugh, *American Medicinal Plants, an Illustrated and Descriptive Guide to Plants Indigenous to and Naturalized in the United States Which Are Used in Medicine*, in Erichsen-Brown, *Medicinal and Other Uses*, p. 226.

69. Kari, *Tanaina Plantlore*, p. 101.

70. Kuhnlein and Turner, *Traditional Plant Foods*, p. 50.

71. Annora Brown, *Old Man's Garden* (Toronto: J. M. Dent and Sons, 1954), p. 140.

72. Wayne Pauly, "Prairie Folklore: Horsetail Tales," *Action: Adult Conservation Team Newsletter*, Dane County Parks, fall 1998, http://www.co.dane.wi. us/parks/adult/action/1998/fall/p6.htm.

73. Ibid.

74. "Horsetail," *Health Library*, Pharmacist's Ultimate Health Corp, 2000, www.healthsoftware. com/users/user1/HORSETAIL.htm.

75. Allen, "Horsetail."

76. Roy Lukes, "Scouring Rushes Were the First Scouring Brushes," *Nature-Wise*, *Door County Advocate*, December 15, 2000, http://www. doorbell.net/lukes/a121500.htm.

77. Jon Krakauer, "Death of an Innocent: How Christopher McCandless Lost His Way in the Wilds," *Outside Magazine*, January 1993, Mariah Media, 2001, www.outsidemag.com/ magazine/0193/9301fdea.html.

78. Pat Holloway, "A Tale of Two Species," *Georgeson Botanical Garden Review*, vol. 6, no. 3 (1997): 7–8, http://www.lter.uaf.edu/salrm/gbg/ Pubs/Review/Rev.pdf.

79. Walshe, *Plants of Quetico*, p. 117.

80. Johnson et al., *Western Boreal*, p. 196.

81. Theodat Gabriel Sagard, *Le gran voyage du pays des Hurons*, in Erichsen-Brown, *Medicinal and Other Uses*, p. 351.

82. Frank G. Speck, "Medicine Practices of the Northeastern Algonquians," in Erichsen-Brown, *Medicinal and Other Uses*, p. 352.

83. Smith, *Ethnobotany of the Ojibwe*, p. 352.

84. "New Orleans Mead," *Canadian Pharmaceutical Journal*, 1876, in Erichsen-Brown, *Medicinal and Other Uses*, p. 352.

85. Dorothy Molter Museum, "Who Was Dorothy Molter?" www.canoecountry.com/dorothy/who.

86. Menihek Trails Group, "March—Can It Be Spring?" March 6, 2001, www.labrador-west. com/menihektrails/marchat.htm.

87. FEIS Online, "*Empetrum nigrum,*" www.fs.fed. us/database/feis/plants/shrub/empnig/ fire_ecology.html.

88. Ibid.

89. Alice Ramirez, "Bunchberries," *Explorations,* December 1996, www.chem.ucla.edu/~alice/ explorations/churchill/botbu.htm.

90. Moerman, *Native American Ethnobotany,* p. 209.

91. Ibid.

92. Judith Colwell, "Day 4," *Northern Lights,* 1995, www.standford.edu/~jcolwell/NL/Day4.html.

93. David Rockwell, *The Nature of North America: A Handbook to the Continent Rocks, Plants, and Animals* (New York: Berkeley Books, 1998), p. 202. There is also more aspen information at http://www.extremescience.com/aspengrove. htm. In the former it's 160 acres and in the latter 200 acres. It may have grown that much in the meantime.

94. Douglas W. Johnson, "Biogeography of Quaking Aspen (*Populus tremuloides*)," San Francisco State University, Department of Geography, http://bss.sfsu.edu/geog/bholzman/ courses/fall99projects/aspen.htm.

95. Rockwell, *Nature of North America,* p. 202.

96. http://co-trading-post.com/sku_html/3473.htm. (This website had expired as of March 2002.)

97. Karl Snyder, "Aspen Fall Colors," Estes Park Online.com, 1999–2001, www.estesparkonline. com/EP-News-Aspen2000.HTML.

98. Johnson, "Biogeography of Quaking Aspen."

99. FEIS Online, "*Populus tremuloides,*" www.fs.fed. us/database/feis/plants/tree/poptre/ value_and_use.html.

100. Johnson et al., *Western Boreal,* p. 38.

101. Henry D. Thoreau, *The Illustrated Maine Woods* (Princeton, N.J.: Princeton University Press, 1974), p. 235.

102. E. Pauline Johnson, *Flint and Feather: The Complete Poems of E. Pauline Johnson* (Toronto: Musson Book Company, 1931), pp. 139–141.

103. Johnson, "Biogeography of Quaking Aspen."

104. Kuhnlein and Turner, *Traditional Plant Foods,* p. 162.

105. Schofield, *Discovering Wild Plants,* p. 236.

106. Kuhnlein and Turner, *Traditional Plant Foods,* p. 163.

107. Kari, *Tanaina Plantlore,* p. 91.

108. Kuhnlein and Turner, *Traditional Plant Foods,* p. 163.

109. Eastman, *Swamp and Bog,* pp. 129–30.

Region 5: The Prairie

1. Sharon Butala, *The Perfection of the Morning: An Apprenticeship in Nature* (Toronto: HarperCollins, 1994), p. 101.

2. Bernard DeVoto, ed., *The Journals of Lewis and Clark* (Boston: Houghton Mifflin, 1953), p. 419.

3. Ian Frazier, *Great Plains* (New York: Farrar Straus Giroux, 1989), p. 196.

4. Daniel Moerman, *Native American Ethnobotany,* (Portland, Ore.: Timber Press, 1998), p. 72.

5. Ibid., p. 521.

6. Ibid., p. 127.

7. Pierre-Jean de Smet, *Life, Letters, and Travels of Father de Smet, 1801–1873,* ed. Hiram Martin Chittenden and Alfred Talbot Richardson (New York: Kraus Reprint Co., 1969), vol. III, p. 906.

8. Moerman, *Native American Ethnobotany,* p. 293.

9. Ibid.

10. Ibid., p. 377.

11. Melvin Randolph Gilmore, *Uses of Plants by the Indians of the Missouri River Region* (Washington, D.C.: U.S. Government Printing Office, 1919), p. 66.

12. John Madson, *Where the Sky Began, Land of the Tallgrass Prairie* (Boston: Houghton Mifflin, 1982), p. 57.

13. FEIS Online, "*Sorghastrum nutans,*" www.fs. fed.us/database/feis/plants/graminoid/sornut/ value_and_use.html.

14. Ibid., "*Spartina pectinata,*" www.fs.fed.us/ database/feis/plants/graminoid/spapec/ botanical_and_ecological_characteristics.html.

15. Madson, *Where the Sky Began,* p. 212.

16. Moerman, *Native American Ethnobotany,* p. 542.

17. Gilmore, *Uses of Plants,* p. 131.

18. Timothy Coffey, *The History and Folklore of Native American Wildflowers* (Boston: Houghton Mifflin, 1993), p. 256.

19. Gilmore, *Uses of Plants,* p. 131.

20. Steven Foster and Varro E. Tyler, *Tyler's Honest Herbal: A Sensible Guide to the Use of Herbs and Related Remedies,* 4th ed. (New York: Haworth Herbal Press, 1999), pp. 143–144.

21. Steven Foster, "Understanding Echinacea," Steven Foster Group, 2000, www.stevenfoster. com/education/monograph/echinacea.html.

22. jps.net/ahaherb/newsletter.

23. Moerman, *Native American Ethnobotany,* p. 229.

24. Jack Sanders, *Hedgemaids and Fairy Candles: The Lives and Lore of North American Wildflowers*

(Camden, Maine: Ragged Mountain Press, 1993), p. 187.

25. Ibid., p. 186.

26. C. Dwight Marsh and A. B. Clawson, "*Eupatorium urticaefolium* as a Poisonous Plant," *Journal of Agricultural Research*, vol. 11, no. 13 (December 24, 1917): 699–707. Journal of Agricultural Research Index, http://preserve.nal.usda.gov:8300/jag/v11/v11i13/110699/110699.htm.

27. Edwin James, *An Account of an Expedition from Pittsburgh to the Rocky Mountains* (Ann Arbor, Mich.: University Microfilms, 1966), vol. I, p. 72.

28. Ana Nez Heatherley, *Healing Plants: A Medicinal Guide to Native North American Plants and Herbs* (Toronto: HarperCollins, 1998), p. 92.

29. Ibid., pp. 92–93.

30. Neill Diboll and Bob Stadyk were both kind enough to talk to me about their great passion for prairies forbs. Diboll has been important in conserving native plant seeds and using them in his nursery; he is often sought out as a consultant on prairie regeneration.

31. Naomi Mathews, "Maryland's Queen of Flowers: Black-eyed Susans," *Garden Guides*, 1999, www.gardenguides.com/articles/maryland.htm.

32. Wilfrid Blunt, *The Compleat Naturalist: A Life of Linnaeus* (London: Collins, 1971), pp. 26, 37.

33. Coffey, *Wildflowers*, p. 273.

34. William Bartram, *Travels through North and South Carolina, Georgia, East and West Florida* . . . (Dublin: Moore, Jones, M'Allister, and Rice, 1793), p. 325.

35. Madson, *Where the Sky Began*, p. 93.

36. Aldo Leopold, *A Sand County Almanac* (New York: Oxford University Press, 1966), p. 45.

37. Ibid., pp. 48–49.

38. Moerman, *Native American Ethnobotany*, p. 531.

39. Gilmore, *Uses of Plants*, p. 132.

40. Moerman, *Native American Ethnobotany*, p. 531.

41. Virgil J. Vogel, *American Indian Medicine* (Norman, Okla.: University of Oklahoma Press, 1970), p. 248.

42. FEIS Online, "*Solidago canadensis*," www.fs.fed.us/database/feis/plants/forb/solcan/value_and_use.html.

43. M. Grieve, *A Modern Herbal: The Medicinal, Culinary, Cosmetic, and Economic Properties* . . . (London: Jonathan Cape, 1931), p. 361.

44. Heatherley, *Healing Plants*, p. 96.

45. Ibid., p. 97.

46. Sanders, *Hedgemaids*, p. 208.

47. Huron Smith, *Ethnobotany of the Meskwaki*, in Charlotte Erichsen-Brown, *Medicinal and Other Uses of North American Plants: A Historical Survey with Special Reference to the Eastern Indian Tribes* (New York: Dover, 1989, originally published 1979), p. 390.

48. Sanders, *Hedgemaids*, pp. 208–209.

49. Inge N. Dobelis, ed., *Magic and Medicine of Plants* (Pleasantville, N.Y.: Reader's Digest, 1986), p. 198.

50. http://forums.gardenweb.com/forums/load/natives/msg0512550330050.html?19.

51. Vogel, *American Indian Medicine*, p. 371.

52. Ibid., p. 287.

53. Diana Beresford-Kroeger, *Bioplanning: A North Temperate Garden* (Kingston, Ont.: Quarry Press, 1999), p. 55.

54. Gilmore, *Uses of Plants*, p. 93.

55. Ibid.

56. A. T. Andreas, *History of the State of Nebraska*, Part 7, "Colonel Fremont's Explorations" (Chicago: Western Historical Company, 1882), Kansas Collection Books, www.ukans.edu/carrie/kancoll/andreas_ne/history/erlyhst-p7.html#fremont.

57. From an interview by Murray Aspden with Julie Hrapko, former curator of botany, Alberta Provincial Museum.

58. Vogel, *American Indian Medicine*, p. 323.

59. Dobelis, *Magic and Medicine of Plants*, p. 336.

60. Moerman, *Native American Ethnobotany*, p. 193.

61. Ibid., p. 321.

62. Henry Youle Hind, *Narrative of the Canadian Red River Exploring Expedition of 1857 and of the Assinniboine and Saskatchewan Exploring Expedition of 1858*, in Barry Kaye and D. W. Moodie, "*Psoralea* Food Resource on the Northern Plains," *Plains Anthropologist*, vol. 23, no. 82 (November 1978): 332.

63. Edwin Thompson Denig, *Five Indian Tribes of the Upper Missouri*, in Kaye and Moodie, "*Psoralea* Food Resource," p. 333.

64. Paul Russell Cutright, *Lewis and Clark: Pioneering Naturalists* (Urbana: University of Illinois Press, 1969), p. 91.

65. Frederick Pursh, *Flora Americae Septentrionalis; or, A Systematic Arrangement and Description of the Plants of North America* (London: White, Cochrane, 1814), vol. II, p. 476.

66. Smet, *Life*, vol. II, p. 906.

67. Annora Brown, *Old Man's Garden* (Toronto: J. M. Dent and Sons, 1954), p. 57.
68. Moerman, *Native American Ethnobotany*, p. 381.
69. Personal letter.
70. Alexander Mackenzie, *Journal of the Voyage to the Pacific*, ed. Walter Sheppe (New York: Dover, 1995), p. 76.
71. Gilmore, *Uses of Plants*, p. 181.
72. Brown, *Old Man's Garden*, p. 12.
73. "Native People's Garden," Devonian Botanic Garden, University of Alberta, 1997–2002, www.discoveredmonton.com/devonian/natvgrdn.html.
74. Daniel Williams Harmon, *A Journal of Voyages and Travels in the Interior of North America*, in Mary Alice Downie and Mary Hamilton, *"And Some Brought Flowers": Plants in a New World* (Toronto: University of Toronto Press, 1980), "Rose."
75. "Wildman" Steve Brill and Evelyn Dean, *Identifying and Harvesting Edible and Medicinal Plants in Wild (and Not So Wild) Places* (New York: Hearst Books, 1994), pp. 158–59.
76. Personal interview conducted by Murray Aspden.
77. Max Braithwaite, *Why Shoot the Teacher?* (Toronto: McClelland and Stewart, 1965), pp. 124–125.
78. Thomas G. Barnes, "Wild about Wildflowers," University of Kentucky, College of Agriculture, 1998, www.ca.uky.edu/agc/pubs/for/for71/for71.htm.

Region 6: The Desert
1. Steven J. Phillips and Patricia Wentworth Comus, eds., *A Natural History of the Sonoran Desert* (Tucson: Arizona–Sonoran Desert Museum Press, 2000), pp. 105–106. This is a superb reference book on the Sonoran Desert.
2. Ibid., p. 112.
3. Mitchel P. McClaran and Thomas R. van Devender, eds., *The Desert Grassland* (Tucson: University of Arizona Press, 1995), p. 244.
4. Ibid., p. 252.
5. Phillips and Comus, *Natural History*, p. 117.
6. McClaran and van Devender, *Desert Grassland*, p. 253.
7. Ruth Kirk, *Desert: The American Southwest* (Boston: Houghton Mifflin, 1973), p. 82.
8. Bill Thomas, *Wild Woodlands: The Old-Growth Forests of America* (Dallas: Taylor, 1992), p. 126.

9. John Alcock, *Sonoran Desert Spring* (Tucson: University of Arizona Press, 1994), p. 131.
10. Gary Paul Nabhan, *The Desert Smells Like Rain: A Naturalist in Papago Indian Country* (San Francisco: North Point Press, 1982), p. 26. This is one of my favorite desert books. Nabhan is a wonderful writer and all his books are must-reads for anyone who cares about ecology.
11. Ruth Underhill, *Papago Indian Religion* (New York: Columbia University Press, 1946), p. 67.
12. Ibid., p. 51.
13. David L. Eppele, "On the Desert," Arizona Cactus, 2001, http://www.arizonacactus.com/story.htm. (Not available online as of March 2002.)
14. Thomas, *Wild Woodlands*, p. 128.
15. Ibid., p. 134.
16. Linda McMillin Pyle, "Saguaro Fruit," *Desert USA*, June 1998, Digital West Media, 2002, http://www.desertusa.com/mag98/june/papr/jun_lil.html.
17. "Cactus Garden," Arizona–Sonoran Desert Museum, 2002, http://www.desertmuseum.org/about/exhibits/cactus_garden.html. This is a wonderful resource for anyone ready to explore the desert.
18. Raymond Turner, Janice E. Bowers, and Tony L. Burgess, *Sonoran Desert Plants: An Ecological Atlas* (Tucson: University of Arizona Press, 1995), p. 317.
19. Philips and Comus, *Natural History*, p. 195.
20. "Senita Cactus," *Plant Walk*, Herbarium, University of Arizona, 1998, www.ag.arizona.edu/arboretum/pwalk/pw13.htm.
21. Park S. Nobel, *Remarkable Agaves and Cacti* (New York: Oxford University Press, 1994), p. 65.
22. Alcock, *Sonoran Desert Spring*, p. 83.
23. Phillips and Comus, *Natural History*, p. 198.
24. Janice Emily Bowers, *The Mountains Next Door* (Tucson: University of Arizona Press, 1991), p. 68.
25. L. S. M. Curtin, *By the Prophet of the Earth: Ethnobotany of the Pima* (Santa Fe: University of Arizona Press: 1949; Tucson: University of Arizona Press, 1997), http://www.uapress.arizona.edu/online.bks/prophet/plants.htm.
26. Neil Luebke, "Cacti: More Than Stickers," *Lore* magazine, 1996; Milwaukee Public Museum, 2000, http://www.mpm.edu/research/botany/botany_lore02.html.

27. Ann Zwinger, *A Desert Country near the Sea: A Natural History of the Cape Region of Baja California* (New York: Harper and Row, 1983), p. 59.
28. Alcock, *Sonoran Desert Spring*, p. 135.
29. Bowers, *Mountains Next Door*, p. 70.
30. Alcock, *Sonoran Desert Spring*, p. 161.
31. Ibid., p. 137.
32. David Jeffery, "Arizona's Suburbs of the Sun," *National Geographic*, vol. 152, no. 4 (October 1977), p. 495.
33. Daniel Moerman, *Native American Ethnobotany*, (Portland, Ore.: Timber Press, 1998), p. 367.
34. Nobel, *Agaves and Cacti*, p. 61.
35. Ibid., p. 47.
36. Edward Abbey, *Desert Solitaire: A Season in the Wilderness* (New York: Ballantine Books, 1968), p. 25.
37. Pedro Casteñeda, *The Journey of Coronado*, trans. George Parker Winship (Ann Arbor, Mich.: University Microfilms, 1966), p. 30.
38. Nobel, *Agaves and Cacti*, p. 56.
39. Peter Fidler, *Journal of a Journey Overland from Buckingham House to the Rocky Mountains . . .* , in Mary Alice Downie and Mary Hamilton, *"And Some Brought Flowers": Plants in a New World* (Toronto: University of Toronto Press, 1980), "Prickly Pear Cactus."
40. Melvin Randolph Gilmore, *Uses of Plants by the Indians of the Missouri River Region* (Washington, D.C.: U.S. Government Printing Office, 1919), p. 104.
41. Phillips and Comus, *Natural History*, p. 211.
42. Yvonne Savio, "Prickly Pear Cactus," *Family Farm Series*, University of California–Davis, June 1989, rev. July 1989, www.sfc.ucdavis.edu/pubs/brochures/pricklypear.html.
43. Bowers, *Mountains Next Door*, p. 81.
44. Phillips and Comus, *Natural History*, p. 214.
45. Gilmore, *Uses of Plants*, pp. 104–105.
46. H. F. Whiting, *Havasupai Habitat: A. F. Whiting's Ethnography of a Traditional Indian Culture*, ed. Steven A. Weber and P. David Seaman (Tucson: University of Arizona Press, 1985), pp. 225–226.
47. Alcock, *Sonoran Desert Spring*, p. 106.
48. Raymond B. Cowles, *Desert Journal: A Naturalist Reflects on Arid California*, with Elna S. Bakker (Berkeley: University of California Press, 1977), p. 189.
49. Ibid., p. 195.
50. Whiting, *Havasupai Habitat*, pp. 109–111.
51. Phillips and Comus, *Natural History*, p. 238.
52. Jack Hill, personal interview, May 1999.
53. Alcock, *Sonoran Desert Spring*, pp. 171–172.
54. Sara St. Antoine, "Ironwood: Carving New Life from Ancient Trees," summer 1994, www.askpubs.com/wood/AncientTrees.htm.
55. Whiting, *Havasupai Habitat*, pp. 49–50.
56. "The 'Real' Ocotillo," Maricopa Center for Learning and Instruction, http://www.mcli.dist.maricopa.edu/ocotillo/oco_real.html.
57. Ruth Corbett Cross, "A Fence That You Water: It Is a Cinch to Grow," *Tucson Citizen*, August 9, 1996, p. 34.
58. Jack Hill, personal interview, May 1999.
59. Howard Scott Gentry, *Agaves of Continental North America* (Tucson: University of Arizona Press, 1982), p. 385.
60. Nobel, *Agaves and Cacti*, p. 144.
61. Wilfred William Robbins, John Peabody Harrington, and Barbara Freire-Marreca, *Ethnobotany of the Tewa Indians*, Smithsonian Institution of the Bureau of American Ethnology Bulletin 55 (Washington, D.C.: U.S. Government Printing Office, 1916), p. 50.
62. Timothy Coffey, *The History and Folklore of North American Wildflowers* (Boston: Houghton Mifflin, 1993), p. 317.
63. Thomas, *Wild Woodlands*, p. 136.
64. FEIS Online, "*Yucca brevifolia*," www.fs.fed.us/database/feis/plants/tree/yucbre/botanical_and_ecological_characteristics.html.
65. Matilda Coxe Stevenson, *Ethnobotany of the Zuni Indians*, Extract from the Thirteenth Annual Report of the Bureau of American Ethnology (Washington, D.C.: U.S. Government Printing Office, 1915), p. 72.
66. Tim Flannery, *The Eternal Frontier: An Ecological History of North America and Its Peoples* (New York: Atlantic Monthly Press, 2001), p. 140.
67. Gary Paul Nabhan, *Gathering the Desert* (Tucson: University of Arizona Press, 1985), p. 19.
68. J. L. Gardea-Torresdey, J. Bibb, and A. Hernandez, "Uptake of Toxic Heavy Metal Ions by Inactivated Cells of *Larrea tridentata* (Creosote Bush)," Abstract from the 1997 Conference on Hazardous Waste Research, Hazardous Substance Research Centers, http://www.engg.ksu.edu/HSRC/97abstracts/p22.html.
69. Nabhan, *Gathering the Desert*, p. 14.
70. Jack Hill, personal interview, May 1999.

71. Nabhan, *Gathering the Desert*, p. 14.
72. Ibid.
73. Ibid., pp. 14–15.
74. Ibid., p. 15.
75. Angela Powers, personal interview, May 1999.
76. Whiting, *Havasupai Habitat*, p. 228.
77. Nabhan, *Gathering the Desert*, p. 18.
78. Ibid., p. 15.
79. Ibid., p. 13.
80. *Ma huang* is first mentioned in the classic Chinese herbal of the Divine Plowman Emperor, Shen-Nong's Ben Cao Jing, which contains a list of 365 herbs from the first century A.D., and is the basis of the modern Chinese materia medica.
81. FEIS Online, *"Encelia farinosa,"* www.fs.fed.us/database/feis/plants/shrub/encfar/botanical_and_ecological_characteristics.html.
82. Alcock, *Sonoran Desert Spring*, pp. 64–65.
83. Kirk, *Desert*, p. 141.
84. Gary Tremper, "The History and Promise of Jojoba," *Armchair World*, 1996, http://www.armchair.com/warp/jojoba1.html.
85. Ibid.
86. Kirk, *Desert*, p. 304.

Region 7: California
1. California Native Plant Society, "Rare Plant Program," CNPS, 1999–2002, quoted from the 5th ed. of CNPS *Inventory*, February 1994, www.cnps.org/rareplants/about_flora.htm.
2. Stephen J. Pyne, *Fire in America: A Cultural History of Wildland and Rural Fire* (Seattle: University of Washington Press, 1997), p. 408.
3. Emory E. Smith, *The Golden Poppy* (San Francisco: Murdoch Press, 1902), p. 9.
4. Mary Austin, "The Land of Journeys' Ending," in Ann Zwinger, ed., *Writing the Western Landscape* (Boston: Beacon Press, 1994), p. 54.
5. "Golden Poppies Goodnight," lyrics by Mary A. Lombard and music by Leo Bruck (1902) in Smith, *Golden Poppy*, p. 192.
6. Jake Brouwer, "The California Poppy," *Echo Mountain Echoes*, vol. 2, no. 1 (spring 1997), http://aaaim.com/echo/v2n1/poppy2.htm.
7. John O'Neill, "California Poppy," 1999, http://www.pyramid.net/gallery/webnew/poppy.html.
8. Thomas C. Blackburn, ed., *December's Child: A Book of Chumash Oral Narratives* (Berkeley: University of California Press, 1975), p. 100.
9. Daniel Moerman, *Native American Ethnobotany* (Portland, Ore.: Timber Press, 1998) p. 228.
10. Timothy Coffey, *The History and Folklore of North American Wildflowers* (Boston: Houghton Mifflin, 1993), p. 29.
11. "California Poppy," California State University, Chico, Meriam Library, special collections, www.csuchico.edu/lbib/spc/netpages/calpoppy.html.
12. J. R. Dunster, "California Poppy Page," Bear's Clover Yosemite, Pottery and Art Page, 2001–2002, www.geocities.com/bearsclover/poppy.html.
13. FEIS Online, *"Pinus edulis,"* www.fs.fed.us/database/feis/plants/tree/pinedu/botanical_and_ecological_characteristics.html.
14. John Muir, *The Mountains of California* (Berkeley, Calif.: Ten Speed Press, 1977), pp. 221–222.
15. "Chapter 8: Seeds, Fruits, and Cones," Non-Wood Forest Products from Conifers (12), Food and Agriculture Organization of the UN, 1995, www.fao.org/docrep/x0453e/X0453e12.htm.
16. Moerman, *Native American Ethnobotany*, pp. 406–408.
17. Jamie Rappaport Clark, "Endangered and Threatened Wildlife and Plants," *Federal Register*, Department of Interior Fish and Wildlife Service, vol. 63, no. 155 (August 12, 1998): 43100–43116, www.cdpr.ca.gov/docs/es/estext/fr081298.txt.
18. Coyote Man, *The Destruction of the People* (Berkeley, Calif.: Brother William Press, 1973), p. 39.
19. Douglass F. Roy, *Silvical Characteristics of Redwood* (Sequoia sempervirens), in James A. Snyder, "The Ecology of *Sequoia sempervirens*," December 1992, http://www.batnet.com/askmar/Redwoods/Masters_Thesis.html.
20. United States Department of the Interior, *The Redwoods: A National Opportunity for Conservation and Alternatives for Action*, in Snyder, "Ecology of *Sequoia sempervirens*."
21. Lynwood Carranco, *Redwood Lumber Industry*, and David Kelly and Gary Braasch, *Secrets of the Old Growth Forest*, in Snyder, "Ecology of *Sequoia sempervirens*."
22. Edward C. Stone, *Ecology of the Redwoods: Inter-relationship of Redwood and its Environment*, in Snyder, "Ecology of *Sequoia sempervirens*."
23. Snyder, "Ecology of *Sequoia sempervirens*."
24. Shaun Walker, "The Luna Tree Sit and Julia Butterfly," www.ottermedia.com/LunaJulia.html.
25. Donald Culross Peattie, *A Natural History of*

Western Trees (Boston: Houghton Mifflin, 1991), p. 6.

26. Ibid., p. 9

27. Maureen Gilmer, "The Redwoods—California Big Trees," *Garden Forum: Plant Profiles*, 1998, 1999, 2000, www.gardenforum.com/redwoods. html.

28. Malcolm Margolin, *The Ohlone Way: Indian Life in the San Francisco–Monterey Bay Area* (Berkeley, Calif.: Heyday Books, 1978), archive. sportserver.com/newsroom/sports/oth/1998/ oth/mor/feat/archive/040498/mor2415.html.

29. Anna Berg, "Sudden Oak Death Marks Demise of a Culture," *City on a Hill Press*, vol. 35, no. 15 (February 15, 2001), www.slugwire.org/weekly/ archives/01Feb15/oak.html. (Website no longer available as of March 2002.)

30. Bruce Pavlik et al., *Oaks of California* (Los Olivos, Calif.: Cachuma Press, 1995).

31. Moerman, *Native American Ethnobotany*, p. 458.

32. Timothy Bradley, Marika Walter, and Marjorie Patrick, "Oak Woodland," *Exploring Ecosystems at Limestone Canyon*, University of California at Irvine, http://compphys.bio.uci.edu/ Limestone/oakwoodland.htm.

33. Coyote Man, *Sun, Moon and Stars* (Berkeley, Calif.: Brother William Press, 1973), p. 43.

34. Bradley, Walter, and Patrick, "Oak Woodland."

35. John Muir, *Mountains of California*, p. 225.

36. Moerman, *Native American Ethnobotany*, pp. 458–459.

37. Bradley, Walter, and Patrick, "Oak Woodland."

38. "Ask Mother," *Mother Earth News*, November 18, 1999, Ogden Publications, 2001–2002, www. motherearthnews.com/askmother/john.shtml.

39. Moerman, *Native American Ethnobotany*, pp. 458–459.

40. Ibid.

41. Blackburn, *December's Child*, pp. 94–95.

42. Philip M. McDonald, "*Quercus douglasii*: Blue Oak," *Silvics of North America: Hardwoods*, vol. 2, Agriculture Handbook 654, USDA Forest Service, www.na.fs.fed.us/spfo/pubs/silvics_manual/ volume_2/quercus/douglasii.htm.

43. "Common Plants of the Blue Oak Savanna," *Stebbins Cold Canyon Reserve*, University of California Natural Reserve System, nrs.ucop. edu/reserves/stebbins/plants/savanna.htm.

44. Francis M. Fultz, *The Elfin Forest of California* Los Angeles: Times-Mirror Press, 1923), www. notfrisco.com/calmem/chaparral/fultz02.html.

45. Moerman, *Native American Ethnobotany*, p. 460.

46. Pyne, *Fire in America*, p. 415.

47. Linda Thomas, "Brush Fire," *Tapestry Rambles*, Women Online Worldwide, 1995–2001, www. wowwomen.com/tapestry/arch_rambles/ brushfire.html.

48. H. H. [Helen Hunt Jackson], *Bits of Travel at Home* (Boston: Roberts Brother, 1878), p. 45.

49. Jack London, *Jack London's California: The Golden Poppy and Other Writings*, ed. Sal Noto (New York: Beaufort Books, 1986), p. 71.

50. Moerman, *Native American Ethnobotany*, p. 86.

51. Maureen Gilmer, "Of California Fire and Natives," *Garden Forum*, 1998, 1999, 2000, www.gardenforum.com/FireAndNatives.html.

52. "Myrtle Wood Tree," The Real Oregon Gift Factory, www.realoregongift.com/myrtletree. htm.

53. P. J. Gladnick, "Promontory Utah and the Nation's Railway Uniting," PageWise, 2001, http://mo.essortment.com/promontory_raic. htm.

54. Michael Ellis, "California Bay Trees," *Footloose Forays*, December 17, 2000, www. footlooseforays.iohome.net/cgi-bin/ Topic.pl?topic=44&public.

55. Brenda Ryan, "California Gold," http:// alchemylitmag.home.mindspring.com/ califgold.html.

56. Moerman, *Native American Ethnobotany*, p. 263.

57. "*Heteromeles arbutifolia*: Christmas Berry," Santa Barbara County Education Office, www.sbceo. k12.ca.us/~mcssb/sbpanda/toyon.html.

58. California Native Plant Society, "Toyon: *Heteromeles arbutifolia*," *Fremontia*, January 1974, vol. 1, no. 4: 23–25, http://lacnps.org/ toyon.html.

59. Arnel Guanlao, "Bay Area Ramblings," *Picture This*, February 12, 2000, www.calphoto.com/ antonio2.htm.

60. John McPhee, "Los Angeles against the Mountains," *The Control of Nature*, Sun Valley Writers' Conference, 2002, www.svwc.com/ McPhee.htm.

61. Moerman, *Native American Ethnobotany*, p. 49.

62. Marc Kummel, "*Adenostoma fasciculatum*: Chamise, Greasewood," *Treebeard's Flora: Woody Plants of the Central Santa Ynez Mountains*, 1998, www.rain.org/~mkummel/flora/adefas.html.

63. Moerman, *Native American Ethnobotany*, p. 93.

64. Sarah Royce, *A Frontier Lady: Recollections of the*

Gold Rush and Early California (New Haven, Conn.: Yale University Press, 1932), pp. 49–50.

65. Jane Strong, "Coyote Bush," *Jane Strong's Nature Guide*, Pelican Network, October 2000, www.pelicannetwork.net/fg/coyote_bush.html.

66. Judith Lowry, "Coyote Bush Defended," *Point Reyes Light*, May 10, 2001, Tomales Bay Publishing, 1995–2002, www.ptreyeslight.com/stories/may10_01/coyote_bush.html.

67. Coyote Man, *Destruction of the People*, p. 10.

68. Jeff Caldwell, "Berries for the Birds," *Birds and Native Plants*, California Native Plant Society, www.stanford.edu/~rawlings/birds.htm.

69. Las Pilitas Nursery, "*Rhamnus californica*," www.laspilitas.com/plants/566.htm.

70. Rebecca Rosen Lum, "'Healing Garden' Protects Redding's Sole Synagogue," *Jewish Bulletin News*, San Francisco Jewish Community Publications, 2002.

71. Christopher and Dolores Lynn Nyerges, "Wild Coffee Alternatives," *Mother Earth News*, August 1999.

72. Moerman, *Native American Ethnobotany*, p. 473.

73. Peter Hacker, "Poison Oak and Ivy," www.dhc.net/~pmhack/POISON.HTM.

74. Julie Carville, "The Oak Woodlands," www.gv.net/~rsthomas/plant/OAK_WOODLANDS.htm.

75. Moerman, *Native American Ethnobotany*, pp. 145–146.

76. "Pinnacles Hiking: 24 March 2001," marmot.net/danm/adv/2001_03_24_Pinnacles_Hike/pages/128_2866_IMG.htm. (Web page no longer available as of March 2002.)

77. George R. Mead, *The Ethnobotany of the California Indians: A Compendium of the Plants, Their Users, and Their Uses* (Greeley, Colo.: Museum of Anthropology, University of Northern Colorado, 1972), p.124.

78. Moerman, *Native American Ethnobotany*, pp. 343–344.

79. Mark McNair, Exeter University, UK, www.ex.ac.uk/~MRMacnai/guttatus.html.

80. Rick Lovett, "'Extinct' Oregon Flower Reappears," *Daily InScight*, June 21, 1999, Academic Press, American Association for the Advancement of Science, 1999, www.academicpress.com/inscight/06211999/graphb.htm.

81. C. F. Saunders, *Western Wild Flowers and Their Stories* (New York: Doubleday Doran, 1933), pp. 116–118.

82. Carol Kaesuk Yoon, "For a New Species, a Few Genes Are All That's Needed," September 5, 1995, *The New York Times*, 1995, www.qeced.net/bio/genbio/exNatSel.htm.

83. Ibid.

84. Charles Jeffries, "Floral Tribute," in *Come Into the Garden: A Treasury of Garden Verse*, comp. Jill Hollis (London: Random Century Group, 1992), p. 18.

85. David Rogers, "The Native Irises (Flags or Fleur-de-lis) of the Santa Lucia Mountains," *The Double Cone Quarterly*, spring 2001, vol. 4, no. 1, www.ventanawild.org/news/se01/irises.html.

86. Coffey, *History and Folklore*, p. 321.

Region 8: Montane

1. Ann Zwinger, *Run, River, Run* (New York: Harper and Row, 1975), p. 13.

2. Kathryn Eberhart, "Tall Bluebells (*Mertensia paniculata*)," Alaska Screensavers, 1997–2002, www.alaskascreensavers.com/gallery/matvalley/bluebell.htm.

3. "In Mountain and Meadow," May 15, 1996, naturalist.org, 1995, 1996, 1998, www.naturalist.org/brow4/96515.php3.

4. Daniel Moerman, *Native American Ethnobotany* (Portland, Ore.: Timber Press, 1998), p. 351.

5. Ibid., p. 525.

6. Nancy J. Turner et al., *Thompson Ethnobotany: Knowledge and Usage of Plants by the Thompson Indians of British Columbia* (Victoria: Royal British Columbia Museum, 1990), p. 206.

7. Diane Rogers, "Herb Growers Find Several Paths to Market," *The Western Producer*, July 27, 2000, www.producer.com/articles/20000727/farm_living/20000727fl01.html.

8. "Leedy's Roseroot," State of Minnesota, Department of Natural Resources, 1993, adapted from Nancy Sather, *Leedy's Roseroot: A Cliffside Glacial Relict* (St. Paul: Minnesota Department of Natural Resources, 1993), www.dnr.state.mn.us/ecological_services/nhnrp/Lroseroot.pdf.

9. Moerman, *Native American Ethnobotany*, p. 525.

10. "Plants Used for Food-Related Purposes," *Traditional Plant Use in the Hazeltons*, Industry Canada, http://collections.ic.gc.ca/hazeltons/food.htm.

11. Linda Kershaw, Andy MacKinnon, and Jim Pojar, *Plants of the Rocky Mountains* (Edmonton, Alt.: Lone Pine, 1998), p. 48.

12. C. S. Rafinesque, *Medical Flora*, in Charlotte

Erichsen-Brown, *Medicinal and Other Uses of North American Plants: A Historical Survey with Special Reference to the Eastern Indian Tribes* (New York: Dover, 1989, originally published 1979), p. 34.

13. Kershaw et al., *Plants of the Rocky Mountains*, p. 48.

14. Henriette Kress, "Gathering Juniper (*Juniperus communis*) Berries," *Henriette's Herbal Homepage*, The Culinary Herblist, January 1996, http://ibiblio.org/herbmed/neat-stuff/junicomm.html.

15. Rod Smith, "In Like Gin," *Private Clubs Magazine Online*, January/February 2001, www.privateclubs.com/archives/2001-jan-feb/wine_inlikegin.htm.

16. "Garden of the Gods and White House Ranch Colorado Springs City Parks," http://www.mindspring.com/~duckgoose/gog.html.

17. Turner et al., *Thompson Ethnobotany*, p. 94.

18. Clark Schreibeis Wildlife Art, "Natural Finish Wood Sculptures," 2001. www.clarkschreibeis.com/wood.htm.

19. Moerman, *Native American Ethnobotany*, p. 586.

20. "Window Mountain Lake," Crowsnest Vacation Creation, http://www.telusplanet.net/public/dtwilson/wmlake.html.

21. Mary Ann Spahr, "Glacier Lily, *Erythronium grandiflorum*," *Northwest Natives*, Rainier Audubon Society, *The Heron Herald Newsletter*, September 1999, http://www.rainieraudubon.org/nn/glacier-lily.htm.

22. Turner et al., *Thompson Ethnobotany*, pp. 123–124.

23. Moerman, *Native American Ethnobotany*, p. 227.

24. Brian Keating and Dee Keating, "Mistaya Lodge," *Notes from the Field*, July 23–26, 1999, http://www.calgaryzoo.ab.ca/zoofaritour/mistaya2.htm.

25. Moerman, *Native American Ethnobotany*, pp. 35–36.

26. Turner et al., *Thompson Ethnobotany*, pp. 97–98; Moerman, *Native American Ethnobotany*, pp. 34–35.

27. Ibid.

28. Albert Kellogg, *Forest Trees of California*, in David Rogers, "An Addendum on the Botanical History of Santa Lucia Fir, *Abies bracteata*, with Excerpts from the Notes and Letters of Early Collectors," *The Double Cone Quarterly*, vol. 1, no. 3 (winter solstice 1998), www.ventanawild.org/news/ws98/slfirs2.html.

29. Charles S. Sargent, *The Silva of North America*, in Rogers, "Addendum on Santa Lucia Fir."

30. David Rogers, "Perfect Pattern of Silvan Perfection on the Symmetrical Plan, the Rare Santa Lucia Fir," *The Double Cone Quarterly*, vol. 1, no. 2 (fall equinox 1998), www.ventanawild.org/news/fe98/slfirs.html.

31. Robert Devine, "Rocky Mountain National Park," *National Geographic Traveler*, July/August 1991, p. 82.

32. Turner et al., *Thompson Ethnobotany*, pp. 100–101; Moerman, *Native American Ethnobotany*, p. 398.

33. April Pettinger, *Native Plants in the Coastal Garden: A Guide for Gardeners in British Columbia and the Pacific Northwest* (Vancouver: Whitecap Books, 1996), p. 69.

34. Turner et al., *Thompson Ethnobotany*, pp. 102–103.

35. Moerman, *Native American Ethnobotany*, p. 405.

36. Rob Curtis, "Sierran Montane Forests," *Bird Habitats*, eNature.com, 2002, http://www.enature.com/habitats/show_sublifezone.asp?sublifezoneID=42.

37. Audrey DeLella Benedict, *A Sierra Club Naturalist's Guide to the Southern Rockies: The Rocky Mountain Regions of Southern Wyoming, Colorado, and Northern New Mexico* (San Francisco: Sierra Club Books, 1991), p. 277.

38. Moerman, *Native American Ethnobotany*, pp. 410–412.

39. "Ponderosa Pine Forest of the Colorado Plateau," *Land Use History of North America*, Northern Arizona University, www.cpluhna.nau.edu/Biota/ponderosa_forest.htm.

40. John Muir, *The Yosemite* (New York: Century, 1912), "Chapter 5: The Trees of the Valley," reprinted in *John Muir Exhibit*, Sierra Club, 2002, www.sierraclub.org/john_muir_exhibit/writings/the_yosemite/chapter_5.html.

41. backpacking.about.com/library/weekly/aa041301a.htm. (Web page not available as of March 2002.)

42. "Bitterroot: *Lewisia rediviva Pursh*," Montana Historical Society Press, http://www.his.state.mt.us/departments/press/bitter_root.html.

43. W. J. Hooker, in Timothy Coffey, *The History and Folklore of North American Wildflowers* (Boston: Houghton Mifflin, 1993), p. 47.

44. Carl A. Geyer, "Notes on the Vegetation and General Character of the Missouri and Oregon Territories, Made During a Botanical Journey

from the State of Missouri, across the South-
Pass of the Rocky Mountains, to the Pacific,
during the years 1843 and 1844," in Coffey,
History and Folklore, p. 46.

45. "Bitterroot: *Lewisia rediviva Pursh*."
46. Kershaw et al., *Plants of the Rocky Mountains*,
p. 122.
47. "Bitterroot: *Lewisia rediviva Pursh*."
48. Ibid.
49. Nancy Turner, interview, June 2, 1999.
50. Gene Bush, "Shooting Stars," *Garden Clippin's*,
Munchkin Nursery and Gardens, 2002, http://
www.munchkinnursery.com/newsletter/
shooting-stars/.
51. Moerman, *Native American Ethnobotany*,
pp. 202–203.
52. James S. Lippincott, "Among the Wild Flowers
of San Diego," *Information Relative to the City of
San Diego, California*, San Diego, Office of the
San Diego Daily Union, 1874, p. 29, in San
Diego Historical Society, www.sandiegohistory.
org/books/city1874/page29.htm.
53. Opal Whiteley, "Along the Road," *Opal
Whiteley's Enchanted Fairyland*, http://www.
liloriole.net/text-road-01.htm.
54. "Studying the Frigid Shootingstar," Berry
Botanical Garden, http://www.berrybot.org/
ar_stdoau.html.
55. John Gerard, *The Herball or General History of
Plants*, in Erichsen-Brown, *Medicinal and Other
Uses*, p. 314.
56. Turner et al., *Thompson Ethnobotany*, p. 248.
57. Kershaw et al., *Plants of the Rocky Mountains*,
p. 198.
58. Barbara Earle and Scott Earle, "Figwort
(Snapdragon) Family: Scrophulariaceae," Idaho
Mountain Wildflowers, 1999/2000, www.
larkspurbooks.com/Scroph1.html.
59. Henry D. Thoreau, *The Journal of Henry D.
Thoreau*, ed. Bradford Torrey and Francis H.
Allen (New York: Dover, 1962), vol. V,
pp. 127–128 [vol. 1, p. 560].
60. Moerman, *Native American Ethnobotany*,
pp. 142–144.
61. J. W. Brewer, K. J. Collyard, and C. E. J. Lott,
"Analysis of Sugars in Dwarf Mistletoe
Nectar," in R. M. Polhill, coord., *The Golden
Bough*, vol. 10, February 1988, The Herbarium,
Royal Botanic Gardens, Kew, www.science.siu.
edu/parasitic-plants/Ant.DM.html.
62. Moerman, *Native American Ethnobotany*, p. 84.

**Region 9: The Tall Trees: The Pacific
Northwest**

1. Sallie Tisdale, *Stepping Westward: The Long
Search for Home in the Pacific Northwest* (New
York: Henry Holt, 1991), p. 50.
2. Chris Maser, *The Redesigned Forest* (Toronto:
Stoddart, 1988), p. 19.
3. Jonathan Raban, *Passage to Juneau: A Sea and Its
Meanings* (New York: Pantheon Books, 1999),
p. 69.
4. John Davies, *Douglas of the Forests* (Seattle:
University of Washington Press, 1980), App. IV.
5. David Douglas, *Journal Kept by David Douglas
during His Travels in North America, 1823–1827*
(London: William Wesley and Son, 1914),
pp. 338–339.
6. Nancy J. Turner, *Plant Technology of First Peoples
in British Columbia* (Vancouver: UBC Press,
1998), p. 96.
7. Ibid., p. 97.
8. Stephen J. Pyne, *Fire in America: A Cultural
History of Wildland and Rural Fire* (Seattle:
University of Washington Press, 1997), p. 337.
9. Audrey Grescoe, *Giants: The Colossal Trees of
Pacific North America* (Vancouver: Raincoast
Books, 1997), p. 41.
10. Jon R. Luoma, *The Hidden Forest: The Biography
of an Ecosystem* (New York: Henry Holt, 1999),
p. 40.
11. Michael J. Caduto and Joseph Bruchac, *Keepers
of Life: Discovering Plants through Native Stories
and Earth Activities for Children* (Saskatoon,
Sask.: Fifth House Publishers, 1994), p. 202.
12. Deanna Kawatski, *Wilderness Mother* (New
York: Lyons and Burford, 1994), p. 188.
13. Turner, *Plant Technology*, pp. 87–88.
14. Ibid., p. 89.
15. Donald Culross Peattie, *A Natural History of
Western Trees* (Boston: Houghton Mifflin,
1991), p. 151.
16. Cameron Young, *The Forests of British Columbia*
(North Vancouver: Whitecap Books, 1985),
p. 66.
17. Turner, *Plant Technology*, p. 98.
18. Ibid.
19. Ibid., p. 99.
20. Ibid.
21. Arthur Plotnik, *The Urban Tree Book: An
Uncommon Field Guide for City and Town* (New
York: Three Rivers Press, 2000), p. 345.
22. Alice C. Fletcher and Francis La Flesche, *The

Omaha Tribe (New York: Johnson Reprint, 1970), pp. 457–458.

23. Melvin Randolph Gilmore, *Uses of Plants by the Indians of the Missouri River Region* (Washington, D.C.: USGPO, 1919), pp. 11–12.

24. Rolf W. Mathewes and Richard J. Hebda, "Holocene History of Cedar and Native Indian Cultures of the North American Pacific Coast," *Science*, vol. 225 (August 17, 1984): 711.

25. Emily Carr, *Klee Wyck* (Toronto: Oxford University Press, 1941), p. 6.

26. Turner, *Plant Technology*, p. 72.

27. Ibid., pp. 72–73.

28. Stephen E. Ambrose, *Undaunted Courage: Meriwether Lewis, Thomas Jefferson, and the Opening of the American West* (New York: Simon & Schuster, 1996), p. 339.

29. "Legends," *Native Online*, 2000, www.nativeonline.com/legends.html#REDCEDAR. This is a really good Web site to get a background on many aspects of West Coast cultures.

30. Daniel Moerman, *Native American Ethnobotany* (Portland, Ore.: Timber Press, 1998), pp. 558–561.

31. Carr, *Klee Wyck*, pp. 27, 29.

32. Peattie, *Western Trees*, p. 223.

33. Ibid., p. 254

34. Turner, *Plant Technology*, p. 67.

35. Ibid., pp. 67, 77.

36. Erna Gunther, *Ethnobotany of Western Washington: The Knowledge and Use of Indigenous Plants by Native Americans* (Seattle: University of Washington Press, 1973), p. 16.

37. Douglas, *Journal*, p. 108.

38. Ibid., p. 221.

39. Turner, *Plant Technology*, p. 128.

40. Ibid.

41. Ruth Kirk, *The Olympic Rain Forest: An Ecological Web* (Seattle: University of Washington Press, 1992), p. 61.

42. Turner, *Plant Technology*, pp. 131–132.

43. Grescoe, *Giants*, pp. 120–121.

44. Ibid., p. 150.

45. Ibid., p. 152.

46. David Tirrell Hellyer, *At the Forest's Edge: Memoir of a Physician-Naturalist* (Seattle: Pacific Search Press, 1985), p. 316.

47. Herbert Eugene Bolton, *Fray Juan Crespi, Missionary Explorer on the Pacific Coast, 1769–1774* (New York: AMS Press, 1871), p. 232.

48. Archibald Menzies, *Menzies' Journal of Vancouver's Voyage, April to October, 1792*, ed. C.F. Newcombe (Victoria, B.C.: William H. Cullin, 1923), p. 20.

49. Young, *Forests of British Columbia*, p. 80.

50. Wallace W. Hansen, "*Arbutus menziesii*— Madrone, Madrona," Native Plants of the Northwest, Native Plant Nursery & Gardens, www.nwplants.com/madrone/.

51. Douglas, *Journal*, p. 102.

52. Ibid., p. 221.

53. Turner, *Plant Technology*, p. 212.

54. Athelstan George Harvey, *Douglas of the Fir: A Biography of David Douglas, Botanist* (Cambridge, Mass.: Harvard University Press, 1947), p. 56.

55. Arthur R. Kruckeberg, *Gardening with Native Plants of the Pacific Northwest: An Illustrated Guide* (Seattle: University of Washington Press, 1982), pp. 67–69. This book was invaluable in my research on plants of the Pacific Northwest.

56. Turner, *Plant Technology*, p. 166.

57. Ibid., pp. 166–167.

58. Kruckeberg, *Gardening with Native Plants*, pp. 67–69.

59. Nalini Nadkarni and Erica Guttman, "Native Plant of the Month: Pacific Dogwood," *Green Screens*, September 1999, www.olywa.net/speech/september99/native.html.

60. Turner, *Plant Technology*, pp. 148–149.

61. Ibid., p. 149.

62. Sydney Eddison, "Good Looks Begin at the Edge," *Fine Gardening*, no. 55 (May/June 1997), p. 26.

63. Turner, *Plant Technology*, pp. 164–165.

64. Richard Hebda, "Natural History: Ocean Spray," *Coastal Grower, Index of Native Plants*, Royal British Columbia Museum, 1995, http://rbcml.rbcm.gov.bc.ca/nh_papers/nativeplants/holodisc.html.

65. Las Pilitas Nursery, "*Holodiscus discolor*," www.laspilitas.com/plants/350.htm.

66. E-mail from John Browne Jr. Originally on the Pacific Northwest Native Plant Society Digest: pnw-natives@tardigrade.net. Every native plant society has a terrific Web site and chat lines where it's possible to pick up masses of really good information on these plants.

67. Nancy J. Turner, *Food Plants of Interior First Peoples* (Vancouver: UBC Press, 1997), p. 67.

68. Ambrose, *Undaunted Courage*, p. 294.

69. Sister M. Alfreda Elsensohn, "A Flora of the Camas Prairie Region in the Vicinity of

Cottonwood, Idaho, Designed for the Use of High School Students," Master's Thesis, Education, University of Idaho, 1939, p. 17, www.lili.org/camas.html.

70. Homer, *Odyssey*, vol. 7, lines 253 ff. Trans. Albert Cook (New York: W.W. Norton Company, 1967), p. 97.

71. Craig Childs, "The Millworker and the Forest," *High Country News*, vol. 31, no. 18 (September 27, 1999), www.hcn.org/servlets/hcn.Article?article_id=5266.

72. William Frederic Badè, *The Life and Letters of John Muir* (Boston: Houghton Mifflin, 1924), p. 71.

73. Carr, *Klee Wyck*, p. 49.

74. Turner, *Plant Technology*, p. 104.

75. Ibid.

76. Ibid.

77. FEIS Online, "*Oplopanax horridus*," www.fs.fed.us/database/feis/plants/shrub/oplhor/botanical_and_ecological_characteristics.html.

78. Leslie A. Viereck and Elbert A. Little Jr., *Alaska Trees and Shrubs* (Washington, D.C.: U.S. Department of Agriculture, 1972; reprinted, Fairbanks: University of Alaska Press, 1994), p. 198.

79. Turner, *Plant Technology*, p. 140.

80. Ibid.

81. Priscilla Russell Kari, *Tanaina Plantlore: Dena'ina K'et'una—An Ethnobotany of the Dena'ina Indians of Southcentral Alaska*, 2nd ed. (Anchorage: National Park Service Alaska Region, 1987), p. 23.

82. Hellyer, *At the Forest's Edge*, p. 319.

83. "Western Sword Fern, Christmas Fern, Sword Holly Fern (*Polystichum munitum*)," Henry Cowell Redwoods State Park, 1997, www.best.com/~dforthof/HenryCowell/wild.cgi/polystichum_munitum.

84. Turner, *Plant Technology*, p. 64.

Region 10: The Tundra

1. John Muir, *The Cruise of the Corwin: Journal of the Arctic Expedition of 1881 in Search of Delong and Jeanette*, ed. William F. Badè (Boston: Houghton Mifflin, 1917), p. 266.

2. Tim Flannery, *The Eternal Frontier: An Ecological History of North America and Its Peoples* (New York: Atlantic Monthly Press, 2001), p. 356.

3. Steven B. Young, *To the Arctic: An Introduction to the Far Northern World* (New York: John Wiley and Sons, 1989), p. 193.

4. Priscilla Russell Kari, *Tanaina Plantlore: Dena'ina K'et'una—An Ethnobotany of the Dena'ina. . . .* (Anchorage: National Park Service Alaska Region, 1987), p. 56.

5. M. L. Fernald, Rhodora article quoted in Charlotte Erichsen-Brown, *Medicinal and Other Uses of North American Plants: A Historical Survey with Special Reference to the Eastern Indian Tribes* (New York: Dover, 1989, originally published 1979), p. 47.

6. E. C. Pielou, *A Naturalist's Guide to the Arctic* (Chicago: University of Chicago Press, 1994), p. 114.

7. John Burroughs, *Alaska, the Harriman Expedition*, in John A. Murray, ed., *A Republic of Rivers: Three Centuries of Nature Writing from Alaska and the Yukon* (New York: Oxford University Press, 1990), p. 119.

8. Kari, *Tanaina Plantlore*, pp. 54–55.

9. Janice J. Schofield, *Discovering Wild Plants: Alaska, Western Canada, the Northwest* (Anchorage: Alaska Northwest Books, 1989), p. 25.

10. Kari, *Tanaina Plantlore*, p. 71.

11. Timothy Coffey, *The History and Folklore of North American Wildflowers* (Boston: Houghton Mifflin, 1993), p. 90.

12. Schofield, *Discovering Wild Plants*, p. 218.

13. Adam F. Szczawinski and Nancy J. Turner, *Wild Green Vegetables of Canada* (Ottawa: National Museum of Natural Sciences, 1980), p. 100.

14. Robert Bylot and William Baffin, *Ventursome Voyage of Robert Bylot and William Baffin*, in Erichsen-Brown, *Medicinal and Other Uses*, p. 420.

15. Michael McRae (who went gathering plants with Janice Schofield in Alaska), "I Am Monkey Flower," *Outside* magazine, July 1997, online in *Outside Online*, Mariah Media, Inc.: 2001, www.outsidemag.com/magazine/0797/9707flower.html.

16. Szczawinski and Turner, *Wild Green Vegetables*, p. 102.

17. Derek Johnson, Linda Kershaw, Andy McKinnon, and Jim Pojar, *Plants of the Western Boreal Forest and Aspen Parkland* (Edmonton, Alta.: Lone Pine Publishing, 1995), p. 111.

18. Aldo Leopold, *A Sand County Almanac* (New York: Oxford University Press, 1966), p. 26.

19. Young, *To the Arctic*, pp. 187–188.

20. Robert Marshall, *Alaska Wilderness*, in Murray, *A Republic of Rivers*, pp. 173–176.

21. Young, *To the Arctic*, pp. 187–188.
22. www.ankn.uaf.edu/sop/sopv4i4.pdf.
23. Schofield, *Discovering Wild Plants*, p. 14.
24. Ibid., p. 95.
25. Richard Hebda, "Sweet Coltsfoot," *Coastal Grower, Index for Native Plants*, Royal British Columbia Museum, 1995, rbcm1.rbcm.gov.bd. ca/nh_papers/nativeplants/petafrig.html.
26. Daniel Moerman, *Native American Ethnobotany* (Portland, Ore.: Timber Press, 1998), p. 388.
27. *Plants for a Future—Species Database*, Cornwall, UK, 1997–2000, www.ibiblio.org/pfaf/cgi-bin/ arr_html? Myosotis+alpestris.
28. John Muir, *Travels in Alaska*, in Mary Durant, *Who Named the Daisy? Who Named the Rose? A Roving Dictionary of North American Wildflowers* (New York: Dodd Mead, 1976), pp. 67–68.
29. Jeannette Mirsky, *To the Arctic! The Story of Northern Exploration from Earliest Times to the Present* (Chicago: University of Chicago Press, 1934, 1970), p. 100.
30. Moerman, *Native American Ethnobotany*, p. 378.
31. Ibid., p. 141.
32. Greg Henry, *Retrospective Growth Analysis in High Arctic Dwarf Shrubs*, University of British Columbia, Geography Department, www.geog. ubc.ca/research/henry1.htm.
33. Kari, *Tanaina Plantlore*, p. 122.
34. Harriet V. Kuhnlein and Nancy J. Turner, *Traditional Plant Foods of Canadian Indigenous Peoples: Nutrition, Botany, and Use* (Philadelphia: Gordon and Breach Science, 1991), p. 82.
35. Schofield, *Discovering Wild Plants*, p. 139.
36. Kari, *Tanaina Plantlore*, p. 132.
37. Pielou, *Naturalist's Guide to the Arctic*, pp. 126–127.
38. Kari, *Tanaina Plantlore*, p. 157.
39. Szczawinski and Turner, *Wild Green Vegetables*, p. 141.
40. E-mail from E. C. Pielou, August 2, 2000.
41. A. E. Porsild, *Emergency Food in Arctic Canada*, in Erichsen-Brown, *Medicinal and Other Uses*, p. 421.
42. Szczawinski and Turner, *Wild Green Vegetables*, p. 140.
43. Pielou, *Naturalist's Guide to the Arctic*, p. 136.
44. D. B. O. Savile, *Arctic Adaptations in Plants* (Canadian Department of Agriculture, 1972), p. 26.
45. Schofield, *Discovering Wild Plants*, p. 100.
46. Murray, *A Republic of Rivers*, p. 82.
47. Nancy J. Turner and Adam Szczawinski, *Edible Wild Fruits and Nuts of Canada*, vol. 3 (Markham, Ont.: Fitzhenry and Whiteside for National Museums of Natural Sciences, 1988), p. 173.
48. Johnson et al., *Plants of the Western Boreal Forest*, p. 115.
49. John Eastman, *The Book of Swamp and Bog: Trees, Shrubs, and Wildflowers of Eastern Freshwater Wetlands* (Mechanicsburg, Penn.: Stackpole Books, 1995), p. 81.
50. Durant, *Who Named the Daisy?*, p. 180.
51. Ibid.
52. Schofield, *Discovering Wild Plants*, p. 242.
53. Kuhnlein and Turner, *Traditional Plant Foods*, p. 263.
54. Moerman, *Native American Ethnobotany*, p. 270.
55. Schofield, *Discovering Wild Plants*, p. 110.
56. "*Lycopodium clavatum*—Running Clubmoss," Forest Capital Development Association, 2000/ 2001, www.borealforest.org/ferns/fern9.htm.
57. Pielou, *Naturalist's Guide to the Arctic*, p. 185.
58. Michael J. Caduto and Joseph Bruchac, *Keepers of Life: Discovering Plants through Native Stories and Earth Activities for Children* (Saskatoon, Sask.: Fifth House Publishers, 1994), p. 91.
59. Durant, *Who Named the Daisy?*, p. 106.
60. Kari, *Tanaina Plantlore*, p. 176.
61. Mary Alice Downie and Mary Hamilton, "*And Some Brought Flowers*": *Plants in a New World* (Toronto: University of Toronto Press, 1980), p. 88.
62. Caduto and Bruchac, *Keepers of Life*, p. 80.
63. Scofield, *Discovering Wild Plants*, p. 221.
64. Pielou, *Naturalist's Guide to the Arctic*, p. 188.
65. Johnson et al., *Plants of the Western Boreal Forest*, p. 333.
66. Kuhnlein and Turner, *Traditional Plant Foods*, p. 38.
67. Mirsky, *To the Arctic!*, pp. 115–117.
68. A. C. Parker, *Iroquois Uses of Maize and Other Food Plants* (Albany: University of the State of New York, 1910; reprinted, Ohsweken, Ont.: Iroquois Publishing and Craft Supplies, 1994), p. 94.
69. Claude-Jean Allouez, *Father Allouez's Journey to Lake Superior, 1665–1667*, in Downie and Hamilton, "*And Some Brought Flowers*," "Tripe de Roche."
70. Samuel Hearne, *A Journey from Prince of Wales's Fort in Hudson's Bay to the Northern Ocean, in the Years 1769, 1770, 1771 and 1772*, in Downie

and Hamilton, *"And Some Brought Flowers,"* "Tripe de Roche."

71. Kuhnlein and Turner, *Traditional Plant Foods*, pp. 38–39.

72. Szczawinski and Turner, *Wild Green Vegetables*, p. 45.

73. James N. Corbridge and William A. Weber, *A Rocky Mountain Lichen Primer* (Niwot, Colo.: University Press of Colorado, 1998), p. 23.

74. Kari, *Tanaina Plantlore*, p. 178.

75. Sylvia Duran Sharnoff, comp., *Bibliographic Database of the Human Uses of Lichens*, www.lichen.com/usetype.html.

76. Moerman, *Native American Botany*, p. 582.

77. Chanchal Cabrera, *A Review of Some Medicinal Plants of the Pacific North West*, Gaia Garden Herbal Dispensary and Clinic, 2000, http://www.gaiagarden.com/articles/pacific_north_west_herbs.html.

78. Nancy J. Turner, *Plant Technology of First Peoples in British Columbia* (Vancouver: UBC Press, 1998), p. 55.

79. Ibid., pp. 55–56.

The Three Sisters: Agriculture

1. Tim Flannery, *The Eternal Frontier: An Ecological History of North America and Its Peoples* (New York: Atlanta Monthly Press, 2001), p. 247.

2. Both William Cronon, *Changes in the Land: Indians, Colonists, and the Ecology of New England* (New York: Farrar Straus Giroux, 1983), p. 43 ff., and Flannery, *Eternal Frontier*, p. 256, have interesting chapters on this.

3. Doug MacCleery, "When Is a Landscape Natural?" *Forest Landowner Magazine*, January 1999, http://www.forestland.org/mag_jan99_mac.html.

4. Henry A. Wallace and Earl N. Bressman, *Corn and Corn Growing*, in Betty Fussell, *The Story of Corn: The Myths and History, the Culture and Agriculture, the Art and Science of America's Quintessential Crop* (New York: North Point Press, 1992), p. 61.

5. Frank Waters, *Book of the Hopi* (New York: Viking, 1963), pp. 7–8.

6. Paul Christoph Mangelsdorf, in Fussell, *Story of Corn*, p. 81.

7. Gary Paul Nabhan, *Enduring Seeds: Native American Agriculture and Wild Plant Conservation* (San Francisco: North Point Press, 1989), p. 53.

8. William H. MacLeish, *The Day before America: Changing the Nature of a Continent* (Boston: Houghton Mifflin, 1994), p. 157.

9. Frances B. King, "American Indian Plant Use: An Overview," in Marsha C. Bol, ed., *The Stars Above, the Earth Below: American Indians and Nature* (Niwot, Colo.: Roberts Rinehart Publishers for Carnegie Museum of Natural History, 1998), p. 175.

10. A. C. Parker, *Iroquois Uses of Maize and Other Food Plants* (Albany: University of the State of New York, 1910; reprinted, Ohsweken, Ont.: Iroquois Publishing and Craft Supplies, 1994), p. 12.

11. Samuel de Champlain, *Voyages of Samuel de Champlain 1604–1608*, in Mary Alice Downie and Mary Hamilton, *"And Some Brought Flowers": Plants in a New World* (Toronto: University of Toronto Press, 1980), "Corn."

12. Walton Galinat, in Fussell, *Story of Corn*, p. 67.

13. King, "American Indian Plant Use," pp. 174–175.

14. Nabhan, *Enduring Seeds*, pp. 184–185.

15. Ibid., p. 53.

16. Cronon, *Changes in the Land*, p. 43.

17. Robert Hendrickson, *Ladybugs, Tiger Lilies, and Wallflowers: A Gardener's Book of Words* (Upper Saddle River, N.J.: Prentice Hall, 1993), p. 48.

18. Melvin Randolph Gilmore, *Uses of Plants by the Indians of the Missouri River Region* (Washington, D.C.: U.S. Government Printing Office, 1919), p. 68.

19. Gilbert L. Wilson, *Buffalo Bird Woman's Garden: Agriculture of the Hidatsa Indians* (St. Paul: Minnesota Historical Society Press, 1987), pp. 13, 22, 23, 36, 37, 38.

20. Albert Yava, *Big Falling Snow: A Tewa-Hopi Indian's Life and Times . . .* (New York: Crown, 1978), p. 5.

21. Dean Saxton and Lucille Saxton, *Legends and Lore of the Papago and Pima Indians* (Tucson: University of Arizona Press, 1973), p. 29.

22. Marsha C. Bol, "Nature as a Model for American Indian Societies: An Overview," in Bol, *Stars Above, Earth Below*, p. 235.

23. Elizabeth Rozin, *Blue Corn and Chocolate* (New York: Alfred A. Knopf, 1992), p. 141.

24. Howard S. Russell, *Indian New England before the Mayflower* (Hanover, N.H.: University Press of New England, 1980), p. 76.

25. Frances Densmore, *Indian Use of Wild Plants for*

Food, Medicine, and Charms, reprint of *Uses of Plants by the Chippewa Indians*, in the Forty-fourth Annual Report of the Bureau of American Ethnology, 1926–1927 (Washington, D.C.: U.S. Government Printing Office, 1928, reprinted, Ohsweken, Ont.: Iroqrafts Ltd., 1993), p. 319.

26. Michael A. Weiner, *Earth Medicine—Earth Foods; Plant Remedies, and Natural Foods of the North American Indians* (New York: Macmillan, 1972), p. 168.

27. Norman Kolpas and Barbara Pool Fenzl, *Southwest the Beautiful Cookbook: Recipes from America's Southwest* (San Francisco: Collins Publishers, 1994), p. 91.

28. Russell, *Before the Mayflower*, p. 76.

29. Nabhan, *Enduring Seeds*, p. 180.

30. Kolpas and Fenzl, *Southwest the Beautiful Cookbook*, p. 91.

31. Hendrickson, *Ladybugs, Tiger Lilies, and Wallflowers*, p. 49.

32. Ibid.

33. Fussell, *Story of Corn*, p. 104.

34. Herman J. Viola and Carolyn Margolis, eds., *Seeds of Change: A Quincentennial Commemoration* (Washington, D.C.: Smithsonian Institution Press, 1991), p. 23.

35. "Echo Plant Information Sheet—Tepary Bean," N. Fort Myers, Fla., Educational Concerns for Hunger Organization, 1999. http://www.echonet.org/tropicalag/plantinfo/Phaseolusacutifolius.pdf.

36. F. W. Waugh, *Iroquois Foods and Food Preparation* (Ottawa: Government Printing Bureau, 1916; facsimile ed., Ottawa: National Museums of Canada, 1973), p. 104.

37. Wilson, *Agriculture of the Hidatsa*, p. 82.

38. Ibid., p. 83.

39. Parker, *Iroquois Uses of Maize*, p. 89.

40. Nabhan, *Enduring Seeds*, pp. 74–75.

41. Parker, *Iroquois Uses of Maize*, p. 38.

42. Bol, *Stars Above, Earth Below*, p. 165.

43. Wilson, *Agriculture of the Hidatsa*, p. 20.

44. "Westward Ho: Frontier Life and Exploration, 1776–1848," *Welcome to America the Bountiful*, University of California at Davis, www.lib.ucdavis.edu/exhibits/food/panel3.html.

45. Ibid.

46. Gilmore, *Uses of Plants*, p. 118.

47. King, "American Indian Plant Use," p. 177.

48. Nabhan, *Enduring Seeds*, p. 50.

49. Ibid., p. 49.

50. Wilson, *Agriculture of the Hidatsa*, p. 75.

51. Ibid., p. 74.

52. Rozin, *Blue Corn and Chocolate*, pp. 142–143.

53. Densmore, *Indian Use of Wild Plants*, p. 319.

54. Virgil J. Vogel, *American Indian Medicine* (Norman: University of Oklahoma Press, 1970), p. 356.

55. Parker, *Iroquois Uses of Maize*, p. 37.

56. Amadeo M. Rea, "Corn Man and Tobacco Woman," in Bol, *Stars Above, Earth Below*, p. 214.

57. Wilson, *Agriculture of the Hidatsa*, p. 69.

58. Ibid., p. 70.

59. "Sunflower Story," Red River Commodities, 2000, www.redriv.com/sunflower_story.html.

60. National Sunflower Association, www.sunflowernsa.com/pubinfo/history/.

61. Wilson, *Agriculture of the Hidatsa*, p. 21.

62. William A. Niering and Nancy C. Olmstead, *National Auduban Society Field Guide to North American Wildflowers: Eastern Region* (New York: Alfred A. Knopf, 1997), p. 383.

63. Charles Bixler Heiser, *The Sunflower* (Norman: University of Oklahoma Press, 1976), pp. 29–35.

64. Annora Brown, *Old Man's Garden* (Toronto: J. M. Dent and Sons, 1954), p. 66.

65. Wilson, *Agriculture of the Hidatsa*, p. 16.

66. Frederick C. Klein, *Sunflower Seeds and Seoul Food*, in Addall, Muze, 1995–2002, www.addall.com/Browse/Detail/1566250757.html.

67. Nabhan, *Enduring Seeds*, p. 130.

68. Lyal Tait, *Tobacco in Canada* (Tillsonburg, Ont.: Flue-Cured Tobacco Growers' Marketing Board, 1968), p. 11.

69. Brown, *Old Man's Garden*, pp. 151–152.

70. John D. Hunter, *Manners and Customs of Several Indian Tribes Located West of the Mississippi*, in Vogel, *American Indian Medicine*, p. 384.

71. Frank G. Speck, "Virginia Indian Folk Lore Plants"; Gladys Tantaquidgeon, "Mohegan Medicinal Practices, Weather-Lore, and Superstition"; and Frank G. Speck, "Medicine Practices of the Northeastern Algonquians", in Vogel, *American Indian Medicine*, p. 384.

72. David I. Bushnell, Jr., "The Choctaw of Bayou Lacomb, St. Tammany Parish, Louisiana," in Vogel, *American Indian Medicine*, p. 384.

73. John R. Swanton, "Religious Beliefs and Medical Practices of the Creek Indians," in Vogel, *American Indian Medicine*, p. 384.

74. Frank G. Speck, "Catawba Medicines and

Curative Practices," in Vogel, *American Indian Medicine*, p. 384.

75. Wilson, *Agriculture of the Hidatsa*, p. 121.
76. Ibid., p. 126.
77. Tait, *Tobacco in Canada*, p. 13.
78. Ibid., p. 18.
79. Richard Dunlop, *Doctors of the American Frontier*, in Vogel, *American Indian Medicine*, p. 384.
80. William O. Douglas, "The People of Cades Cove," in Vogel, *American Indian Medicine*, p. 384.
81. Edward P. Claus, *Gathercoal and Wirth Pharmacognosy*, in Vogel, *American Indian Medicine*, p. 384.
82. Sheila Connor, *New England Natives* (Cambridge, Mass.: Harvard University Press, 1994), p. 207.
83. Maude Kegg, *Portage Lake: Memories of an Ojibwe Childhood*, in Paula Giese, "Wild Rice—Mahnoomin," North American Indian Resources, 1995, www.kstrom.net/isk/food/wildrice.html.
84. "Groundnuts," in J. A. Duke, *Herbalbum: An Anthology of Varicose Verse* (Laurel, Md.: J. Medrow, 1985).

Bibliography

Abbey, Edward. *Beyond the Wall: Essays from the Outside*. New York: Holt, Rinehart, and Winston, 1984.

———. *Cactus Country*. New York: Time-Life Books, 1973.

———. *Desert Solitaire: A Season in the Wilderness*. New York: Ballantine Books, 1968.

Abram, David. *The Spell of the Sensuous: Perception and Language in a More-Than-Human World*. New York: Vintage Books, 1997.

ACC Public Information Office. "The Tree That Owns Itself." *ACC Online*. Unified Government of Athens–Clarke County, 1999–2002. www. athensclarkecounty.com/tour/tour10.htm.

Adair, Mary J. *Prehistoric Agriculture in the Central Plains*. Lawrence: University of Kansas, 1988.

Adams, Dennis. Information Services Coordinator. *Spanish Moss: Its Nature, History, and Uses*. Beaufort County (South Carolina) Public Library. http://www. bcgov.org/bftlib/spanish.htm.

Agle, A. E. [One Feather]. "About Pipestone." *One Feather, Spirit and Sacred Pipe Carver*. http://my peoplepc.com/member/cherlyn/onefeather/id3.html.

Alaska Magazine Editorial Staff. *The Alaska-Yukon Wild Flowers Guide*. Ed. Helen A. White et al. Anchorage: Alaska Northwest Publishing Company, 1974.

Alcock, John. *Sonoran Desert Spring*. Tucson: University of Arizona Press, 1994.

———. *Sonoran Desert Summer*. Tucson: University of Arizona Press, 1990.

Allen, Carla. "Horsetail." *The Vine*, June 5, 1998. Reprinted from *The Yarmouth Vanguard*. www.klis.com/scove/060598.htm.

Allison, Mead A. "Can the Loss of Texas Coastal Wetlands Be Halted . . . or Reversed?" *Quarterdeck*, vol. 5, no. 2 (summer 1997). Texas A & M University Department of Oceanography. www.ocean.tamu.edu/Quarterdeck/AD5.2/allison.html.

Ambrose, Stephen E. *Undaunted Courage: Meriwether Lewis, Thomas Jefferson, and the Opening of the American West*. New York: Simon and Schuster, 1996.

American Chestnut Foundation. "Research and Restoration." http://chestnut.acf.org/r_r.htm.

Ames, Guy, and Steve Driver. *Sustainable Pecan Production*. Appropriate Technology Transfer for Rural Areas. www.attra.org/attra-pub/pecan.html#pecanculture.

Amos, William H., and Stephen H. Amos. *Atlantic and Gulf Coasts (Audubon Society Nature Guide)*. New York: Alfred A. Knopf, 1985.

Andreas, A. T. *History of the State of Nebraska*, Part 7, "Colonel Frémont's Explorations." Chicago: Western Historical Company, 1882. Kansas Collection Books. www.ukans.edu/carrie/kancoll/andreas_ne/history/erlyhst-p7.html#fremont.

Angle, Paul M., comp. *Prairie State: Impressions of Illinois, 1673–1967, by Travelers and Other Observers*. Chicago: University of Chicago, 1968.

Anttila, Carina K. et al. "Greater Male Fitness of a Rare Invader (*Spartina alterniflora*, Poaceae) Threatens a Common Native (*Spartina foliosa*) with Hybridization." *American Journal of Botany*, vol. 85, no. 11 (November 1998): 1597–1601.

Apperson, G. L. *The Social History of Smoking*. London: M. Secker, 1914.

Archer, Sellers G., and Clarence E. Bunch. *The American Grass Book: A Manual of Pasture and Range Practices*. Norman: University of Oklahoma Press, 1953.

Arizona–Sonoran Desert Museum. "Cactus Garden." 2002. http://www.desertmuseum.org/about/exhibits/cactus_garden.html.

Armstrong, W. P. "Major Types of Chemical Compounds in Plants and Animals." *Wayne's Word*. WOLFFIA, Inc. http://waynesword.palomar.edu/chemid1.htm.

Assiniwi, Bernard. *Survival in the Bush*. Toronto: Copp Clark, 1972.

Audubon, John James. *The Birds of America*. New York: Dover, 1967.

Austin, Mary. *The Land of Journeys' Ending*. Tucson: University of Arizona Press, 1983.

Austin, Richard L. *Wild Gardening: Strategies and Procedures Using Native Plantings*. Scarborough, Ont.: Prentice Hall Canada, 1986.

Axelrod, Daniel I. "Rise of the Grassland Biome, Central North America." *The Botanical Review*, vol. 51, no. 2 (1985): 163–201. backpacking.about.com/library/weekly/aa041301a.htm.

Badè, William Frederic. *The Life and Letters of John Muir*. Boston: Houghton Mifflin, 1924.

Badger, Curtis J. *Salt Tide: Cycles and Currents of Life along the Coast*. Harrisburg, Penn.: Stackpole Books, 1993.

Bagust, Harold. *The Gardener's Dictionary of Horticultural Terms*. London: Cassell Publishers, 1992.

Baker, Herbert G. *Plants and Civilization*. Belmont, Calif.: Wadsworth, 1965.

Baker, Mary Francis. *Florida Wild Flowers: An Introduction to the Florida Flora*. New York: Macmillan, 1938.

Balick, Michael J., and Paul Alan Cox. *Plants, People, and Culture: The Science of Ethnobotany*. New York: Scientific American Library, 1996.

Barbour, George M. *Florida for Tourists, Invalids, and Settlers: Containing Practical Information Regarding Climate, Soil, and Productions; Cities, Towns, and People; The Culture of the Orange and Other Tropical Fruits; Farming and Gardening; Scenery and Resorts; Sport; Routes of Travel, etc, etc.* Gainesville: University of Florida Press, 1964. Facsimile of 1882 ed.

Barbour, Michael G., and William Dwight Billings, eds. *North American Terrestrial Vegetation.* Cambridge: Cambridge University Press, 1988.

Barnes, Ruth A., comp. *I Hear America Singing: An Anthology of Folk Poetry.* Chicago: John C. Winston, 1937.

Barnes, Thomas G. "Wild about Wildflowers." University of Kentucky, College of Agriculture, 1998. www.ca.uky.edu/agc/pubs/for/for71/for71.htm.

Barr, Claude A. *Jewels of the Plains: Wildflowers of the Great Plains, Grasslands, and Hills.* Minneapolis: University of Minnesota Press, 1983.

Bartram, John, and William Bartram. *John and William Bartram's America: Selections from the Writings of the Philadelphia Naturalists.* Ed. Helen Gere Cruickshank. New York: Devin-Adair, 1957.

Bartram, William. *Travels and Other Writings: Travels through North and South Carolina, Georgia, East and West Florida, Travels in Georgia and Florida, 1773–74—A Report to Dr. John Fothergill, Miscellaneous Writings.* New York: Library of America, 1996.

———. *Travels of William Bartram.* Ed. Mark van Doren. New York: Dover, 1928.

———. *Travels through North and South Carolina, Georgia, East and West Florida, the Cherokee Country, the Extensive Territories of the Muscogulges or Creek Confederacy, and the Country of the Chactaws: Containing an Account of the Soil and Natural Productions of those Regions; Together with Observations on the Manners of the Indians.* Dublin: Moore, Jones, M'Allister, and Rice, 1793.

Beardy, Flora, and Robert Coutts, eds. *Voices from Hudson Bay: Cree Stories from York Factory.* Montreal and Kingston: McGill–Queen's University Press, 1996.

Beaubien, Elisabeth. "Western Trillium." *Plantwatch.* University of Alberta, Devonian Botanic Garden: 1997–2002. www.devonian.ualberta.ca/pwatch/westtr.htm#NAMES.

Benedict, Audrey DeLella. *A Sierra Club Naturalist's Guide to the Southern Rockies: The Rocky Mountain Regions of Southern Wyoming, Colorado, and Northern New Mexico.* San Francisco: Sierra Club Books, 1991.

Bennett, Ethel Hume Patterson, ed. *New Harvesting: Contemporary Canadian Poetry, 1918–1938.* Toronto: Macmillan Company of Canada, 1938.

Benyus, Janine M. *Biomimicry: Innovation Inspired by Nature.* New York: Quill (William Morrow), 1997.

Beresford-Kroeger, Diana. *Bioplanning: A North Temperate Garden.* Kingston, Ont.: Quarry Press, 1999.

Berg, Anna. "Sudden Oak Death Marks Demise of a Culture." *City on a Hill Press*, vol. 35, no. 15 (February 2001). www.slugwire.org/weekly/archives/01Feb15/oak.html.

Berger, John J. *The Sierra Club Guide to Understanding Forests.* San Francisco: Sierra Club Books, 1998.

Berkeley, Edmund, and Dorothy Smith Berkeley. *John Clayton, Pioneer of American Botany.* Chapel Hill: University of North Carolina Press, 1963.

———. *The Life and Travels of John Bartram from Lake Ontario to the River St. John.* Tallahassee: University Presses of Florida, 1982.

Bernhardt, Peter. *Wily Violets and Underground Orchids: Revelations of a Botanist.* First edition. New York: William Morrow and Company, 1989.

Bernstein, Lisa. "Time Capsule 2000." Cape Publishing, 1998–2002. www.capemay.com/lisa.htm.

Berry Botanical Garden. "Studying the Frigid Shootingstar." http://www.berrybot.org/ar_stdoau.html.

Bickel, Karl August. *The Mangrove Coast, the Story of the West Coast of Florida.* New York: Coward-McCann, 1942.

Bird, Isabella L. *A Lady's Life in the Rocky Mountains.* New York: G. P. Putnam's Sons, 1879.

Blackburn, Thomas C., ed. *December's Child: A Book of Chumash Oral Narratives.* Berkeley: University of California Press, 1975.

Blume, Martha. "Getting to Know the New Birds of the Bay." *Bay Weekly Online.* March 1–7, 2001. www.bayweekly.com/year01/issue9_9/lead9_9.html.

Blunt, Wilfrid. *The Compleat Naturalist: A Life of Linnaeus.* London: Collins, 1971.

Boas, Franz. *Kwakiutl Ethnography.* Ed. Helen Codere. Chicago: University of Chicago Press, 1966.

Bol, Marsha C., ed. *The Stars Above, the Earth Below: American Indians and Nature.* Niwot, Colo.: Roberts Rinehart Publishers for Carnegie Museum of Natural History, 1998.

Bolger, William C. "Elizabeth C. White." *History of the Whitesbog Cranberry and Blueberry Plantation, Burlington County, New Jersey.* March 12, 1997. http://www.whitesbog.org/elizabethwhite.html.

Bolton, Herbert Eugene. *Fray Juan Crespi, Missionary Explorer on the Pacific Coast, 1769–1774.* New York: AMS Press, 1971.

Bonnycastle, R.H.G. [Richard Henry Gardyne]. *A Gentleman Adventurer: The Arctic Diaries of R.H.G. Bonnycastle.* Ed. Heather Robertson. Toronto: Lester and Orpen Dennys, 1984.

Bonta, M. "Restoring our Tallgrass Prairie." *American Horticulturist,* vol. 70 (1991): 10–18.

Boorman, Sylvia. *Wild Plums in Brandy: A Cookery Book of Wild Foods.* Expanded edition. New York: McGraw-Hill, 1969.

Bossu, M. *Travels through That Part of North America Formerly Called Louisiana.* Trans. J. R. Forster. London: T. Davies, 1771.

Botkin, Daniel B. *Discordant Harmonies: A New Ecology for the Twenty-First Century.* New York: Oxford University Press, 1990.

Bourinot, Arthur Stanley. *Ottawa Lyrics; and, Verses for Children.* Ottawa: Graphic Publishers, 1929.

Bowden, Charles. *The Secret Forest.* Albuquerque: University of New Mexico Press, 1993.

Bowers, Janice Emily. *A Sense of Place: The Life and Work of Forrest Shreve.* Tucson: University of Arizona Press, 1988.

———. *The Mountains Next Door.* Tucson: University of Arizona Press, 1991.

Boyd, Robert. "Strategies of Indian Burning in the Willamette Valley." *Canadian Journal of Anthropology,* vol. 5 (1986): 65–86.

———, ed. *Indians, Fire, and the Land in the Pacific Northwest.* Corvallis: Oregon State University Press, 1999.

Boyle, David. *Uncle Jim's Canadian Rhymes for Family and Kindergarten Use.* Toronto: Musson Book Co., 1908.

Bradley, Timothy, Marika Walter, and Marjorie Patrick. "Oak Woodland." *Exploring Ecosystems at Limestone Canyon.* University of California at Irvine. http://compphys.bio.uci.edu/limestone/oakwoodland.htm.

Bragg, T. B., and James L. Stubbendieck, eds. *Prairie Pioneers: Ecology, History, and Culture.* Proceedings of the 11th North American Prairie Conference. Lincoln: University of Nebraska, 1989.

Braithwaite, Max. *Why Shoot the Teacher?* Toronto: McClelland and Stewart, 1965.

Braun, E. Lucy. *Deciduous Forests of Eastern North America.* New York: Free Press, 1950, 1974.

Brill, "Wildman" Steve, and Evelyn Dean. *Identifying and Harvesting Edible and Medicinal Plants in Wild (and Not So Wild) Places.* New York: Hearst Books, 1994.

Britton, Nathaniel Lord, and Addison Brown. *An Illustrated Flora of the Northern United States and Canada: From Newfoundland to the Parallel of* the Southern Boundary of Virginia and from the Atlantic Ocean Westward to the 102d Meridian. Second edition, revised and enlarged. New York: Dover Publications, 1970.

Broadfoot, Barry. *The Pioneer Years, 1895–1914: Memories of Settlers Who Opened the West.* Toronto: Doubleday Canada, 1976.

Brooks, Maurice. *The Appalachians.* Boston: Houghton Mifflin, 1965.

Broome, Tom. "The Coonties of Florida." The Cycad Jungle. cycadjungle.8m.com/cycajungle/The%20Coontie%20of%20Florida.htm.

Brouwer, Jake. "The California Poppy." *Echo Mountain Echoes,* vol. 2, no. 1 (spring 1997). http://aaaim.com/echo/v2nl/poppy2.htm.

Brower, David, ed. *Gentle Wilderness: The Sierra Nevada.* New York: Sierra Club and Ballantine Books, 1968.

Brown, Annora. *Old Man's Garden.* Toronto: J. M. Dent and Sons, 1954.

Brown, David E. *The Grizzly in the Southwest: Documentary of an Extinction.* First edition. Norman: University of Oklahoma Press, 1985.

Brown, Lauren. *Grasslands.* New York: Alfred A. Knopf, 1985.

Brown, Sam. "The Tribe's First Christmas." www.kumeyaay.org/Stories3.htm.

Bryson, Bill. *A Walk in the Woods: Rediscovering America on the Appalachian Trail.* Toronto: Doubleday Canada, 1997.

Buchmann, Stephen L., and Gary Paul Nabhan. *The Forgotten Pollinators.* Washington, D.C.: Shearwater Books, 1996.

Buckles, Mary Parker. *Margins: A Naturalist Meets Long Island Sound.* New York: North Point Press, 1997.

Bucuvalas, Tina, Peggy A. Bulger, and Stetson Kennedy. *South Florida Folklife.* Jackson: University Press of Mississippi, 1994.

"Building a Defensive Line to Tackle Phragmites." *University of Delaware Sea Grant Reporter,* vol. 19, no. 1 (2000). Graduate College of Marine Studies, University of Delaware. http://www.ocean.udel.edu/publications/Newsletter/reporter/special00/00.html#3.

Burkhardt, Frederick, et al., eds. *The Correspondence of Charles Darwin.* Volume 8, *1860.* Cambridge: Cambridge University Press, 1993.

Burkill, I. H. *A Dictionary of Economic Products of the Malay Peninsula.* 2 vols. London: Crown Agents for the Colonies, 1935.

Burroughs, John. *Signs and Seasons.* Cambridge, Mass.: Riverside Press, 1886.

Bush, Gene. "Shooting Stars." *Garden Clippin's.* Munchkin Nursery and Gardens, 2002. http://www.munchkinnursery.com/newsletter/shooting-stars/.

Butala, Sharon. *The Perfection of the Morning: An Apprenticeship in Nature.* Toronto: HarperCollins, 1994.

———. *Wild Stone Heart: An Apprentice in the Fields.* Toronto: HarperFlamingo Canada, 2000.

Butler, Sir William Francis. *The Wild Northland: Being the Story of a Winter Journey, with Dogs, across Northern North America.* New York. A. S. Barnes, 1904.

Cabrera, Chanchal. *A Review of Some Medicinal Plants of the Pacific North West.* Gaia Garden Herbal Dispensary and Clinic, 2000. http://www.gaiagarden.com/articles/pacific_north_west_herbs.html.

Caduto, Michael J., and Joseph Bruchac. *Keepers of Life: Discovering Plants through Native Stories and Earth Activities for Children.* Saskatoon, Sask.: Fifth House Publishers, 1994.

Caldwell, Jeff. "Berries for the Birds." *Birds and Native Plants.* California Native Plant Society. www.stanford.edu/~rawlings/birds.htm.

California Native Plant Society. "Rare Plant Program." CNPS, 1999–2002. www.cnps.org/rareplants/about.flora.htm.

———. "Toyon: *Heteromeles arbutifolia.*" *Fremontia,*

vol. 1, no. 4 (January 1974): 23–25. http://lacnps.org/toyon.html.

California State University, Chico. "California Poppy." Meriam Library Special collections. www.csuchio.edu/lbib/spc/netpages/calpoppy.html.

Canadian Department of Fisheries and Oceans. "By the Sea." Module 10 Coastal Bogs. http://www.gfc.dfo.ca/habitat/imaget/anzip.htm.

Cannings, Richard J., and Sydney G. Cannings. *British Columbia: A Natural History*. Vancouver: Greystone Books, 1996.

———. *Mountains and Northern Forests*. Vancouver: Greystone Books, 1998.

Cape Cod Cranberry Growers Association. "The History of Cranberry Production." www.cranberries.org/thecranberry/tbody.html.

Carman, Bliss. *Poems*. Toronto: McClelland and Stewart Ltd., 1931.

Carr, Emily. *Klee Wyck*. Toronto: Oxford University Press, 1941.

Carr, Patrick. *Sunshine States: Wild Times and Extraordinary Lives in the Lands of Gators, Guns, and Grapefruit*. New York: Doubleday, 1990.

Carson, Rachel. *The Sea Around Us*. New York: Oxford University Press, 1961.

Carville, Julie. "The Oak Woodlands." www.gv.net/.rsthomas/plant/OAK_WOODLANDS.htm.

Case, Frederick W., Jr., and Roberta B. Case. *Trilliums*. Portland, Ore.: Timber Press, 1997.

Casselman, Bill. *Canadian Garden Words*. Toronto: Little, Brown Canada, 1997.

Casteñeda, Pedro. *The Journey of Coronado*. Trans. George Parker Winship. Ann Arbor, Mich.: University Microfilms, 1966.

Caswell, Maryanne. *Pioneer Girl*. Toronto: McGraw-Hill Co. of Canada, 1964.

Catesby, Mark. *The Natural History of Carolina, Florida, and the Bahama Islands: Containing the Figures of Birds, Fishes, Serpents, Insects, and Plants: Particularly the Forest-Trees, Shrubs, and Other Plants, Not Hitherto Described, or Very Incorrectly Figured by Authors*. London: Author, 1731–1743.

Chapman, V. J. *Seaweeds and Their Uses*. With chapters by D. J. Chapman. Third edition. London: Chapman and Hall, 1980.

Chapman, William K., and Alan E. Bessette. *Trees and Shrubs of the Adirondacks*. Utica, N.Y.: North Country Books, 1990.

Childs, Craig. "The Millworker and the Forest." *High Country News*, vol. 31, no. 18 (September 27, 1999). www.hcn.org/servlets/hcn.Article?article id=5266.

Clark, Jamie Rappaport. "Endangered and Threatened Wildlife and Plants." *Federal Register*. Department of Interior Fish and Wildlife Service, vol. 63, no. 155 (August 12, 1998): 43100–43116. www.cdpr.ca.gov/docs/es/estext/fr081298.txt.

Clark Schreibeis Wildlife Art. www.clarkschreibeis.com/wood.htm.

Clemons, Elizabeth Cameron. *Tide Pools and Beaches*. New York: Knopf, 1964.

Coats, Alice M. *Garden Shrubs and Their Histories*. New York: Simon and Schuster, 1992.

———. *The Plant Hunters: Being a History of the Horticultural Pioneers, Their Quests, and Their Discoveries from the Renaissance to the Twentieth Century*. New York: McGraw-Hill, 1969.

Cody, William J. *Flora of the Yukon Territory*. Ottawa: National Research Press, 1996.

Coffey, Timothy. *The History and Folklore of North American Wildflowers*. Boston: Houghton Mifflin, 1993.

Collins, Scott L., and Linda L. Wallace, eds. *Fire in North American Tallgrass Prairies*. First edition. Norman, Okla.: University of Oklahoma Press, 1990.

Colombo, John Robert, ed. *Songs of the Indians.* Ottawa: Oberon Press, 1983.

Colston Burrell, C. "Environmental Concerns of Harvesting Wild Materials." Vol. 2 of a report commissioned by the Saskatchewan Dried Flowers Association. AgriCarta, 1999. www.aginfonet.com/ aglibrary/content/sk_driedflower/environmental_ concerns.html.

———, guest ed. *Ferns: Wild Things Make a Comeback in the Garden.* Brooklyn: Brooklyn Botanic Garden Publication, 1994.

Colwell, Judith. "Day 4." *Northern Lights*, 1995. www.stanford.edu/~jcolwell/NL/Day4.html.

Connor, Sheila. *New England Natives.* Cambridge, Mass.: Harvard University Press, 1994.

Corbridge, James N., and William A. Weber. *A Rocky Mountain Lichen Primer.* Niwot, Colo.: University Press of Colorado, 1998.

Corner, E. J. H. *The Natural History of Palms.* London: Weidenfeld and Nicolson, 1966.

Costello, David F. *The Prairie World.* Minneapolis: University of Minnesota Press, 1969, 1980.

Coulombe, Deborah. *The Seaside Naturalist: A Guide to Nature Study at the Seashore.* In cooperation with the University of New Hampshire. Englewood Cliffs, N.J.: Prentice Hall, 1984.

Cowles, Raymond B. *Desert Journal: A Naturalist Reflects on Arid California.* With Elna S. Bakker. Berkeley: University of California Press, 1977.

Cox, J. *Landscaping with Nature: Using Nature's Design to Plan Your Yard.* Emmaus, Penn.: Rodale Press, 1991.

Coyote Man. *The Destruction of the People.* Berkeley, Calif.: Brother William Press, 1973.

———. *Sun, Moon, and Stars.* Berkeley, Calif.: Brother William Press, 1973.

Craighead, Frank C. *The Trees of South Florida.* Coral Gables: University of Miami Press, 1971.

Craighead, John Johnson, Frank C. Craighead, Jr., and Ray Davies. *A Field Guide to Rocky Mountain Wildflowers from Northern Arizona and New Mexico to British Columbia.* Boston: Houghton Mifflin, 1963.

Cronon, William. *Changes in the Land: Indians, Colonists, and the Ecology of New England.* New York: Farrar Straus Giroux, 1983.

———, ed. *Uncommon Ground: Toward Reinventing Nature.* New York: W. W. Norton, 1995.

Crosby, Alfred W. *The Columbian Exchange: Biological and Cultural Consequences of 1492.* Westport: Greenwood Publishing Co., 1972.

———. *Ecological Imperialism: The Biological Expansion of Europe, 900–1900.* Cambridge: Cambridge University Press, 1986.

Cross, Ruth Corbett. "A Fence That You Water: It Is a Cinch to Grow." *Tucson Citizen*, August 9, 1996.

Cunningham, Isabel Shipley. *Frank N. Meyer, Plant Hunter in Asia.* Ames: Iowa State University Press, 1984.

Curtin, L. S. M. *By the Prophet of the Earth: Ethnobotany of the Pima.* Santa Fe: University of Arizona Press, 1949; Tucson: University of Arizona Press, 1997. http://www.uapress.arizona.edu/online. bks/prophet/plants.htm.

Curtis, Rob. "Sierran Montane Forests." *Bird Habitats.* eNature.com, 2002. http://www.enature.com/habitats/ show_sublifezone.asp?sublifezoneID=42.

Cutright, Paul Russell. *Lewis and Clark: Pioneering Naturalists.* Urbana: University of Illinois Press, 1969.

Cutter, Donald C. *Malaspina and Galiano: Spanish Voyages to the Northwest Coast, 1791 and 1792.* Vancouver: Douglas and McIntyre, 1991.

Daigle, Cheryl, and Jim Dow. "Life Within the Rockweed." *Quoddy Tides*, September 22, 2000. www. cobscook.org/ART9-rock2.htm.

Dale, Arlene. *Comparative Wood Anatomy of Some Shrubs Native to the Northern Rocky Mountains.*

Ogden, Utah: Intermountain Forest and Range Experiment Station, 1968.

Dallman, Peter R. *Plant Life in the World's Mediterranean Climates: California, Chile, South Africa, Australia, and the Mediterranean Basin.* Berkeley: University of California Press, 1998.

Daniel, Glenda, and Jerry Sullivan. *A Sierra Club Naturalist's Guide to the North Woods of Michigan, Wisconsin, and Minnesota.* San Francisco: Sierra Club Books, 1981.

Darwin, Charles. *Voyage of the Beagle: Charles Darwin's Journal of Researches.* Ed. Janet Browne and Michael Neve. London: Penguin Books, 1989.

Davenport-Hines, R. P. T. *Gothic: Four Hundred Years of Excess, Horror, Evil, and Ruin.* New York: North Point Press, 1999.

Davies, John. *Douglas of the Forests: The North American Journals of David Douglas.* Seattle: University of Washington Press, 1980.

Davis, Mary Byrd, ed. *Eastern Old-Growth Forests: Prospects for Rediscovery and Recovery.* Washington, D.C.: Island Press, 1996.

Davis, Wade. *One River: Explorations and Discoveries in the Amazon Rain Forest.* New York: Touchstone, 1996.

Day, Leslie. "The City Naturalist—Sweet Gum Tree." NY West Side Site: The 79th Street Boat Basin Flora and Fauna Society, 1996. www.nysite.com/nature/flora/sweetgum/htm.

———. *The Serpent and the Rainbow: A Harvard Scientist's Astonishing Journey into the Secret Societies of Haitian Voodoo, Zombies, and Magic.* New York: Touchstone, 1997.

De Angulo, Jaime. *Coyote Man and Old Doctor Loon.* San Francisco: Turtle Island Foundation, 1973.

Deland, Margaret. *Florida Days.* Boston: Little, Brown, 1889.

Dellenbaugh, Frederick S. *A Canyon Voyage; The Narrative of the Second Powell Expedition Down the Green-Colorado River from Wyoming, and the Explorations on Land, in the Years 1871 and 1872.* New Haven, Conn.: Yale University Press, 1962.

Densmore, Frances. *Indian Use of Wild Plants for Food, Medicine, and Charms.* Reprint of *Uses of Plants by the Chippewa Indians* from the 44th annual report of the U.S. Bureau of American Ethnology, 1926–1927. Washington, D.C.: U.S. Government Printing Office, 1928. Reprinted, Ohsweken, Ont.: Iroqrafts Ltd., 1993.

Dent, Huntley. *The Feast of Santa Fe: Cooking of the American Southwest.* New York: Simon and Schuster, 1985.

Devine, Robert. *Alien Invasion: America's Battle with Non-Native Animals and Plants.* Washington, D.C.: National Geographic Society, 1998.

———. "The Cheatgrass Problem." *The Atlantic Monthly*, vol. 271, no. 5 (1993): 40–45.

———. "Rocky Mountain National Park." *National Geographic Traveler*, July/August 1991, pp. 72–85.

Devonian Botanic Garden. "Native People's Garden." University of Alberta, 1997–2002. www.discoveredmonton.com/devonian/natvgrdn.html.

DeVoto, Bernard, ed. *The Journals of Lewis and Clark.* Boston: Houghton Mifflin, 1953.

Diamond, Jared. *Guns, Germs, and Steel: The Fates of Human Societies.* New York: W.W. Norton, 1997.

Dickinson, Jonathan. *Jonathan Dickinson's Journal, or, God's Protecting Providence, Being the Narrative of a Journey from Port Royal in Jamaica to Philadelphia between August 23, 1696 and April 1, 1697.* Ed. Evangeline Walker Andrews and Charles McLean Andrews. New Haven, Conn: Yale University Press, 1945.

Dictionary of Ecology and Environmental Science. Ed. Henry W. Art. New York: Henry Holt, 1983.

Diggs, George M., Jr., Barney L. Lipscomb, and Robert J. O'Kennon. "Introduction." *Shinners and Mahler's Illustrated Flora of North Central Texas.* Part of the *Illustrated Texas Floras Project*, The Austin College Center for Environmental Studies and the

Botanical Research Institute of Texas, 1999. http://artemis.austinc.edu/acad/bio/gdiggs/introduction.html.

Dirr, Michael A. *A Manual of Woody Landscape Plants: Their Identification, Ornamental Characteristics, Culture, Propagation, and Uses*. 3rd ed. Champaign, Ill.: Stipes, 1983.

Diver, Steve, and Guy Ames. "Sustainable Pecan Production, Horticulture Production Guide." http://www.attra.org/attra-pub/pecan.html#pecanculture.

Dix, R. L. *A History of Biotic and Climatic Changes within the North American Grassland*. Oxford: Blackwell Scientific Publications, 1964.

Dobelis, Inge N., ed. *Magic and Medicine of Plants*. Pleasantville, N.Y.: Reader's Digest, 1986.

Dobyns, Henry F. *Prehistoric Indian Occupation within the Eastern Area of the Yuman Complex: A Study in Applied Archaeology*. New York: Garland, 1974.

Doll, Jerry. "Hemp Dogbane Biology and Management." University of Wisconsin, Department of Agronomy Publication Agronomy Advice. Fd. Crops. 33.0. ipcm.wisc.edu/uw_weeds/articles/hempdog.htm.

Dorothy Molter Museum. "Who Was Dorothy Molter?" www.canoecountry.com/dorothy/who.

———. *Journal Kept by David Douglas during His Travels in North America, 1823–1827*. London: William Wesley and Son, 1914.

Douglas, Marjory Stoneman. *The Everglades: River of Grass*. New York: Rinehart and Company, 1947.

———. *Nine Florida Stories*. Jacksonville: University of North Florida Press, 1990.

———. *River in Flood and Other Florida Stories*. Gainesville: University Press of Florida, 1998.

Downie, Mary Alice, and Mary Hamilton. *"And Some Brought Flowers": Plants in a New World*. Toronto: University of Toronto Press, 1980.

Downie, Mary Alice, and Barbara Robertson. *The*

Well-Filled Cupboard: Everyday Pleasures of the Canadian Home and Garden, with Over 200 Recipes. Toronto: Lester and Orpen Dennys, 1987.

Driver, Clive. "Ending at the West End." *Provincetown Banner*, December 5, 1996. www.provincetownbanner.com/history/12/5/1996/1.

Duke, James A. *Handbook of Edible Weeds*. Boca Raton, Fla.: CRC Press, 1992.

———. *Handbook of Energy Crops*. Unpublished, in Purdue University Center for New Crops and Plant Products, "*Rhizophora mangle* L.," New Crop: The New Crop Resource Online Program, 1983. www.hort.purdue.edu/newcrop/duke_energy/Rhizophora_mangle.html.

———. *Handbook of Energy Crops*. Unpublished in Purdue University Center for New Crops and Plant Products, "*Zizania aquatica* L.," New Crop: The New Crop Resource Online Program, 1983. www.hort.purdue.edu/newcrop/duke_energy/Zizania_aquatica.html.

———. *Herbalbum: An Anthology of Varicose Verse*. Laurel, Md.: J. Medrow, 1985.

Duke, James A., and K. K. Wain, *Medicinal Plants of the World*. 3 vols. 1981. Computer index.

Duncan, Frances. *Mary's Garden and How It Grew*. New York: Century, 1904.

Duncan, Wilbur H., and Marion B. Duncan. *The Smithsonian Guide to Seaside Plants of the Gulf and Atlantic Coasts from Louisiana to Massachusetts, Exclusive of Lower Peninsular Florida*. Washington, D.C.: Smithsonian Institution Press, 1987.

———. *Trees of the Southeastern United States*. Athens: University of Georgia Press, 1988.

Dunmire, William W., and Gail D. Tierney. *Wild Plants and Native Peoples of the Four Corners*. Santa Fe: Museum of New Mexico Press, 1997.

Dunster, J. R. "California Poppy Page." Bear's Clover Yosemite, Pottery and Art Page, Bearsclover, 2001–2002. www.geocities.com/bearsclover/poppy.html.

Dunwich, Gerina. Wicca Candle Magick. www. crystalforest1.homestead.com/oilrecipes.html.

Durant, Mary. *Who Named the Daisy? Who Named the Rose? A Roving Dictionary of North American Wildflowers*. 1st paperback ed. New York: Dodd Mead, 1976.

Earle, Alice Morse. *Old Time Gardens, Newly Set Forth*. New York: Macmillan, 1901.

Earle, Barbara, and Scott Earle. "Figwort (Snapdragon) Family: Scrophulariaceae." Idaho Mountain Wildflowers, 1999/2000. www. larkspurbooks.com/Scrophl.html.

Eastman, John. *The Book of Forest and Thicket: Trees, Shrubs, and Wildflowers of Eastern North America*. Mechanicsburg, Penn: Stackpole Books, 1992.

————. *The Book of Swamp and Bog: Trees, Shrubs, and Wildflowers of Eastern Freshwater Wetlands*. Mechanicsburg, Penn: Stackpole Books, 1995.

Eberhart, Kathryn. "Tall Bluebells (*Mertensia paniculata*)." Alaska Screensavers, 1997–2002. www. alaskascreensavers.com/gallery/matvalley/bluebell.htm.

ECHO. "Echo Plant Information Sheet—Tepary Bean." N. Fort Myers, Fla.: Educational Concerns for Hunger Organization, 1999. http://www.echonet. org/tropicalag/plantinfo/Phaseolusacutifolius.pdf.

Eddison, Sydney. "Good Looks Begin at the Edge." *Fine Gardening*, no. 55 (May/June 1997): 26.

Egan, Timothy. *The Good Rain: Across Time and Terrain in the Pacific Northwest*. New York: Alfred A. Knopf, 1990.

Egerton-Warburton, Louise. Chicago Botanic Garden, Scientific Programs, 1997–2001. www. chicago-botanic.org/research/conservation/egerton/ html.

Eisenberg, Evan. *The Ecology of Eden*. Toronto: Random House of Canada, 1998.

Elias, Thomas S. *The Complete Trees of North America: Field Guide and Natural History*. New York: Van Nostrand Reinhold, 1980.

Elliott, Doug. *Wild Roots: A Forager's Guide to the Edible and Medicinal Roots, Tubers, Corms, and Rhizomes of North America*. Rochester, Vt.: Healing Arts Press, 1995.

Ellis, Brian "Fox." "The Cottonwood." *Science and Children*, vol. 38, no. 4 (January 2001): 42–46.

Ellis, Gerry, and Karen Kane. *America's Rain Forest*. Minocqua, Wis.: North Word Press, 1991.

Ellis, Michael. "California Bay Trees." *Footloose Forays*. December 17, 2000. www.footlooseforays. iohome.net/cgi-bin/Topic.pl?topic=44&public.

Ellis, Shona, et al. *"Diospyros virginiana." An Ethnobotany of the UBC Arboretum*. March 30, 1995, web adaptation August 1998. http://botany.ubc.ca/ arboretum/UBC030.HTM.

Elmore, Francis Hapgood. *Ethnobotany of the Navajo*. New York: AMS Press, 1978.

Elsensohn, Sister M. Alfreda. *A Flora of the Camas Prairie Region in the Vicinity of Cottonwood, Idaho, Designed for the Use of High School Students*. Master's thesis, education. University of Idaho, 1939. www. lili.org/camas.html.

Emboden, William A. *Bizarre Plants: Magical, Monstrous, Mythical*. New York: Macmillan, 1974.

Emerson, Ralph Waldo. *Poems*. Boston: Houghton Mifflin, 1911.

Eppele, David L. "On the Desert," Arizona Cactus: 2001. http://www.arizonacactus.com/story.htm.

Erichsen-Brown, Charlotte. *Medicinal and Other Uses of North American Plants: A Historical Survey with Special Reference to the Eastern Indian Tribes*. New York: Dover 1989, originally published 1979.

Estes, James R., Ronald J. Tyrl, and Jere N. Brunken, eds. *Grasses and Grasslands: Systematics and Ecology*. Norman: University of Oklahoma Press, 1982.

"Everglades Plant Diversity." Andrews University. www.biol.andrews.edu/everglades/plant/plant.htm.

Eyre, F. H., ed. *Forest Cover Types of the United States*

and Canada. Washington, D.C.: Society of American Foresters, 1980.

Facciola, Stephen. *Cornucopia: A Source Book of Edible Plants*. Vista, Calif.: Kampong Publications, 1990.

Fagan, Brian M. *The Great Journey: The Peopling of Ancient America*. New York: Thames and Hudson, 1987.

Farney, D. "Can the Tallgrass Prairies Be Saved?" *National Geographic*, vol. 157, no. 1 (1980): 37–61.

Faull, J. H., ed. *The Natural History of the Toronto Region*. Toronto: Canadian Institute, 1913.

Felter, Harvey Wickes, and John Uri Lloyd. "*Corydalis.*—Turkey-Corn." *King's American Dispensatory*. Henriette's Herbal Homepage, 2000–2002. http://www.ibiblio.org/herbmed/ eclectic/kings/dicentra-cana.html.

———. "*Veratrum viride.*" *King's American Dispensatory*. Henriette's Herbal Homepage, 2000–2002. http://www.ibiblio.org/herbmed/ eclectic/kings/veratrum-viri.html.

Ferguson, Mary, and R. M. Saunders. *Wildflowers through the Seasons*. Toronto: Key Porter Books, 1995.

Fielding, Raymond R. *Shrubs of Nova Scotia: A Guide to Native Shrubs, Small Trees, and Woody Vines*. Halifax: Nimbus Publishing and the Nova Scotia Department of Natural Resources, 1998.

Fisher, John. *The Origins of Garden Plants*. London: Constable, 1982.

Fisher, Ron. *Heartland of a Continent: America's Plains and Prairies*. Washington, D.C.: National Geographic Society, 1991.

Fitzharris, Tim. *Forests: A Journey into North America's Vanishing Wilderness*. Toronto: Stoddart, 1991.

———. *The Wild Prairie: A Natural History of the Western Plains*. Toronto: Oxford University Press, 1983.

Fitzharris, Tim, and John Livingston. *Canada: A Natural History*. Markham: Penguin Books Canada, 1988.

Flannery, Tim. *The Eternal Frontier: An Ecological History of North America and Its Peoples*. New York: Atlantic Monthly Press, 2001.

Fletcher, Alice C., and Francis La Flesche. *The Omaha Tribe*. New York: Johnson Reprint, 1970.

Flying Fish & Company. "Wildlife." www. flyingfishnco.com/wildlife.htm.

Food and Agriculture Organization of the UN. "Seeds, Fruits, and Cones." Non-Wood Forest Products from Conifers, 1995. www.fao.org/docrep/ X0453e/X0453e12.htm.

Forest Trees of Florida. Seventeenth edition. Bob Crawford, Commissioner of Agriculture. Tallahassee: Florida Division of Forestry, 1997.

Foss, Krista. "Study Finds Ginseng Reduces Blood Sugar." *Globe and Mail*. April 10, 2000, p. A5.

Foster, John, Dick Harrison, and I. S. MacLaren, eds. *Buffalo*. Edmonton: University of Alberta Press, 1992.

Foster, Steven. "Understanding Echinacea." Steven Foster Group, 2000. www.stevenfoster.com/ education/monograph/echinacea.html.

Foster, Steven, and J. A. Duke. *A Field Guide to Medicinal Plants, Eastern and Central North America*. Boston: Houghton Mifflin, 1990.

Foster, Steven, and Varro E. Tyler. *Tyler's Honest Herbal: A Sensible Guide to the Use of Herbs and Related Remedies*. 4th ed. New York: Haworth Herbal Press, 1999.

Fowler, Cary, and Pat Mooney. *Shattering: Food, Politics, and the Loss of Genetic Diversity*. Tucson: University of Arizona Press, 1990.

Fox, J. F. "Alteration and Coexistence of Tree Species." *American Naturalist*, vol. 111 (1977): 69–89.

Fox, William T. *At the Sea's Edge: An Introduction to*

Coastal Oceanography for the Amateur Naturalist. Englewood Cliffs, N.J.: Prentice Hall, 1983.

Frazer, Sir James. *The Golden Bough: A Study in Magic and Religion.* Ware, Hertfordshire: Wordsworth Reference, 1993.

Frazier, Ian. *Great Plains.* New York: Farrar Straus Giroux, 1989.

Freeman, Linda. "The Smithsonian Institution." Mount Shasta Companion, 2001. www.siskiyous.edu/shasta/art/smith

Frémont, J. C. *The Exploring Expedition to the Rocky Mountains, Oregon, and California.* Buffalo: G. H. Derby, 1850.

Frost, Robert. *The Poetry of Robert Frost.* Ed. Edward Connery Lathem. New York: Holt, Rinehart, and Winston, 1969.

————. *You Come Too: Favorite Poems for Young Readers.* New York: Holt, Rinehart, and Winston, 1959.

Fuller, Margaret. *Mountains: A Natural History and Hiking Guide.* New York: John Wiley & Sons, 1989.

Fultz, Francis M. *The Elfin Forest of California.* Los Angeles: Times-Mirror Press, 1923. http://www.notfrisco.com/calmem/chaparral/fultz02.html.

Fussell, Betty. *The Story of Corn: The Myths and History, the Culture and Agriculture, the Art and Science of America's Quintessential Crop.* New York: North Point Press, 1992.

Gadd, Ben. *Handbook of the Canadian Rockies.* Jasper, Alta.: Corax Press, 1986.

Gaertner, E. E. "Breadstuff from Fir (*Abies balsamea*)." *Economic Botany,* vol. 26, no. 1 (1970): 69–72.

Gale, Bonnie. "American Willow Growers Network." *English Basketry Willows.* October 3, 2000. http://www.msu.edu/user/shermanh/galeb/awgndesc.htm.

Gannett, Lewis. *Cream Hill: Discoveries of a Week-end Countryman.* New York: Viking Press, 1949.

Garcés, Francisco Tomas Hermenegildo. *On the Trail of a Spanish Pioneer: The Diary and Itinerary of Francisco Garcés (Missionary Priest) in His Travels through Sonora, Arizona, and California, 1775–1776.* New York: F. P. Harper, 1900.

Gardea-Torresdey, J. L., J. Bibb, and A. Hernandez. "Uptake of Toxic Heavy Metal Ions by Inactivated Cells of *Larrea tridentata* (Creosote Bush)." Abstracts from the 1997 Conference on Hazardous Waste Research, Hazardous Substance Research Centers. http://www.engg.ksu.edu/HSRC/97abstracts/p22.html.

"Garden of the Gods and White House Ranch Colorado Springs City Parks." http://www.mindspring.com/~duckgoose/gog.html.

Garden Web. http://forums.gardenweb.com/forums/load/natives/msg0512550330050.html.

Garrison, George A., et al. *Vegetation and Environmental Features of Forest and Range Ecosystems.* Washington, D.C.: Forest Service, U.S. Department of Agriculture, 1977.

Garvin, John William, ed. *Canadian Verse for Boys and Girls.* Toronto: Nelson, 1930.

Gayton, Don. *The Wheatgrass Mechanism: Science and Imagination in the Western Canadian Landscape.* Saskatoon, Sask.: Fifth House Publishers, 1990.

Gentry, Howard Scott. *Agaves of Continental North America.* Tucson: University of Arizona Press, 1982.

Gibbons, Euell. *Stalking the Blue-Eyes Scallop: Field Guide Edition.* New York: David McKay, Company, 1970.

————. *Stalking the Healthful Herbs.* New York: David McKay, 1966.

————. *Stalking the Wild Asparagus.* New York: David McKay, 1962.

Gibbons, Whit. "Hey Kids, Put Down the Video Games and Play Outside." *Online Athens.* July 23, 2000. www.onlineathens.com/stories/072400/opi_0724000003.shtml.

Giese, Paula. "Wild Rice—Mahnoomin." North

American Indian Resources. 1995. www.kstrom.net/isk/food/wildrice.htm.

Gill, Edward Anthony Wharton. *A Manitoba Chore Boy: The Experiences of a Young Emigrant Told from His Letters*. London: Religious Tract Society, 1912.

Gillaspie, Angela. "Poke Salet: The Versatile Veggie." *Momma's Southern Stuff*. August 1998. http://home.talkcity.com/KudzuKorner/gnats-themomma/sostuf.html.

Gillies, Craille Maguire. "Could, Should, Wood." *Globe and Mail*, March 24, 2001, p. R23.

Gilmer, Maureen. "Of California Fire and Natives." *Garden Forum*. Maureen Gilmer, 1998, 1999, 2000. www.gardenforum.com/FireAndNatives.html.

———. "The Redwoods—California Big Trees." *Garden Forum: Plant Profiles*. Maureen Gilmer, 1998, 1999, 2000. www.gardenforum.com/redwoods.html.

Gilmore, Melvin Randolph. *Uses of Plants by the Indians of the Missouri River Region*. Washington, D.C.: U.S. Government Printing Office, 1919.

Gladnick, P. J. "Promontory Utah and the Nation's Railway Uniting." PageWise, Inc., 2001. http://mo.essortment.com/promontory_raic.htm.

Gleason, Henry A., and Arthur Cronquist. *Manual of the Vascular Plants of Northeastern United States and Adjacent Canada*. Second edition. New York: New York Botanical Garden, 1991.

Godfrey, Robert K. *Trees, Shrubs, and Woody Vines of Northern Florida and Adjacent Georgia and Alabama*. Athens: University of Georgia Press, 1988.

Goldstein, Natalie. *Rebuilding Prairies and Forests*. Chicago: Children's Press, 1994.

Goodrich, Jennie, Claudia Lawson, and Vana Parrish Lawson. *Kashaya Pomo Plants*. Los Angeles: American Indian Studies Center, University of California, 1980.

Gosse, Philip Henry. *The Canadian Naturalist; A Series of Conversations on the Natural History of Lower Canada*. London: John Van Voorst, 1840.

Goudie, Elizabeth. *Woman of Labrador*. Toronto: Peter Martin, 1973.

Goulson, Cary F. *Seventeenth-Century Canada: Source Studies*. Toronto: Macmillan of Canada, 1970.

Grahame, John D., and Thomas D. Sisk, eds. *Canyons, Cultures and Environmental Change: An Introduction to the Land-use History of the Colorado Plateau*. http://www.cpluhna.nau.edu/.

Gray, Asa. *Gray's Manual of Botany: A Handbook of the Flowering Plants and Ferns of the Central and Northeastern United States and Adjacent Canada*. Eighth edition. Largely rewritten and expanded by Merritt Lyndon Fernald. New York: American Book Co., 1950.

Gray, Charlotte. *Sisters in the Wilderness: The Lives of Susanna Moodie and Catharine Parr Traill*. Toronto: Viking, 1999.

Great Plains Flora Association. *Flora of the Great Plains*. Coordinated by Ronald L. McGregor. Ed. T. M. Barkley et al. Lawrence: University Press of Kansas, 1986.

Gregg, Josiah. *Commerce of the Prairies, or, The Journal of a Santa Fé Trader: During Eight Expeditions across the Great Western Prairies, and a Residence of Nearly Nine Years in Northern Mexico*. Ed. Max L. Moorhead. Norman, Okla.: University of Oklahoma Press, 1954.

Grescoe, Audrey. *Giants: The Colossal Trees of Pacific North America*. Vancouver: Raincoast Books, 1997.

Gribbin, John R., and M. Gribbin. *Children of the Ice: Climate and Human Origins*. Oxford: Basil Blackwell, 1990.

Grieve, M. *A Modern Herbal: The Medicinal, Culinary, Cosmetic, and Economic Properties, Cultivation, and Folk-lore of Herbs, Grasses, Fungi, Shrubs, and Trees with All Their Modern Scientific Uses*. London: Jonathan Cape, 1931.

Grove, Frederick Philip. *Over Prairie Trails*. Intro. Malcolm Ross. Toronto: McClelland and Stewart, 1970.

Guanlao, Arnel. "Bay Area Ramblings." *Picture This*, February 12, 2000. www.calphoto.com/antonio2.htm.

Gunther, Erna. *Ethnobotany of Western Washington: The Knowledge and Use of Indigenous Plants by Native Americans*. Seattle: University of Washington Press, 1973.

H. H. [Helen Hunt Jackson]. *Bits of Travel at Home*. Boston: Roberts Brother, 1878.

Hacker, Peter. "Poison Oak and Ivy." www.dhc.net/~pmhack/POISON.HTM.

Hale, Mason E. *Lichen Handbook: A Guide to the Lichens of Eastern North America*. Washington, D.C.: Smithsonian Institution, 1961.

Hale, Sarah Josepha. *Flora's Interpreter; Or, The American Book of Flowers and Sentiments*. Boston: [no publisher given], 1832.

Hall, Barbara. "A Jewel of a Weed." *Weeds and Wild Things*. Suite101.com, August 25, 1998. http://www.suite101.com/article.cfm/weeds_and_wild_things/10044.

———. "Poison I Eye-Eye-Eye-Eye VEE." *Weeds and Wild Things*. Suite101.com, February 23, 1998. http://www.suite101.com/article.cfm/weeds_and_wild_things/5847.

Hallworth, Beryl, and C. C. Chinnappa. *Plants of Kananaskis Country in the Rocky Mountains of Alberta*. Edmonton: University of Alberta Press, 1997.

Hamel, Paul B., and Mary U. Chiltoskey. *Cherokee Plants: Their Uses—A 400 Year History*. Sylva, N.C.: Herald Publishing, 1975.

Hammond, George Peter. *Coronado's Seven Cities*. Albuquerque: United States Coronado Exposition Commission, 1940.

Hammond, George Peter, and Agapito Rey. *Narratives of the Coronado Expedition, 1540–1542*. New York: AMS Press, 1977.

———. *New Mexico in 1602: Juan de Montoya's Relation of the Discovery of New Mexico*. Albuquerque: Quivira Society, 1938.

———. *The Rediscovery of New Mexico, 1580–1594: The Explorations of Chamuscado, Espejo, Castana de Sosa, Morlete, and Leyva de Bonilla and Humaña*. Albuquerque: University of New Mexico Press, 1966.

Hansen, Wallace W. "*Arbutus menziesii*—Madrone, Madrona." Native Plants of the Northwest, Native Plant Nursery and Gardens, 2002. www.nwplants.com/madrone/.

Hardy, Campbell. *Forest Life in Acadie: Sketches of Sport and Natural History in the Lower Provinces of the Canadian Dominion*. London: Chapman and Hall, 1869.

Harrington, H. D. *Edible Native Plants of the Rocky Mountains*. Albuquerque: University of New Mexico Press, 1967.

Harris, John. *Northern Regions; Or, Uncle Richard's Relation of Captain Parry's Voyages for the Discovery of a North-West Passage, and Franklin's and Cochrane's Overland Journeys to Other Parts of the World*. New York: O. A. Roorbach, 1827.

Harris, Mark S. "Herbs Used in Period and How They Were Used." *Stefan's Florilegium*, January 19, 2001. www.florilegium.org/files/PLANTS/herbs-msg.text.

Hartmann, William K. *Desert Heart: Chronicles of the Sonoran Desert*. Tucson: Fisher Books, 1989.

Hartzell, Hal, Jr. *The Yew Tree: A Thousand Whispers—Biography of a Species*. Eugene, Ore.: Hulogosi, 1991.

Harvey, Athelstan George. *Douglas of the Fir: A Biography of David Douglas, Botanist*. Cambridge, Mass.: Harvard University Press, 1947.

Haviland, William A., and Marjory W. Power. *The Original Vermonters: Native Inhabitants, Past and Present*. Hanover, N.H.: University Press of New England, 1981.

Hawke, David J. *Wetlands: A Habitat in Peril*. Toronto: Stoddart/Boston Mills Press, 1994.

Healey, B. J. *A Gardener's Guide to Plant Names*. New York: Charles Scribner's Sons, 1972.

Hearne, Samuel. *A Journey from Prince of Wales's Fort, in Hudson's Bay, to the Northern Ocean: Undertaken by Order of the Hudson's Bay Company for the Discovery of Copper Mines, a North West Passage, &c. in the Years 1769, 1770, 1771, and 1772.* London: A. Strahan and T. Cadell, 1795. Republished, Edmonton, Alta.: M. G. Hurtig, 1971.

Heatherley, Ana Nez. *Healing Plants: A Medicinal Guide to Native North American Plants and Herbs.* Toronto: HarperCollins, 1998.

Hebda, Richard. "Natural History: Ocean Spray." *Coastal Grower. Index for Native Plants,* Royal British Columbia Museum, 1995. http://rbcm1.rbcm.gov.bc.ca/nh_papers/nativeplants/holodisc.html.

———. "Soapberry (*Shepherdia canadensis*)." *Coastal Grower. Index for Native Plants,* Royal British Columbia Museum, 1995. http://rbcm1.rbcm.gov.bc.ca/nh_papers/nativeplants/shepherd.html.

———. "Sweet Coltsfoot (*Petasites frigidus*)." *Coastal Grower. Index for Native Plants,* Royal British Columbia Museum, 1995. http://rbcm1.rbcm.gov.bc.ca/nh_papers/nativeplants/petafrig.html.

Hedrick, Ulysses Prentiss. *A History of Horticulture in America to 1860.* New York: Oxford University Press, 1950.

Heinrich, Bernd. *The Trees in My Forest.* New York: Cliff Street Books, 1997.

Heiser, Charles Bixler. *The Sunflower.* Norman: University of Oklahoma Press, 1976.

Hellyer, David Tirrell. *At the Forest's Edge: Memoir of a Physician-Naturalist.* Seattle: Pacific Search Press, 1985.

Helmig, Gabriele. "Aboriginal Place Names in British Columbia." *Dreamspeaker.* Indian and Northern Affairs Canada, spring 2000. www.ainc-inac.gc.ca/nr/nwlts/drm/s2000/apn-e.html.

Hendrickson, Robert. *Ladybugs, Tiger Lilies, and Wallflowers: A Gardener's Book of Words.* Upper Saddle River, N.J.: Prentice Hall, 1993.

Henry Cowell Redwoods State Park. "Western Sword Fern, Christmas Fern, Sword Holly Fern (*Polystichum munitum*)." www.best.com/~dforthof/HenryCowell/wild.cgi/polystichum_munitum.

Henry, Greg. *Retrospective Growth Analysis in High Arctic Dwarf Shrubs.* University of British Columbia, Geography Department. www.geog.ubc.ca/research/henry1.htm.

Herbst, Josephine. *New Green World.* New York: Hastings House Publishers, 1954.

Hiaasen, Carl. *Tourist Season.* New York: Putnam's Sons, 1986.

Hiatt, Melana. "Bulrush (*Scirpus* species)." Edible Wild Kitchen. www.ediblewild.com/bill.html.

Hickman, Pamela. *The Kids Canadian Tree Book.* Toronto: Kids Can Press, 1995.

Hill, Julia Butterfly. *The Legacy of Luna: The Story of a Tree, a Woman, and the Struggle to Save the Redwoods.* San Francisco: Harper San Francisco, 2000.

Hillman, R. B. "Red Maple—*Acer rubrum.*" September 18, 2001. web.vet.cornell.edu/CVM/HANDOUTS/plants/MAPLE.html.

Hillyer, Robert. *Collected Poems.* New York: Alfred A. Knopf, 1961.

Hitchcock, C. Leo. *Vascular Plants of the Pacific Northwest.* Seattle: University of Washington Press, 1955.

Hitchcock, C. Leo, and Arthur Cronquist. *Flora of the Pacific Northwest: An Illustrated Manual.* Seattle: University of Washington Press, 1973.

Hoagland, Edward. *Balancing Acts: Essays.* New York: Lyons Press, 1999.

Holbrook, Florence. *The Book of Nature Myths.* Boston: Houghton Mifflin, 1902.

Hollis, Jill, comp. *Come Into the Garden: A Treasury of Garden Verse.* London: Random Century Group, 1992.

Holloway, Pat. "A Tale of Two Species." *Georgeson Botanical Garden Review,* vol. 6, no. 3 (1997): 6–8.

Homer. *Odyssey*. Trans. Albert Cook. New York: W.W. Norton & Company, 1967.

Hosie, R. C. *Native Trees of Canada*. Seventh edition. Ottawa: Canadian Forestry Service, Department of Fisheries and Forestry, 1969.

Houk, Rose. *Great Smoky Mountains National Park: A Natural History Guide*. Boston: Houghton Mifflin, 1993.

"How Pussy Willows Got Their Name." http://www.mooncrystal.com/~gypsy/cattrivia/pwillow.html.

Hubbard, Mina. *A Woman's Way through Unknown Labrador*. New York: McClure, 1908. Reprinted, St. John's, Nfld.: Breakwater Books, 1981.

Hultén, Eric. *Flora of Alaska and Neighboring Territories: A Manual of the Vascular Plants*. Stanford: Stanford University Press, 1968.

Hummel, Monte, ed. *Endangered Species: The Future for Canada's Wilderness*. Toronto: Key Porter Books, 1989.

Humphrey, Harry Baker. *Makers of North American Botany*. New York: Ronald Press, 1961.

Humphreys, Jay. "Building Better Sea Oats." *Fathom Magazine*, spring 1996. Florida Sea Grant College Program, 1998–2000. www.flseagrant.org/science/library/fathom_magazine/volume-8_issue-1/sea_oats.htm.

Hutchens, Alma R. *A Handbook of Native American Herbs*. Boston: Shambhala, 1992.

Hutchison, Isobel Wylie. *North to the Rime-Ringed Sun: Being the Record of an Alaskan-Canadian Journey Made in 1933–34*. London: Blackie & Son, 1934.

Hyams, Edward. *Plants in the Service of Man: 10,000 Years of Domestication*. New York: J. B. Lippincott, 1971.

Industry Canada. *Traditional Plant Use in the Hazeltons*. Schoolnet Digital Collection Program. http://collections.ic.gc.ca/hazeltons/food.htm.

Ingersoll, Ernest. *Crest of the Continent: A Record of a Summer's Ramble in the Rocky Mountains and Beyond*. Chicago: R. R. Donnelley & Sons, 1885.

"Ironweeds." *GardenWeb*. The Virtual Mirror, Inc. http://forums.gardenweb.com/forum/load/natives/msg0512550330050.html.

Irvine, F. R. *Woody Plants of Ghana*. London: Oxford University Press, 1961.

Isern, Tom. "Plains Folk: Plum Full." *News for North Dakotans*, September 24, 1998. Agriculture Communication, North Dakota State University. http://www.ext.nodak.edu/extnews/newsrelease/1998/092498/20plains.htm.

James, Edwin. *An Account of an Expedition from Pittsburgh to the Rocky Mountains*. Ann Arbor, Mich.: University Microfilms, 1966.

Jeffery, David. "Arizona's Suburbs of the Sun." *National Geographic*, vol. 152, no. 4 (October 1977): 486–517.

Jensen, Jens. *Siftings*. Baltimore: Johns Hopkins University Press, 1990.

Jepson, Willis Linn. *The Jepson Manual: Higher Plants of California*. Ed. James C. Hickman. Berkeley: University of California Press, 1993.

Jewett, Sarah Orne. "Marsh Rosemary." *Atlantic Monthly*, vol. 57, no. 343 (May 1886): 590–601.

Joern, Anthony, and Kathleen H. Keeler, eds. *The Changing Prairie: North American Grasslands*. New York: Oxford University Press, 1995.

Johnson, Alfred E., ed. *Archaic Prehistory on the Prairie-Plains Border*. Lawrence: University of Kansas, 1980.

Johnson, Derek, Linda Kershaw, Andy MacKinnon, and Jim Pojar. *Plants of the Western Boreal Forest and Aspen Parkland*. Edmonton, Alta.: Lone Pine Publishing, 1995.

Johnson, Douglas W. "Biogeography of Quaking Aspen (*Populus tremuloides*)." San Francisco State University, Department of Geography. http://bss.

sfsu.edu/geog/bholzman/courses/fall99projects/aspen.htm.

Johnson, E. Pauline. *Flint and Feather: The Complete Poems of E. Pauline Johnson.* Toronto: Musson Book Co., 1931.

Johnson, Jon D., and Gorden Ekuan. "Hybrid Poplar Research Program." Washington State University. http://www.puyallup.wsu.edu/poplar/.

Johnson, Lorraine. *Grow Wild! Native-Plant Gardening in Canada and Northeastern United States.* Toronto: Random House of Canada, 1998.

————. *The Ontario Naturalized Garden: The Complete Guide to Using Native Plants.* Vancouver: Whitecap Books, 1995.

Johnson, Paul. *A History of the American People.* London: Phoenix Giant, 1997.

Jones, Ann. *Guide to America's Outdoors: Middle-Atlantic.* Washington, D.C.: National Geographic, 2001.

Jones, M. *Dr. Kane, the Arctic Here: A Narrative of His Adventures and Explorations in the Polar Regions—A Book for Boys.* London: T. Nelson, 1879.

Jorgensen, Neil. *A Guide to New England's Landscape.* Barre, Mass.: Barre Publishers, 1971.

————. *A Sierra Club Naturalist's Guide to Southern New England.* San Francisco: Sierra Club Books, 1978.

Josselyn, John. *John Josselyn, Colonial Traveler: A Critical Edition of Two Voyages to New England.* Ed. Paul J. Lindholdt. Hanover, N.H.: University Press of New England, 1988.

Kappel-Smith, Diana. *Desert Time: A Journey through the American Southwest.* First edition. Boston: Little, Brown, 1992.

Kari, Priscilla Russell. *Tanaina Plantlore: Dena'ina K'et'una—An Ethnobotany of the Dena'ina Indians of Southcentral Alaska.* Second edition. Anchorage: National Park Service Alaska Region, 1987.

Kartesz, John T. *A Synonymized Checklist of the Vascular Flora of the United States, Canada, and Greenland.* Second edition. Portland: Timber Press, 1994.

Kavasch, E. Barrie. *Native Harvests: Recipes and Botanicals of the American Indian.* New York: Vintage Books, 1979.

Kawatski, Deanna. *Wilderness Mother.* New York: Lyons and Burford, 1994.

Kaye, Barry, and D. W. Moodie. "*Psoralea* Food Resource on the Northern Plains." *Plains Anthropologist*, vol. 23, no. 82 (November 1978): 329–336.

Keating, Brian, and Dee Keating. "Mistaya Lodge." *Notes from the Field*, July 23–26, 1999. http://www.calgaryzoo.ab.ca/zoofaritour/mistaya2.htm.

Keeler, Harriet. *Our Native Trees and How to Identify Them: A Popular Study of Their Habits and Their Peculiarities.* Toronto: McClelland, Goodchild, and Stewart, 1900.

Keen, George Gillet, and Sarah Pamela Williams. *Cracker Times and Pioneer Lives: The Florida Reminiscences of George Gillet Keene and Sarah Pamela Williams.* Ed. James M. Denham and Canter Brown Jr. Columbia: University of South Carolina Press, 2000.

Keller, Jack. "Wild Plums." The Winemaking Home Page. http://winemaking.jackkeller.net/plums.asp.

Kelly, David, and Gary Braasch. *Secrets of the Old Growth Forest.* Salt Lake City: Peregrine Smith Books, 1988.

Kendall, George Wilkins. *Narrative of the Texan Santa Fé Expedition: Comprising a Description of a Tour through Texas and across the Great Southwestern Prairies the Camanche and Caygüa Hunting-Grounds, with an Account of the Sufferings from Want of Food, Losses from Hostile Indians, and Final Capture of the Texans, and Their March, as Prisoners, to the City of Mexico.* New York: Harper and Bros., 1844.

Kennedy, Des. *An Ecology of Enchantment: A Year in a Country Garden.* Toronto: HarperCollins, 1998.

Kennedy, William. *Texas: The Rise, Progress, and Prospects of the Republic of Texas*. London: Hastings, 1841.

Kershaw, Linda, Andy MacKinnon, and Jim Pojar. *Plants of the Rocky Mountains*. Edmonton, Alta.: Lone Pine, 1998.

Ketchum, Richard M. *The Secret Life of the Forest*. New York: American Heritage Press, 1970.

Kindscher, Kelly. *Edible Wild Plants of the Prairie: An Ethnobotanical Guide*. Lawrence: University Press of Kansas, 1987.

————. *Medicinal Wild Plants of the Prairie: An Ethnobotancial Guide*. Lawrence: University Press of Kansas, 1992.

King, James E. "Late Quaternary Vegetational History of Illinois." *Ecological Monographs*, vol. 51, no. 1 (1981): 43–62.

Kirk, Ruth. *Desert: The American Southwest*. Boston: Houghton Mifflin, 1973.

————. *The Olympic Rain Forest: An Ecological Web*. Seattle: University of Washington Press, 1992.

Klein, Frederick C. *Sunflower Seeds and Seoul Food*. In Addall, Muze Inc., 1995–2002. www.addall.com/Browse/Detail/1566250757.html.

Klinka, K., et al. *Indicator Plants of Coastal British Columbia*. Vancouver: University of British Columbia Press, 1989.

Knobel, Edward. *Field Guide to the Grasses, Sedges, and Rushes of the United States*. New York: Dover, 1977.

Kolpas, Norman, and Barbara Pool Fenzl. *Southwest the Beautiful Cookbook: Recipes from America's Southwest*. San Francisco: Collins Publishers, 1994.

Kozloff, Eugene N. *Plants and Animals of the Pacific Northwest: An Illustrated Guide to the Natural History of Western Oregon, Washington, and British Columbia*. Vancouver: J.J. Douglas, 1976.

Krakauer, Jon. "Death of an Innocent: How Christopher McCandless Lost His Way in the Wilds." *Outside Magazine*, January 1993. Mariah Media, Inc., 2001. www.outsidemag.com/magazine/0193/9301fdea.htm.

Kraulis, J. A. *The Rocky Mountains: Crest of a Continent*. Toronto: Key Porter Books, 1986.

Krech, Shepard, III. *The Ecological Indian: Myth and History*. New York: W. W. Norton, 1999.

Kress, Henriette. "Gathering Juniper (*Juniperus communis*) Berries." Henriette's Herbal Homepage. The Culinary Herbalist, January 1996. http://ibiblio.org/herbmed/neat-stuff/junicomm.html.

Kricher, John C. *A Field Guide to Ecology of Eastern Forests, North America*. Boston: Houghton Mifflin, 1988.

Kroeber, A. L., and E. W. Gifford, comps. *Karok Myths*. Ed. Grace Buzaljko. Berkeley: University of California Press, 1980.

Kruckeberg, Arthur R. *Gardening with Native Plants of the Pacific Northwest: An Illustrated Guide*. Seattle: University of Washington Press, 1982.

Kucera, C. L., and R. E. McDermott. "Sugar Maple–Basswood Studies in the Forest-Prairie Transition of Central Missouri." *American Midland Naturalist*, vol. 54, no. 2 (1955): 495–503.

Küchler, A. W. *Manual to Accompany the Map of Potential Vegetation of the Conterminus United States*. New York: American Geographical Society, 1964.

Kuhnlein, Harriet V., and Nancy J. Turner. *Traditional Plant Foods of Canadian Indigenous Peoples: Nutrition, Botany, and Use*. Philadelphia: Gordon and Breach Science, 1991.

Kummel, Marc. "*Adenostoma fasciculatum*: Chamise, Greasewood." *Treebeard's Flora: Woody Plants of the Central Santa Ynez Mountains*, 1998. www.rain.org/~mkummel/flora/adefas/html.

Lamb, L. "A Tall-Grass Prairie in a Small Backyard: Adventurous Gardener Evokes a Bygone Landscape." *Fine Gardening*, May–June 1990, 47–52.

Lamb, L., and G. Rhynard. *Plants of Carolinian*

Canada. Don Mills: Federation of Ontario Naturalists, 1994.

Lampman, Archibald. *The Poems of Archibald Lampman.* Ed. Duncan Campbell Scott. Third edition. Toronto: Morang and Co., 1905.

Landon, Fred, ed. *The Diary of Benjamin Lundy, Written during His Journey through Upper Canada, January 1832.* Toronto: Ontario Historical Society Papers and Records, 1922.

Landry, Peter. "Blupete's Wildflowers of Nova Scotia." June 2001. http://www.blupete.com/Nature/Wildflowers/Wild.htm.

Lanier, Sidney. *Florida: Its Scenery, Climate, and History.* Gainesville: University of Florida Press, 1973. Facsimile of 1875 ed.

———. *Poems.* New York: Charles Scribner's Sons, 1928.

Lanner, Ronald M. *The Piñon Pine: A Natural and Cultural History.* Reno: University of Nevada Press, 1981.

Las Pilitas Nursery. *"Holodiscus discolor."* www.laspilitas.com/plants/350.htm.

———. *"Rhamnus californica."* www.laspilitas.com/plants/566.htm.

Lauriault, Jean. *Identification Guide to the Trees of Canada.* Richmond Hill, Ont.: Fitzhenry and Whiteside, 1989.

Lawrence, Elizabeth. *Gardening for Love: The Marketing Bulletins.* Durham, N.C.: Duke University Press, 1987.

———. *Through the Garden Gate.* Ed. Bill Neal. Chapel Hill: University of North Carolina Press, 1990.

Lawrence, R. D. *The Natural History of Canada.* Toronto: Key Porter, 1988.

Leak, W. B. "Age Distribution in Virgin Red Spruce and Northern Hardwoods." *Ecology,* vol. 56, no. 6 (1975): 1451–1454.

Leduc, M. A. *The Explorer's Guide to Boreal Forest Plants.* Cobalt, Ont.: Highway Book Shop, 1997.

Legasy, Karen L. *Forest Plants of Northeastern Ontario.* Edmonton, Alta.: Lone Pine, 1995.

"Legends." *Native Online,* 2000. www.nativeonline.com/legends.html#REDCEDAR.

Lemmon, Kenneth. *The Golden Age of Plant Hunters.* London: Phoenix House, 1968.

Leopold, Aldo. *A Sand County Almanac.* New York: Oxford University Press, 1966.

Leshuk, Dave. "John Muir's Wisconsin Days." *John Muir Exhibit,* Sierra Club. Reprinted from *Wisconsin Natural Resources,* vol. 12, no. 3 (May/June 1988). http://www.sierraclub.org/john_muir_exhibit/life/muir_wisconsin_dave_leshuk.html.

Lewington, Anna, and Edward Parker. *Ancient Trees: Trees That Live for a Thousand Years.* London: Collins and Brown, 1999.

Lippincott, James S. "Among the Wild Flowers of San Diego." *Information Relative to the City of San Diego, California.* San Diego: Office of the San Diego Daily Union, 1874, in San Diego Historical Society. www.sandiegohistory.org/books/city1874/page29.htm.

Littlejohn, Bruce, and Weyland Drew. *Superior: The Haunted Shore.* Toronto: FireflyBooks, 1994.

Little, Charles E. *The Dying of the Trees: The Pandemic in America's Forests.* New York: Viking, 1995.

Little, Elbert Luther. *National Audubon Society Field Guide to North American Trees: Eastern Region.* New York: Alfred A. Knopf, 1980.

Littleflower's the Medicine of North American Plants. Little Flower Publications. www.geocities.com/RodeoDrive/Mall/4992/poison_ivy.html.

London, Jack. *Jack London's California: The Golden Poppy and Other Writings.* Ed. Sal Noto. New York: Beaufort Books, 1986.

Long, Robert W., and Olga Lakela. *A Flora of*

Tropical Florida: A Manual of the Seed Plants and Ferns of Southern Peninsula Florida. Coral Gables: University of Miami Press, 1971.

Longfellow, Henry Wadsworth. *The Works of Henry Wadsworth Longfellow.* Boston: Houghton & Mifflin, 1886.

Lopez, Barry. *Arctic Dreams: Imagination and Desire in a Northern Landscape.* New York: Charles Scribner's Sons, 1986.

———. *Crossing Open Ground.* New York: Vintage Books, 1989.

———. *Giving Birth to Thunder, Sleeping with His Daughter: Coyote Builds North America.* Kansas City: Sheed Andrews and McMeel, 1977.

Lovett, Rick. "'Extinct' Oregon Flower Reappears." *Daily InSight,* June 21, 1999. Academic Press, American Association for the Advancement of Science, 1999. www.academicpress/com/inscight/06211999/graphb.htm.

Lowdermilk, Walter Clay. *Conquest of the Land through 7,000 Years.* Washington, D.C.: U.S. Department of Agriculture, Soil Conservation Service, 1953.

Lower, A. R. M. *The North American Assault on the Canadian Forest: A History of the Lumber Trade between Canada and the United States.* Toronto: Ryerson Press, 1938.

Lowry, Judith. "Coyote Bush Defended." *Point Reyes Light,* May 10, 2001. Tomales Bay Publishing, 1995–2002. www.ptreyeslight.com/stories/may10_01/coyote_bush.html.

Lubbock Avalanche-Journal. "Bill May Make Saw Palmetto Florida Crop." Lubbock Online.com. *Lubbock-Avalanche Journal,* May 6, 1997. http://www.lubbockonline.com/news/040797/billmay.htm.

Luebke, Neil. "Cacti—More Than Stickers." *Lore* magazine, 1996. Milwaukee Public Museum, 2000. http://www.mpm.edu/research/botany/botany_lore02.html.

Lukes, Roy. "Scouring Rushes Were the First Scouring Brushes." *Nature-Wise, Door County*

Advocate, December 15, 2000. http://www.doorbell.net/lukes/a121500.htm.

Lum, Rebecca Rosen. "'Healing Garden' Protects Redding's Sole Synagogue." *Jewish Bulletin News.* San Francisco Jewish Community Publications, 2002.

Luoma, Jon R. *The Hidden Forest: The Biography of an Ecosystem.* New York: Henry Holt, 1999.

Luther, Ken. "The Company of Weeds." *Saturday Night Magazine,* July/August 1996. www.sheridanc.on.ca/~kem/weeds/weeds.htm.

Lynch, Jeremiah. *Three Years in the Klondike.* Ed. Dale L. Morgan. Chicago: R. R. Donnelley, 1967.

Lynch, Wayne. *Married to the Wind.* North Vancouver: Whitecap Books, 1984.

Lyons, Janet, and Sandra Jordan. *Walking the Wetlands: A Hiker's Guide to Common Plants and Animals of Marshes, Bogs, and Swamps.* New York: John Wiley and Sons, 1989.

MacCleery, Doug. "When Is a Landscape Natural?" *Forest Landowner Magazine,* January 1999. http://www.forestland.org/mag_jan99_mac.html.

Macfarlane, J. *The Cook Not Mad: Or, Rational Cookery.* Ed. Roy Abrahamson. Reprint of 1831 edition with additional recipes and explanatory notes. Toronto: Cherry Tree Press, 1972.

Mackenzie, Alexander. *Journal of the Voyage to the Pacific.* Ed. Walter Sheppe. New York: Dover, 1995.

MacLeish, William H. *The Day before America: Changing the Nature of a Continent.* Boston: Houghton Mifflin, 1994.

MacLeod, Salome. *Memories of Beach Point and Cape Bear.* Charlottetown, P.E.I.: Author, 1967.

MacMahon, James A. *Deserts.* New York: Alfred A. Knopf, 1985.

Macnair, Mark. "*Mimulus guttatus* Complex." January 22, 2001. http://www.ex.ac.uk/~MRMacnai/guttatus.html.

Madson, John. *Tall Grass Prairie*. Helena: Falcon Press, 1993.

———. *Where the Sky Began, Land of the Tallgrass Prairie*. Boston: Houghton Mifflin, 1982.

Maguire, Sarah. *Flora Poetica: The Chatto Book of Botanical Verse*. London: Chatto and Windus, 2001.

Mair, Charles. *Dreamland and Other Poems, Tecumseh: A Drama*. Toronto: University of Toronto Press, 1974.

Maissurow, D. K. "The Role of Fire in the Perpetuation of Virgin Forests of Northern Wisconsin." *Journal of Forestry*, vol. 39, no. 2 (1941): 201–207.

Malin, James C. *History and Ecology: Studies of the Grassland*. Ed. Robert P. Swierenga. Lincoln: University of Nebraska Press, 1984.

Manley, Sean, and Robert Manley. *Beaches: Their Lives, Legends, and Lore*. Philadelphia: Chilton, 1968.

Manners, Robert A. *An Ethnological Report on the Hualapai Indians of Arizona*. New York: Garland, 1974.

Manning, R. *Grassland: The History, Biology, Politics, and Promise of the American Prairie*. New York: Penguin Books, 1995, 1997.

Margolin, Malcolm. *The Ohlone Way: Indian Life in the San Francisco–Monterey Bay Area*. Berkeley: Heyday Books, 1978. archive.sportserver.com/ newsroom/sports/oth/1998/oth/mor/feat/archive/ 040498/mor2415.html.

Maricopa Center for Learning and Instruction. "The 'Real' Ocotillo." http://www.mcli.dist.maricopa.edu/ ocotillo/oco_real.html.

Marinelli, Janet. *Stalking the Wild Amaranth: Gardening in the Age of Extinction*. New York: Henry Holt, 1998.

Marquardt, William H., ed. *Culture and Environment in the Domain of the Calusa*. Gainesville: Institute of Archaeology and Paleoenvironmental Studies, University of Florida, 1992.

Marsh, C. Dwight, and A. B. Clawson. "*Eupatorium urticaefolium* as a Poisonous Plant." *Journal of Agricultural Research*, vol. 11, no. 13 (December 24, 1917): 699–716.

Martin, Laura C. *The Wildflower Meadow Book: A Gardener's Guide*. Chester, Conn.: Globe Pequot Press, 1990.

Martin, W. Keble. *The Concise British Flora in Colour*. London: Ebury Press, 1965.

Martinez, Raymond J. *The Story of Spanish Moss and Its Relatives*. Home Publications, 1959, Communities Online, Inc., 1997. www.communityonline.com/ local/culture/spanishmoss.htm.

Maryland Department of Natural Resources, Forest Service. "Quiet Giant, the Wye Oak." http://www. dnr.state.md.us/forests/trees/giant.html.

Maser, Chris. *Forest Primeval: The Natural History of an Ancient Forest*. San Francisco: Sierra Club Books, 1994.

———. *The Redesigned Forest*. Toronto: Stoddart, 1988.

Mason, Marjorie Rohde. *One Hundred Years of Grasslands: An Historical Survey of the Settlement and Development of the Stockgrowing Industry of the Grasslands of Southern Saskatchewan*. North Battleford, Sask.: Turner-Warwick, 1993.

Mathewes, Rolf W., and Richard Hebda. "Holocene History of Cedar and Native Indian Cultures of the North American Pacific Coast." *Science*, vol. 225 (August 17, 1984): 711.

Mathews, Anne. *Where the Buffalo Roam*. New York: Grove Weidenfeld, 1992.

Mathews, Naomi. "Maryland's Queen of Flowers: Black-eyed Susans." *Garden Guides*, 1999. www. gardenguides.com/articles/maryland.htm.

Maud, Ralph. *A Guide to B. C. Indian Myth and Legend: A Short History of Myth-Collecting and a Survey of Published Texts*. Vancouver: Talonbooks, 1982.

McAllister, Ian. *The Great Bear Rainforest: Canada's Forgotten Coast*. San Francisco: Sierra Club Books, 1997.

McCarthy, James. *A Papago Traveler: The Memories of James McCarthy*. Tucson: Sun Tracks and University of Arizona Press, 1985.

McClaran, Mitchel P., and Thomas R. van Devender, eds. *The Desert Grassland*. Tucson: University of Arizona Press, 1995.

McDonald, Philip M. "*Quercus douglasii:* Blue Oak." *Silvics of North America: Hardwoods*, vol. 2. Agriculture Handbook 654, USDA Forest Service. www.na.fs.fed.us/spfo/pubs/silvics_manual/volume_2/quercus/douglasii.htm.

McFeeley, Dan. "Added Contributions to Bee Tidbits." *Some Tidbits about Bees and Honey from Prehistory to Present*. April 6, 1999. http://www.chebucto.ns.ca/~ag151/addendum.html.

McGrath, Judy Waldner. *Dyes from Lichens and Plants: A Canadian Dyer's Guide*. Toronto: Van Nostrand Reinhold, 1977.

McHugh, Tom. *The Time of the Buffalo*. New York: Alfred A. Knopf, 1972.

McKibben, Bill. *The End of Nature*. New York: Anchor Books, 1990.

McMinn, Howard E., Evelyn Maino, and H. W. Shepherd. *An Illustrated Manual of Pacific Coast Trees, with Lists of Trees Recommended for Various Uses on the Pacific Coast*. Berkeley: University of California Press, 1967.

McMurtry, Larry. *Sacajawea's Nickname: Essays on the American West*. New York: New York Review Books, 2001.

McNulty, Tim. *In Blue Mountain Dusk*. Seattle: Broken Moon Press, 1992.

McPhee, John. *In Suspect Terrain*. New York: Farrar Straus and Giroux, 1983.

———. "Los Angeles against the Mountains." *The Control of Nature*. Sun Valley Writers' Conference, 2002. www.svwc.com/McPhee.htm.

McQueen, Cyrus B. *Field Guide to the Peat Mosses of Boreal North America*. Hanover, N.H.: University Press of New England, 1990.

McRae, Michael. "I Am Monkey Flower." *Outside Magazine*, July 1997. *Outside Online*, Mariah Media, Inc., 2001. www.outsidema.com/magazine/0797/9707flower.html.

Mead, George R. *The Ethnobotany of the California Indians: A Compendium of the Plants, Their Users, and Their Uses*. Greeley, Colo.: Museum of Anthropology, University of Northern Colorado, 1972.

"Medicine of North American Plants." http://www.geocities.com/littleflowers_medicinal_plants/sarsaparilla.htm.

Medsger, Oliver. *Edible Wild Plants*. New York: Macmillan, 1957.

Meehan, Thomas. *The Native Flowers and Ferns of the United States: In Their Botanical, Horticultural, and Popular Aspects*. Boston: L. Prang and Company, 1878.

Menihek Trails Group. "March—Can It Be Spring." March 6, 2001. www.labrador-west.com/menihektrails/marchat.htm.

Menzies, Archibald. *Menzies' Journal of Vancouver's Voyage, April to October, 1792*. Ed. C. F. Newcombe. Victoria, B.C.: William H. Cullin, 1923.

Merilees, Bill. *The New Gardening for Wildlife: A Guide for Nature Lovers*. Vancouver: Whitecap Books, 2000.

Michaux, François André. *The North American Sylva; or, A Description of the Forest Trees of the United States, Canada, and Nova Scotia Considered Particularly with Respect to Their Use in the Arts and Their Introduction into Commerce, to Which Is Added a Description of the Most Useful of the European Forests Trees*. Trans. A. L. Hillhouse. Paris: C. d'Hautel, 1819.

Mikkelson, Martha. "Picking Pin Cherries." 1998. http://www.dwave.net/~martha/pincher.htm.

Millay, Edna St. Vincent. *Second April*. New York: Mitchell Kennerley, 1921.

Miller, Howard A., and Samuel H. Lamb. *Oaks of*

North America. Happy Camp, Calif.: Naturegraph Publishers, 1985.

Mirov, Nicholas T., and Jean Hasbrouck. *The Story of Pines*. Bloomington: Indiana University Press, 1976.

Mirsky, Jeannette. *Elisha Kent Kane and the Seafaring Frontier*. Ed. Oscar Handlin. Boston: Little, Brown, 1954.

———. *To the Arctic! The Story of Northern Exploration from Earliest Times to the Present*. Chicago: University of Chicago Press, 1934, 1970.

Moerman, Daniel. *Native American Ethnobotany*. Portland, Ore.: Timber Press, 1998.

Mohlenbrock, Robert H. *Where Have All the Wildflowers Gone? A Region-by-Region Guide to Threatened or Endangered U.S. Wildflowers*. New York: Macmillan, 1983.

Montana Historical Society Press. "Bitterroot: *Lewisia rediviva Pursh*." http://www.his.state.mt.us/departments/press/bitter_root.html.

Moodie, Susanna. *Life in the Clearings*. Ed. Robert L. McDougall. Toronto: Macmillan of Canada, 1959.

———. *Roughing It in the Bush*. London: Virago Press, 1986.

Moon, William Least Heat. *PrairyErth: A Deep Map*. Boston: Houghton Mifflin, 1991.

Moore, Michael. *Medicinal Plants of the Desert and Canyon West*. Santa Fe: Museum of New Mexico Press, 1989.

———. *Medicinal Plants of the Mountain West*. Santa Fe: Museum of New Mexico Press, 1979.

Moorwood, William. *Traveler in a Vanished Landscape: The Life and Times of David Douglas*. New York: Clarkson N. Potter, 1973.

Morgan, John P., and Douglas R. Collicutt. *Restoring Canada's Native Prairies: A Practical Manual*. Argyle, Man.: Prairie Habitats, 1995.

Morgan, Ted. *Wilderness at Dawn: The Settling of the*

North American Continent. New York: Simon and Schuster, 1993.

Morin, N.D., ed. *Flora of North America: North of Mexico*. New York: Oxford University Press, 1971.

Morison, Samuel Eliot. *The European Discovery of America: The Northern Voyages, A.D. 500–1600*. New York: Oxford University Press, 1971.

Morley, Jim. *Muir Woods: The History, Sights, and Seasons of the Famous Redwood Forest near San Francisco: A Pictorial Guide*. Berkeley: Howell-North Books, 1968.

Morris, Alton C, ed. *Folksongs of Florida*. Gainesville: University of Florida Press, 1950.

Morrissey, Jennifer. "Awed by Cottonwoods." *Green Line*, vol. 11, no. 1 (spring 1999). http://coloradoriparian.org/GreenLine/V11-1/Cottonwoods.html.

Morss, Elisabeth W. *Herbs of a Rhyming Gardener*. Boston: Branden Press, 1971.

Morton, Julia F. *Atlas of Medicinal Plants of Middle America: Bahamas to Yucatan*. Springfield, Ill.: C. C. Thomas, 1981.

———. "Can the Red Mangrove Provide Food, Feed, and Fertilizer?" *Economic Botany* 19 (1965): 113–123.

Muench, David. *Desert Images: An American Landscape*. New York: Harcourt Brace Jovanovich, 1979.

Muir, John. *The Cruise of the Corwin: Journal of the Arctic Expedition of 1881 in Search of DeLong and the Jeanette*. Ed. William Frederic Badè. Boston: Houghton Mifflin, 1917.

———. *The Mountains of California*. Berkeley, Calif.: Ten Speed Press, 1977.

———. *South of Yosemite*. Ed. Frederic R. Gunsky. Garden City, N.Y.: American Museum of Natural History, Natural History Press, 1968.

———. *The Yosemite*. New York: Century, 1912. *John*

Muir Exhibit, Sierra Club, 2002. www.sierraclub.org/ john_muir_exhibit/writings/the_yosemite.

Munz, Philip A. *California Mountain Wildflowers*. Berkeley: University of California Press, 1963.

Munz, Philip A., and David D. Keck. *A California Flora*. Berkeley: University of California Press, 1959.

Murphy, Louise Mary. *Sweet Canada: Twelve Bird Songs and a Round*. Toronto: Longmans, Green, 1923.

Murray, John A., ed. *A Republic of Rivers: Three Centuries of Nature Writing from Alaska and the Yukon*. New York: Oxford University Press, 1990.

Musgrave, Toby, Will Musgrave, and Chris Gardner. *The Plant Hunters: Two Hundred Years of Adventure and Discovery Around the World*. New York: Sterling, 1998.

"Myrtle Wood Tree." The Real Oregon Gift Factory. www.realoregongift.com/myrtletree/htm.

Nabhan, Gary Paul. *Cultures of Habitat: On Nature, Culture, and Story*. Washington, D.C.: Counterpoint, 1997.

———. *The Desert Smells Like Rain: A Naturalist in Papago Indian Country*. San Francisco: North Point Press, 1982.

———. *Enduring Seeds: Native American Agriculture and Wild Plant Conservation*. San Francisco: North Point Press, 1989.

———. *Gathering the Desert*. Tucson: University of Arizona Press, 1985.

Nabhan, Gary Paul, and Caroline Wilson. *Canyons of Color: Utah's Slickrock Wildlands*. San Francisco: HarperCollinsWest, 1995.

Nadkarni, Nalini, and Erica Guttman. "Native Plant of the Month: Pacific Dogwood." *Green Screens*, September 1999. www.olywa.net/speech/ september99/native.html.

Nairne, Thomas. *Nairne's Muskhogean Journals: The 1708 Expedition to the Mississippi River*. Ed. Alexander

Moore. Jackson: University Press of Mississippi, 1988.

Nash, Thomas H., ed. *Lichen Biology*. Cambridge: Cambridge University Press, 1996.

National Forum on BioDiversity. *Biodiversity*. Ed. E. O. Wilson. Washington, D.C.: National Academy Press, 1988.

National Pecan Shellers Association. "Pecans, So Good for You." http://www.ilovepecans.org/.

"National Register of Big Trees." American Forests. www.americanforests.org/resources/bigtrees.

National Sunflower Association. www.sunflowernsa. com/pubinfo/history/.

Naturalist.org. "In Mountain and Meadow— Journal." May 15, 1996. Naturalist.org, 1995, 1996, 1998. http://www.naturalist.org/brow4/ 96515.php3.

"Neighborhood Pride, Childhood Memories Make the Town." *SouthCoast Today*, June 30, 2000. The Standard-Times, 2001. www.s-t.com/daily/06-00/06-30-00/a061o030.htm.

Newman, Peter C. *Company of Adventurers, Volume 2: Caesars of the Wilderness*. Markham, Ont.: Viking, 1987.

Nickrent, Daniel. "Ant-Dwarf Mistletoe Relationships." *The Golden Bough*, vol. 10 (February 1988). Department of Biology, Southern Illinois University at Carbondale. http://www.science.siu. edu/parasitic-plants/Ant.DM.html.

Niering, William A. *Wetlands: An Audubon Society Nature Guide*. New York: Alfred A. Knopf, 1985.

Niering, William A., and Nancy C. Olmstead. *National Audubon Society Field Guide to North American Wildflowers: Eastern Region*. New York: Alfred A. Knopf, 1997.

Nightingale, Marie. *Out of Old Nova Scotia Kitchens: A Collection of Traditional Recipes of Nova Scotia and the Stories of the People Who Cooked Them*. Fifth edition. Halifax: Petheric Press, 1973.

Nobel, Park S. *Remarkable Agaves and Cacti*. New York: Oxford University Press, 1994.

Noble, David Grant, ed. *The Hohokam: Ancient People of the Desert*. Santa Fe, N.M.: School of American Research Press, 1991.

Norse, Elliott A. *Ancient Forests of the Pacific Northwest*. Washington, D.C.: Island Press, 1990.

North American Prairie Conference (6th, 1978, Ohio State University). *The Prairie Peninsula—in the "Shadow" of Transeau: Proceedings of the Sixth North American Prairie Conference, the Ohio State University, Columbus, Ohio, 12–17 August 1978*. Ed. Ronald L. Stuckey and Karen J. Reese. Columbus: College of Biological Sciences, Ohio State University, 1981.

North American Prairie Conference (9th, 1984, Moorhead, Minn.). *The Prairie—Past, Present, and Future: Proceedings of the Ninth North American Prairie Conference, Held July 29 to August 1, 1984, Moorhead, Minnesota*. Ed. Gary K. Clambey and Richard H. Pemble. Fargo, N.D.: Tri-College University Center for Environmental Studies, 1986.

North American Prairie Conference (13th, 1992, Windsor, Ont.). *Proceedings of the Thirteenth North American Prairie Conference: Spirit of the Land, Our Prairie Legacy: Held 6–9 August 1992, Windsor, Ontario, Canada*. Chaired by Lloyd O. W. Burridge. Ed. Robert G. Wickett, et al. Windsor, Ont.: Department of Parks and Recreation, 1994.

North, Marianne. *A Vision of Eden: The Life and Work of Marianne North*. Exeter: Webb and Bower, 1980.

Nyerges, Christopher. *Guide to Wild Foods and Useful Plants*. Chicago: Chicago Review Press, 1999.

———. "Wild Coffee Alternatives." *Mother Earth News*, August/September 1999, pp. 10–12.

Odum, Howard T., and Katherine Carter Ewel, eds. *Cypress Swamps*. Gainesville: University Press of Florida, 1985.

Oesterreicher, Michel. *Pioneer Family: Life on Florida's 20th Century Frontier*. Tuscaloosa: University of Alabama Press, 1996.

Oldham, Michael J. *Distribution and Status of Vascular Plants of Southwestern Ontario*. Ontario Ministry of Natural Resources, Aylmer District, 1993.

———. *Natural Heritage Resources of Ontario: Rare Vascular Plants*. Peterborough: Natural Heritage Information Centre, Ontario Ministry of Natural Resources, 1996.

Olson, Steven. *The Prairie in Nineteenth-Century American Poetry*. Norman, Okla.: University of Oklahoma Press, 1994.

O'Neill, John. "California Poppy." 1999. http://www.pyramid.net/gallery/webnew/poppy.html.

Oosting, Henry John. *The Study of Plant Communities: An Introduction to Plant Ecology*. Second edition. San Francisco: W. H. Freeman, 1956.

Orchid Safari. www.geocities.com/~marylois/index.html.

Orlean, Susan. *The Orchid Thief*. New York: Random House, 1998.

Osborn, Stellanova. *Beside the Cabin*. Sault Ste. Marie, Mich.: Northwoods Press, 1957.

Oswalt, Wendell H. *Alaskan Eskimos*. Scranton, Penn.: Chandler, 1967.

Oxford Companion to Gardens. Ed. Sir Geoffrey Jellicoe, et al. Oxford: Oxford University Press, 1986.

Pacific Northwest Native Plant Society Digest. pnw-natives@tardigrade.net.

Packard, Stephen, and Cornelia F. Mutel, eds. *The Tallgrass Restoration Handbook for Prairies, Savannas, and Woodlands*. Washington: Island Press, 1997.

Pallardy, S. G., T. A. Nigh, et al. "Changes in Forest Composition in Central Missouri: 1969–1982." *American Midland Naturalist*, vol. 120, no. 2 (1988): 380–390.

Parker, A. C. *Iroquois Uses of Maize and Other Food Plants*. Albany: University of the State of New York,

1910. Reprinted, Ohsweken, Ont: Iroquois Publishing and Craft Supplies, 1994.

Parker, G. R., et al. "Tree Dynamics in an Old-Growth, Deciduous Forest." *Forest Ecology and Management*, vol. 11, nos. 1 & 2 (1985): 31–57.

Parry, Caroline, ed. *Eleanora's Diary: The Journals of a Canadian Pioneer Girl*. Richmond Hill, Ont.: Scholastic Canada, 1994.

Parsons, Frances Theodora [Mrs. William Starr Dana]. *How to Know the Wild Flowers: A Guide to the Names, Haunts, and Habits of Our Common Wild Flowers*. New York: Dover, 1963.

Pauly, Wayne, "Prairie Folklore: Horsetail Tails." *Action: Adult Conservation Team Newsletter*: Dane County Parks, fall 1998. http://www.co.dane.wi.us/parks/adult/action/1998/fall/p6.htm.

Pavlik, Bruce, et al. *Oaks of California*. Los Olivos, Calif.: Cachuma Press, 1995.

Peattie, Donald Culross. *A Natural History of Trees of Eastern and Central North America*. Boston: Houghton Mifflin, 1991.

———. *A Natural History of Western Trees*. Boston: Houghton Mifflin, 1991.

Pérez de Luxán, Diego. *Expedition into New Mexico Made by Antonio de Espejo, 1582–1583, as Revealed in the Journal of Diego Pérez de Luxán, a Member of the Party*. Trans. George Peter Hammond and Agapito Rey. Los Angeles: Quivira Society, 1929.

Peters, Paige. "Fruits, Roots, and Shoots of the Northwoods." *Northbound*, vol. 11, no. 1 (spring 2001): 3. Trees for Tomorrow, http://www.treesfortomorrow.com/download/WILDEDIB.pdf.

Peterson, Roger Tory, and Margaret McKenny. *A Field Guide to Wildflowers of Northeastern and North-Central North America: A Visual Approach Arranged by Color, Form, and Detail*. Boston: Houghton Mifflin, 1968.

Petersons, Andris. "The Beach Pea of Salmon Cove." *Gazette*, vol. 32, no. 21 (July 13, 2000). Memorial University of Newfoundland. www.mun.ca/univrel/gazette/1999–2000/July, 13/research.html.

Petrides, George A., and Roger Tory Peterson. *A Field Guide to Western Trees (Peterson Guide # 44)*. New York: Houghton Mifflin, 1998.

Pettinger, April. *Native Plants in the Coastal Garden: A Guide for Gardeners in British Columbia and the Pacific Northwest*. Vancouver: Whitecap Books, 1996.

Pharmacist's Ultimate Health, Health Library. "Horsetail." Pharmacist's Ultimate Health Corp., 2000. www.healthsoftware.com/users/user1/HORSETAIL.htm.

Phillips, Steven J., and Patricia Wentworth Comus, eds. *A Natural History of the Sonoran Desert*. Tucson: Arizona–Sonora Desert Museum Press, 2000.

Pielou, E. C. *After the Ice Age: The Return of Life to Glaciated North America*. Chicago: University of Chicago Press, 1991.

———. *Fresh Water*. Chicago: University of Chicago Press, 1998.

———. *A Naturalist's Guide to the Arctic*. Chicago: University of Chicago Press, 1994.

———. *The World of Northern Evergreens*. Ithaca, N.Y.: Comstock Publishing Associates, 1988.

Pitcher, J. S., and J. S. McKnight. "Black Willow—*Salix nigra*." *Trees of Western North Carolina*. Reprinted online with permission of the USDA Forest Service. wildwnc.org/trees/Salix_nigra.html.

Plants for a Future—Species Database. Cornwall, UK, 1997–2000. www.ibiblio.org/pfaf/.

Platt, Rutherford Hayes. *The Great American Forest*. Englewood Cliffs, N.J.: Prentice Hall, 1965.

Plotkin, Mark J. *Tales of a Shaman's Apprentice: An Ethnobotanist Searches for New Medicines in the Amazon Rain Forest*. New York: Penguin Books, 1994.

Plotnik, Arthur. *The Urban Tree Book: An Uncommon Field Guide for City and Town*. New York: Three Rivers Press, 2000.

Pojar, Jim, and Andy MacKinnon, eds. *Plants of the Pacific Northwest Coast: Washington, Oregon, British*

Columbia, and Alaska. Redmond: Lone Pine Publishing, 1994.

Polhill, R. M., coord. *The Golden Bough*, vol. 10, February 1988. Kew: The Herbarium, Royal Botanic Gardens. www.science.siu.edu/parasitic-plants/Ant. DM.html.

Pollan, Michael. *The Botany of Desire: A Plant's-Eye View of the World*. New York: Random House, 2001.

Polunin, Nicholas Vladimir. *Circumpolar Arctic Flora*. Oxford: Clarendon Press, 1959.

Porsild, A. E. *Illustrated Flora of the Canadian Arctic Archipelago*. Ottawa: National Museums of Canada, 1964.

———. *Rocky Mountain Wildflowers*. Ottawa: National Museum of Natural Sciences, National Museums of Canada and Parks Canada, Depatment of Indian and Northern Affairs, 1974.

Potyondi, Barry. *In Palliser's Triangle: Living in the Grasslands, 1850–1930*. Saskatoon, Sask.: Purich Publishing, 1995.

Powell, John Wesley. *Down the Colorado: Diary of the First Trip through the Grand Canyon, 1869*. New York: E. P. Dutton, 1969.

Pratt, E. J. *Complete Poems*. Toronto: University of Toronto Press, 1989.

Pratt, Mara L. *Fairyland of Flowers*. Boston: Educational Publishers, 1890.

Pressley, Benjamin. "Conquering the Darkness: Primitive Lighting Methods." *Bulletin of Primitive Technology*, no. 12 (fall 1996). http://www.hollowtop. com/spt_html/lighting.htm.

"Project Profile: Balmville, N.Y., Historic Tree Merits Extraordinary Measures." Originally printed in *Arbor Age Magazine*, January 1996. http://www. acrtinc.com/balmtree.html.

Purchis, George. "Growing with the Tide." *Tidal Tales*. Newsletter of the Hayward Shoreline Interpretive Center. Hayward, Calif.: Naturalist's Library, May/June 1998. www.hard.dst.ca.us/ hayshore/library/marsh5_98.htm.

Pursh, Frederick. *Flora Americae Septentrionalis; or, A Systematic Arrangement and Description of the Plants of North America*. London: White, Cochrane, 1814.

Pyle, Linda McMillin. "Saguaro Fruit." *Desert USA*, June 1998. Digital West Media, Inc., 2002. http://www.desertusa.com/mag98/june/papr/jun_ lil.html.

Pyne, Stephen J. *Fire in America: A Cultural History of Wildland and Rural Fire*. Seattle: University of Washington Press, 1997.

Pynn, Larry. *Last Stands: A Journey through North America's Vanishing Ancient Rainforests*. Vancouver: New Star Books, 1999.

Raban, Jonathan. *Passage to Juneau: A Sea and Its Meanings*. New York: Pantheon Books, 1999.

Ramirez, Alice. "Bunchberries." *Explorations*, December 1996. www.chem.ucla.edu/~alice/ explorations/churchill/botbu.htm.

Rawlings, Marjorie Kinnan. *Cross Creek*. Toronto: Reginald Saunders, 1942.

Reading, Jeff. *Eating Smoke: A Review of Non-Traditional Use of Tobacco among Aboriginal People*. Ottawa: Health Canada, 1996.

Red River Commodities, Inc. "Sunflower Story." 2000.www.redriv.com/sunflower_story.html.

Reed, Monique. "Sedge Meadow." *Lick Creek Park*. Herbarium, Department of Biology, Texas A&M University. 1996. www.csdl.tamu.edu/FLORA/LCP/ LCPINTRO.HTML.

Reid, Kenneth C. "*Psoralea esculenta* as a Prairie Resource." *Plains Anthropologist*, vol. 22 (1978): 321–325.

Robbins, Wilfred William, John Peabody Harrington, and Barbara Freire-Marreco. *Ethnobotany of the Tewa Indians*. Smithsonian Institution, Bureau of American Ethnology Bulletin

55. Washington, D.C.: U.S. Government Printing Office, 1916.

Roberts, June Carver. *Born in the Spring: A Collection of Spring Wildflowers.* Athens, Oh.: Ohio University Press, 1976.

Robichaux, Robert H., ed. *Ecology of Sonoran Desert Plants and Plant Communities.* Tucson: University of Arizona Press, 1999.

Robuck, O. Wayne. *The Common Plants of the Muskegs of Southeast Alaska.* U.S. Department of Agriculture, 1985.

Rockwell, David. *The Nature of North America: A Handbook to the Continent Rocks, Plants, and Animals.* New York: Berkeley Books, 1998.

Rogers, David. "An Addendum on the Botanical History of Santa Lucia Fir, *Abies bracteata*, with Excerpts from the Notes and Letters of Early Collectors." *The Double Cone Quarterly*, vol. 1, no. 3 (winter solstice 1998). www.ventanawild.org/news/ws98/slfirs2.html.

———. "The Native Irises (Flags or Fleur-de-lis) of the Santa Lucia Mountains." *The Double Cone Quarterly*, vol. 4, no. 1 (spring 2001). www.ventanawild.org/news/se01/irises.html.

———. "Perfect Pattern of Silvan Perfection on the Symmetrical Plan, the Rare Santa Lucia Fir." *The Double Cone Quarterly*, vol. 1, no. 2 (fall equinox 1998). www.ventanawild.org/news/fe98/slfirs.html.

Rogers, Diane. "Herb Growers Find Several Paths to Market." *The Western Producer*, July 27, 2000. www.producer.com/articles/20000727/farm_living/20000727fl01.html.

Rook, Earl J. S. *Flora, Fauna, Earth, and Sky . . . The Natural History of the Northwoods.* 1999. www.rook.org/earl/bwca/nature/index.html.

Rowe, Stan. *Home Place: Essays on Ecology.* Edmonton, Alta.: NuWest Publishers, 1992.

Royce, Sarah. *A Frontier Lady: Recollections of the Gold Rush and Early California.* New Haven, Conn.: Yale University Press, 1932.

Rozin, Elisabeth. *Blue Corn and Chocolate.* New York: Alfred A. Knopf, 1992.

Rudloe, Jack. *The Wilderness Coast: Adventures of a Gulf Coast Naturalist.* New York: E. P. Dutton, 1988.

Runesson, Ulf. "*Lycopodium clavatum*—Running Clubmoss." Forest Capital Development Association, 2000/2001. www.borealforest.org/ferns/fern9.htm.

Runkel, Sylvan T., and Alvin F. Bull. *Wildflowers of Indiana Woodlands.* Ames: Iowa State University Press, 1994.

Rupp, Rebecca. *Red Oaks and Black Birches: The Science and Lore of Trees.* Pownal, Vt.: Storey Communications, 1990.

Russell, Howard S. *Indian New England before the Mayflower.* Hanover, N.H.: University Press of New England, 1980.

Rutsch, Edward S. *Smoking Technology of the Aborigines of the Iroquois Area of New York State.* Rutherford, N.J.: Fairleigh Dickinson University Press, 1973.

Ryan, Brenda. "California Gold." http://alchemylitmag.home.mindspring.com/califgold.html.

Salim, Emil, and Ola Ullsten, co-chairmen. *Our Forests, Our Future.* Report of the World Commission on Forests and Sustainable Development. Cambridge: Cambridge University Press, 1999.

Sanders, Jack. *Hedgemaids and Fairy Candles: The Lives and Lore of North American Wildflowers.* Camden, Maine: Ragged Mountain Press, 1993.

Sanft, Heather. "How to Plant a Living Lattice Fence." *East Coast Gardener*, vol. 3, no. 2 (March 1999). http://www.klis.com/fundy/ecg/fence3-2.htm.

Sather, Nancy. *Leedy's Roseroot: A Cliffside Glacial Relict.* St. Paul: Minnesota Department of Natural Resources, 1993. Adapted to digital document "Leedy's Roseroot," by State of Minnesota, Department of Natural Resources, 1993. http://www.dnr.state.mn.us/ecological_services/nhnrp/Lroseroot.pdf.

Sauleda, Ruben P. "The Orchids of Florida." The Everglades. www.members.tripod.com/~rsauleda/.

Saunders, C. F. *Western Wild Flowers and Their Stories*. New York: Doubleday Doran, 1933.

Savile, D. B. O. *Artic Adaptations in Plants*. Canadian Department of Agriculture, 1972.

Savio, Yvonne. "Prickly Pear Cactus." *Family Farm Series*. University of California–Davis, June 1989, rev. July 1989. www.sfc.ucdavis.edu/pubs/brochures/pricklypear.html.

Saxton, Dean, and Lucille Saxton. *Legends and Lore of the Papago and Pima Indians*. Tucson: University of Arizona Press, 1973.

Schama, Simon. *Landscape and Memory*. Toronto: Random House of Canada, 1995.

Schofield, Janice J. *Discovering Wild Plants: Alaska, Western Canada, the Northwest*. Anchorage: Alaska Northwest Books, 1989.

Schultes, Richard Evans, and Albert Hofmann. *Plants of the Gods: Origins of Hallucinogenic Use*. New York: A. van der Marck Editions, 1979.

Schwaab, Eugene L., and Jacqueline Bull. *Travels in the Old South, Volume II*. Lexington, Kentucky: University Press of Kentucky.

Schwartz, Randall. *Carnivorous Plants*. Ed. Deborah Leavy. New York: Praeger Publishers, 1974.

Scott, Richard W. *The Alpine Flora of the Rocky Mountains, Volume I. The Middle Rockies*. Salt Lake City: University of Utah Press, 1995.

Sebesta, Lou. "Balmville Tree." NYS Department of Environmental Conservation. Reprinted from *New York State Conservationist*, April 1999. www.dec.state.nv.us/website/dpae/cons/balmvilletree/html.

Seeger, Ruth Porter. *American Folk Songs for Children in Home, School, and Nursery School: A Book for Children, Parents, and Teachers*. Garden City, N.Y.: Doubleday, 1948.

Senate Subcommittee on the Boreal Forest. *Competing Realities: The Boreal Forest at Risk*. Report of the Sub-Committee on Boreal Forest of the Standing Senate Committee on Agriculture and Forestry. Chair of the Subcommittee, The Honourable Nicholas W. Taylor. June 1999. http://www.parl.gc.ca/36/1/parlbus/commbus/senate/com-e/bore-e/rep-e/rep09jun99part2-e.htm.

Service, Robert. *The Best of Robert Service*. Philadelphia: Running Press, 1983.

Sharnoff, Sylvia Duran, comp. *Bibliographic Database of the Human Uses of Lichens*. www.lichen.com/usetype.html.

Shulman, Alix Kates. *Drinking the Rain*. New York: Farrar Straus Giroux, 1995.

Siry, Joseph V. *Marshes of the Ocean Shore: Development of an Ecological Ethic*. College Station: Texas A&M University Press, 1984.

Small, John Kunkel. *Ferns of the Southeastern States*. Facsimile of 1938 edition. New York: Hafner Publishing, 1964.

Smet, Pierre-Jean de. *Life, Letters, and Travels of Father de Smet, 1801–1873*. 4 vols. Ed. Hiram Martin. Chittenden and Alfred Talbot Richardson. New York: Kraus Reprint Co., 1969.

Smith, Emory E. *The Golden Poppy*. San Francisco: Murdoch Press, 1902.

Smith, Huron. *Ethnobotany of the Menomini Indians*. Bulletin of the Public Museum of the City of Milwaukee, vol. 4, no. 1. Milwaukee: Board of Trustees, Public Museum, 1923.

Smith, Richard M. *Wild Plants of America: A Select Guide for the Naturalist and Traveler*. New York: John Wiley, 1989.

Smith, Rod. "In Like Gin." *Private Clubs Magazine Online*. January/February 2001. www.privateclubs.com/archives/2001-jan0feb/wine_inlikegin.htm.

Snyder, James A. "The Ecology of *Sequoia semper-virens*." December 1992. askmar, 1996. http://www.batnet.com/askmar/Redwoods/Masters_Thesis.html.

Snyder, Karl. "Aspen Fall Colors." Estes Park Online. com, 1999–2001. http://www. estesparkonline.com/EP-News-Aspen2000.HTML.

Spahr, Mary Anne. "Glacier Lily, *Erythronium grandiflorum*." *Northwest Natives.* Rainier Audubon Society, *The Herald Heron Newsletter*, September 1999. http://www.rainieraudubon.org/nn/glacier-lily.htm.

Speck, Frank G. "Virginia Indian Folk Lore Plants." American Philosophical Society, 1941. Unpublished manuscript.

Spongberg, Stephen A. *A Reunion of Trees: The Discovery of Exotic Plants and Their Introduction into North American and European Landscapes.* Cambridge, Mass.: Harvard University Press, 1990.

St. Antoine, Sara. "Ironwood: Carving New Life from Ancient Trees." Ask!! Distinctive Gifts and Ironwood, summer 1994. http://www.askpubs.com/wood/AncientTrees.htm.

Steele, Richard M. "Plant Hunting: Privilege Pleasure Posterity." *The Rosebay Newsletter*, vol. 27 (spring 2000). http://rosebay.org/chapterweb/rosebay/plant_hunting_in_nova_scotia.htm.

Stegner, Wallace. *Beyond the Hundredth Meridian: John Wesley Powell and the Second Opening of the West.* New York: Penguin, 1992.

Sternberg, Guy, and Jim Wilson. *Landscaping with Native Trees: The Northeast, Midwest, Midsouth, and Southeast Edition.* Shelburne, Vt.: Chapters Publishing Ltd., 1995.

Stevenson, Matilda Coxe. *Ethnobotany of the Zuni Indians.* Extract from the Thirteenth Annual Report of the Bureau of American Ethnology. Washington, D.C.: U.S. Government Printing Office, 1915.

Stewart, John. *An Account of PEI in 1806: Forest Trees and Other Vegetable Products.* http://www3.pei.sympatico.ca/garyschneider/peiacc/pei4forest.html.

Stienstra, Tom. "Outwest: Land of Ishi Inspires Wonder and Awe." Nando.net, 1998. Scripps-McClatchy Western, 1998. http://archive.

sportserver.com/newsroom/sports/oth/1998/oth/mor/feat/archive/040498/mor2415.html.

Stowe, Harriet Beecher. *Palmetto Leaves.* Gainesville: University Press of Florida, 1968. Facsimile of 1873 ed.

Strickland, Major [Samuel]. *Twenty-Seven Years in Canada West: Or the Experience of an Early Settler.* Ed. Agnes Strickland. Two volumes in one. Edmonton, Alta.: M. G. Hurtig, 1970.

Strong, Donald R. and Mei-Yin Wu. "American Insects Feed on English Cordgrass." *Biocentral News and Information*, vol. 20 no. 2 (June 1999). http://pest.cabweb.org/Journals/BNI/Bni20-2/gennews.htm.

Strong, Frances. *All the Year Round: A Nature Reader.* Boston: Ginn and Company, 1905.

Strong, Jane. "Coyote Bush." *Jane Strong's Nature Guide.* Pelican Network, October 2000. www.pelicannetwork.net/fg/coyote_bush/coyote_bush.html.

Suzuki, David, and Holly Dressel. *From Naked Ape to Superspecies: A Personal Perspective on Humanity and the Global Eco-Crisis.* Toronto: Stoddart Publishing, 1999.

Swanton, John R. *Social and Religious Beliefs and Usages of the Chickasaw Indians.* U.S. Bureau of American Ethnology. Forty-fourth annual report, 126/27. Washington, D.C. 1928. 169–273.

Swerdlow, Joel L. *Nature's Medicine, Plants That Heal: A Chronicle of Mankind's Search for Healing Plants through the Ages.* Washington, D.C.: National Geographic, 2000.

Szczawinski, Adam F., and Nancy J. Turner. *Wild Green Vegetables of Canada.* Ottawa: National Museum of Natural Sciences, 1980.

Tait, Lyal. *The Petuns Tobacco Indians of Canada.* Port Burwell, Ont.: Erie, 1971.

———. *Tobacco in Canada.* Tillsonburg, Ont.: Flue-Cured Tobacco Growers' Marketing Board, 1968.

Tanner, John. *A Narrative of the Captivity and Adventures of John Tanner, U.S. Interpreter at the Sault de Ste. Marie, during Thirty Years Residence among the*

Indians, in the Interior of North America. New York: G. & C. & H. Carvill, 1830.

Taylor, Patricia A. *Easy Care Native Plants: A Guide to Selecting and Using Beautiful American Flowers, Shrubs, and Trees in Gardens and Landscapes*. New York: Henry Holt and Co., 1996.

Taylor, Raymond L. *Plants of Colonial Days*. Mineola, N.Y.: Dover, 1996.

Taylor's Guide to Annuals. Ed. Gordon DeWolf Jr. Boston: Houghton Mifflin, 1986.

Taylor's Guide to Natural Gardening. Ed. Roger Holmes. Boston: Houghton Mifflin, 1993.

Taylor's Guide to Perennials. Ed. Gordon DeWolf Jr. Boston: Houghton Mifflin, 1986.

Taylor's Guide to Shrubs. Ed. Gordon DeWolf Jr. Boston: Houghton Mifflin, 1987.

Teal, John, and Mildred Teal. *Life and Death of the Salt Marsh*. Boston: Little, Brown, 1969.

Teit, James. *Traditions of the Thompson River Indians of British Columbia*. Boston and New York: Houghton Mifflin, 1898.

Tekulsky, Mathew. *The Hummingbird Garden: Turning Your Garden, Window Box, or Backyard into a Beautiful Home for Hummingbirds*. New York: Crown, 1990.

Thomas, Alfred Barnaby. *After Coronado: Spanish Exploration Northeast of New Mexico, 1696–1727*. Second edition. Norman: University of Oklahoma Press, 1966.

Thomas, Bill. *Wild Woodlands: The Old-Growth Forests of America*. Dallas: Taylor, 1992.

Thomas, Linda. "Brush Fire." *Tapestry Rambles*. Women Online Worldwide, 1995–2001. www. wowwomen.com/tapestry/arch_rambles/brushfire/html.

Thomson, John W. *American Arctic Lichens. I. The Macrolichens*. New York: Columbia University Press, 1984.

Thoreau, Henry D. *Cape Cod*. Ed. Joseph J. Moldenhauer. Princeton, N.J.: Princeton University Press, 1988.

———. *The Illustrated Maine Woods*. Princeton, N.J.: Princeton University Press, 1974.

———. *The Journal of Henry D. Thoreau*. Ed. Bradford Torrey and Francis H. Allen. 14 vols. bound as two. New York: Dover, 1962.

———. *The Major Essays of Henry David Thoreau*. Ed. Richard Dillman. New York: Whitston, 2001.

———. *Walden*. Ed. Stephen Allen Fender. Oxford: Oxford University Press, 1997.

Time-Life Books, ed. *Grasslands and Tundra*. Alexandria: Time-Life Books, 1987.

Tiner, Ralph W. *A Field Guide to Coastal Wetland Plants of the Northeastern United States*. Amherst: University of Massachusetts Press, 1987.

———. *In Search of Swampland: A Wetland Sourcebook and Field Guide*. New Brunswick, N.J.: Rutgers University Press, 1998.

Tisdale, Sallie. *Stepping Westward: The Long Search for Home in the Pacific Northwest*. New York: H. Holt, 1991.

Townsend, John Kirk. *Narrative of a Journey across the Rockies to the Columbia River*. Philadelphia: Henry Perkins, 1839. www.xmission.com/~drudy/mtman/html/townsend.html.

Traill, Catherine Parr. *The Backwoods of Canada: Being Letters from the Wife of an Emigrant Officer*. Toronto: McClelland and Stewart, 1969, 1971 (first published 1846).

———. *Canadian Settler's Guide*. Toronto: McClelland and Stewart, 1969 (first published 1855).

———. *Canadian Wild Flowers*. Montreal: John Lovell, 1868. Facsimile edition, Toronto: Coles, 1972.

———. *Female Emigrant's Guide, and Hints on Canadian Housekeeping*. Toronto: Maclear, 1854.

———. *Stories of the Canadian Forest, or, Little Mary and Her Nurse*. New York: C. S. Francis, 1859.

Travels in the Great Western Prairies, the Anahuac and Rocky Mountains, and in the Oregon Territory: An 1839 Wagon Train Journal. American Biography Series, Thomas Jefferson Farnham, 1843.

Trelawny, John G. *Wildflowers of the Yukon and Northwestern Canada Including Adjacent Alaska*. Sidney, B.C.: Gray's Publishing, 1983.

Tremper, Gary. "The History and Promise of Jojoba." *Armchair World*, 1996. http://www.armchair.com/warp/jojoba1.html.

Tripple Brook Farm. "Asimina." www.tripplebrookfarm.com/iplants/Asimina.html.

Trollope, Frances Milton. *Domestic Manners of the Americans*. 4th ed. 2 vols. London: Whittaker, Treacher, 1832.

Tryon, Alice F., and Rolla M. Tryon. *Ferns and Allied Plants with Special Reference to Tropical America*. New York: Springer-Verlag, 1982.

Turner, John W. *American Arctic Lichens. 2. The Microlichens*. Madison: University of Wisconsin Press, 1997.

Turner, Nancy J. *Food Plants of Coastal First Peoples*. Vancouver: UBC Press, 1995.

———. *Food Plants of Interior First Peoples*. Vancouver: UBC Press, 1997.

———. *Plant Technology of First Peoples in British Columbia*. Vancouver: UBC Press, 1998.

Turner, Nancy J., and Adam Szczawinski. *Edible Wild Fruits and Nuts of Canada*. Volume 3. Markham, Ont.: Fitzhenry and Whiteside for National Museums of Natural Sciences, 1988.

———. *Wild Coffee and Tea Substitutes of Canada*. Ottawa: National Museums of Canada, 1978.

Turner, Nancy J., et al., *Thompson Ethnobotany: Knowledge and Usage of Plants by the Thompson Indians of British Columbia*. Victoria: Royal British Columbia Museum, 1990.

Turner, Raymond, Janice E. Bowers, and Tony L. Burgess. *Sonoran Desert Plants: An Ecological Atlas*. Tucson: University of Arizona Press, 1995.

Twain, Mark. *Adventures of Huckleberry Finn*. Oxford: Oxford University Press, 1999.

———. *Roughing It*. Pleasantville, N.Y.: Reader's Digest Association, 1994.

U.S. Department of Agriculture. "*Platanus x acerfolia*: 'Columbia' and 'Liberty.' " *U.S. National Arboretum Plant Introduction*. Floral and Nursery Plants Research Unit, November 1999. www.usna.usda.gov/Newinto/platanus/pdf.

U.S. Department of Agriculture, Forest Service, Rocky Mountain Research Station, Fire Sciences Laboratory. Fire Effects Information System, July 2001. http://www.fs.fed.us/database/feis/_[2001].

U.S. Fish and Wildlife Service. "Endangered species facts: Leedy's Roseroot." http://midwest.fws.gov/Endangered/plants/leedysro.html.

Underhill, Ruth. *Papago Indian Religion*. New York: Columbia University Press, 1946.

University of Arizona Arboretum. "Plant Walk." http://www.ag.arizona.edu/arboretum/pwalk/pw13.htm.

University of California, Natural Reserve System. "Common Plants of the Blue Oak Savanna." Stebbins Cold Canyon Reserve. http://nrs.ucop.edu/reserves/stebbins/plants/savanna.htm.

University of California at Davis. "Westward Ho: Frontier Life and Exploration, 1776–1848." *Welcome to America the Bountiful*. www.lib.ucdavis.edu/exhibits/food/panel3/html.

University of Connecticut Cooperative Extension Forestry. "Why Is Swamp Thing Taking Over the Forests?" http://www.canr.uconn.edu/ces/forest/redmaple.htm.

Ursin, Michael J. *Life In and Around the Salt Marshes: A Handbook of Plant and Animal Life In and Around the Temperate Atlantic Coastal Marshes.* New York: Crowell, 1972.

Vance, F. R., J. R. Jowsey, et al. *Wildflowers across the Prairies: With a New Section on Grasses, Sedges, and Rushes.* Vancouver: Greystone Books, 1999.

Vankat, John L. *The Natural Vegetation of North America.* New York: John Wiley and Sons, 1979.

Vansittart, Jane, ed. *Lifelines: The Stacey Letters, 1836–1858.* London: Peter Davies, 1976.

Verrazzano, Giovanni da. "Translation of the Cellère Codex." Trans. Susan Tarrow. *The Voyages of Giovanni da Verrazzano, 1524–1528.* Ed. Lawrence C. Wroth. New Haven, Conn.: Yale University Press, 1970. http://bc.barnard. columbia.edu/~/gordis/earlyAC/documents/ verrazan.htm.

Viereck, Leslie A., and Elbert L. Little Jr. *Alaska Trees and Shrubs.* Washington, D.C.: U.S. Department of Agriculture, 1972. Reprinted, Fairbanks, University of Alaska Press, 1994.

Vileisis, Ann. *Discovering the Unknown Landscape: A History of America's Wetlands.* Washington: Island Press, 1997.

Viola, Herman J., and Carolyn Margolis, eds. *Seeds of Change: A Quincentennial Commemoration.* Washington, D.C.: Smithsonian Institution Press, 1991.

Vitt, Dale, Janet E. Marsh, and Robin B. Bovey. *Mosses, Lichens, and Ferns of Northwest North America.* Edmonton, Alta.: Lone Pine Publishing, 1988.

Vogel, Virgil J. *American Indian Medicine.* Norman, Okla.: University of Oklahoma Press, 1970.

Walbran, John T. *British Columbia Coast Names, Their Origin and History.* Ottawa: Government Printing Bureau, 1909. Reprinted, Vancouver: Library's Press, 1971.

Walker, Shaun. "The Luna Tree Sit and Julia Butterfly." www.ottermedia.com/LunaJulia.html.

Walshe, Shan. *Plants of Quetico and the Ontario Shield.* Toronto: University of Toronto Press, 1980.

Ward, Barbara, and René Dubos. *Only One Earth: The Care and Maintenance of a Small Planet.* New York: W. W. Norton, 1983.

Wasielewski, Jeff. "Everybody, Gumbo Limbo Mon!" *Dade Florida Native Plant Society Newsletter*, February 2001. http://www.fnps.org/dade/ pastnewslets/newslet102.html.

Wasowski, Sally, and Andy Wasowski. *Native Gardens for Dry Climates.* New York: Clarkson Potter, 1995.

"Watch the Weeds." Fishtalk Radio, 2000–2001. www.fishtalkradio.net/watch_the_weeds.htm.

Waters, Frank. *Book of the Hopi.* New York: Viking, 1963.

Waugh, F. W. *Iroquois Foods and Food Preparation.* Ottawa: Government Printing Bureau, 1916. Facsimile ed. Ottawa: National Museums of Canada, 1973.

Weiner, Michael A. *Earth Medicine—Earth Foods: Plant Remedies, Drugs, and Natural Foods of the North American Indians.* New York: Macmillan, 1972.

Welker, Glenn. "Lakota (Sioux): The Origin of the Prairie Rose." Native American Bedtime Stories, 1996–2001. www.the-office.com/bedtime-story/ sioux-prairierose.htm.

West, George Arbor. *Tobacco, Pipes, and Smoking Customs of the American Indians.* Westport: Greenwood Press, 1970.

Wheeler, David. *The Penguin Book of Garden Writing.* London: Penguin Books, 1996.

Wherry, Edgar T. *The Fern Guide: Northeastern and Midland United States and Adjacent Canada.* Garden City, N.Y.: Doubleday and Company, 1961.

Whiteley, Opal. "Along the Road." *Opal Whiteley's Enchanted Fairyland.* http://www.liloriole.net/text- road-01.htm.

Whitesbog Preservation Trust. "Discover Historic Whitesbog Village." Burlington County Library. www.whitesbog.org.

Whiting, Alfred. *Havasupai Habitat: A. F. Whiting's Ethnography of a Traditional Indian Culture.* Ed. Stephen Weber and David Seaman. Tucson: University of Arizona Press, 1985.

Whitman, Walt. *Leaves of Grass.* Toronto: Musson Book Co., 1900.

Whittier, John Greenleaf. *The Complete Poetical Works of John Greenleaf Whittier.* Cambridge edition. Ed. Horace E. Scudder. Boston: Houghton, Mifflin, 1894.

Whittle, Tyler. *The Plant Hunters: Being an Examination of Collecting with an Account of the Careers and the Methods of a Number of Those Who Have Searched the World for Wild Plants.* New York: PAJ Publications, 1988.

Wick, Nancy. "Gardener Works to Preserve History of Campus Trees." *University Week*, March 3, 2001. University of Washington. http://depts.washington. edu/uweek/archives/2001.03. MAR_29/_article6. html.

Wilder, Laura Ingalls. *Little House on the Prairie.* New York. Harper and Row, 1935.

Willard, Terry. *Edible and Medicinal Plants of the Rocky Mountains and Neighbouring Territories.* Calgary: Wild Rose College of Natural Healing, 1992.

Williams, David, and Catherine Hart. "American Sweetgum—*Liquidambar styraciflua*." Illinois Plant, Landscape, and Nursery Technology. University of Illinois Extension. http://www.extension.uiuc.edu/ IPLANT/plant_select/arboretum_trees/ Sweetgum.htm.

Williams, William Carlos. *The Collected Poems of William Carlos Williams.* Two volumes. New York: New Directions, 1988.

Wilson, Edward O. *The Future of Life.* New York: Alfred A. Knopf, 2002.

Wilson, Gilbert L. *Buffalo Bird Woman's Garden: Agriculture of the Hidatsa Indians.* St. Paul: Minnesota Historical Society Press, 1987.

Wood, William. *New England's Prospect.* 1634. Republished, Boston: John Wilson and Son, 1865.

Woodex Bearing Company, Inc. "Some Other Examples of Wood Bearing Use, from Our Readers." www.woodex-meco.com/ history1.htm.

Wright, Mabel Osgood. *The Friendship of Nature: A New England Chronicle of Birds and Flowers.* Baltimore: Johns Hopkins University Press, 1999.

Wunderlin, Richard P. *Guide to the Vascular Plants of Florida.* Gainesville: University Press of Florida, 1998.

Yava, Albert. *Big Falling Snow: A Tewa-Hopi Indian's Life and Times and the History and Traditions of His People.* New York: Crown, 1978.

Yoon, Carol Kaesuk. "For a New Species, a Few Genes Are All That's Needed." September 5, 1995. *The New York Times*, 1995. http://www.qeced.net/ bio/genbio/exNatSel.htm.

Young, Cameron. *The Forests of British Columbia.* North Vancouver: Whitecap Books, 1985.

Young, Delbert A. *According to Hakluyt: Tales of Adventure and Exploration.* Toronto: Clarke, Irwin, 1973.

Young, Steven B. *To the Arctic: An Introduction to the Far Northern World.* New York: John Wiley and Sons, 1989.

Zatz, Arline. *New Jersey's Special Places.* Countryman Press, 1998. Adapted in "Hidden New Jersey: Island Beach, Part 2," *Destinations*, GORP, Countryman Press. http://www.gorp.com/gorp/publishers/ countryman/hik_njs1htm.

Zona, Scott. *Flora Neotropics, Monograph 71, Roystonea (Arecaceae: Arecoideae).* New York: New York Botanical Garden, 1996.

Zwinger, Ann. *A Desert Country near the Sea: A Natural History of the Cape Region of Baja California.* New York: Harper and Row, 1983.

———. *The Mysterious Lands: An Award-Winning Naturalist Explores the Four Great Deserts of the Southwest.* Tucson: University of Arizona Press, 1989.

———. *Run, River, Run.* New York: Harper and Row, 1975.

———, ed. *Writing the Western Landscape.* Boston: Beacon Press, 1994.

Botanical References

Armitage, Allan M. *Herbaceous Perennial Plants: A Treatise on Their Identification, Culture, and Garden Attributes*. Second edition. Champaign: Stipes Publishing, 1997.

Botanica: The Illustrated A–Z of Over 10,000 Garden Plants and How to Cultivate Them. Managing eds. Susan Page and Margaret Olds. Vancouver: Raincoast Books, 1998.

Budd, Archibald, and Keith F. Best. *Wild Plants of the Canadian Prairies*. Ottawa: Research Branch, Canada Department of Agriculture, 1964. Reprinted 1974.

CalFlora database. www.calflora.org.

California Native Plant Society. www.cnps.org.

Capon, Brian. *Botany for Gardeners: An Introduction and Guide*. Portland, Ore.: Timber Press, 1990.

Columbia Encyclopedia. Sixth edition. Columbia University Press, 2001. www.bartleby.com/65/.

Dendrology at Virginia Tech. www.cnr.vt.edu/dendro/dendrology/syllabus/biglist_frame.htm.

Department of Biology, Herbarium, Texas A&M University. www.csdl.tamu.edu/FLORA.

DesertUSA.www.desertusa.com.

Dirr, Michael A. *A Manual of Woody Landscape Plants, Their Identification, Ornamental Characteristics, Culture, Propagation, and Uses*. Third edition. Champaign, Ill.: Stipes, 1983.

FEIS Online. Fire Effects Information System, USDA Forest Service. www.fs.fed.us/database/feis.

Flora of North America Online Database. http://hua.huh.harvard.edu/cgi-bin/Flora/flora.pl?FLORA_ID=12395.

Floridata. http://www.floridata.com/main_fr.cfm?state=ref_contents&viewsrc=lists/contents.cfm.

Harcourt Academic Press Dictionary of Science and Technology. www.harcourt.com/dictionary.

Hosie, R. C. *Native Trees of Canada*. Eighth edition. Markham, Ont.: Fitzhenry and Whiteside Ltd. in cooperation with the Canadian Forestry Service (Environment Canada), 1990.

International Plant Names Index. www.ipni.org.

Jones, Samuel B., Jr., and Leonard E. Foote. *Gardening with Native Wild Flowers*. Portland, Ore.: Timber Press, 1997.

Krüssmann, Gerd. *Manual of Cultivated Conifers*. Second edition. Ed. Hans-Dieter Warda. Portland, Ore.: Timber Press, 1995.

Mickel, John T. *Ferns for American Gardens*. New York: Macmillan, 1994.

New Royal Horticultural Society Dictionary of Gardening. Ed. Anthony Huxley, Mark Griffiths, and Margot Levy. London: Macmillan Reference Ltd., 1999.

Newmaster, Steven G., Allan G. Harris, and Linda J. Kershaw. *Wetland Plants of Ontario*. Vancouver: Lone Pine Publishing, 1997.

Plants for a Future. database. University of Leeds. http://www.comp.leeds.ac.ok/pfaf/.

Reader's Digest A–Z Encyclopedia of Garden Plants. Ed. Christopher Brickell, Trevor Cole, and Judith D. Zuk. Montreal: Reader's Digest Association (Canada), 1997.

Southwest Missouri State University, Department of Biology, Ozarks Regional Herbarium. http://biology.smsu.edu/Herbarium.

USDA Forest Service. www.fs.fed.us.

USGS Western Wetland Flora. www.npwrc.usgs.gov/resource/othrdata/westflor/westflor.htm.

Index

Page numbers in *italics* refer to illustrations.

Eragrostis lehmanniana, 338
Ericaceae, 58–62, *61*, 63–67, *64,*
    65, 70–71, 92–96, *93, 95,*
    175–76, *175*, 261–63, *261*,
    416–18, *417*, 449–50,
    497–504, *498, 499, 501,*
    529–31, *529*, 542, *542*
Eriophorum spp., 536–37, *536*
Eryngium:
    aquaticum, 315
    yuccifolium, 315, *315*
Erythronium:
    americanum, 104–5, 450
    grandiflorum, 450–52, *450*
    grandiflorum var. *candidum*, 452
Eschscholzia californica, 397–99,
    397
Eupatorium:
    maculatum, 301–2, *301*
    perfoliatum, 301
    urticifolium (syn. *E. rugosum*),
        302
Euphorbia esula, 334, 442
evening primrose, 112–13, *112*

Fabaceae, 47–49, *48*, 96–97,
    201–2, *201*, 269–70, 317–25,
    317, 319, 320, 323, 324,
    360–69, *361, 362, 367, 369,*
    434–35, 567–69
Fagaceae, 15–19, *15, 17*, 23–30,
    131–37, *131, 135*, 410–16
Fagus grandifolia, 17–19, *17*, 31
fairy-slipper, 514–16, *515*
fairy spud, 121, *121*
false banana, 168–70, *169*
false cypress, Nootka, 489–91,
    489
farewell-summer, 82–84, *83*
fern, *71*
    Christmas, 522
    licorice, 493
    ostrich, 92
    resurrection, 237–38, *238*
    sword, 522, *522*
    Western bracken, 92
Ferocactus:
    cylindraceus, 349–50
    wislizenii, 348–49, *349*
fescue, tall, 334
Festuca elatior, 334
feverbush, 168
Ficus:
    aurea, 234–36, *235*
    sycomorus, 156
fig, strangler, 234–36, *235*

filbert:
    American, 19
    European, 19
fir:
    balsam, 256–57, *256*
    bristlecone, 455–56
    Douglas, 442, 457, 473–78,
        475
    grand, 454–55, 505
    Santa Lucia, 455–56
    sub-alpine, 442, 452–54
    white, 453–54
fireweed, 259–61, *260*
flag:
    blue, 176–77, *177*
    poison, 176–77, *177*
    sweet, 178, 270
flagroot, 178
flannelbush, 430–31
Florida, 221–41, *221*
Florida moss, 147–49, *149*
flower-of-a-day, 90–91, *90*
foamberry, 277, *277*
forget-me-not, Alpine, 538–40,
    538
Fouquieriaceae, 369–71, *371*
Fouquieria splendens, 369–71, *371*
four-o'clock plant, 112–13, *112*
franklinia, 144–46, *145*
Franklinia alatamaha, 144–46,
    145
fremontia, California, 430–31
Fremontodendron californicum (syn.
    Fremontia californica),
    430–31
Fritillaria camschatcensis, 452,
    543–44, *544*
Fucaceae, 213–14, 215–17, *216*
Fucus vesiculosus, 215–17, *216*
Fumariaceae, 97–100, *98*, 516,
    516
fungi, 558–59
fungus, bracket, 559

galleta, big, 391
Gaultheria:
    procumbens, 41, 94–96, *95*, 463
    shallon, 499–500, *499*
gayfeather, Kansas, 304, *304*
Gaylussacia:
    baccata, 71
    brachycera, 71
    dumosa, 70–71
Geum triflorum, 329, *331*
Gigartinaceae, 214–15, *215*
gin berry, 447–48

ginger, wild, 80–82, *81*
ginseng:
    Alaskan, *see* devil's club
    American, 79–80, *79*
girasole, 302–3, *303*
glasswort, *208*, 209
Glechoma hederacea, 334
Gleditsia triacanthos, 42
glycine, 74, 96–97
goat nut, 389–91, *390*
goldenrod, 310–12, *311*
    stiff (rigid), 312
    sweet, 312
golden slipper, 115–16, *115*
goldthread, 182–83
    three-leaved, 182–83
gopher berry, 70–71
gourd:
    bottle, 570
    ornamental, 569–71
    wild, 570
grama:
    blue, 288–89
    sideoats, 287–89, *288*
grape:
    dull Oregon, 506
    Oregon, 505–6, *506*
grass:
    blue grama, 288–89
    buffalo, 289–90, *289*
    buffel, 338
    cord, *see* cordgrass
    cotton, 536–37, *536*
    dune, *202*, 203
    Indian, 292, *292*
    June, 290–91
    Lehmann love, 338
    needle, 207
    prairie June, 290–91
    scurvy, 534
    sea, 204–5
    shave, *see* Equisetum, Equisetum
        spp.
    slough, 292, 294–95
    switch, 291–92, *291*
    tall panic, 291–92, *291*
    whitlow, 534–35, *535*
    wood, 292, *292*
    wrack, 204–5, *204*
grassbur, 203–4
grass-of-Parnassus, Northern,
    548–49, *549*
greasewood, 382–84, *383*,
    423–24
ground ivy, 334
groundnut, 74, 96–97

groundsel bush, 208
groundsel tree, 208
grouseberry, 449–50
Guaiacum sanctum, 234
gumbo-limbo, 223, 233, *233*
gum guaiacum, 234
gum plant, 307–9, *308*
Gymnocladus dioicus, 41–43, *42*

hackmatack, 247–49, *248*
Halesia carolina, 52–53, *52*
Hamamelidaceae, 62–63, *63*,
    160–62, *161*
Hamamelis virginiana, 62–63, *63*
hay:
    marsh, 207
    salt-meadow, 207
haybell, 126–28, *127*
hazelnut:
    American, 19, 62
    beaked, 19
heal-all, 128–29, *128*
heath, beach, 199
heather:
    Arctic bell, 542, *542*, 546–47
    beach, 199
    false, 199
    white, 542, *542*, 545
Hedera, 130
Hedysarum:
    alpinum, 269, 270
    mackenzii (syn. *H. boreale*),
        269–70
Helianthus, Helianthus spp.,
    571–72
    annuus, 571
    anomalus, 572
    giganteus, 572
    maximiliani, 572
    strumosus, 572
    tuberosus, 302–3, *303*
hellebore:
    Eurasian, 179
    false, 179, 181
    swamp, 179, 181
Helleborus, 179
hemlock:
    Canadian, 44–46, *45*
    Eastern, 44–46, *45*
    Western, 481–83, *481*
hemp, Indian, 78
Hepatica:
    acutiloba, 124
    americana, 124–25, *125*
    round-lobed, 124–25, *125*
Hesperis matronalis, 334

Hesperoyucca whipplei:
    var. *caespitosa*, 380
    var. *intermedia*, 380
    var. *parishii*, 380
    var. *percursa*, 380
Heteromeles arbutifolia (syn.
    Photinia arbutifolia), 422–23
hickory, 19–23
    bird's-eye, 20–21, *20*
    bitternut, 22–23
    broom, 22
    butternut, 22
    Carolina, 20–21, *20*
    pignut, 22
    shagbark, 20–21, *20*
    swamp, 22–23
Hilaria rigida (syn. *Pleuraphis
    rigida*), 391
Hippocastanaceae, 15, 49
hog apple, 62, 87–89, *88*
holly, American, 167–68, *167*
Holodiscus:
    discolor, 510–11, *510*
    dumosus, 510
honey mushrooms, 25
honeysuckle, *xv*, 123–24, *123*
hop hornbeam, Eastern, 54–55,
    55
hopniss, 74, 96–97
horse chestnut, 15
horsetail, 266–69, *266*
    common, 266, 268
huckleberry:
    black, 71
    box, 71
    bush, 70–71
    dwarf, 70–71
    red, 500–501, *501*
Hudsonia tomentosa, 199
Huperzia selago, 552
Hydrophyllaceae, 100–101, *100*
Hydrophyllum virginianum,
    100–101, *100*
Hypericaceae, 189
Hypericum:
    perforatum, 189
    pyramidatum, 189
    virginicum (syn. *Triadenum
        virginicum*), 189

ice plant, 110–11, *111*
Ilex:
    opaca, 167–68, *167*
    paraguayensis, 168
    verticillata, 168
Impatiens capensis, 84–85, *85*

Indian cucumber, 105–6
Indian cup, 309–10, *309*
Indian ginger, 80–82, *81*
Indian gum, 309–10, *309*
Indian hemp, 78
Indian paint, 117–18, *117*
Indian pipe, 110–11, *111*
Indian poke, 179, 181
Indian potato, 269, 270
Indian tea, 386–87
Indian turnip, 72–75, *73*
indigo, 320–21
    blue false, 319–20, *319*
Indigofera, 320
inkweed, 118–21, *119*
inside-out flower, 507
Inula, 443
Ipomoea pandurata, 91
Iridaceae, 176–77, *177*, 438–39,
    439
iris:
    coast, 438–39, *439*
    Douglas, 438–39, *439*
    Marin, 439
    Thompson's, 439
Iris:
    douglasiana, 438–39, *439*
    versicolor, 176–77, *177*
    virginica, 177
Irish moss, 214–15, *215*
ironweed, 312–13, *313*, 315
ironwood, 54–55, *55*
    desert, 366–68, *367*, 417
ivy, 130
    American, 129–30
    Boston, 129
    five-fingered, 129–30
    ground, 334

Jack-in-the-Pulpit, 72–75, *73*
    Northern, 72
Jerusalem artichoke, 302–3, *303*
jewelweed, spotted, 84–85, *85*
Joe-Pye weed, 301–2, *301*
    spotted, 301–2, *301*
Johnny-jump-up, 128–29, *128*
joint grass, 266–69, *266*
joint pine, 385–87, *386*
jointweed, 181–82
jojoba, 389–91, *390*
Joshua tree, 376–78, *376*
Judas tree, 58
Juglandaceae, 19–23, *20*
Juncaceae, 207–8
Juncus roemerianus, 207–8
Juneberry, 183–84, *183*

maximum, 60–61, *61*
periclymenoides (syn. *R. nudiflorum*), 59, 62
prinophyllum, 62
Rhodymeniaceae, 217–18
Rhus, Rhus spp., 68–69, *68*
   aromatica, 68–69
   copallina, 69
   glabra, 69
   ovata, 429–30, *430*
   typhina, 69
   vernix, 69
rice, wild, 575–77
rice root, 452, 543, *544*
ripgut, 292, 294–95
riverbush, 162–64, *162*
Robinia pseudoacacia, 47–49, *48*
rocket:
   dame's, 334
   sea, 197–98, *197*, 199
rock lily, *xv*, 123–24, *123*
rock tripe, 556–57
rockweed, 215–17, *216*
   bladder, 215–17, *216*
   knotted, 213–14
Rosa:
   acicularis, 329, 331–32, *331*
   arkansana, 331
   woodsii, 331
Rosaceae, 56–57, *56*, 183–84,
   183, 257–59, *258*, 329,
   331–32, *331*, 422–24,
   465–67, *466*, 508–11, *509*,
   510, 546–48, *547*
rose:
   cattle, 332
   prairie, 329, 331–32, *331*
   prickly, 329, 331–32, *331*
rosebay rhododendron, 60–61,
   61
   California, 501–4, *501*
roseroot, 446–47, *446*
   Leedy's, 447, 467
Roystonea:
   elata, 226–28, *226*
   regia, 228
Rubiaceae, 162–64, *162*
Rubus:
   acaulis (syn. *R. arcticus* subsp.
     acaulis), 548
   chamaemorus, 548
   spectabilis, 508–10, *509*, 548
Rudbeckia hirta, 306–7, *306*

Sabal palmetto, 228–31, *229*, 234
sage, coastal, 424–26, *425*

sagebrush, California (coast),
   424–26, *425*
Sagittaria latifolia, 170–72, *170*
saguaro, 339–45, *339*
sahuaro, 339–45, *339*
St. John's wort:
   common, 189
   great, 189
   marsh, 189
Saint-Mary's-seal, 106–8, *107*
salal, 499–500, *499*
Salicaceae, 13–15, 157–60,
   273–75, *273*, 527–29, *528*
Salicornia spp., *208*, 209
Salix, Salix spp., 157–59
   arctica, 158, 527–29, *528*
   babylonica, 158
   discolor, 160
   exigua, 356
   nigra, 158, 159–60
salmonberry, 508–10, *509*, 548
Salsola pestifer, 333–34
saltbush:
   crested, 199
   four-wing, 384–85, *385*
   hoary, 384–85, *385*
Sambucus pubens (syn. *Sambucus
   racemosa* spp. *pubens*), 507–8,
   508
sandbur, 203–4
sand food, 388, 391, 393
sandspur, 203–4
Sanguinaria canadensis, 117–18,
   117
Sarcobatus, 383
Sarraceniaceae, 193–94, *193*, *194*
Sarracenia purpurea, 193, *193*
sarsaparilla:
   American, 270–71
   false, 270–71
   wild, 270–71
sassafras, 53–54, *53*
Sassafras albidum, 53–54, *53*
sawgrass, 240–41, *241*
Saxifraga, Saxifraga spp., 549–50
   bronchialis, 550
   caespitosa, 550
   flagellaris, 550
   oppositifolia, 550
Saxifragaceae, 548–50, *549*
saxifrage, 549–50
   purple, 550
   spotted, 550
   tufted, 550
Schizachyrium scoparium, 287, *287*
Schlumbergera, 355

Scirpus, Scirpus spp., 173–74, *173*
   acutus, 174
   maritimus, 174
   validus, 174
Scrophulariaceae, 185–86, *185*,
   186, 332–33, *333*, 435–38,
   436, 467–68, *467*, *550*, 551
sea lavender, 210–11, *210*
sea lettuce, 218, *219*
sea myrtle, 208
sea oats, 196–97, *196*
sea rocket, 197–98, *197*, 199
sea whistle, 213–14
sedge, *163*, 263–66
   bristle-stalked, 265
   cotton, 536–37, *536*
   golden, 266
   hairy-fruited (slender), 266
   hay, 265
   inland, 265
   lakeshore, 264
   large, 264
   Norway, 266
   water, 264
Sedum:
   integrifolium subsp. *leedyi*, 447,
     467
   integrifolium (syn. *Rhodiola
     integrifolia*), 446–47, *446*
   lanceolatum, 447
senita, 346
sequoia, giant, 153, 407–9, *407*
Sequoiadendron giganteum, 153,
   407–9, *407*
Sequoia sempervirens, 153, 403–6,
   403
Serenoa repens, 146–47, *146*
serviceberry, 183–84, *183*
shadbush, 183–84, *183*
shad scale, 384–85, *385*
Shepherdia canadensis, 277, *277*
shindagger, 375
shoestring, 317–18, *317*
shooting star, 462–63, *463*
   frigid, 463
   pretty, 462
   tall mountain, 462
Silene acaulis, 540–41, *540*
silphium, cut-leaved, 307–9, *308*
Silphium:
   laciniatum, 307–9, *308*
   perfoliatum, 309–10, *309*
silver bell, 52–53, *52*
   Carolina, 52–53, *52*
Simmondsiaceae, 389–91, *390*
Simmondsia chinensis, 389–91, *390*